LONDON MATHEMATICAL SOCIETY LECTURE NOTE SERIES

Managing Editor: Professor M. Reid, Mathematics Institute,
University of Warwick, Coventry CV4 7AL. United Kingdom

The titles below are available from booksellers, or from Cambridge University Press at www.cambridge.org/mathematics.

211 Groups '93 Galway / St Andrews I, C.M. CAMPBELL *et al*
212 Groups '93 Galway / St Andrews II, C.M. CAMPBELL *et al*
213 General theory of Lie groupoids and Lie algebroids, K.C.H. MACKENZIE
214 Generalised Euler–Jacobi inversion formula and asymptotics beyond all orders, V. KOWALENKO *et al*
215 Number theory: Paris 1992-3, S. DAVID (ed)
216 Stochastic partial differential equations, A. ETHERIDGE (ed)
217 Quadratic forms with applications to algebraic geometry and topology, A. PFISTER
218 Surveys in combinatorics, 1995, P. ROWLINSON (ed)
220 Algebraic set theory, A. JOYAL & I. MOERDIJK
221 Harmonic approximation, S.J. GARDINER
222 Advances in linear logic, J.-Y. GIRARD, Y. LAFONT & L. REGNIER (eds)
223 Analytic semigroups and semilinear initial boundary value problems, K. TAIRA
224 Computability, enumerability, unsolvability, S.B. COOPER, T.A. SLAMAN & S.S. WAINER (eds)
225 A mathematical introduction to string theory, S. ALBEVERIO *et al*
226 Novikov conjectures, index theorems and rigidity I, S.C. FERRY, A. RANICKI & J. ROSENBERG (eds)
227 Novikov conjectures, index theorems and rigidity II, S.C. FERRY, A. RANICKI & J. ROSENBERG (eds)
228 Ergodic theory of $\mathbf{Z}^d$ actions, M. POLLICOTT & K. SCHMIDT (eds)
229 Ergodicity for infinite dimensional systems, G. DA PRATO & J. ZABCZYK
230 Prolegomena to a middlebrow arithmetic of curves of genus 2, J.W.S. CASSELS & E.V. FLYNN
231 Semigroup theory and its applications, K.H. HOFMANN & M.W. MISLOVE (eds)
232 The descriptive set theory of Polish group actions, H. BECKER & A.S. KECHRIS
233 Finite fields and applications, S. COHEN & H. NIEDERREITER (eds)
234 Introduction to subfactors, V. JONES & V.S. SUNDER
235 Number theory: Séminaire de théorie des nombres de Paris 1993–94, S. DAVID (ed)
236 The James forest, H. FETTER & B.G. DE BUEN
237 Sieve methods, exponential sums, and their applications in number theory, G.R.H. GREAVES *et al* (eds)
238 Representation theory and algebraic geometry, A. MARTSINKOVSKY & G. TODOROV (eds)
240 Stable groups, F.O. WAGNER
241 Surveys in combinatorics, 1997, R.A. BAILEY (ed)
242 Geometric Galois actions I, L. SCHNEPS & P. LOCHAK (eds)
243 Geometric Galois actions II, L. SCHNEPS & P. LOCHAK (eds)
244 Model theory of groups and automorphism groups, D.M. EVANS (ed)
245 Geometry, combinatorial designs and related structures, J.W.P. HIRSCHFELD *et al* (eds)
246 p-Automorphisms of finite p-groups, E.I. KHUKHRO
247 Analytic number theory, Y. MOTOHASHI (ed)
248 Tame topology and O-minimal structures, L. VAN DEN DRIES
249 The atlas of finite groups - ten years on, R.T. CURTIS & R.A. WILSON (eds)
250 Characters and blocks of finite groups, G. NAVARRO
251 Gröbner bases and applications, B. BUCHBERGER & F. WINKLER (eds)
252 Geometry and cohomology in group theory, P.H. KROPHOLLER, G.A. NIBLO & R. STÖHR (eds)
253 The q-Schur algebra, S. DONKIN
254 Galois representations in arithmetic algebraic geometry, A.J. SCHOLL & R.L. TAYLOR (eds)
255 Symmetries and integrability of difference equations, P.A. CLARKSON & F.W. NIJHOFF (eds)
256 Aspects of Galois theory, H. VÖLKLEIN, J.G. THOMPSON, D. HARBATER & P. MÜLLER (eds)
257 An introduction to noncommutative differential geometry and its physical applications (2nd edition), J. MADORE
258 Sets and proofs, S.B. COOPER & J.K. TRUSS (eds)
259 Models and computability, S.B. COOPER & J. TRUSS (eds)
260 Groups St Andrews 1997 in Bath I, C.M. CAMPBELL *et al* (eds)
261 Groups St Andrews 1997 in Bath II, C.M. CAMPBELL *et al* (eds)
262 Analysis and logic, C.W. HENSON, J. IOVINO, A.S. KECHRIS & E. ODELL
263 Singularity theory, W. BRUCE & D. MOND (eds)
264 New trends in algebraic geometry, K. HULEK, F. CATANESE, C. PETERS & M. REID (eds)
265 Elliptic curves in cryptography, I. BLAKE, G. SEROUSSI & N. SMART
267 Surveys in combinatorics, 1999, J.D. LAMB & D.A. PREECE (eds)
268 Spectral asymptotics in the semi-classical limit, M. DIMASSI & J. SJÖSTRAND
269 Ergodic theory and topological dynamics of group actions on homogeneous spaces, M.B. BEKKA & M. MAYER
271 Singular perturbations of differential operators, S. ALBEVERIO & P. KURASOV
272 Character theory for the odd order theorem, T. PETERFALVI. Translated by R. SANDLING
273 Spectral theory and geometry, E.B. DAVIES & Y. SAFAROV (eds)
274 The Mandelbrot set, theme and variations, T. LEI (ed)
275 Descriptive set theory and dynamical systems, M. FOREMAN, A.S. KECHRIS, A. LOUVEAU & B. WEISS (eds)
276 Singularities of plane curves, E. CASAS-ALVERO
277 Computational and geometric aspects of modern algebra, M. ATKINSON *et al* (eds)
278 Global attractors in abstract parabolic problems, J.W. CHOLEWA & T. DLOTKO
279 Topics in symbolic dynamics and applications, F. BLANCHARD, A. MAASS & A. NOGUEIRA (eds)
280 Characters and automorphism groups of compact Riemann surfaces, T. BREUER
281 Explicit birational geometry of 3-folds, A. CORTI & M. REID (eds)
282 Auslander–Buchweitz approximations of equivariant modules, M. HASHIMOTO
283 Nonlinear elasticity, Y.B. FU & R.W. OGDEN (eds)
284 Foundations of computational mathematics, R. DEVORE, A. ISERLES & E. SÜLI (eds)
285 Rational points on curves over finite fields, H. NIEDERREITER & C. XING
286 Clifford algebras and spinors (2nd Edition), P. LOUNESTO
287 Topics on Riemann surfaces and Fuchsian groups, E. BUJALANCE, A.F. COSTA & E. MARTÍNEZ (eds)
288 Surveys in combinatorics, 2001, J.W.P. HIRSCHFELD (ed)

289 Aspects of Sobolev-type inequalities, L. SALOFF-COSTE
290 Quantum groups and Lie theory, A. PRESSLEY (ed)
291 Tits buildings and the model theory of groups, K. TENT (ed)
292 A quantum groups primer, S. MAJID
293 Second order partial differential equations in Hilbert spaces, G. DA PRATO & J. ZABCZYK
294 Introduction to operator space theory, G. PISIER
295 Geometry and integrability, L. MASON & Y. NUTKU (eds)
296 Lectures on invariant theory, I. DOLGACHEV
297 The homotopy category of simply connected 4-manifolds, H.-J. BAUES
298 Higher operands, higher categories, T. LEINSTER (ed)
299 Kleinian groups and hyperbolic 3-manifolds, Y. KOMORI, V. MARKOVIC & C. SERIES (eds)
300 Introduction to Möbius differential geometry, U. HERTRICH-JEROMIN
301 Stable modules and the D(2)-problem, F.E.A. JOHNSON
302 Discrete and continuous nonlinear Schrödinger systems, M.J. ABLOWITZ, B. PRINARI & A.D. TRUBATCH
303 Number theory and algebraic geometry, M. REID & A. SKOROBOGATOV (eds)
304 Groups St Andrews 2001 in Oxford I, C.M. CAMPBELL, E.F. ROBERTSON & G.C. SMITH (eds)
305 Groups St Andrews 2001 in Oxford II, C.M. CAMPBELL, E.F. ROBERTSON & G.C. SMITH (eds)
306 Geometric mechanics and symmetry, J. MONTALDI & T. RATIU (eds)
307 Surveys in combinatorics 2003, C.D. WENSLEY (ed.)
308 Topology, geometry and quantum field theory, U.L. TILLMANN (ed)
309 Corings and comodules, T. BRZEZINSKI & R. WISBAUER
310 Topics in dynamics and ergodic theory, S. BEZUGLYI & S. KOLYADA (eds)
311 Groups: topological, combinatorial and arithmetic aspects, T.W. MÜLLER (ed)
312 Foundations of computational mathematics, Minneapolis 2002, F. CUCKER *et al* (eds)
313 Transcendental aspects of algebraic cycles, S. MÜLLER-STACH & C. PETERS (eds)
314 Spectral generalizations of line graphs, D. CVETKOVIĆ, P. ROWLINSON & S. SIMIĆ
315 Structured ring spectra, A. BAKER & B. RICHTER (eds)
316 Linear logic in computer science, T. EHRHARD, P. RUET, J.-Y. GIRARD & P. SCOTT (eds)
317 Advances in elliptic curve cryptography, I.F. BLAKE, G. SEROUSSI & N.P. SMART (eds)
318 Perturbation of the boundary in boundary-value problems of partial differential equations, D. HENRY
319 Double affine Hecke algebras, I. CHEREDNIK
320 L-functions and Galois representations, D. BURNS, K. BUZZARD & J. NEKOVÁŘ (eds)
321 Surveys in modern mathematics, V. PRASOLOV & Y. ILYASHENKO (eds)
322 Recent perspectives in random matrix theory and number theory, F. MEZZADRI & N.C. SNAITH (eds)
323 Poisson geometry, deformation quantisation and group representations, S. GUTT *et al* (eds)
324 Singularities and computer algebra, C. LOSSEN & G. PFISTER (eds)
325 Lectures on the ricci flow, P. TOPPING
326 Modular representations of finite groups of Lie type, J.E. HUMPHREYS
327 Surveys in combinatorics 2005, B.S. WEBB (ed)
328 Fundamentals of hyperbolic manifolds, R. CANARY, D. EPSTEIN & A. MARDEN (eds)
329 Spaces of Kleinian groups, Y. MINSKY, M. SAKUMA & C. SERIES (eds)
330 Noncommutative localization in algebra and topology, A. RANICKI (ed)
331 Foundations of computational mathematics, Santander 2005, L.M. PARDO, A. PINKUS, E. SÜLI & M.J. TODD (eds)
332 Handbook of tilting theory, L. ANGELERI HÜGEL, D. HAPPEL & H. KRAUSE (eds)
333 Synthetic differential geometry (2nd Edition), A. KOCK
334 The Navier–Stokes equations, N. RILEY & P. DRAZIN
335 Lectures on the combinatorics of free probability, A. NICA & R. SPEICHER
336 Integral closure of ideals, rings, and modules, I. SWANSON & C. HUNEKE
337 Methods in Banach space theory, J.M.F. CASTILLO & W.B. JOHNSON (eds)
338 Surveys in geometry and number theory, N. YOUNG (ed)
339 Groups St Andrews 2005 I, C.M. CAMPBELL, M.R. QUICK, E.F. ROBERTSON & G.C. SMITH (eds)
340 Groups St Andrews 2005 II, C.M. CAMPBELL, M.R. QUICK, E.F. ROBERTSON & G.C. SMITH (eds)
341 Ranks of elliptic curves and random matrix theory, J.B. CONREY, D.W. FARMER, F. MEZZADRI & N.C. SNAITH (eds)
342 Elliptic cohomology, H.R. MILLER & D.C. RAVENEL (eds)
343 Algebraic cycles and motives I, J. NAGEL & C. PETERS (eds)
344 Algebraic cycles and motives II, J. NAGEL & C. PETERS (eds)
345 Algebraic and analytic geometry, A. NEEMAN
346 Surveys in combinatorics, 2007, A. HILTON & J. TALBOT (eds)
347 Surveys in contemporary mathematics, N. YOUNG & Y. CHOI (eds)
348 Transcendental dynamics and complex analysis, P.J. RIPPON & G.M. STALLARD (eds)
349 Model theory with applications to algebra and analysis I, Z. CHATZIDAKIS, D. MACPHERSON, A. PILLAY & A. WILKIE (eds)
350 Model theory with applications to algebra and analysis II, Z. CHATZIDAKIS, D. MACPHERSON, A. PILLAY & A. WILKIE (eds)
351 Finite von Neumann algebras and masas, A.M. SINCLAIR & R.R. SMITH
352 Number theory and polynomials, J. MCKEE & C. SMYTH (eds)
353 Trends in stochastic analysis, J. BLATH, P. MÖRTERS & M. SCHEUTZOW (eds)
354 Groups and analysis, K. TENT (ed)
355 Non-equilibrium statistical mechanics and turbulence, J. CARDY, G. FALKOVICH & K. GAWEDZKI
356 Elliptic curves and big Galois representations, D. DELBOURGO
357 Algebraic theory of differential equations, M.A.H. MACCALLUM & A.V. MIKHAILOV (eds)
359 Moduli spaces and vector bundles, L. BRAMBILA-PAZ, S.B. BRADLOW, O. GARCÍA-PRADA & S. RAMANAN (eds)
361 Words: Notes on verbal width in groups, D. SEGAL
363 Foundations of computational mathematics, Hong Kong 2008, F. CUCKER, A. PINKUS & M.J. TODD (eds)
364 Partial differential equations and fluid mechanics, J.C. ROBINSON & J.L. RODRIGO (eds)
365 Surveys in combinatorics 2009, S. HUCZYNSKA, J.D. MITCHELL & C.M. RONEY-DOUGAL (eds)
366 Highly oscillatory problems, B. ENGQUIST, A. FOKAS, E. HAIRER & A. ISERLES (eds)

London Mathematical Society Lecture Note Series: 362

Differential Tensor Algebras and their Module Categories

R. BAUTISTA, L. SALMERÓN AND R. ZUAZUA
Universidad Nacional Autónoma de México

CAMBRIDGE
UNIVERSITY PRESS

CAMBRIDGE
UNIVERSITY PRESS

University Printing House, Cambridge CB2 8BS, United Kingdom

One Liberty Plaza, 20th Floor, New York, NY 10006, USA

477 Williamstown Road, Port Melbourne, VIC 3207, Australia

314-321, 3rd Floor, Plot 3, Splendor Forum, Jasola District Centre, New Delhi - 110025, India

103 Penang Road, #05-06/07, Visioncrest Commercial, Singapore 238467

Cambridge University Press is part of the University of Cambridge.

It furthers the University's mission by disseminating knowledge in the pursuit of education, learning and research at the highest international levels of excellence.

www.cambridge.org
Information on this title: www.cambridge.org/9780521757683

First published 2009

A catalogue record for this publication is available from the British Library

Library of Congress Cataloging in Publication data
Bautista, R., 1943–
Differential tensor algebras and their module categories / R. Bautista, L. Salmerón, and R. Zuazua.
p. cm. – (London Mathematical Society lecture note series ; 359)
Includes bibliographical references and index.
ISBN 978-0-521-75768-3 (pbk.)
1. Tensor algebra. 2. Representations of algebras. 3. Categories (Mathematics) I. Salmerón, L. II. Zuazua, R. III. Title. IV. Series.
QA200.B38 2009
512′.57 – dc22 2009014316

ISBN 978-0-521-75768-3 Paperback

Contents

Preface

This monograph is concerned with the notions of ditalgebras (an acronym for "differential tensor algebras") and the study of their categories of modules. It involves reduction techniques which have proved to be very useful in the development of the theory of representation of finite-dimensional algebras. Our aim has been to present in a systematic, elementary and self-contained as possible way some of the main results obtained with these methods. They were originally introduced by the Kiev School in representation theory of algebras, in an attempt to formalize and generalize matrix problems methods.

The presentation given here has many common features with the original one of A. V. Roiter and M. Kleiner [46], in terms of differential graded categories, as well as with the formulation given by Y. Drozd [28] (and further developed by W. Crawley-Boevey [19] and [20]), in terms of bocses. It is clear that some applications of these techniques, notably in the study of coverings in representation theory of algebras, will require the categorical formulation of the theory, as suggested in [30]. However, for the sake of simplicity, we preferred to work here in the more concrete ring theoretical context of ditalgebras. We assume from the reader some familiarity with the basics of representation theory of algebras and homological algebra (including the basics of the theory of additive categories with exact structures), which can be obtained from the first sections of [29], [47] and [3] (respectively, [32] and [27]).

In the representation theory of finite-dimensional algebras, the notions of finite, tame and wild representation type play a central role. An algebra is of finite representation type if it has finitely many pairwise non-isomorphic indecomposable modules. It is of wild representation type, or simply wild, if it contains the problem of finding a normal form for pairs of square matrices over a field under simultaneous conjugation by a non-singular matrix. Finally, it is of tame representation type, or simply tame, if the pairwise non-isomorphic indecomposable modules in each dimension can be described by a finite number of one-parameter families. For precise definitions, see Sections 22 and 27. This monograph includes a fresh point of view of well-known facts on tame and wild ditalgebras, on tame and wild algebras, and on their modules. But there are also some new results and some new proofs.

We will review, for instance, Drozd's Tame and Wild Theorem, stating that a finite-dimensional algebra over an algebraically closed field is either tame or wild, but not both. We review also Crawley's Theorem on the existence of generic modules for tame finite-dimensional algebras over an algebraically closed field, and his Structure Theorem for its Auslander–Reiten quiver.

Our approach presents a formal alternative to the use of bocses with underlying additive categories and pull-back reduction constructions. This is replaced by the use of some special dual basis and what we call "reduction by an admissible module". This approach permits to perform explicit calculations with a reasonable effort. As an illustration of this, Section 24 includes a more conceptual proof of the fact that critical bocses are wild than the original proof of Drozd (see [19]) or than some of its subsequent simplifications (see [49]), where an explicit bimodule which produces wildness is exhibited.

The presentation given here of the reduction by an admissible module is more general than the one given in [6]. We believe that this approach has some promising potential since it provides a systematic treatment of a wide variety of reductions.

Let us comment on one interesting new result proved in Section 31. Let K denote a field extension of our ground field k. As usual, if Λ is some k-algebra, Λ-Mod denotes the category of left Λ-modules. We consider the induced K-algebra $\Lambda^K = \Lambda \otimes_k K$. Recall that the *endolength* of a Λ-module M is by definition the length of the right $\mathrm{End}_\Lambda(M)^{op}$-module M. A *generic* Λ-module is an indecomposable Λ-module with finite endolength and not finitely generated over Λ.

We will prove that if Λ is a finite-dimensional algebra over an algebraically closed field k and the induced algebra Λ^K is not wild, then every generic Λ^K-module is *rationally induced* from a generic Λ-module. More precisely, any generic Λ^K-module is of the form $G \otimes_{k(t)} K(t)$, where G is some generic Λ-module equipped with a natural structure of Λ-$k(t)$-bimodule. This is related to the study in [37], where it is shown that the extension of a field to its algebraic closure preserves generic tameness.

Now assume that k is algebraically closed and let $K = k(t)$, the rational function field of k. It has been proved in [11] that Λ^K is of finite representation type if and only if every indecomposable Λ^K-module is induced from a Λ-module. In Section 31, we show that Λ^K is not wild if and only if every generic Λ^K-module is rationally induced from a generic Λ-module. Our proof is derived from the compatibility of the scalar extension with reduction operations and a "scalar extended version" of Crawley's article [20] on the existence and description of generic modules for tame algebras Λ over an algebraically closed field k.

We have included a series of exercises in order to illustrate and enrich the content of these notes. As usual, some of them contain parts of various research works. We have added reference paragraphs at the end of some sections, where we tried to provide fair recognition of previous work on the subject from which these notes developed.

R. Bautista, L. Salmerón and R. Zuazua

1

t-algebras and differentials

From now on k denotes a fixed ground field. Whenever we consider a k-algebra or a bimodule, we always assume that the field k acts centrally on them. We start with some basic notions and notation, and some elementary remarks.

Definition 1.1. *We say that the k-algebra T is freely generated by the pair (A, V) if A is a k-subalgebra of T, V is a A-A-subbimodule of T, and the following universal property is satisfied: for any k-algebra B, any morphism of k-algebras $A \xrightarrow{\phi_0} B$ and any morphism of A-A-bimodules $V \xrightarrow{\phi_1} B$, where the structure of A-A-bimodule of B is obtained by restriction through ϕ_0, there exists a unique morphism of k-algebras $T \xrightarrow{\phi} B$, which extends both ϕ_0 and ϕ_1.*

Definition 1.2. *Consider a k-algebra A and any A-A-bimodule V. For $i \geq 2$, we write $V^{\otimes i}$ for the tensor product $V \otimes_A V \otimes_A \cdots \otimes_A V$ of i copies of V, and set $V^{\otimes 0} = A$ and $V^{\otimes 1} = V$. The vector space $T_A(V) = \bigoplus_{i=0}^{\infty} V^{\otimes i}$ admits a natural structure of k-algebra with product determined by the canonical isomorphism $V^{\otimes i} \otimes_A V^{\otimes j} \longrightarrow V^{\otimes(i+j)}$. The algebra $T_A(V)$ is called the tensor algebra of V over A.*

Lemma 1.3. *Consider a k-algebra A and any A-A-bimodule V. Then:*

(1) The tensor algebra $T_A(V)$ is freely generated by (A, V).
(2) If the algebra T is freely generated by (A, V), the morphism $T_A(V) \longrightarrow T$ determined by the inclusions of A and V in T is an isomorphism.
(3) Assume that T is an algebra freely generated by (A, V), and that $V = V' \bigoplus V''$ is a bimodule decomposition of the A-A-bimodule V. Then, the subalgebra A' of T generated by $A \cup V'$ is freely generated by (A, V') and T is freely generated by $(A', A'V''A')$.

Proof. (1) and (2) are easy to show. We show (3): let $\pi_1 : V \longrightarrow V'$ be the projection and consider the algebra morphism $T \xrightarrow{\pi} T_A(V')$ determined by the inclusion of A in $T_A(V')$ and π_1. Consider also the morphism $T_A(V') \xrightarrow{\sigma} T$ determined by the inclusion of A and V' in T. Then, clearly, $\pi\sigma$ is the identity map, and the restriction of σ to its image provides the isomorphism $T_A(V') \cong A'$. Now, consider the morphism $T_{A'}(A'V''A') \xrightarrow{\phi} T$ determined by the inclusions of A' and $A'V''A'$ in T. Consider also the morphism of A-A-bimodules $V \xrightarrow{\psi_1} T_{A'}(A'V''A')$, which maps each $v' \in V'$ onto $v' \in A'$, and each $v'' \in V''$ onto $v'' \in A'V''A'$. Then, the morphism $T \xrightarrow{\psi} T_{A'}(A'V''A')$ determined by the inclusion of A in A' and ψ_1 is an inverse for ϕ. □

Definition 1.4. *We say that T is a graded k-algebra if T is a k-algebra which admits a vector space decomposition $T = \oplus_{i \geq 0}[T]_i$ such that $[T]_i[T]_j \subseteq [T]_{i+j}$, for all i, j. Thus, $[T]_0$ is a subalgebra of T and each $[T]_i$ is a $[T]_0$-subbimodule of T. The elements $a \in [T]_i$ are called homogeneous of degree i, and we write* $\deg(a) = i$.

We say that T is a t-algebra if T is a graded k-algebra and T is freely generated by the pair $([T]_0, [T]_1)$.

Remark 1.5. *Consider a k-algebra A and any A-A-bimodule V:*

(1) The tensor algebra $T_A(V)$ with its standard grading given by $[T_A(V)]_i = V^{\otimes i}$ is a graded algebra.
(2) If T is a t-algebra, and we make $A = [T]_0$ and $V = [T]_1$, then there is an isomorphism of graded k-algebras $T_A(V) \cong T$ if we consider the standard grading on $T_A(V)$. In particular, for each n, the product of n elements induces an isomorphism

$$[T]_1 \otimes_{[T]_0} [T]_1 \otimes_{[T]_0} \cdots \otimes_{[T]_0} [T]_1 \xrightarrow{\cong} [T]_n.$$

We often identify both bimodules.

Definition 1.6. *Assume T is a graded k-algebra. Then we say that δ is a differential on T if $\delta : T \to T$ is a linear transformation such that $\delta([T]_i) \subseteq [T]_{i+1}$, for all i, and δ satisfies Leibniz rule: $\delta(ab) = \delta(a)b + (-1)^{\deg(a)} a\delta(b)$, for all homogeneous elements $a, b \in T$.*

Remark 1.7. *If T is a graded k-algebra and δ is a differential on T, then:*

(1) By induction, we obtain the following formula, for any homogeneous elements $t_1, t_2, \ldots, t_n \in T$

$$\delta(t_1 t_2 \cdots t_n) = \sum_{i=1}^{n} (-1)^{(\sum_{j=1}^{i-1} \deg(t_j))} t_1 t_2 \cdots t_{i-1} \delta(t_i) t_{i+1} t_{i+2} \cdots t_n.$$

(2) *The linear map* $\delta^2 : T \longrightarrow T$ *satisfies* $\delta^2(ab) = \delta^2(a)b + a\delta^2(b)$, *for any homogeneous elements* $a, b \in T$. *Again, by induction, we obtain the following formula, for any homogeneous elements* $t_1, \ldots, t_n \in T$

$$\delta^2(t_1 t_2 \cdots t_n) = \sum_{i=1}^{n} t_1 t_2 \cdots t_{i-1} \delta^2(t_i) t_{i+1} t_{i+2} \cdots t_n.$$

(3) *From (1), (2) and (1.5), we obtain that if* T *is a t-algebra, the differential* δ *and its square* δ^2 *are determined by their values on* $A = [T]_0$ *and on* $V = [T]_1$. *In particular, we can also derive that* $\delta^2(A) = 0$ *and* $\delta^2(V) = 0$ *imply* $\delta^2 = 0$.

Lemma 1.8. *Let* T *be a t-algebra. Denote* $A = [T]_0$ *and* $V = [T]_1$. *Assume we have a pair of linear maps* $\delta_0 : A \longrightarrow [T]_1$ *and* $\delta_1 : V \longrightarrow [T]_2$ *such that* $\delta_0(ab) = \delta_0(a)b + a\delta_0(b)$, $\delta_1(av) = \delta_0(a)v + a\delta_1(v)$ *and* $\delta_1(va) = \delta_1(v)a - v\delta_0(a)$, *for* $a, b \in A$ *and* $v \in V$. *Then, these maps extend uniquely to a differential* $\delta : T \longrightarrow T$.

Proof. Since T is a t-algebra, freely generated by (A, V), we may assume that $T = T_A(V)$, with its standard grading. We shall define a linear map δ_n from each of the direct summands $V^{\otimes n}$ of T to T. We use the same symbols δ_0 and δ_1 to denote their compositions with the inclusions to T. Then, for $n \geq 2$, define δ_n by the formula

$$\delta_n(v_1 \otimes \cdots \otimes v_n) = \sum_{i=1}^{n} (-1)^{(i-1)} v_1 v_2 \cdots v_{i-1} \delta_1(v_i) v_{i+1} v_{i+2} \cdots v_n,$$

where each $v_i \in V$. See Remark (1.7). This formula yields a well-defined linear map $\delta_n : V^{\otimes n} \longrightarrow T$, because $\delta_1(av_i) = \delta_0(a)v_i + a\delta_1(v_i)$ and $\delta_1(v_{i-1}a) = \delta_1(v_{i-1})a - v_{i-1}\delta_0(a)$, for any $a \in A$. Then, there is a linear map $\delta : T \longrightarrow T$ which extends all these maps δ_n. It is clear that $\delta([T]_i) \subseteq [T]_{i+1}$. It remains to show that δ satisfies Leibniz rule. By assumption, δ already satisfies Leibniz rule for products of the form ab, av and va, with $a, b \in A$ and $v \in V$. From this and the definition of δ, it follows that δ satisfies Leibniz rule for products of the form aw and wa, with $a \in A$ and $w \in V^{\otimes n}$. To finish our proof, it is enough to show that given $u_n = \otimes_{s=1}^{n} v_s$ and $w_m = \otimes_{r=1}^{m} v'_r$, with $v_s, v'_r \in V$, then $\delta(u_n \otimes w_m) = \delta(u_n) \otimes w_m + (-1)^{\deg(u_n)} u_n \otimes \delta(w_m)$. This is a straightforward calculation using the definition of δ. □

2

Ditalgebras and modules

In this section, we introduce the basic objects studied in these notes. Namely, ditalgebras and their categories of modules. Its study constitutes a natural generalization of the theory of algebras and their categories of modules. At the same time, it has proved to be a useful tool in establishing some deep results in representation theory of algebras.

Definition 2.1. *A differential t-algebra or ditalgebra $\mathcal{A}$ is by definition a pair $\mathcal{A} = (T, \delta)$, where T is a t-algebra and δ is a differential on T satisfying $\delta^2 = 0$.*

A morphism of ditalgebras $\phi : (T, \delta) \to (T', \delta')$ is a morphism of k-algebras $\phi : T \to T'$, satisfying $\phi([T]_i) \subseteq [T']_i$, for all i, and $\delta'\phi = \phi\delta$.

Clearly, we can consider the category of ditalgebras over k, where the morphisms are composed as maps.

Definition 2.2. *The category of modules (or representations) of the ditalgebra $\mathcal{A} = (T, \delta)$, denoted by $\mathcal{A}$-Mod, is defined as follows. Denote by $A = A_{\mathcal{A}} = [T]_0$, a k-subalgebra of T, and by $V = V_{\mathcal{A}} = [T]_1$, an A-A-subbimodule of T. The objects of $\mathcal{A}$-Mod are all the A-modules. Given $M, N \in \mathcal{A}$-Mod, a morphism $f : M \to N$ in $\mathcal{A}$-Mod is a pair $f = (f^0, f^1)$, with $f^0 \in \mathrm{Hom}_k(M, N)$ and $f^1 \in \mathrm{Hom}_{A\text{-}A}(V, \mathrm{Hom}_k(M, N))$ satisfying that*

$$af^0(m) = f^0(am) + f^1(\delta(a))(m),$$

for any $a \in A$ and $m \in M$. The Hom-space in this category is denoted by $\mathrm{Hom}_{\mathcal{A}}(M, N)$. Given $f \in \mathrm{Hom}_{\mathcal{A}}(M, N)$ and $g \in \mathrm{Hom}_{\mathcal{A}}(N, L)$ in $\mathcal{A}$-Mod, consider the composition of morphisms of A-A-bimodules

$$V \otimes_A V \xrightarrow{g^1 \otimes f^1} \mathrm{Hom}_k(N, L) \otimes_A \mathrm{Hom}_k(M, N) \xrightarrow{\pi} \mathrm{Hom}_k(M, L),$$

where the last morphism is induced by composition. Since T is a t-algebra, we can identify $V \otimes_A V$ with $[T]_2$. Then the composition gf is defined, for any

$v \in V$, by the following

$$
\begin{aligned}
(gf)^0 &= g^0 f^0;\\
(gf)^1(v) &= g^0 f^1(v) + g^1(v) f^0 + \pi(g^1 \otimes f^1)(\delta(v)).
\end{aligned}
$$

The full subcategory of $\mathcal{A}$-Mod consisting of all finite-dimensional objects will be denoted by $\mathcal{A}$-mod.

Proposition 2.3. *Given a ditalgebra $\mathcal{A}$, the definition above indeed gives rise to a k-category $\mathcal{A}$-Mod.*

Proof. First we see that $gf = ((gf)^0, (gf)^1)$ is indeed a morphism. Clearly, $(gf)^0 \in \mathrm{Hom}_k(M, L)$. Let us verify that $(gf)^1 \in \mathrm{Hom}_{A\text{-}A}(V, \mathrm{Hom}_k(M, L))$. For this, take $v \in V$, $a \in A$ and $m \in M$, then

$$
\begin{aligned}
[(gf)^1(av)](m) &= [g^0 f^1(av)](m) + [g^1(av) f^0](m)\\
&\quad + [\pi(g^1 \otimes f^1)(\delta(av))](m)\\
&= ag^0[f^1(v)(m)] - g^1(\delta(a))[f^1(v)(m)] + [ag^1(v) f^0](m)\\
&\quad + [\pi(g^1 \otimes f^1)(\delta(a)v + a\delta(v))](m)\\
&= ag^0[f^1(v)(m)] - g^1(\delta(a))[f^1(v)(m)] + [ag^1(v) f^0](m)\\
&\quad + [g^1(\delta(a)) f^1(v)](m) + a[\pi(g^1 \otimes f^1)(\delta(v))](m)\\
&= ag^0[f^1(v)(m)] + [ag^1(v) f^0](m)\\
&\quad + a[\pi(g^1 \otimes f^1)(\delta(v))](m)\\
&= a[(gf)^1(v)](m).
\end{aligned}
$$

Now, take $a \in A$, $v \in V$ and $m \in M$, then

$$
\begin{aligned}
[(gf)^1(va)](m) &= [g^0 f^1(va)](m) + [g^1(va) f^0](m)\\
&\quad + [\pi(g^1 \otimes f^1)(\delta(va))](m)\\
&= [g^0(f^1(v)a)](m) + (g^1(v)a)[f^0(m)]\\
&\quad + [\pi(g^1 \otimes f^1)(\delta(v)a - v\delta(a))](m)\\
&= g^0[f^1(v)(am)] + (g^1(v))[af^0(m)]\\
&\quad + [\pi(g^1 \otimes f^1)(\delta(v)a - v\delta(a))](m)\\
&= g^0[f^1(v)(am)] + (g^1(v))[f^0(am) + f^1(\delta(a))(m)]\\
&\quad + [\pi(g^1 \otimes f^1)(\delta(v)a)](m) - [g^1(v) f^1(\delta(a))](m)\\
&= g^0[f^1(v)(am)] + (g^1(v))[f^0(am)]\\
&\quad + [\pi(g^1 \otimes f^1)(\delta(v)a)](m)\\
&= [g^0 f^1(v)](am) + [g^1(v) f^0](am)\\
&\quad + [\pi(g^1 \otimes f^1)(\delta(v))](am)\\
&= [(gf)^1(v)](am)\\
&= [(gf)^1(v)a](m).
\end{aligned}
$$

Finally, $gf \in \mathrm{Hom}_{\mathcal{A}}(M, L)$, because, for $a \in A$ and $m \in M$, we have

$$\begin{aligned}
[(gf)^0](am) &= [g^0 f^0](am) \\
&= g^0[af^0(m) - f^1(\delta(a))(m)] \\
&= g^0[af^0(m)] - g^0[f^1(\delta(a))](m) \\
&= ag^0[f^0(m)] - g^1(\delta(a))[f^0(m)] - g^0[f^1(\delta(a))](m) \\
&= [a(g^0 f^0)](m) \\
&\quad - [g^0 f^1(\delta(a)) + g^1(\delta(a)) f^0 + \pi(g^1 \otimes f^1)(\delta^2(a))][m] \\
&= [a(g^0 f^0)](m) - (gf)^1(\delta(a))[m] \\
&= [a(gf)^0](m) - (gf)^1(\delta(a))[m].
\end{aligned}$$

Clearly, for each $M \in \mathcal{A}$-Mod, the morphism $I_M = (I_M, 0)$ plays the role of an identity. Now we show that the composition is associative (it is clearly bilinear). Consider the morphisms $M \xrightarrow{f} N$, $N \xrightarrow{g} L$ and $L \xrightarrow{h} K$ in $\mathcal{A}$-Mod.

It is clear that $[h(gf)]^0 = [(hg)f]^0$. In order to show that $[h(gf)]^1 = [(hg)f]^1$, having in mind our identification $V \otimes_A V = [T]_2$, consider $v \in V$ and let $\delta(v) = \sum_a u_a \otimes w_a$, with $u_a, w_a \in V$. Assume that, for each a

$$\delta(u_a) = \sum_b u^1_{ab} \otimes u^2_{ab} \quad \text{and} \quad \delta(w_a) = \sum_c w^1_{ac} \otimes w^2_{ac}.$$

Then

$$\begin{aligned}
[h(gf)]^1(v) &= h^0(gf)^1(v) + h^1(v)(gf)^0 + \textstyle\sum_a h^1(u_a)(gf)^1(w_a) \\
&= h^0[g^0 f^1(v) + g^1(v) f^0 + \textstyle\sum_a g^1(u_a) f^1(w_a)] + h^1(v) g^0 f^0 \\
&\quad + \textstyle\sum_a h^1(u_a)[g^0 f^1(w_a) + g^1(w_a) f^0 + \sum_c g^1(w^1_{ac}) f^1(w^2_{ac})],
\end{aligned}$$

and

$$\begin{aligned}
[(hg)f]^1(v) &= (hg)^0 f^1(v) + (hg)^1(v) f^0 + \textstyle\sum_a (hg)^1(u_a) f^1(w_a) \\
&= h^0 g^0 f^1(v) + [h^0 g^1(v) + h^1(v) g^0 + \textstyle\sum_a h^1(u_a) g^1(w_a)] f^0 \\
&\quad + \textstyle\sum_a [h^0 g^1(u_a) + h^1(u_a) g^0 + \sum_b h^1(u^1_{ab}) g^1(u^2_{ab})] f^1(w_a).
\end{aligned}$$

Then, we have to show that

$$\sum_{a,c} h^1(u_a) g^1(w^1_{ac}) f^1(w^2_{ac}) = \sum_{a,b} h^1(u^1_{ab}) g^1(u^2_{ab}) f^1(w_a).$$

We have

$$\sum_a \delta(u_a) w_a + (-1)^{\deg(u_a)} u_a \delta(w_a) = \delta\left(\sum_a u_a \otimes w_a\right) = \delta^2(v) = 0,$$

which implies the following equality in $V \otimes_A V \otimes_A V = [T]_3$ (we use again that T is a t-algebra)

$$\sum_{a,c} u_a \otimes w^1_{ac} \otimes w^2_{ac} = \sum_{a,b} u^1_{ab} \otimes u^2_{ab} \otimes w_a.$$

Then, the following composition applied to the last equality gives the desired result

$$V \otimes_A V \otimes_A V \xrightarrow{h^1 \otimes g^1 \otimes f^1} \mathrm{Hom}_k(L,K) \otimes_A \mathrm{Hom}_k(N,L) \otimes_A \mathrm{Hom}_k(M,N)$$
$$\xrightarrow{\pi} \mathrm{Hom}_k(M,K).$$

□

Lemma 2.4. *Any morphism of ditalgebras $\phi : \mathcal{A} \longrightarrow \mathcal{A}'$ induces, by restriction, a functor $F_\phi : \mathcal{A}'$-Mod$\longrightarrow \mathcal{A}$-Mod. To give the explicit formula on morphisms, denote by $A = A_\mathcal{A}$ and $V = V_\mathcal{A}$, and with A' and V' the corresponding objects for the ditalgebra $\mathcal{A}'$; consider also the morphisms $\phi_0 : A \rightarrow A'$ and $\phi_1 : V \rightarrow V'$, induced by ϕ. Then, if $M \in \mathcal{A}'$-Mod, $F_\phi(M)$ is the A-module obtained from M by restriction of scalars through ϕ_0. The receipt on morphisms is given, for any $(f^0, f^1) \in \mathrm{Hom}_{\mathcal{A}'}(M,N)$, by $F_\phi(f^0, f^1) = (f^0, f^1\phi_1)$.*

If ϕ is surjective, then F_ϕ is faithful and injective on objects. Moreover, if $\phi' : \mathcal{A}' \longrightarrow \mathcal{A}''$ is another morphism of ditalgebras, then $F_{\phi'\phi} = F_\phi F_{\phi'}$.

Proof. We first show that $F_\phi(f^0, f^1) \in \mathrm{Hom}_\mathcal{A}(F_\phi(M), F_\phi(N))$, whenever we have $(f^0, f^1) \in \mathrm{Hom}_{\mathcal{A}'}(M,N)$. For $m \in M$ and $a \in A$, we have

$$\begin{aligned} F_\phi(f)^0(am) &= f^0(am) \\ &= f^0[\phi_0(a)m] \\ &= \phi_0(a)f^0(m) - f^1(\delta'(\phi_0(a)))[m] \\ &= af^0(m) - f^1\phi_1(\delta(a))[m] \\ &= aF_\phi(f)^0[m] - F_\phi(f)^1(\delta(a))[m]. \end{aligned}$$

In order to show that F_ϕ preserves the composition, take $f \in \mathrm{Hom}_{\mathcal{A}'}(M,N)$ and $g \in \mathrm{Hom}_{\mathcal{A}'}(N,L)$. Therefore, $[F_\phi(gf)]^0 = (gf)^0 = g^0 f^0 = [F_\phi(g)F_\phi(f)]^0$. Moreover, for $v \in V$ with $\delta(v) = \sum_i v_i^1 \otimes v_i^2$, we have $\delta'(\phi(v)) = \phi\delta(v) = \sum_i \phi(v_i^1) \otimes \phi(v_i^2)$. Therefore

$$\begin{aligned} \left[F_\phi(g)F_\phi(f)\right]^1(v) &= F_\phi(g)^0 F_\phi(f)^1(v) + F_\phi(g)^1(v)F_\phi(f)^0 \\ &\quad + \pi(F_\phi(g)^1 \otimes F_\phi(f)^1)(\delta(v)) \\ &= g^0 f^1(\phi(v)) + g^1(\phi(v))f^0 + \textstyle\sum_i g^1(\phi(v_i^1))f^1(\phi(v_i^2)) \\ &= g^0 f^1(\phi(v)) + g^1(\phi(v))f^0 + \pi(g^1 \otimes f^1)(\delta'(\phi(v))) \\ &= (gf)^1(\phi(v)) \\ &= \left[F_\phi(gf)\right]^1(v). \end{aligned}$$

We have seen that $F_\phi(gf) = F_\phi(g)F_\phi(f)$. Clearly, F_ϕ preserves identities. □

Remark 2.5. *Whenever $\mathcal{A}$ is a ditalgebra, there is a canonical embedding*

$$L = L_\mathcal{A} : A\text{-Mod} \longrightarrow \mathcal{A}\text{-Mod},$$

which is the identity on objects, and satisfies $L(f^0) = (f^0, 0)$, *whenever* $f^0 \in \mathrm{Hom}_A(M, N)$. *As a consequence,* $\mathcal{A}$-Mod *is an additive category: Given* M_1, M_2 *in* $\mathcal{A}$-Mod, *their direct sum in* $\mathcal{A}$-Mod *is* $\{M_i \xrightarrow{(\sigma_i,0)} M_1 \oplus M_2 \xrightarrow{(\pi_i,0)} M_i\}_{i=1,2}$, *where* $\{M_i \xrightarrow{\sigma_i} M_1 \oplus M_2 \xrightarrow{\pi_i} M_i\}_{i=1,2}$ *is the direct sum of* M_1 *and* M_2 *in* A-Mod.

As in any additive category, each morphism $f \in \mathrm{Hom}_{\mathcal{A}}(M_1 \oplus M_2, N_1 \oplus N_2)$ *can be written as a matrix* $f = \begin{pmatrix} f_{11} & f_{12} \\ f_{21} & f_{22} \end{pmatrix}$ *of morphisms* $f_{ji} = (\pi_j, 0) f (\sigma_i, 0) : M_i \longrightarrow N_j$. *In our case, the explicit description of* f_{ji} *is given by* $f^0_{ji} = \pi_j f^0 \sigma_i$, *and, for* $v \in V$, $f^1_{ji}(v) = \pi_j f^1(v)\sigma_i$. *And, as usual, if* $f \in \mathrm{Hom}_{\mathcal{A}}(M_1, N_1)$ *and* $g \in \mathrm{Hom}_{\mathcal{A}}(M_2, N_2)$, *with* $f \oplus g$, *we denote the direct sum morphism* $\begin{pmatrix} f & 0 \\ 0 & g \end{pmatrix} \in \mathrm{Hom}_{\mathcal{A}}(M_1 \oplus M_2, N_1 \oplus N_2)$.

Remark 2.6. *If* Λ *is a k-algebra, the associated regular ditalgebra is* $\mathcal{R}_\Lambda = (\Lambda, 0)$, *where* $[\Lambda]_0 = \Lambda$. *Clearly* $\mathcal{R}_\Lambda$-Mod $\cong \Lambda$-Mod. *Given a ditalgebra* $\mathcal{A} = (T, \delta)$, *the projection* $T \longrightarrow [T]_0 = A$ *determines a morphism of ditalgebras* $\pi : \mathcal{A} \longrightarrow \mathcal{R}_A$. *The induced functor* $F_\pi : A$-Mod$\longrightarrow \mathcal{A}$-Mod *is the canonical embedding* $L_{\mathcal{A}}$ *of (2.5).*

Definition 2.7. *Let* $\mathcal{A} = (T, \delta)$ *be any ditalgebra. Then, its opposite ditalgebra* $\mathcal{A}^{op} = (T^{op}, \delta^{op})$ *is constituted by the opposite algebra* T^{op} *of* T, *where* $v \cdot^{op} u = uv$, *with the same grading and* $\delta^{op}(t) = (-1)^{\deg(t)}\delta(t)$, *for any homogeneous* $t \in T$. *In particular,* $\delta(a) = \delta^{op}(a)$, *for any* $a \in A$, *and* $\delta^{op}(v) = -\sum_i v_i \cdot^{op} u_i$, *whenever* $\delta(v) = \sum_i u_i v_i$ *with* $v, v_i, u_i \in V$.

Proposition 2.8. *Let* $\mathcal{A}$ *be any ditalgebra. Then, the functor* $D = \mathrm{Hom}_A(-, k) :$ k-Mod$\longrightarrow k$-Mod *permits to define a contravariant k-functor*

$$D : \mathcal{A}\text{-Mod} \longrightarrow \mathcal{A}^{op}\text{-Mod}$$

where, given $f = (f^0, f^1) \in \mathrm{Hom}_{\mathcal{A}}(M, N)$, $D(f) := (D(f^0), -D(f^1))$, *with* $D(f^1)[v] = D(f^1[v]) \in \mathrm{Hom}_k(D(N), D(M))$, *for any* $v \in V$. *Moreover, if we restrict ourselves to finite-dimensional modules, we obtain a duality*

$$D : \mathcal{A}\text{-mod} \longrightarrow \mathcal{A}^{op}\text{-mod} \quad \textit{with} \quad D^2 \cong Id.$$

Proof. We first show that $D(f) \in \mathrm{Hom}_{\mathcal{A}^{op}}(D(N), D(M))$. It is clear that $D(f^0) \in \mathrm{Hom}_k(D(N), D(M))$. We have that $A_{\mathcal{A}^{op}} = [T^{op}]_0 = A^{op}$ and $V^{op} := V_{\mathcal{A}^{op}} = [T^{op}]_1$ has the same underlying vectorspace V, but its A^{op}-bimodule action is related to the A-bimodule action on V by the formula $a^{op} v b^{op} = bva$, where we agree to write a^{op} for the element a of A when considered as an element of A^{op}. If we consider left A^{op}-modules as right A-modules, then

$\mathrm{Hom}_{A^{op}\text{-}A^{op}}(V^{op}, \mathrm{Hom}_k(D(N), D(M))) = \mathrm{Hom}_{A\text{-}A}(V, \mathrm{Hom}_k(D(N), D(M)))$. We want to show that $D(f^1) \in \mathrm{Hom}_{A\text{-}A}(V, \mathrm{Hom}_k(D(N), D(M)))$. Let $v \in V$, $a \in A$, $\phi \in D(N)$ and $m \in M$, then

$$\begin{aligned}\left(D(f^1)[va]\right)(\phi)[m] &= \phi(f^1(va)[m]) \\ &= \phi((f^1(v)a)[m]) \\ &= \phi(f^1(v)[am]) \\ &= (\phi f^1(v))[am] \\ &= (\phi f^1(v))a[m] \\ &= \left[\left(D(f^1)[v]\right)(\phi)\right]a[m] \\ &= \left[\left(D(f^1)[v]\right)a\right](\phi)[m].\end{aligned}$$

Moreover

$$\begin{aligned}\left(D(f^1)[av]\right)(\phi)[m] &= \phi(f^1(av)[m]) \\ &= \phi((af^1(v))[m]) \\ &= \phi(af^1(v)[m]) \\ &= (\phi a)[f^1(v)[m]] \\ &= (\phi a)f^1(v)[m] \\ &= \left[\left(D(f^1)[v]\right)(\phi a)\right][m] \\ &= a\left[D(f^1)[v]\right](\phi)[m].\end{aligned}$$

Now we verify that $D(f)$ is a morphism, which means that, for $a \in A$ and $\phi \in D(N)$, $([D(f^0)](\phi))a = (D(f^0))(\phi a) - D(f^1)[\delta(a)](\phi)$. Take $m \in M$, then

$$\begin{aligned}\left[\left([D(f^0)](\phi)\right)a\right][m] &= \left(\phi f^0\right)a[m] \\ &= (\phi f^0)[am] \\ &= \phi(f^0[am]) \\ &= \phi\left(af^0[m] - f^1(\delta(a))[m]\right) \\ &= (\phi a)(f^0[m]) - \left(\phi f^1(\delta(a))\right)[m] \\ &= \left(D(f^0)\right)(\phi a)[m] - D(f^1)[\delta(a)](\phi)[m].\end{aligned}$$

It is clear that D preserves identities; now we check that D reverses composition. Take $g \in \mathrm{Hom}_{\mathcal{A}}(M, N)$ and $f \in \mathrm{Hom}_{\mathcal{A}}(N, L)$. We have to verify that each component of $D(fg)$ and $D(g)D(f)$ coincide. This is clear for the first one. Take $v \in V^{op}$, then

$$\begin{aligned}-D(fg)^1(v) &= -D\left[f^0g^1(v) + f^1(v)g^0 + \pi(f^1 \otimes g^1)\delta(v)\right] \\ &= -D(g^1(v))D(f^0) - D(g^0)D(f^1(v)) - D\left[\pi(f^1 \otimes g^1)\delta(v)\right].\end{aligned}$$

Moreover

$$\begin{aligned}[D(g)D(f)]^1(v) = &-D(g^0)D(f^1(v)) - D(g^1(v))D(f^0) \\ &+ \widehat{\pi}(D(g^1) \otimes D(f^1))\delta^{op}(v).\end{aligned}$$

We are referring to the following maps

$$V \xrightarrow{\delta} [T]_2 = V \otimes_A V, \qquad V^{op} \xrightarrow{\delta^{op}} [T^{op}]_2 = V^{op} \otimes_{A^{op}} V^{op},$$

$$V \otimes_A V \xrightarrow{f^1 \otimes g^1} \mathrm{Hom}_k(N, L) \otimes_A \mathrm{Hom}_k(M, N) \xrightarrow{\pi} \mathrm{Hom}_k(M, L), \text{ and}$$

$$V^{op} \otimes_{A^{op}} V^{op} \xrightarrow{D(g^1)\otimes D(f^1)} \mathrm{Hom}_k(DN, DM) \otimes_{A^{op}} \mathrm{Hom}_k(DL, DN) \xrightarrow{\widehat{\pi}} \mathrm{Hom}_k(DL, DM),$$

where π and $\widehat{\pi}$ denote composition. Assume that $\delta(v) = \sum_i u_i \otimes v_i$, with $v, u_i, v_i \in V$, then $\delta^{op}(v) = -\sum_i v_i \otimes u_i$. Hence

$$\begin{aligned} -D\left[\pi(f^1 \otimes g^1)\delta(v)\right] &= -D\left[\textstyle\sum_i f^1(u_i)g^1(v_i)\right] \\ &= -\textstyle\sum_i D(g^1)(v_i)D(f^1)(u_i) \\ &= \widehat{\pi}(D(g^1) \otimes D(f^1))\delta^{op}(v). \end{aligned}$$

Then, we have a contravariant k-functor $D : \mathcal{A}\text{-Mod} \longrightarrow \mathcal{A}^{op}\text{-Mod}$. Now consider its restriction to finite-dimensional modules $D : \mathcal{A}\text{-mod} \longrightarrow \mathcal{A}^{op}\text{-mod}$. Let us show an isomorphism of functors $\eta : Id \longrightarrow D^2$. We can consider the standard duality $D : A\text{-mod} \longrightarrow A^{op}\text{-mod}$ and we know that the evaluation map $\eta^0_M : M \longrightarrow D^2M$ yields an isomorphism of functors $\eta^0 : Id_{A\text{-mod}} \longrightarrow D^2$. Then, we have an isomorphism of functors $\eta : Id_{\mathcal{A}\text{-mod}} \longrightarrow D^2$, defined by $\eta_M = (\eta^0_M, 0)$, for each object $M \in \mathcal{A}$-mod. □

Exercise 2.9. *Let $\{M_i\}_{i\in I}$ be a family of modules of the ditalgebra $\mathcal{A}$ and write $A := A_{\mathcal{A}}$. Consider the product $P := \prod_i M_i$ of the family in A-Mod, with associated projections $\{\pi^0_i : P \longrightarrow M_i\}_{i\in I}$, and the coproduct $Q := \coprod_i M_i$ of the family in A-Mod, with associated injections $\{\sigma^0_i : M_i \longrightarrow Q\}_{i\in I}$. Show that:*

(1) P is the product of the family $\{M_i\}_{i\in I}$ in the category $\mathcal{A}$-Mod, with associated projections $\{\pi_i := (\pi^0_i, 0) : P \longrightarrow M_i\}_{i\in I}$.
(2) Q is the coproduct of the family $\{M_i\}_{i\in I}$ in the category $\mathcal{A}$-Mod, with associated injections $\{\sigma_i := (\sigma^0_i, 0) : M_i \longrightarrow Q\}_{i\in I}$.
(3) Therefore, for $M, N \in \mathcal{A}$-Mod, we have canonical isomorphisms

$$\begin{cases} \mathrm{Hom}_{\mathcal{A}}(\coprod_i M_i, N) \cong \prod_{i\in I} \mathrm{Hom}_{\mathcal{A}}(M_i, N), & \text{given by } g \mapsto (g\sigma_i), \text{ and} \\ \mathrm{Hom}_{\mathcal{A}}(M, \prod_i M_i) \cong \prod_{i\in I} \mathrm{Hom}_{\mathcal{A}}(M, M_i), & \text{given by } f \mapsto (\pi_i f). \end{cases}$$

Exercise 2.10. *Let $\mathcal{A} = (T, \delta)$ be any k-ditalgebra and write $A = [T]_0$ and $V = [T]_1$. Given $M, N \in \mathcal{A}$-Mod, we have canonical morphisms of k-algebras $\psi_M : A \longrightarrow \mathrm{End}_k(M)$ and $\psi_N : A \longrightarrow \mathrm{End}_k(N)$, given by the corresponding A-module structures on the k-vector spaces M and N. For any pair $g = (g^0, g^1) \in \mathrm{Hom}_k(M, N) \times \mathrm{Hom}_{A\text{-}A}(V, \mathrm{Hom}_k(M, N))$, we can consider* *the map* $\psi_g : A \longrightarrow \mathrm{End}_k(M \oplus N)$ *defined, for* $a \in A$, *by the recipe*

$$\psi_g(a) := \begin{pmatrix} \psi_M(a) & 0 \\ \psi_N(a)g^0 - g^0\psi_M(a) - g^1(\delta(a)) & \psi_N(a) \end{pmatrix}.$$

(1) Show that ψ_g *is a morphism of algebras. Moreover, notice that* $\psi_0 = \psi_g$ *iff* $g \in \mathrm{Hom}_{\mathcal{A}}(M, N)$.

(2) Let $\eta : \mathcal{A} \longrightarrow \mathcal{A}'$ *be a morphism of ditalgebras, denote by* $\eta_0 : A \longrightarrow A'$ *and* $\eta_1 : V \longrightarrow V'$ *the canonical restrictions of* η, *and let* $F_\eta : \mathcal{A}'\text{-Mod} \longrightarrow \mathcal{A}\text{-Mod}$ *be the restriction functor. Assume that* η_0 *is an epimorphism of algebras and that, for any* $M, N \in \mathcal{A}'\text{-Mod}$, η_1 *induces a surjection (resp. injection)*

$$\mathrm{Hom}_{A'\text{-}A'}(V', \mathrm{Hom}_k(M, N)) \xrightarrow{\eta_1^*} \mathrm{Hom}_{A\text{-}A}(V, \mathrm{Hom}_k(F_\eta(M), F_\eta(N))).$$

Show that F_η *is a full (resp. faithful) functor.*

(3) Prove Silver's theorem: assume that $\eta : R \longrightarrow S$ *is a morphism of algebras, then,* η *is an epimorphism of algebras iff the restriction functor* $F_\eta : S\text{-Mod} \longrightarrow R\text{-Mod}$ *is a full and faithful functor.*

Hints: For (2), notice that $\psi_M\eta_0 = \psi_{F_\eta(M)}$. *Then, for any morphism* $f \in \mathrm{Hom}_{\mathcal{A}}(F_\eta(M), F_\eta(N))$ *with* $f^1 = \eta_1^*(g^1)$, *show that* $g := (f^0, g^1)$ *satisfies* $\psi_g\eta_0 = \psi_0\eta_0$.

Exercise 2.11. *A quasiditalgebra* $\mathcal{A}$ *is by definition a pair* $\mathcal{A} = (T, \delta)$, *where* T *is a t-algebra and* δ *is a differential on* T *such that there is an element* $d \in [T]_2$, *with* $\delta(d) = 0$, *satisfying* $\delta^2(u) = du - ud$, *for all homogeneous* $u \in T$. *Let us write* $A = A_{\mathcal{A}} = [T]_0$ *and* $V = V_{\mathcal{A}} = [T]_1$. *The category of modules of the quasiditalgebra* $\mathcal{A}$, *denoted by* $\mathcal{A}$-Mod, *has as objects all the* A*-modules. Given* $M, N \in \mathcal{A}$-Mod, *a morphism* $f : M \to N$ *in* $\mathcal{A}$-Mod *is a pair* $f = (f^0, f^1)$, *with* $f^0 \in \mathrm{Hom}_k(M, N)$ *and* $f^1 \in \mathrm{Hom}_{A\text{-}A}(V, \mathrm{Hom}_k(M, N))$ *satisfying that*

$$af^0(m) = f^0(am) + f^1(\delta(a))(m),$$

for any $a \in A$ *and* $m \in M$. *The* Hom*-space in this category is denoted by* $\mathrm{Hom}_{\mathcal{A}}(M, N)$. *Given* $f \in \mathrm{Hom}_{\mathcal{A}}(M, N)$ *and* $g \in \mathrm{Hom}_{\mathcal{A}}(N, L)$, *consider the composition of morphisms of A-A-bimodules*

$$V \otimes_A V \xrightarrow{g^1 \otimes f^1} \mathrm{Hom}_k(N, L) \otimes_A \mathrm{Hom}_k(M, N) \xrightarrow{\pi} \mathrm{Hom}_k(M, L),$$

where the last morphism is induced by composition. Since T *is a t-algebra, we can identify* $V \otimes_A V$ *with* $[T]_2$. *Then the composition* gf *is defined, for any*

$v \in V$, by the following

$$\begin{aligned}(gf)^0 &= g^0 f^0 - \pi(g^1 \otimes f^1)(d);\\(gf)^1(v) &= g^0 f^1(v) + g^1(v) f^0 + \pi(g^1 \otimes f^1)(\delta(v)).\end{aligned}$$

*Show that $\mathcal{A}$-*Mod *is indeed a k-category.*

Exercise 2.12. *A morphism of quasiditalgebras $\phi : \mathcal{A} = (T, \delta) \longrightarrow \mathcal{A}' = (T', \delta')$ is by definition a morphism $\phi : T \longrightarrow T'$ of graded algebras such that $\delta'\phi = \phi\delta$ and $\phi(d) = d'$. Show that any such morphism induces, by restriction, a functor $F_\phi : \mathcal{A}'$-*Mod$\longrightarrow\mathcal{A}$-Mod. *To give the explicit formula on morphisms, denote by $A = A_{\mathcal{A}}$ and $V = V_{\mathcal{A}}$, and with A' and V' the corresponding objects for the quasiditalgebra $\mathcal{A}'$; consider also the morphisms $\phi_0 : A \to A'$ and $\phi_1 : V \to V'$, induced by ϕ. Then, if $M \in \mathcal{A}'$-*Mod*, $F_\phi(M)$ is the A-module obtained from M by restriction of scalars through ϕ_0. The receipt on morphisms is given by $F_\phi(f^0, f^1) = (f^0, f^1\phi_1)$.*

References. The idea to interpret a wide class of matrix problems as problems of representations of differential graded categories was first proposed by A. V. Roiter and M. Kleiner in [46]. The purpose was to consider an interesting class of problems for which it was possible to give a formulation generalizing the representations of quivers introduced by P. Gabriel (see [31]). The idea was to extend to a formulation based on representations of a new type of combinatorial objects called bigraphs. The language was further developed by other specialists of different countries (see the bibliography). This monograph was particularly influenced by the developments made by the Kiev School in representation theory (see [46] and [28]), by M. Butler and W. Crawley-Boevey, in England (see [16], [19] and [20]), and the Mexican group (see [6]). The approach we present here is an improvement of the line proposed in [6].

3

Bocses, ditalgebras and modules

Here, we recall the notion of a bocs and its category of modules. Normal bocses and ditalgebras are formulations of the same mathematical object. There is a natural correspondence between them, which we describe with detail in the following.

Definition 3.1. *A bocs $\mathcal{B}$ is a quadruple $\mathcal{B} = (A, U, \mu, \epsilon)$, where A is any k-algebra, U is an A-A-bimodule, $\mu : U \longrightarrow U \otimes_A U$ and $\epsilon : U \longrightarrow A$ are morphisms of A-A-bimodules such that ϵ is surjective and the following diagrams commute*

$$\begin{array}{ccccc} U \otimes_A U & \xleftarrow{\mu} & U & \xrightarrow{\mu} & U \otimes_A U \\ \downarrow{\scriptstyle 1\otimes\epsilon} & & \| & & \downarrow{\scriptstyle \epsilon\otimes 1} \\ U \otimes_A A & \xrightarrow{\cong} & U & \xleftarrow{\cong} & A \otimes_A U, \end{array}$$

$$\begin{array}{ccccc} U & \xrightarrow{\mu} & U \otimes_A U & \xrightarrow{1\otimes\mu} & U \otimes_A (U \otimes_A U) \\ \| & & & & \downarrow{\scriptstyle \cong} \\ U & \xrightarrow{\mu} & U \otimes_A U & \xrightarrow{\mu\otimes 1} & (U \otimes_A U) \otimes_A U. \end{array}$$

μ is called the comultiplication, ϵ is the counit and $V = \operatorname{Ker} \epsilon$ is the kernel of the bocs $\mathcal{B}$. The commutativity of the second diagram will be recalled by saying that μ is coassociative. The bocs is called normal iff there is an element $w \in U$ such that $\mu(w) = w \otimes w$ and $\epsilon(w) = 1$. In this case, w is called a grouplike of the bocs.

A morphism of normal bocses $\psi : \mathcal{B} \longrightarrow \mathcal{B}'$ is a pair $\psi = (\psi_0, \psi_1)$, where $\psi_0 : A \longrightarrow A'$ is a morphism of algebras and $\psi_1 : U \longrightarrow U'$ is a morphism of A-A-bimodules, where U' is considered as an A-A-bimodule by restriction using ψ_0. Moreover, it is required that $\psi_1(w) = w'$, and that the following *diagram is commutative*

$$\begin{array}{ccccccc}
U \otimes_A U & & \xleftarrow{\qquad\qquad\mu\qquad\qquad} & & U & \xrightarrow{\epsilon} & A \\
{\scriptstyle \psi_1 \otimes \psi_1}\downarrow & & & & {\scriptstyle \psi_1}\downarrow & & \downarrow{\scriptstyle \psi_0} \\
U' \otimes_A U' & \xrightarrow{\tau} & U' \otimes_{A'} U' & \xleftarrow{\mu'} & U' & \xrightarrow{\epsilon'} & A',
\end{array}$$

where τ is the canonical map.

The category of modules (or representations) of the bocs $\mathcal{B}$ has as objects the left A-modules and the space of morphisms is defined by $\mathrm{Hom}_{\mathcal{B}}(M, N) = \mathrm{Hom}_A(U \otimes_A M, N)$. *If $f \in \mathrm{Hom}_{\mathcal{B}}(M, N)$ and $g \in \mathrm{Hom}_{\mathcal{B}}(N, L)$, their composition gf in $\mathcal{B}$-Mod is given by the following composition in A-Mod*

$$U \otimes_A M \xrightarrow{\mu \otimes 1} U \otimes_A U \otimes_A M \xrightarrow{1 \otimes f} U \otimes_A N \xrightarrow{g} L.$$

It is not hard to show that $\mathcal{B}$-Mod is indeed a k-category.

Lemma 3.2. *Given a normal bocs $\mathcal{B} = (A, U, \mu, \epsilon)$, with grouplike w and kernel V, we can associate the ditalgebra $\mathcal{A}(\mathcal{B}) = (T, \delta)$, where $T = T_A(V)$, with its standard grading, and δ is the differential obtained from the linear maps $\delta_0 : A \longrightarrow V = [T]_1$ and $\delta_1 : V \longrightarrow V \otimes_A V = [T]_2$ defined by the rules $\delta_0(a) = aw - wa$ and $\delta_1(v) = \mu(v) - w \otimes v - v \otimes w$ using (1.8).*

If $\psi : \mathcal{B} \longrightarrow \mathcal{B}'$ is a morphism of normal bocses, then $\mathcal{A}(\psi) : \mathcal{A}(\mathcal{B}) \longrightarrow \mathcal{A}(\mathcal{B}')$ is the morphism of ditalgebras defined by $\psi_0 : A \longrightarrow A'$ and the restriction $\psi_1' : V \longrightarrow V'$ of ψ_1 to the kernels V and V' of the corresponding bocses.

Proof. We have that $V \otimes_A V \subseteq U \otimes_A U$, because ϵ splits as a morphism of left A-modules and as a morphism of right A-modules. Then, notice that whenever $\epsilon(w) = 1$ and $u \in U$, with $\mu(u) = \sum_i u_i^1 \otimes u_i^2$, we have, from the first diagram in the definition of a bocs, that

$$\mu(u) - u \otimes w - w \otimes u + w\epsilon(u) \otimes w = \sum_i [u_i^1 - w\epsilon(u_i^1)] \otimes [u_i^2 - \epsilon(u_i^2)w].$$

Thus, for $v \in V$, indeed $\mu(v) - v \otimes w - w \otimes v \in V \otimes_A V$. It is easy to see that we can really apply (1.8). From (1.7), we know that, in order to show that $\delta^2 = 0$, it is enough to verify that $\delta^2(A) = 0$ and $\delta^2(V) = 0$.

If $a \in A$, then

$$\begin{aligned}
\delta^2(a) &= \delta(aw - wa) \\
&= \mu(aw - wa) - w \otimes (aw - wa) - (aw - wa) \otimes w \\
&= a\mu(w) - \mu(w)a + w \otimes wa - aw \otimes w \\
&= [w \otimes w - \mu(w)]a - a[w \otimes w - \mu(w)] \\
&= 0.
\end{aligned}$$

Now, for $v \in V$, we have $\delta(v) = \mu(v) - w \otimes v - v \otimes w \in [T]_2$ can be written in the form $\delta(v) = \sum_i u_i \otimes v_i$, with $u_i, v_i \in V$. Then

$$\begin{aligned}\textstyle\sum_i \mu(u_i) \otimes v_i &= (\mu \otimes 1)(\textstyle\sum_i u_i \otimes v_i)\\ &= (\mu \otimes 1)(\mu(v) - w \otimes v - v \otimes w)\\ &= (\mu \otimes 1)\mu(v) - \mu(v) \otimes w - \mu(w) \otimes v.\end{aligned}$$

Similarly, we have $\sum_i u_i \otimes \mu(v_i) = (1 \otimes \mu)\mu(v) - v \otimes \mu(w) - w \otimes \mu(v)$. Thus

$$\begin{aligned}\delta^2(v) &= \textstyle\sum_i (\delta(u_i) \otimes v_i - u_i \otimes \delta(v_i))\\ &= \textstyle\sum_i [(\mu(u_i) - w \otimes u_i - u_i \otimes w) \otimes v_i\\ &\quad -u_i \otimes (\mu(v_i) - w \otimes v_i - v_i \otimes w)]\\ &= \textstyle\sum_i \mu(u_i) \otimes v_i - \sum_i w \otimes u_i \otimes v_i - \sum_i u_i \otimes \mu(v_i)\\ &\quad + \textstyle\sum_i u_i \otimes v_i \otimes w\\ &= [(\mu \otimes 1)\mu(v) - \mu(v) \otimes w - \mu(w) \otimes v]\\ &\quad - w \otimes [\mu(v) - w \otimes v - v \otimes w]\\ &\quad - [(1 \otimes \mu)\mu(v) - v \otimes \mu(w) - w \otimes \mu(v)]\\ &\quad + [\mu(v) - w \otimes v - v \otimes w] \otimes w\\ &= 0,\end{aligned}$$

because $(\mu \otimes 1)\mu = (1 \otimes \mu)\mu$ and $\mu(w) = w \otimes w$. □

Lemma 3.3. *Given a ditalgebra $\mathcal{A} = (T, \delta)$, take $A = [T]_0$ and $V = [T]_1$. Then, we can associate a bocs $\mathcal{B}(\mathcal{A}) = (A, U, \mu, \epsilon)$, where the components are constructed as follows. The A-A-bimodule U has underlying vector space $U = A \times V$, where the action of A is given by the formulas*

$$\begin{aligned}(b, v)a &= (ba, va)\\ a(b, v) &= (ab, av + \delta(a)b),\end{aligned}$$

for $a, b \in A$ and $v \in V$. If we denote by $w = (1, 0)$, then $wA = A \times \{0\}$ and, if we identify V with $\{0\} \times V$, we have $U = wA \oplus V$ as right A-modules. Moreover, $\delta(a) = aw - wa$ and, hence, $U = Aw \oplus V$, as left A-modules. By definition

$$\begin{aligned}\mu(wa + v) &= w \otimes wa + \delta(v) + v \otimes w + w \otimes v\\ \epsilon(wa + v) &= a,\end{aligned}$$

for $a \in A$ and $v \in V$. Then, $\mathcal{B}(\mathcal{A})$ is a normal bocs with grouplike w and kernel V.

If $\psi : \mathcal{A} \longrightarrow \mathcal{A}'$ is a morphism of ditalgebras, consider the restrictions of ψ, to the homogeneous component of degree zero, $\psi_0 : A \longrightarrow A'$, and to the homogeneous component of degree 1, $\psi_1 : V \longrightarrow V'$. Then, define $\overline{\psi}_1$:

$U \longrightarrow U'$ by $\overline{\psi}_1(wa + v) = w'\psi_0(a) + \psi_1(v)$, for all $a \in A$ and $v \in V$. Then, $\mathcal{B}(\psi) := \overline{\psi} = (\psi_0, \overline{\psi}_1) : \mathcal{B}(\mathcal{A}) \longrightarrow \mathcal{B}(\mathcal{A}')$ is a morphism of normal bocses.

Proof. The verifications are straightforward. □

Remark 3.4. *If we denote by* Ditalg *and* Nbocs *the categories of ditalgebras and normal bocses respectively, the associations described in (3.2) and (3.3), yield an equivalence of k-categories $\mathcal{B}$:* Ditalg$\longrightarrow$Nbocs *with quasi-inverse $\mathcal{A}$:* Nbocs$\longrightarrow$Ditalg.

Proposition 3.5. *Let $\mathcal{A}$ be a ditalgebra and $\mathcal{B}(\mathcal{A})$ be the corresponding normal bocs with grouplike w. Then, there is an isomorphism of k-categories*

$$F : \mathcal{B}(\mathcal{A})\text{-Mod} \longrightarrow \mathcal{A}\text{-Mod}$$

satisfying $F(M) = M$, for any A-module M, and, for any $f \in \mathrm{Hom}_{\mathcal{B}(\mathcal{A})}(M, N)$, $v \in V$ and $m \in M$

$$\begin{aligned} F(f)^0[m] &= f(w \otimes m) \\ F(f)^1(v)[m] &= f(v \otimes m). \end{aligned}$$

Proof. The verifications are straightforward. We just describe the inverse $G : \mathcal{A}\text{-Mod} \longrightarrow \mathcal{B}(\mathcal{A})\text{-Mod}$ of F. We have, $G(M) = M$, for any $M \in \mathcal{A}$-Mod and given $f \in \mathrm{Hom}_{\mathcal{A}}(M, N)$, $G(f) : U \otimes_A M \longrightarrow N$ is defined by $G(f)[(wa + v) \otimes m] = f^0(am) + f^1(v)[m]$, for any $a \in A$, $v \in V$ and $m \in M$. □

Exercise 3.6. *Given a bocs $\mathcal{B} = (A, U, \mu, \epsilon)$, with kernel V, choose $w \in U$ with $\epsilon(w) = 1$. Show that we can associate the quasiditalgebra $\mathcal{A}(\mathcal{B}) = (T, \delta)$, where $T = T_A(V)$, with its standard grading, and δ is the differential obtained from the linear maps $\delta_0 : A \longrightarrow V = [T]_1$ and $\delta_1 : V \longrightarrow V \otimes_A V = [T]_2$ defined by the rules $\delta_0(a) = aw - wa$ and $\delta_1(v) = \mu(v) - w \otimes v - v \otimes w$ using (1.8). Here, the required element $d \in [T]_2$, for the formula involving δ^2, is $d = w \otimes w - \mu(w)$. See (2.11).*

Exercise 3.7. *Given a quasiditalgebra $\mathcal{A} = (T, \delta)$, with $d \in [T]_2$ such that $\delta(d) = 0$ and $\delta^2(u) = du - ud$, take $A = [T]_0$ and $V = [T]_1$. Then, we can associate a bocs $\mathcal{B}(\mathcal{A}) = (A, U, \mu, \epsilon)$, where the components are constructed as follows. The A-A-bimodule U has underlying vector space $U = A \times V$, where the action of A is given by the formulas*

$$\begin{aligned} (b, v)a &= (ba, va) \\ a(b, v) &= (ab, av + \delta(a)b), \end{aligned}$$

for $a, b \in A$ and $v \in V$. If we denote by $w = (1, 0)$, then $wA = A \times \{0\}$ and, if we identify V with $\{0\} \times V$, we have $U = wA \oplus V$ as right A-modules.

Moreover, $\delta(a) = aw - wa$ and, hence, $U = Aw \oplus V$, as left A-modules. By definition

$$\mu(wa + v) = w \otimes wa + \delta(v) + v \otimes w + w \otimes v - da$$
$$\epsilon(wa + v) = a,$$

for $a \in A$ and $v \in V$. Then, $\mathcal{B}(\mathcal{A})$ is a bocs with kernel V.

References. A. V. Roiter proposed in [45] an alternative to the approach to matrix problems through differential graded categories. It was supported by the new notion of a *bocs* (an acronym for "bimodule over a category with coalgebra structure") and its category of representations. There, the natural correspondence between normal bocses and differential graded categories is established. We have presented this in the ring theoretical context of our notes. Although the bocs language is elegant and in some sense natural, we preferred to use in our notes the ditalgebra formulation because it seems to lend itself to easier explicit concrete calculations.

4

Layered ditalgebras

Even though the category of modules makes sense for arbitrary ditalgebras, its main properties can be derived only for special types of ditalgebras. A relevant aspect of the underlying t-algebra, for such special ditalgebras, is presented in this section.

Lemma 4.1. *Let T be an algebra freely generated by the pair (R, W). Assume W admits an R-R-bimodule decomposition $W = W_0 \bigoplus W_1$. Then:*

(1) T admits a unique structure of a graded k-algebra such that the elements of R and W_0 are homogeneous of degree 0, and those of W_1 are homogeneous of degree 1. Denote by $A = [T]_0$ and $V = [T]_1$.

(2) With the given grading, T is a t-algebra.

(3) A is freely generated by the pair (R, W_0).

(4) The product of T determines an isomorphism $A \otimes_R W_1 \otimes_R A \longrightarrow V$ of A-A-bimodules.

Proof. By (1.3), the inclusions of R and W in T induce an isomorphism $T_R(W) \cong T$, and the first item is clear from this. It also becomes clear that with this grading, A is the subalgebra of T generated by R and W_0, and that $V = AW_1A$. Then, (2) and (3) follow from (1.3).

Let $\widehat{T}$ denote the tensor algebra of $A \otimes_R W_1 \otimes_R A$ over A, and consider the morphism $\mu : \widehat{T} \longrightarrow T$ induced by the inclusion of A in T and the product $A \otimes_R W_1 \otimes_R A \longrightarrow T$. Consider also the morphism $\phi : T \longrightarrow \widehat{T}$ induced by the inclusion of R in $\widehat{T}$, and the R-R-bimodule morphism $\phi_1 : W \longrightarrow \widehat{T}$ which maps each $w_0 \in W_0$ onto w_0 and each $w_1 \in W_1$ onto $1 \otimes w_1 \otimes 1$. Then, ϕ turns out to be an inverse for μ. □

We shall consider t-algebras with this additional combinatorial information supplied by R and $W = W_0 \oplus W_1$.

Definition 4.2. *Assume R is a k-algebra and W is an R-R-bimodule endowed with a bimodule decomposition $W = W_0 \oplus W_1$. Then, we say that the t-algebra T has layer (R, W), if we are in the situation of last lemma. That is, iff $R, W_0 \subseteq [T]_0$, $W_1 \subseteq [T]_1$, and T is freely generated by the pair (R, W).*

Remark 4.3. *Let T be a t-algebra with layer (R, W) and assume δ is a differential on T. Then:*

(1) $\delta(R) = 0$ iff $\delta : T \longrightarrow T$ is a morphism of R-R-bimodules.
(2) If $\delta(R) = 0$, then δ is determined by its values on W; moreover, in this case, $\delta^2 = 0$ iff $\delta^2(W) = 0$ (see (1.7)).

Lemma 4.4. *Let T be a t-algebra with layer (R, W). Then, any morphism δ : $W \longrightarrow T$ of R-R-bimodules such that $\delta(W_0) \subseteq [T]_1$ and $\delta(W_1) \subseteq [T]_2$ extends uniquely to a differential $\widehat{\delta} : T \longrightarrow T$ with $\widehat{\delta}(R) = 0$.*

Proof. Since T is freely generated by (R, W), we may assume that $T = T_R(W)$. We shall define a morphism of R-R-bimodules $\widehat{\delta}_n$ from each of the direct summands $W^{\otimes n} = \oplus_{d_1,\ldots,d_n \in \{0,1\}} W_{d_1} \otimes \cdots \otimes W_{d_n}$ of T to T. Take $\widehat{\delta}_0 := 0, \widehat{\delta}_1 := \delta$ and define $\widehat{\delta}_n$ by the formula

$$\widehat{\delta}_n\,(w_1 \otimes \cdots \otimes w_n) = \sum_{i=1}^{n} \left((-1)^{\left(\sum_{j=1}^{i-1} \deg(w_j)\right)} w_1 \cdots w_{i-1}\delta(w_i)w_{i+1} \cdots w_n \right),$$

where each $w_i \in W_{d_i}$. See remarks (1.7). This formula defines a morphism of R-R-bimodules $\widehat{\delta}_n : W^{\otimes n} \longrightarrow T$, because δ is a morphism of R-R-bimodules. Then, there is a morphism $\widehat{\delta} : T \longrightarrow T$ of R-R-bimodules which extends all these maps $\widehat{\delta}_n$. Since, by definition $\deg\left(\otimes_{i=1}^n w_i\right) = \sum_i \deg(w_i)$, the extra assumption on δ guarantees that $\widehat{\delta}([T]_i) \subseteq [T]_{i+1}$. It remains to show that $\widehat{\delta}$ satisfies Leibniz rule. Since $\widehat{\delta}$ is a morphism of R-R-bimodules with $\widehat{\delta}(R) = 0$, it is enough to show that given $u_n = \otimes_{s=1}^n w_s$ and $v_m = \otimes_{r=1}^m w'_r$, with $w_s, w'_r \in W_0 \cup W_1$, then $\widehat{\delta}(u_n \otimes v_m) = \widehat{\delta}(u_n) \otimes v_m + (-1)^{\deg(u_n)} u_n \otimes \widehat{\delta}(v_m)$. This is a straightforward calculation, using the definition of $\widehat{\delta}$. □

Definition 4.5. *Assume R is a k-algebra and W is an R-R-bimodule endowed with a bimodule decomposition $W = W_0 \oplus W_1$. A ditalgebra $\mathcal{A} = (T, \delta)$ has layer (R, W) iff its underlying t-algebra T has layer (R, W) and $\delta(R) = 0$. A ditalgebra equipped with a layer is called a layered ditalgebra.*

Notation 4.6. *Given a ditalgebra $\mathcal{A} = (T, \delta)$ with layer (R, W), we can consider the ditalgebra $\mathcal{A}^t = (T_R(W), \underline{\delta})$, where the differential $\underline{\delta}$ is obtained from δ by conjugation by the canonical isomorphism $\theta : T_R(W) \longrightarrow T$. We call $\mathcal{A}^t$ the tensor realization of the ditalgebra $\mathcal{A}$. Of course $\mathcal{A}^t$, as well as the canonical* *isomorphism of ditalgebras $\theta_{\mathcal{A}} = \theta : \mathcal{A}^t \to \mathcal{A}$, depends on the chosen layer of $\mathcal{A}$. Once this layer is fixed, we often identify $\mathcal{A}$ with $\mathcal{A}^t$ through $\theta_{\mathcal{A}}$. We just have to be careful when we deal with different layers of a ditalgebra.*

Whenever we have the ditalgebra $\mathcal{A}$ with layer (R, W), we usually adopt the following notation: $T = T_{\mathcal{A}} \cong T_R(W)$, $A = A_{\mathcal{A}} = [T]_0 \cong T_R(W_0)$ and $V = V_{\mathcal{A}} = [T]_1 \cong A \otimes_R W_1 \otimes_R A$.

Lemma 4.7. *Assume that $\mathcal{A}$ is a ditalgebra with layer (R, W) and differential δ. Take $M, N \in \mathcal{A}$-Mod. Then:*

(1) Whenever $(f^0, f^1) \in \mathrm{Hom}_{\mathcal{A}}(M, N)$, we have that $f^0 \in \mathrm{Hom}_R(M, N)$.
(2) If $f = (f^0, f^1) \in \mathrm{Hom}_R(M, N) \times \mathrm{Hom}_{A\text{-}A}(V, \mathrm{Hom}_k(M, N))$, then f is a morphism in $\mathcal{A}$-Mod iff

$$wf^0(m) = f^0(wm) + f^1(\delta(w))(m),$$

for each $w \in W_0$ and $m \in M$.

Proof. (1) follows from the definition of morphism in $\mathcal{A}$-Mod using that $\delta(R) = 0$. For the second part, recall that $A \cong T_R(W_0)$ is generated as an R-module by R and the finite products $w = w_1 w_2 \cdots w_n$, where $w_1, \ldots, w_n \in W_0$ and $n \geq 1$. Since f^1 is a morphism of R-R-bimodules, it will be enough to show that $wf^0(m) = f^0(wm) + f^1(\delta(w))(m)$, for such products w of length $n \geq 1$ and all $m \in M$. This is true for $n = 1$ by assumption, so assume $n \geq 2$, that it holds for such products of length $n - 1$ and take $w' = w_2 w_3 \cdots w_n$, then

$$\begin{aligned} f^1(\delta(w))[m] &= f^1\left(\delta(w_1 w')\right)[m] \\ &= f^1\left(\delta(w_1)w' + w_1\delta(w')\right)[m] \\ &= f^1\left(\delta(w_1)w'\right)[m] + f^1\left(w_1\delta(w')\right)[m] \\ &= f^1(\delta(w_1))[w'm] + w_1 f^1\left(\delta(w')\right)[m] \\ &= w_1 f^0[w'm] - f^0[w_1 w'm] + w_1\left(w' f^0[m] - f^0[w'm]\right) \\ &= w_1 w' f^0[m] - f^0[w_1 w'm] \\ &= wf^0[m] - f^0[wm]. \end{aligned}$$

□

Remark 4.8. *Assume that A is freely generated by the pair (R, U). Any left A-module M has associated a representation $A \longrightarrow \mathrm{End}_k(M)$ of A in the vector space M, which is determined by its restrictions to R and U. Hence, by an R-module M and by an action $\phi : U \otimes_R M \longrightarrow M$ of U on M. The procedure can be reversed: Fix an R-module M and consider the corresponding R-module structure map $R \xrightarrow{\phi_0} \mathrm{End}_k(M)$. Then, for any morphism $\phi_1 : U \otimes_R M \longrightarrow M$ in R-Mod, the space M can be endowed with an A-module structure determined by ϕ_0 and ϕ_1. Let us be more precise: consider the natural isomorphism*

$\mathrm{Hom}_R(U \otimes_R M, M) \xrightarrow{\eta} \mathrm{Hom}_{R\text{-}R}(U, \mathrm{Hom}_k(M, M))$. *Then, the image* $\eta(\phi_1)$ *is an* R-R*-bimodule morphism* $U \longrightarrow \mathrm{End}_k(M)$*, which, together with* ϕ_0 *determine, using that* A *is freely generated by* R *and* U*, an algebra morphism* $A \longrightarrow \mathrm{End}_k(M)$*. Hence, a structure of left* A*-module on* M.

Remark 4.9. *Assume that* R *is a* k*-subalgebra of the algebra* A*,* U *is an* R-R*-bimodule and* $M, N \in A$-Mod*. Then, we have an isomorphism*

$$\mathrm{Hom}_{A\text{-}A}(A \otimes_R U \otimes_R A, \mathrm{Hom}_k(M, N)) \xrightarrow{\beta} \mathrm{Hom}_{R\text{-}R}(U, \mathrm{Hom}_k(M, N)),$$

given by $\beta(f)[u] = f(1 \otimes u \otimes 1)$*, with* $\beta^{-1}(h)[a \otimes u \otimes a'] = ah(u)a'$*, which is natural in* N *and* M.

Assume that $\mathcal{A}$ *is a ditalgebra with layer* (R, W) *and consider the associated* R-R*-bimodule decomposition* $W = W_0 \oplus W_1$*. Whenever* $f = (f^0, f^1) \in \mathrm{Hom}_{\mathcal{A}}(M, N)$*, we know that* $f^1 \in \mathrm{Hom}_{A\text{-}A}(AW_1A, \mathrm{Hom}_k(M, N))$ *is determined by its restriction to* W_1.

Conversely, if $f^0 \in \mathrm{Hom}_R(M, N)$ *and* $f^1 \in \mathrm{Hom}_{R\text{-}R}(W_1, \mathrm{Hom}_k(M, N))$ *are such that* $wf^0(m) = f^0(wm) + \hat{f}^1(\delta(w))[m]$*, for* $w \in W_0$ *and* $m \in M$*, we obtain, from (4.1) and (4.7), that* $(f^0, \hat{f}^1) \in \mathrm{Hom}_{\mathcal{A}}(M, N)$*, where* $\hat{f}^1$ *is obtained by extension from* f^1 *using* β*. When there is no danger of confusion, abusing the language, we shall use the same symbol to denote* f^1 *and its extension* $\hat{f}^1$*. We have derived the following result.*

Lemma 4.10. *Assume that* $\mathcal{A}$ *is a ditalgebra with layer* (R, W) *and consider the associated* R-R*-bimodule decomposition* $W = W_0 \oplus W_1$*. Whenever we have* $M, N \in R$-Mod*, we can consider the vector space*

$$\mathrm{Phom}_{R\text{-}W}(M, N) = \mathrm{Hom}_R(M, N) \times \mathrm{Hom}_{R\text{-}R}(W_1, \mathrm{Hom}_k(M, N)).$$

Then, given $M, N \in \mathcal{A}$-Mod*, there is an exact sequence*

$$0 \longrightarrow \mathrm{Hom}_{\mathcal{A}}(M, N) \xrightarrow{\sigma} \mathrm{Phom}_{R\text{-}W}(M, N) \xrightarrow{\partial} \mathrm{Hom}_R(W_0 \otimes_R M, N),$$

where $\sigma(g^0, g^1) = (g^0, g^1)$ *and* $\partial(f^0, f^1)[w \otimes m] = f^0(wm) - wf^0(m) + f^1(\delta(w))[m]$*, for* $w \in W_0$ *and* $m \in M$.

Therefore, the choice of a layer (R, W) for a ditalgebra $\mathcal{A}$ permits some control of $\mathcal{A}$-Mod using R-Mod and R-R-bimodule morphisms involving W_0 and W_1.

References. The word "layer" is used in our notes with a different meaning than the one attached to it by Crawley-Boevey in [19] or [20]. Our notion generalizes his proposal and allows a more general detailed analysis of properties of the category of modules according to restrictions imposed on the layer.

Both formulations propose to enrich with an additional algebraic-combinatorial structure, namely a layer, a given ditalgebra (or bocs), with the aim to permit an understanding of the associated module category. They include the possibility of a layer determined by some bigraph or marked bigraph as in the original definitions of free bocses, used by Roiter in [45], or of a semifree bocs, used by Drozd in [28].

5
Triangular ditalgebras

For many interesting ditalgebras $\mathcal{A}$, the category of $\mathcal{A}$-modules is not abelian. However, the fact that $\mathcal{A}$ satisfies the following triangularity condition permits to initiate the study of $\mathcal{A}$-Mod supported by elementary induction arguments.

Definition 5.1. *We say that the layer (R, W) of a ditalgebra $\mathcal{A} = (T, \delta)$ is triangular iff:*

(1) There is a sequence of R-R-subbimodules $0 = W_0^0 \subseteq W_0^1 \subseteq \cdots \subseteq W_0^r = W_0$ such that $\delta(W_0^{i+1}) \subseteq A_i W_1 A_i$, for all $i \in [0, r-1]$, where A_i denotes the R-subalgebra of A generated by W_0^i.
(2) There is a sequence of R-R-subbimodules $0 = W_1^0 \subseteq W_1^1 \subseteq \cdots \subseteq W_1^s = W_1$ such that $\delta(W_1^{i+1}) \subseteq AW_1^i AW_1^i A$, for all $i \in [0, s-1]$.

If, moreover, each subbimodule W_0^i is a direct summand of W_0^{i+1}, for $i \in [0, r-1]$, the layer (R, W) is called additive triangular. A ditalgebra $\mathcal{A}$ is called triangular (resp. additive triangular) iff it admits a triangular (resp. additive triangular) layer.

Our definition of triangularity imposes a restriction on the bimodule W_1, which is not present in Crawley's formulation of the subject (see [19] and [20]). This point of view permits to show, for a wide class of ditalgebras $\mathcal{A}$, that the category $\mathcal{A}$-Mod has nice properties, as we shall see through this section.

Remark 5.2. *Assume $\mathcal{A}$ is an additive triangular ditalgebra and adopt the notation of last definition. Then, we can derive from (1.3) and (4.1) that, for all $i \in [1, r]$:*

(1) A_i is freely generated by the pair (R, W_0^i).
(2) The product $A_i \otimes_R W_1 \otimes_R A_i \longrightarrow A_i W_1 A_i$ is an isomorphism.

Lemma 5.3. *Assume $\mathcal{A}$ is a ditalgebra with triangular layer (R, W), adopt the notation of (5.1) and assume (1) and (2) of (5.2). Suppose that $M, N \in R$-Mod and $(f^0, f^1) \in \mathrm{Hom}_R(M, N) \times \mathrm{Hom}_{R\text{-}R}(W_1, \mathrm{Hom}_k(M, N))$. Then:*

(1) If f^0 is a retraction in R-Mod and N is the underlying R-module of an object $N \in \mathcal{A}$-Mod, then there is a structure of left A-module on M such that $(f^0, \hat{f}^1) \in \mathrm{Hom}_{\mathcal{A}}(M, N)$, where $\hat{f}^1$ denotes the unique extension of f^1 to V.
(2) If f^0 is a section in R-Mod and M is the underlying R-module of an object $M \in \mathcal{A}$-Mod, then there exists a structure of left A-module on N such that $(f^0, \hat{f}^1) \in \mathrm{Hom}_{\mathcal{A}}(M, N)$, where $\hat{f}^1$ denotes the unique extension of f^1 to V.

Proof. We shall prove only (1), since the proof of (2) is similar. Therefore, we assume that f^0 admits a right inverse $g^0 \in \mathrm{Hom}_R(N, M)$.

We will define inductively, for $j \in [1, r]$, morphisms $\phi_j : W_0^j \otimes_R M \longrightarrow M$ of R-modules, using the next formula

$$(E_j): \qquad \phi_j(w \otimes m) = g^0[wf^0(m)] - g^0[f^1_{j-1}(\delta(w))(m)],$$

for $w \in W_0^j$ and $m \in M$, where $f^1_{j-1} : A_{j-1}W_1A_{j-1} \longrightarrow \mathrm{Hom}_k(M, N)$ is the unique extension of f^1. We agree that $f^1_0 = f^1$.

Notice that at each step j of the induction, once ϕ_j has already been defined, we obtain a structure of left A_j-module for M from (5.2)(1) and (4.8). Then, condition (2) of (5.2) and the isomorphism of (4.9) guarantee the existence of the extension $f^1_j \in \mathrm{Hom}_{A_j\text{-}A_j}(A_jW_1A_j, \mathrm{Hom}_k(M, N))$ of f^1.

For $j = 1$, $w \in W_0^1$ and, from the triangularity of the layer, we get $\delta(w) \in W_1$. Thus, $f^1(\delta(w))$ is defined and the formula (E_1) defines a morphism ϕ_1. Assuming ϕ_j defined, we noticed before that f^1_j is defined too, and, again from the triangularity of the layer, we know that $\delta(w) \in A_jW_1A_j$, for $w \in W_0^{j+1}$. Then, formula (E_{j+1}) defines a morphism ϕ_{j+1} of R-modules.

The morphism ϕ_r determines the desired left A-module structure for M, because $f^0g^0 = 1_N$ and formula (E_r) imply that $(f^0, f^1_r) \in \mathrm{Hom}_{\mathcal{A}}(M, N)$ as shown in (4.7). □

In the proof of last lemma, the filtration of W_1 required in the definition of the triangular layer was irrelevant. The next result shows an elementary application which requires only this second condition.

Lemma 5.4. *Let $\mathcal{A}$ be a triangular ditalgebra, take $M \in \mathcal{A}$-Mod and $f = (f^0, f^1) \in \mathrm{End}_{\mathcal{A}}(M)$. Then, f is nilpotent iff f^0 is so.*

Proof. Adopt the notation of (5.1) for the triangular ditalgebra $\mathcal{A}$. It is clear that the nilpotency of f implies the same property for f^0. In order to prove the converse, let us write $f^{(n)} = \big((f^{(n)})^0, (f^{(n)})^1\big)$, for any natural number n, and assume that $(f^{(m)})^0 = (f^0)^m = 0$. For $w_1 \in W_1^1$, we have $\delta(w_1) = 0$, and then

$$(f^{(2m)})^1(w_1) = (f^{(m)})^0(f^{(m)})^1(w_1) + (f^{(m)})^1(w_1)(f^{(m)})^0 = 0.$$

If we have, for $t \in \mathbb{N}$, that $(f^{(tm)})^1(W_1^j) = 0$, then $(f^{([t+1]m)})^1(W_1^{j+1}) = 0$. Indeed, if $w \in W_1^{j+1}$, by triangularity, $\delta(w) = \sum_i v_i^1 v_i^2$, with $v_i^1, v_i^2 \in AW_1^j A$; hence

$$\begin{aligned}(f^{([t+1]m)})^1(w) &= (f^{(tm)})^0(f^{(m)})^1(w) + (f^{(tm)})^1(w)(f^{(m)})^0 \\ &\quad + \textstyle\sum_i (f^{(tm)})^1(v_i^1)(f^{(m)})^1(v_i^2) \\ &= \textstyle\sum_i (f^{(tm)})^1(v_i^1)(f^{(m)})^1(v_i^2) \\ &= 0.\end{aligned}$$

Thus, $f^{((s+1)m)} = 0$, and f is nilpotent. □

Definition 5.5. *We say that the ditalgebra $\mathcal{A}$ is a Roiter ditalgebra iff it admits a triangular layer (R, W) and the following property is satisfied: for any $M, N \in R$-*Mod*, any isomorphism $f^0 \in \mathrm{Hom}_R(M, N)$ and any morphism $f^1 \in \mathrm{Hom}_{R\text{-}R}(W_1, \mathrm{Hom}_k(M, N))$, if one of M or N has a structure of left A-module, then the other one admits also a structure of left A-module such that $(f^0, \hat{f}^1) : M \longrightarrow N$ is a morphism in $\mathcal{A}$-*Mod*.*

From (5.3), additive triangular ditalgebras are Roiter ditalgebras. As we shall see, this property plays an important role in the understanding of the category $\mathcal{A}$-Mod and it is preserved under important constructions of new ditalgebras from old ones.

Recall that an algebra R is called *semisimple* iff the left R-module R is semisimple (equivalently, any left R-module is semisimple see [45] or [23]). We obtain many Roiter ditalgebras from the following.

Lemma 5.6. *Assume that $\mathcal{A}$ is a ditalgebra with triangular layer (R, W) such that R is semisimple. Then, $\mathcal{A}$ is a Roiter ditalgebra.*

Proof. We can assume that the underlying t-algebra T of $\mathcal{A}$ is $T = T_R(W)$ and consider the R-R-bimodule filtration $0 = W_0^0 \subseteq W_0^1 \subseteq \cdots \subseteq W_0^r = W_0$ of W_0 provided by (5.1)(1). By (5.3), it will be enough to verify conditions (1) and (2) of (5.2). Let us denote by $*$ the product in the algebra T; given R-R-subbimodules U and V of T, denote by $U * V$ the subbimodule of T generated by the products $u * v$, with $u \in U$ and $v \in V$; define $U^{*0} := R$ and $U^{*1} := U$; finally, define recursively $U^{*n} := U^{*(n-1)} * U$, for $n \geq 2$. Then, the subalgebra A_i of T generated by R and W_0^i is just $A_i = \oplus_{n\geq 0}(W_0^i)^{*n} \subseteq \oplus_{n\geq 0}(W_0)^{\otimes n} = A$.

Therefore, in order to show that A_i is freely generated by the pair (R, W_0^i), it will be enough to show that, for $n \geq 2$, the product map $\pi_n : (W_0^i)^{\otimes n} \longrightarrow T$ is injective, and, therefore, $(W_0^i)^{\otimes n} \cong (W_0^i)^{*n}$, and $A_i \cong T_R(W_0^i)$. Now, if we already know, by induction hypothesis, that π_{n-1} is injective, then π_n coincides with the following composition of injective maps (recall that R is semisimple!)

$$\begin{array}{ccccc} (W_0^i)^{\otimes n} & & & & T \\ \cong \downarrow & & & & \uparrow \\ (W_0^i)^{\otimes(n-1)} \otimes_R W_0^i & \longrightarrow & (W_0)^{\otimes(n-1)} \otimes_R W_0^i & \longrightarrow & (W_0)^{\otimes(n-1)} \otimes_R W_0. \end{array}$$

Similarly, we have $A_i \otimes_R W_1 \otimes_R A_i = \oplus_{n,m}[(W_0^i)^{*n} \otimes_R W_1 \otimes_R (W_0^i)^{*m}]$ and $A_i * W_1 * A_i = \oplus_{n,m}[(W_0^i)^{*n} * W_1 * (W_0^i)^{*m}] \subseteq \oplus_{n,m}[(W_0)^{\otimes n} \otimes_R W_1 \otimes_R (W_0)^{\otimes m}] \subseteq T$. Thus, in order to show that (5.2)(2) holds, we have to show that the product map $(W_0^i)^{\otimes n} \otimes_R W_1 \otimes_R (W_0^i)^{\otimes m} \longrightarrow (W_0)^{\otimes n} \otimes_R W_1 \otimes_R (W_0)^{\otimes m} \subseteq T$ is injective. This follows, as before, from the fact that left and right R-modules are flat. □

Lemma 5.7. *Assume that $\mathcal{A}$ is a Roiter ditalgebra with triangular layer (R, W). Then, for any morphism $f = (f^0, f^1) : M \longrightarrow N$ in $\mathcal{A}$-Mod, we have:*

(1) If f^0 is a retraction in R-Mod, there exists a morphism $h = (h^0, h^1) : M' \longrightarrow M$ in $\mathcal{A}$-Mod with h^0 isomorphism and $(fh)^1 = 0$.
(2) If f^0 is a section in R-Mod, there exists a morphism $g = (g^0, g^1) : N \longrightarrow N'$ in $\mathcal{A}$-Mod with g^0 isomorphism and $(gf)^1 = 0$.

Proof. We only prove (1). Adopt the notation of (5.1) and notice that the proof follows by induction from the following.

Claim. Given $i \in [1, s]$, if $f^1(W_1^{i-1}) = 0$, then there exists a morphism $h = (h^0, h^1) : M' \longrightarrow M$ in $\mathcal{A}$-Mod with h^0 isomorphism and $(fh)^1(W_1^i) = 0$.

Proof of the claim: let $t^0 \in \mathrm{Hom}_R(N, M)$ be a right inverse for f^0 in R-Mod. Denote by M' the underlying R-module of M. Define, for $w \in W_1$, $h^1(w) := -t^0 f^1(w) \in \mathrm{Hom}_k(M', M)$. Then, $h^1 \in \mathrm{Hom}_{R\text{-}R}(W_1, \mathrm{Hom}_k(M', M))$. Since $\mathcal{A}$ is a Roiter ditalgebra, we know that M' admits a left A-module structure such that $h = (1_M, h^1)$ is a morphism in $\mathcal{A}$-Mod. Since $f^1(W_1^{i-1}) = 0$, then $h^1(W_1^{i-1}) = 0$. Now, if $w \in W_1^i$, we obtain from the triangularity condition that $\delta(w) = \sum_j a_j w_j b_j w'_j c_j$, for some $a_j, b_j, c_j \in A$ and $w_j, w'_j \in W_1^{i-1}$. Then, $(fh)^1(w) = f^0 h^1(w) + f^1(w) 1_M + \sum_j a_j f^1(w_j) b_j h^1(w'_j) c_j$. Therefore, $(fh)^1(w) = -f^0 t^0 f^1(w) + f^1(w) 1_M = 0$, and the claim is proved. □

Lemma 5.8. *Assume that $\mathcal{A}$ is a Roiter ditalgebra with triangular layer (R, W) and consider any morphism $f = (f^0, f^1) : M \longrightarrow N$ in $\mathcal{A}$-Mod. Then, f is an isomorphism in $\mathcal{A}$-Mod iff f^0 is an isomorphism in R-Mod.*

Proof. It is clear that if f is an isomorphism in $\mathcal{A}$-Mod, then f^0 is an isomorphism in R-Mod. Conversely, assume that f^0 is an isomorphism in R-Mod. By (5.7), there is morphism $g = (g^0, g^1) : N \longrightarrow N'$ with $(gf)^1 = 0$ and g^0 isomorphism in R-Mod. This implies that $g^0 f^0 : M \longrightarrow N'$ is an A-module isomorphism, hence $gf = (g^0 f^0, 0)$ is an isomorphism in $\mathcal{A}$-Mod, and so f is a section in $\mathcal{A}$-Mod. Dually, we show that f is a retraction in $\mathcal{A}$-Mod. □

Lemma 5.9. *Assume that $\mathcal{A}$ is a Roiter ditalgebra with layer (R, W) such that $W_0 = 0$, and consider any morphism $f = (f^0, f^1) : M \longrightarrow N$ in $\mathcal{A}$-Mod. Then, f is a section (resp. retraction) in $\mathcal{A}$-Mod iff f^0 is a section (resp. retraction) in R-Mod.*

Proof. Assume that f^0 is a section in R-Mod. By (5.7), there is morphism $g = (g^0, g^1) : N \longrightarrow N'$ with $(gf)^1 = 0$ and g^0 isomorphism. This implies that $g^0 f^0$ is a section in R-Mod and, since $W_0 = 0$, it is also a section in A-Mod. Then, $gf = (g^0 f^0, 0)$ has a left inverse in $\mathcal{A}$-Mod, so f is a section in $\mathcal{A}$-Mod. Similarly, we prove that f is a retraction in $\mathcal{A}$-Mod when f^0 is a retraction in R-Mod. □

Definition 5.10. *In an additive category $\mathcal{C}$ a non-zero object M is called indecomposable iff $M \cong M_1 \oplus M_2$ in $\mathcal{C}$ implies that $M_1 = 0$ or $M_2 = 0$. We say that N is a direct summand of M in $\mathcal{C}$ iff $M \cong N \oplus N'$ in $\mathcal{C}$, for some $N' \in \mathcal{C}$. A Krull–Schmidt category $\mathcal{C}$ is an additive k-category, where each object is a finite direct sum of indecomposable objects in $\mathcal{C}$ with local endomorphism rings.*

Let us recall some essential properties of Krull–Schmidt categories (see, for instance, [32](3.3))

Lemma 5.11. *If $\mathcal{C}$ is a Krull–Schmidt category, then we have:*

(1) All idempotents split in $\mathcal{C}$. This means that, for any $M \in \mathcal{C}$ and any idempotent $e \in \mathrm{End}_{\mathcal{C}}(M)$, there exists an isomorphism $h : M \longrightarrow M_1 \oplus M_2$ in $\mathcal{C}$ such that the following diagram commutes

$$\begin{array}{ccc} M & \xrightarrow{h} & M_1 \oplus M_2 \\ {\scriptstyle e}\downarrow & & \downarrow\begin{pmatrix} 0 & 0 \\ 0 & 1 \end{pmatrix} \\ M & \xrightarrow{h} & M_1 \oplus M_2. \end{array}$$

Observe that idempotents split in $\mathcal{C}$ iff, for any $M \in \mathcal{C}$ and any idempotent $e \in \mathrm{End}_{\mathcal{C}}(M)$, there exist morphisms $N \xrightarrow{\sigma} M$ and $M \xrightarrow{\pi} N$ in $\mathcal{C}$ such that $\sigma\pi = e$ and $\pi\sigma = 1_N$.

(2) *Whenever $M \cong \bigoplus_{i=1}^n d_i M_i$ in $\mathcal{C}$, where the M_i are non-isomorphic indecomposables and $n, d_i \geq 1$, the isoclasses of the objects M_i, as well as their multiplicities d_i, are determined uniquely.*

Lemma 5.12. *If $\mathcal{A}$ is a Roiter ditalgebra, then idempotents split in $\mathcal{A}$-Mod. As a consequence of this, $M \in \mathcal{A}$-Mod is indecomposable iff $\mathrm{End}_{\mathcal{A}}(M)$ admits no non-trivial idempotents. In particular, for any $M \in \mathcal{A}$-Mod with $\dim_k \mathrm{End}_{\mathcal{A}}(M)$ finite, we have that M is indecomposable iff $\mathrm{End}_{\mathcal{A}}(M)$ is local.*

Proof. Fix $M \in \mathcal{A}$-Mod and take an idempotent $e = (e^0, e^1) \in \mathrm{End}_{\mathcal{A}}(M)$. Observe first that if $e^1 = 0$, then e^0 is an idempotent of $\mathrm{End}_A(M)$; hence, there is an isomorphism $M \xrightarrow{h^0} M_1 \oplus M_2$ in A-Mod such that $h^0 e^0 (h^0)^{-1} = \begin{pmatrix} 0 & 0 \\ 0 & 1 \end{pmatrix}$. Applying the embedding functor $L_{\mathcal{A}}$ to this, we obtain a splitting for e in $\mathcal{A}$-Mod.

To finish the proof, it will be enough to show the existence of an isomorphism $h : M' \longrightarrow M$ in $\mathcal{A}$-Mod with $(h^{-1}eh)^1 = 0$. Adopt the notation of (5.1) and notice that the proof follows by induction from the following.

Claim. Given $i \in [1, s]$, if $e^1(W_1^{i-1}) = 0$, then there exists an isomorphism $h : M' \longrightarrow M$ in $\mathcal{A}$-Mod with $(h^{-1}eh)^1(W_1^i) = 0$.

Proof of the claim: e idempotent implies that e^0 is an idempotent too. From $e^1(W_1^{i-1}) = 0$ and the triangularity we obtain, for $w \in W_1^i$, that $e^1(w) = (e^2)^1(w) = e^0 e^1(w) + e^1(w)e^0$. Hence, $e^0 e^1(w) e^0 = 0$. Now, for $w \in W_1$, define $h^1(w) := e^1(w) f^0$, where $f^0 \in \mathrm{Hom}_R(M, M)$ will be specified in a moment. Then, clearly $h^1(W_1^{i-1}) = 0$ and $h^1 \in \mathrm{Hom}_{R\text{-}R}(W_1, \mathrm{Hom}_k(M, M))$. Since $\mathcal{A}$ is a Roiter ditalgebra, there is a left A-module structure on the underlying R-module M' of M such that $h = (1_M, h^1) : M' \longrightarrow M$ is a morphism in $\mathcal{A}$-Mod. From (5.8), we know that h is an isomorphism in $\mathcal{A}$-Mod. Let $g := h^{-1} : M \longrightarrow M'$. Then, $g^0 = 1_M$ and, since $(gh)^1 = 0$, we obtain for $w \in W_1^i$ that $g^1(w) = -h^1(w)$. Then, by triangularity, we have for $w \in W_1^i$

$$\begin{aligned}(geh)^1(w) &= g^0(eh)^1(w) + g^1(w)(eh)^0 \\ &= e^0 h^1(w) + e^1(w) - h^1(w)e^0 \\ &= e^0 e^1(w) f^0 + e^1(w) - e^1(w) f^0 e^0 \\ &= e^0 e^1(w) f^0 + e^0 e^1(w) + e^1(w) e^0 - e^1(w) f^0 e^0.\end{aligned}$$

Since $e^0 e^1(w) e^0 = 0$, choosing $f^0 := 2e^0 - 1_M$ we obtain that $(geh)^1(w) = 0$, as we wanted. □

The next statement follows from (5.12) and some basic results on additive categories (see, for instance, [32](3.3)). The assumption on W_1 guarantees that Hom-spaces between finite-dimensional $\mathcal{A}$-modules are again finite dimensional.

Theorem 5.13. *Let $\mathcal{A}$ be a Roiter ditalgebra with layer (R, W) such that the R-R-bimodule W_1 is finitely generated. Then, $\mathcal{A}$-mod is a Krull–Schmidt category.*

Lemma 5.14. *Let $\mathcal{A}$ be a Roiter ditalgebra. Suppose $M \xrightarrow{f} E \xrightarrow{g} N$ is a pair of composable morphisms in $\mathcal{A}$-Mod with $gf = 0$ and*

$$0 \longrightarrow M \xrightarrow{f^0} E \xrightarrow{g^0} N \longrightarrow 0$$

is a split exact sequence of left R-modules. Then, there is an isomorphism $h : E' \longrightarrow E$ in $\mathcal{A}$-Mod such that $(gh)^1 = 0$ and $(h^{-1}f)^1 = 0$.

Proof. By (5.7), we may assume that $f^1 = 0$. Once again, we adopt the notation of (5.1). The proof will follow by induction from the following.

Claim. Given $i \in [1, s]$, if $f^1 = 0$ and $g^1(W_1^{i-1}) = 0$, then there exists an isomorphism $h : E' \longrightarrow E$ in $\mathcal{A}$-Mod with $(h^{-1}f)^1 = 0$ and $(gh)^1(W_1^i) = 0$.

Proof of the claim: First follow the argument in the proof of the claim in (5.7), where using a right inverse t^0 for g^0 we constructed an isomorphism $h = (1_E, h^1) : E' \longrightarrow E$ in $\mathcal{A}$-Mod, where E' has the same underlying R-module that E has, $h^1(w) = -t^0 g^1(w)$, for $w \in W_1$, and $(gh)^1(W_1^i) = 0$. We will show that $(h^{-1}f)^1 = 0$. For this, we first notice that $(gf)^1 = 0$ implies $0 = (gf)^1(w) = g^1(w)f^0$, for any $w \in W_1$. Thus, $h^1(w)f^0 = 0$, for any $w \in W_1$, too. Since $f^1 = 0$, we know that $f^0 \in \mathrm{Hom}_A(M, E)$. Then, if $a, b \in A$, $w \in W_1$ and $m \in M$, we have $\left(h^1(awb)f^0\right)[m] = \left(ah^1(w)b\right)\left[f^0[m]\right] = ah^1(w)\left[bf^0[m]\right] = a\left(h^1(w)f^0[bm]\right) = 0$. Thus, $h^1(v)f^0 = 0$ holds for any $v \in AW_1A = V$. Now, if $u := h^{-1}$, we get that $u^0 = 1_E$. Moreover, if $w \in W_1$ and we write $\delta(w) = \sum_j v_j v'_j$, with $v_j, v'_j \in AW_1A$, then $0 = (uh)^1(w) = h^1(w) + u^1(w) + \sum_j u^1(v_j)h^1(v'_j)$. Then, $(uf)^1(w) = u^0 f^1(w) + u^1(w)f^0 + \sum_j u^1(v_j)f^1(v'_j) = u^1(w)f^0 = [-h^1(w) - \sum_j u^1(v_j)h^1(v'_j)]f^0 = 0$. Which implies that $(uf)^1 = 0$, as wanted. □

Exercise 5.15. *Let $\mathcal{A}$ be a triangular ditalgebra with layer (R, W) and let $f : M \longrightarrow N$ be a morphism in $\mathcal{A}$-Mod. Show that f is an epimorphism (resp. a monomorphism) in $\mathcal{A}$-Mod whenever f^0 is an epimorphism (resp. a monomorphism) in R-Mod.*

Exercise 5.16. *Let $R = k \times k \times k$ and consider the canonical decomposition of the unit $1 = e_1 + e_2 + e_3$ of R. Consider also the R-R-bimodules $W_0 := (ke_2 \otimes_k ke_1) \oplus (ke_3 \otimes_k ke_1)$, $W_1 := ke_3 \otimes_k ke_2$ and $W = W_0 \oplus W_1$. Make $\alpha := e_2 \otimes e_1 \in W_0$, $\beta := e_3 \otimes e_1 \in W_0$ and $v := e_3 \otimes e_2 \in W_1$. Thus, $W_0 := R\alpha R \oplus R\beta R$ and $W_1 := RvR$. Consider the t-algebra $T := T_R(W)$ and the R-R-bimodule morphism $\delta : W \to T$ defined by $\delta(a\alpha + b\beta + cv) := bv\alpha$, for $a, b, c \in k$. Extend this map to a differential $\delta : T \to T$ using (4.4). Show that $\mathcal{A} = (T, \delta)$ is a ditalgebra with triangular layer (R, W). Consider: the $\mathcal{A}$-module M defined by $e_1M = 0$, $e_2M = k$ and $e_3M = 0$; the $\mathcal{A}$-module N defined by $e_1N = 0$, $e_2N = 0$ and $e_3N = k$; and the morphism $f = (f^0, f^1) : M \longrightarrow N$ in $\mathcal{A}$-*Mod *defined by $f^0 = 0$ and $f^1(v) = 1_k$. Show that f is a monomorphism which is not a kernel in $\mathcal{A}$-*Mod*. Thus, $\mathcal{A}$-*Mod *is not an abelian category.*

Exercise 5.17. *Prove statement (2) of Lemma (5.3).*

Exercise 5.18. *Let $\mathcal{A}$ be a triangular ditalgebra with layer (R, W) and assume $M \in \mathcal{A}$-*Mod*. Then, for any nilpotent ideal I of $\mathrm{End}_R(M)$, the subset $\widetilde{I} := \{f \in \mathrm{End}_{\mathcal{A}}(M) \mid f^0 \in I\}$ is a nilpotent ideal of $\mathrm{End}_{\mathcal{A}}(M)$. If, moreover, $W_0 = 0$, then $I \times \mathrm{Hom}_{R\text{-}R}(W_1, \mathrm{Hom}_k(M, M)) \subseteq \widetilde{I}$. Thus, if $W_0 = 0$, then $\widetilde{0} = \{0\} \times \mathrm{Hom}_{R\text{-}R}(W_1, \mathrm{Hom}_k(M, M))$ is a nilpotent ideal of $\mathrm{End}_{\mathcal{A}}(M)$.*

Exercise 5.19. *Let $R = k$ and $W := (k \otimes_k k) \times (k \otimes_k k)$. Make $\alpha := (1 \otimes 1, 0)$, $v := (0, 1 \otimes 1)$, $W_0 := R\alpha R$ and $W_1 := RvR$. Thus, we have an R-R-bimodule decomposition $W = W_0 \oplus W_1$. Consider the t-algebra $T := T_R(W)$ and the R-R-bimodule morphism $\delta : W \to T$ defined by $\delta(a\alpha + bv) := \alpha v\alpha$, for $a, b \in k$. Extend this map to a differential $\delta : T \to T$ using (4.4). Show that $\mathcal{A} = (T, \delta)$ is a ditalgebra with (non-triangular) layer (R, W). Show there exist $M \in R$-*Mod*, $N \in \mathcal{A}$-*Mod *and $f \in \mathrm{Phom}_{R\text{-}W}(M, N)$, with f^0 an R-isomorphism, such that M admits no structure of $\mathcal{A}$-module with $f \in \mathrm{Hom}_{\mathcal{A}}(M, N)$.*

Exercise 5.20. *Let $R = k$ and $W := (k \otimes_k k) \times (k \otimes_k k)$. Make $\alpha := (1 \otimes 1, 0)$, $v := (0, 1 \otimes 1)$, $W_0 := R\alpha R$ and $W_1 := RvR$. Thus, we have an R-R-bimodule decomposition $W = W_0 \oplus W_1$. Consider the t-algebra $T := T_R(W)$ and the R-R-bimodule morphism $\delta : W \to T$ defined by $\delta(a\alpha + bv) := a\alpha v - bv^2$, for $a, b \in k$. Extend this map to a differential $\delta : T \to T$ using (4.4). Show that $\mathcal{A} = (T, \delta)$ is a ditalgebra with (non-triangular) layer (R, W). Show there exist a morphism $f : M \to N$ in $\mathcal{A}$-*Mod*, which is not an isomorphism, but f^0 is an isomorphism in R-*Mod*. Finally, show an idempotent endomorphism in $\mathcal{A}$-*mod *which does not split.*

Exercise 5.21. *Let $R = k$ and $W := (k \otimes_k k)$. Make $v := 1 \otimes 1$, $W_0 := 0$ and $W_1 := RvR$. Thus, we have an R-R-bimodule decomposition $W = W_0 \oplus W_1$. Consider the t-algebra $T := T_R(W)$ and the R-R-bimodule morphism $\delta : W \to T$ defined by $\delta(av) := -av^2$, for $a \in k$. Extend this map to a differential $\delta : T \to T$ using (4.4). Show that $\mathcal{A} = (T, \delta)$ is a ditalgebra with (non-triangular) layer (R, W). Notice that, $\mathcal{A}$ satisfies (5.3)(1) and (5.3)(2). Show a non-nilpotent endomorphism $f = (f^0, f^1)$ in* $\mathcal{A}$-mod *with f^0 nilpotent (see (5.4)). Show there exist a morphism $f : M \to N$ in* $\mathcal{A}$-Mod, *which is not an isomorphism, while f^0 is an isomorphism in* R-Mod.

References. The original definition of triangularity of a free bocs involved the existence of a "height" map $\mathbb{h} : \mathbb{B}_0 \cup \mathbb{B}_1 \longrightarrow \mathbb{N}$ on the union of the set of solid arrows and the set of dotted arrows, such that the differential of any arrow t could be expressed by means of arrows t' with $\mathbb{h}(t') < \mathbb{h}(t)$ (see [46] and [45]). The arrows are the generators of the bimodules W_0 and W_1 of the layer, as we will make precise later on. In Crawley's formulation (see [19]) this partial ordering of the arrows is only required on solid arrows. Here, we require the existence of two independent partial orders, determined by the filtrations of W_0 and W_1, as was suggested in [6]. As we remarked before, the extra filtration on W_1 permits to guarantee nice algebraic properties for the category of modules. The independence of the partial orders makes it easier to follow their independent impact on the study of the category of modules.

The importance of the condition required in (5.5) was made explicit in [45]. We propose the name of *Roiter ditalgebra* as a tribute to the importance of the seminal work of Professor A. V. Roiter in this line of representation theory.

Lemma (5.7) was first remarked by Ovsienko in [43]; Lemmas (5.8) and (5.12) are taken from [6], but they evolved from [46].

6

Exact structures in $\mathcal{A}$-Mod

In this section, we present a natural exact structure for the category of modules over a Roiter ditalgebra $\mathcal{A}$. We will recall some terminology and general concepts of exact category theory as needed.

Definition 6.1. *If $\mathcal{C}$ is an additive k-category where idempotents split, a pair (s, d) of composable morphisms $M \xrightarrow{s} E \xrightarrow{d} N$ in $\mathcal{C}$ is called exact iff s is the kernel of d and d is the cokernel of s. A morphism of exact pairs from $M \xrightarrow{s} E \xrightarrow{d} N$ to $M' \xrightarrow{s'} E' \xrightarrow{d'} N'$ is a triple of morphisms (u, v, w) in $\mathcal{C}$, making the following diagram commutative*

$$\begin{array}{ccccc} M & \xrightarrow{s} & E & \xrightarrow{d} & N \\ {\scriptstyle u}\downarrow & & {\scriptstyle v}\downarrow & & {\scriptstyle w}\downarrow \\ M' & \xrightarrow{s'} & E' & \xrightarrow{d'} & N'. \end{array}$$

If u, v and w are isomorphisms, (u, v, w) is an isomorphism of exact pairs. If there is such an isomorphism with $u = 1_M$ and $w = 1_N$, the exact pairs are called equivalent.

Once we have fixed a class of exact pairs $\mathcal{E}$ closed under isomorphisms, if $(s, d) \in \mathcal{E}$, s is called an inflation and d is called a deflation.

Lemma 6.2. *Assume that $\mathcal{A}$ is a Roiter ditalgebra. Then, whenever*

$$0 \longrightarrow M \xrightarrow{f^0} E \xrightarrow{g^0} N \longrightarrow 0$$

is an exact sequence in A-Mod, we have that $M \xrightarrow{(f^0,0)} E \xrightarrow{(g^0,0)} N$ is an exact pair in $\mathcal{A}$-Mod.

Proof. Write $f = (f^0, 0)$ and $g = (g^0, 0)$. We only prove that $f = \operatorname{Ker} g$ (the equality $g = \operatorname{Cok} f$ is proved dually). So suppose that $Z \xrightarrow{t} E$ satisfies $gt = 0$ in $\mathcal{A}$-Mod. Then, $g^0 t^0 = 0$ and $g^0 t^1(v) = 0$, for all $v \in V$. Then, there exist a unique $h^0 \in \mathrm{Hom}_R(Z, M)$ with $f^0h^0 = t^0$ and, for each $v \in V$, there exists a unique $h^1(v) \in \mathrm{Hom}_k(Z, M)$ with $f^0h^1(v) = t^1(v)$. Since f^0 is an injective morphism of A-modules, we get that $h^1 \in \mathrm{Hom}_{A\text{-}A}(V, \mathrm{Hom}_k(Z, M))$. Moreover, for $a \in A$ and $z \in Z$

$$f^0(ah^0(z)) = at^0(z) = t^0(az) + t^1(\delta(a))(z) = f^0\left[h^0(az) + h^1(\delta(a))(z)\right].$$

Using again that f^0 is injective, we get that $h = (h^0, h^1) \in \mathrm{Hom}_{\mathcal{A}}(Z, M)$. Clearly, h is unique such that $fh = t$ in $\mathcal{A}$-Mod. □

Definition 6.3. *Assume that $\mathcal{C}$ is an additive k-category and $\mathcal{E}$ is a class of exact pairs in $\mathcal{C}$ closed under isomorphisms. Then, $\mathcal{E}$ is called an exact structure on $\mathcal{C}$ iff:*

(1) The composition of deflations is a deflation.
(2) For each morphism $f : Z' \longrightarrow Z$ and each deflation $d : Y \longrightarrow Z$, there exists a morphism $f' : Y' \longrightarrow Y$ and a deflation $d' : Y' \longrightarrow Z'$ such that $df' = fd'$.
(3) Identities are deflations. If gf is a deflation, then so is g.
(3)^{op} Identities are inflations. If gf is a inflation, then so is f.

A pair $(\mathcal{C}, \mathcal{E})$, as above, is called an exact category. A functor $F : \mathcal{C} \longrightarrow \mathcal{C}'$, where the categories $\mathcal{C}$ and $\mathcal{C}'$ are endowed with exact structures $\mathcal{E}$ and $\mathcal{E}'$, respectively, is called exact iff it maps $\mathcal{E}$ into $\mathcal{E}'$. Given an exact category $(\mathcal{C}, \mathcal{E})$, an object $P \in \mathcal{C}$ (resp. $I \in \mathcal{C}$) is called $\mathcal{E}$-projective (resp. $\mathcal{E}$-injective) iff $\mathrm{Hom}_{\mathcal{C}}(P, -) : \mathcal{C} \longrightarrow k\text{-Mod}$ is an exact functor (resp. $\mathrm{Hom}_{\mathcal{C}}(-, I) : \mathcal{C}^{op} \longrightarrow k\text{-Mod}$ is an exact functor).

We will provide a unified context for the exact structures on $\mathcal{A}$-Mod we are going to consider.

Definition 6.4. *Assume that $\mathcal{A}$ is a Roiter ditalgebra with layer (R, W) and let S be a k-subalgebra of R. Consider the class $\mathcal{E}_{\mathcal{A},S}$ of composable pairs of morphisms $M \xrightarrow{f} E \xrightarrow{g} N$ such that there is a commutative diagram in $\mathcal{A}$-Mod*

$$\begin{array}{ccccc} M & \xrightarrow{f} & E & \xrightarrow{g} & N \\ \downarrow \cong & & \downarrow \cong & & \downarrow \cong \\ M' & \xrightarrow{(\varphi^0,0)} & E' & \xrightarrow{(\psi^0,0)} & N', \end{array}$$

such that $0 \longrightarrow M' \xrightarrow{\varphi^0} E' \xrightarrow{\psi^0} N' \longrightarrow 0$ is a split exact sequence in S-Mod. We shall say that the Roiter ditalgebra $\mathcal{A}$ is S-acceptable iff, for any morphism $s : X \longrightarrow Y$ in $\mathcal{A}$-Mod such that $s^0 : X \longrightarrow Y$ is a section or a retraction in

S-Mod, *there are isomorphisms* $r : X' \longrightarrow X$ *and* $t : Y \longrightarrow Y'$ *in* $\mathcal{A}$-Mod *such that* $(tsr)^1 = 0$.

We will establish that $\mathcal{E}_{\mathcal{A},S}$ is an exact structure of $\mathcal{A}$-Mod, for any S-acceptable $\mathcal{A}$. In this section, we will concentrate our attention on the following family of examples. In Section 32, we consider another type of example.

Remark 6.5. *If $\mathcal{A}$ is a Roiter ditalgebra with layer (R, W), by Lemma (5.7), $\mathcal{A}$ is R-acceptable.*

Lemma 6.6. *Let $\mathcal{A}$ be an S-acceptable ditalgebra. Then, $\mathcal{E}_{\mathcal{A},S}$ is a class of exact pairs in* $\mathcal{A}$-Mod *closed under isomorphisms. Moreover, if we consider morphisms $f : M \longrightarrow E$ and $g : E \longrightarrow N$ in* $\mathcal{A}$-Mod, *then:*

(1) $M \xrightarrow{f} E \xrightarrow{g} N$ *is a conflation iff $gf = 0$ and the sequence* $0 \longrightarrow M \xrightarrow{f^0} E \xrightarrow{g^0} N \longrightarrow 0$ *is a split exact sequence in* S-Mod.
(2) $g : E \longrightarrow N$ *is a deflation iff* $g^0 : E \longrightarrow N$ *is a retraction in* S-Mod.
(3) $f : M \longrightarrow E$ *is an inflation iff* $f^0 : M \longrightarrow E$ *is a section in* S-Mod.

Proof. We will prove (1) and, at the same time, that $\mathcal{E}_{\mathcal{A},S}$ is a class of exact pairs in $\mathcal{A}$-Mod. If $M \xrightarrow{f} E \xrightarrow{g} N \in \mathcal{E}_{\mathcal{A},S}$, there is a commutative diagram in $\mathcal{A}$-Mod

$$\begin{array}{ccccc} M & \xrightarrow{f} & E & \xrightarrow{g} & N \\ \downarrow{\scriptstyle t} & & \downarrow{\scriptstyle s} & & \downarrow{\scriptstyle r} \\ M' & \xrightarrow{(\varphi^0,0)} & E' & \xrightarrow{(\psi^0,0)} & N', \end{array}$$

such that r, s, t are isomorphisms and $0 \longrightarrow M' \xrightarrow{\varphi^0} E' \xrightarrow{\psi^0} N' \longrightarrow 0$ is a split exact sequence in S-Mod. In particular, we have the commutative diagram in S-Mod

$$\begin{array}{ccccc} M & \xrightarrow{f^0} & E & \xrightarrow{g^0} & N \\ \downarrow{\scriptstyle t^0} & & \downarrow{\scriptstyle s^0} & & \downarrow{\scriptstyle r^0} \\ M' & \xrightarrow{\varphi^0} & E' & \xrightarrow{\psi^0} & N', \end{array}$$

where r^0, s^0, t^0 are isomorphisms. Then, $0 \longrightarrow M \xrightarrow{f^0} E \xrightarrow{g^0} N \longrightarrow 0$ is a split exact sequence in S-Mod.

Conversely, if $M \xrightarrow{f} E \xrightarrow{g} N$ is a pair of composable morphisms in $\mathcal{A}$-Mod such that $gf = 0$ and

$$0 \longrightarrow M \xrightarrow{f^0} E \xrightarrow{g^0} N \longrightarrow 0$$

is a split exact sequence in S-Mod, then, since $\mathcal{A}$ is S-acceptable, there is a commutative square in $\mathcal{A}$-Mod

$$\begin{array}{ccc} E & \xrightarrow{g} & N \\ {\scriptstyle s}\downarrow & & \downarrow{\scriptstyle r} \\ E' & \xrightarrow{(\psi^0,0)} & N', \end{array}$$

where r, s are isomorphisms. It follows that ψ^0 is a retraction in S-Mod. Now, consider the kernel $\varphi^0 : M' \longrightarrow E'$ of the morphism $\psi^0 : E' \longrightarrow N'$ in A-Mod. Since $gf = 0$, from (6.2), we know that there is an induced morphism of $\mathcal{A}$-modules $t : M \longrightarrow M'$ such that the following diagram commutes

$$\begin{array}{ccccc} M & \xrightarrow{f} & E & \xrightarrow{g} & N \\ {\scriptstyle t}\downarrow & & {\scriptstyle s}\downarrow & & \downarrow{\scriptstyle r} \\ M' & \xrightarrow{(\varphi^0,0)} & E' & \xrightarrow{(\psi^0,0)} & N', \end{array}$$

where the lower row is an exact pair in $\mathcal{A}$-Mod. In particular, we have a commutative diagram

$$\begin{array}{ccccc} M & \xrightarrow{f^0} & E & \xrightarrow{g^0} & N \\ {\scriptstyle t^0}\downarrow & & {\scriptstyle s^0}\downarrow & & \downarrow{\scriptstyle r^0} \\ M' & \xrightarrow{\varphi^0} & E' & \xrightarrow{\psi^0} & N', \end{array}$$

which implies that t^0 is an isomorphism. Since $\mathcal{A}$ is a Roiter ditalgebra, t is an isomorphism too and $M \xrightarrow{f} E \xrightarrow{g} N$ is an exact pair of $\mathcal{A}$-Mod, which belongs to $\mathcal{E}_{A,S}$.

(2) If $g : E \longrightarrow N$ is a deflation, clearly we know that g^0 is a retraction in S-Mod. Moreover, if we start from a morphism $g : E \longrightarrow N$ in $\mathcal{A}$-Mod such that $g^0 : E \longrightarrow N$ is a retraction in S-Mod, since $\mathcal{A}$ is acceptable there is a commutative square in $\mathcal{A}$-Mod

$$\begin{array}{ccc} E & \xrightarrow{g} & N \\ {\scriptstyle s}\downarrow & & \downarrow{\scriptstyle r} \\ E' & \xrightarrow{(\psi^0,0)} & N', \end{array}$$

where r, s are isomorphisms. Then, as before, we consider the kernel φ^0 of ψ^0 in A-Mod and construct a commutative square in $\mathcal{A}$-Mod

$$\begin{array}{ccccc} M & \xrightarrow{f} & E & \xrightarrow{g} & N \\ \downarrow{\scriptstyle t} & & \downarrow{\scriptstyle s} & & \downarrow{\scriptstyle r} \\ M' & \xrightarrow{(\varphi^0,0)} & E' & \xrightarrow{(\psi^0,0)} & N', \end{array}$$

where t is an isomorphism, so g is a deflation.

The proof of (3) is dual to the proof of (2). □

Proposition 6.7. *If $\mathcal{A}$ is an S-acceptable ditalgebra, then $\mathcal{E}_{\mathcal{A},S}$ is an exact structure on $\mathcal{A}$-Mod.*

Proof. From (6.6), items (1), (3) and $(3)^{op}$ of Definition (6.3) are clear. Let us prove the remaining one (2). Take a morphism $f : Z' \longrightarrow Z$ and any deflation $d : Y \longrightarrow Z$. Consider the morphism $(f, d) : Z' \bigoplus Y \longrightarrow Z$ in $\mathcal{A}$-Mod. Since d is a deflation, d^0 is a retraction in S-Mod and, therefore, so is (f^0, d^0). Then, $Z' \bigoplus Y \xrightarrow{(f,d)} Z$ is a deflation and its kernel has the form $\begin{pmatrix} d' \\ -f' \end{pmatrix} : Y' \longrightarrow Z' \bigoplus Y$. Therefore, we have a pull-back diagram in S-Mod

$$\begin{array}{ccc} Y' & \xrightarrow{(d')^0} & Z' \\ {\scriptstyle (f')^0}\downarrow & & \downarrow{\scriptstyle f^0} \\ Y & \xrightarrow{d^0} & Z. \end{array}$$

Since d^0 is a retraction in S-Mod, the same is true for $(d')^0$. Therefore, d' is a deflation, proving our statement. □

Corollary 6.8. *If $\mathcal{A}$ is a Roiter ditalgebra with layer (R, W), then $\mathcal{E}_{\mathcal{A}} := \mathcal{E}_{\mathcal{A},R}$ is an exact structure on $\mathcal{A}$-Mod.*

Remark 6.9. *If Λ is a k-algebra and $\mathcal{R}$ is its regular ditalgebra (which has layer $(\Lambda, 0)$), then the identification of $\mathcal{R}$-Mod with Λ-Mod maps the exact structure $\mathcal{E}_{\mathcal{R},k1}$ onto the standard exact structure of Λ-Mod.*

Remark 6.10. *Let $(\mathcal{C}, \mathcal{E})$ be an exact category. Then, we know that:*

(0) *In any commutative diagram in $\mathcal{C}$*

$$\begin{array}{ccccc} \xi : N & \xrightarrow{s} & E & \xrightarrow{d} & M \\ {\scriptstyle u}\downarrow & & {\scriptstyle v}\downarrow & & {\scriptstyle w}\downarrow \\ \xi' : N' & \xrightarrow{s'} & E' & \xrightarrow{d'} & M, \end{array}$$

where $\xi, \xi' \in \mathcal{E}$, if u and v are isomorphisms, so is v.

(1) *Assume* $\xi : N \xrightarrow{s} E \xrightarrow{d} M$ *is an exact pair of* $\mathcal{E}$ *and* $N \xrightarrow{f} N'$ *is a morphism in* $\mathcal{C}$. *Then, there is a commutative diagram*

$$\begin{array}{ccccc} \xi : N & \xrightarrow{s} & E & \xrightarrow{d} & M \\ \downarrow{\scriptstyle f} & & \downarrow & & \| \\ \xi' : N' & \xrightarrow{s'} & E' & \xrightarrow{d'} & M \end{array}$$

in $\mathcal{C}$, *where* $\xi' \in \mathcal{E}$ *and the first square is a pushout diagram. We fix one of such diagrams and write* $f\xi := \xi'$.

$(1)^{op}$ *Assume* $\xi : N \xrightarrow{s} E \xrightarrow{d} M$ *is an exact pair of* $\mathcal{E}$ *and* $M' \xrightarrow{g} M$ *is a morphism in* $\mathcal{C}$. *Then, there is a commutative diagram*

$$\begin{array}{ccccc} \xi' : N & \xrightarrow{s'} & E' & \xrightarrow{d'} & M' \\ \| & & \downarrow & & \downarrow{\scriptstyle g} \\ \xi : N & \xrightarrow{s} & E & \xrightarrow{d} & M \end{array}$$

in $\mathcal{C}$, *where* $\xi' \in \mathcal{E}$ *and the second square is a pullback diagram. We fix one of such diagrams and write* $\xi g := \xi'$.

(2) *Factorization property: assume that we have a morphism*

$$\begin{array}{ccccc} \xi : N & \xrightarrow{s} & E & \xrightarrow{p} & M \\ \downarrow{\scriptstyle f} & & \downarrow{\scriptstyle h} & & \downarrow{\scriptstyle g} \\ \overline{\xi} : \overline{N} & \xrightarrow{\overline{s}} & \overline{E} & \xrightarrow{\overline{p}} & \overline{M} \end{array}$$

of exact pairs in $\mathcal{E}$, *then there is a commutative diagram*

$$\begin{array}{ccccc} \xi : N & \xrightarrow{s} & E & \xrightarrow{p} & M \\ \downarrow{\scriptstyle f} & & \downarrow{\scriptstyle h'} & & \| \\ f\xi : \overline{N} & \longrightarrow & E' & \longrightarrow & M \\ \| & & \downarrow{\scriptstyle h''} & & \downarrow{\scriptstyle g} \\ \overline{\xi} : \overline{N} & \xrightarrow{\overline{s}} & \overline{E} & \xrightarrow{\overline{p}} & \overline{M}, \end{array}$$

with $h''h' = h$. *The dual statement holds for* $\overline{\xi}g$.

(3) *Whenever there is a commutative diagram*

$$\begin{array}{ccccc} \xi : N & \xrightarrow{s} & E & \xrightarrow{d} & M \\ \downarrow{\scriptstyle f} & & \downarrow & & \| \\ \widehat{\xi} : \widehat{N} & \xrightarrow{\widehat{s}} & \widehat{E} & \xrightarrow{\widehat{d}} & M \end{array}$$

in $\mathcal{C}$, *where* $\xi, \widehat{\xi} \in \mathcal{E}$, *we have that* $f\xi$ *and* $\widehat{\xi}$ *are equivalent.*

(3)^{op} *Whenever there is a commutative diagram*

$$\begin{array}{ccccc} \widehat{\xi}: N & \xrightarrow{\widehat{s}} & \widehat{E} & \xrightarrow{\widehat{d}} & \widehat{M} \\ \| & & \downarrow & & \downarrow{\scriptstyle g} \\ \xi: N & \xrightarrow{s} & E & \xrightarrow{d} & M \end{array}$$

in $\mathcal{C}$, where $\xi, \widehat{\xi} \in \mathcal{E}$, we have that ξg and $\widehat{\xi}$ are equivalent.

Proposition 6.11. *If $(\mathcal{C}, \mathcal{E})$ is an exact category, there is a k-bifunctor*

$$\mathrm{Ext}_{\mathcal{E}}(-, ?) : \mathcal{C}^{op} \times \mathcal{C} \longrightarrow k\text{-Mod}.$$

It is defined by paraphrasing the standard construction for module categories using Baer sums. Thus, given $M, N \in \mathcal{C}$, $\mathrm{Ext}_{\mathcal{E}}(M, N)$ is by definition the class of equivalence classes $[\xi]$ of exact pairs $\xi : N \xrightarrow{s} E \xrightarrow{d} M$ in $\mathcal{E}$. The sum of two such classes is given by the formula $[\xi] + [\xi'] = [(\nabla(\xi \bigoplus \xi'))\Delta]$, where Δ and ∇ are the diagonal $M \xrightarrow{\Delta} M \bigoplus M$ and codiagonal $N \bigoplus N \xrightarrow{\nabla} N$ morphisms in $\mathcal{C}$.

If $N \xrightarrow{f} N'$ and $M' \xrightarrow{g} M$ are morphisms in $\mathcal{C}$, we have that the linear map $\mathrm{Ext}_{\mathcal{E}}(M, f) : \mathrm{Ext}_{\mathcal{E}}(M, N) \longrightarrow \mathrm{Ext}_{\mathcal{E}}(M, N')$ is given by $\mathrm{Ext}_{\mathcal{E}}(M, f)[\xi] = [f\xi]$, and the linear map $\mathrm{Ext}_{\mathcal{E}}(g, N) : \mathrm{Ext}_{\mathcal{E}}(M, N) \longrightarrow \mathrm{Ext}_{\mathcal{E}}(M', N)$ is given by $\mathrm{Ext}_{\mathcal{E}}(g, N)[\xi] = [\xi g]$.

Now, we concentrate our attention on the exact category $(\mathcal{A}\text{-Mod}, \mathcal{E}_{\mathcal{A}})$, where $\mathcal{A}$ is a given Roiter ditalgebra. For simplicity, we will write $\mathrm{Ext}_{\mathcal{A}} := \mathrm{Ext}_{\mathcal{E}_{\mathcal{A},R}}$. For this case, from (5.14), we have the following special fact.

Remark 6.12. *Assume that $\mathcal{A}$ is a Roiter ditalgebra. Then, every conflation $M \xrightarrow{f} E \xrightarrow{g} N$ in $\mathcal{E}_{\mathcal{A}}$ is equivalent to some conflation $M \xrightarrow{(s,0)} E \xrightarrow{(d,0)} N$ in $\mathcal{E}_{\mathcal{A}}$, such that $0 \longrightarrow M \xrightarrow{s} E \xrightarrow{d} N \longrightarrow 0$ is an exact sequence in A-Mod which splits when restricted to R.*

Proposition 6.13. *Let $\mathcal{A}$ be a Roiter ditalgebra with layer (R, W) and consider the exact structure $\mathcal{E}_{\mathcal{A}}$ on $\mathcal{A}$-Mod. Then, there is an exact sequence of vector spaces*

$$0 \longrightarrow \mathrm{Hom}_{\mathcal{A}}(M, N) \xrightarrow{\sigma_{\mathcal{A}}} \mathrm{Phom}_{R\text{-}W}(M, N) \xrightarrow{\partial_{\mathcal{A}}} \mathrm{Hom}_R(W_0 \otimes_R M, N)$$
$$\xrightarrow{\psi_{\mathcal{A}}} \mathrm{Ext}_{\mathcal{A}}(M, N) \longrightarrow 0.$$

Moreover, if we denote by $U = U_{\mathcal{A}} : \mathcal{A}\text{-Mod} \longrightarrow R\text{-Mod}$ the forgetful functor, defined by $U(f^0, f^1) = f^0$, and $L = L_{\mathcal{A}} : A\text{-Mod} \longrightarrow \mathcal{A}\text{-Mod}$ is the canonical embedding, then $\mathrm{Hom}_R(W_0 \otimes_R UL(M), UL(N)) \xrightarrow{\psi_{\mathcal{A}}} \mathrm{Ext}_{\mathcal{A}}(L(M), L(N))$ is a natural morphism in the variables $M, N \in A\text{-Mod}$.

Proof. First notice that, since A is freely generated by (R, W_0), we know from (4.8) that any A-module M is uniquely determined by an R-module M and its action morphism $\phi_M \in \mathrm{Hom}_R(W_0 \otimes_R M, M)$. Given $M, N \in A$-Mod and $\gamma \in \mathrm{Hom}_R(W_0 \otimes_R M, N)$, we can construct a new left A-module $N \oplus_\gamma M$ as follows: its underlying R-module is the direct sum $N \oplus M$ of the R-modules N and M; its action morphism is the composition

$$W_0 \otimes_R (N \oplus M) \cong (W_0 \otimes_R N) \oplus (W_0 \otimes_R M) \xrightarrow{\begin{pmatrix} \phi_N & \gamma \\ 0 & \phi_M \end{pmatrix}} N \oplus M.$$

With matrix notation, for $n \in N$, $m \in M$ and $w \in W_0$, this means that $w \begin{pmatrix} n \\ m \end{pmatrix} = \begin{pmatrix} wn + \gamma(w \otimes m) \\ wm \end{pmatrix}$. If $s^0 = \begin{pmatrix} 1 \\ 0 \end{pmatrix} : N \longrightarrow N \bigoplus M$ (resp. $p^0 = (0\ 1) : N \bigoplus M \longrightarrow M$) is the canonical injection (resp. projection) in R-Mod, then $N \xrightarrow{s^0} N \bigoplus_\gamma M \xrightarrow{p^0} M$ is an exact sequence in A-Mod. Conversely, if $M, N, E \in A$-Mod and E has underlying R-module $N \oplus M$ and $N \xrightarrow{s^0} E \xrightarrow{p^0} M$ is an exact sequence in A-Mod, then $E = N \bigoplus_\gamma M$ for some morphism $\gamma \in \mathrm{Hom}_R(W_0 \otimes_R M, N)$.

With this notation in mind, we define $\psi(\gamma) \in \mathrm{Ext}_{\mathcal{A}}(M, N)$ as the equivalence class of

$$\psi_\gamma \quad : \quad N \xrightarrow{(s^0,0)} N \oplus_\gamma M \xrightarrow{(p^0,0)} M.$$

Let us show that ψ is surjective. Consider an exact pair $\xi : N \xrightarrow{f} E \xrightarrow{g} M$ in $\mathcal{E}_{\mathcal{A}}$. From (6.12), we may assume that $f^1 = 0$, $g^1 = 0$ and that the exact sequence $0 \longrightarrow N \xrightarrow{f^0} E \xrightarrow{g^0} M \longrightarrow 0$ is a split sequence in R-Mod. Then, there is an isomorphism h^0 in R-Mod such that the following diagram

$$\begin{array}{ccccc} N & \xrightarrow{f^0} & E & \xrightarrow{g^0} & M \\ \| & & \downarrow{\scriptstyle h^0} & & \| \\ N & \xrightarrow{s^0} & N \oplus M & \xrightarrow{p^0} & M \end{array}$$

commutes in R-Mod. Since $\mathcal{A}$ is a Roiter ditalgebra, the R-module $N \bigoplus M$ admits a structure of an A-module $\widehat{N \bigoplus M}$ such that $(h^0, 0) \in \mathrm{Hom}_{\mathcal{A}}(E, \widehat{N \bigoplus M})$. Then, $s^0 = h^0 f^0$ and $p^0 = g^0 (h^0)^{-1}$ are morphisms of A-modules. Hence, by our previous observation, $\widehat{N \bigoplus M} = N \oplus_\gamma M$, for some $\gamma \in \mathrm{Hom}_R(W_0 \otimes_R M, N)$. Now, we just have to embed the last diagram with L_A in $\mathcal{A}$-Mod to realize that $\psi(\gamma) = [\xi]$.

Now we prove the exactness of the sequence of our proposition assuming that ψ is additive. Suppose that $\psi(\gamma) = 0$. Then, there exists an isomorphism

h in $\mathcal{A}$-Mod such that the following diagram commutes in $\mathcal{A}$-Mod

$$\begin{array}{ccccc} \psi_0 : N & \xrightarrow{(s^0,0)} & N \oplus_0 M & \xrightarrow{(p^0,0)} & M \\ \| & & \downarrow h & & \| \\ \psi_\gamma : N & \xrightarrow{(s^0,0)} & N \oplus_\gamma M & \xrightarrow{(p^0,0)} & M. \end{array}$$

The commutativity implies that $s^0 = h^0 s^0$, $p^0 h^0 = p^0$ and, for each $v \in V$, $h^1(v)s^0 = 0$ and $p^0 h^1(v) = 0$. Hence

$$h^0 = \begin{pmatrix} 1 & g^0 \\ 0 & 1 \end{pmatrix} \quad \text{and} \quad h^1(v) = \begin{pmatrix} 0 & g^1(v) \\ 0 & 0 \end{pmatrix},$$

for some $(g^0, g^1) \in \mathrm{Phom}_{R\text{-}W}(M, N)$. The condition

$$0 = wh^0 \begin{pmatrix} n \\ m \end{pmatrix} - h^0 \left(w \begin{pmatrix} n \\ m \end{pmatrix} \right) - h^1(\delta(w)) \begin{pmatrix} n \\ m \end{pmatrix},$$

for $w \in W_0$, $n \in N$ and $m \in M$, is equivalent to the equality

$$\gamma(w \otimes m) = g^0(wm) - wg^0(m) + g^1(\delta(w))[m].$$

This means that $\partial_A(g^0, g^1) = \gamma$. The argument can be reversed to construct from an element $(g^0, g^1) \in \mathrm{Phom}_{R\text{-}W}(M, N)$, making $\gamma := \partial_A(g^0, g^1)$, an isomorphism $h = (h^0, h^1)$ with the matricial description given before, which makes the last diagram commutative. Then $\mathrm{Im}\, \partial_A = \mathrm{Ker}\, \psi_A$, and the sequence of our statement is exact.

For the proof of the naturality of ψ, assume that $f^0 : N \longrightarrow N'$ is a morphism of A-Mod, write $f = L(f^0) = (f^0, 0)$ and take $\gamma \in \mathrm{Hom}_R(W_0 \otimes_R M, N)$. Then, consider the morphism $h := (f^0 \oplus 1_M, 0) : N \oplus_\gamma M \longrightarrow N' \oplus_{f^0\gamma} M$ in $\mathcal{A}$-Mod. Since the following diagram commutes in $\mathcal{A}$-Mod

$$\begin{array}{ccccc} \psi_\gamma \;: N & \xrightarrow{(s^0,0)} & N \oplus_\gamma M & \xrightarrow{(p^0,0)} & M \\ f \downarrow & & \downarrow h & & \| \\ \psi_{f^0\gamma} : N' & \xrightarrow{(s^0,0)} & N' \oplus_{f^0\gamma} M & \xrightarrow{(p^0,0)} & M, \end{array}$$

we obtain that $f\psi_\gamma$ and $\psi_{f^0\gamma}$ are equivalent exact pairs and we have shown the naturality of ψ in the second variable.

Now, if we have a morphism $g^0 : M' \longrightarrow M$ in A-Mod and an element $\gamma \in \mathrm{Hom}_R(W_0 \otimes_R M, N)$, then $\gamma(1 \otimes g^0) \in \mathrm{Hom}_R(W_0 \otimes_R M', N)$. Make $g = L(g^0) = (g^0, 0)$ and consider the morphism $h = (1_N \oplus g^0, 0) : N \oplus_{\gamma(1\otimes g^0)}$

$M' \longrightarrow N \oplus_\gamma M$ in $\mathcal{A}$-Mod. Thus, the following diagram commutes in $\mathcal{A}$-Mod

$$\begin{array}{ccccc} \psi_{\gamma(1\otimes g^0)} : N & \xrightarrow{(s^0,0)} & N \oplus_{\gamma(1\otimes g^0)} M' & \xrightarrow{(p^0,0)} & M' \\ \| & & \downarrow h & & \downarrow g \\ \psi_\gamma \quad : N & \xrightarrow{(s^0,0)} & N \oplus_\gamma M & \xrightarrow{(p^0,0)} & M. \end{array}$$

Then, $\psi_\gamma g$ is equivalent to $\psi_{\gamma(1\otimes g^0)}$ and we have proved that ψ is natural in the first variable too.

Now, we show that ψ is additive. Assume that $\gamma, \lambda \in \mathrm{Hom}_R(W_0 \otimes_R M, N)$ and denote by $\gamma \hat{\oplus} \lambda$ the composition

$$W_0 \otimes_R (M \oplus M) \cong (W_0 \otimes_R M) \oplus (W_0 \otimes_R M) \xrightarrow{\gamma \oplus \lambda} N \oplus N.$$

Then we have $\gamma + \lambda = \nabla^0 \left(\gamma \hat{\oplus} \lambda\right) \left(1 \otimes \Delta^0\right)$. Notice that the map

$$h^0 : (N \oplus N) \oplus_{\gamma \hat{\oplus} \lambda} (M \oplus M) \longrightarrow \left(N \oplus_\gamma M\right) \oplus (N \oplus_\lambda M)$$

given by $(n, n', m, m') \mapsto (n, m, n', m')$ is a morphism of $\mathcal{A}$-modules. Moreover, we have an equivalence of exact pairs $(1_{N\oplus N}, (h^0, 0), 1_{M\oplus M}) : \psi_{\gamma \hat{\oplus} \lambda} \longrightarrow \psi_\gamma \bigoplus \psi_\lambda$. Then, by the naturality of ψ proved before, we have

$$\begin{aligned} [\psi_{(\gamma+\lambda)}] = \left[\psi_{\nabla^0(\gamma\hat{\oplus}\lambda)(1\otimes\Delta^0)}\right] &= \left[\left(\nabla \psi_{(\gamma\hat{\oplus}\lambda)}\right) \Delta\right] \\ &= \left[\left(\nabla \left(\psi_\gamma \bigoplus \psi_\lambda\right)\right) \Delta\right] = [\psi_\gamma] + [\psi_\lambda]. \end{aligned}$$

Finally, the linearity of ψ follows from the k-linearity of U and L, the naturality proved above and the fact that the bifunctors B appearing as domain or codomain of ψ are k-bilinear, hence satisfy $cb = B(c1_M, N)(b)$, for $c \in k$ and $b \in B(M, N)$. □

Remark 6.14. *Assume that the algebra A is freely generated by the pair (R, W_0). Given $M, N \in A$-Mod, denote by $\mathrm{Ext}_{A,R}(M, N)$ the subspace of $\mathrm{Ext}_A(M, N)$ consisting of the equivalence classes of exact pairs $N \longrightarrow E \longrightarrow M$ in A-Mod which split when restricted to R-Mod. With the notation used in last proof, we can define a map*

$$\mathrm{Hom}_R(W_0 \otimes_R M, N) \xrightarrow{\psi_A} \mathrm{Ext}_{A,R}(M, N)$$

with recipe $\gamma \mapsto [N \xrightarrow{s^0} N \oplus_\gamma M \xrightarrow{p^0} M]$. Using the same argument as before, we can prove that ψ_A is surjective and is natural in both variables when considered as a natural transformation $\mathrm{Hom}_R(W_0 \otimes_R U(M), U(N)) \xrightarrow{\psi_A} \mathrm{Ext}_{A,R}(M, N)$, where now $U : A\text{-Mod} \longrightarrow R\text{-Mod}$ is the forgetful functor.

Given a Roiter ditalgebra $\mathcal{A}$ with layer (R, W), the morphism $\psi_{\mathcal{A}}$ coincides with the composition

$$\mathrm{Hom}_R(W_0 \otimes_R M, N) \xrightarrow{\psi_A} \mathrm{Ext}_{A,R}(M, N) \xrightarrow{\rho} \mathrm{Ext}_{\mathcal{A}}(M, N),$$

where the last map ρ is induced by the canonical embedding $L_{\mathcal{A}}$ (and is surjective by (6.12)). The transformation $\mathrm{Ext}_{A,R}(M, N) \xrightarrow{\rho} \mathrm{Ext}_{\mathcal{A}}(L(M), L(N))$, *is natural in $M, N \in A$-*Mod*. Moreover, $\psi_{\mathcal{A}} = \psi_A$ when we identify $\mathcal{A}$-*Mod *with A-*Mod *for a Roiter ditalgebra $\mathcal{A}$ with layer (R, W) where $W_1 = 0$.*

Lemma 6.15. *Let T be an algebra freely generated by the pair (R, W). Then, the following sequence of T-T-bimodules is exact*

$$0 \longrightarrow T \otimes_R W \otimes_R T \xrightarrow{d} T \otimes_R T \xrightarrow{m} T \longrightarrow 0,$$

where m is determined by the product of T and $d(1 \otimes w \otimes 1) = 1 \otimes w - w \otimes 1$, for $w \in W$.

Proof. We are considering the extension d of the morphism in $\mathrm{Hom}_{R\text{-}R}(W, T \otimes_R T)$ defined by $w \mapsto 1 \otimes w - w \otimes 1$. It is clear that m is surjective and that $md = 0$. We can assume that $T = T_R(W)$ and consider the standard grading on T. To prove the injectivity of d, let us identify $T \otimes_R T$ with $\bigoplus_{i=0}^{\infty} \left(W^{\otimes i} \otimes_R T\right)$. If we have $d\left(\sum_{i=1}^n \beta_i \otimes w_i \otimes \alpha_i\right) = 0$, where β_i, α_i are homogeneous elements of T and $w_i \in W$, then $\sum_i \beta_i \otimes w_i\alpha_i = \sum_i \beta_i w_i \otimes \alpha_i$ in $T \otimes_R T$. If β_i has maximal degree $\deg(\beta_i) = m$, then $\deg(\beta_i w_i) = m + 1$. Therefore, $\sum_j \beta_j w_j \otimes \alpha_j = 0$, where j runs through the indexes i such that $\deg(\beta_i) = m$. Then, $\sum_j \beta_j \otimes w_j \otimes \alpha_j = 0$. Indeed, the product in T induces a section $T \otimes_R W \xrightarrow{\mu} T$ in R-Mod, hence $T \otimes_R W \otimes_R T \xrightarrow{\mu \otimes 1} T \otimes_R T$ is injective. Repeating this trick, in a finite number of steps, we obtain that $\sum_{i=1}^n \beta_i \otimes w_i \otimes \alpha_i = 0$, and d is injective.

Now, assume that $m\left(\sum_{i=1}^n \beta_i \otimes \alpha_i\right) = 0$, where β_i, α_i are homogeneous elements of T. We can assume that each one of them is either a finite product of elements of W or an element of R. We want to show that the argument in last formula belongs to $\mathrm{Im}\, d$. Notice that in $T \otimes_R T$, we have the equality

$$\beta \otimes w\gamma = \beta(1 \otimes w - w \otimes 1)\gamma + \beta w \otimes \gamma,$$

for any $\beta, \gamma \in T$ and $w \in W$. Then, an easy induction argument shows that $\sum_{i=1}^n \beta_i \otimes \alpha_i = \Theta + \sum_{i=1}^n \beta_i' \otimes \alpha_i'$, where $\Theta \in \mathrm{Im}\, d$ and $\alpha_i' \in R$. Then, $\sum_i \beta_i' \otimes \alpha_i' = \sum_i \beta_i'\alpha_i' \otimes 1 = 0$, because $0 = \sum_i \beta_i\alpha_i = \sum_i \beta_i'\alpha_i'$, and we are done. □

Lemma 6.16. *Let $\mathcal{A}$ be a ditalgebra with layer (R, W). Then, for any $M \in R$-Mod and $N \in \mathcal{A}$-Mod, there is an isomorphism*

$$\mathrm{Hom}_{\mathcal{A}}(A \otimes_R M, N) \xrightarrow{\eta} \mathrm{Hom}_R(M, N) \times \mathrm{Hom}_{R\text{-}R}(W_1, \mathrm{Hom}_k(A \otimes_R M, N)).$$

It is given by $\eta(f^0, f^1) = (f_0^0, f^1)$, where f_0^0 is the composition $M \cong R \otimes_R M \longrightarrow N$ of the canonical isomorphism with the restriction of $f^0 : A \otimes_R M \longrightarrow N$ to $R \otimes_R M$. If we denote by $L = L_{\mathcal{A}} : A\text{-Mod} \longrightarrow \mathcal{A}\text{-Mod}$ the canonical embedding, then this isomorphism

$$\mathrm{Hom}_{\mathcal{A}}(L(A \otimes_R M), L(N)) \xrightarrow{\eta}$$
$$\mathrm{Hom}_R(M, N) \times \mathrm{Hom}_{R\text{-}R}(W_1, \mathrm{Hom}_k(A \otimes_R M, N))$$

is natural in the variables $M \in R$-Mod and $N \in A$-Mod.

Proof. If $f = (f^0, f^1) \in \mathrm{Hom}_{\mathcal{A}}(A \otimes_R M, N)$, then from the definition of morphism in $\mathcal{A}$-Mod we have the formula

$$f^0(a \otimes m) = af_0^0(m) - f^1(\delta(a))[1 \otimes m],$$

for $m \in M$ and $a \in A$. Hence, f is determined by the pair of maps (f_0^0, f^1) and η is injective. Now, given a pair of maps $(f_0^0, f^1) \in \mathrm{Hom}_R(M, N) \times \mathrm{Hom}_{R\text{-}R}(W_1, \mathrm{Hom}_k(A \otimes_R M, N))$, the last formula defines a morphism of R-modules $f^0 : A \otimes_R M \longrightarrow N$ such that (f^0, f^1) is a morphism in $\mathcal{A}$-Mod, and η is surjective. The naturality of η on M and N is a straightforward verification. □

Proposition 6.17. *Let $\mathcal{A}$ be a Roiter ditalgebra with layer (R, W). Then, the projective objects relative to the associated exact structure $\mathcal{E}_{\mathcal{A}}$ are the direct summands in $\mathcal{A}$-Mod of the objects $A \otimes_R M$, where $M \in R$-Mod. Moreover, every $M \in \mathcal{A}$-Mod admits the following conflation in $\mathcal{E}_{\mathcal{A}}$*

$$A \otimes_R W_0 \otimes_R M \xrightarrow{(d^0,0)} A \otimes_R M \xrightarrow{(m^0,0)} M,$$

where m^0 is determined by the action of A on M and $d^0(1 \otimes w \otimes m) = 1 \otimes wm - w \otimes m$, for $w \in W_0$ and $m \in M$. Thus $\mathcal{A}$-Mod with the exact structure $\mathcal{E}_{\mathcal{A}}$ has enough projectives and it has global projective dimension 1.

Proof. We first show that $A \otimes_R M$ is $\mathcal{E}_{\mathcal{A}}$-projective, for any $M \in R$-Mod. Consider any conflation $N' \xrightarrow{f} E \xrightarrow{g} N$ in $\mathcal{E}_{\mathcal{A}}$. We only have to show that the following induced sequence is exact

$$0 \longrightarrow \mathrm{Hom}_{\mathcal{A}}(A \otimes_R M, N') \longrightarrow \mathrm{Hom}_{\mathcal{A}}(A \otimes_R M, E)$$
$$\longrightarrow \mathrm{Hom}_{\mathcal{A}}(A \otimes_R M, N) \longrightarrow 0.$$

By (6.12), we may assume that $g^1 = 0$, $f^1 = 0$, and the sequence of A-modules

$$0 \longrightarrow N' \xrightarrow{f^0} E \xrightarrow{g^0} N \longrightarrow 0$$

is exact and splits when restricted to R. Then, applying to this split exact sequence the functor $\text{Hom}_R(M, -) \times \text{Hom}_{R\text{-}R}(W_1, \text{Hom}_k(A \otimes_R M, -))$ yields an exact sequence which is isomorphic, by (6.16), to the sequence whose exactness we wanted to show.

To finish the proof, recall that A is freely generated by the pair (R, W_0). Then, (6.15) gives us an exact sequence

$$0 \longrightarrow A \otimes_R W_0 \otimes_R A \longrightarrow A \otimes_R A \longrightarrow A \longrightarrow 0,$$

which splits as an exact sequence of R-A-bimodules. Tensor this sequence by the right with the left A-module M. Then, apply the embedding functor $L_{\mathcal{A}}$, to obtain the wanted conflation in $\mathcal{E}_{\mathcal{A}}$. □

Remark 6.18. *Let A be an algebra freely generated by (R, W_0). Consider the exact structure $\mathcal{E}_{A\text{-}R}$ on A-Mod formed by the exact sequences in A-Mod which split when restricted to R-Mod. Then, the sequence of (6.15) is a conflation of $\mathcal{E}_{A\text{-}R}$. Notice that the existence of the isomorphism $\text{Hom}_A(A \otimes_R M, N) \cong \text{Hom}_R(M, N)$, which is natural in the variables $N \in A$-Mod and $M \in R$-Mod, implies that the A-modules of the form $A \otimes_R M$ are $\mathcal{E}_{A\text{-}R}$-projectives. Then, the long exact sequence for the induced conflation $A \otimes_R W_0 \otimes_R M \longrightarrow A \otimes_R M \longrightarrow M$ in $\mathcal{E}_{A\text{-}R}$ gives*

$$0 \to \text{Hom}_A(M, N) \to \text{Hom}_A(A \otimes_R M, N) \to \text{Hom}_A(A \otimes_R W_0 \otimes_R M, N) \to \text{Ext}_{A\text{-}R}(M, N) \to 0,$$

or equivalently

$$0 \to \text{Hom}_A(M, N) \to \text{Hom}_R(M, N) \to \text{Hom}_R(W_0 \otimes_R M, N) \to \text{Ext}_{A\text{-}R}(M, N) \to 0.$$

This is the sequence of (6.13), when $W_1 = 0$. It is interesting to notice that when $\mathcal{A}$ is a Roiter ditalgebra with layer (R, W) and $W_1 \neq 0$, the previous argument applied to ($\mathcal{A}$-Mod, $\mathcal{E}_{\mathcal{A}}$), using (6.17) does not yield the svelter sequence of (6.13).

Definition 6.19. *We say that a ditalgebra $\mathcal{A}$ has semisimple layer iff the algebra R in the layer (R, W) of $\mathcal{A}$ is semisimple.*

Proposition 6.20. *Let $\mathcal{A}$ be a Roiter ditalgebra. Then, every projective (resp. injective) A-module is an $\mathcal{E}_{\mathcal{A}}$-projective (resp. $\mathcal{E}_{\mathcal{A}}$-injective) $\mathcal{A}$-module. If,* *furthermore, $\mathcal{A}$ is a Roiter ditalgebra with semisimple layer, then every $\mathcal{E}_{\mathcal{A}}$-projective (resp. $\mathcal{E}_{\mathcal{A}}$-injective) $\mathcal{A}$-module is a direct summand in $\mathcal{A}$-Mod of some projective (resp. injective) A-module.*

Proof. Assume $P \in A$-Mod is a projective A-module and consider any conflation $Q \longrightarrow E \longrightarrow P$ in $\mathcal{E}_{\mathcal{A}}$. From (6.12), this conflation is equivalent to one of the form $Q \xrightarrow{(f^0,0)} E' \xrightarrow{(g^0,0)} P$. The last conflation splits because P is a projective A-module. Hence the first conflation also splits. The argument for injectives is dual.

Now, assume that $\mathcal{A}$ is a Roiter ditalgebra with semisimple layer (R, W) and take any $\mathcal{E}_{\mathcal{A}}$-projective $\mathcal{A}$-module M. Consider an exact sequence in A-Mod of the form

$$0 \longrightarrow K \xrightarrow{f^0} P \xrightarrow{g^0} M \longrightarrow 0,$$

where P is a projective A-module. Since R is semisimple, the exact pair $K \xrightarrow{(f^0,0)} P \xrightarrow{(g^0,0)} M$ lies in $\mathcal{E}_{\mathcal{A}}$. It splits because M is $\mathcal{E}_{\mathcal{A}}$-projective. Thus, M is a direct summand of P in $\mathcal{A}$-Mod. The argument for injectives is dual. □

Exercise 6.21. *An exact pair $\zeta : N \xrightarrow{f} E \xrightarrow{g} M$ in the additive category $\mathcal{C}$ is split iff there exist morphisms $s : M \longrightarrow E$ and $p : E \longrightarrow N$ such that $gs = 1_M$, $pf = 1_N$ and $1_E = fp + sg$. Show that the following statements are equivalent:*

(1) ζ is split;
(2) g is a retraction;
(3) f is a section;
(4) ζ is equivalent to the exact pair $\zeta_0 : N \xrightarrow{(1_N,0)^t} N \oplus M \xrightarrow{(0,1_M)} M$.

If $\mathcal{C} = \mathcal{A}$-Mod, where $\mathcal{A}$ is a Roiter ditalgebra, ζ_0 represents the zero element in the space $\mathrm{Ext}_{\mathcal{A}}(M, N)$. It coincides with the image under the canonical embedding L_A : A-Mod$\longrightarrow\mathcal{A}$-Mod of the trivial exact sequence

$$0 \longrightarrow N \longrightarrow N \oplus M \longrightarrow M \longrightarrow 0.$$

Exercise 6.22. *Let $\mathcal{A}$ be a ditalgebra with layer (R, W) such that W_1 is a finitely generated R-R-bimodule. Consider a non-split exact pair $N \xrightarrow{\sigma} E \xrightarrow{\pi} M$ in $\mathcal{A}$-mod. Show that*

$$\dim_k \mathrm{End}_{\mathcal{A}}(E) < \dim_k \mathrm{End}_{\mathcal{A}}(M \oplus N).$$

Hints: Consider the linear map $\eta : \mathrm{Hom}_{\mathcal{A}}(M, N) \longrightarrow \mathrm{Hom}_{\mathcal{A}}(E, E)$, given by $g \mapsto \sigma g \pi$, and the map $\rho : \mathrm{End}_{\mathcal{A}}(E) \longrightarrow \mathrm{Hom}_{\mathcal{A}}(N, M)$, given by $f \mapsto \pi f \sigma$. Write $H_0 := \mathrm{Im}\, \eta$ and $H_1 := \mathrm{Ker}\, \rho$. Notice that η identifies $\mathrm{Hom}_{\mathcal{A}}(M, N)$

with H_0 *and that* $\mathrm{End}_{\mathcal{A}}(E)/H_1$ *embeds into* $\mathrm{Hom}_{\mathcal{A}}(N, M)$*. Notice also that* H_1/H_0 *embeds into* $\mathrm{End}_{\mathcal{A}}(N) \times \mathrm{End}_{\mathcal{A}}(M)$ *through the linear map* $f \mapsto (f_N, f_M)$*, which associates to each* $f \in H_1$ *the pair defined by the formulas* $f\sigma = \sigma f_N$ *and* $\pi f = f_M \pi$*. All this implies that* $\dim \mathrm{End}_{\mathcal{A}}(E) \leq \dim \mathrm{End}_{\mathcal{A}}(N \oplus M)$*. If we had equality, the previous embeddings would be isomorphisms. The element* $f \in H_1$ *mapped to the pair* $(1_N, 0)$ *by the last isomorphism gives a left inverse for* σ*.*

Exercise 6.23. *Let* $\mathcal{A}$ *is a Roiter ditalgebra with layer* (R, W) *such that* W_1 *is a finitely generated* R*-*R*-bimodule. For* $d \in \mathbb{N}$*, let* $\mathcal{A}\text{-mod}_d$ *denote the class of* d*-dimensional* $\mathcal{A}$*-modules. Take* $M \in \mathcal{A}\text{-mod}_d$ *with minimal* $\dim_k \mathrm{End}_{\mathcal{A}}(M)$ *and consider its decomposition* $M = M_1 \oplus \cdots \oplus M_s$ *as a direct sum of indecomposables. Show that* $\mathrm{Ext}_{\mathcal{A}}(M_i, M_j) = 0$*, for all* $i \neq j$*.*

Hint: Assume $0 \neq [\xi] \in \mathrm{Ext}_{\mathcal{A}}(M_i, M_j)$*, consider the canonical projection* $\pi_i : \oplus_{r \neq j} M_r \longrightarrow M_i$ *and the non-split pullback conflation* $\xi\pi_i$*. Then, apply (6.22) to it and obtain a contradiction.*

Exercise 6.24. *Assume that* $\mathcal{A}$ *and* $\mathcal{B}$ *are Roiter ditalgebras and that* $G : \mathcal{A}\text{-Mod} \longrightarrow \mathcal{B}\text{-Mod}$ *is a full and faithful functor. Suppose that* $\zeta : N \xrightarrow{f} E \xrightarrow{g} M$ *is a composable pair in* $\mathcal{A}$*-Mod such that*

$$G(\zeta) : G(N) \xrightarrow{G(f)} G(E) \xrightarrow{G(g)} G(M)$$

is a split conflation of $\mathcal{E}_{\mathcal{B}}$*. Show that* ζ *is a split conflation in* $\mathcal{E}_{\mathcal{A}}$*.*

Exercise 6.25. *Let* $\mathcal{A}$ *be a Roiter ditalgebra and* $M \xrightarrow{f} E \xrightarrow{g} N \in \mathcal{E}_{\mathcal{A}}$*. From the general theory of categories with exact structures, we have the exact sequence*

$$\begin{aligned} 0 \longrightarrow \mathrm{Hom}_{\mathcal{A}}(Z, M) &\xrightarrow{f^*} \mathrm{Hom}_{\mathcal{A}}(Z, E) \xrightarrow{g^*} \mathrm{Hom}_{\mathcal{A}}(Z, N) \\ \xrightarrow{\partial} \mathrm{Ext}^1_{\mathcal{A}}(Z, M) &\xrightarrow{f'} \mathrm{Ext}^1_{\mathcal{A}}(Z, E) \xrightarrow{g'} \mathrm{Ext}^1_{\mathcal{A}}(Z, N), \end{aligned}$$

for any $Z \in \mathcal{A}$*-Mod. Use (6.13) to show that in fact* g' *is surjective. Formulate and prove the dual statement.*

References. The exact structure $\mathcal{E}_{\mathcal{A}}$ on the category of modules over a Roiter ditalgebra was first introduced by S. A. Ovsienko in [43] and then studied in [6]. Proposition (6.13) was recovered from the last reference.

7
Almost split conflations in $\mathcal{A}$-Mod

In this section, we show that the exact category ($\mathcal{A}$-mod, $\mathcal{E}_{\mathcal{A}}$) has almost split conflations, under certain finiteness conditions for the Roiter ditalgebra $\mathcal{A}$. We will also require that $\mathcal{A}$ has semisimple layer, as in (6.19).

We will first recall some terminology and some basic facts on the theory of almost split conflations in a *Krull–Schmidt exact category* $(\mathcal{C}, \mathcal{E})$, that is a Krull–Schmidt category $\mathcal{C}$, as in (5.10), endowed with an exact structure $\mathcal{E}$. Their proofs are similar to those required for the case of finitely generated modules over a finite-dimensional algebra. We will apply these results for the particular case $\mathcal{A}$-mod, see (5.13), and later in some other contexts.

Definition 7.1. *Assume $(\mathcal{C}, \mathcal{E})$ is an exact category. A morphism $g : E \longrightarrow N$ in $\mathcal{C}$ is right almost split iff it is not a retraction and, for every morphism $t : Z \longrightarrow N$, which is not a retraction, there is a morphism $\bar{t} : Z \longrightarrow E$ such that $g\bar{t} = t$. The morphism $f : M \longrightarrow E$ in $\mathcal{C}$ is left almost split iff it is not a section and, for every morphism $t : M \longrightarrow Z$, which is not a section, there is a morphism $\bar{t} : E \longrightarrow Z$ such that $\bar{t}f = t$.*

Lemma 7.2. *Assume $(\mathcal{C}, \mathcal{E})$ is a Krull–Schmidt exact category. The existence of a right almost split morphism $g : E \longrightarrow N$ in $\mathcal{C}$ implies that N is an idecomposable object. Dually, if $f : M \longrightarrow E$ is left almost split, M is indecomposable.*

Proof. The right ideal $g\mathrm{Hom}_{\mathcal{C}}(N, E)$ is the unique maximal proper right ideal of $\mathrm{End}_{\mathcal{C}}(N)$. Hence this last algebra is local, and N is indecomposable. The other statement is dual. □

Definition 7.3. *The exact category $(\mathcal{C}, \mathcal{E})$ has right almost split morphisms iff for any indecomposable $N \in \mathcal{C}$, there exists a right almost split morphism $g : E \longrightarrow N$ in $\mathcal{C}$. $(\mathcal{C}, \mathcal{E})$ has left almost split morphisms iff for any indecomposable $M \in \mathcal{C}$, there exists a left almost split morphism $f : M \longrightarrow E$ in $\mathcal{C}$. A morphism $h : M \longrightarrow N$ in $\mathcal{C}$ is called right minimal iff the existence of any factorization*

$h = ht$ in $\mathcal{C}$, entails that t is an isomorphism. The morphism h is called left minimal iff the existence of any factorization $h = th$ in $\mathcal{C}$, entails that t is an isomorphism.

Notice that if $g : E \longrightarrow N$ and $g' : E' \longrightarrow N$ are right minimal almost split morphisms, then there is an isomorphism $s : E \longrightarrow E'$ such that $g's = g$.

Proposition 7.4. *If N is an object in the Krull–Schmidt exact category at which a right almost split morphism ends, then there exists a minimal right almost split morphism $E \to N$.*

Proof. It is not hard to see that any right almost split morphism $E \to N$ in $\mathcal{A}$-mod with minimal number of indecomposable direct summands is a right minimal almost split morphism. □

Lemma 7.5. *Assume that $g : E \longrightarrow N$ is a right almost split morphism and $g_1 : E_1 \longrightarrow N$ is a right minimal almost split morphism in a Krull–Schmidt exact category $(\mathcal{C}, \mathcal{E})$. Then we have the following commutative diagram in $\mathcal{C}$*

$$\begin{array}{ccc} E & \xrightarrow{g} & N \\ \downarrow{\cong} & & \| \\ E_1 \oplus E_2 & \xrightarrow{(g_1,0)} & N. \end{array}$$

Proof. It follows from the fact that idempotents split in $\mathcal{C}$. □

Lemma 7.6. *Assume that $M \xrightarrow{f} E \xrightarrow{g} N$ is a conflation of the Krull–Schmidt exact category $(\mathcal{C}, \mathcal{E})$, where M is indecomposable and g is a right almost split morphism, then g is a minimal right almost split morphism.*

Proof. Choose a minimal right almost split morphism ending at N, which exists by (7.4), and compare it with g using (7.5). □

Lemma 7.7. *Assume that $g : E \longrightarrow N$ is a morphism in the Krull–Schmidt exact category $(\mathcal{C}, \mathcal{E})$. Then, g is a right almost split morphism iff g is not a retraction and, for any non-isomorphism $h : Z \longrightarrow N$, where Z is indecomposable, there exists $\overline{h} : Z \longrightarrow E$ such that $g\overline{h} = h$.*

Lemma 7.8. *Assume that $M \xrightarrow{f} E \xrightarrow{g} N$ is a conflation in the Krull–Schmidt exact category $(\mathcal{C}, \mathcal{E})$, where g is a minimal right almost split morphism, then M is indecomposable.*

Proof. First notice that the following holds. If $M \xrightarrow{f} E$ is a morphism in $\mathcal{C}$ and $M = \oplus_i M_i \xrightarrow{\pi_i} M_i$ are the canonical projections into the indecomposable direct summands of M, then f is a section iff each π_i factors through f.

Choose i such that π_i does not factor through f. Then, in the following pushout diagram

$$\begin{array}{rcccc} \xi : & M & \xrightarrow{f} & E & \xrightarrow{g} & N \\ & \downarrow{\scriptstyle \pi_i} & & \downarrow{\scriptstyle t} & & \| \\ \pi_i\xi : & M_i & \xrightarrow{f'} & E' & \xrightarrow{g'} & N, \end{array}$$

the conflation $\pi_i\xi$ does not split. It follows that g' is a right almost split morphism, and, from (7.6), it is a minimal right almost split morphism. Hence, t is an isomorphism and the conflations ξ and $\pi_i\xi$ are isomorphic. □

Theorem 7.9. *Assume that $\xi : M \xrightarrow{f} E \xrightarrow{g} N$ is a conflation in the Krull–Schmidt exact category $(\mathcal{C}, \mathcal{E})$. Then, the following statements are equivalent:*

(1) f is a left almost split morphism and g is a right almost split morphism;
(2) f is a minimal left almost split morphism;
(3) g is a minimal right almost split morphism;
(4) M is indecomposable and g is a right almost split morphism;
(5) N is indecomposable and f is a left almost split morphism.

Proof. From (7.7), we know that g is right almost split iff $\xi \neq 0$ and $\xi r = 0$, for any indecomposable X and any non-isomorphism $r : X \longrightarrow N$. Dually, f is left almost split iff $\xi \neq 0$ and $r\xi = 0$, for any indecomposable X and any non-isomorphism $r : M \longrightarrow X$.

Now, assume g is a minimal right almost split morphism. Thus, $\xi \neq 0$, and, from (7.8), M is indecomposable. Take an indecomposable X and a non-isomorphism $r : M \longrightarrow X$. Suppose that $r\xi \neq 0$ is a non-trivial conflation, then there exist morphisms t and s making the following diagram commutative

$$\begin{array}{rcccc} 0 \neq r\xi : & X & \longrightarrow & E' & \longrightarrow & N \\ & \downarrow{\scriptstyle t} & & \downarrow{\scriptstyle s} & & \| \\ \xi : & M & \xrightarrow{f} & E & \xrightarrow{g} & N \\ & \downarrow{\scriptstyle r} & & \downarrow & & \| \\ r\xi : & X & \longrightarrow & E' & \longrightarrow & N. \end{array}$$

Thus, the conflations $(rt)r\xi$ and $r\xi$ are equivalent. Since r is not an isomorphism, neither is rt. Hence, rt is a nilpotent element in the local algebra $\mathrm{End}_{\mathcal{C}}(X)$. Thus, $r\xi = 0$, a contradiction. Hence, f is left almost split, from the dual of (7.6), since N is indecomposable, we obtain that f is minimal left almost split. We have shown that (3) implies (2).

Dually, (2) implies (3). Assuming (3) (or (2)), by (7.8) we know that N is indecomposable. Hence, by (7.6), (5) is equivalent to (3) (or (2)). Dually, (4) is equivalent to (2) (or (3)). Statement (2) (and (3)) imply (1), which, by (7.2), implies (5). The cycle of equivalences is now closed. □

Definition 7.10. *A conflation $M \xrightarrow{f} E \xrightarrow{g} N$ in the Krull–Schmidt exact category $(\mathcal{C}, \mathcal{E})$ is an almost split conflation iff it satisfies any of the equivalent conditions of last theorem.*

We immediately obtain the following:

Proposition 7.11. *Assume $M \xrightarrow{f} E \xrightarrow{g} N$ and $M' \xrightarrow{f'} E' \xrightarrow{g'} N'$ are almost split conflations in the Krull–Schmidt exact category $(\mathcal{C}, \mathcal{E})$. Then $M \cong M'$ iff $N \cong N'$. Moreover, any of them is equivalent to the existence of a commutative diagram of the form*

$$\begin{array}{ccccc} M & \xrightarrow{f} & E & \xrightarrow{g} & N \\ \downarrow\cong & & \downarrow\cong & & \downarrow\cong \\ M' & \xrightarrow{f'} & E' & \xrightarrow{g'} & N'. \end{array}$$

Now, we focus our attention on the category of modules of a Roiter ditalgebra $\mathcal{A}$ and give a series of results which will contribute to the proof of the main theorem (7.18).

Lemma 7.12. *Let $\mathcal{A}$ be a Roiter ditalgebra and make $\mathbb{E} = \mathrm{End}_{\mathcal{A}}(A)^{op}$. Then, the functor*

$$F = \mathrm{Hom}_{\mathcal{A}}(A, -) : \mathcal{A}\text{-Mod} \longrightarrow \mathbb{E}\text{-Mod}$$

is an exact faithful functor. If $\mathcal{A}$ has semisimple layer and A is left noetherian, then the restriction of F to the full subcategory of $\mathcal{A}$-Mod consisting of the finitely generated A-modules is full and faithful.

Proof. Consider the layer (R, W) of $\mathcal{A}$. Since $A \cong A \otimes_R R$, by (6.17), A is a projective object relative to the exact structure $\mathcal{E}_{\mathcal{A}}$. Hence, F is an exact functor.

Assume now that $h \in \mathrm{Hom}_{\mathcal{A}}(M, N)$ satisfies that $F(h) = 0$. This means that $hg = 0$, for any $g \in \mathrm{Hom}_{\mathcal{A}}(A, M)$, or equivalently, $h^0 g^0 = 0$ and, for any $v \in V$, $h^1(v)g^0 + h^0 g^1(v) + \sum_i h^1(u_i)g^1(v_i) = 0$, where $\delta(v) = \sum_i u_i \otimes v_i$. We want to show that $h = 0$. Given $m_0 \in M$, there is a morphism of A-modules $g^0 : A \longrightarrow M$ such that $g^0(a) = am_0$. Let $g = (g^0, 0)$, thus $g \in \mathrm{Hom}_{\mathcal{A}}(A, M)$, and $h^0 g^0 = 0$ and $h^1(v)g^0 = 0$. Evaluating at the identity element $1 \in A$, we obtain $h^0(m_0) = 0$ and $h^1(v)[m_0] = 0$. Thus, $h = 0$ and F is a faithful functor.

Assume A is left noetherian and M, N are finitely generated A-modules. We want to verify that $F : \mathrm{Hom}_{\mathcal{A}}(M, N) \longrightarrow \mathrm{Hom}_{\mathbb{E}}(F(M), F(N))$ is surjective. It is clear that $F : \mathrm{Hom}_{\mathcal{A}}(A, A) \longrightarrow \mathrm{Hom}_{\mathbb{E}}(F(A), F(A))$ is an isomorphism. Using that F is additive, for any natural numbers n and n', we obtain isomorphisms

$$F : \mathrm{Hom}_{\mathcal{A}}(A^n, A^{n'}) \longrightarrow \mathrm{Hom}_{\mathbb{E}}(F(A^n), F(A^{n'})).$$

In our case, there are exact sequences in A-Mod of the form

$$0 \longrightarrow K \xrightarrow{t^0} A^n \xrightarrow{p^0} M \longrightarrow 0 \text{ and } 0 \longrightarrow K' \xrightarrow{v^0} A^m \xrightarrow{q^0} K \longrightarrow 0.$$

Then, defining $s^0 := t^0 q^0$ and using (6.2) and the fact that R is semisimple, we obtain that their image under the canonical embedding in $\mathcal{A}$-Mod give the $\mathcal{E}_{\mathcal{A}}$-projective resolution

$$A^m \xrightarrow{s=(s^0,0)} A^n \xrightarrow{p=(p^0,0)} M,$$

and then, applying F, we get the exact sequence of $\mathbb{E}$-modules

$$F(A^m) \xrightarrow{F(s)} F(A^n) \xrightarrow{F(p)} F(M) \longrightarrow 0,$$

which is a projective resolution of $F(M)$ in $\mathbb{E}$-Mod. If we consider for the $\mathcal{A}$-module N the corresponding construction $A^{m'} \xrightarrow{s'} A^{n'} \xrightarrow{p'} N$, then any morphism $f \in \mathrm{Hom}_{\mathbb{E}}(F(M), F(N))$ can be lifted as follows

$$\begin{array}{ccccccc} F(A^m) & \xrightarrow{F(s)} & F(A^n) & \xrightarrow{F(p)} & F(M) & \longrightarrow & 0 \\ \downarrow{\scriptstyle f_2} & & \downarrow{\scriptstyle f_1} & & \downarrow{\scriptstyle f} & & \\ F(A^{m'}) & \xrightarrow{F(s')} & F(A^{n'}) & \xrightarrow{F(p')} & F(N) & \longrightarrow & 0. \end{array}$$

From the previous statements, $f_2 = F(h_2)$ and $f_1 = F(h_1)$ for some $h_1 \in \mathrm{Hom}_{\mathcal{A}}(A^n, A^{n'})$ and $h_2 \in \mathrm{Hom}_{\mathcal{A}}(A^m, A^{m'})$. Since $F(s'h_2) = F(s')f_2 = f_1 F(s) = F(h_1 s)$, we have $s'h_2 = h_1 s$, and hence there exists $h \in \mathrm{Hom}_{\mathcal{A}}(M, N)$ such that the following diagram commutes in $\mathcal{A}$-Mod

$$\begin{array}{ccccc} A^m & \xrightarrow{s} & A^n & \xrightarrow{p} & M \\ \downarrow{\scriptstyle h_2} & & \downarrow{\scriptstyle h_1} & & \downarrow{\scriptstyle h} \\ A^{m'} & \xrightarrow{s'} & A^{n'} & \xrightarrow{p'} & N. \end{array}$$

Applying F to this diagram, we obtain that $F(h) = f$, and F is full. □

Remark 7.13. *Let $\mathcal{A}$ be a Roiter ditalgebra and make $\mathbb{E} = \mathrm{End}_{\mathcal{A}}(A)^{op}$. If we denote, for $b \in A$, $\rho_b : A \longrightarrow A$ the right multiplication by b, then we have the following canonical embedding of A in $\mathbb{E}$: $A \xrightarrow{\cong} \mathrm{End}_A(A)^{op} \longrightarrow \mathrm{End}_{\mathcal{A}}(A)^{op} = \mathbb{E}$ where $b \mapsto \hat{\rho}_b = (\rho_b, 0)$. The algebra $\mathbb{E}$*

*becomes an A-A-bimodule. In particular, for $h \in \mathbb{E}$ and $b \in A$, we have $hb = \hat{\rho}_b h$, where the right product is composition in $\mathcal{A}$-*Mod.

Lemma 7.14. *Let $\mathcal{A}$ be a Roiter ditalgebra and make $\mathbb{E} = \mathrm{End}_{\mathcal{A}}(A)^{op}$. Then, for each $M \in A$-*Mod*, there is a linear map $\sigma_M : \mathbb{E} \otimes_A M \longrightarrow \mathrm{Hom}_{\mathcal{A}}(A, M)$ given by $\sigma(f \otimes m) = f_m$, where $f_m^0(a) = f^0(a)m$ and $f_m^1(v)[a] = f^1(v)[a]m$. It defines a natural transformation of functors from A-*Mod *to $\mathbb{E}$-*Mod

$$\mathbb{E} \otimes_A M \xrightarrow{\sigma_M} \mathrm{Hom}_{\mathcal{A}}(A, L(M)),$$

*where $L = L_{\mathcal{A}} : A$-*Mod $\to \mathcal{A}$-Mod *is the canonical embedding functor. Moreover, if $\mathcal{A}$ has semisimple layer, A is left noetherian and M is finitely generated, then σ_M is an isomorphism.*

Proof. For the last claim, notice that $\sigma_A : \mathbb{E} \otimes_A A \to \mathbb{E}$ is just the isomorphism given by multiplication. From the additivity of the functors $\mathbb{E} \otimes_A -$ and $\mathrm{Hom}_{\mathcal{A}}(A, L(-))$, we obtain that σ_{A^n} is an isomorphism too. Finally, if we consider a free resolution $A^m \to A^n \to M \to 0$ of M in A-Mod, we can apply the exact functors $\mathbb{E} \otimes_A -$ and $\mathrm{Hom}_{\mathcal{A}}(A, L(-))$ to it, and then σ_M is an isomorphism because the following diagram commutes

$$\begin{array}{ccccccc}
\mathbb{E} \otimes_A A^m & \longrightarrow & \mathbb{E} \otimes_A A^n & \longrightarrow & \mathbb{E} \otimes_A M & \longrightarrow & 0 \\
\downarrow{\scriptstyle \sigma_{A^m}} & & \downarrow{\scriptstyle \sigma_{A^n}} & & \downarrow{\scriptstyle \sigma_M} & & \\
\mathrm{Hom}_{\mathcal{A}}(A, A^m) & \longrightarrow & \mathrm{Hom}_{\mathcal{A}}(A, A^n) & \longrightarrow & \mathrm{Hom}_{\mathcal{A}}(A, M) & \longrightarrow & 0.
\end{array}$$

□

Thus, in the following and until the end of this section, we fix a Roiter ditalgebra $\mathcal{A}$ such that $\mathcal{A}$-mod is a Krull–Schmidt category. For the main result, we will also require that $\mathcal{A}$ is finite dimensional in the following sense.

Definition 7.15. *A layered k-ditalgebra $\mathcal{A}$ with layer (R, W) is called finite dimensional iff A and W_1 are finite dimensional over k.*

Definition 7.16. *Given a full subcategory $\mathcal{H}$ of a module category* $\mathbb{E}$-mod *and $M \in \mathbb{E}$-*mod*, a morphism $p : H(M) \longrightarrow M$, where $H(M) \in \mathcal{H}$ is a right $\mathcal{H}$-approximation of M iff, for any $h : Z \longrightarrow M$ in* $\mathbb{E}$-Mod *with $Z \in \mathcal{H}$, there exists a morphism $\overline{h} : Z \longrightarrow H(M)$ such that $p\overline{h} = h$.*

The subcategory $\mathcal{H}$ is called contravariantly finite in $\mathbb{E}$-mod *iff every $M \in$* $\mathbb{E}$-mod *admits a right $\mathcal{H}$-approximation.*

Lemma 7.17. *Let $\mathcal{A}$ be a finite-dimensional Roiter ditalgebra and consider the finite-dimensional algebra $\mathbb{E} = \mathrm{End}_{\mathcal{A}}(A)^{op}$. Then, the full subcategory $\mathcal{I}$ of*

$\mathbb{E}$-mod *formed by the* $\mathbb{E}$*-modules isomorphic to modules of the form* $\mathbb{E} \otimes_A M$, *for some* $M \in A$-mod, *is contravariantly finite in* $\mathbb{E}$-mod.

Proof. Given $M \in \mathbb{E}$-mod, our assumptions imply that $\mathbb{E}$ is a left finite-dimensional A-module, hence $M \in A$-mod. Consider the product morphism $p : \mathbb{E} \otimes_A M \longrightarrow M$ and assume $h : \mathbb{E} \otimes_A N \longrightarrow M$ is a morphism of $\mathbb{E}$-modules. Then, there is a morphism $\overline{h} : \mathbb{E} \otimes_A N \longrightarrow \mathbb{E} \otimes_A M$ such that $\overline{h}(f \otimes n) = f \otimes h(1 \otimes n)$, for $n \in N$ and $f \in \mathbb{E}$. Then, $p\overline{h} = h$. □

Theorem 7.18. *Let $\mathcal{A}$ be a finite-dimensional Roiter ditalgebra with semisimple layer. Then,* $(\mathcal{A}\text{-mod}, \mathcal{E}_{\mathcal{A}})$ *has almost split conflations, that means the following: each indecomposable object* $N \in \mathcal{A}$-mod *admits a minimal right almost split morphism* $g : E \longrightarrow N$ *and a minimal left almost split morphism* $f : N \longrightarrow G$*; moreover, for any non-*$\mathcal{E}_{\mathcal{A}}$*-projective indecomposable object* N *in* $\mathcal{A}$-mod, *there exist an almost split conflation* $M \xrightarrow{f} E \xrightarrow{g} N$*, and, for any non-*$\mathcal{E}_{\mathcal{A}}$*-injective indecomposable object* M *in* $\mathcal{A}$-mod, *there exist an almost split conflation* $M \xrightarrow{f} E \xrightarrow{g} N$.

Proof. We first show that each indecomposable object $N \in \mathcal{A}$-mod admits a minimal right almost split morphism $g : E \longrightarrow N$. From (7.12) and (7.14), we know there is a full and faithful functor $F = \mathrm{Hom}_{\mathcal{A}}(A, -) : \mathcal{A}\text{-mod} \longrightarrow \mathbb{E}\text{-mod}$, and we can identify the $\mathbb{E}$-module $F(M)$ with $\mathbb{E} \otimes_A M$, for each $M \in \mathcal{A}$-mod. From our assumptions, we know that the k-algebra $\mathbb{E} = \mathrm{End}_{\mathcal{A}}(A)^{op}$ is finite dimensional and, from the general representation theory for artin algebras, we know that the category $\mathbb{E}$-mod has almost split sequences. Since the object N is indecomposable in the Krull–Schmidt category $\mathcal{A}$-mod, see (5.13), it has a local endomorphism ring $\mathrm{End}_{\mathcal{A}}(N) \cong \mathrm{End}_{\mathbb{E}}(F(N))$; hence, $F(N)$ is an indecomposable object in $\mathbb{E}$-mod. Thus, the indecomposable $\mathbb{E}$-module $F(N)$ admits a minimal right almost split morphism $d : E \longrightarrow F(N)$ in $\mathbb{E}$-mod. Consider the full subcategory $\mathcal{I}$ of $\mathbb{E}$-mod of Lemma (7.17), and the right approximation $F(E) = \mathbb{E} \otimes_A E \xrightarrow{p} E$ constructed in its proof. Then, the composition $dp : F(E) \longrightarrow F(N)$ is the image under F of a morphism $g : E \longrightarrow N$ in $\mathcal{A}$-mod. We claim that g is a right almost split morphism. Indeed, if g was a retraction, $F(g) = dp$ would be a retraction, and so would d, which is impossible. Now, if $h : Z \longrightarrow N$ is a morphism in $\mathcal{A}$-mod, which is not a retraction, then $F(h)$ is not a retraction, and hence it factors through d, say $F(h) = dt$, where $t \in \mathrm{Hom}_{\mathbb{E}}(F(Z), E)$. Since p is the right approximation of E in $\mathcal{I}$, there exists $\overline{t} \in \mathrm{Hom}_{\mathbb{E}}(F(Z), F(E))$ such that $p\overline{t} = t$. Take $\overline{h} \in \mathrm{Hom}_{\mathcal{A}}(Z, E)$ with $F(\overline{h}) = \overline{t}$. Since F is faithful, $g\overline{h} = h$, and we have shown that g is right almost split. From (7.4), we know there exists a minimal right almost split morphism $g' : E' \longrightarrow N$.

Now, if furthermore N is not an $\mathcal{E}_{\mathcal{A}}$-projective, there exists a non-trivial deflation $d' : L \longrightarrow N$ in $\mathcal{A}$-mod. Using that g' is right almost split, we can factor d' through g', and derive directly from the definition of an exact structure that g' is a deflation too. Hence, there exists a conflation $M \xrightarrow{f'} E \xrightarrow{g'} N$ in $\mathcal{E}_{\mathcal{A}}$, where g' is a minimal right almost split morphism. By (7.9), this is an almost split conflation.

The remaining part of the theorem can be proved using the duality D of (2.8). Indeed, (R^{op}, W^{op}) is a triangular layer for $\mathcal{A}^{op}$ with R^{op} semisimple. By (5.6), $\mathcal{A}^{op}$ is a Roiter ditalgebra. Since $D : \mathcal{A}^{op}\text{-mod} \longrightarrow \mathcal{A}\text{-mod}$ maps $\mathcal{E}_{\mathcal{A}^{op}}$ onto $\mathcal{E}_{\mathcal{A}}$, and non $\mathcal{E}_{\mathcal{A}^{op}}$-projective modules onto non $\mathcal{E}_{\mathcal{A}}$-injectives, we can transform the right-hand part of the statement of the theorem for $\mathcal{A}^{op}$-mod into the left-hand part of the statement of the theorem for $\mathcal{A}$-mod. □

Proposition 7.19. *Let $\mathcal{A}$ be a finite-dimensional Roiter ditalgebra with semisimple layer. Assume that N is not $\mathcal{E}_{\mathcal{A}}$-projective and denote by $S(N, M)$ the set of equivalence classes of almost split conflations and the trivial one in* $\mathrm{Ext}_{\mathcal{A}}(N, M)$. *Then, $S(N, M)$ is the socle of the right* $\mathrm{End}_{\mathcal{A}}(N)$*-module* $\mathrm{Ext}_{\mathcal{A}}(N, M)$ *and is also the socle of the left* $\mathrm{End}_{\mathcal{A}}(M)$*-module* $\mathrm{Ext}_{\mathcal{A}}(N, M)$. *They are both simple and generated by any previously fixed almost split conflation $M \xrightarrow{f} E \xrightarrow{g} N$.*

Proof. Assume that $\xi : M \xrightarrow{f} E \xrightarrow{g} N$ is an almost split conflation and take any non-trivial conflation $\xi' : M \xrightarrow{f'} E' \xrightarrow{g'} N$. Then, there exist morphisms s and t such that the following diagram commutes

$$\begin{array}{ccccc} \xi' : M & \xrightarrow{f'} & E' & \xrightarrow{g'} & N \\ \downarrow t & & \downarrow s & & \| \\ \xi : M & \xrightarrow{f} & E & \xrightarrow{g} & N. \end{array}$$

Hence, $[t\xi'] = [\xi]$. It follows that $\mathrm{socExt}_{\mathcal{A}}(N, M) \subseteq \mathrm{End}_{\mathcal{A}}(M)\xi$. Indeed, for $\xi' \in \mathrm{socEnd}_{\mathcal{A}}(N, M)$, if t was not invertible, ξ would split; thus, t is invertible and ξ' is equivalent to $t^{-1}\xi$. Recall that the algebra $\mathrm{End}_{\mathcal{A}}(M)$ is local, with radical formed by the non-invertible morphisms $r : M \longrightarrow M$. From the dual of (7.7), $r\xi = 0$, for any such morphism r; hence, $\mathrm{End}_{\mathcal{A}}(M)\xi \subseteq \mathrm{socExt}_{\mathcal{A}}(N, M)$. As before, $S(N, M) \subseteq \mathrm{End}_{\mathcal{A}}(M)\xi$, and given $f \in \mathrm{End}_{\mathcal{A}}(M)$, we have $f\xi \in S(N, M)$, because $f\xi$ is an almost split conflation if f is an isomorphism and it is trivial otherwise. The rest of the statement is obtained dually. □

Exercise 7.20. *Let $\mathcal{A}$ be a Roiter ditalgebra with layer (R, W), where R is a basic k-algebra of finite representation type. Denote by $Y \in R$-mod the direct* *sum of a complete set of representatives of the non-isomorphic indecomposable finite-dimensional R-modules, make* $P := A \otimes_R Y$ *and* $\mathbb{E} := \mathrm{End}_A(P)^{op}$. *Show that the functor*

$$F = \mathrm{Hom}_A(P, -) : \mathcal{A}\text{-Mod} \longrightarrow \mathbb{E}\text{-Mod}$$

is an exact faithful functor. If A is finite dimensional, its restriction to the full subcategory of $\mathcal{A}$-Mod *consisting of the finitely generated A-modules is full and faithful.*

Hint: Adapt the proof of (7.12). For this, use that A is a direct summand of P to show that F is faithful. Then, given $M \in \mathcal{A}$-mod, *use (6.17) to produce a projective resolution in* $\mathcal{A}$-mod *of the form*

$$Q_2 \xrightarrow{(s^0,0)} Q_1 \xrightarrow{(p^0,0)} M,$$

where Q_1, Q_2 *are finite direct sums of direct summands of P.*

References. The existence of almost split conflations for bocses, under some finiteness conditions, was independently proved by W. L. Burt and M. C. R. Butler in [16], and by R. Bautista and M. M. Kleiner in [7]. The proofs are different, but they both use M. Auslander and S. O. Smalø's work [4], on almost split sequences in subcategories. The proof given here is different and does not require the use of Auslander–Smalø theorem.

8
Quotient ditalgebras

In this section, we present some basic construction procedures that yield new ditalgebras $\mathcal{A}^z$ from a given one $\mathcal{A}$. Each one of these constructions $\mathcal{A} \mapsto \mathcal{A}^z$ has an associated functor $F_z : \mathcal{A}^z\text{-Mod} \longrightarrow \mathcal{A}\text{-Mod}$ which relates the corresponding module categories. We usually work with layered ditalgebras. However, it is convenient to work sometimes in a slightly more general context. We will first discuss the notions of ideals and quotient ditalgebras.

Definition 8.1. *Assume that T is a k-algebra freely generated by the pair (B, U), as in (1.1). Then, an ideal I of T is called (B, U)-compatible iff the ideal $I_B := I \cap B$ of B and the B-B-subbimodule $I_U := I \cap U$ of U satisfy $I = \langle I_B \cup I_U \rangle$. Here, $\langle I_B \cup I_U \rangle$ denotes the ideal of the algebra T generated by $I_B \cup I_U$.*

Remark 8.2. *Recall that a homogeneous ideal I of a graded k-algebra $T = \oplus_{i \geq 0}[T]_i$ is by definition an ideal of T generated by homogeneous elements (equivalently, for any $t \in I$, the homogeneous component t_i of t also belongs to I, for all $i \geq 0$). Thus, in this case, $I = \oplus_{i \geq 0} I \cap [T]_i$ and the quotient T/I admits a canonical structure of graded k-algebra, where $[T/I]_i = ([T]_i + I)/I$.*

Remark 8.3. *If T is a k-algebra freely generated by the pair (B, U), then:*

(1) Assume that I is a (B, U)-compatible ideal of T and consider the canonical grading of T determined by the isomorphism $T \cong T_B(U)$, then I is a homogeneous ideal of T.

(2) If I_B is an ideal of B and I_U is a B-B-subbimodule of U satisfying the relation $I_B U + U I_B \subseteq I_U$, then $I := \langle I_B \cup I_U \rangle$ is a (B, U)-compatible ideal of T, with $I_B = I \cap B$ and $I_U = I \cap U$.

Proposition 8.4. *Assume that T is a k-algebra freely generated by the pair (B, U) and let I be a (B, U)-compatible ideal of T. Consider the canonical projection $\eta : T \longrightarrow T/I$. Then, T/I is freely generated by the pair $(\eta(B), \eta(U))$. Moreover, the natural morphism of algebras $\pi : T \longrightarrow T_{B/I_B}(U/I_U)$ induces an isomorphism*

$$T_{B/I_B}(U/I_U) \xrightarrow{\overline{\eta}} T/I$$

such that $\overline{\eta}\pi = \eta$, thus identifying $\eta(B)$ with B/I_B and $\eta(U)$ with U/I_U.

Proof. Since $I_B U + U I_B \subseteq I_U$ holds, indeed U/I_U is a B/I_B-bimodule and the tensor algebra $T_{B/I_B}(U/I_U)$ is defined. By the universal property of T, the canonical projections $B \longrightarrow B/I_B$ and $U \longrightarrow U/I_U$ determine a morphism of algebras $\pi : T \longrightarrow T_{B/I_B}(U/I_U)$. Recall that I is an ideal of T containing I_B and I_U, thus the restriction $\eta_0 : B \longrightarrow T/I$ of η maps I_B to zero and hence induces a morphism of algebras $\overline{\eta}_0 : B/I_B \longrightarrow T/I$. Similarly, the restriction $\eta_1 : U \longrightarrow T/I$ maps I_U to zero and hence induces a morphism $\overline{\eta}_1 : U/I_U \longrightarrow T/I$. By the universal property of the tensor algebra, there is a morphism of algebras $\overline{\eta} : T_{B/I_B}(U/I_U) \longrightarrow T/I$ such that $\overline{\eta}\pi = \eta$. This implies that $\mathrm{Ker}\,\pi \subseteq \mathrm{Ker}\,\eta = I$. Moreover, π is a morphism of algebras such that $\pi(I_B) = 0$ and $\pi(I_U) = 0$. Thus, since $I = \langle I_B \cup I_U \rangle$, we also have that $I \subset \mathrm{Ker}\,\pi$. Then, $\overline{\eta}$ is an isomorphism and the proposition follows from this. □

Definition 8.5.

(1) Given a t-algebra T, as in (1.4), a t-ideal I of T is any $([T]_0, [T]_1)$-compatible ideal of T. From (8.4), we know that in this case, T/I is a t-algebra with the grading determined in the quotient by the homogeneous ideal I of T. Thus, the projection map $T \longrightarrow T/I$ is a morphism of graded algebras.

(2) Given a ditalgebra $\mathcal{A} = (T, \delta)$, an ideal I of $\mathcal{A}$ is a t-ideal of T such that $\delta(I) \subseteq I$. In this case, the differential δ induces a differential $\overline{\delta} : T/I \longrightarrow T/I$ and the pair $\mathcal{A}/I := (T/I, \overline{\delta})$ is a ditalgebra: the quotient ditalgebra of $\mathcal{A}$ modulo the ideal I. The projection map $\eta : T \longrightarrow T/I$ is a morphism of ditalgebras $\eta : \mathcal{A} \longrightarrow \mathcal{A}/I$.

Lemma 8.6. *Assume that $\mathcal{A} = (T, \delta)$ is some ditalgebra and make $A := [T]_0$ and $V := [T]_1$. Then:*

(1) If I is an ideal of $\mathcal{A}$, then the ideal $I_A := I \cap A$ of A and the A-A-subbimodule $I_V := I \cap V$ of V satisfy that $I_A V + V I_A + \delta(I_A) \subseteq I_V$.

(2) If I_A is any ideal of A, then $I_V := I_A V + V I_A + \delta(I_A)$ is an A-A-subbimodule of V, and $I := \langle I_A \cup I_V \rangle$ is an ideal of the ditalgebra $\mathcal{A}$.

Proof. The first part is clear. For the second, we first use (8.3) to guarantee that I is a t-ideal of T. Then, notice that $\delta(I_A) \subseteq I_V$ and that, by Leibniz rule, we first get that $\delta(I_V) \subseteq I$ and then $\delta(I) \subseteq I$. □

Lemma 8.7. *Suppose that $I_1 \subseteq I_2$ are ideals of the ditalgebra $\mathcal{A}$. Then, I_2/I_1 is an ideal of the quotient ditalgebra $\mathcal{A}/I_1$ and there is an isomorphism of ditalgebras θ such that the following diagram commutes*

$$\begin{array}{ccc} \mathcal{A} & \xrightarrow{\eta_1} & \mathcal{A}/I_1 \\ {\scriptstyle \eta_2}\downarrow & & \downarrow{\scriptstyle \eta} \\ \mathcal{A}/I_2 & \xrightarrow{\theta} & (\mathcal{A}/I_1)/(I_2/I_1), \end{array}$$

where η_1, η_2, η are the canonical projections.

Proof. Assume that $\mathcal{A} = (T, \delta)$, thus $\mathcal{A}/I_1 = (T/I_1, \overline{\delta})$. Clearly, I_2/I_1 is an ideal of the algebra T/I_1. We have to see that I_2/I_1 is a t-ideal of T/I_1. This means that the equality $I_2/I_1 = \langle (I_2/I_1) \cap [T/I_1]_0 \cup (I_2/I_1) \cap [T/I_1]_1 \rangle$ holds. This is clear because $[T/I_1]_0 = ([T]_0 + I_1)/I_1$, $[T/I_1]_1 = ([T]_1 + I_1)/I_1$ and $I_2 = \langle I_2 \cap [T]_0 \cup I_2 \cap [T]_1 \rangle$. Finally, $\delta(I_2) \subseteq I_2$ implies that $\overline{\delta}(I_2/I_1) \subseteq I_2/I_1$. Hence, I_2/I_1 is an ideal of the ditalgebra $\mathcal{A}/I_1$. Noether theorem for algebras guarantees the existence of an isomorphism $\theta : T/I_2 \longrightarrow (T/I_1)/(I_2/I_1)$ such that $\eta\eta_1 = \theta\eta_2$, which is a morphism of ditalgebras. □

Definition 8.8. *Let $\mathcal{A} = (T, \delta)$ be a ditalgebra and make $A := [T]_0$ and $V := [T]_1$. Then, if S is a subset of A, the ideal I of $\mathcal{A}$ generated by S is the ideal $I = \langle I_A \cup I_V \rangle$ of $\mathcal{A}$, where I_A is the ideal of the algebra A generated by S and $I_V = I_A V + V I_A + \delta(I_A)$.*

Notice that, in the context of last definition, I is the smallest ideal J of the algebra T such that $S \subseteq J$ and $\delta(S) \subseteq J$. Thus, the ideal of $\mathcal{A}$ generated by S coincides with the ideal of T generated by $S \cup \delta(S)$.

Remark 8.9. *Let $\mathcal{A} = (T, \delta)$ be a ditalgebra and denote by $A = [T]_0$ and $V = [T]_1$. Consider an ideal I of $\mathcal{A}$ generated by some subset S of A. Write $\overline{A} := A/I_A$, $\overline{V} := V/(I_A V + \delta(I_A) + V I_A)$ and consider the canonical morphism $\pi : T \longrightarrow T_{\overline{A}}(\overline{V})$, induced by the projections $\pi_0 : A \longrightarrow \overline{A}$ and $\pi_1 : V \longrightarrow \overline{V}$. Then, from (8.4), applied to the algebra T freely generated by (A, V), we know that there is an isomorphism $\overline{\eta} : T_{\overline{A}}(\overline{V}) \longrightarrow T/I$ such that $\overline{\eta}\pi = \eta$, where $\eta : \mathcal{A} \longrightarrow \mathcal{A}/I$ is the canonical projection to the quotient ditalgebra $\mathcal{A}/I$. Then, with the help of this isomorphism, we can transfer the differential of $\mathcal{A}/I$ to a differential $\overline{\delta}$ on the tensor algebra $\overline{T} := T_{\overline{A}}(\overline{V})$, and get a nice realization of $\mathcal{A}/I$. Notice that $\overline{\delta}$ can also be obtained as the differential produced using*

Lemma (1.8) from the linear maps $\overline{\delta}_0 : \overline{A} \longrightarrow [\overline{T}]_1$ *and* $\overline{\delta}_1 : \overline{V} \longrightarrow [\overline{T}]_2$ *induced by the composition* $\pi\delta$, *when restricted respectively to* A *and* V. *Write* $\overline{\mathcal{A}} := (\overline{T}, \overline{\delta})$. *In practice, it is often nicer to work with the ditalgebra morphism* $\pi : \mathcal{A} \longrightarrow \overline{\mathcal{A}}$ *than with the canonical projection* $\eta : \mathcal{A} \longrightarrow \mathcal{A}/I$.

Lemma 8.10. *Consider any ideal* I *of the ditalgebra* $\mathcal{A}$ *generated by some subset* S *of* A. *Then, the canonical projection* $\eta : \mathcal{A} \longrightarrow \mathcal{A}/I$ *induces by restriction a functor* $F_\eta : \mathcal{A}/I\text{-Mod} \longrightarrow \mathcal{A}\text{-Mod}$, *which is full and faithful. The image of* F_η *is formed by the objects of* $\mathcal{A}$-Mod *annihilated by* S *(or equivalently, by the ideal* I_A *of* A *generated by* S*).*

Proof. Adopt the notation from last remark. Then, we can replace the morphism η by the morphism π. Let us show that $F_\pi : \overline{\mathcal{A}}\text{-Mod} \longrightarrow \mathcal{A}\text{-Mod}$ is full and faithful. We already know from (2.4) that F_π is a faithful functor. Now, fix $M, N \in \overline{\mathcal{A}}$-Mod and take any $f = (f^0, f^1) \in \mathrm{Hom}_{\mathcal{A}}(F_\pi(M), F_\pi(N))$. We started with an action of $\overline{A}$ on the vector space M, which determines the object $F_\pi(M)$, that is an action of A on the same vectorspace M such that $I_A M = 0$. Since the ideal I_A annihilates both $F_\pi(M)$ and $F_\pi(N)$, we have that $f^1(I_A V + V I_A) = 0$. Moreover, given $a \in I_A$ and $m \in M$, the formula $af^0(m) = f^0(am) + f^1(\delta(a))(m)$ implies that $f^1(\delta(a)) = 0$. Hence, we know that $f^1(I_A V + \delta(I_A) + V I_A) = 0$ and there exists a morphism $g^1 \in \mathrm{Hom}_{\overline{A}\text{-}\overline{A}}(\overline{V}, \mathrm{Hom}_k(M, N)) = \mathrm{Hom}_{A\text{-}A}(\overline{V}, \mathrm{Hom}_k(F_\pi(M), F_\pi(N)))$ such that $g^1\pi_1 = f^1$. Define $g^0 = f^0$ and $g = (g^0, g^1)$. From the definition of morphism in $\mathcal{A}$-Mod, we have that for any $a \in A$ and $m \in F_\pi(M)$, $af^0(m) = f^0(am) + f^1(\delta(a))[m]$. Denote $\overline{a} = \pi(a)$, for $a \in A$. Then, we can rewrite the last formula as $\overline{a}g^0(m) = g^0(\overline{a}m) + g^1(\overline{\delta}(\overline{a}))[m]$ because, $g^1(\overline{\delta}(\overline{a}))[m] = g^1(\overline{\delta}\pi(a))[m] = g^1(\pi\delta(a))[m] = f^1(\delta(a))[m]$. This shows that $g \in \mathrm{Hom}_{\overline{\mathcal{A}}}(M, N)$, and it is now clear that $F_\pi(g) = f$. □

Definition 8.11. *Given two layered ditalgebras* $\mathcal{A}$ *and* $\mathcal{A}'$ *with layers* (R, W) *and* (R', W') *respectively, a morphism of ditalgebras* $\phi : \mathcal{A} \longrightarrow \mathcal{A}'$ *is called compatible with the layers (or a morphism of layered ditalgebras) iff* $\phi(R) \subseteq R'$ *and* $\phi(W_i) \subseteq W'_i$, *for* $i \in \{0, 1\}$.

We observe that, in this case, the corresponding morphism of t-algebras $\phi : T \longrightarrow T'$ is completely determined by its restrictions $\phi_r : R \longrightarrow R'$ and $\phi_b : W \longrightarrow W'$. Moreover, any pair (ϕ_r, ϕ_b), where $\phi_r : R \longrightarrow R'$ is a morphism of k-algebras and $\phi_b : W \longrightarrow W'$ is a morphism of R-R-bimodules with $\phi_b(W_i) \subseteq W'_i$ for $i \in \{0, 1\}$, extends uniquely to a morphism of t-algebras $\phi : T \longrightarrow T'$.

Whenever $\phi : \mathcal{A} \longrightarrow \mathcal{A}'$ is a morphism compatible with the layers, we have a commutative square

$$\begin{array}{ccc} \mathcal{A}'\text{-Mod} & \xrightarrow{F_\phi} & \mathcal{A}\text{-Mod} \\ \Big\downarrow U_{\mathcal{A}'} & & \Big\downarrow U_{\mathcal{A}} \\ R'\text{-Mod} & \xrightarrow{F_{\phi_r}} & R\text{-Mod}, \end{array}$$

where $U_{\mathcal{A}}$ is the forgetful functor and F_{ϕ_r} is the restriction functor associated to the algebra morphism ϕ_r. Moreover, this compatibility condition permits to study the action of the functor F_ϕ on a morphism $f \in \mathrm{Hom}_{\mathcal{A}'}(M, N)$ when considered as an element of $\mathrm{Phom}_{R'\text{-}W'}(M, N)$.

Lemma 8.12. *Assume we have a morphism of ditalgebras $\phi : \mathcal{A} \longrightarrow \mathcal{A}'$ which is compatible with given layers (R, W) of $\mathcal{A}$ and (R', W') of $\mathcal{A}'$. Assume furthermore, that ϕ induces surjective maps $R \longrightarrow R'$ and $W \longrightarrow W'$. Then, if (R, W) is triangular, so is (R', W').*

Proof. Assume the triangularity of (R, W) and adopt the notation of definition (5.1). Then, consider the sequences of R'-R'-bimodules $0 = \phi(W_0^0) \subseteq \phi(W_0^1) \subseteq \cdots \subseteq \phi(W_0^r) = W_0'$ and $0 = \phi(W_1^0) \subseteq \phi(W_1^1) \subseteq \cdots \subseteq \phi(W_1^s) = W_1'$. By definition, A_i denotes the R-subalgebra of A generated by W_0^i. If A_i' denotes the R'-subalgebra of A' generated by $\phi(W_0^i)$, then clearly $A_i' = \phi(A_i)$. Now it is clear that $\delta'\phi(W_0^{i+1}) = \phi\delta(W_0^{i+1}) \subseteq \phi(A_i W_1 A_i) = A_i' W_1' A_i'$, for all $i \in [0, r-1]$, and that $\delta'\phi(W_1^{i+1}) = \phi\delta(W_1^{i+1}) \subseteq \phi(AW_1^i AW_1^i A) = A'\phi(W_1^i)A'\phi(W_1^i)A'$, for all $i \in [0, s-1]$. □

Definition 8.13. *Given a ditalgebra $\mathcal{A} = (T, \delta)$ with layer (R, W), an ideal I of $\mathcal{A}$ is compatible with the layer iff I is an (R, W)-compatible ideal of T.*

Remark 8.14. *If I is a homogeneous ideal of the t-algebra T, which is compatible with a layer of T, then I is a t-ideal of T.*

Proposition 8.15. *Assume $\mathcal{A}$ is a layered ditalgebra, let I be an ideal of $\mathcal{A}$ compatible with its layer (R, W) and consider the canonical projection $\mathcal{A} \xrightarrow{\eta} \mathcal{A}/I$. Then, the quotient ditalgebra $\mathcal{A}/I$ admits the layer $(\eta(R), \eta(W))$, with $\eta(W) = \eta(W_0) \oplus \eta(W_1)$, thus η is a morphism of layered ditalgebras. Moreover, by last lemma, $\mathcal{A}/I$ is triangular whenever $\mathcal{A}$ is so.*

The ditalgebra $\mathcal{A}/I$ admits the following natural realization: consider the ideal $I_R := I \cap R$, the R-R-subbimodules $I_{W_j} := I \cap W_j$ of W_j, for $j \in \{0, 1\}$, the algebra $R^q := R/I_R$, the R^q-R^q-bimodules $W_j^q := W_j/I_{W_j}$, $j \in \{0, 1\}$, and form $W^q = W_0^q \oplus W_1^q$. Then, the canonical isomorphism

$$T_{R^q}(W^q) = T_{R/I_R}(W/[I \cap W]) \xrightarrow{\overline{\eta}} T/I$$

described in (8.4) permits to transfer the differential of $\mathcal{A}/I$ to a differential δ^q on the tensor algebra $T^q := T_{R^q}(W^q)$. Then, we have an isomorphism of *layered ditalgebras*

$$\mathcal{A}^q = (T^q, \delta^q) \xrightarrow{\overline{\eta}} (T/I, \overline{\delta}) = \mathcal{A}/I.$$

We shall often identify $\mathcal{A}/I$ *with* $\mathcal{A}^q$, *through* $\overline{\eta}$, *and we denote by* F^q *the functor* $\mathcal{A}^q$-Mod$\longrightarrow\mathcal{A}$-Mod *induced by the canonical morphism* $\mathcal{A} \rightarrow \mathcal{A}^q$. *Thus, the ditalgebra* $\mathcal{A}^q$ *has layer* (R^q, W^q), *which is triangular whenever* (R, W) *is so. The functor* F^q *is full and faithful by (8.10). Its image is formed by the objects in* $\mathcal{A}$-Mod *annihilated by the ideal* $I_A := I \cap A$ *of* A, *which is generated by* I_R *and* I_{W_0}.

In the following lemmas, we specialize the last statement to several types of ideals of layered ditalgebras.

Lemma 8.16. *Assume* $\mathcal{A} = (T, \delta)$ *is a ditalgebra with layer* (R, W) *and let* I_R *be an ideal of the algebra* R. *Then, the ideal* I *of* $\mathcal{A}$ *generated by* I_R *is compatible with the given layer.*

The layered ditalgebra $\mathcal{A}/I$ *admits the following natural realization: consider the algebra* $R^s = R/I_R$, *the* R^s-R^s-*bimodules* $W_j^s = W_j/(I_R W_j + W_j I_R)$, *for* $j \in \{0, 1\}$, *and form* $W^s = W_0^s \oplus W_1^s$. *Then, there is a canonical isomorphism* $T^s := T_{R^s}(W^s) \xrightarrow{\overline{\eta}} T/I$ *of t-algebras. We can translate the differential* $\overline{\delta}$ *of* $\mathcal{A}/I$ *to a differential* δ^s *on* T^s *using* $\overline{\eta}$. *Therefore, we have an isomorphism of layered ditalgebras*

$$\mathcal{A}^s = (T^s, \delta^s) \xrightarrow{\overline{\eta}} (T/I, \overline{\delta}) = \mathcal{A}/I.$$

We shall often identify $\mathcal{A}/I$ *with* $\mathcal{A}^s$ *through* $\overline{\eta}$.

We say that the layered ditalgebra $\mathcal{A}^s$ *is obtained from* $\mathcal{A}$ *by factoring out the ideal* I_R *and we denote by* $F^s : \mathcal{A}^s$-Mod$\longrightarrow\mathcal{A}$-Mod *the functor induced by the canonical morphism* $\mathcal{A} \rightarrow \mathcal{A}^s$. *Thus, the ditalgebra* $\mathcal{A}^s$ *has layer* (R^s, W^s), *which is triangular whenever* (R, W) *is so. The functor* F^s *is full and faithful; its image consists of the objects in* $\mathcal{A}$-Mod *annihilated by* I_R.

Proof. If we write $A = [T]_0$ and $V = [T]_1$, we have the ideal $I_A = AI_RA$ of A, the A-A-subbimodule $I_V = I_A V + V I_A + \delta(I_A) = I_A V + V I_A$ of V and $I = \langle I_A \cup I_V \rangle$. Since I_A and $I_V = AI_RAV + VAI_RA$ are contained in $\langle I \cap R \rangle$, we have $I = \langle (I \cap R) \cup (I \cap W) \rangle$, and the ideal I is compatible with the layer (R, W) of $\mathcal{A}$. Then, we can apply (8.15). With the notation introduced there, we have that $R^q = R/I_R = R^s$ and $I_{W_j} = I_R W_j + W_j I_R$, for $j \in \{0, 1\}$. Thus, the quotient ditalgebra $\mathcal{A}/I$ admits the realization $\mathcal{A}^s = \mathcal{A}^q$ described in (8.15). Thus, our assertions on $\mathcal{A}^s$ and F^s hold. □

Lemma 8.17. *Assume* $\mathcal{A} = (T, \delta)$ *is a ditalgebra with layer* (R, W). *Assume that* $e \in R$ *is an idempotent of* R *satisfying that* $eR(1 - e) = 0 = (1 - e)Re$.

Consider the ideal I_R of R generated by $1-e$. Then, the ditalgebra $\mathcal{A}^s$ obtained from $\mathcal{A}$ by factoring out the ideal I_R, described in (8.16), is a layered ditalgebra. It admits the following natural realization: consider the algebra $R^d = eRe$, the R^d-R^d-bimodules $W_j^d = eW_je$, for $j \in \{0,1\}$, and form $W^d = W_0^d \oplus W_1^d$. Then, our assumption on e guarantees that there are canonical isomorphisms $R^s \xrightarrow{\gamma_r} R^d$ and $W^s \xrightarrow{\tau_b} W^d$, induced by the morphism of algebras $R \longrightarrow R^d$ and the morphism $W \longrightarrow W^d$ of R-R-bimodules, given by the receipt $x \mapsto exe$.

There is an induced isomorphism of t-algebras $T^s \xrightarrow{\gamma} T^d := T_{R^d}(W^d)$. We can translate the differential δ^s of $\mathcal{A}^s$ to a differential δ^d on T^d using γ. Therefore, we have an isomorphism of layered ditalgebras

$$\mathcal{A}^s = (T^s, \delta^s) \xrightarrow{\gamma} (T^d, \delta^d) = \mathcal{A}^d.$$

We shall often identify $\mathcal{A}^s$ with $\mathcal{A}^d$ through γ.

We call $\mathcal{A}^d$ the ditalgebra obtained from $\mathcal{A}$ by deletion of the idempotent $1-e$. Associated to this construction, we have the canonical projection $\mathcal{A} \xrightarrow{\eta} \mathcal{A}^d$, and the full and faithful functor $F^d = F_\eta : \mathcal{A}^d\text{-Mod} \longrightarrow \mathcal{A}\text{-Mod}$ whose image consists of the objects annihilated by the element $1-e$. The layer of $\mathcal{A}^d$ is $(R^d, W^d) = (eRe, eWe)$, which will be triangular whenever (R, W) is so.

Lemma 8.18. *Assume $\mathcal{A} = (T, \delta)$ is a ditalgebra with layer (R, W) and that W_0' is an R-R-subbimodule of W_0 such that $\delta(W_0') \subseteq W_1$. Consider the ideal I of $\mathcal{A}$ generated by W_0'. Then, I is an ideal of $\mathcal{A}$ compatible with the layer.*

The layered ditalgebra $\mathcal{A}/I$, admits the following natural realization: consider the algebra $R^p = R$, the R-R-bimodules $W_0^p = W_0/W_0'$, $W_1^p = W_1/\delta(W_0')$, and form $W^p = W_0^p \oplus W_1^p$. Then, there is a canonical isomorphism $T^p := T_{R^p}(W^p) \xrightarrow{\overline{\eta}} T/I$ of t-algebras. We can translate the differential $\overline{\delta}$ of $\mathcal{A}/I$ to a differential δ^p on T^p using $\overline{\eta}$. Therefore, we have an isomorphism of layered ditalgebras

$$\mathcal{A}^p = (T^p, \delta^p) \xrightarrow{\overline{\eta}} (T/I, \overline{\delta}) = \mathcal{A}/I.$$

We shall often identify $\mathcal{A}/I$ with $\mathcal{A}^p$ through $\overline{\eta}$.

We say that the layered ditalgebra $\mathcal{A}^p$ is obtained from $\mathcal{A}$ by factoring out the subbimodule W_0' and we denote by $F^p : \mathcal{A}^p\text{-Mod} \longrightarrow \mathcal{A}\text{-Mod}$ the functor induced by the canonical morphism $\mathcal{A} \longrightarrow \mathcal{A}^p$. Thus, the ditalgebra $\mathcal{A}^p$ has layer (R^p, W^p), which is triangular whenever (R, W) is so. The functor F^p is full and faithful and its image consists of the objects in $\mathcal{A}$-Mod annihilated by W_0'.

Proof. If we write $A = [T]_0$ and $V = [T]_1$, we have the ideal $I_A = AW_0'A$ of A, the A-A-subbimodule $I_V = I_AV + VI_A + \delta(I_A)$ of V and $I = \langle I_A \cup I_V \rangle$. Then, $I_A = I \cap A = AW_0'A$ and $I_V = AW_1AW_0'A + AW_0'AW_1A +$

$\delta(AW'_0A)$. It follows that $I_R = I \cap R = 0$, $I_{W_0} = I \cap W_0 = W'_0$ and $I_{W_1} = I \cap W_1 = \delta(W'_0)$. Then, $I = \langle I_R \cup I_W \rangle$, where $I_W = I_{W_0} \oplus I_{W_1}$, and I is an ideal of $\mathcal{A}$ compatible with the layer (R, W). Then, we can apply (8.15). With the notation introduced there, we have that the quotient ditalgebra $\mathcal{A}/I$ admits the realization $\mathcal{A}^q = \mathcal{A}^p$. Thus, our assertions on $\mathcal{A}^p$ and F^p hold. □

Lemma 8.19. *Assume $\mathcal{A} = (T, \delta)$ is a ditalgebra with layer (R, W). Assume that we have the following decompositions of R-R-bimodules $W_0 = W'_0 \oplus W''_0$ and $W_1 = \delta(W'_0) \oplus W''_1$. Consider the layered ditalgebra $\mathcal{A}^p$ obtained from $\mathcal{A}$ factoring out the subbimodule W'_0, as in the last lemma. We can construct another realization of the layered ditalgebra $\mathcal{A}^p$ as follows. Consider the algebra $R^r = R$, the R^r-R^r-bimodules $W^r_0 = W''_0$, $W^r_1 = W''_1$, and form $W^r = W^r_0 \oplus W^r_1$. Then, there are canonical isomorphisms $R^p \xrightarrow{Id} R^r$ and $W^p \xrightarrow{\tau_b} W^r$. There is an induced isomorphism of t-algebras $T^p \xrightarrow{\gamma} T^r := T_{R^r}(W^r)$. We can translate the differential δ^p of T^p to a differential δ^r on T^r using γ. Therefore, we have an isomorphism of layered ditalgebras*

$$\mathcal{A}^p = (T^p, \delta^p) \xrightarrow{\gamma} (T^r, \delta^r) = \mathcal{A}^r.$$

We shall often identify $\mathcal{A}^p$ with $\mathcal{A}^r$ through γ.

We call $\mathcal{A}^r$ the ditalgebra obtained from $\mathcal{A}$ by regularization of the bimodule W'_0. Associated to this construction, we have the canonical projection $\mathcal{A} \xrightarrow{\eta} \mathcal{A}^r$, and the the full and faithful functor $F^r = F_\eta : \mathcal{A}^r\text{-Mod} \longrightarrow \mathcal{A}\text{-Mod}$ where the image consists of the objects annihilated by W'_0. The layer of $\mathcal{A}^r$ is $(R^r, W^r) = (R, W''_0 \oplus W''_1)$. It will be triangular whenever (R, W) is so.

Assume that $\mathcal{A}$ is a Roiter ditalgebra and make $K := W'_0 \cap \operatorname{Ker} \delta$. Then an object $M \in \mathcal{A}\text{-Mod}$ is isomorphic to some object in the image of F^r iff $KM = 0$. Thus, if δ restricts to an injection $\delta : W'_0 \longrightarrow \delta(W'_0)$ and $\mathcal{A}$ is a Roiter ditalgebra, the functor F^r is dense.

Proof. We verify the last statement. Assume $M \in \mathcal{A}\text{-Mod}$ satisfies $KM = 0$, and let us show the existence of an isomorphism $\widehat{M} \xrightarrow{f} M$ in $\mathcal{A}\text{-Mod}$, where $\widehat{M}$ is annihilated by W'_0.

Consider the action map $\psi \in \operatorname{Hom}_{R\text{-}R}(W_0, \operatorname{Hom}_k(M, M))$ of W_0 on the A-module M, and denote by ψ' its restriction to W'_0. The equality $KM = 0$ implies the existence of a morphism $\delta(W'_0) \xrightarrow{g} \operatorname{Hom}_k(M, M)$ of R-R-bimodules with $g\delta = \psi'$. Consider the morphism $f^1 = \begin{pmatrix} g & 0 \end{pmatrix} \in \operatorname{Hom}_{R\text{-}R}(W_1, \operatorname{Hom}_k(M, M))$, recall that $W_1 = \delta(W'_0) \oplus W''_1$. Since $\mathcal{A}$ is a Roiter ditalgebra, there exists an object $\widehat{M} \in \mathcal{A}\text{-Mod}$ with underlying R-module M, and such that $f = (Id_M, f^1) \in \operatorname{Hom}_{\mathcal{A}}(\widehat{M}, M)$ is an isomorphism. Therefore, if $w \in W'_0, m \in M$ and we denote by $w \cdot m$ the action of A on $\widehat{M}$, we have $w \cdot m = wm - f^1(\delta(w))[m] = wm - \psi(w)[m] = wm - wm = 0$. Hence, W'_0 annihilates $\widehat{M}$, and we are done.

The fact that $M \cong N$ in $\mathcal{A}$-Mod with $W_0'N = 0$ implies that $KM = 0$ follows from the definition of morphism. □

Lemma 8.20. *Assume $\mathcal{A} = (T, \delta)$ is a ditalgebra with layer (R, W), and that there is an R-R-bimodule decomposition $W_0 = W_0' \oplus W_0''$ with $\delta(W_0') = 0$. Then, we can consider another layer for the same ditalgebra $\mathcal{A}$, namely (R^a, W^a), where R^a is the subalgebra of T generated by R and W_0', and, by definition, $W_0^a = R^a W_0'' R^a$ and $W_1^a = R^a W_1 R^a$. Lemma (1.3) guarantees that T is freely generated by R^a and W^a, and $\delta(W_0') = 0$ implies that indeed (R^a, W^a) is a layer for $\mathcal{A}$. We keep the notation of $\mathcal{A}$ for our ditalgebra with its original layer, and we denote by $\mathcal{A}^a$ the same ditalgebra equipped with this new layer. We say that $\mathcal{A}^a$ is the ditalgebra obtained from $\mathcal{A}$ by absorption of the subbimodule W_0'. We will consider the identity map $\eta : T \longrightarrow T$ as a morphism of ditalgebras $\eta : \mathcal{A} \longrightarrow \mathcal{A}^a$, which is clearly an isomorphism. Notice, however, that η and η^{-1} are not compatible with the given layers. The induced functor $F^a = F_\eta : \mathcal{A}^a$-Mod$\longrightarrow \mathcal{A}$-Mod is an isomorphism of categories. Moreover, the layer (R^a, W^a) of $\mathcal{A}^a$ is triangular, whenever the layer (R, W) of $\mathcal{A}$ is so.*

Proof. Assume the triangularity of (R, W) and adopt the notation of definition (5.1). Let $\pi : W_0 = W_0' \oplus W_0'' \longrightarrow W_0''$ be the canonical projection. Then, consider the sequences of R^a-R^a-bimodules $0 = R^a\pi(W_0^0)R^a \subseteq R^a\pi(W_0^1)R^a \subseteq \cdots \subseteq R^a\pi(W_0^r)R^a = R^a\pi(W_0)R^a = R^a W_0'' R^a = W_0^a$ and $0 = R^a W_1^0 R^a \subseteq R^a W_1^1 R^a \subseteq \cdots \subseteq R^a W_1^s R^a = W_1^a$. By definition, A_i denotes the R-subalgebra of A generated by W_0^i, and A_i^a denotes the R^a-subalgebra of $A^a = A$ generated by $R^a\pi(W_0^i)R^a$. Since $R \subseteq R^a$ and $W_0^i \subseteq R^a + \pi(W_0^i)$, we know that $A_i \subseteq A_i^a$, and hence $\delta(R^a\pi(W_0^{i+1})R^a) = R^a\delta(\pi(W_0^{i+1}))R^a \subseteq R^a A_i W_1 A_i R^a \subseteq A_i^a R^a W_1 R^a A_i^a$, for all $i \in [0, r-1]$, and $\delta(R^a W_1^{i+1} R^a) = R^a\delta(W_1^{i+1})R^a \subseteq R^a A W_1^i A W_1^i A R^a \subseteq A^a R^a W_1^i R^a A^a R^a W_1^i R^a A^a$, for all $i \in [0, s-1]$. □

Exercise 8.21. *Assume that $\mathcal{A} = (T, \delta)$ is some ditalgebra and make $A = [T]_0$ and $V = [T]_1$. Suppose that $S \subseteq V$ satisfies that $\delta(S) \subseteq I$, where I is the ideal of T generated by S. Show that I is an ideal of $\mathcal{A}$. Consider the A-A-subbimodule I_V of V generated by S and the t-algebra $\widehat{T} := T_A(V/I_V)$. Then, δ induces a differential $\widehat{\delta}$ on $\widehat{T}$ such that $\mathcal{A}/I \cong (\widehat{T}, \widehat{\delta})$. Show that the restriction functor $F_\eta : \mathcal{A}/I$-Mod$\longrightarrow \mathcal{A}$-Mod, associated to the canonical projection $\eta : \mathcal{A} \longrightarrow \mathcal{A}/I$, is faithful, but not necessarily full.*

Exercise 8.22. *Assume $\mathcal{A} = (T, \delta)$ is any ditalgebra. An ideal I of $\mathcal{A}$ is called triangular iff it is generated by some subset H of $A = [T]_0$, for which there is a filtration of subsets $\{0\} = H_0 \subseteq H_1 \subseteq \cdots \subseteq H_\ell = H$ such that*

$$\delta(H_i) \subseteq AH_{i-1}V + VH_{i-1}A, \text{ for all } i \in [1, \ell].$$

*Here, $AH_{i-1}V$ is the A-A-subbimodule of T generated by the products ahv, where $a \in A$, $h \in H_{i-1}$ and $v \in V$, and $VH_{i-1}A$ is defined similarly. Write $I_A = I \cap A$ and show that $M \cong N$ in $\mathcal{A}$-*Mod *and $I_A M = 0$ implies that $I_A N = 0$.*

Hint: Show by induction on $i \in [0, \ell]$ that $H_i N = \{0\}$.

Exercise 8.23. *Show that it is not true in general that, given a ditalgebra $\mathcal{A} = (T, \delta)$ and $M \cong N$ in $\mathcal{A}$-*Mod*, we necessarily have that the annihilators of the A-modules M and N coincide (here, $A = [T]_0$).*

Exercise 8.24. *Assume that $\mathcal{A}$ is a ditalgebra and that I is a triangular ideal of $\mathcal{A}$, as in (8.22). Write $I_A = I \cap A$ and denote by $(\mathcal{A}, I)$-*Mod *the full subcategory of $\mathcal{A}$-*Mod *formed by all $\mathcal{A}$-modules with $I_A M = 0$. By (8.10), the canonical projection $\eta : \mathcal{A} \longrightarrow \mathcal{A}/I$ induces an equivalence of categories $\mathcal{A}/I$-*Mod $\simeq$ *$(\mathcal{A}, I)$-*Mod.

If $\psi : \mathcal{A} \longrightarrow \mathcal{A}'$ is a morphism of ditalgebras and I is a triangular ideal of $\mathcal{A}$, with associated filtration $\{0\} = H_0 \subseteq H_1 \subseteq \cdots \subseteq H_\ell = H$, then the subset $\psi(H)$ of A' generates a triangular ideal I' of $\mathcal{A}'$ with associated filtration $\{0\} = \psi(H_0) \subseteq \psi(H_1) \subseteq \cdots \subseteq \psi(H_\ell) = \psi(H)$. If $\eta' : \mathcal{A}' \longrightarrow \mathcal{A}'/I'$ is the corresponding canonical projection, the morphism ψ induces a morphism of ditalgebras $\overline{\psi} : \mathcal{A}/I \longrightarrow \mathcal{A}'/I'$ such that $\overline{\psi}\eta = \eta'\psi$. Consider the induced commutative diagram

$$\begin{array}{ccc} \mathcal{A}'\text{-Mod} & \xrightarrow{F_\psi} & \mathcal{A}\text{-Mod} \\ {\scriptstyle F_{\eta'}}\big\uparrow & & \big\uparrow{\scriptstyle F_\eta} \\ \mathcal{A}'/I'\text{-Mod} & \xrightarrow{F_{\overline{\psi}}} & \mathcal{A}/I\text{-Mod}. \end{array}$$

*Show that whenever $M \in \mathcal{A}/I$-*Mod *and $N \in \mathcal{A}'$-*Mod *satisfy that $F_\psi(N) \cong F_\eta(M)$, then there is $N' \in \mathcal{A}'/I'$-*Mod *with $F_{\overline{\psi}}(N') \cong M$. Thus, if F_ψ is dense, so is $F_{\overline{\psi}}$ and it induces a dense functor*

$$\overline{F}_\psi : (\mathcal{A}', I')\text{-Mod} \longrightarrow (\mathcal{A}, I)\text{-Mod}.$$

References. The notion of the ideal of a ditalgebra has evolved from the notion of a "regular ideal" introduced in [46], where some form of the basic contructions exhibited in this section can be traced back. Such is the case of regularization or factoring out some part of W_0 annihilated by the differential. The general presentation given here, although elementary, is very useful, as we shall see later.

9

Frames and Roiter ditalgebras

In this section, we explore necessary conditions under which a functor between module categories of layered ditalgebras $F : \mathcal{A}'\text{-Mod} \longrightarrow \mathcal{A}\text{-Mod}$ reflects the property of being a Roiter ditalgebra.

Definition 9.1. *Let the ditalgebras $\mathcal{A}$ and $\mathcal{A}'$ be layered by (R, W) and (R', W') respectively. Then, a functor $F : \mathcal{A}'\text{-Mod} \longrightarrow \mathcal{A}\text{-Mod}$ is framed by the pair $(F_\circ, \xi_F)$ iff:*

(1) $F_\circ : R'\text{-Mod} \longrightarrow R\text{-Mod}$ *is a functor;*

(2) $\xi_F : \mathrm{Phom}_{R'\text{-}W'}(M', N') \longrightarrow \mathrm{Phom}_{R\text{-}W}(F_\circ(M'), F_\circ(N'))$ *is a natural transformation in M' and N' in* $R'\text{-Mod}$*;*

(3) $U_{\mathcal{A}}F(M') = F_\circ U_{\mathcal{A}'}(M')$*, for all* $M' \in \mathcal{A}'\text{-Mod}$*;*

(4) for any $M', N' \in \mathcal{A}'\text{-Mod}$, the following square commutes

$$\begin{array}{ccc} \mathrm{Phom}_{R'\text{-}W'}(M', N') & \xrightarrow{\xi_F} & \mathrm{Phom}_{R\text{-}W}(F_\circ(M'), F_\circ(N')) \\ \uparrow{\scriptstyle \sigma'} & & \uparrow{\scriptstyle \sigma} \\ \mathrm{Hom}_{\mathcal{A}'}(M', N') & \xrightarrow{F} & \mathrm{Hom}_{\mathcal{A}}(F(M'), F(N')). \end{array}$$

Here, $U_{\mathcal{A}} : \mathcal{A}\text{-Mod} \longrightarrow R\text{-Mod}$ denotes the functor that maps each object M onto its underlying R-module and each morphism (f^0, f^1) onto f^0. When there is no danger of confusion, we write simply M instead of $U_{\mathcal{A}}(M)$. We will also identify f and $\sigma(f)$.

Lemma 9.2. *Assume that the ditalgebras $\mathcal{A}$ and $\mathcal{A}'$ admit triangular layers (R, W) and (R', W') respectively, and that the functor $F : \mathcal{A}'\text{-Mod} \longrightarrow \mathcal{A}\text{-Mod}$ is framed by the pair $(F_\circ, \xi_F)$. Then, the following conditions all together imply that $\mathcal{A}'$ is a Roiter ditalgebra whenever $\mathcal{A}$ is so:*

(1) F is full.

(2) ξ_F is injective.

(3) *If* $\xi_F(f^0, f^1) = (g^0, g^1)$ *and* f^0 *is bijective, so is* g^0.

(4) *Let* $M' \in R'$-Mod *and* $\widehat{M} \in \mathcal{A}$-Mod *satisfy* $U_{\mathcal{A}}(\widehat{M}) = F_\circ(M')$; *take* $N' \in \mathcal{A}'$-Mod *and suppose that the pair* $\xi_F(f^0, f^1) \in \text{Phom}_{R\text{-}W}(F_\circ(M'), F_\circ(N'))$ *is an isomorphism* $\widehat{M} \to F(N')$ *in* $\mathcal{A}$-Mod. *Then, there exists an object* $\widehat{M}' \in \mathcal{A}'$-Mod *with* $F(\widehat{M}') = \widehat{M}$ *and* $U_{\mathcal{A}'}(\widehat{M}') = M'$.

$(4)^{op}$ *Let* $M' \in R'$-Mod *and* $\widehat{M} \in \mathcal{A}$-Mod *satisfy* $U_{\mathcal{A}}(\widehat{M}) = F_\circ(M')$; *take* $N' \in \mathcal{A}'$-Mod *and suppose that the pair* $\xi_F(f^0, f^1) \in \text{Phom}_{R\text{-}W}(F_\circ(N'), F_\circ(M'))$ *is an isomorphism* $F(N') \to \widehat{M}$ *in* $\mathcal{A}$-Mod. *Then, there exists an object* $\widehat{M}' \in \mathcal{A}'$-Mod *with* $F(\widehat{M}') = \widehat{M}$ *and* $U_{\mathcal{A}'}(\widehat{M}') = M'$.

Proof. Assume that $M' \in R'$-Mod and $N' \in \mathcal{A}'$-Mod. Make $N := F(N')$. Consider an isomorphism $f^0 : M' \longrightarrow N'$ in R'-Mod and a morphism of bimodules $f^1 \in \text{Hom}_{R'\text{-}R'}(W_1', \text{Hom}_k(M', N'))$. From (3), we know that the first component of $(g^0, g^1) = \xi_F(f^0, f^1)$ is an isomorphism of R-modules. Since $\mathcal{A}$ is a Roiter ditalgebra, there exists a left A-module $\widehat{M}$ with underlying R-module $M := F_\circ(M')$, such that $g = (g^0, g^1) : \widehat{M} \longrightarrow N$ is a morphism in $\mathcal{A}$-Mod. By (5.8) we know that (g^0, g^1) is an isomorphism in $\mathcal{A}$-Mod. Condition (4) implies that $\widehat{M} = F(\widehat{M}')$ and $U_{\mathcal{A}'}(\widehat{M}') = M'$, for some $\widehat{M}' \in \mathcal{A}'$-Mod. Now, since F is a full functor, there exists $h = (h^0, h^1) \in \text{Hom}_{\mathcal{A}'}(\widehat{M}', N')$ with $F(h) = g \in \text{Hom}_{\mathcal{A}}(F(\widehat{M}'), F(N'))$. Since F is framed by $(F_\circ, \xi_F)$, we have that $\xi_F(h^0, h^1) = \xi_F(f^0, f^1)$. The injectivity of ξ_F implies that $(f^0, f^1) = h$ is a morphism.

The other requirement of (5.5) is verified dually, using condition $(4)^{op}$. □

Proposition 9.3. *Assume that the layered ditalgebra* $\mathcal{A}^z$ *is obtained from the layered ditalgebra* $\mathcal{A}$, *by one of the basic operations: factoring out an ideal of* R *(which includes the possibility of deletion of an idempotent), factoring out a subbimodule of* W_0 *(which includes the possibility of regularization) or absorption, as described in (8.16), (8.18) and (8.20), respectively. Then, if the original* $\mathcal{A}$ *is a Roiter ditalgebra, so is* $\mathcal{A}^z$.

Proof. In every possible case $z \in \{s, p, a\}$, we have a morphism of ditalgebras $\eta : \mathcal{A} \longrightarrow \mathcal{A}^z$, which induces a full and faithful functor $F_\eta : \mathcal{A}^z\text{-Mod} \longrightarrow \mathcal{A}\text{-Mod}$. Consider the restriction functor $F_\circ : R^z\text{-Mod} \longrightarrow R\text{-Mod}$ associated to the morphism $R \longrightarrow R^z$ induced by η. Given $M', N' \in R^z$-Mod, define

$$\begin{array}{c}\text{Hom}_{R^z}(M', N') \times \text{Hom}_{R^z\text{-}R^z}(W_1^z, \text{Hom}_k(M', N')) \\ \Big\downarrow \xi_F \\ \text{Hom}_R(M', N') \times \text{Hom}_{R\text{-}R}(W_1, \text{Hom}_k(M', N')),\end{array}$$

by $\xi_F(f^0, f^1) = (F_\circ(f^0), g^1)$, where $g^1(w) = f^1(\eta[w])$, for any $w \in W_1$. It is easy to see that the pair $(F_\circ, \xi_F)$ is a frame for the functor $F = F_\eta$.

Since F_0 is faithful and $\eta(W_1)$ generates W_1^z as an R^z-R^z-bimodule, ξ_F is injective. Condition (3) of Lemma (9.2) is clear because the first component of $\xi_F(f^0, f^1)$ is $F_\circ(f^0)$ and $F_\circ$ is a functor.

Let us verify condition (4) of the last lemma. Adopt the notation given there, with $\mathcal{A}' = \mathcal{A}^z$, and make $\xi_F(f^0, f^1) = (g^0, g^1)$. In case $z = s$, since I_R annihilates $F(N')$ which is isomorphic as an R-module to $\widehat{M}$, we also have $I_R\widehat{M} = 0$, and we can take $\widehat{M}'$ to be the corresponding $A^s = A/\langle I_R\rangle$-module with $\overline{a}m' = am'$, for $a \in A$ and $m' \in M'$, where $\overline{a} = \eta(a)$. In case $z = p$, we know that W_0' annihilates $F(N')$ and we claim that it annihilates $\widehat{M}$ too. Indeed, by assumption, $w'g^0(m') = g^0(w'm') + g^1(\delta(w'))[m']$, for $m' \in M'$ and $w' \in W_0'$. The left-hand term is zero and, by the definition of ξ_F, $g^1(\delta(w')) = f^1(\eta(\delta(w'))) = f^1(\delta^p(\eta(w'))) = 0$. Hence, since g^0 is bijective, we also have $w'm' = 0$, as claimed. Then we can take $\widehat{M}'$ to be the corresponding $A^p = A/\langle W_0'\rangle$-module with $\overline{a}m' = am'$, as before. Finally, in case $z = a$, put $\widehat{M}' = F^{-1}(\widehat{M})$. It remains to show that $U_{\mathcal{A}^a}(\widehat{M}') = M'$. First notice that the isomorphism $g^0 : U_{\mathcal{A}}(\widehat{M}) \longrightarrow F(N')$ of R-modules is in fact an isomorphism of R^a-modules because $\delta(R^a) = 0$ and (g^0, g^1) is a morphism. Thus, $rm' = (g^0)^{-1}(rg^0(m'))$, for $r \in R^a$ and $m' \in U_{\mathcal{A}}(\widehat{M}')$. But we also know that f^0 is an isomorphism of R^a-modules, because as linear maps $f^0 = g^0$ and g^0 is invertible. Thus, the structure of M' as an R^a-module is determined by the structure of N' by the same formula $rm' = (g^0)^{-1}(rg^0(m'))$, for $r \in R^a$. The condition $(4)^{op}$ of (9.2) is proved dually.

Finally, the triangularity of $\mathcal{A}^z$ has already been shown in (8.16), (8.18) and (8.20). □

Remark 9.4. *Given a Roiter ditalgebra $\mathcal{A}$, every layered ditalgebra $\mathcal{A}'$, which is isomorphic to $\mathcal{A}$ as a layered ditalgebra (i.e. with an isomorphism ϕ such that ϕ and ϕ^{-1} are compatible with the given layers), is a Roiter ditalgebra too.*

Exercise 9.5. *Let $\mathcal{A} = (T, \delta)$ be a ditalgebra with layer (R, W). Assume that $S \subseteq W_1$ is such that $\delta(S) \subseteq ASV + VSA$, and consider the ideal I of T generated by S. By (8.21), the ideal I is an ideal of $\mathcal{A}$. Show that I is compatible with the layer (R, W) and derive, from (8.15), that the quotient $\mathcal{A}/I$ is layered. Moreover, we can identify $\mathcal{A}/I$ with $(T_R(W_0 \oplus [W_1/RSR]), \widehat{\delta})$. From, (8.12), if $\mathcal{A}$ is triangular, then $\mathcal{A}/I$ is triangular. Show that $\mathcal{A}/I$ is a Roiter ditalgebra, whenever $\mathcal{A}$ is so.*

Exercise 9.6. *Assume that $\mathcal{A}$ and $\mathcal{B}$ are Roiter k-ditalgebras and consider any exact functor $F : \mathcal{B}\text{-Mod} \longrightarrow \mathcal{A}\text{-Mod}$. Thus, it maps conflations in $\mathcal{E}_{\mathcal{B}}$ onto conflations in $\mathcal{E}_{\mathcal{A}}$, and it induces a natural transformation of k-bifunctors $F^* : \mathrm{Ext}_{\mathcal{B}}(-,?) \longrightarrow \mathrm{Ext}_{\mathcal{A}}(F(-), F(?))$. Show that F^* is a monomorphism when F is full and faithful.*

Exercise 9.7. *Assume $\mathcal{A} = (T, \delta)$ is a Roiter ditalgebra with layer (R, W). Let W_0' be a direct summand of the R-R-bimodule W_0 such that $\delta(W_0') = 0$. Consider the subalgebra R^a of T generated by R and W_0'. Thus, R^a is freely generated by R and W_0'. Then, we have the exact functor $G : \mathcal{A}\text{-Mod} \longrightarrow R^a\text{-Mod}$, which maps each morphism $f = (f^0, f^1)$ of $\mathcal{A}$-Mod onto f^0. Show that it induces an epimorphism of bifunctors*

$$G^* : \mathrm{Ext}_{\mathcal{A}}(-,?) \longrightarrow \mathrm{Ext}_{R^a,R}(-,?).$$

Hint: Consider the commutative diagram

$$\begin{array}{ccccccc}
\mathrm{Phom}_{R\text{-}W}(M,N) & \xrightarrow{\partial_{\mathcal{A}}} & \mathrm{Hom}_R(W_0 \otimes_R M, N) & \xrightarrow{\psi_{\mathcal{A}}} & \mathrm{Ext}_{\mathcal{A}}(M,N) & \rightarrow & 0 \\
\downarrow{\scriptstyle \pi} & & \downarrow{\scriptstyle i^*} & & \downarrow{\scriptstyle G^*} & & \\
\mathrm{Hom}_R(M,N) & \xrightarrow{\partial_{R^a}} & \mathrm{Hom}_R(W_0' \otimes_R M, N) & \xrightarrow{\psi_{R^a}} & \mathrm{Ext}_{R^a\text{-}R}(M,N) & \rightarrow & 0,
\end{array}$$

where the first exact row is given by (6.13); the second one is also given by (6.13) but now applied to the ditalgebra $\mathcal{A}'$ with layer (R, W'), with $W_1' = 0$ (see (6.14)); π is the projection in the first component, and i^* is induced by the section $i : W_0' \longrightarrow W_0$. Thus G^* is surjective because $\psi_{R^a} i^*$ is so.

Exercise 9.8. *Let $\mathcal{A}$ be a Roiter ditalgebra with layer (R, W). Let W_0' be a direct summand of the R-R-bimodule W_0 such that $\delta(W_0') = 0$, and let $\mathcal{A}^a$ be the Roiter ditalgebra with layer (R^a, W^a) obtained from $\mathcal{A}$ by absorption of W_0', as in (8.20). Notice that the corresponding equivalence $F_a : \mathcal{A}^a\text{-Mod} \longrightarrow \mathcal{A}\text{-Mod}$ is exact. Moreover, if $G : \mathcal{A}\text{-Mod} \longrightarrow R^a\text{-Mod}$ is the functor introduced in (9.7), then the following exact sequence of bifunctors is induced*

$$0 \longrightarrow \mathrm{Ext}_{\mathcal{A}^a}(-,?) \xrightarrow{F_a^*} \mathrm{Ext}_{\mathcal{A}}(F_a(-), F_a(?)) \xrightarrow{G^*} \mathrm{Ext}_{R^a\text{-}R}(F_a(-), F_a(?)) \longrightarrow 0.$$

Exercise 9.9. *Assume that the ditalgebra $\mathcal{A}^s$ is obtained from the Roiter ditalgebra $\mathcal{A}$, with layer (R, W), by factoring out an ideal I_R of R, as in (8.16). Notice that the corresponding full and faithful functor $F_s : \mathcal{A}^s\text{-Mod} \longrightarrow \mathcal{A}\text{-Mod}$ is exact. Moreover, it induces an isomorphism of bifunctors*

$$\mathrm{Ext}_{\mathcal{A}^s}(-,?) \xrightarrow{F_s^*} \mathrm{Ext}_{\mathcal{A}}(F_s(-), F_s(?)).$$

Exercise 9.10. *Assume $\mathcal{A} = (T, \delta)$ is a Roiter ditalgebra with layer (R, W). Assume that we have the following decompositions of R-R-bimodules $W_0 = W_0' \oplus W_0''$ and $W_1 = \delta(W_0') \oplus W_1''$. Consider the ditalgebra $\mathcal{A}^r$ obtained from $\mathcal{A}$ by regularization of the bimodule W_0', as in (8.19). Notice that the corresponding full and faithful functor $F_r : \mathcal{A}^r\text{-Mod} \longrightarrow \mathcal{A}\text{-Mod}$ is exact. Thus there is a monomorphism of bifunctors*

$$\mathrm{Ext}_{\mathcal{A}^r}(-, ?) \xrightarrow{F_r^*} \mathrm{Ext}_{\mathcal{A}}(F_r(-), F_r(?)).$$

Moreover, if $\delta_{|W_0'}$ is injective, then F_r^* is an isomorphism.

References. This brief section, technical in nature, is paraphrased from [44], where the notion of *frame* is introduced. (9.8) and (9.10) were reported in [6].

10
Product of ditalgebras

Now we introduce the canonical construction of product of ditalgebras and describe its category of modules, which will be useful later.

Lemma 10.1. *Assume that T_1 and T_2 are k-algebras. Then, the cartesian product $T_1 \times T_2$, with the componentwise product, is a k-algebra. Moreover:*

(1) If T_1 and T_1 are freely generated by the pairs (A_1, V_1) and (A_2, V_2), respectively, then the k-algebra $T_1 \times T_2$ is freely generated by the pair $(A_1 \times A_2, V_1 \times V_2)$.

(2) If T_1 and T_2 are graded k-algebras, then $T_1 \times T_2$ has the natural grading $[T_1 \times T_2]_i = [T_1]_i \times [T_2]_i$, for all i. With this grading, the product $T_1 \times T_2$ of two t-algebras T_1 and T_2 is a t-algebra: the product of the t-algebras T_1 and T_2.

(3) If T_1 and T_2 are layered t-algebras with layers (R_1, W^1) and (R_2, W^2), respectively, then the product t-algebra $T_1 \times T_2$ admits the natural layer (R, W), where $R = R_1 \times R_2$, $W_i = W_i^1 \times W_i^2$, for $i \in \{0, 1\}$. We call $T_1 \times T_2$ the product layered t-algebra of T_1 by T_2.

Proof.

(1) Write $A := A_1 \times A_2$ and $V := V_1 \times V_2$. We can consider the morphism of k-algebras $T_A(V) \xrightarrow{\varphi} T_1 \times T_2$ determined by the inclusion maps $A \to T_1 \times T_2$ and $V \to T_1 \times T_2$. Let us show that φ is bijective. The vector space injections of A_i in A and of V_i in V determine linear maps $T_i \xrightarrow{\psi_i} T_A(V)$, which preserve the product. Then, the map $T_1 \times T_2 \xrightarrow{\psi} T_A(V)$, with receipt $\psi(t_1, t_2) = \psi_1(t_1) + \psi_2(t_2)$ is an inverse for φ.

(2) If we assume that $T_1 = \oplus_i [T_1]_i$ and $T_2 = \oplus_j [T_2]_j$, then we have $T_1 \times T_2 = \sum_i [T_1]_i \times [T_2]_i$, where the last sum is direct. The fact that $T_1 \times T_2$ is a t-algebra if T_1 and T_2 are so follows now from (1).
(3) Follows from (1) too.

□

Definition 10.2. *Given the ditalgebras $\mathcal{A}_1 = (T_1, \delta_1)$ and $\mathcal{A}_2 = (T_2, \delta_2)$, the product ditalgebra of $\mathcal{A}_1$ by $\mathcal{A}_2$, denoted by $\mathcal{A}_1 \times \mathcal{A}_2$, is the ditalgebra (T, δ), where T is the product of the t-algebras T_1 and T_2, and $\delta = \delta_1 \times \delta_2$. The projections $T_1 \times T_2 \xrightarrow{\pi_i} T_i$ determine morphisms of ditalgebras $\mathcal{A}_1 \times \mathcal{A}_2 \xrightarrow{\pi_i} \mathcal{A}_i$, for $i \in \{1, 2\}$.*

If $\mathcal{A}_1$ and $\mathcal{A}_2$ are layered ditalgebras, with associated layers (R_1, W^1) and (R_2, W^2), then the product ditalgebra $\mathcal{A}_1 \times \mathcal{A}_2$ admits the natural layer (R, W) described in (10.1). With this layer, $\mathcal{A}_1 \times \mathcal{A}_2$ is called the product layered ditalgebra of $\mathcal{A}_1$ by $\mathcal{A}_2$. The canonical projections $\pi_i : \mathcal{A}_1 \times \mathcal{A}_2 \longrightarrow \mathcal{A}_i$ are morphisms of layered ditalgebras.

Lemma 10.3. *Assume that $\mathcal{A}_1$ and $\mathcal{A}_2$ are any ditalgebras. Then, given any other ditalgebra $\mathcal{A}'$ and any pair of functors $F_1 : \mathcal{A}_1\text{-Mod} \longrightarrow \mathcal{A}'\text{-Mod}$ and $F_2 : \mathcal{A}_2\text{-Mod} \longrightarrow \mathcal{A}'\text{-Mod}$, there is a functor*

$$\mathcal{A}_1\text{-Mod} \times \mathcal{A}_2\text{-Mod} \xrightarrow{F} \mathcal{A}'\text{-Mod},$$

such that, for any morphisms $f_1 \in \mathrm{Hom}_{\mathcal{A}_1}(M_1, N_1)$ and $f_2 \in \mathrm{Hom}_{\mathcal{A}_2}(M_2, N_2)$, the morphism $F(f_1, f_2) \in \mathrm{Hom}_{\mathcal{A}'}(F_1(M_1) \oplus F_2(M_2), F_1(N_1) \oplus F_2(N_2))$ is defined by $F(f_1, f_2) = F_1(f_1) \oplus F_2(f_2)$. More explicitely

$$F(f_1, f_2)^0 = \begin{pmatrix} F_1(f_1)^0 & 0 \\ 0 & F_2(f_2)^0 \end{pmatrix}$$

and, for $v' \in V' = [T']_1$

$$F(f_1, f_2)^1(v') = \begin{pmatrix} F_1(f_1)^1(v') & 0 \\ 0 & F_2(f_2)^1(v') \end{pmatrix}.$$

We shall denote this functor F by $F_1 \oplus F_2$.

The canonical projections $(\mathcal{A}_1 \times \mathcal{A}_2) \xrightarrow{\pi_i} \mathcal{A}_i$ induce the functors $\mathcal{A}_i\text{-Mod} \xrightarrow{F_{\pi_i}} (\mathcal{A}_1 \times \mathcal{A}_2)\text{-Mod}$, which produce an equivalence of categories

$$\mathcal{A}_1\text{-Mod} \times \mathcal{A}_2\text{-Mod} \xrightarrow{F_{\pi_1} \oplus F_{\pi_2}} (\mathcal{A}_1 \times \mathcal{A}_2)\text{-Mod}.$$

Proof. It is easy to see that $F = F_1 \oplus F_2$ is indeed a well-defined functor.

We write, as above, $\mathcal{A}_1 = (T_1, \delta_1)$, $\mathcal{A}_2 = (T_2, \delta_2)$ and $\mathcal{A} = \mathcal{A}_1 \times \mathcal{A}_2 = (T, \delta)$, with $T = T_1 \times T_2$ and $\delta = \delta_1 \times \delta_2$. Make $F_i = F_{\pi_i}$, for $i \in \{1, 2\}$, and consider the functor $F = F_1 \oplus F_2$.

Write $A_1 = [T_1]_0$, $A_2 = [T_2]_0$ and $A = [T]_0$; and $V_1 = [T_1]_1$, $V_2 = [T_2]_1$ and $V = [T]_1$. Consider the decomposition of the unit $1 = e_1 + e_2$ of A as a sum of orthogonal central idempotents corresponding to the decomposition $A = A_1 \times A_2$.

Notice that, given $M \in \mathcal{A}$-Mod, each space $e_i M$ has a natural structure of A_i-module. Indeed, for instance, $e_1 M$ is clearly an A_1-module with the action $a_1 m_1 := (a_1, 0)m_1$, where $a_1 \in A_1$ and $m_1 \in e_1 M$.

We have that any $M \in \mathcal{A}$-Mod can be decomposed as the vector space internal direct sum $M = e_1 M \oplus e_2 M$. Take any morphism $g \in \mathrm{Hom}_{\mathcal{A}}(M, N)$. Since $g : M \longrightarrow N$ is a morphism of $\mathcal{A}$-modules and $\delta(e_i) = 0$, $g^0(e_i m) = e_i g^0(m)$, for all $m \in M$, and g^0 splits as the direct sum of its restrictions $g_i^0 : e_i M \longrightarrow e_i N$, for $i \in \{1, 2\}$. Moreover, $g^1 : V \longrightarrow \mathrm{Hom}_k(M, N)$ is a morphism of A-A-bimodules and, for $v \in V$, the i, j component of the 2×2 matrix of linear maps $g^1(v) : e_1 M \oplus e_2 M \longrightarrow e_1 N \oplus e_2 N$ is precisely $g^1(e_j v e_i) = e_j g^1(v) e_i : e_i M \longrightarrow e_j N$. Since $e_j V e_i = \delta_{i,j} V_i$, for $i, j \in \{1, 2\}$, we have $g^1(e_j v e_i) = 0$, for $i \neq j$. For $v_1 \in V_1$ and $v_2 \in V_2$ define $g_1^1(v_1) := g^1[(v_1, 0)]$ and $g_2^1(v_2) := g^1[(0, v_2)]$. Then, $g_i^1 \in \mathrm{Hom}_{A_i\text{-}A_i}(V_i, \mathrm{Hom}_k(e_i M, e_i N))$. We claim that $g_i = (g_i^0, g_i^1) \in \mathrm{Hom}_{\mathcal{A}_i}(e_i M, e_i N)$. Indeed, for instance, for $a_1 \in A_1$ and $m_1 \in e_1 M$, we have

$$\begin{aligned} a_1 g_1^0(m_1) - g_1^0(a_1 m_1) &= (a_1, 0) g^0(m_1) - g^0((a_1, 0)m_1) \\ &= g^1(\delta[(a_1, 0)])[m_1] \\ &= g^1((\delta_1[a_1], 0))[m_1] \\ &= g_1^1(\delta_1(a_1))[m_1]. \end{aligned}$$

Thus, we can define, for $i \in \{1, 2\}$, the functor $G_i : \mathcal{A}\text{-Mod} \longrightarrow \mathcal{A}_i\text{-Mod}$, such that $G_i(M) = e_i M$ and, given $g \in \mathrm{Hom}_{\mathcal{A}}(M, N)$, $G_i(g) = g_i$, as described above. We claim that the functor $G := (G_1, G_2) : \mathcal{A}\text{-Mod} \longrightarrow \mathcal{A}_1\text{-Mod} \times \mathcal{A}_2\text{-Mod}$ is a quasi-inverse for F.

Indeed, we have the natural isomorphism $\psi : FG \longrightarrow 1_{\mathcal{A}\text{-Mod}}$, given for each $M \in \mathcal{A}$-Mod, by the morphism $\psi_M = (\psi_M^0, 0)$, where $\psi_M^0 : FG(M) = F_1(e_1 M) \oplus F_2(e_2 M) \longrightarrow M$ is the map defined by $\psi_M^0(e_1 m, e_2 m') = e_1 m + e_2 m'$. Likewise, we have that GF is isomorphic to the identity functor on the product category $\mathcal{A}_1\text{-Mod} \times \mathcal{A}_2\text{-Mod}$. □

Exercise 10.4. *If $\mathcal{A}_1$ and $\mathcal{A}_2$ are Roiter ditalgebras, so is its product $\mathcal{A}_1 \times \mathcal{A}_2$. In this case, with the notation of the proof of (10.3), shows that there is an isomorphism of bifunctors*

$$\mathrm{Ext}_{\mathcal{A}_1 \times \mathcal{A}_2}(-, ?) \cong \mathrm{Ext}_{\mathcal{A}_1}(G_1(-), G_1(?)) \oplus \mathrm{Ext}_{\mathcal{A}_2}(G_2(-), G_2(?)).$$

11

Hom-tensor relations and dual basis

In this section, we recall some hom-tensor relations and introduce maps μ, λ and ρ which will play an important role in the construction, in the next section, of ditalgebras obtained by reduction using admissible modules.

The following statement is well known:

Lemma 11.1. *A right S-module U is projective iff there is a dual basis for U, this means a collection $\{(u_i, \lambda_i) \mid i \in I\}$ such that $u_i \in U$ and $\lambda_i \in U^* = \mathrm{Hom}_S(U, S)$, for all $i \in I$, and, for any $u \in U$, the following equality holds: $u = \sum_{i \in I} u_i \lambda_i(u)$, where $\lambda_i(u) = 0$, for almost all $i \in I$.*

If U_S is finitely generated and projective, such dual basis can be chosen with finite index set I. Starting with a left S-module U, we have the corresponding dual statement, using its left dual ${}^*U = \mathrm{Hom}_S(U, S)$.

Lemma 11.2. *Assume that the module U_S admits a finite dual basis $\{(u_i, \lambda_i) \mid i \in I\}$. For each $i \in I$, let $d_i \in {}^*(U^*)$ be given by $d_i(\lambda) = \lambda(u_i)$, for any $\lambda \in U^*$. Then $\{(\lambda_i, d_i) \mid i \in I\}$ is a finite dual basis for U^*.*

Proof. We know that $u_i \in U$ and $\lambda_i \in U^* = \mathrm{Hom}_S(U, S)$, for all $i \in I$. In order to define d_i on U^*, we had to fix a $\lambda \in \mathrm{Hom}_S(U, S)$. Thus, $d_i(\lambda) = \lambda(u_i) \in S$; then for $s \in S$, $d_i(s\lambda) = (s\lambda)(u_i) = s(\lambda(u_i)) = sd_i(\lambda)$. Hence, indeed, $d_i \in {}^*(U^*)$. We have to show that $\lambda = \sum_{i \in I} d_i(\lambda)\lambda_i$, for $\lambda \in U^* = \mathrm{Hom}_S(U, S)$. For a fixed $u \in U$, we have $\lambda(u) = \lambda\left(\sum_{i \in I} u_i \lambda_i(u)\right) = \sum_{i \in I} \lambda(u_i)\lambda_i(u) = \left[\sum_{i \in I} d_i(\lambda)\lambda_i\right](u)$. □

Lemma 11.3. *Given the bimodules ${}_TW_S$ and ${}_RU_S$, there is a natural morphism of T-R-bimodules, $W \otimes_S U^* \xrightarrow{\phi} \mathrm{Hom}_S(U, W)$ such that $\phi(w \otimes \lambda)(u) = w\lambda(u)$, for all $w \in W$, $\lambda \in U^*$ and $u \in U$. Moreover, if U_S admits a finite dual basis, $\{(u_i, \lambda_i) \mid i \in I\}$, then ϕ is an isomorphism and its inverse* *satisfies*

$$\phi^{-1}(h) = \sum_{i \in I} h(u_i) \otimes \lambda_i,$$

for each $h \in \mathrm{Hom}_S(U, W)$.

Proof. It is easy to show that ϕ is well defined and a natural morphism of bimodules. Take $h \in \mathrm{Hom}_S(U, W)$ and $u \in U$, then

$$\phi\left(\sum_{i \in I} h(u_i) \otimes \lambda_i\right)(u) = \sum_{i \in I} h(u_i)\lambda_i(u) = h\left(\sum_{i \in I} u_i \lambda_i(u)\right) = h(u).$$

Moreover, given $\lambda \in U^*$ and $w \in W$, we have

$$\sum_{i \in I} w\lambda(u_i) \otimes \lambda_i = w \otimes \left(\sum_{i \in I} \lambda(u_i)\lambda_i\right) = w \otimes \lambda.$$

□

Lemma 11.4. *Consider the bimodules* ${}_TA_S$ *and* ${}_SB_R$, *then there is a natural morphism of R-T-bimodules* $\psi : B^* \otimes_S A^* \longrightarrow (A \otimes_S B)^*$, *given by the prescription* $\psi(g \otimes f)[a \otimes b] = g(f(a)b)$. *If* A_S *admits a finite dual basis* $\{(a_i, \lambda_i) \mid i \in I\}$, *then* ψ *is an isomorphism and its inverse satisfies*

$$\psi^{-1}(h) = \sum_{i \in I} \rho_i(h) \otimes \lambda_i, \text{ where } \rho_i(h)[b] = h(a_i \otimes b),$$

for each $h \in (A \otimes_S B)^*$.

Proof. It is easy to see that ψ is a well-defined and natural morphism of bimodules. Take $h \in \mathrm{Hom}_R(A \otimes_S B, R)$, $a \in A$ and $b \in B$, then we have $\psi\psi^{-1}(h)[a \otimes b] = \psi(\sum_i \rho_i(h) \otimes \lambda_i)[a \otimes b] = \sum_i \rho_i(h)[\lambda_i(a)b] = \sum_i h(a_i \otimes \lambda_i(a)b) = h(a \otimes b)$. Now, let $g \in B^*$ and $f \in A^*$. Recall that (a_i, λ_i) is a dual basis for A_S and, therefore, (λ_i, d_i), as defined in (11.2), is a dual basis for ${}_SA^*$. Thus, $f = \sum_i d_i(f)\lambda_i$. Then, for $b \in B$

$$\rho_i(\psi(g \otimes f))[b] = g(f(a_i)b) = (gf(a_i))[b] = (gd_i(f))[b],$$

which means that $\rho_i(\psi(g \otimes f)) = gd_i(f)$. Then

$$\psi^{-1}\psi(g \otimes f) = \sum_i \rho_i(\psi(g \otimes f)) \otimes \lambda_i = \sum_i gd_i(f) \otimes \lambda_i$$
$$= g \otimes \sum_i d_i(f)\lambda_i = g \otimes f.$$

□

The proof of the following two lemmas is straightforward.

Lemma 11.5. *Consider the bimodules ${}_TA_S$, ${}_SB_R$ and ${}_RC_L$. Then the following diagram is commutative (where the morphisms ψ are defined in (11.4))*

$$\begin{array}{ccc} C^*\otimes_R B^*\otimes_S A^* & \xrightarrow{\psi\otimes Id} & (B\otimes_R C)^*\otimes_S A^* \\ {\scriptstyle Id\otimes\psi}\downarrow & & \downarrow{\scriptstyle\psi} \\ C^*\otimes_R (A\otimes_S B)^* & \xrightarrow{\psi} & (A\otimes_S B\otimes_R C)^*. \end{array}$$

Lemma 11.6. *Given the bimodules ${}_TE_R$, ${}_SF_L$ and ${}_TG_L$, and the morphisms of bimodules $m_1 : A\otimes_S B \longrightarrow E$, $m_2 : B\otimes_R C\longrightarrow F$, $m_3 : A\otimes_S F\longrightarrow G$, and $m_4 : E\otimes_R C\longrightarrow G$ such that the following diagram commutes*

$$\begin{array}{ccc} A\otimes_S B\otimes_R C & \xrightarrow{m_1\otimes Id} & E\otimes_R C \\ {\scriptstyle Id\otimes m_2}\downarrow & & \downarrow{\scriptstyle m_4} \\ A\otimes_S F & \xrightarrow{m_3} & G, \end{array}$$

then the following diagram commutes (where ψ is defined in (11.4))

$$\begin{array}{ccccc} G^* & \xrightarrow{m_4^*} & (E\otimes_R C)^* & \xleftarrow{\psi} & C^*\otimes_R E^* \\ {\scriptstyle m_3^*}\downarrow & & \downarrow{\scriptstyle (m_1\otimes Id)^*} & & \downarrow{\scriptstyle Id\otimes m_1^*} \\ (A\otimes_S F)^* & \xrightarrow{(Id\otimes m_2)^*} & (A\otimes_S B\otimes_R C)^* & \xleftarrow{\psi} & C^*\otimes_R (A\otimes_S B)^* \\ {\scriptstyle\psi}\uparrow & & \uparrow{\scriptstyle\psi} & & \uparrow{\scriptstyle Id\otimes\psi} \\ F^*\otimes_S A^* & \xrightarrow{m_2^*\otimes Id} & (B\otimes_R C)^*\otimes_S A^* & \xleftarrow{\psi\otimes Id} & C^*\otimes_R B^*\otimes_S A^*. \end{array}$$

Definition 11.7. *The pair (S, P) is called a splitting for the k-algebra Γ iff S is a subalgebra of Γ, P is an ideal of Γ, the module P_S admits a finite dual basis, and there is an S-S-bimodule decomposition: $\Gamma = S\oplus P$. In this case, the product of Γ induces a morphism of S-S-bimodules $m : P\otimes_S P\longrightarrow P$, which we call the multiplication of P. The associativity in Γ implies that the following diagram is commutative*

$$\begin{array}{ccc} P\otimes_S P\otimes_S P & \xrightarrow{m\otimes Id} & P\otimes_S P \\ \downarrow{\scriptstyle Id\otimes m} & & \downarrow{\scriptstyle m} \\ P\otimes_S P & \xrightarrow{m} & P. \end{array}$$

We have the morphism of S-S-bimodules $\mu := \psi^{-1}m^* : P^*\longrightarrow P^*\otimes_S P^*$, where the upper right star is taken over the algebra S. The map μ is called the comultiplication of P.

From (11.6), we know that the following diagram commutes

$$\begin{array}{ccc} P^* & \xrightarrow{\mu} & P^* \otimes_S P^* \\ \downarrow{\scriptstyle \mu} & & \downarrow{\scriptstyle Id\otimes\mu} \\ P^* \otimes_S P^* & \xrightarrow{\mu\otimes Id} & P^* \otimes_S P^* \otimes_S P^*. \end{array}$$

We refer to this property by saying that μ is coassociative.

Lemma 11.8. *Assume we have a splitting (S, P) for the k-algebra Γ, with comultiplication $P^* \xrightarrow{\mu} P^* \otimes_S P^*$. Suppose that $\{(p_j, \gamma_j) \mid j \in J\}$ is a finite dual basis for the right S-module P. Then, for any $\gamma \in P^*$, we have*

$$\mu(\gamma) = \sum_{i,j} \gamma(p_i p_j)\gamma_j \otimes \gamma_i.$$

Proof. Take $\gamma \in P^*$. Then $\mu(\gamma) = \sum_j \gamma_j^1 \otimes \gamma_j^2$ is the unique element of $P^* \otimes_S P^*$ such that $\sum_j \gamma_j^1(\gamma_j^2(p^1)p^2) = \gamma(p^1 p^2)$, for all $p^1, p^2 \in P$. Indeed, using that $\mu = \psi^{-1}m^*$, we have

$$\begin{aligned} \gamma(p^1 p^2) &= \gamma(m[p^1 \otimes p^2]) \\ &= m^*(\gamma)[p^1 \otimes p^2] \\ &= \psi\mu(\gamma)[p^1 \otimes p^2] \\ &= \psi\left(\textstyle\sum_j \gamma_j^1 \otimes \gamma_j^2\right)[p^1 \otimes p^2] \\ &= \textstyle\sum_j \gamma_j^1\left(\gamma_j^2(p^1)p^2\right). \end{aligned}$$

Then, the lemma follows from the equalities

$$\begin{aligned} \textstyle\sum_{i,j} \gamma(p_i p_j)\gamma_j[\gamma_i(p^1)p^2] &= \textstyle\sum_{i,j} \gamma\left(p_i p_j \gamma_j[\gamma_i(p^1)p^2]\right) \\ &= \textstyle\sum_i \gamma\left(p_i \gamma_i(p^1)p^2\right) \\ &= \gamma\left(p^1 p^2\right). \end{aligned}$$

□

Definition 11.9. *Assume that B is a k-algebra and Γ is an algebra with splitting (S, P). A vector space X is said to be adapted to the pair (B, Γ) iff:*

(1) X is a left B-module;
(2) X is a right Γ-module;
(3) with the restriction of the action given in (1) and (2), X is a B-S-bimodule;
(4) X_S admits a finite dual basis.

Definition 11.10. *Associated to a space X adapted to the pair (B, Γ), where Γ has splitting (S, P), we have two natural morphisms of right S-modules, which* *keep relevant information on the right action of P on X, namely*

$$m_\ell : X \otimes_S P \longrightarrow X, \text{ given by } m_\ell(x \otimes p) = xp, \text{ and}$$
$$m_r : X \longrightarrow \mathrm{Hom}_S(P, X), \text{ given by } m_r(x)[p] = xp.$$

They are morphisms of B-S-bimodules iff X is a B-Γ-bimodule. Combined with the isomorphisms given in (11.3) and (11.4): $\psi : P^* \otimes_S X^* \longrightarrow (X \otimes_S P)^*$ and $\phi : X \otimes_S P^* \longrightarrow \mathrm{Hom}_S(P, X)$, they give rise to the following morphisms of S-modules

$$\lambda := \psi^{-1} m_\ell^* : X^* \longrightarrow P^* \otimes_S X^*, \text{ and}$$
$$\rho := \phi^{-1} m_r : X \longrightarrow X \otimes_S P^*.$$

Their receipts satisfy: $\lambda(\nu) = \sum_t \gamma_t \otimes \nu_t$ is the unique element of $P^* \otimes_S X^*$ such that $\sum_t \gamma_t[\nu_t(x)p] = \nu[xp]$, for all $x \in X$ and $p \in P$; and $\rho(x) = \sum_t x_t \otimes \gamma_t$ is the unique element of $X \otimes_S P^*$ such that $\sum_t x_t\gamma_t(p) = xp$, for all $x \in X$ and $p \in P$.

Lemma 11.11. *Assume X is adapted to the pair (B, Γ) and Γ has splitting (S, P). Consider finite dual basis $\{(x_i, \nu_i) \mid i \in I\}$ for X_S, and $\{(p_j, \gamma_j) \mid j \in J\}$ for P_S. Then the morphisms λ and ρ, defined before, can be described by the formulas:*

(1) For any $x \in X$, $\quad \rho(x) = \sum_{j \in J} x p_j \otimes \gamma_j$.
(2) For any $\nu \in X^*$, $\quad \lambda(\nu) = \sum_{i \in I} \sum_{j \in J} \nu(x_i p_j)\gamma_j \otimes \nu_i$.

Proof. The first formula follows from the description of ϕ^{-1} given in (11.3). For the second formula, recall that $\nu \in X^*$; $x_i \in X$ and $\nu_i \in X^*$; $p_j \in P$ and $\gamma_j \in P^*$. Then, from the definition of λ and the receipt for ψ^{-1} given in (11.4), we have

$$\lambda(\nu) = \psi^{-1} m_\ell^*(\nu) = \psi^{-1}(\nu m_\ell) = \sum_{i \in I} \rho_i(\nu m_\ell) \otimes \nu_i,$$

where $\rho_i(\nu m_\ell) \in P^*$. From (11.2), we then know that

$$\rho_i(\nu m_\ell) = \sum_{j \in J} [\rho_i(\nu m_\ell)]\,(p_j)\,\gamma_j = \sum_{j \in J} \nu(x_i p_j)\gamma_j.$$

□

The morphisms ρ and λ are compatible with the comultiplication μ of P, in the following sense.

Lemma 11.12. *The following diagrams commute*

$$\begin{array}{ccc} X^* & \xrightarrow{\lambda} & P^* \otimes_S X^* \\ \lambda\downarrow & & \downarrow\mu\otimes Id \\ P^* \otimes_S X^* & \xrightarrow{Id\otimes\lambda} & P^* \otimes_S P^* \otimes_S X^* \end{array} \qquad \begin{array}{ccc} X & \xrightarrow{\rho} & X \otimes_S P^* \\ \rho\downarrow & & \downarrow Id\otimes\mu \\ X \otimes_S P^* & \xrightarrow{\rho\otimes Id} & X \otimes_S P^* \otimes_S P^*. \end{array}$$

Proof. The following diagram is commutative

$$\begin{array}{ccc} X \otimes_S P \otimes_S P & \xrightarrow{Id\otimes m} & X \otimes_S P \\ m_\ell\otimes Id\downarrow & & \downarrow m_\ell \\ X \otimes_S P & \xrightarrow{m_\ell} & X. \end{array}$$

Then, from (11.6), we obtain the commutativity of the first diagram. In order to show the commutativity of the second diagram, take $x \in X$. Then, from (11.8) and (11.11), we get

$$\begin{aligned} (1 \otimes \mu)\rho(x) &= (1 \otimes \mu)\left[\textstyle\sum_j xp_j \otimes \gamma_j\right] \\ &= \textstyle\sum_j xp_j \otimes \mu(\gamma_j) \\ &= \textstyle\sum_j xp_j \otimes \left[\sum_{i,t} \gamma_j(p_i p_t)\gamma_t \otimes \gamma_i\right] \\ &= \textstyle\sum_{i,t,j} xp_j\gamma_j(p_i p_t) \otimes \gamma_t \otimes \gamma_i \\ &= \textstyle\sum_{i,t} x\left[\sum_j p_j\gamma_j(p_i p_t)\right] \otimes \gamma_t \otimes \gamma_i \\ &= \textstyle\sum_j \left[\sum_t xp_j p_t \otimes \gamma_t\right] \otimes \gamma_j \\ &= \textstyle\sum_j \rho(xp_j) \otimes \gamma_j \\ &= (\rho \otimes 1)\left[\textstyle\sum_j xp_j \otimes \gamma_j\right] \\ &= (\rho \otimes 1)\rho(x). \end{aligned}$$

□

Definition 11.13. *Assume X is a space adapted to the pair (B, Γ), where Γ has splitting (S, P). Consider finite dual basis $\{(x_i, \nu_i) \mid i \in I\}$ for X_S, and $\{(p_j, \gamma_j) \mid j \in J\}$ for P_S. Then, given $\nu \in X^*$, $x \in X$ and $b \in B$, we consider the following element of P^**

$$\gamma_{\nu,x}^b := \sum_j \nu(b[xp_j] - [bx]p_j)\gamma_j.$$

Observe that, for $p \in P$, we obtain $\gamma_{\nu,x}^b(p) = \nu(b[xp] - [bx]p)$, by replacing $p = \sum_j p_j\gamma_j(p)$ at the right-hand side of the equation.

In the following lemmas, we keep the notation and hypothesis of last definition. Since X is not always a B-Γ-bimodule, the morphisms λ and ρ are not in general morphisms of B-modules. The following lemmas will measure the deviation from this property.

Lemma 11.14. *Consider the morphism $X^* \xrightarrow{\lambda} P^* \otimes_S X^*$, as defined in (11.10), $\nu \in X^*$ and $b \in B$, then*

$$\lambda(\nu b) - \lambda(\nu)b = \sum_i \gamma^b_{\nu,x_i} \otimes \nu_i.$$

Proof. Having in mind (11.11)(2), it will be enough to verify the equality $\lambda(\nu)b = \sum_{i,j} \nu([bx_i]p_j)\gamma_j \otimes \nu_i$. We have the isomorphism $\psi : P^* \otimes_S X^* \longrightarrow (X \otimes_S P)^*$. Then, for $x \in X$ and $p \in P$, we have, from (11.11)

$$\begin{aligned}
\psi\left(\textstyle\sum_{i,j} \nu([bx_i]p_j)\gamma_j \otimes \nu_i\right)[x \otimes p] &= \textstyle\sum_{i,j} \nu([bx_i]p_j)\gamma_j(\nu_i(x)p) \\
&= \nu\left(\textstyle\sum_{i,j}[bx_i]p_j\gamma_j(\nu_i(x)p)\right) \\
&= \nu\left(\textstyle\sum_i [bx_i](\nu_i(x)p)\right) \\
&= \nu\left(\left[\textstyle\sum_i bx_i\nu_i(x)\right]p\right) \\
&= \nu([bx]p) \\
&= \nu\left(\left[\textstyle\sum_i x_i\nu_i(bx)\right]p\right) \\
&= \nu\left(\textstyle\sum_i x_i\,[\nu_i(bx)p]\right) \\
&= \nu\left(\textstyle\sum_{i,j} x_i\left[p_j\gamma_j\,[\nu_i(bx)p]\right]\right) \\
&= \nu\left(\textstyle\sum_{i,j}(x_ip_j)\gamma_j\,[\nu_i(bx)p]\right) \\
&= \textstyle\sum_{i,j} \nu(x_ip_j)\gamma_j\,[\nu_i(bx)p] \\
&= \textstyle\sum_{i,j}\left[\nu(x_ip_j)\gamma_j\right][(\nu_ib)(x)p] \\
&= \psi\left[\textstyle\sum_{i,j} \nu(x_ip_j)\gamma_j \otimes \nu_ib\right](x \otimes p) \\
&= \psi\,[\lambda(\nu)b]\,(x \otimes p).
\end{aligned}$$

Then our claims follow. □

Lemma 11.15. *Consider the morphism $X \xrightarrow{\rho} X \otimes_S P^*$, as defined in (11.10), $x \in X$ and $b \in B$, then*

$$\rho(bx) - b\rho(x) = -\sum_i x_i \otimes \gamma^b_{\nu_i,x}.$$

Proof. We just have to recall, from (11.11), that

$$b\rho(x) = \sum_j b[xp_j] \otimes \gamma_j = \sum_{i,j} x_i\nu_i(b[xp_j]) \otimes \gamma_j,$$

while, $\rho(bx) = \sum_j [bx]p_j \otimes \gamma_j = \sum_{i,j} x_i\nu_i([bx]p_j) \otimes \gamma_j$, and take the difference. □

Definition 11.16. *Assume the bimodule ${}_BX_S$ is projective finitely generated as a right S-module. Then we know that the morphism $X \otimes_S X^* \xrightarrow{\phi} \mathrm{Hom}_S(X, X)$, as defined in Lemma (11.3), is an isomorphism. Suppose that $\phi^{-1}(1_X) =$*

$\sum_{i \in I} x_i \otimes \nu_i$, with $x_i \in X$ and $\nu_i \in X^*$. Then $\{(x_i, \nu_i) \mid i \in I\}$ is a finite dual basis for X_S. Any finite dual basis for X_S appears as before.

Therefore, since ϕ is a morphism of B-B-bimodules, we obtain, for any $b \in B$

$$b\left(\sum_{i \in I} x_i \otimes \nu_i\right) = \left(\sum_{i \in I} x_i \otimes \nu_i\right) b.$$

Then, given any bimodule ${}_B U_B$ and $n \geq 2$, there is a natural morphism of S-S-bimodules

$$X^* \otimes_B U^{\otimes n} \otimes_B X \xrightarrow{\sigma_n} (X^* \otimes_B U \otimes_B X)^{\otimes n},$$

where $\sigma_n(\nu \otimes w_1 \otimes w_2 \otimes \cdots \otimes w_n \otimes x)$ is given by

$$\nu \otimes w_1 \otimes \left(\sum_{i \in I} x_i \otimes \nu_i\right) \otimes w_2 \otimes \left(\sum_{i \in I} x_i \otimes \nu_i\right) \otimes w_3 \otimes \cdots$$
$$\otimes \left(\sum_{i \in I} x_i \otimes \nu_i\right) \otimes w_n \otimes x,$$

whenever $w_1, \ldots, w_n \in U$, $\nu \in X^*$ and $x \in X$. We can extend the last definition to the case $n = 1$ taking σ_1 as the identity map. We call σ_n the canonical insertion morphism. It does not depend on the choice of the dual basis.

References. In this section, we started some analysis needed for the construction of a new ditalgebra $\mathcal{A}^X$ from a given one $\mathcal{A}$ using some special kind of module X over a subditalgebra of $\mathcal{A}$. This construction is described in the next section and generalizes the one given in [6]. The generalization is very convenient because it permits us to perform some "reduction operations" in one basic step of type $\mathcal{A} \mapsto \mathcal{A}^X$, as we shall see later.

12

Admissible modules

The notion of admissible modules introduced in this section plays a central role in our text. Given an admissible module X over a proper subditalgebra of a layered ditalgebra $\mathcal{A}$, we will construct a new layered ditalgebra $\mathcal{A}^X$ and an associated functor $F^X : \mathcal{A}^X\text{-Mod}\longrightarrow\mathcal{A}\text{-Mod}$. In the following five sections, we will discuss their basic properties.

Definition 12.1. *Let $\mathcal{A} = (T, \delta)$ be any ditalgebra with layer (R, W). Assume we have R-R-bimodule decompositions $W_0 = W_0' \oplus W_0''$ and $W_1 = W_1' \oplus W_1''$. Consider the subalgebra T' of T generated by R and $W' = W_0' \oplus W_1'$. Then, T' is freely generated by R and W' (see (3) of Lemma (1.3)). Let us write $B := [T']_0$, which is freely generated by the pair (R, W_0'), and let us assume furthermore that $\delta(W_0') \subseteq BW_1'B$ and $\delta(W_1') \subseteq BW_1'BW_1'B$. Then, the differential δ on T restricts to a differential δ' on the t-algebra T' and we obtain a new ditalgebra $\mathcal{A}' = (T', \delta')$ with layer (R, W'). A layered ditalgebra $\mathcal{A}'$ is called a proper subditalgebra of $\mathcal{A}$ if it is obtained from an R-R-bimodule decomposition of W, as we have just described.*

Lemma 12.2. *Assume that $\mathcal{A} = (T, \delta)$ is a ditalgebra with layer (R, W), and let $\mathcal{A}' = (T', \delta')$ be a proper subditalgebra of $\mathcal{A}$ associated to the R-R-bimodule decompositions $W_0 = W_0' \oplus W_0''$ and $W_1 = W_1' \oplus W_1''$. Make $A = [T]_0$ and $B := [T']_0$. Then, we have the following:*

(1) The algebra $A = [T]_0$ is freely generated by the pair $(B, BW_0''B)$.
(2) The t-algebra T has layer $(B, \underline{W})$, where $\underline{W} = \underline{W}_0 \oplus \underline{W}_1$, with $\underline{W}_0 = BW_0''B$ and $\underline{W}_1 = BW_1B$.
(3) There are isomorphisms of B-B-bimodules $B \otimes_R W_0'' \otimes_R B \longrightarrow \underline{W}_0$ and $B \otimes_R W_1 \otimes_R B \longrightarrow \underline{W}_1$ induced by the product of T.

(4) There is a B-B-bimodule decomposition $\underline{W}_1 = \underline{W}'_1 \oplus \underline{W}''_1$, where $\underline{W}'_1 = BW'_1B$ and $\underline{W}''_1 = BW''_1B$.

Proof. Statements (1) and (2) follow from (1.3). For (3) and (4), recall that the product map $A \otimes_R W_1 \otimes_R A \longrightarrow V = AW_1A$ is an isomorphism. Since $W_0 = W'_0 \oplus W''_0$, the inclusion of B in A is a split R-R-bimodule morphism. Then, $B \otimes_R W_1 \otimes_R B \longrightarrow BW_1B$ is an isomorphism too. Then, the decomposition $B \otimes_R W_1 \otimes_R B = (B \otimes_R W'_1 \otimes_R B) \oplus (B \otimes_R W''_1 \otimes_R B)$ is mapped onto the decomposition $BW_1B = BW'_1B \oplus BW''_1B$. By Lemma (4.1)(4), applied to A, which is a freely generated algebra over the pair $(R, W'_0 \oplus W''_0)$, we obtain that $B \otimes_R W''_0 \otimes_R B \longrightarrow \underline{W}_0$ is an isomorphism of B-B-bimodules. □

The utility of the following elementary lemma will be clear in Section 22.

Lemma 12.3. *Assume that $\mathcal{A}' = (T', \delta')$ is a proper subditalgebra of the layered ditalgebra $\mathcal{A} = (T, \delta)$. Write, $A := [T]_0$ and $B := [T']_0$. Then, there is a commutative diagram of functors*

$$\begin{array}{ccc} B\text{-Mod} & \xrightarrow{F_\pi} & A\text{-Mod} \\ \downarrow{\scriptstyle L_{\mathcal{A}'}} & & \downarrow{\scriptstyle L_{\mathcal{A}}} \\ \mathcal{A}'\text{-Mod} & \xleftarrow{F_r} & \mathcal{A}\text{-Mod}, \end{array}$$

where F_r is the restriction functor induced by the inclusion morphism of ditalgebras $r : \mathcal{A}' \longrightarrow \mathcal{A}$ and F_π is the restriction functor associated to the morphism of algebras $\pi : A \longrightarrow B$ induced by the identity of R and the projection morphism of R-R-bimodules $W \longrightarrow W'$. Moreover, if $\mathcal{A}$ is a Roiter ditalgebra with layer (R, W) and $\mathcal{A}'$ admits a triangular layer (R, W'), then $\mathcal{A}'$ is a Roiter ditalgebra.

Proof. The commutativity of the diagram is clear because $\pi r_0 = I_B$, where $r_0 : B \longrightarrow A$ is the restriction of r. Assume that $\mathcal{A}$ is a Roiter ditalgebra with layer (R, W) and that $\mathcal{A}'$ admits a triangular layer (R, W'). Let us show that $\mathcal{A}'$ is a Roiter ditalgebra. Let $f^0 : M \longrightarrow N$ be an isomorphism in R-Mod and $f^1 \in \mathrm{Hom}_{R\text{-}R}(W'_1, \mathrm{Hom}_k(M, N))$. Assume that M admits a structure of an $\mathcal{A}'$-module and consider the $\mathcal{A}$-module $F_\pi(M)$. Consider also the morphism $\widehat{f^1} \in \mathrm{Hom}_{R\text{-}R}(W'_1 \oplus W''_1, \mathrm{Hom}_k(M, N))$ with components f^1 and 0. Since $\mathcal{A}$ is a Roiter ditalgebra, there is a structure of an $\mathcal{A}$-module $\widehat{N}$ on the R-module N such that $\widehat{f} := (f^0, \widehat{f^1}) : F_\pi(M) \longrightarrow \widehat{N}$ is a morphism in $\mathcal{A}$-Mod. Then, $(f^0, f^1) = F_r(\widehat{f}) : M = F_rF_\pi(M) \longrightarrow F_r(\widehat{N})$ is a morphism in $\mathcal{A}'$-Mod. A similar argument can be applied if the additional $\mathcal{A}$-module structure is provided on the R-module N. Thus, $\mathcal{A}'$ is a Roiter ditalgebra. □

Definition 12.4. *Let $\mathcal{A}$ be any ditalgebra with layer (R, W). Given any module $X \in \mathcal{A}$-Mod, the algebra $\Gamma = \mathrm{End}_{\mathcal{A}}(X)^{op}$ acts by the right on X by the rule $xp = p^0(x)$, whenever $p = (p^0, p^1) \in \Gamma$ and $x \in X$. We say that X is an admissible $\mathcal{A}$-module iff the following conditions are satisfied:*

(1) Γ admits a splitting (S, P);
(2) the right S-module X admits a finite dual basis; and
(3) the morphisms f in S are of the form $f = (f^0, 0)$.

Definition 12.5. *Assume that $\mathcal{A}'$ is a proper subditalgebra of the layered ditalgebra $\mathcal{A}$ and that X is an admissible $\mathcal{A}'$-module. Adopt the notation of (12.1). Consider the associated splitting (S, P) of $\Gamma = \mathrm{End}_{\mathcal{A}'}(X)^{op}$. Writing $B = [T']_0$, we have that X is adapted to the pair (B, Γ), as in (11.9), and the last section applies to this situation.*

Define, for $x \in X$, $w \in BW_1'B$ and $\nu \in X^*$, the element $\gamma^w_{\nu,x} \in P^*$, such that, for $p \in P$

$$\gamma^w_{\nu,x}(p) = \nu(p^1(w)[x]).$$

It is easy to see that indeed, $\gamma^w_{\nu,x} \in P^*$. Notice that, these elements contain the elements $\gamma^{\delta(b)}_{\nu,x} = \gamma^b_{\nu,x}$ defined in (11.13). Indeed, by assumption, $\delta(b) \in BW_1'B$ and, since p is a morphism in $\mathcal{A}'$-Mod, we have that $\nu(b[xp] - [bx]p) = \nu(bp^0(x) - p^0(bx)) = \nu\left(p^1(\delta(b))[x]\right)$.

In the following, we fix a proper subditalgebra $\mathcal{A}'$ of $\mathcal{A}$ and an admissible $\mathcal{A}'$-module X, together with the notation introduced in this section. Moreover, $(x_i, \nu_i)_{i\in I}$ and $(p_j, \gamma_j)_{j\in J}$ are finite dual basis for the right S-modules X and P, respectively.

Lemma 12.6. *Consider the comultiplication $P^* \xrightarrow{\mu} P^* \otimes_S P^*$ of P, $\nu \in X^*$, $x \in X$, $w \in BW_1'B$. Since $\mathcal{A}'$ is a proper subditalgebra of $\mathcal{A}$, $\delta(w) = \sum_r u_r v_r$, with $u_r, v_r \in BW_1'B$. Then we have*

$$\mu(\gamma^w_{\nu,x}) = \sum_{i,j} \nu(x_i p_j)\gamma_j \otimes \gamma^w_{\nu_i,x} + \sum_j \gamma^w_{\nu,xp_j} \otimes \gamma_j + \sum_{r,i} \gamma^{u_r}_{\nu,x_i} \otimes \gamma^{v_r}_{\nu_i,x}.$$

Proof. By definition μ is the composition $P^* \xrightarrow{m^*} (P \otimes_S P)^* \xrightarrow{\psi^{-1}} P^* \otimes_S P^*$, as in (11.7). Let us denote by $\xi_1 = \sum_{i,j} \nu(x_i p_j)\gamma_j \otimes \gamma^w_{\nu_i,x}$, $\xi_2 = \sum_j \gamma^w_{\nu,xp_j} \otimes \gamma_j$ and $\xi_3 = \sum_{r,i} \gamma^{u_r}_{\nu,x_i} \otimes \gamma^{v_r}_{\nu_i,x}$, the summands of the right-hand side of the formula of the lemma. We have to show that $\psi(\mu(\gamma^w_{\nu,x})) = \psi(\xi_1 + \xi_2 + \xi_3)$. For $p, q \in P$, having in mind that the product of pq in Γ is the opposite of the composition

$q \circ p$ in the category $\mathcal{A}'$-Mod, we have

$$\begin{aligned}
\psi(\mu(\gamma^w_{\nu,x}))[p\otimes q] &= m^*(\gamma^w_{\nu,x})[p\otimes q]\\
&= \gamma^w_{\nu,x}[pq]\\
&= \nu((q\circ p)^1(w)[x])\\
&= \nu\left(q^0(p^1(w)[x]) + q^1(w)[p^0(x)]\right.\\
&\quad \left.+\textstyle\sum_r q^1(u_r)\left[p^1(v_r)[x]\right]\right)\\
&= \nu\left((p^1(w)[x])q + q^1(w)[xp]\right.\\
&\quad \left.+\textstyle\sum_r q^1(u_r)\left[p^1(v_r)[x]\right]\right).
\end{aligned}$$

Moreover

$$\begin{aligned}
\psi(\xi_1)(p\otimes q) &= \textstyle\sum_{i,j} \nu(x_i p_j)\gamma_j(\gamma^w_{\nu_i,x}(p)q)\\
&= \textstyle\sum_{i,j} \nu\left([x_i p_j]\gamma_j(\gamma^w_{\nu_i,x}(p)q)\right)\\
&= \textstyle\sum_i \nu(x_i[\sum_j p_j\gamma_j(\gamma^w_{\nu_i,x}(p)q)])\\
&= \textstyle\sum_i \nu\left(x_i\left[\gamma^w_{\nu_i,x}(p)q\right]\right)\\
&= \textstyle\sum_i \nu\left(\left[x_i\gamma^w_{\nu_i,x}(p)\right]q\right)\\
&= \textstyle\sum_i \nu\left(\left[x_i\nu_i(p^1(w)[x])\right]q\right)\\
&= \nu\left(\left[p^1(w)[x]\right]q\right),
\end{aligned}$$

we also have that

$$\begin{aligned}
\psi(\xi_2)(p\otimes q) &= \textstyle\sum_j \gamma^w_{\nu,xp_j}\left(\gamma_j(p)q\right)\\
&= \textstyle\sum_j \nu\left((\gamma_j(p)q)^1(w)[xp_j]\right)\\
&= \textstyle\sum_j \nu\left(q^1(w)\left[xp_j\gamma_j(p)\right]\right)\\
&= \nu\left(q^1(w)[xp]\right),
\end{aligned}$$

and we have

$$\begin{aligned}
\psi(\xi_3)(p\otimes q) &= \textstyle\sum_{r,i} \gamma^{u_r}_{\nu,x_i}\left(\gamma^{v_r}_{\nu_i,x}(p)q\right)\\
&= \textstyle\sum_{r,i} \gamma^{u_r}_{\nu,x_i}\left(\nu_i\left(p^1(v_r)[x]\right)q\right)\\
&= \textstyle\sum_{r,i} \nu\left[\left(\nu_i\left(p^1(v_r)[x]\right)q\right)^1(u_r)[x_i]\right]\\
&= \textstyle\sum_{r,i} \nu\left[q^1(u_r)[x_i\nu_i\left(p^1(v_r)[x]\right)]\right]\\
&= \textstyle\sum_r \nu\left[q^1(u_r)[p^1(v_r)[x]]\right].
\end{aligned}$$

The claim follows from this. □

Definition 12.7. *Assume X is an admissible $\mathcal{A}'$-module, where $\mathcal{A}'$ is a proper subditalgebra of the layered ditalgebra $\mathcal{A}$. Having in mind the above notation, we define the S-S-bimodule $W^X = W_0^X \oplus W_1^X$ by the formulas*

$$W_0^X = X^*\otimes_B \underline{W}_0 \otimes_B X \quad \text{and} \quad W_1^X = \left([X^*\otimes_B \underline{W}_1\otimes_B X]\oplus P^*\right)/\mathcal{L},$$

where $\mathcal{L}$ is the S-S-subbimodule of $[X^*\otimes_B \underline{W}_1\otimes_B X]\oplus P^*$ generated by the elements $\gamma^w_{\nu,x} - \nu\otimes w\otimes x$, where $\nu\in X^*$, $x\in X$ and $w\in BW_1'B$. Consider *the t-algebra $T^X := T_S(W^X)$, with layer (S, W^X). The canonical projection*

$$\widehat{W}^X := \big([X^* \otimes_B \underline{W}_0 \otimes_B X] \oplus [X^* \otimes_B \underline{W}_1 \otimes_B X] \oplus P^*\big) \xrightarrow{\eta} W^X$$

extends to an algebra morphism $\widehat{T}^X := T_S(\widehat{W}^X) \xrightarrow{\eta} T_S(W^X) = T^X$. For the sake of simplicity, we shall often use the same symbol t to denote an element $t \in \widehat{T}^X$ and its image $\eta(t) \in T^X$.

Lemma 12.8. *Assume X is an admissible $\mathcal{A}'$-module, where $\mathcal{A}'$ is a proper subditalgebra of the layered ditalgebra $\mathcal{A} = (T, \delta)$. Consider the t-algebra T^X defined in (12.7). Recall that T is freely generated by B and $\underline{W}$, as in (12.2). In particular, $T = B \oplus \langle \underline{W} \rangle$, where $\langle \underline{W} \rangle$ is the two-sided ideal of T generated by $\underline{W}$. Then, for each $(u, v) \in D(X) := [X^* \times X] \cup [(P^* \otimes_S X^*) \times X] \cup [X^* \times (X \otimes_S P^*)]$, there is a linear map*

$$\sigma_{u,v} : T \longrightarrow T^X$$

such that: for $w_1, \ldots, w_n \in \underline{W}$, we have $\sigma_{u,v}(w_1 w_2 \cdots w_n)$ given by

$$\sum_{i_1, i_2, \ldots, i_{n-1}} u \otimes w_1 \otimes x_{i_1} \otimes \nu_{i_1} \otimes w_2 \otimes x_{i_2} \otimes \nu_{i_2} \otimes w_3 \otimes \cdots \otimes x_{i_{n-1}} \otimes \nu_{i_{n-1}} \otimes w_n \otimes v;$$

and, for $b \in B$

$$\begin{cases} \sigma_{\nu,x}(b) = \nu(bx) & \text{if } \nu \in X^*, x \in X; \\ \sigma_{u,x}(b) = \sum_r \varphi_r \psi_r(bx) & \text{if } x \in X, u = \sum_r \varphi_r \otimes \psi_r,\ \varphi_r \in P^*, \psi_r \in X^*; \\ \sigma_{\nu,v}(b) = \sum_s \nu(by_s)\varphi'_s & \text{if } \nu \in X^*, v = \sum_s y_s \otimes \varphi'_s,\ y_s \in X, \varphi'_s \in P^*. \end{cases}$$

Moreover, the following statements hold:

(1) For $x \in X$, $\nu \in X^*$ and any homogeneous $t \in T$, we have

$$\deg \sigma_{\nu,x}(t) = \deg t.$$

(2) For any $x \in X$ and $\nu \in X^*$, we have

$$\begin{cases} \sigma_{u,x} = \sum_r \varphi_r \sigma_{\psi_r,x} & \text{if } u = \sum_r \varphi_r \otimes \psi_r,\ \varphi_r \in P^*, \psi_r \in X^*; \\ \sigma_{\nu,v} = \sum_s \sigma_{\nu,y_s} \varphi'_s & \text{if } v = \sum_s y_s \otimes \varphi'_s,\ y_s \in X, \varphi'_s \in P^*. \end{cases}$$

In particular, for any $x \in X$, $\nu \in X^*$ and $t \in T$, we have

$$\begin{cases} \sigma_{\lambda(\nu),x}(t) = \sum_{i,j} \nu(x_i p_j)\gamma_j \sigma_{\nu_i,x}(t) \quad \text{and} \\ \sigma_{\nu,\rho(x)}(t) = \sum_j \sigma_{\nu,xp_j}(t)\gamma_j. \end{cases}$$

(3) For each $(u, v) \in D(X)$ and $t, t' \in T$, we have

$$\sigma_{u,v}(tt') = \sum_i \sigma_{u,x_i}(t)\sigma_{v_i,v}(t').$$

(4) The map $\sigma_{v,x}$ is S-linear in each one of the variables $v \in X^*$ and $x \in X$; given $x \in X$ and $v \in X^*$, the map $\sigma_{u,x}$ is S-linear in the variable $u \in P^* \otimes_S X^*$ and the map $\sigma_{v,v}$ is S-linear in the variable $v \in X \otimes_S P^*$.

Proof. The map $\sigma_{u,v}$ is well defined on $\langle \underline{W} \rangle$ because, from (12.2), the multiplication maps $\underline{W}^{\otimes n} \longrightarrow \langle \underline{W} \rangle$ induce an isomorphism $\langle \underline{W} \rangle \cong \oplus_{n \geq 1} \underline{W}^{\otimes n}$ and we have the equality $b(\sum_i x_i \otimes v_i) = (\sum_i x_i \otimes v_i)b$, for any $b \in B$. With the notation in the statement of the lemma, the maps $\sigma_{u,x}$ and $\sigma_{v,v}$ are well defined on B by the formulas given there because we have well-defined maps: $\theta_{x,b} : P^* \otimes_S X^* \longrightarrow P^*$, with $\varphi \otimes \psi \mapsto \varphi\psi(bx)$, and $\theta'_{v,b} : X \otimes_S P^* \longrightarrow P^*$, with $x \otimes \varphi \mapsto v(bx)\varphi$.

To visualize these maps, notice that the insertion morphisms of (11.16)

$$X^* \otimes_B \underline{W}^{\otimes n} \otimes_B X \xrightarrow{\sigma_n} (X^* \otimes_B \underline{W} \otimes_B X)^{\otimes n}$$

induce morphisms of S-S-bimodules $X^* \otimes_B \langle \underline{W} \rangle \otimes_B X \xrightarrow{\sigma} \langle X^* \otimes_B \underline{W} \otimes_B X \rangle \hookrightarrow \widehat{T}^X$, and $\sigma(v \otimes t \otimes x) = \sigma_{v,x}(t)$, for any homogeneous element $t \in \langle \underline{W} \rangle \subseteq T$, $v \in X^*$ and $x \in X$. Therefore, $\deg \sigma_{v,x}(t) = \deg t$.

In general, if 1 denotes the identity map on P^*, we have that $\sigma_{u,v}(t) = \xi(u \otimes t \otimes v)$, where ξ is σ, $1 \otimes \sigma$, or $\sigma \otimes 1$, whenever $u \in X^*$ and $v \in X$, $u \in P^* \otimes_S X^*$ and $v \in X$, or $u \in X^*$ and $v \in X \otimes_S P^*$, respectively.

(2) The equalities hold evaluating at $t \in B$ by definition, and they clearly hold evaluating at each $t = w_1 w_2 \cdots w_n$ with $w_1, w_2, \ldots, w_n \in \underline{W}$. Then, they hold evaluating at any arbitrary $t \in T$.

(3) Fix any $u = v \in X^*$ and $v = x \in X$. Then, (3) is clear for any homogeneous $t, t' \in \langle \underline{W} \rangle$. Assume that $t \in B$ and $t' = w_1 w_2 \cdots w_n$ with $w_i \in \underline{W}$, then

$$\begin{aligned}
\sigma_{v,x}(tt') &= \textstyle\sum_{i_1,\ldots,i_{n-1}} v \otimes tw_1 \otimes x_{i_1} \otimes v_{i_1} \otimes \cdots \otimes v_{i_{n-1}} \otimes w_n \otimes x \\
&= \textstyle\sum_{i_1,\ldots,i_{n-1}} vt \otimes w_1 \otimes x_{i_1} \otimes v_{i_1} \otimes \cdots \otimes v_{i_{n-1}} \otimes w_n \otimes x \\
&= \textstyle\sum_{i_1,\ldots,i_{n-1},i} vt(x_i)v_i \otimes w_1 \otimes x_{i_1} \otimes v_{i_1} \otimes \cdots \otimes v_{i_{n-1}} \otimes w_n \otimes x \\
&= \textstyle\sum_{i_1,\ldots,i_{n-1},i} v(tx_i)v_i \otimes w_1 \otimes x_{i_1} \otimes v_{i_1} \otimes \cdots \otimes v_{i_{n-1}} \otimes w_n \otimes x \\
&= \textstyle\sum_i \sigma_{v,x_i}(t)\sigma_{v_i,x}(t').
\end{aligned}$$

Assume that $t = w_1 w_2 \cdots w_n$, with $w_i \in \underline{W}$, and $t' \in B$ then

$$\begin{aligned}\sigma_{v,x}(tt') &= \textstyle\sum_{i_1,\ldots,i_{n-1}} v \otimes w_1 \otimes x_{i_1} \otimes v_{i_1} \otimes \cdots \otimes v_{i_{n-1}} \otimes w_n t' \otimes x \\ &= \textstyle\sum_{i_1,\ldots,i_{n-1}} v \otimes w_1 \otimes x_{i_1} \otimes v_{i_1} \otimes \cdots \otimes v_{i_{n-1}} \otimes w_n \otimes t'x \\ &= \textstyle\sum_{i_1,\ldots,i_{n-1},i} v \otimes w_1 \otimes x_{i_1} \otimes v_{i_1} \otimes \cdots \otimes v_{i_{n-1}} \otimes w_n \otimes x_i v_i(t'x) \\ &= \textstyle\sum_i \sigma_{v,x_i}(t)\sigma_{v_i,x}(t').\end{aligned}$$

Assume that $t, t' \in B$ then

$$\begin{aligned}\sigma_{v,x}(tt') = v(tt'x) = v(\sum_i tx_i v_i(t'x)) &= \sum_i v(tx_i)v_i(t'x) \\ &= \sum_i \sigma_{v,x_i}(t)\sigma_{v_i,x}(t').\end{aligned}$$

Formula (3) holds for arbitrary $t, t' \in T$, and the fixed $u = v \in X^*$ and $v = x \in X$. The general case follows from this. Indeed, fix $t, t' \in T$. If $u = \sum_r \varphi_r \otimes \psi_r$ and $v = x \in X$, then

$$\begin{aligned}\sigma_{u,x}(tt') = \sum_r \varphi_r \sigma_{\psi_r,x}(tt') &= \sum_{r,i} \varphi_r \sigma_{\psi_r,x_i}(t)\sigma_{v_i,x}(t') \\ &= \sum_i \sigma_{u,x_i}(t)\sigma_{v_i,x}(t').\end{aligned}$$

For the other possible form of the pair u, v, the argument is similar.

(4) If $v, v' \in X^*$, $x \in X$, $q \in S$, the equality $\sigma_{qv+v',x} = q\sigma_{v,x} + \sigma_{v',x}$ is easily verified evaluating first at $t \in B$, then at elements of the form $t = w_1 w_2 \cdots w_n$ with $w_1, w_2, \ldots, w_n \in \underline{W}$. If $q \in S$, $u = \sum_r \varphi_r \otimes \psi_r$ and $u' = \sum_s \varphi'_s \otimes \psi'_s$, where $\varphi_r, \varphi'_s \in P^*$ and $\psi_r, \psi'_s \in X^*$, then

$$\sigma_{qu+u',x} = \sum_r q\varphi_r \sigma_{\psi_r,x} + \sum_s \varphi'_s \sigma_{\psi'_s,x} = q\sigma_{u,x} + \sigma_{u',x}.$$

Then, $\sigma_{u,x}$ is S-linear in $u \in P^* \otimes_S X^*$. The S-linearity of $\sigma_{v,x}$ in $x \in X$ and the S-linearity of $\sigma_{v,v}$ in $v \in X \otimes_S P^*$ are obtained similarly. □

Lemma 12.9. *There is a morphism $\widehat{\delta} : [X^* \otimes_B \underline{W} \otimes_B X] \oplus P^* \longrightarrow T^X$ of S-S-bimodules, satisfying $\widehat{\delta}(\gamma) = \mu(\gamma)$, where $\gamma \in P^*$ and $\mu : P^* \longrightarrow P^* \otimes_S P^*$ is the comultiplication of P, and*

$$\widehat{\delta}(v \otimes w \otimes x) = \lambda(v) \otimes w \otimes x + \sigma_{v,x}(\delta(w)) + (-1)^{\deg w+1} v \otimes w \otimes \rho(x),$$

for $w \in \underline{W}_0 \cup \underline{W}_1$, $v \in X^*$ and $x \in X$. Moreover, $\widehat{\delta}(\mathcal{L}) = 0$ so $\widehat{\delta}$ induces a morphism of S-S-bimodules $W^X \longrightarrow T^X$, which can be extended, using (4.4), to a differential δ^X on the t-algebra T^X such that $\mathcal{A}^X := (T^X, \delta^X)$ is a ditalgebra with layer (S, W^X).

Proof. For $r \in \{0,1\}$, consider the maps $\widehat{\delta} : X^* \times \underline{W}_r \times X \longrightarrow T^X$ defined by

$$\widehat{\delta}(\nu, w, x) = \lambda(\nu) \otimes w \otimes x + \sigma_{\nu,x}(\delta(w)) + (-1)^{\deg w+1} \nu \otimes w \otimes \rho(x).$$

We verify first that this is a B-balanced map. From (11.14) and (12.8), we have

$$\begin{aligned}
\widehat{\delta}(\nu b, w, x) &= \lambda(\nu b) \otimes w \otimes x + \sigma_{\nu b,x}(\delta(w)) + (-1)^{\deg w+1} \nu b \otimes w \otimes \rho(x) \\
&= \lambda(\nu) \otimes bw \otimes x + \textstyle\sum_i \gamma^b_{\nu,x_i} \otimes \nu_i \otimes w \otimes x + \sigma_{\nu,x}(b\delta(w)) \\
&\quad + (-1)^{\deg w+1} \nu \otimes bw \otimes \rho(x) \\
&= \lambda(\nu) \otimes bw \otimes x + \textstyle\sum_i \nu \otimes \delta(b) \otimes x_i \otimes \nu_i \otimes w \otimes x \\
&\quad + \sigma_{\nu,x}(b\delta(w)) + (-1)^{\deg w+1} \nu \otimes bw \otimes \rho(x) \\
&= \lambda(\nu) \otimes bw \otimes x + \sigma_{\nu,x}(\delta(b)w) \\
&\quad + \sigma_{\nu,x}(b\delta(w)) + (-1)^{\deg w+1} \nu \otimes bw \otimes \rho(x) \\
&= \lambda(\nu) \otimes bw \otimes x + \sigma_{\nu,x}(\delta(bw)) \\
&\quad + (-1)^{\deg w+1} \nu \otimes bw \otimes \rho(x) \\
&= \widehat{\delta}(\nu, bw, x).
\end{aligned}$$

Similarly, from (11.15) and (12.8), we obtain

$$\begin{aligned}
\widehat{\delta}(\nu, w, bx) &= \lambda(\nu) \otimes w \otimes bx + \sigma_{\nu,bx}(\delta(w)) \\
&\quad + (-1)^{\deg w+1} \nu \otimes w \otimes \rho(bx) \\
&= \lambda(\nu) \otimes wb \otimes x + \sigma_{\nu,x}(\delta(w)b) \\
&\quad + (-1)^{\deg w+1} \nu \otimes wb \otimes \rho(x) \\
&\quad + (-1)^{\deg w} \nu \otimes w \otimes \textstyle\sum_i x_i \otimes \gamma^b_{\nu_i,x} \\
&= \lambda(\nu) \otimes wb \otimes x + \sigma_{\nu,x}(\delta(w)b) \\
&\quad + (-1)^{\deg w+1} \nu \otimes wb \otimes \rho(x) \\
&\quad + \textstyle\sum_i (-1)^{\deg w} \nu \otimes w \otimes x_i \otimes \nu_i \otimes \delta(b) \otimes x \\
&= \lambda(\nu) \otimes wb \otimes x + \sigma_{\nu,x}(\delta(w)b) + \sigma_{\nu,x}((-1)^{\deg w} w\delta(b)) \\
&\quad + (-1)^{\deg w+1} \nu \otimes wb \otimes \rho(x) \\
&= \lambda(\nu) \otimes wb \otimes x + \sigma_{\nu,x}(\delta(wb)) \\
&\quad + (-1)^{\deg w+1} \nu \otimes wb \otimes \rho(x) \\
&= \widehat{\delta}(\nu, wb, x).
\end{aligned}$$

Now, recall that ρ is a morphism of right S-modules, λ is a morphism of left S-modules and $\sigma_{\nu,x}$ is S-linear in ν and x. Then, there is a well-defined morphism of S-S-bimodules $\widehat{\delta} : X^* \otimes_B \underline{W} \otimes_B X \longrightarrow T^X$, and we can extend this to a morphism of S-S-bimodules as in the statement of this lemma, making $\widehat{\delta}(\gamma) = \mu(\gamma)$, for $\gamma \in P^*$.

Now, to show that $\widehat{\delta}(\mathcal{L}) = 0$, it is enough to see that $\widehat{\delta}(\gamma^w_{\nu,x} - \nu \otimes w \otimes x) = 0$, for each $\nu \in X^*$, $w \in BW'_1B$ and $x \in X$. Assume that $\delta(w) = \sum_r u_r v_r$, with

$u_r, v_r \in BW_1'B$. Using (11.11), (12.6) and (12.8), we have

$$\begin{aligned}\widehat{\delta}(\gamma^w_{v,x} - v \otimes w \otimes x) &= \mu(\gamma^w_{v,x}) - \lambda(v) \otimes w \otimes x - \sigma_{v,x}(\delta(w)) \\ &\quad - v \otimes w \otimes \rho(x) \\ &= \sum_{i,j} v(x_i p_j)\gamma_j \otimes \gamma^w_{v_i,x} + \sum_j \gamma^w_{v,xp_j} \otimes \gamma_j \\ &\quad + \sum_{r,i} \gamma^{u_r}_{v,x_i} \otimes \gamma^{v_r}_{v_i,x} - \lambda(v) \otimes w \otimes x \\ &\quad - \sum_r \sigma_{v,x}(u_r v_r) - v \otimes w \otimes \rho(x) \\ &= \sum_{i,j} v(x_i p_j)\gamma_j \otimes \gamma^w_{v_i,x} + \sum_j \gamma^w_{v,xp_j} \otimes \gamma_j \\ &\quad + \sum_{r,i} \gamma^{u_r}_{v,x_i} \otimes \gamma^{v_r}_{v_i,x} - \sum_{i,j} v(x_i p_j)\gamma_j \otimes v_i \otimes w \otimes x \\ &\quad - \sum_{i,r} \sigma_{v,x_i}(u_r)\sigma_{v_i,x}(v_r) - \sum_j v \otimes w \otimes xp_j \otimes \gamma_j \\ &= 0.\end{aligned}$$

Then, $\widehat{\delta}$ induces a morphism of S-S-bimodules $W^X \xrightarrow{\widehat{\delta}} T^X$ onto the t-algebra T^X with layer (S, W^X). Since $\deg \widehat{\delta}(w) = \deg(w) + 1$, for $w \in W_0^X \cup W_1^X$, from (4.4) we can extend $\widehat{\delta}$ to a differential $\delta^X : T^X \longrightarrow T^X$, with $\delta^X(S) = 0$. In order to show that $(\delta^X)^2 = 0$, it will be enough to see that $(\delta^X)^2(W^X) = 0$. For this, we prove first the following:

Claim. For any homogeneous element $t \in \langle \underline{W} \rangle \subseteq T$, $v \in X^*$ and $x \in X$, the following formula holds

$$\delta^X(\sigma_{v,x}(t)) = \sigma_{\lambda(v),x}(t) + \sigma_{v,x}(\delta(t)) + (-1)^{\deg t+1}\sigma_{v,\rho(x)}(t).$$

Proof of claim: Indeed, our claim is true for any homogeneous $t \in \underline{W}$ by definition of δ^X. Suppose that $t \in \underline{W}^n$ and our claim proved for elements in $\underline{W}^{n-1}$. We may assume that $t = wt'$, where $w \in \underline{W}$ and $t' \in \underline{W}^{n-1}$ are homogeneous elements. Then, from Leibniz rule for δ^X and δ, and from the formula (12.8)(3), we have

$$\begin{aligned}\delta^X[\sigma_{v,x}(t)] &= \delta^X[\textstyle\sum_i \sigma_{v,x_i}(w)\sigma_{v_i,x}(t')] \\ &= \textstyle\sum_i \delta^X[\sigma_{v,x_i}(w)]\sigma_{v_i,x}(t') + (-1)^{\deg w} \sum_i \sigma_{v,x_i}(w)\delta^X[\sigma_{v_i,x}(t')] \\ &= \textstyle\sum_i \sigma_{\lambda(v),x_i}(w)\sigma_{v_i,x}(t') + \sum_i \sigma_{v,x_i}(\delta(w))\sigma_{v_i,x}(t') \\ &\quad +(-1)^{\deg w+1} \textstyle\sum_i \sigma_{v,\rho(x_i)}(w)\sigma_{v_i,x}(t') \\ &\quad +(-1)^{\deg w} \textstyle\sum_i \sigma_{v,x_i}(w)\sigma_{\lambda(v_i),x}(t') \\ &\quad +(-1)^{\deg w} \textstyle\sum_i \sigma_{v,x_i}(w)\sigma_{v_i,x}(\delta(t')) \\ &\quad +(-1)^{\deg w+\deg t'+1} \textstyle\sum_i \sigma_{v,x_i}(w)\sigma_{v_i,\rho(x)}(t') \\ &= \sigma_{\lambda(v),x}(wt') + \sigma_{v,x}(\delta(w)t') \\ &\quad +(-1)^{\deg w+1} \textstyle\sum_i \sigma_{v,\rho(x_i)}(w)\sigma_{v_i,x}(t') \\ &\quad +(-1)^{\deg w} \textstyle\sum_i \sigma_{v,x_i}(w)\sigma_{\lambda(v_i),x}(t') \\ &\quad +(-1)^{\deg w}\sigma_{v,x}(w\delta(t')) \\ &\quad +(-1)^{\deg w+\deg t'+1}\sigma_{v,\rho(x)}(wt')\end{aligned}$$

$$\begin{aligned}
&= \sigma_{\lambda(\nu),x}(t) + \sigma_{\nu,x}(\delta(t)) + (-1)^{\deg t+1}\sigma_{\nu,\rho(x)}(t) \\
&\quad + (-1)^{\deg w+1} \textstyle\sum_i \sigma_{\nu,\rho(x_i)}(w)\sigma_{\nu_i,x}(t') \\
&\quad + (-1)^{\deg w} \textstyle\sum_i \sigma_{\nu,x_i}(w)\sigma_{\lambda(\nu_i),x}(t') \\
&= \sigma_{\lambda(\nu),x}(t) + \sigma_{\nu,x}(\delta(t)) + (-1)^{\deg t+1}\sigma_{\nu,\rho(x)}(t).
\end{aligned}$$

The last equality follows from the formula

$$\sum_i \rho(x_i) \otimes \nu_i - \sum_i x_i \otimes \lambda(\nu_i) = 0,$$

which is a consequence of (11.11). Then, our claim has been established.

Now take $\nu \in X^*$, $w \in \underline{W}_0 \cup \underline{W}_1$ and $x \in X$. We have to compute each summand in the expression

$$\begin{aligned}
(\delta^X)^2[\sigma_{\nu,x}(w)] = \delta^X[\sigma_{\lambda(\nu),x}(w)] + \delta^X[\sigma_{\nu,x}(\delta(w))] \\
+ (-1)^{\deg w+1}\delta^X[\sigma_{\nu,\rho(x)}(w)].
\end{aligned}$$

Assume that $\lambda(\nu) = \sum_t \gamma_t' \otimes \nu_t'$, then, from Leibniz rule and (11.12), we have

$$\begin{aligned}
\delta^X[\sigma_{\lambda(\nu),x}(w)] &= \textstyle\sum_t \delta^X(\gamma_t') \otimes \nu_t' \otimes w \otimes x - \sum_t \gamma_t' \otimes \lambda(\nu_t') \otimes w \otimes x \\
&\quad - \textstyle\sum_t \gamma_t' \otimes \sigma_{\nu_t',x}(\delta(w)) \\
&\quad - (-1)^{\deg w+1} \textstyle\sum_t \gamma_t' \otimes \nu_t' \otimes w \otimes \rho(x) \\
&= (\mu \otimes 1)\lambda(\nu) \otimes w \otimes x - (1 \otimes \lambda)\lambda(\nu) \otimes w \otimes x \\
&\quad - \textstyle\sum_t \gamma_t' \otimes \sigma_{\nu_t',x}(\delta(w)) \\
&\quad - (-1)^{\deg w+1} \textstyle\sum_t \gamma_t' \otimes \nu_t' \otimes w \otimes \rho(x) \\
&= -\sigma_{\lambda(\nu),x}(\delta(w)) - (-1)^{\deg w+1}\lambda(\nu) \otimes w \otimes \rho(x).
\end{aligned}$$

Our claim implies that

$$\begin{aligned}
\delta^X[\sigma_{\nu,x}(\delta(w))] &= \sigma_{\lambda(\nu),x}(\delta(w)) + \sigma_{\nu,x}(\delta^2(w)) + (-1)^{\deg w+2}\sigma_{\nu,\rho(x)}(\delta(w)) \\
&= \sigma_{\lambda(\nu),x}(\delta(w)) + (-1)^{\deg w}\sigma_{\nu,\rho(x)}(\delta(w)).
\end{aligned}$$

Assume that $\rho(x) = \sum_t x_t' \otimes \gamma_t'$, then, from Leibniz rule and (11.12), we have

$$\begin{aligned}
\delta^X[\sigma_{\nu,\rho(x)}(w)] &= \textstyle\sum_t \sigma_{\lambda(\nu),x_t'}(w) \otimes \gamma_t' + \sum_t \sigma_{\nu,x_t'}(\delta(w)) \otimes \gamma_t' \\
&\quad + (-1)^{\deg w+1} \textstyle\sum_t \sigma_{\nu,\rho(x_t')}(w) \otimes \gamma_t' \\
&\quad + (-1)^{\deg w} \textstyle\sum_t \sigma_{\nu,x_t'}(w) \otimes \mu(\gamma_t') \\
&= \lambda(\nu) \otimes w \otimes \rho(x) + \sigma_{\nu,\rho(x)}(\delta(w)) \\
&\quad + (-1)^{\deg w+1}\nu \otimes w \otimes (\rho \otimes 1)\rho(x) \\
&\quad + (-1)^{\deg w}\nu \otimes w \otimes (1 \otimes \mu)\rho(x) \\
&= \lambda(\nu) \otimes w \otimes \rho(x) + \sigma_{\nu,\rho(x)}(\delta(w)).
\end{aligned}$$

Finally, adding all these terms, we obtain

$$\begin{aligned}(\delta^X)^2[\sigma_{\nu,x}(w)] &= -\sigma_{\lambda(\nu),x}(\delta(w)) - (-1)^{\deg w+1}\lambda(\nu)\otimes w\otimes \rho(x)\\ &\quad +\sigma_{\lambda(\nu),x}(\delta(w)) + (-1)^{\deg w}\sigma_{\nu,\rho(x)}(\delta(w))\\ &\quad +(-1)^{\deg w+1}[\lambda(\nu)\otimes w\otimes \rho(x) + \sigma_{\nu,\rho(x)}(\delta(w))]\\ &= 0.\end{aligned}$$

If $\gamma \in P^*$ and $\mu(\gamma) = \sum_t \gamma_t' \otimes \gamma_t''$, then, from the coassociativity of μ, we have that

$$\begin{aligned}(\delta^X)^2(\gamma) &= \delta^X(\textstyle\sum_t \gamma_t' \otimes \gamma_t'')\\ &= \textstyle\sum_t \left[\mu(\gamma_t')\otimes \gamma_t'' - \gamma_t' \otimes \mu(\gamma_t'')\right]\\ &= [(\mu\otimes 1) - (1\otimes \mu)]\mu(\gamma)\\ &= 0.\end{aligned}$$

From the above arguments, we obtain $(\delta^X)^2 = 0$. □

Proposition 12.10. *Assume X is an admissible $\mathcal{A}'$-module, where $\mathcal{A}'$ is a proper subditalgebra of the layered ditalgebra $\mathcal{A}$. Consider the layered ditalgebra $\mathcal{A}^X = (T^X, \delta^X)$ constructed in last lemma. Then, with the notation fixed above, there is a functor $F^X : \mathcal{A}^X$-Mod$\longrightarrow \mathcal{A}$-Mod defined as follows.*

Given $M \in \mathcal{A}^X$-Mod, let us denote by $*$ the left action of $T^X = T_S(W^X)$ on it. Then $F^X(M)$ is obtained from the B-module $X \otimes_S M$ considering the action of A on it given as follows. Recall that A is freely generated by B and $\underline{W}_0$, then to define an A-module structure on the B-module $X\otimes_S M$ we have to give a morphism of left B-modules $\underline{W}_0 \otimes_B X \otimes_S M \longrightarrow X \otimes_S M$. Consider the composition

$$\begin{aligned}\underline{W}_0 \otimes_B X\otimes_S M &\longrightarrow \operatorname{End}_S(X)\otimes_B \underline{W}_0 \otimes_B X \otimes_S M\\ &\xrightarrow{\phi^{-1}\otimes 1} X\otimes_S (X^*\otimes_B \underline{W}_0 \otimes_B X)\otimes_S M \xrightarrow{1\otimes *} X\otimes_S M,\end{aligned}$$

where the first map satisfies $w\otimes x\otimes m \mapsto 1\otimes w\otimes x\otimes m$. Thus, for $w \in \underline{W}_0$

$$w\cdot (x\otimes m) = \sum_i x_i \otimes (\nu_i \otimes w\otimes x) * m.$$

Then, having in mind (12.8), for any $a \in A$, we have the formula

$$a\cdot (x\otimes m) = \sum_i x_i \otimes \sigma_{\nu_i,x}(a) * m.$$

If $f = (f^0, f^1) \in \operatorname{Hom}_{\mathcal{A}^X}(M,N)$, then $F^X(f)$ is defined by

$$\begin{aligned}(F^X(f))^0[x\otimes m] &= x\otimes f^0(m) + \textstyle\sum_j xp_j \otimes f^1(\gamma_j)[m]\\ (F^X(f))^1(w)[x\otimes m] &= \textstyle\sum_i x_i \otimes f^1(\nu_i \otimes w\otimes x)[m],\end{aligned}$$

where $x \in X$, $m \in M$ and $w \in W_1$. Moreover, for $v \in V$, $x \in X$ and $m \in M$, we have

$$(F^X(f))^1(v)[x \otimes m] = \sum_i x_i \otimes f^1(\sigma_{v_i,x}(v))[m].$$

Proof. In order to verify the formula giving the action of A on $X \otimes_S M$, first notice that an easy induction and (12.8)(3) show that it holds for $w = w_1 w_2 \cdots w_t$ with $w_i \in \underline{W}_0$. For $b \in B$, $b \cdot (x \otimes m) = bx \otimes m = \sum_i x_i v_i(bx) \otimes m = \sum_i x_i \otimes v_i(bx) * m = \sum_i x_i \otimes \sigma_{v_i,x}(b) * m$. Since, A is freely generated by B and $\underline{W}_0$, we get the desired formula.

Now, we show that $F^X(f) \in \mathrm{Phom}_{R\text{-}W}(F^X(M), F^X(N))$. In order to show that the map $g^0 : X \times M \longrightarrow X \otimes_S N$ given by $g^0(x,m) = x \otimes f^0(m) + \sum_j xp_j \otimes f^1(\gamma_j)[m]$ is S-balanced notice that $\sum_j [xs]p_j \otimes \gamma_j = \sum_j xp_j \otimes \gamma_j s$, for any $x \in X$ and $s \in S$, because ρ is a morphism of right S-modules. Therefore

$$\begin{aligned} g^0(xs, m) &= xs \otimes f^0(m) + \textstyle\sum_j [xs]p_j \otimes f^1(\gamma_j)(m) \\ &= x \otimes sf^0(m) + \textstyle\sum_j xp_j \otimes f^1(\gamma_j s)(m) \\ &= x \otimes f^0(sm) + \textstyle\sum_j xp_j \otimes f^1(\gamma_j)(sm) \\ &= g^0(x, sm), \end{aligned}$$

which implies that $(F^X(f))^0$ is well defined. It is clearly an R-linear map. Obviously, $(F^X(f))^1$ is well defined. It is a morphism of R-R-bimodules because, as we remarked before, $r(\sum_i x_i \otimes v_i) = (\sum x_i \otimes v_i)r$, for any $r \in R$. Now, we show the following:

Claim. For $v \in V$, $x \in X$ and $m \in M$,

$$(F^X(f))^1(v)[x \otimes m] = \sum_i x_i \otimes f^1(\sigma_{v_i,x}(v))[m].$$

Proof of the claim: From (12.2), T has layer $(B, \underline{W})$, thus $V = A'\underline{W}_1 A'$, where A' is freely generated by B and $\underline{W}_0$. Thus, we can assume that $v = w_1' \cdots w_t' w w_1'' \cdots w_s''$, with $w_i', w_j'' \in \underline{W}_0$ and $w \in \underline{W}_1$. Then, the claim follows by an easy induction argument from the following three cases.

Case 1: $v = bwb'$, for $w \in W_1$ and $b, b' \in B$.

Here, since $b(\sum_i x_i \otimes v_i) = (\sum_i x_i \otimes v_i)b$, we have

$$\begin{aligned} \left(F^X(f)\right)^1(v)[x \otimes m] &= \left(F^X(f)\right)^1(bwb')[x \otimes m] \\ &= b\left(F^X(f)\right)^1(w)[b'x \otimes m] \\ &= b\left(\textstyle\sum_i x_i \otimes f^1(v_i \otimes w \otimes b'x)[m]\right) \\ &= \textstyle\sum_i x_i \otimes f^1(v_i b \otimes w \otimes b'x)[m] \\ &= \textstyle\sum_i x_i \otimes f^1(v_i \otimes bwb' \otimes x)[m] \\ &= \textstyle\sum_i x_i \otimes f^1(\sigma_{v_i,x}(v))[m]. \end{aligned}$$

Case 2: $v = wv'$, for $w \in \underline{W}_0$ and $v' \in V$, where v' satisfies the claim. Here, we have

$$\begin{aligned}
\left(F^X(f)\right)^1(v)[x \otimes m] &= \left(F^X(f)\right)^1(wv')[x \otimes m] \\
&= w\left(F^X(f)\right)^1(v')[x \otimes m] \\
&= \textstyle\sum_i w \cdot \left(x_i \otimes f^1(\sigma_{v_i,x}(v'))[m]\right) \\
&= \textstyle\sum_{i,j} x_j \otimes (v_j \otimes w \otimes x_i) * f^1(\sigma_{v_i,x}(v'))[m] \\
&= \textstyle\sum_{i,j} x_i \otimes \sigma_{v_i,x_j}(w) * f^1(\sigma_{v_j,x}(v'))[m] \\
&= \textstyle\sum_{i,j} x_i \otimes f^1(\sigma_{v_i,x_j}(w)\sigma_{v_j,x}(v'))[m] \\
&= \textstyle\sum_i x_i \otimes f^1(\sigma_{v_i,x}(v))[m].
\end{aligned}$$

Case 3: $v = v'w$, for $w \in \underline{W}_0$ and $v' \in V$, where v' satisfies the claim. Here, we have

$$\begin{aligned}
\left(F^X(f)\right)^1(v)[x \otimes m] &= \left(F^X(f)\right)^1(v'w)[x \otimes m] \\
&= \left(F^X(f)\right)^1(v')[w \cdot (x \otimes m)] \\
&= \left(F^X(f)\right)^1(v')[\textstyle\sum_j x_j \otimes (v_j \otimes w \otimes x) * m] \\
&= \textstyle\sum_{i,j} x_i \otimes f^1(\sigma_{v_i,x_j}(v'))[(v_j \otimes w \otimes x) * m] \\
&= \textstyle\sum_{i,j} x_i \otimes f^1(\sigma_{v_i,x_j}(v')\sigma_{v_j,x}(w))[m] \\
&= \textstyle\sum_i x_i \otimes f^1(\sigma_{v_i,x}(v))[m].
\end{aligned}$$

Our claim is proved. To show that $F^X(f) \in \mathrm{Hom}_{\mathcal{A}}(F^X(M), F^X(N))$, we have to show that

$$a(F^X(f))^0(x \otimes m) = (F^X(f))^0(a[x \otimes m]) + (F^X(f))^1(\delta(a))[x \otimes m],$$

for any $x \in X$, $m \in M$ and $a \in A$. Since A is freely generated by B and $\underline{W}_0 = BW_0''B$, it is enough to show that this formula holds for $a = b \in B$ and $a = w \in \underline{W}_0$, as in (4.7). For $b \in B$, we have

$$\begin{aligned}
b(F^X(f))^0(x \otimes m) &= bx \otimes f^0(m) + \textstyle\sum_j b[xp_j] \otimes f^1(\gamma_j)[m] \\
&= bx \otimes f^0(m) + \textstyle\sum_j [bx]p_j \otimes f^1(\gamma_j)[m] \\
&\quad + \textstyle\sum_j (b[xp_j] - [bx]p_j) \otimes f^1(\gamma_j)[m] \\
&= (F^X(f))^0(bx \otimes m) \\
&\quad + \textstyle\sum_i x_i \otimes f^1(\sum_j v_i(b[xp_j] - [bx]p_j)\gamma_j)[m] \\
&= (F^X(f))^0(bx \otimes m) + \textstyle\sum_i x_i \otimes f^1(\gamma^b_{v_i,x})[m] \\
&= (F^X(f))^0(bx \otimes m) + \textstyle\sum_i x_i \otimes f^1(v_i \otimes \delta(b) \otimes x)[m] \\
&= (F^X(f))^0(bx \otimes m) + (F^X(f))^1(\delta(b))[x \otimes m],
\end{aligned}$$

because $\gamma_{\nu,x}^b = \nu \otimes \delta(b) \otimes x$ in T^X. Now, for $w \in \underline{W}_0$, we have

$$\begin{aligned} w(F^X(f))^0(x \otimes m) &= \textstyle\sum_i x_i \otimes (\nu_i \otimes w \otimes x) * f^0(m) \\ &\quad + \textstyle\sum_{i,j} x_i \otimes (\nu_i \otimes w \otimes xp_j) * f^1(\gamma_j)[m] \\ &= \textstyle\sum_i x_i \otimes (\nu_i \otimes w \otimes x) * f^0(m) \\ &\quad + \textstyle\sum_{i,j} x_i \otimes f^1(\nu_i \otimes w \otimes xp_j \otimes \gamma_j)[m] \\ &= \textstyle\sum_i x_i \otimes (\nu_i \otimes w \otimes x) * f^0(m) \\ &\quad + \textstyle\sum_i x_i \otimes f^1(\nu_i \otimes w \otimes \rho(x))[m], \end{aligned}$$

because f^1 is a morphism of left $[T^X]_0$-modules and (11.11). Moreover, using again (11.11) and that f^1 is a morphism of right $[T^X]_0$-modules, we have

$$\begin{aligned} F^X(f)^0(w(x \otimes m)) &= (F^X(f))^0\big[\textstyle\sum_i x_i \otimes (\nu_i \otimes w \otimes x) * m\big] \\ &= \textstyle\sum_i x_i \otimes f^0((\nu_i \otimes w \otimes x) * m) \\ &\quad + \textstyle\sum_{i,j} x_i p_j \otimes f^1(\gamma_j)((\nu_i \otimes w \otimes x) * m) \\ &= \textstyle\sum_i x_i \otimes f^0((\nu_i \otimes w \otimes x) * m) \\ &\quad + \textstyle\sum_{i,j} x_i p_j \otimes f^1(\gamma_j \otimes \nu_i \otimes w \otimes x)[m] \\ &= \textstyle\sum_i x_i \otimes f^0((\nu_i \otimes w \otimes x) * m) \\ &\quad + \textstyle\sum_i x_i \otimes f^1(\lambda(\nu_i) \otimes w \otimes x)[m]. \end{aligned}$$

Now, we compute $\Delta := w(F^X(f))^0(x \otimes m) - (F^X(f))^0(w(x \otimes m))$, using our previous claim

$$\begin{aligned} \Delta &= \textstyle\sum_i x_i \otimes \big[(\nu_i \otimes w \otimes x) * f^0(m) - f^0((\nu_i \otimes w \otimes x) * m)\big] \\ &\quad + \textstyle\sum_i x_i \otimes \big[f^1(\nu_i \otimes w \otimes \rho(x))[m] - f^1(\lambda(\nu_i) \otimes w \otimes x)[m]\big] \\ &= \textstyle\sum_i x_i \otimes f^1(\delta^X(\nu_i \otimes w \otimes x))[m] \\ &\quad + \textstyle\sum_i x_i \otimes \big[f^1(\nu_i \otimes w \otimes \rho(x))[m] - f^1(\lambda(\nu_i) \otimes w \otimes x)[m]\big] \\ &= \textstyle\sum_i x_i \otimes f^1[\lambda(\nu_i) \otimes w \otimes x + \sigma_{\nu_i,x}(\delta(w)) \\ &\quad + (-1)^{\deg w+1} \nu_i \otimes w \otimes \rho(x)][m] \\ &\quad + \textstyle\sum_i x_i \otimes \big[f^1(\nu_i \otimes w \otimes \rho(x))[m] - f^1(\lambda(\nu_i) \otimes w \otimes x)[m]\big] \\ &= \textstyle\sum_i x_i \otimes f^1[\sigma_{\nu_i,x}(\delta(w))][m] \\ &= (F^X(f))^1[\delta(w)](x \otimes m). \end{aligned}$$

Therefore, $F^X(f)$ is a morphism in $\mathcal{A}$-Mod.

Now, assume $f : M \longrightarrow N$ and $g : N \longrightarrow L$ are morphisms in $\mathcal{A}^X$-Mod. We will prove that $F^X(gf) = F^X(g)F^X(f)$.

Take $x \in X$ and $m \in M$. Suppose that $\delta^X(\gamma_j) = \mu(\gamma_j) = \sum_s \gamma_{j,s}^1 \otimes \gamma_{j,s}^2$, for any j. From (11.12) and (11.11), we know that

$$\sum_{i,j} xp_i p_j \otimes \gamma_j \otimes \gamma_i = (\rho \otimes 1)\rho(x) = (1 \otimes \mu)\rho(x) = \sum_{j,s} xp_j \otimes \gamma_{j,s}^1 \otimes \gamma_{j,s}^2.$$

Then, we have

$$\begin{aligned}(F^X(g)F^X(f))^0(x \otimes m) &= (F^X(g))^0\left((F^X(f))^0(x \otimes m)\right)\\ &= (F^X(g))^0\big(x \otimes f^0(m) + \textstyle\sum_j xp_j \otimes f^1(\gamma_j)[m]\big)\\ &= x \otimes g^0 f^0(m) + \textstyle\sum_j xp_j \otimes g^0 f^1(\gamma_j)[m]\\ &\quad + \textstyle\sum_j xp_j \otimes g^1(\gamma_j)[f^0(m)]\\ &\quad + \textstyle\sum_{j,t} xp_j p_t \otimes g^1(\gamma_t) f^1(\gamma_j)[m]\\ &= x \otimes g^0 f^0(m) + \textstyle\sum_j xp_j \otimes (gf)^1(\gamma_j)[m]\\ &= F^X(gf)^0(x \otimes m).\end{aligned}$$

This means that $(F^X(g)F^X(f))^0 = F^X(gf)^0$. Now, let $w \in W_1$ and assume that $\delta(w) = \sum_s v_s^1 \otimes v_s^2$, with $v_s^1, v_s^2 \in V$. Using again (12.8), we get, for each $i \in I$ and $\nu_i \otimes w \otimes x \in W_1^X$

$$\begin{aligned}\delta^X(\nu_i \otimes w \otimes x) &= \sigma_{\lambda(\nu_i),x}(w) + \sigma_{\nu_i,x}(\delta(w)) + (-1)^{\deg w+1}\sigma_{\nu_i,\rho(x)}(w)\\ &= \textstyle\sum_{r,t} \nu_i(x_r p_t)\gamma_t \otimes \nu_r \otimes w \otimes x\\ &\quad + \textstyle\sum_{r,s} \sigma_{\nu_i,x_r}(v_s^1) \otimes \sigma_{\nu_r,x}(v_s^2)\\ &\quad + \textstyle\sum_r \nu_i \otimes w \otimes xp_r \otimes \gamma_r.\end{aligned}$$

Therefore

$$\begin{aligned}&(F^X(g)F^X(f))^1(w)[x \otimes m]\\ &\quad = F^X(g)^0 F^X(f)^1(w)[x \otimes m]\\ &\qquad + F^X(g)^1(w) F^X(f)^0[x \otimes m]\\ &\qquad + \textstyle\sum_s F^X(g)^1(v_s^1) F^X(f)^1(v_s^2)[x \otimes m]\\ &\quad = (F^X(g))^0\left(\textstyle\sum_i x_i \otimes f^1(\nu_i \otimes w \otimes x)[m]\right)\\ &\qquad + F^X(g)^1(w)\big(x \otimes f^0(m) + \textstyle\sum_j xp_j \otimes f^1(\gamma_j)[m]\big)\\ &\qquad + \textstyle\sum_{s,i} F^X(g)^1(v_s^1)\left(x_i \otimes f^1(\sigma_{\nu_i,x}(v_s^2))[m]\right)\\ &\quad = \textstyle\sum_i x_i \otimes g^0 f^1(\nu_i \otimes w \otimes x)[m]\\ &\qquad + \textstyle\sum_{i,j} x_i p_j \otimes g^1(\gamma_j)[f^1(\nu_i \otimes w \otimes x)(m)]\\ &\qquad + \textstyle\sum_i x_i \otimes g^1(\nu_i \otimes w \otimes x)[f^0(m)]\\ &\qquad + \textstyle\sum_{i,j} x_i \otimes g^1(\nu_i \otimes w \otimes xp_j)[f^1(\gamma_j)(m)]\\ &\qquad + \textstyle\sum_{i,r,s} x_r \otimes g^1(\sigma_{\nu_r,x_i}(v_s^1))[f^1(\sigma_{\nu_i,x}(v_s^2))(m)]\\ &\quad = \textstyle\sum_i x_i \otimes g^0 f^1(\nu_i \otimes w \otimes x)[m]\\ &\qquad + \textstyle\sum_i x_i \otimes g^1(\nu_i \otimes w \otimes x)[f^0(m)]\\ &\qquad + \textstyle\sum_i x_i \otimes \pi[g^1 \otimes f^1]\left(\delta^X(\nu_i \otimes w \otimes x)\right)[m]\\ &\quad = \textstyle\sum_i x_i \otimes (gf)^1(\nu_i \otimes w \otimes x)[m]\\ &\quad = (F^X(gf))^1(w)[x \otimes m].\end{aligned}$$

Hence, $(F^X(g)F^X(f))^1 = (F^X(gf))^1$, and F^X preserves composition. It is clear that F^X preserves identities and k-linear combinations, thus F^X is a k-functor. □

Exercise 12.11. *In the context of Definition (12.7), (12.8) and (12.9), assume that* $\delta(W_0') = 0$. *Show that, for any homogeneous* $t \in T$, *we have*

$$\delta^X(\sigma_{\nu,x}(t)) = \sigma_{\lambda(\nu),x}(t) + \sigma_{\nu,x}(\delta(t)) + (-1)^{\deg t+1}\sigma_{\nu,\rho(x)}(t).$$

Hint: Use the claim of the proof of (12.9) and (11.11). Then, notice that, for $\nu \in X^*$ *and* $x \in X$, *the element* $\gamma_{\nu,x} \in P^*$, *defined by* $\gamma_{\nu,x}(p) = \nu(xp)$ *satisfies that* $\gamma_{\nu,xs} = (\gamma_{\nu,x})s$, *for* $s \in S$. *This implies the formula* $\sum_j \nu(xp_j)\gamma_j s = \sum_j \nu((xs)p_j)\gamma_j$.

References. As we remarked in the preface to these notes, the use of the construction of $\mathcal{A}^X$ from $\mathcal{A}$ substitutes the use of the amalgam or pushout construction indicated in [28] or [19]. With its help, we can also avoid the use of additive ground categories; thus we remain closer to the original approach of Roiter and Kleiner in [46], where we can work directly with points and arrows generating a ditalgebra, as will be clear later.

13
Complete admissible modules

Given a layered ditalgebra $\mathcal{A}$ and an admissible module X over a proper subditalgebra $\mathcal{A}'$ of $\mathcal{A}$, we constructed in the previous section a new layered ditalgebra $\mathcal{A}^X$ and an associated functor $F^X : \mathcal{A}^X\text{-Mod} \longrightarrow \mathcal{A}\text{-Mod}$. Now, we start the study of F^X by examining sufficient conditions on X which guarantee that F^X is a full and faithful functor. We keep the notations of last section.

Remark 13.1. *Any layered ditalgebra $\mathcal{A}$ can be considered as a proper subditalgebra of itself. Namely, if (R, W) is the layer of $\mathcal{A}$, then $\mathcal{A}$ is the proper subditalgebra associated to the trivial R-R-bimodule decompositions $W_0 = W_0' \oplus 0$ and $W_1 = W_1' \oplus 0$. Given any admissible $\mathcal{A}$-module X, whenever we refer to the functor $F^X : \mathcal{A}^X\text{-Mod} \longrightarrow \mathcal{A}\text{-Mod}$, we mean the previous constructions of $\mathcal{A}^X$ and F^X, using the trivial decomposition mentioned above.*

Definition 13.2. *Given a layered ditalgebra $\mathcal{A}$, an admissible $\mathcal{A}$-module X is called complete iff the functor $F^X : \mathcal{A}^X\text{-Mod} \longrightarrow \mathcal{A}\text{-Mod}$ is full and faithful.*

Lemma 13.3. *Let $\mathcal{A}$ be a ditalgebra with layer (R, W). Assume that X is an admissible $\mathcal{A}$-module, where $\Gamma = \mathrm{End}_{\mathcal{A}}(X)^{op}$ admits the splitting (S, P). Assume, furthermore, that X is finite dimensional over the ground field k, that W_1 is a finitely generated R-R-bimodule, and that the algebra S is semisimple. Then, X is a complete admissible $\mathcal{A}$-module.*

Proof. Let us recall some notation: Here, $\mathcal{A} = (T, \delta)$, $B = [T]_0$ and T has layer $(B, \underline{W})$, with $\underline{W}_0 = B0B = 0$ and $\underline{W}_1 = BW_1B = [T]_1$. Then, the ditalgebra $\mathcal{A}^X = (T^X, \delta^X)$ has layer (S, W^X), where $W_0^X = 0$ and

$$W_1^X = \left([X^* \otimes_B \underline{W}_1 \otimes_B X] \oplus P^*\right) / \mathcal{L},$$

where $\mathcal{L}$ is the S-S-subbimodule of $[X^* \otimes_B \underline{W}_1 \otimes_B X] \oplus P^*$ generated by the elements $\gamma_{\nu,x}^{w} - \nu \otimes w \otimes x$, where $x \in X$, $\nu \in X^*$ and $w \in BW_1B$.

We need to prove that, for $M, N \in \mathcal{A}^X$-Mod, the map

$$\mathrm{Hom}_{\mathcal{A}^X}(M,N) \xrightarrow{F^X} \mathrm{Hom}_{\mathcal{A}}(F^X(M), F^X(N))$$

is an isomorphism. For this, we will define an inverse map

$$\mathrm{Hom}_{\mathcal{A}}(F^X(M), F^X(N)) \xrightarrow{G} \mathrm{Hom}_{\mathcal{A}^X}(M,N).$$

Since S is a semisimple k-algebra, the left S-module N is projective and we can consider a (possibly infinite) dual basis $\{(n_t, \eta_t) \mid t \in \tau\}$ for it. Then, for each linear map $g \in \mathrm{Hom}_k(X \otimes_S M, X \otimes_S N)$, $m \in M$ and $t \in \tau$, we can consider the linear endomorphism $g_{t,m} \in \mathrm{End}_k(X)$ given by

$$g_{t,m}(x) = \sigma(1 \otimes \eta_t) g(x \otimes m),$$

for any $x \in X$. Here, $\sigma : X \otimes_S S \longrightarrow X$ is the canonical isomorphism. Since X is finite dimensional, for a fixed m, there are only finitely many t with $g_{t,m} \neq 0$.

Then, we can consider, for each $t \in \tau$ and $m \in M$, the map

$$\mathrm{Hom}_{\mathcal{A}}(F^X(M), F^X(N)) \longrightarrow \mathrm{End}_{\mathcal{A}}(X)$$

such that $f \mapsto f_{t,m}$, where $f^0_{t,m}$ is the linear map associated to f^0 as above, and $f^1_{t,m}(w) = f^1(w)_{t,m}$, for any $w \in \underline{W}_1$. It is easy to see that $f^0_{t,m} \in \mathrm{Hom}_R(X,X)$ and $f^1_{t,m} \in \mathrm{Hom}_{B\text{-}B}(\underline{W}_1, \mathrm{Hom}_k(X,X))$. Then, for $w \in W_0$ and $x \in X$

$$\begin{aligned} w f^0_{t,m}(x) - f^0_{t,m}(wx) &= w\sigma(1 \otimes \eta_t) f^0(x \otimes m) - \sigma(1 \otimes \eta_t) f^0(wx \otimes m) \\ &= \sigma(1 \otimes \eta_t) w f^0(x \otimes m) - \sigma(1 \otimes \eta_t) f^0(wx \otimes m) \\ &= \sigma(1 \otimes \eta_t) f^1(\delta(w))(x \otimes m) \\ &= f^1_{t,m}(\delta(w))(x). \end{aligned}$$

Therefore, $f_{t,m}$ is indeed an endomorphism of X in $\mathcal{A}$-Mod. Since W_1 is finitely generated and X is finite dimensional, we also have that $f_{t,m}$ is zero for almost all t. We claim that: $f_{t,sm} = s f_{t,m}$, for $t \in \tau$, $m \in M$ and $s \in S$. Indeed, given $s \in S$ and $h \in \Gamma$, the product sh in Γ means the composition $h \circ s = (h^0, h^1) \circ (s^0, 0)$ in $\mathrm{End}_{\mathcal{A}}(X)$. Thus, for $w \in \underline{W}_1$ and $x \in X$, we have

$$\begin{cases} (sh)^0(x) = h^0 s^0(x) = h^0(xs), \text{ and} \\ (sh)^1(w)[x] = h^1(w) s^0(x) = h^1(w)[xs]. \end{cases}$$

Then, for $h = f_{t,m}$, we have: $(s f_{t,m})^0(x) = f^0_{t,m}(xs) = \sigma(1 \otimes \eta_t) f^0(xs \otimes m)$ $= f^0_{t,sm}(x)$ and $(s f_{t,m})^1(w)[x] = f^1_{t,m}(w)[xs] = \sigma(1 \otimes \eta_t) f^1(w)[xs \otimes m] =$ $f^1_{t,sm}(w)[x]$, as claimed. For later use, it is also convenient to notice the following. Clearly, $\mathrm{Hom}_B(X,X)$ and $\mathrm{Hom}_{B\text{-}B}(\underline{W}_1, \mathrm{Hom}_k(X,X))$ are S-S-bimodules, where $(sh^0s')[x] = h^0(xs)s'$ and $(sh^1s')(w)[x] = (h^1(w)[xs])s'$, for any $h^0 \in \mathrm{Hom}_B(X,X)$ and $h^1 \in \mathrm{Hom}_{B\text{-}B}(\underline{W}_1, \mathrm{Hom}_k(X,X))$, $s, s' \in S$, $x \in X$ and $w \in \underline{W}_1$. Then, for $h \in \Gamma$ and $s, s' \in S$, we have $shs' = (sh^0s', sh^1s')$.

Consider the canonical projections $\pi_S : \Gamma \longrightarrow S$ and $\pi_P : \Gamma \longrightarrow P$. Then, for $f \in \mathrm{Hom}_{\mathcal{A}}(F^X(M), F^X(N))$ we can define $G(f) \in \mathrm{Hom}_{\mathcal{A}^X}(M,N)$ by the following, for $m \in M$, $w \in \underline{W}_1$, $x \in X$, $\nu \in X^*$ and $\gamma \in P^*$

$$\begin{aligned}(G(f))^0[m] &= \textstyle\sum_t \pi_S(f_{t,m})n_t \\ (G(f))^1(\nu \otimes w \otimes x)[m] &= \sigma(\nu \otimes 1)\left[f^1(w)(x \otimes m)\right] \\ (G(f))^1(\gamma)[m] &= \textstyle\sum_t \gamma[\pi_P(f_{t,m})]n_t,\end{aligned}$$

where, again, σ denotes the canonical isomorphism $S \otimes_S N \to N$.

Then, $G(f)^0 \in \mathrm{Hom}_k(M,N)$. Let us verify that $G(f)^1$ is well defined. Consider the map $\widehat{G(f)}^1 : [X^* \otimes_B \underline{W}_1 \otimes_B X] \oplus P^* \longrightarrow \mathrm{Hom}_k(M,N)$ given by

$$\widehat{G(f)}^1[\nu \otimes w \otimes x + \gamma][m] = \sigma(\nu \otimes 1)\left[f^1(w)(x \otimes m)\right] + \sum_t \gamma[\pi_P(f_{t,m})]n_t.$$

Notice that, for $s, s' \in S$ and $\gamma \in P^*$, we have

$$\begin{aligned}\widehat{G(f)}^1\left[s\gamma s'\right][m] &= \textstyle\sum_t (s\gamma s')\pi_P[f_{t,m}]n_t \\ &= s\textstyle\sum_t \gamma\left(s'\pi_P[f_{t,m}]\right)n_t \\ &= s\textstyle\sum_t \gamma\left(\pi_P[f_{t,s'm}]\right)n_t \\ &= s\left(\widehat{G(f)}^1\right)[\gamma][s'm] \\ &= \left[s\left(\widehat{G(f)}^1[\gamma]\right)s'\right][m].\end{aligned}$$

Moreover, for $s, s' \in S$, $\nu \in X^*$, $w \in \underline{W}_1$ and $x \in X$

$$\begin{aligned}\widehat{G(f)}^1\left[s\nu \otimes w \otimes xs'\right][m] &= \sigma(s\nu \otimes 1)\left[f^1(w)(xs' \otimes m)\right] \\ &= s\sigma(\nu \otimes 1)\left[f^1(w)(x \otimes s'm)\right] \\ &= \left(s\left[\widehat{G(f)}^1(\nu \otimes w \otimes x)\right]s'\right)[m].\end{aligned}$$

Then, $\widehat{G(f)}^1$ is a morphism of S-S-bimodules. It remains to prove that the following equality holds

$$\widehat{G(f)}^1(\gamma^w_{\nu,x}) = \widehat{G(f)}^1(\nu \otimes w \otimes x).$$

To show this, recall again that X is admissible and, therefore, the morphisms s in S have the form $s = (s^0, 0)$. This implies that the element $p_{t,m} := \pi_P(f_{t,m})$ of P satisfies that $p^1_{t,m} = f^1_{t,m}$. Notice also that, for $\theta \in X \otimes_S N$ and $\nu \in X^*$, we have $\sigma(\nu \otimes 1)[\theta] = \sum_t \nu(\sigma(1 \otimes \eta_t)[\theta])\, n_t$. Then

$$\begin{aligned}
\widehat{G(f)}^1(\gamma^w_{\nu,x})[m] &= \textstyle\sum_t \gamma^w_{\nu,x}[\pi_P(f_{t,m})]n_t \\
&= \textstyle\sum_t \nu\left(p^1_{t,m}(w)[x]\right) n_t \\
&= \textstyle\sum_t \nu\left(f^1_{t,m}(w)[x]\right) n_t \\
&= \textstyle\sum_t \nu\left(\sigma(1 \otimes \eta_t) f^1(w)[x \otimes m]\right) n_t \\
&= \sigma(\nu \otimes 1)\left[f^1(w)(x \otimes m)\right] \\
&= \widehat{G(f)}^1(\nu \otimes w \otimes x)[m].
\end{aligned}$$

It follows that $G(f) \in \mathrm{Hom}_{\mathcal{A}^X}(M, N)$, because the layer $(S, \underline{W}^X)$ of $\mathcal{A}^X$ satisfies $\underline{W}^X_0 = 0$.

Now, we verify that G is indeed an inverse for F^X. For this, notice that the right S-module Γ admits the dual basis $\{(1_X, \pi_S)\} \cup \{(p_j, \gamma_j \pi_P) \mid j \in J\}$. Observe also that, for $\theta \in X \otimes_S N$, we have $\theta = \sum_t \sigma(1 \otimes \eta_t)[\theta] \otimes n_t$. Then

$$\begin{aligned}
(F^X(G(f)))^0[x \otimes m] &= x \otimes G(f)^0(m) + \textstyle\sum_j xp_j \otimes G(f)^1(\gamma_j)[m] \\
&= \textstyle\sum_t x \otimes \pi_S(f_{t,m})n_t + \sum_{t,j} xp_j \otimes \gamma_j[\pi_P(f_{t,m})]n_t \\
&= \textstyle\sum_t x\left[1\pi_S(f_{t,m}) + \sum_j p_j\gamma_j[\pi_P(f_{t,m})]\right] \otimes n_t \\
&= \textstyle\sum_t x f_{t,m} \otimes n_t \\
&= \textstyle\sum_t f^0_{t,m}(x) \otimes n_t \\
&= \textstyle\sum_t \sigma(1 \otimes \eta_t)[f^0(x \otimes m)] \otimes n_t \\
&= f^0(x \otimes m).
\end{aligned}$$

We also have

$$\begin{aligned}
(F^X(G(f)))^1(w)[x \otimes m] &= \textstyle\sum_i x_i \otimes G(f)^1(\nu_i \otimes w \otimes x)[m] \\
&= \textstyle\sum_i x_i \otimes \sigma(\nu_i \otimes 1)\left[f^1(w)(x \otimes m)\right] \\
&= f^1(w)(x \otimes m).
\end{aligned}$$

Hence, $F^X G = Id$.

Assume that $h \in \mathrm{Hom}_{\mathcal{A}^X}(M, N)$, we claim that, in $\mathrm{End}_{\mathcal{A}}(X)$

$$F^X(h)_{t,m} = 1_X\eta_t(h^0(m)) + \sum_j p_j\eta_t(h^1(\gamma_j)[m]).$$

That is

$$\begin{aligned}
&\left(F^X(h)^0_{t,m}, F^X(h)^1_{t,m}\right) \\
&\quad = \left(1^0_X\eta_t(h^0(m)) + \sum_j p^0_j\eta_t(h^1(\gamma_j)[m]), \sum_j p^1_j\eta_t(h^1(\gamma_j)[m])\right).
\end{aligned}$$

Let us first show the equality of the first components, so take $x \in X$

$$\begin{aligned}\left(F^X(h)_{t,m}\right)^0[x] &= \sigma(1 \otimes \eta_t)[F^X(h)^0(x \otimes m)] \\ &= \sigma(1 \otimes \eta_t)\left[x \otimes h^0(m) + \sum_j xp_j \otimes h^1(\gamma_j)[m]\right] \\ &= x\eta_t(h^0(m)) + \sum_j xp_j\eta_t(h^1(\gamma_j)[m]) \\ &= \left[1_X^0\eta_t(h^0(m)) + \sum_j p_j^0\eta_t(h^1(\gamma_j)[m])\right](x).\end{aligned}$$

For the other components, we have, for $w \in W_1$ and $x \in X$

$$\begin{aligned}\left(F^X(h)_{t,m}\right)^1(w)[x] &= \sigma(1 \otimes \eta_t)\left[F^X(h)^1(w)[x \otimes m]\right] \\ &= \sum_i \sigma(1 \otimes \eta_t)\left[x_i \otimes h^1(\nu_i \otimes w \otimes x)[m]\right] \\ &= \sum_i x_i\eta_t\left(h^1(\nu_i \otimes w \otimes x)[m]\right).\end{aligned}$$

Since h is a morphism in $\mathcal{A}^X$-Mod, h^1 satisfies $h^1(\gamma_{\nu,x}^w - \nu \otimes w \otimes x) = 0$, for $w \in BW_1B$, $x \in X$ and $\nu \in X^*$, therefore

$$\begin{aligned}h^1(\nu \otimes w \otimes x) &= h^1(\gamma_{\nu,x}^w) \\ &= \sum_j h^1\left(\gamma_{\nu,x}^w(p_j)\gamma_j\right) \\ &= \sum_j \gamma_{\nu,x}^w(p_j)h^1\left(\gamma_j\right) \\ &= \sum_j \nu\left(p_j^1(w)[x]\right)h^1(\gamma_j).\end{aligned}$$

Consider the last expression for $\nu = \nu_i$, evaluate it on $m \in M$, and then apply η_t to the result to obtain

$$\eta_t\left(h^1(\nu_i \otimes w \otimes x)[m]\right) = \eta_t\left(\sum_j \nu_i\left(p_j^1(w)[x]\right)h^1(\gamma_j)[m]\right).$$

Then, multiply x_i by this element of S and sum over i, to get

$$\begin{aligned}\sum_i x_i\eta_t\left(h^1(\nu_i \otimes w \otimes x)[m]\right) &= \sum_i x_i\eta_t\left(\sum_j \nu_i\left(p_j^1(w)[x]\right)h^1(\gamma_j)[m]\right) \\ &= \sum_{i,j} x_i\nu_i\left(p_j^1(w)[x]\right)\eta_t\left(h^1(\gamma_j)[m]\right) \\ &= \sum_j p_j^1(w)[x]\eta_t\left(h^1(\gamma_j)[m]\right).\end{aligned}$$

Thus, our claim holds. Let us see how to use this claim to end our proof

$$(G(F^X(h)))^0[m] = \sum_t \pi_S(F^X(h)_{t,m})n_t = \sum_t \eta_t(h^0(m))n_t = h^0[m].$$

Moreover

$$\begin{aligned}(G(F^X(h)))^1[\gamma](m) &= \textstyle\sum_t \gamma[\pi_P(F^X(h)_{t,m})]n_t \\ &= \textstyle\sum_t \gamma\left(\sum_j p_j\eta_t(h^1(\gamma_j)[m])\right) n_t \\ &= \textstyle\sum_{t,j} \gamma(p_j)\eta_t\left(h^1(\gamma_j)[m]\right) n_t \\ &= \textstyle\sum_j \gamma\left(p_j\right) h^1(\gamma_j)[m] \\ &= h^1\left(\sum_j \gamma(p_j)\gamma_j\right)[m] \\ &= h^1[\gamma](m).\end{aligned}$$

Finally

$$\begin{aligned}G(F^X(h))^1[\nu\otimes w\otimes x](m) &= \sigma(\nu\otimes 1)(F^X(h))^1(w)[x\otimes m] \\ &= \textstyle\sum_i \sigma(\nu\otimes 1)\left[x_i\otimes h^1(\nu_i\otimes w\otimes x)(m)\right] \\ &= \textstyle\sum_i \nu(x_i)h^1(\nu_i\otimes w\otimes x)(m) \\ &= h^1\left(\sum_i \nu(x_i)\nu_i\otimes w\otimes x\right)(m) \\ &= h^1\left(\nu\otimes w\otimes x\right)[m].\end{aligned}$$

□

Remark 13.4. *Assume X is an admissible $\mathcal{A}'$-module, where $\mathcal{A}' = (T', \delta')$ is a proper subditalgebra of the layered ditalgebra $\mathcal{A} = (T, \delta)$. As we pointed out before, in (13.1), we can perform the construction of two different layered ditalgebras $\mathcal{A}^X$ and $\mathcal{A}'^X$, and two functors $F^X : \mathcal{A}^X$-Mod$\longrightarrow\mathcal{A}$-Mod and $F'^X : \mathcal{A}'^X$-Mod$\longrightarrow\mathcal{A}'$-Mod, following the prescriptions of (12.9) and (12.10). They can be related as follows. Let us recall that $\mathcal{A}'$ is determined by R-R-bimodule decompositions $W_0 = W_0' \oplus W_0''$ and $W_1 = W_1' \oplus W_1''$ of the layer (R, W) of $\mathcal{A}$. By construction, T' is the subalgebra of T generated by R and $W' = W_0' \oplus W_1'$, $\mathcal{A}'$ has layer (R, W'), and the inclusion $r : T' \longrightarrow T$ yields a morphism of ditalgebras $r : \mathcal{A}' \longrightarrow \mathcal{A}$. Denote $B = [T']_0$, then T has layer $(B, \underline{W})$, with $\underline{W}_0 = BW_0''B$ and $\underline{W}_1 = BW_1B$. Moreover, if $\Gamma = \mathrm{End}_{\mathcal{A}'}(X)^{op}$ has splitting (S, P), the layer of $\mathcal{A}^X$ has the form (S, W^X), where*

$$W_0^X = X^* \otimes_B \underline{W}_0 \otimes_B X \quad \textit{and} \quad W_1^X = \left([X^* \otimes_B \underline{W}_1 \otimes_B X] \oplus P^*\right)/\mathcal{L},$$

and $\mathcal{L}$ is the S-S-subbimodule of $[X^* \otimes_B \underline{W}_1 \otimes_B X] \oplus P^*$ generated by the elements $\gamma^w_{\nu,x} - \nu\otimes w\otimes x$, with $w \in BW_1'B$, $x \in X$ and $\nu \in X^*$.

Now, for the construction of $\mathcal{A}'^X$, we are considering the trivial R-R-bimodule decompositions $W_0' = W_0' \oplus 0$ and $W_1' = W_1' \oplus 0$ of the layer (R, W') of $\mathcal{A}'$. T' has layer $(B, \underline{W}')$, with $\underline{W}_0' = B0B = 0$ and $\underline{W}_1' = BW_1'B$. Then, the ditalgebra $\mathcal{A}'^X = (T'^X, \delta'^X)$ has layer (S, W'^X), where $W_0'^X = 0$ and

$$W_1'^X = \left([X^* \otimes_B \underline{W}_1' \otimes_B X] \oplus P^*\right)/\mathcal{L}',$$

where $\mathcal{L}'$ is the S-S-subbimodule of $[X^* \otimes_B \underline{W}'_1 \otimes_B X] \oplus P^*$ *generated by the elements* $\gamma^w_{\nu,x} - \nu \otimes w \otimes x$, *with* $w \in BW'_1B$, $x \in X$ *and* $\nu \in X^*$.

The decomposition $\underline{W}_1 = \underline{W}'_1 \oplus \underline{W}''_1$ *of (12.2) determines the decomposition*

$$X^* \otimes_B \underline{W}_1 \otimes_B X = \left[X^* \otimes_B \underline{W}'_1 \otimes_B X\right] \oplus \left[X^* \otimes_B \underline{W}''_1 \otimes_B X\right].$$

The differences $\gamma^w_{\nu,x} - \nu \otimes w \otimes x$ *lie in* $\left[X^* \otimes_B \underline{W}'_1 \otimes_B X\right] \oplus P^*$. *Thus,* $\mathcal{L} = \mathcal{L}'$ *and* $W^X_1 = W'^X_1 \oplus \left[X^* \otimes_B \underline{W}''_1 \otimes_B X\right]$, *as S-S-bimodules. Then, the inclusion of* W'^X *in* W^X *determines a morphism of ditalgebras* $\hat{r} : \mathcal{A}'^X \longrightarrow \mathcal{A}^X$. *With this notation in mind, the following diagram is commutative*

$$\begin{array}{ccc} \mathcal{A}^X\text{-Mod} & \xrightarrow{F^X} & \mathcal{A}\text{-Mod} \\ {\scriptstyle F_{\hat{r}}}\downarrow & & \downarrow{\scriptstyle F_r} \\ \mathcal{A}'^X\text{-Mod} & \xrightarrow{F'^X} & \mathcal{A}'\text{-Mod}, \end{array}$$

where F_r *and* $F_{\hat{r}}$ *are the restriction functors associated to* r *and* $\hat{r}$, *respectively.*

Proposition 13.5. *Assume* X *is a complete admissible* $\mathcal{A}'$*-module, where* $\mathcal{A}'$ *is a proper subditalgebra of the layered ditalgebra* $\mathcal{A}$. *Then, the associated functor* $F^X : \mathcal{A}^X\text{-Mod} \longrightarrow \mathcal{A}\text{-Mod}$ *is full and faithful.*

Proof. Adopt the notation of (13.4) and take $M, N \in \mathcal{A}^X\text{-Mod}$. We want to show that

$$\mathrm{Hom}_{\mathcal{A}^X}(M, N) \xrightarrow{F^X} \mathrm{Hom}_{\mathcal{A}}(F^X(M), F^X(N))$$

is an isomorphism. We first show that it is injective. Assume that $F^X(f) = 0$, then $0 = F_r F^X(f) = F'^X F_{\hat{r}}(f)$ and, by assumption, $F_{\hat{r}}(f) = 0$. Then, $f^0 = 0$ and $f^1(W'^X_1) = 0$. It remains to prove that $f^1\left[X^* \otimes_B \underline{W}''_1 \otimes_B X\right] = 0$. But, for $w \in W_1$, $x \in X$, $\nu \in X^*$ and $m \in M$

$$\begin{aligned} f^1(\nu \otimes w \otimes x)[m] &= \textstyle\sum_i \nu(x_i) f^1(\nu_i \otimes w \otimes x)[m] \\ &= \sigma(\nu \otimes 1)\left(\textstyle\sum_i x_i \otimes f^1(\nu_i \otimes w \otimes x)[m]\right) \\ &= \sigma(\nu \otimes 1)\left(F^X(f)^1(w)[x \otimes m]\right) \\ &= 0. \end{aligned}$$

Now we show that F^X is full. Take $g \in \mathrm{Hom}_{\mathcal{A}}(F^X(M), F^X(N))$, consider $g_| = F_r(g) \in \mathrm{Hom}_{\mathcal{A}'}(F'^X F_{\hat{r}}(M), F'^X F_{\hat{r}}(N))$ and take an element $h \in \mathrm{Hom}_{\mathcal{A}'^X}(F_{\hat{r}}(M), F_{\hat{r}}(N))$ such that $F'^X(h) = g_|$. Then, we can define $f \in \mathrm{Phom}_{S\text{-}W^X}(M, N)$ as $f = (h^0, [h^1, h_1])$, where $h_1 : \left[X^* \otimes_B \underline{W}''_1 \otimes_B X\right] \to \mathrm{Hom}_k(M, N)$ is the morphism of S-S-bimodules satisfying

$$h_1(\nu \otimes w \otimes x)[m] = \sigma(\nu \otimes 1)\left[g^1(w)[x \otimes m]\right],$$

where $\sigma : S \otimes_S N \to N$ is the canonical isomorphism. We just have to show that $f \in \mathrm{Hom}_{\mathcal{A}^X}(M, N)$. For this, the following statement will be useful.

Claim. $$f^1(\sigma_{\nu,x}(v))[m] = \sigma(\nu \otimes 1)\left[g^1(v)[x \otimes m]\right],$$

for $x \in X$, $\nu \in X^*$, $v \in V = [T]_1$, $m \in M$, and $\sigma_{\nu,x} : \langle \underline{W} \rangle \longrightarrow T^X$ is the map defined in (12.8).

Proof of the claim: Recall that $V \subseteq \langle \underline{W} \rangle = \bigoplus_n \underline{W}^n$. Moreover, $V = \bigoplus_n \left(\underline{W}^n \cap V\right)$. Then, it is enough to show the claim for each element $v \in \underline{W}^n \cap V$. We shall do this by induction on the number n.

If $n = 1$, $v \in \underline{W} \cap V = \underline{W}_1 = BW_1'B \oplus BW_1''B$. Our statement holds by definition of f^1 if $v \in BW_1''B$. If $v = w \in BW_1'B$, since $F'^X(h) = g_|$, we have

$$\begin{aligned}
f^1(\sigma_{\nu,x}(w))[m] &= h^1(\nu \otimes w \otimes x)[m] \\
&= \textstyle\sum_i \nu(x_i) h^1(\nu_i \otimes w \otimes x)[m] \\
&= \sigma(\nu \otimes 1)\left(\textstyle\sum_i x_i \otimes h^1(\nu_i \otimes w \otimes x)[m]\right) \\
&= \sigma(\nu \otimes 1) F'^X(h)^1(w)[x \otimes m] \\
&= \sigma(\nu \otimes 1)[g^1(w)[x \otimes m]].
\end{aligned}$$

Then, the claim is clear for $n = 1$.

Now assume that $v = w_1 w_2 \cdots w_n$, with $w_i \in \underline{W}$ and $n \geq 2$. In the following, deg denotes the degree map on the graded t-algebra T, we make $w = w_1$ and $w' = w_2 \cdots w_n$. We consider two cases.

If $\deg w = 0$, then $\deg w' = 1$. Write $g^1(w')[x \otimes m] = \sum_q x^q \otimes n^q$. Then, using our induction hypothesis, we obtain

$$\begin{aligned}
f^1(\sigma_{\nu,x}(v))[m] &= f^1(\sigma_{\nu,x}(ww'))[m] \\
&= \textstyle\sum_i f^1(\sigma_{\nu,x_i}(w)\sigma_{\nu_i,x}(w'))[m] \\
&= \textstyle\sum_i \sigma_{\nu,x_i}(w) * f^1(\sigma_{\nu_i,x}(w'))[m] \\
&= \textstyle\sum_i \sigma_{\nu,x_i}(w) * \sigma(\nu_i \otimes 1) g^1(w')[x \otimes m] \\
&= \textstyle\sum_i \sigma_{\nu,x_i}(w) * \sum_q \nu_i(x^q) n^q \\
&= \textstyle\sum_q \sigma_{\nu,\sum_i x_i \nu_i(x^q)}(w) * n^q \\
&= \textstyle\sum_q \sigma_{\nu,x^q}(w) * n^q \\
&= \textstyle\sum_{i,q} \sigma_{\nu(x_i)\nu_i,x^q}(w) * n^q \\
&= \sigma(\nu \otimes 1)\left[\textstyle\sum_{i,q} x_i \otimes \sigma_{\nu_i,x^q}(w) * n^q\right] \\
&= \sigma(\nu \otimes 1) w \cdot \left[\textstyle\sum_q x^q \otimes n^q\right] \\
&= \sigma(\nu \otimes 1) w g^1(w')[x \otimes m] \\
&= \sigma(\nu \otimes 1) g^1(ww')[x \otimes m] \\
&= \sigma(\nu \otimes 1) g^1(v)[x \otimes m].
\end{aligned}$$

If $\deg w = 1$, then $\deg w' = 0$. Then, by our induction hypothesis, we obtain

$$\begin{aligned} f^1(\sigma_{v,x}(v))[m] &= f^1(\sigma_{v,x}(ww'))[m] \\ &= \textstyle\sum_i f^1(\sigma_{v,x_i}(w)\sigma_{v_i,x}(w'))[m] \\ &= \textstyle\sum_i f^1(\sigma_{v,x_i}(w))(\sigma_{v_i,x}(w'))[m] \\ &= \textstyle\sum_i f^1(\sigma_{v,x_i}(w))[(\sigma_{v_i,x}(w')) * m] \\ &= \textstyle\sum_i \sigma(v \otimes 1)g^1(w)[x_i \otimes (\sigma_{v_i,x}(w')) * m] \\ &= \sigma(v \otimes 1)g^1(w)[(\textstyle\sum_i x_i \otimes (\sigma_{v_i,x}(w')) * m] \\ &= \sigma(v \otimes 1)g^1(w)[w'(x \otimes m)] \\ &= \sigma(v \otimes 1)g^1(ww')[x \otimes m]. \end{aligned}$$

This finishes our induction. We know that, for $m \in M$ and $x \in X$, the following formula holds

$$g^0(x \otimes m) = x \otimes h^0(m) + \sum_j xp_j \otimes h^1(\gamma_j)[m].$$

Then, for $w \in \underline{W}_0$

$$\begin{aligned} g^0(w \cdot (x \otimes m)) &= g^0(\textstyle\sum_i x_i \otimes (v_i \otimes w \otimes x) * m) \\ &= \textstyle\sum_i x_i \otimes h^0[(v_i \otimes w \otimes x) * m] \\ &\quad + \textstyle\sum_{i,j} x_i p_j \otimes h^1(\gamma_j)[(v_i \otimes w \otimes x) * m]. \end{aligned}$$

With this in mind, our claim, and the fact $\sum_i x_i \otimes \sigma(v_i \otimes 1)[\theta] = \theta$, for all $\theta \in X \otimes_S N$, we obtain

$$\begin{aligned} \textstyle\sum_i x_i \otimes f^1(\sigma_{v_i,x}[\delta(w)])[m] &= \textstyle\sum_i x_i \otimes \sigma(v_i \otimes 1)g^1(\delta(w))[x \otimes m] \\ &= g^1(\delta(w))[x \otimes m] \\ &= wg^0(x \otimes m) - g^0(w(x \otimes m)) \\ &= \textstyle\sum_i x_i \otimes (v_i \otimes w \otimes x) * f^0(m) \\ &\quad + \textstyle\sum_{i,j} x_i \otimes (v_i \otimes w \otimes xp_j) * f^1(\gamma_j)[m] \\ &\quad - \textstyle\sum_i x_i \otimes f^0[(v_i \otimes w \otimes x) * m] \\ &\quad - \textstyle\sum_{i,j} x_i p_j \otimes f^1(\gamma_j)[(v_i \otimes w \otimes x) * m]. \end{aligned}$$

Moreover, if $\Delta_i := f^1(v_i \otimes w \otimes \rho(x))[m] - f^1(\lambda(v_i) \otimes w \otimes x)[m]$, then

$$\begin{aligned} \Delta_i &= f^1(\textstyle\sum_j v_i \otimes w \otimes xp_j \otimes \gamma_j)[m] \\ &\quad - f^1(\textstyle\sum_{r,j} v_i(x_r p_j)\gamma_j \otimes v_r \otimes w \otimes x)[m] \\ &= \textstyle\sum_j (v_i \otimes w \otimes xp_j) f^1(\gamma_j)[m] \\ &\quad - \textstyle\sum_{r,j} v_i(x_r p_j) f^1(\gamma_j)(v_r \otimes w \otimes x)[m] \\ &= \textstyle\sum_j (v_i \otimes w \otimes xp_j) f^1(\gamma_j)[m] \\ &\quad - \textstyle\sum_{r,j} v_i(x_r p_j) f^1(\gamma_j)[(v_r \otimes w \otimes x) * m]. \end{aligned}$$

With the last two series of equalities in mind, compute $\sum_i x_i \otimes \Delta_i$ and derive the equality

$$\begin{aligned}\textstyle\sum_i x_i \otimes f^1(\sigma_{v_i,x}(\delta(w)))[m] = \textstyle\sum_i x_i \otimes f^1(v_i \otimes w \otimes \rho(x))[m] \\ -\textstyle\sum_i x_i \otimes f^1(\lambda(v_i) \otimes w \otimes x)[m] \\ +\textstyle\sum_i x_i \otimes (v_i \otimes w \otimes x) * f^0(m) \\ -\textstyle\sum_i x_i \otimes f^0[(v_i \otimes w \otimes x) * m].\end{aligned}$$

This is equivalent to

$$\sum_i x_i \otimes f^1(\delta^X(v_i \otimes w \otimes x))[m] = \sum_i x_i \otimes \big((v_i \otimes w \otimes x) * f^0[m] \\ - f^0[(v_i \otimes w \otimes x) * m]\big).$$

Applying $\sigma(v \otimes 1)$ to the last equality, we finally get

$$f^1(\delta^X(v \otimes w \otimes x))[m] = (v \otimes w \otimes x) * f^0[m] - f^0[(v \otimes w \otimes x) * m].$$

Thus, $f \in \mathrm{Hom}_{\mathcal{A}^X}(M, N)$ and $F^X(f) = g$. □

Remark 13.6. *Assume X is a (not necessarily complete) admissible $\mathcal{A}'$-module, where $\mathcal{A}'$ is a proper subditalgebra of the layered ditalgebra $\mathcal{A}$. The argument of last proof shows that, for $M, N \in \mathcal{A}^X$-Mod, with the notations of (13.4)*

$$\mathrm{Hom}_{\mathcal{A}^X}(M, N) \xrightarrow{F^X} \mathrm{Hom}_{\mathcal{A}}(F^X(M), F^X(N))$$

is an isomorphism whenever

$$\mathrm{Hom}_{\mathcal{A}'^X}(F_{\hat{P}}(M), F_{\hat{P}}(N)) \xrightarrow{F'^X} \mathrm{Hom}_{\mathcal{A}}(F'^X F_{\hat{P}}(M), F'^X F_{\hat{P}}(N))$$

is an isomorphism.

14

Bimodule filtrations and triangular admissible modules

In this section, we show that, for any initial subditalgebra $\mathcal{A}'$ of a triangular ditalgebra $\mathcal{A}$ (see (14.8)), assuming some triangularity condition on the admissible $\mathcal{A}'$-module X, the associated ditalgebra $\mathcal{A}^X$ is triangular. In the proof of this, we have to make some special choice of dual basis, compatible with the required filtrations, in order to handle the maps μ, λ, ρ and δ^X.

Definition 14.1. *An R-S-bimodule filtration $\mathcal{F}(U)$, of length ℓ of the R-S-bimodule U, is a sequence of R-S-subbimodules*

$$0 = U_0 \subseteq U_1 \subseteq \cdots \subseteq U_{\ell-1} \subseteq U_\ell = U.$$

The height associated to such a filtration of U is the map $\mathbb{h} : U \longrightarrow [0, \ell]$ defined by $\mathbb{h}(u) = j$ iff $u \in U_j \setminus U_{j-1}$, for $j \in [1, \ell]$, and $\mathbb{h}(0) = 0$. Thus, for $u \in U$, we have that $\mathbb{h}(u) \leq t$ iff $u \in U_t$.

The filtration is called right additive (resp. left additive) iff, as right S-modules (resp. left R-modules), U_j is a direct summand of U_{j+1}, for all $j < \ell$. It is called additive iff U_j is a direct summand of U_{j+1} as an R-S-bimodule, for all $j < \ell$.

A family of generators $Z = \{u_i \mid i \in I\}$ of the right S-module U is called compatible with the filtration $\mathcal{F}(U)$ iff each right S-module U_j is generated by $Z \cap U_j$, for all $j \in [0, \ell]$.

Definition 14.2. *Assume that we have an R-S-bimodule filtration $\mathcal{F}(U)$ of length ℓ of the R-S-bimodule U, $0 = U_0 \subseteq U_1 \subseteq \cdots \subseteq U_{\ell-1} \subseteq U_\ell = U$, then the dual filtration $\mathcal{F}(U)^*$ of $\mathcal{F}(U)$ is the following filtration of the S-R-bimodule U^**

$$0 = [U^*]_0 \subseteq [U^*]_1 \subseteq \cdots \subseteq [U^*]_{\ell-1} \subseteq [U^*]_\ell = U^*,$$

where $[U^*]_j$ *equals by definition the following* S-R-*subbimodule of* U^*

$$[U^*]_j = \{\lambda \in U^* \mid \lambda(U_{\ell-j}) = 0\}.$$

Definition 14.3. *Assume that we have an* R-S-*bimodule filtration* $\mathcal{F}(U)$ *of length* ℓ *of the* R-S-*bimodule* U, $0 = U_0 \subseteq U_1 \subseteq \cdots \subseteq U_{\ell-1} \subseteq U_\ell = U$. *If* U *is projective as a right* S-*module, a finite dual basis* $\{(u_i, \lambda_i) \mid i \in I\}$ *for* U *is compatible with the given filtration iff*

$$\mathbb{h}(u_i) + \mathbb{h}(\lambda_i) = \ell + 1, \quad \text{for all } i \in I.$$

Here the height of each element u_i *is taken relative to the given filtration* $\mathcal{F}(U)$, *while the height of* λ_i *is taken relative to the dual filtration* $\mathcal{F}(U)^*$ *of* U^*.

Lemma 14.4. *Assume that we have a right additive* R-S-*bimodule filtration* $\mathcal{F}(U)$ *of length* ℓ *of the* R-S-*bimodule* U, $0 = U_0 \subseteq U_1 \subseteq \cdots \subseteq U_{\ell-1} \subseteq U_\ell = U$. *Assume that the* S-*module* U *is finitely generated and projective, then* U_S *admits a finite dual basis compatible with* $\mathcal{F}(U)$. *It is constructed as follows. By assumption,* $U_j = T_j \oplus \cdots \oplus T_1$, *where* $U_j = U_{j-1} \oplus T_j$. *Consider, for each* j, *the canonical projection* $\pi_j : U \longrightarrow T_j$ *and consider a finite dual basis* $\{(u_i, \lambda_i) \mid i \in I(j)\}$ *of the projective right* S-*module* T_j *(with* $\lambda_i \neq 0$ *and* $u_i \neq 0$, *for all* i*). Denote by* $\widehat{\lambda}_i$ *the composition* $U \xrightarrow{\pi_j} T_j \xrightarrow{\lambda_i} S$, *for all* $i \in I(j)$. *Thus,* $u_i \in U$ *and* $\widehat{\lambda}_i \in U^*$, *for* $i \in I(j)$. *If we make* $I := \bigcup_{j\in[1,\ell]} I(j)$, *then* $\{(u_i, \widehat{\lambda}_i) \mid i \in I\}$ *is a finite dual basis for* U_S *compatible with* $\mathcal{F}(U)$.

Proof. Consider $u \in U$. Therefore, $u = \sum_j v_j$, with $v_j \in T_j$. Then, we get $u = \sum_{j\in[1,\ell]} \left(\sum_{i\in I(j)} u_i\lambda_i(v_j)\right) = \sum_{j\in[1,\ell]} \sum_{i\in I(j)} u_i\widehat{\lambda}_i(u)$. Then, $\{(u_i, \widehat{\lambda}_i) \mid i \in I\}$ is indeed a right finite dual basis of U.

Now fix $i \in I$, say with $i \in I(j)$, then we have that $\mathbb{h}(u_i) = j$. We need to show that $\mathbb{h}(\widehat{\lambda}_i) = \ell + 1 - j$. Since $\widehat{\lambda}_i = \lambda_i\pi_j$, we have $\widehat{\lambda}_i(U_{j-1}) = \widehat{\lambda}_i(T_1 \oplus \cdots \oplus T_{j-1}) = 0$. Since $\lambda_i \neq 0$, $\widehat{\lambda}_i(U_j) = \lambda_i(T_j) \neq 0$. Make $t := \ell + 1 - j$, then we have seen that $\widehat{\lambda}_i(U_{\ell-t}) = 0$, but $\widehat{\lambda}_i(U_{\ell-t+1}) \neq 0$. Then $\widehat{\lambda}_i \in [U^*]_t \setminus [U^*]_{t-1}$. Hence, $\mathbb{h}(\widehat{\lambda}_i) = t = \ell + 1 - j$. □

Lemma 14.5. *Assume that we have a* R-S-*bimodule filtration* $\mathcal{F}(U)$ *of the* R-S-*bimodule* U. *Assume that the right* S-*module* U *has a finite dual basis* $\{(u_i, \lambda_i) \mid i \in I\}$ *compatible with* $\mathcal{F}(U)$. *Then:*

(1) $\{u_i \mid i \in I\}$ *is a family of right generators of* U *compatible with* $\mathcal{F}(U)$.
(2) $\{\lambda_i \mid i \in I\}$ *is a family of left generators of* U^* *compatible with* $\mathcal{F}(U)^*$.

Proof. We have our filtrations $\mathcal{F}(U)$ and $\mathcal{F}(U)^*$, with length ℓ

$$0 = U_0 \subseteq U_1 \subseteq \cdots \subseteq U_{\ell-1} \subseteq U_\ell = U, \text{ and}$$
$$0 = [U^*]_0 \subseteq [U^*]_1 \subseteq \cdots \subseteq [U^*]_{\ell-1} \subseteq [U^*]_\ell = U^*.$$

By assumption, $\mathbb{h}(u_i) + \mathbb{h}(\lambda_i) = \ell + 1$, for all $i \in I$. We first show the following *claim*, for a fixed $i \in I$:

(1′) If $u \in U$ satisfies $\mathbb{h}(u) < \mathbb{h}(u_i)$, then $\lambda_i(u) = 0$; and
(2′) If $\lambda \in U^*$ satisfies $\mathbb{h}(\lambda) < \mathbb{h}(\lambda_i)$, then $\lambda(u_i) = 0$.

We first show (1′): Suppose $\mathbb{h}(\lambda_i) = j$, then λ_i lies in $[U^*]_j \setminus [U^*]_{j-1}$, which means that $\lambda_i(U_{\ell-j}) = 0$ and $\lambda_i(U_{\ell-j+1}) \neq 0$. Since $j = \mathbb{h}(\lambda_i) = \ell + 1 - \mathbb{h}(u_i)$, we have $\lambda_i(U_{\mathbb{h}(u_i)-1}) = 0$. But, U_t consists of all elements u in U with $\mathbb{h}(u) \leq t$. In particular, if $u \in U$ satisfies $\mathbb{h}(u) < \mathbb{h}(u_i)$, then $\lambda_i(u) = 0$, as claimed.

Now we show (2′): Suppose $\mathbb{h}(\lambda) = j$. Then $\lambda \in [U^*]_j$, and $\lambda(U_{\ell-j}) = 0$. Now, $\mathbb{h}(\lambda) < \mathbb{h}(\lambda_i)$ means that $\mathbb{h}(u_i) = \ell + 1 - \mathbb{h}(\lambda_i) \leq \ell - \mathbb{h}(\lambda) = \ell - j$, and hence $\lambda(u_i) = 0$, as claimed.

We already know that $u_i \in U$ and $\{u_i \mid i \in I\}$ generate the right S-module U. Denote by I^j the set of indexes $i \in I$ with $u_i \in U_j$, for all $j \in [0, \ell]$. In order to prove (1), we have to show that each U_j is right generated by $\{u_i \mid i \in I^j\}$. If $u \in U_j$, since we have a right dual basis for U, we have from (1′)

$$u = \sum_{i \in I} u_i \lambda_i(u) = \sum_{i \in I \mid j \geq \mathbb{h}(u) \geq \mathbb{h}(u_i)} u_i \lambda_i(u) = \sum_{i \in I^j} u_i \lambda_i(u).$$

We also know, from (11.2), that $\lambda_i \in U^*$ and $\{\lambda_i \mid i \in I\}$ generate the left S-module U^*. Denote by I^j the set of indexes $i \in I$ with $\lambda_i \in [U^*]_j$, for all $j \in [0, \ell]$. In order to prove (2), we have to show that each $[U^*]_j$ is left generated by $\{\lambda_i \mid i \in I^j\}$. If $\lambda \in [U^*]_j$, since we have a left dual basis for U^*, we have from (2′)

$$\lambda = \sum_{i \in I} \lambda(u_i)\lambda_i = \sum_{i \in I \mid j \geq \mathbb{h}(\lambda) \geq \mathbb{h}(\lambda_i)} \lambda(u_i)\lambda_i = \sum_{i \in I^j} \lambda(u_i)\lambda_i.$$

□

Definition 14.6. *Assume that we have an admissible $\mathcal{A}$-module X, for the layered ditalgebra $\mathcal{A}$. Assume that (S, P) is the splitting associated to the algebra $\Gamma = \mathrm{End}_{\mathcal{A}}(X)^{op}$ and $B = [T]_0$. Then, we say that X is triangular iff:*

(1) We have a right additive filtration of S-S-subbimodules of P

$$0 = P^{(\ell_P+1)} \subseteq P^{(\ell_P)} \subseteq \cdots \subseteq P^{(1)} = P,$$

such that $P^{(i)}P^{(j)} \subseteq P^{(i+j)}$ *for all* $i, j \in [1, \ell_P]$ *with* $i + j \leq \ell_P$, *and* $P^{(i)}P^{(j)} = 0$, *otherwise.*

(2) *We have a right additive filtration of* B-S-*subbimodules of* X

$$\mathcal{F}(X): \quad 0 = X_0 \subseteq X_1 \subseteq \cdots \subseteq X_{\ell_X} = X,$$

such that $X_j P \subseteq X_{j-1}$, *for all* $j \in [1, \ell_X]$.

Lemma 14.7. *Assume that we have a triangular admissible* $\mathcal{A}$-*module* X, *for the layered ditalgebra* $\mathcal{A}$. *Assume that* (S, P) *is the splitting associated to the algebra* $\Gamma = \text{End}_{\mathcal{A}}(X)^{op}$. *Reorder the indexes of the sequence of* S-S-*subbimodules of* P *given in the last definition, making* $P_j := P^{(\ell_P - j+1)}$, *for all* $j \in [0, \ell_P]$, *to obtain a right additive* S-S-*bimodule filtration*

$$\mathcal{F}(P): \quad 0 = P_0 \subseteq P_1 \subseteq \cdots \subseteq P_{\ell_P - 1} \subseteq P_{\ell_P} = P.$$

Consider a dual basis $\{(p_j, \gamma_j) \mid j \in J\}$ *for the right* S-*module* P *compatible with* $\mathcal{F}(P)$. *Consider also a dual basis* $\{(x_i, \nu_i) \mid i \in I\}$ *for the right* S-*module* X *compatible with the filtration* $\mathcal{F}(X)$ *of* X *given in last definition. Consider the heights associated to the filtrations* $\mathcal{F}(X)$, $\mathcal{F}(X)^*$ *and* $\mathcal{F}(P)^*$. *Then:*

(1) $\mathbb{h}(xp) < \mathbb{h}(x)$, *for any* $0 \neq x \in X$ *and* $p \in P$.
(2) $\nu(x_i p) \neq 0$ *implies that* $\mathbb{h}(\nu_i) < \mathbb{h}(\nu)$, *for any* $i \in I$, $p \in P$ *and* $\nu \in X^*$.
(3) $\gamma(p_i p_j) \neq 0$ *implies* $\mathbb{h}(\gamma_i) + \mathbb{h}(\gamma_j) \leq \mathbb{h}(\gamma)$, *for any* $\gamma \in P^*$ *and* $i, j \in J$.

Proof.

(1) By definition of triangular admissible module, multiplication of a non-zero element $x \in X$ by an element $p \in P$ decreases the height.
(2) Assume $r = \mathbb{h}(\nu)$ and $\nu(x_i p) \neq 0$. If $x_i \in X_{\ell_X - r+1}$, then $x_i p \in X_{\ell_X - r}$ and then, using that $\nu \in [X^*]_r$, we obtain the contradiction $\nu(x_i p) = 0$. Then, $\mathbb{h}(x_i) > \ell_X - r + 1$. Since the dual basis of X is compatible with the filtration, this is equivalent to the fact $\mathbb{h}(\nu_i) < r$, as claimed.
(3) Assume $m = \mathbb{h}(\gamma)$ and $\gamma(p_i p_j) \neq 0$. Since $\gamma \in [P^*]_m$, $p_i p_j \notin P_{\ell_P - m}$. Then, $\mathbb{h}(p_i p_j) \geq \ell_P - m + 1$. Make $\mathbb{h}(p_i) = u$ and $\mathbb{h}(p_j) = v$. Then

$$p_i p_j \in P_u P_v = P^{(\ell_P - u+1)} P^{(\ell_P - v+1)} \subseteq P^{(2\ell_P - u - v+2)} = P_{u+v-\ell_P - 1}.$$

Since the dual basis of P is compatible with the filtration $\mathcal{F}(P)$, we have the equalities $u = \ell_P + 1 - \mathbb{h}(\gamma_i)$ and $v = \ell_P + 1 - \mathbb{h}(\gamma_j)$. Add these equalities and substitute in the following: $\ell_P - m + 1 \leq \mathbb{h}(p_i p_j) \leq u + v - \ell_P - 1 = \ell_P + 1 - \mathbb{h}(\gamma_i) - \mathbb{h}(\gamma_j)$. Hence, $m \geq \mathbb{h}(\gamma_i) + \mathbb{h}(\gamma_j)$, as claimed.

□

Definition 14.8. *Assume that $\mathcal{A} = (T, \delta)$ is a ditalgebra with triangular layer (R, W). Then, we have a sequence of R-R-subbimodules $0 = W_0^0 \subseteq W_0^1 \subseteq \cdots \subseteq W_0^{\ell_0} = W_0$ such that $\delta(W_0^{s+1}) \subseteq A_s W_1 A_s$, for all $s \in [0, \ell_0 - 1]$, where A_s denotes the R-subalgebra of A generated by W_0^s. There is also a sequence of R-R-subbimodules $0 = W_1^0 \subseteq W_1^1 \subseteq \cdots \subseteq W_1^{\ell_1} = W_1$ such that $\delta(W_1^{s+1}) \subseteq AW_1^s AW_1^s A$, for all $s \in [0, \ell_1 - 1]$.*

Assume that $\mathcal{A}'$ is a proper subditalgebra of $\mathcal{A}$ associated to the R-R-bimodule decompositions $W_0 = W_0' \oplus W_0''$ and $W_1 = W_1' \oplus W_1''$. Then, $\mathcal{A}'$ is called an initial subditalgebra of $\mathcal{A}'$ iff we have $W_0' = W_0^{\ell_0'}$ and $W_1' = W_1^{\ell_1'}$, for some $\ell_0' \in [0, \ell_0]$ and $\ell_1' \in [0, \ell_1]$.

Lemma 14.9. *Assume that $\mathcal{A}'$ is an initial subditalgebra of the triangular ditalgebra $\mathcal{A}$. With the notations above, from (12.2), we know that $\mathcal{A}'$ inherits a triangular structure for his layer (R, W') from the triangular structure of the layer (R, W) of $\mathcal{A}$. Indeed, the filtrations $0 = W_0^0 \subseteq W_0^1 \subseteq \cdots \subseteq W_0^{\ell_0'} = W_0'$ and $0 = W_1^0 \subseteq W_1^1 \subseteq \cdots \subseteq W_1^{\ell_1'} = W_1'$, constitute a triangular structure for the layer (R, W').*

For $r \in \{0, 1\}$, denote by $\ell_r'' = \ell_r - \ell_r'$ and $[W_r'']^s := W_r'' \cap W_r^{\ell_r'+s}$, for all $s \in [0, \ell_r'']$. Then, we have the R-R-bimodule filtration: $0 = [W_r'']^0 \subseteq [W_r'']^1 \subseteq \cdots \subseteq [W_r'']^{\ell_r''} = W_r''$, satisfying that $W_r^{\ell_r'+s} = W_r' \oplus [W_r'']^s$, for all $s \in [0, \ell_r'']$. We can also write $[W_r']^s = W_r^s$, for $s \in [0, \ell_r']$.

Proof. We only prove that $\delta(W_1^{s+1}) \subseteq BW_1^s BW_1^s B$, for all $s \in [0, \ell_1' - 1]$. Since A is freely generated by B and $BW_0''B$, we have that $A = B \oplus C$, where C is the ideal of A generated by W_0''. Since T is freely generated by R and $W_0' \oplus W_0'' \oplus W_1' \oplus W_1''$, we have $BW_1'BW_1'B \cap \overline{C} = 0$, where $\overline{C}$ is the ideal of T generated by W_0''. We also know that $\delta(W_1^{s+1}) \subseteq AW_1^s AW_1^s A$, because (R, W) is a triangular layer for $\mathcal{A}$, and $\delta(W_1') \subseteq BW_1'BW_1'B$, because $\mathcal{A}'$ is a proper subditalgebra of A. Then, if $s + 1 \leq \ell_1'$ and $w \in W_1^{s+1}$, we have $\delta(w) = t + t'$, with $t' \in \overline{C}$ and $t \in BW_1^s BW_1^s B$. Thus, $t' = 0$ and we are done. □

Theorem 14.10. *Assume that X is an admissible triangular $\mathcal{A}'$-module, where $\mathcal{A}'$ is an initial subditalgebra of the triangular ditalgebra $\mathcal{A}$. Then, the ditalgebra $\mathcal{A}^X = (T^X, \delta^X)$ admits a triangular layer (S, W^X).*

Proof. Adopt the notations of (14.9) and write $\underline{W}_0^s = B[W_0'']^s B$, for $s \in [0, \ell_0'']$, and $\underline{W}_1^s = B[W_1'']^s B$, for $s \in [0, \ell_1'']$. Consider the S-S-bimodule filtration $\mathcal{F}(P)$: $0 = P_0 \subseteq P_1 \subseteq \cdots \subseteq P_{\ell_P - 1} \subseteq P_{\ell_P} = P$, and the B-S-bimodule filtration $\mathcal{F}(X)$: $0 = X_0 \subseteq X_1 \subseteq \cdots \subseteq X_{\ell_X - 1} \subseteq X_{\ell_X} = X$, as in (14.7). We want to show that (S, W^X) is triangular. Take

$$[W_0^X]^m := \sum_{r+2\ell_X s+t\leq m} [X^*]_r\widehat{\otimes}_B \underline{W}_0^s \widehat{\otimes}_B X_t, \quad \text{for } m \in [0, 2\ell_X(\ell_0''+1)];$$

$$[W_1^X]^m := [P^*]_m, \text{ for } m \in [0, \ell_P], \text{ and}$$

$$[W_1^X]^{\ell_P+m} := \sum_{r+2\ell_X s+t\leq m} \left([X^*]_r\widehat{\otimes}_B \underline{W}_1^s \widehat{\otimes}_B X_t\right) \oplus P^*, \quad \text{for } m \in [0, 2\ell_X(\ell_1''+1)].$$

Here, we again abuse the language, since we mean the image of these spaces in T^X. The symbol $H'\widehat{\otimes}_R G'$ means the image of the canonical morphism $H' \otimes_R G' \longrightarrow H \otimes_R G$, when H' and G' are R-submodules of H and G respectively. Then, we have the S-S-bimodule filtrations

$$0 = [W_0^X]^0 \subseteq [W_0^X]^1 \subseteq \cdots \subseteq [W_0^X]^{2\ell_X(\ell_0''+1)} = W_0^X$$
$$0 = [W_1^X]^0 \subseteq [W_1^X]^1 \subseteq \cdots \subseteq [W_1^X]^{\ell_P+2\ell_X(\ell_1''+1)} = W_1^X.$$

For the last equality, notice that a generator of type $\nu \otimes w \otimes x \in W_1^X = [(X^* \otimes BW_1B \otimes X) \oplus P^*]/\mathcal{L}$ with $\nu \in X^*$, $w \in W_1 = W_1' \oplus W_1''$, $x \in X$, decomposes as $\nu \otimes w \otimes x = \nu \otimes w' \otimes x + \nu \otimes w'' \otimes x$, with $w' \in W_1'$ and $w'' \in W_1''$, and, therefore, $\nu \otimes w \otimes x = \nu \otimes w'' \otimes x + \gamma_{\nu,x}^{w'}$.

In the following, we consider heights with respect to the filtrations $\mathcal{F}(X)$ and $\mathcal{F}(P)$ (and their duals), and with respect to the filtrations of W_0'' and W_1''.

First of all notice that $[W_0^X]^m$ is generated as an S-S-bimodule by the elements $\nu \otimes w \otimes x$, with $\nu \in X^*$, $w \in W_0''$ and $x \in X$, such that $\mathbb{h}(\nu) + 2\ell_X\mathbb{h}(w) + \mathbb{h}(x) \leq m$. Indeed, any element of $[W_0^X]^m$ is a finite sum of elements of the form $\nu \otimes \underline{w} \otimes x$, with $\nu \in [X^*]_r$, $\underline{w} \in B[W_0'']^s B$ and $x \in X_t$, with $r + 2\ell_X s + t \leq m$. But $\underline{w} = \sum_q b_q w_q b_q'$, with $w_q \in [W_0'']^s$ and $b_q, b_q' \in B$. Then, $\nu \otimes \underline{w} \otimes x = \sum_q \nu \otimes b_q w_q b_q' \otimes x = \sum_q \nu b_q \otimes w_q \otimes b_q' x$, where $\mathbb{h}(\nu b_q) \leq \mathbb{h}(\nu) \leq r$, $\mathbb{h}(w_q) \leq s$ and $\mathbb{h}(b_q' x) \leq \mathbb{h}(x) \leq t$. Thus, $\mathbb{h}(\nu b_q) + 2\ell_X\mathbb{h}(w_q) + \mathbb{h}(b_q' x) \leq m$.

Similarly, $[W_1^X]^m$ is generated, for $m > \ell_P$, as an S-S-bimodule by P^* and the elements $\nu \otimes w \otimes x$, with $\nu \in X^*$, $w \in W_1''$ and $x \in X$, such that $\mathbb{h}(\nu) + 2\ell_X\mathbb{h}(w) + \mathbb{h}(x) \leq m$.

In order to handle the comultiplication μ and the maps λ and ρ involved in the definition of the differential δ^X, we fix dual basis of X and P which are compatible with the filtrations $\mathcal{F}(X)$ and $\mathcal{F}(P)$, as in (14.7).

Step 1: We first show that $\delta^X([W_0^X]^{m+1}) \subseteq A_m^X W_1^X A_m^X$, where A_m^X is the subalgebra of A^X generated by S and $[W_0^X]^m$, for all m such that $0 \leq m \leq 2\ell_X(\ell_0''+1) - 1$.

Choose a generator $\nu \otimes w \otimes x$ of $[W_0^X]^{m+1}$, with $x \in X$, $w \in W_0''$ and $\nu \in X^*$, with $\mathrm{lh}(\nu) + 2\ell_X \mathrm{lh}(w) + \mathrm{lh}(x) \leq m + 1$. We will show that each summand of

$$\delta^X(\nu \otimes w \otimes x) = \lambda(\nu) \otimes w \otimes x + \sigma_{\nu,x}(\delta(w)) - \nu \otimes w \otimes \rho(x)$$

lies in $A_m^X W_1^X A_m^X$.

Since $\lambda(\nu) = \sum_{i,j} \nu(x_i p_j)\gamma_j \otimes \nu_i$, the first summand is $\sum_{i,j} \nu(x_i p_j)\gamma_j \otimes \nu_i \otimes w \otimes x \in P^*[W_0^X]^m \subseteq W_1^X A_m^X$, because from (14.7), $\mathrm{lh}(\nu_i) + 2\ell_X \mathrm{lh}(w) + \mathrm{lh}(x) < \mathrm{lh}(\nu) + 2\ell_X \mathrm{lh}(w) + \mathrm{lh}(x) \leq m + 1$, whenever $\nu(x_i p_j) \neq 0$.

Since $\rho(x) = \sum_j x p_j \otimes \gamma_j$, the third summand is $-\sum_j \nu \otimes w \otimes x p_j \otimes \gamma_j \in [W_0^X]^m P^* \subseteq A_m^X W_1^X$, because from (14.7), $\mathrm{lh}(\nu) + 2\ell_X \mathrm{lh}(w) + \mathrm{lh}(x p_j) < \mathrm{lh}(\nu) + 2\ell_X \mathrm{lh}(w) + \mathrm{lh}(x) \leq m + 1$, whenever $x \neq 0$.

Denote by $\underline{A}_s$ the subalgebra of A generated by B and $[W_0'']^s$, for $0 \leq s \leq \ell_0''$. Then, the subalgebra $A_{\ell_0'+s}$ of A generated by R and $W_0^{\ell_0'+s}$ coincides with $\underline{A}_s$. The triangularity condition for W_0 implies that $\delta([W_0'']^{s+1}) \subseteq \underline{A}_s W_1 \underline{A}_s$.

Assume that x, w and ν are not zero, and denote by $s + 1 = \mathrm{lh}(w)$, thus $2\ell_X \mathrm{lh}(w) < m + 1$. From the above, we know that $\delta(w) = \sum_q a_q w_q a'_q$, with $a_q, a'_q \in \underline{A}_s$ and $w_q \in W_1$. Then, by (12.8)

$$\sigma_{\nu,x}(\delta(w)) = \sum_{i,j} \sigma_{\nu,x_i}(a_q)\sigma_{\nu_i,x_j}(w_q)\sigma_{\nu_j,x}(a'_q).$$

The middle factors $\sigma_{\nu_i,x_j}(w_q)$ are in W_1^X. To conclude that $\sigma_{\nu,x}(\delta(w)) \in A_m^X W_1^X A_m^X$, we will see that $\sigma_{\nu,x}(a) \in A_m^X$, for $a \in \underline{A}_s$. We may assume that a has the form $a = b_0 w_1 b_1 w_2 \cdots b_{n-1} w_n b_n$, with $b_u \in B$ and $w_u \in [W_0'']^s$. Then, $\sigma_{\nu,x}(a)$ is a sum of products of elements of the form $\sigma_{\nu_i,x_j}(b\widehat{w}b')$, with $i, j \in I$, $b, b' \in B$ and $\widehat{w} \in [W_0'']^s$. Again, $\sigma_{\nu_i,x_j}(b\widehat{w}b') = \nu_i b \otimes \widehat{w} \otimes b' x_j$, and $\mathrm{lh}(\nu_i b) + 2\ell_X \mathrm{lh}(\widehat{w}) + \mathrm{lh}(b' x_j) \leq \mathrm{lh}(\nu_i) + 2\ell_X \mathrm{lh}(\widehat{w}) + \mathrm{lh}(x_j) \leq 2\ell_X + 2\ell_X \mathrm{lh}(\widehat{w}) = 2\ell_X(\mathrm{lh}(\widehat{w}) + 1) \leq 2\ell_X \mathrm{lh}(w) < m + 1$, which implies that $\nu_i b \otimes w \otimes b' x_j \in [W_0^X]^m$.

Step 2: Now assume that $\gamma \in P^* \cap [W_1^X]^{m+1}$, with $m \in [0, \ell_P + 2\ell_X(\ell_1'' + 1) - 1]$. Let us show that $\delta^X(\gamma) = \mu(\gamma) \in A^X[W_1^X]^m A^X[W_1^X]^m A^X$.

From (11.8), we know that $\mu(\gamma) = \sum_{i,j} \gamma(p_i p_j)\gamma_j \otimes \gamma_i$. Then, for any $m \in [\ell_P, \ell_P + 2\ell_X(\ell_1'' + 1) - 1]$, we have $P^* \subseteq [W_1^X]^m$, and $\mu(\gamma) \in P^* P^* \subseteq A^X[W_1^X]^m A^X[W_1^X]^m A^X$. Now, if $m \in [0, \ell_P - 1]$, we have $\gamma \in [W_1^X]^{m+1} = [P^*]_{m+1}$. From (14.7), $\gamma(p_i p_j) \neq 0$ implies $\mathrm{lh}(\gamma_i) + \mathrm{lh}(\gamma_j) \leq \mathrm{lh}(\gamma) \leq m + 1$, and $\mathrm{lh}(\gamma_i)$ and $\mathrm{lh}(\gamma_j)$ are both smaller than m. Then, $\mu(\gamma) \in [P^*]_m[P^*]_m \subseteq A^X[W_1^X]^m A^X[W_1^X]^m A^X$.

Step 3: Finally, we show that $\delta^X([W_1^X]^{m+1}) \subseteq A^X[W_1^X]^m A^X[W_1^X]^m A^X$, for all $m \in [0, \ell_P + 2\ell_X(\ell_1'' + 1) - 1]$.

Notice that for $m \in [0, \ell_P - 1]$, step 3 follows from step 2. So assume that $m \in [\ell_P, \ell_P + 2\ell_X(\ell_1'' + 1) - 1]$. Then, again from step 2, it will be enough to choose a generator of the form $\nu \otimes w \otimes x$ of $[W_1^X]^{m+1}$, with $x \in X$, $w \in W_1''$ and $\nu \in X^*$, with $\mathbb{h}(\nu) + 2\ell_X \mathbb{h}(w) + \mathbb{h}(x) \leq m + 1$. We will show that each of the summands in

$$\delta^X(\nu \otimes w \otimes x) = \lambda(\nu) \otimes w \otimes x + \sigma_{\nu,x}(\delta(w)) + \nu \otimes w \otimes \rho(x),$$

lies in $A^X[W_1^X]^m A^X[W_1^X]^m A^X$.

Since $\lambda(\nu) = \sum_{i,j} \nu(x_i p_j)\gamma_j \otimes \nu_i$, the first summand is $\sum_{i,j} \nu(x_i p_j)\gamma_j \otimes \nu_i \otimes w \otimes x \in P^*[W_1^X]^m \subseteq A^X[W_1^X]^m A^X[W_1^X]^m A^X$, because from (14.7), $\mathbb{h}(\nu_i) + 2\ell_X \mathbb{h}(w) + \mathbb{h}(x) < \mathbb{h}(\nu) + 2\ell_X \mathbb{h}(w) + \mathbb{h}(x) \leq m + 1$, whenever $\nu(x_i p_j) \neq 0$.

Since $\rho(x) = \sum_j x p_j \otimes \gamma_j$, the third summand is $\sum_j \nu \otimes w \otimes x p_j \otimes \gamma_j \in [W_1^X]^m P^* \subseteq A^X[W_1^X]^m A^X[W_1^X]^m A^X$, because from (14.7), $\mathbb{h}(\nu) + 2\ell_X \mathbb{h}(w) + \mathbb{h}(x p_j) < \mathbb{h}(\nu) + 2\ell_X \mathbb{h}(w) + \mathbb{h}(x) \leq m + 1$, whenever $x \neq 0$.

Suppose that w, ν and x are not zero, and denote by $s + 1 = \mathbb{h}(w)$, thus $2\ell_X \mathbb{h}(w) < m + 1$. By assumption, $\delta(w) = \sum_q a_q u_q b_q v_q c_q$, with $a_q, b_q, c_q \in A$ and $u_q, v_q \in W_1^{\ell_1' + s}$. Then

$$\sigma_{\nu,x}(\delta(w)) = \sum_{i,j,r,t} \sigma_{\nu,x_i}(a_q)\sigma_{\nu_i,x_j}(u_q)\sigma_{\nu_j,x_r}(b_q)\sigma_{\nu_r,x_t}(v_q)\sigma_{\nu_t,x}(c_q).$$

The factors $\sigma_{\nu,x_i}(a_q)$, $\sigma_{\nu_j,x_r}(b_q)$ and $\sigma_{\nu_t,x}(c_q)$ are in A^X. To conclude that $\sigma_{\nu,x}(\delta(w)) \in A^X[W_1^X]^m A^X[W_1^X]^m A^X$, we will see that $\sigma_{\nu,x}(u) \in [W_1^X]^m$, for $u \in W_1^{\ell_1' + s}$. We may assume that $u = u' + u''$, with $u' \in W_1'$ and $u'' \in [W_1'']^s$. Then, $\sigma_{\nu,x}(u) = \gamma_{\nu,x}^{u'} + \nu \otimes u'' \otimes x$, where clearly $\gamma_{\nu,x}^{u'} \in P^* \subseteq [W_1^X]^m$ and $\nu \otimes u'' \otimes x = \sum_{i,j} \nu(x_i)\nu_i \otimes u'' \otimes x_j \nu_j(x)$, where each $\nu_i \otimes u'' \otimes x_j \in [W_1^X]^m$ because, as before, $\mathbb{h}(\nu_i) + 2\ell_X \mathbb{h}(u'') + \mathbb{h}(x_j) \leq 2\ell_X + 2\ell_X \mathbb{h}(u'') = 2\ell_X(\mathbb{h}(u'') + 1) \leq 2\ell_X(s + 1) < m + 1$. This implies that $\sigma_{\nu,x}(u) \in [W_1^X]^m$. □

Exercise 14.11. *Assume that X is a triangular admissible $\mathcal{A}'$-module, where $\mathcal{A}'$ is a proper subditalgebra of the layered ditalgebra $\mathcal{A}$. Assume furthermore that $\delta(W_0') = 0$. For $\nu \in X^*$ and $x \in X$, we consider the linear map $\sigma_{\nu,x} : T \longrightarrow T^X$ defined in (12.8). Let I be a triangular ideal of $\mathcal{A}$, as in (8.22), with associated filtration $\{0\} = H_0 \subseteq H_1 \subseteq \cdots \subseteq H_\ell = H$. Show that:*

(1) The ideal I^X of $\mathcal{A}^X$ generated by

$$H^X := \{\sigma_{\nu,x}(h) \mid \nu \in X^*, x \in X \text{ and } h \in H\}$$

is a triangular ideal of $\mathcal{A}^X$ *with associated filtration* $\{0\} = H_0^X \subseteq H_1^X \subseteq \cdots \subseteq H_{2\ell_X(\ell+1)}^X = H^X$, *where* ℓ_X *is the length of the filtration associated to* X, *see (14.6), and for* $m \in [0, 2\ell_X(\ell+1)]$

$$H_m^X := \{\sigma_{\nu,x}(h) \mid \nu \in X^*, x \in X, h \in H \text{ and } \mathbb{h}(\nu) + 2\ell_X \mathbb{h}(h) + \mathbb{h}(x) \leq m\}.$$

The heights are naturally taken relative to the filtrations of X, *or its dual filtration of* X^*, *or to the given set filtration of* H.

(2) *The ideal* $I_{A^X}^X = I^X \cap A^X$ *of* $\mathcal{A}^X$ *is also generated by the set*

$$\{\sigma_{\nu,x}(a) \mid \nu \in X^*, x \in X \text{ and } a \in I \cap A\}.$$

(3) *The functor* $F^X : \mathcal{A}^X\text{-Mod} \longrightarrow \mathcal{A}\text{-Mod}$ *induces a functor*

$$F^X : (\mathcal{A}^X, I^X)\text{-Mod} \longrightarrow (\mathcal{A}, I)\text{-Mod}.$$

Hints: To show the triangularity of I^X *using the given filtration, adapt the proof of step 1 in the proof of (14.10), using (12.11). (2) follows from (12.8)(3). For (3), use the formula for the action of* A *on* $F^X(M)$ *in (12.10).*

15

Free bimodule filtrations and free ditalgebras

In this section, we introduce the notion of a free triangular ditalgebra $\mathcal{A}$ and give sufficient conditions on an admissible module X which yield a free triangular ditalgebra $\mathcal{A}^X$. We start with some preliminary definitions and lemmas.

Definition 15.1. *Assume that we have an R-S-bimodule filtration*

$$\mathcal{F}(U)\colon\ 0 = U_0 \subseteq U_1 \subseteq \cdots \subseteq U_{\ell-1} \subseteq U_\ell = U$$

of the R-S-bimodule U. Let E be any k-algebra containing S as a subalgebra. Then, if $\mathcal{F}(U)$ is right additive, we will consider its right extended filtration

$$\mathcal{F}(U)\otimes_S E\colon\ 0 = U_0\otimes_S E \subseteq U_1\otimes_S E \subseteq \cdots \subseteq U_{\ell-1}\otimes_S E \subseteq U_\ell\otimes_S E = U\otimes_S E.$$

We often write $[U\otimes_S E]_m = U_m\otimes_S E$, for $m\in[0,\ell]$. Similarly, if G is a k-algebra containing R and $\mathcal{F}(U)$ is left additive filtration, we will consider the corresponding left extended filtration $G\otimes_R \mathcal{F}(U)$, which is a G-S-bimodule filtration of $G\otimes_R U$. Finally, for E and G as before, if $\mathcal{F}(U)$ is additive, we can also consider the two-sided extended filtration $G\otimes_R\mathcal{F}(U)\otimes_S E$.

Lemma 15.2. *Assume that $\mathcal{F}(U)$ is a right additive R-S-bimodule filtration of the bimodule U, where S is a subalgebra of the k-algebra E which admits a left S-module decomposition $E = S\oplus P$ in S-*Mod*. The last condition holds, for instance if S is semisimple or if E is an algebra freely generated by a pair (S, W). Then, we have $\mathbb{h}(u\otimes 1) = \mathbb{h}(u)$, for $u\in U$. Here, the height maps correspond to the filtration $\mathcal{F}(U)$ and its associated right extension $\mathcal{F}(U)\otimes_S E$.*

Proof. Assume that $\mathbb{h}(u) = r$. By assumption, $U_r = U_{r-1}\oplus V$, for some right S-module V, and $u = u' + v$, with $u'\in U_{r-1}$ and $0\neq v\in V$. Assume that $\mathbb{h}(u\otimes 1) < r$, then $v\otimes 1 = u\otimes 1 - u'\otimes 1 \in (V\otimes_S E)\cap(U_{r-1}\otimes_S E) = 0$.

Our assumptions imply that the canonical map $\theta : V \cong V \otimes_S S \longrightarrow U \otimes_S E$ is injective and, therefore, $v = 0$; a contradiction. □

Definition 15.3. *If $\mathcal{F}(U)$ and $\mathcal{F}(V)$ are R-S-bimodule filtrations of the same length ℓ, then a morphism $\tau : \mathcal{F}(U) \longrightarrow \mathcal{F}(V)$ is a morphism $\tau : U \longrightarrow V$ of R-S-bimodules such that $\tau(U_i) \subseteq V_i$, for all $i \in [0, \ell]$. It is an isomorphism iff $\tau_i(U_i) = V_i$ for all i. When there is such an isomorphism, as usual, we write $\mathcal{F}(U) \cong \mathcal{F}(V)$.*

Lemma 15.4. *Assume that S is a semisimple subalgebra of the k-algebra T. Suppose that we have an R-S-bimodule filtration $\mathcal{F}(U)$ of length ℓ of the R-S-bimodule U, which is right finitely generated over S. Then:*

(1) The morphism $\tau : T \otimes_S U^* \longrightarrow (U \otimes_S T)^*$, determined by the rule $\tau(t \otimes \lambda)[u \otimes t'] = t\lambda(u)t'$, is an isomorphism of T-R-bimodule filtrations $T \otimes_S (\mathcal{F}(U))^* \xrightarrow{\cong} (\mathcal{F}(U) \otimes_S T)^*$.

(2) For any right finite dual basis $\{(u_i, \lambda_i) \mid i \in I\}$ for U_S compatible with the filtration $\mathcal{F}(U)$, we have the right finite dual basis for $(U \otimes_S T)_T$

$$\{(u_i \otimes 1, \tau(1 \otimes \lambda_i)) \mid i \in I\},$$

which is compatible with the right additive R-T-bimodule filtration $\mathcal{F}(U) \otimes_S T$ of $U \otimes_S T$.

Proof. Consider the canonical isomorphism $\rho : T \longrightarrow T^* = \mathrm{End}_T(T)$ which maps t onto the left multiplication by t. Then, the composition $\tau = \psi(\rho \otimes 1) : T \otimes_S U^* \longrightarrow (U \otimes_S T)^*$, where ψ is as in (11.4), is an isomorphism such that $\tau(t \otimes \lambda)[u \otimes t'] = t\lambda(u)t'$. Notice that the first dual is taken over S, while the second one is taken over T.

We first show that τ is a morphism of filtrations. Take $m \in [1, \ell]$ and a typical generator $t \otimes \lambda$ in $[T \otimes_S U^*]_m = T \otimes_S [U^*]_m$, with $t \in T$ and $\lambda \in [U^*]_m$. Thus, $\lambda(U_{\ell-m}) = 0$. Assume that $u \in U_{\ell-m}$ and $t' \in T$, then $\tau(t \otimes \lambda)[u \otimes t'] = t\lambda(u)t' = 0$, which implies that $\tau(t \otimes \lambda)$ maps $U_{\ell-m} \otimes T = [U \otimes_S T]_{\ell-m}$ to zero. Hence, $\tau(t \otimes \lambda) \in [(U \otimes_S T)^*]_m$, as claimed.

If we have a right finite dual basis $\{(u_i, \lambda_i) \mid i \in I\}$ for U, then for any $u \in U$ and $t \in T$, we have that $u \otimes t = \sum_i u_i\lambda_i(u) \otimes t = \sum_i (u_i \otimes 1)\lambda_i(u)t = \sum_i (u_i \otimes 1)\tau(1 \otimes \lambda_i)[u \otimes t]$. This implies that $\{(u_i \otimes 1, \tau(1 \otimes \lambda_i)) \mid i \in I\}$ is a right finite dual basis for $(U \otimes_S T)_T$.

Now we observe that $\mathbb{h}(\lambda) = \mathbb{h}(\tau(1 \otimes \lambda))$, for any $\lambda \in U^*$. Indeed, notice that, for $r \in [1, \ell]$ and $u \in U_r$, $\tau(1 \otimes \lambda)[u \otimes 1] = \lambda(u)$. Hence, $\lambda(U_r) = 0$ iff $\tau(1 \otimes \lambda)(U_r \otimes T) = 0$. Thus, $\lambda \in [U^*]_m \setminus [U^*]_{m-1}$ iff $\lambda(U_{\ell-m}) = 0$ and $\lambda(U_{\ell-m+1}) \neq 0$. This is equivalent to the fact that $\tau(1 \otimes \lambda)(U_{\ell-m} \otimes T) = 0$ and $\tau(1 \otimes \lambda)(U_{\ell-m+1} \otimes T) \neq 0$. That is $\tau(1 \otimes \lambda) \in [(U \otimes T)^*]_m \setminus [(U \otimes T)^*]_{m-1}$.

Now, under the assumptions of item (2), we know from (15.2) that $\mathbb{h}(u_i) = \mathbb{h}(u_i \otimes 1)$, for all i. Therefore, by the above observation, $\mathbb{h}(\tau(1 \otimes \lambda_i)) + \mathbb{h}(u_i \otimes 1) = \mathbb{h}(\lambda_i) + \mathbb{h}(u_i) = \ell + 1$. Then, the dual basis $\{(u_i \otimes 1, \tau(1 \otimes \lambda_i)) \mid i \in I\}$ is compatible with the right additive R-T-filtration $\mathcal{F}(U) \otimes_S T$ of $U \otimes_S T$.

Finally, from (14.5), we know that the set $\{\tau(1 \otimes \lambda_i) \mid \mathbb{h}(\tau(1 \otimes \lambda_i)) \leq m\}$ generates $[(U \otimes_S T)^*]_m$ as a left T-module, for $m \in [1, \ell]$. Clearly, $1 \otimes \lambda_i \in T \otimes [U^*]_m$, for any λ_i with $\mathbb{h}(\tau(1 \otimes \lambda_i)) = \mathbb{h}(\lambda_i) \leq m$. Then, $\tau([T \otimes_S U^*]_m) = [(U \otimes_S T)^*]_m$. Thus, τ is an isomorphism of T-R-filtrations. □

Definition 15.5. *Assume that we have an R-S-bimodule filtration*

$$\mathcal{F}(U)\colon\ 0 = U_0 \subseteq U_1 \subseteq \cdots \subseteq U_{\ell-1} \subseteq U_\ell = U$$

of the R-S-bimodule U and that R_0 is a subalgebra of R. Then, $\mathcal{F}(U)$ is called left R_0-free iff there exists a left additive R_0-S-bimodule filtration

$$\mathcal{F}(\widetilde{U})\colon\ 0 = \widetilde{U}_0 \subseteq \widetilde{U}_1 \subseteq \cdots \subseteq \widetilde{U}_{\ell-1} \subseteq \widetilde{U}_\ell = \widetilde{U}$$

of some R_0-S-bimodule $\widetilde{U}$ such that $R \otimes_{R_0} \mathcal{F}(\widetilde{U}) \cong \mathcal{F}(U)$.

Similarly, if S_0 is a subalgebra of S, $\mathcal{F}(U)$ is called right S_0-free iff there exists a right additive R-S_0-bimodule filtration $\mathcal{F}(\widetilde{U})$ such that $\mathcal{F}(\widetilde{U}) \otimes_{S_0} S \cong \mathcal{F}(U)$. Finally, for R_0 and S_0 as above, we say that the filtration $\mathcal{F}(U)$ is R_0-S_0-free iff there exists an additive R_0-S_0-bimodule filtration $\mathcal{F}(\widetilde{U})$ such that $R \otimes_{R_0} \mathcal{F}(\widetilde{U}) \otimes_{S_0} S \cong \mathcal{F}(U)$.

The following will explain the term "free" chosen in the last definition. We only examine the R_0-R_0-free case, since the others are similar.

Definition 15.6. *Suppose that R_0 is a subalgebra of the k-algebra R. Then, an R-R-bimodule W is called freely generated by its R_0-R_0-subbimodule $\widetilde{W}$ over R_0 iff each morphism $\varphi : \widetilde{W} \longrightarrow V$ of R_0-R_0-bimodules, where V is an R-R-bimodule, extends in a unique way to a morphism of R-R-bimodules $\overline{\varphi} : W \longrightarrow V$.*

Lemma 15.7. *Suppose that R_0 is a subalgebra of the k-algebra R. Let W be an R-R-bimodule and $\widetilde{W}$ be an R_0-R_0-subbimodule of W. Then:*

(1) The bimodule W is freely generated by $\widetilde{W}$ over R_0 iff the product map $\rho : R \otimes_{R_0} \widetilde{W} \otimes_{R_0} R \longrightarrow W$, which sends each generator $r \otimes w \otimes r'$ onto rwr', is an isomorphism of R-R-bimodules.

(2) Assume that R_0 is isomorphic to a finite product of copies of the ground field k and let $1 = \sum_{i=1}^{n} e_i$ be the corresponding decomposition of the unit

of the algebra R_0 as a sum of central primitive orthogonal idempotents. Then, W is freely generated by some (resp. finitely generated) $\widetilde{W}$ over R_0 iff $W \cong \bigoplus_{\alpha\in A} Re_{t(\alpha)} \otimes_k e_{s(\alpha)}R$, where A is a (resp. finite) set and $\alpha \mapsto (s(\alpha), t(\alpha))$ is the receipt of some map $A \to [1,n]^2$.

Proof.

(1) Assume that W is freely generated by $\widetilde{W}$ over R_0 and consider the morphism of R-R-bimodules ρ as in the statement of the lemma. Consider also the morphism of R_0-bimodules $\psi : \widetilde{W} \longrightarrow R \otimes_{R_0} \widetilde{W} \otimes_{R_0} R$ which maps each $w \in \widetilde{W}$ onto $1 \otimes w \otimes 1$. Then, by assumption, there exists a morphism of R-bimodules $\overline{\psi} : W \longrightarrow R \otimes_{R_0} \widetilde{W} \otimes_{R_0} R$ such that $\overline{\psi}(w) = 1 \otimes w \otimes 1$, for $w \in \widetilde{W}$. Thus, $\overline{\psi}\rho(r \otimes w \otimes r') = \overline{\psi}(rwr') = r\overline{\psi}(w)r' = r \otimes w \otimes r'$, for $r, r' \in R$ and $w \in \widetilde{W}$. Then, $\overline{\psi}\rho = id$. Moreover, the maps $\rho\overline{\psi}$ and id_W both extend the inclusion map $\widetilde{W} \longrightarrow W$ and, by the uniqueness property in the definition of freely generated bimodule, they must coincide.

Now assume that ρ is an isomorphism of R-bimodules and let $\varphi : \widetilde{W} \longrightarrow V$ be a morphism of R_0-bimodules, where V is an R-bimodule. Consider the map $\overline{\varphi} : R \otimes_{R_0} \widetilde{W} \otimes_{R_0} R \longrightarrow V$ defined by $\overline{\varphi}(r \otimes w \otimes r') = r\varphi(w)r'$, for $r, r' \in R$ and $w \in \widetilde{W}$. Then, the morphism of R-bimodules $\overline{\overline{\varphi}} = \overline{\varphi}\rho^{-1}$ extends φ.

(2) First, we claim that all the R_0-R_0-bimodules $\widetilde{W}$ are of the form $\widetilde{W} \cong \bigoplus_{\alpha\in A} R_0e_{t(\alpha)} \otimes_k e_{s(\alpha)}R_0$, where A is a set and $\alpha \mapsto (s(\alpha), t(\alpha))$ is the receipt of some map $A \to [1,n]^2$. Indeed, $\widetilde{W}$ is a left $R_0 \otimes_k R_0^{op}$-module, and $R_0 \otimes_k R_0^{op}$ is a semisimple finite-dimensional k-algebra, which decomposes as $R_0 \otimes_k R_0^{op} = \bigoplus_{i,j} R_0e_i \otimes_k e_jR_0^{op}$. Since, $R_0e_i \cong k \cong e_jR_0^{op}$, the simple $R_0 \otimes_k R_0^{op}$-modules are of the form $R_0e_i \otimes_k e_jR_0^{op}$, and our claim follows.

Now, notice that the product map is an isomorphism of left R-modules $R \otimes_{R_0} R_0e_j \cong Re_j$, for all j. Therefore, $W \cong R \otimes_{R_0} \widetilde{W} \otimes_{R_0} R \cong R \otimes_{R_0} \left[\bigoplus_{\alpha\in A} R_0e_{t(\alpha)} \otimes_k e_{s(\alpha)}R_0\right] \otimes_{R_0} R \cong \bigoplus_{\alpha\in A} Re_{t(\alpha)} \otimes_k e_{s(\alpha)}R$.

□

From this lemma, we immediately obtain the following:

Corollary 15.8. *Suppose that R_0 is a subalgebra of the k-algebra R. Then:*

(1) For any R_0-R_0-bimodule $\widetilde{W}$, the R-R-bimodule $R \otimes_{R_0} \widetilde{W} \otimes_{R_0} R$ is freely generated by the image of the map $\widetilde{W} \xrightarrow{\tau} R \otimes_{R_0} \widetilde{W} \otimes_{R_0} R$ given by $\tau(w) = 1 \otimes w \otimes 1$, which we denote by $1 \otimes \widetilde{W} \otimes 1$.

(2) Given any isomorphism of R-R-bimodules $\varphi : W \longrightarrow V$, if W is freely generated by $\widetilde{W}$ over R_0, then V is freely generated by $\varphi(\widetilde{W})$ over R_0.

(3) *If W is an R-R-bimodule freely generated by its R_0-R_0-subbimodule $\widetilde{W}$, which admits an R_0-R_0-bimodule decomposition $\widetilde{W} = \widetilde{W}' \oplus \widetilde{W}''$, then $W' := R\widetilde{W}'R$ and $W'' := R\widetilde{W}''R$ are freely generated by $\widetilde{W}'$ and $\widetilde{W}''$, respectively.*

(4) *Assume W is an R-R-bimodule and $\mathcal{F}(W)$ is an R-R-bimodule filtration. Then, $\mathcal{F}(W)$ is an R_0-R_0-free bimodule filtration iff there exists an R_0-R_0-subbimodule $\widetilde{W}$ of W and an additive R_0-R_0-bimodule filtration $\mathcal{F}(\widetilde{W})$ such that the product map $\rho : R \otimes_{R_0} \widetilde{W} \otimes_{R_0} R \longrightarrow \mathcal{F}(W)$ induces an isomorphism of filtrations $R \otimes_{R_0} \mathcal{F}(\widetilde{W}) \otimes_{R_0} R \longrightarrow \mathcal{F}(W)$.*

Definition 15.9. *Given two filtrations $\mathcal{F}(W')$ and $\mathcal{F}(W'')$ of the R-S-bimodules W' and W'', with lengths ℓ' and ℓ'', respectively, their direct sum $\mathcal{F}(W') \oplus \mathcal{F}(W'')$ is the filtration of length $\ell = \ell' + \ell''$ of the R-S-bimodule $W = W' \oplus W''$ defined by $W_s := W'_s$, for all $s \in [0, \ell']$, and $W_{\ell'+s} := W' \oplus W''_s$, for all $s \in [0, \ell'']$.*

Remark 15.10. *Assume we are given the filtration $\mathcal{F}(W)$ of length ℓ of an R-S-bimodule W, which admits a decomposition as a direct sum of bimodules $W = W' \oplus W''$ such that $W' = W_{\ell'}$, for some $\ell' \in [0, \ell]$. Then, $W'_s := W_s$, for $s \in [0, \ell']$, constitutes a filtration $\mathcal{F}(W)_{W'}$ of W'; and $W''_s := W_{\ell'+s} \cap W''$, for $s \in [0, \ell'']$, where $\ell'' = \ell - \ell'$, constitutes a filtration $\mathcal{F}(W)^{W''}$ of W''. Moreover, $\mathcal{F}(W) = \mathcal{F}(W)_{W'} \oplus \mathcal{F}(W)^{W''}$.*

Remark 15.11. *Given a subalgebra R_0 of the k-algebra R. If $\mathcal{F}(W')$ and $\mathcal{F}(W'')$ are R_0-R_0-free filtrations, then so is $\mathcal{F}(W') \oplus \mathcal{F}(W'')$.*

Remark 15.12. *In the following, we will occasionally find the assumption on a pair of semisimple k-algebras R_0 and S_0 requiring that $R_0 \otimes_k S_0^{op}$ is a semisimple algebra too. This is satisfied, for instance, if both algebras R_0 and S_0 are finite products of copies of the field k, or if both algebras R_0 and S_0 are finite dimensional and the field k is perfect (see [23](7.8)).*

Lemma 15.13. *Assume that R_0 is a subalgebra of the k-algebra R, such that $R_0 \otimes_k R_0^{op}$ is a semisimple algebra, and let $\mathcal{F}(W) = \mathcal{F}(W') \oplus \mathcal{F}(W'')$ be an equality of R-R-bimodule filtrations. Then, if $\mathcal{F}(W)$ is an R_0-R_0-free bimodule filtration, so are the filtrations $\mathcal{F}(W') = \mathcal{F}(W)_{W'}$ and $\mathcal{F}(W'')$.*

Proof. By assumption, we know the existence of an additive R_0-R_0-bimodule filtration $\mathcal{F}(\widetilde{W})$, for some R_0-R_0-subbimodule $\widetilde{W}$ of W, such that the product map $\rho : R \otimes_{R_0} \widetilde{W} \otimes_{R_0} R \longrightarrow W$ determines an isomorphism of R-R-bimodule filtrations $R \otimes_{R_0} \mathcal{F}(\widetilde{W}) \otimes_{R_0} R \xrightarrow{\cong} \mathcal{F}(W)$. Denote by ℓ, ℓ' and ℓ'' the lengths of $\mathcal{F}(W)$, $\mathcal{F}(W')$ and $\mathcal{F}(W'')$, respectively. Then, $\ell = \ell' + \ell''$ and $W_{\ell'} = W'$.

Make $\widetilde{W}' := \widetilde{W}_{\ell'}$ and notice that $\mathcal{F}(\widetilde{W}') := \mathcal{F}(\widetilde{W})_{\widetilde{W}'}$ is an additive R_0-R_0-bimodule filtration and there is an isomorphism of filtrations $R \otimes_{R_0} \mathcal{F}(\widetilde{W}') \otimes_{R_0} R \xrightarrow{\cong} \mathcal{F}(W')$ induced by the product map. Thus, $\mathcal{F}(W')$ is an R_0-R_0-free filtration.

Denote by $\pi'' : W \longrightarrow W''$ the projection map and define, for $s \in [0, \ell'']$, $\widetilde{W}''_s := \pi''(\widetilde{W}_{\ell'+s})$. Then, we have an R_0-R_0-bimodule filtration $\mathcal{F}(\widetilde{W}'')$ of the R_0-R_0-subbimodule $\widetilde{W}'' = \pi''(\widetilde{W})$ of W''. We will show that the product map induces an isomorphism of filtrations $\rho'' : R \otimes_{R_0} \mathcal{F}(\widetilde{W}'') \otimes_{R_0} R \longrightarrow \mathcal{F}(W'')$. Consider the exact commutative diagram

$$\begin{array}{ccccc}
R \otimes \widetilde{W}' \otimes R & \xrightarrow{1\otimes i\otimes 1} & R \otimes \widetilde{W}_{\ell'+s} \otimes R & \xrightarrow{1\otimes \nu\otimes 1} & R \otimes \left[\widetilde{W}_{\ell'+s}/\widetilde{W}'\right] \otimes R \\
\downarrow{\scriptstyle \rho'} & & \downarrow{\scriptstyle \rho} & & \downarrow{\scriptstyle \overline{\rho}} \\
W' & \longrightarrow & W_{\ell'+s} & \xrightarrow{\pi''} & W''_s .
\end{array}$$

Here, $\widetilde{W}' \xrightarrow{i} \widetilde{W}_{\ell'+s}$ is the inclusion, $\widetilde{W}_{\ell'+s} \xrightarrow{\nu} \widetilde{W}_{\ell'+s}/\widetilde{W}'$ is the canonical projection, π'' and $1 \otimes \nu \otimes 1$ are surjective, and $\overline{\rho}$ is the map induced by the isomorphisms ρ and ρ'. Then, $\overline{\rho}$ is an isomorphism of R-R-bimodules and W''_s is freely generated by $\overline{\rho}(1 \otimes \left[\widetilde{W}_{\ell'+s}/\widetilde{W}'\right] \otimes 1) = \pi''(\widetilde{W}_{\ell'+s})$. This implies that $\rho'' : R \otimes_{R_0} \widetilde{W}''_s \otimes_{R_0} R \longrightarrow W''_s$ is an isomorphism, for all $s \in [0, \ell'']$. Thus, ρ'' is an isomorphism of filtrations, as we wanted to show. Finally, $\mathcal{F}(\widetilde{W}'')$ is additive because $R_0 \otimes_k R_0^{op}$ is semisimple. □

Definition 15.14. *Suppose that $\mathcal{A}$ is a layered ditalgebra with triangular layer (R, W) and that R_0 is a subalgebra of R. Then, the ditalgebra $\mathcal{A}$ (or its layer (R, W)) is called R_0-free iff the filtration $\mathcal{F}(W_0)$ realizing the triangularity of the layer at degree zero (see condition (1) of definition (5.1)) is R_0-R_0-free. The ditalgebra $\mathcal{A}$ (or its layer (R, W)) is called completely R_0-free iff both filtrations $\mathcal{F}(W_0)$ and $\mathcal{F}(W_1)$ realizing the triangularity of the layer (see definition (5.1)) are R_0-R_0-free.*

Definition 15.15. *Suppose that X is a triangular admissible $\mathcal{A}$-module, of the ditalgebra $\mathcal{A}$ with layer (R, W). Denote by (S, P) the splitting of the algebra $\Gamma = \mathrm{End}_{\mathcal{A}}(X)^{op}$. Assume that S_0 and R_0 are subalgebras of S and R, respectively. Then, X is called $[R_0, S_0]$-free iff the filtration of R_0-S-bimodules $\mathcal{F}'(X)$, which is obtained from $\mathcal{F}(X)$ by considering each term as an R_0-S-bimodule, is right S_0-free (see (15.5)). In the last requirement, $\mathcal{F}(X)$ is the filtration of B-S-bimodules realizing the triangularity of the module X (see condition (2) of definition (14.6)). The object X is called completely $[R_0, S_0]$-free iff X is $[R_0, S_0]$-free and the filtration $\mathcal{F}(P)$ of P is S_0-S_0-free (see (14.7)).*

Proposition 15.16. *Let $\mathcal{A}$ be a ditalgebra with an R_0-free triangular layer (R, W), where R_0 is a semisimple subalgebra of R such that $R_0 \otimes_k R_0^{op}$ is a semisimple algebra. Consider an initial subditalgebra $\mathcal{A}'$ of $\mathcal{A}$ and an admissible $\mathcal{A}'$-module X; denote by (S, P) the associated splitting of the algebra $\Gamma = \mathrm{End}_{\mathcal{A}'}(X)^{op}$. Furthermore, assume that S_0 is a semisimple subalgebra of S such that $S_0 \otimes_k S_0^{op}$ is a semisimple algebra, and that X is a triangular admissible $[R_0, S_0]$-free $\mathcal{A}'$-module. Then, the ditalgebra $\mathcal{A}^X = (T^X, \delta^X)$ admits an S_0-free triangular layer (S, W^X).*

Proof. Suppose that the initial subditalgebra $\mathcal{A}'$ of $\mathcal{A}$ is associated to the R-R-bimodule decompositions $W_0 = W_0' \oplus W_0''$ and $W_1 = W_1' \oplus W_1''$.

Since $\mathcal{A}$ is R_0-free, by (15.10) and (15.13), we know that that $\mathcal{F}(W_0)^{W_0''}$ is an R_0-R_0-free filtration, where $\mathcal{F}(W_0)$ is the filtration realizing the triangularity of the layer of $\mathcal{A}$ at degree zero (see (5.1)). Then, there is an additive R_0-R_0-bimodule filtration $\mathcal{F}(\widetilde{W}_0'')$ such that the product map induces an isomorphism of R-R-bimodule filtrations

$$R \otimes_{R_0} \mathcal{F}(\widetilde{W}_0'') \otimes_{R_0} R \xrightarrow{\cong} \mathcal{F}(W_0'') := \mathcal{F}(W_0)^{W_0''}.$$

Let us adopt the notations of the proof of (14.10), where we had $[W_0'']^s = W_0'' \cap W_0^{\ell_0'+s}$ and $\underline{W}_0^s = B[W_0'']^s B$, for $s \in [0, \ell_0'']$, where ℓ_0' and ℓ_0'' are the lengths of the filtrations $\mathcal{F}(W_0)_{W_0'}$ and $\mathcal{F}(W_0'')$. Then, we have the filtration $\mathcal{F}(W_0^X)$ of the S-S-bimodule W_0^X given by

$$[W_0^X]^m := \sum_{r+2\ell_X s+t \leq m} [X^*]_r \widehat{\otimes}_B \underline{W}_0^s \widehat{\otimes}_B X_t, \quad \text{for } m \in [0, 2\ell_X(\ell_0''+1)].$$

We have $\underline{W}_0^s = B[W_0'']^s B \cong B \otimes_R [W_0'']^s \otimes_R B$ because the filtration $\mathcal{F}(W_0'')$ is additive and, from (12.2), we know that $B \otimes_R W_0'' \otimes_R B \cong BW_0''B = \underline{W}_0$. Each $[X^*]_r \otimes_B \underline{W}_0^s \otimes_B X_t$ can be identified with its image $[X^*]_r \widehat{\otimes}_B \underline{W}_0^s \widehat{\otimes}_B X_t$ under the canonical map in $W_0^X = X^* \otimes_B \underline{W}_0 \otimes_B X$. Indeed, consider the isomorphism $\eta_{r,t}^s$ given by the composition of the following isomorphisms

$$\begin{aligned}
[X^*]_r \otimes_B \underline{W}_0^s \otimes_B X_t &\cong [X^*]_r \otimes_B \big(B \otimes_R [W_0'']^s \otimes_R B\big) \otimes_B X_t \\
&\cong [X^*]_r \otimes_B \big(B \otimes_R R \otimes_{R_0} [\widetilde{W}_0'']^s \otimes_{R_0} R \otimes_R B\big) \otimes_B X_t \\
&\cong [X^*]_r \otimes_{R_0} [\widetilde{W}_0'']^s \otimes_{R_0} X_t,
\end{aligned}$$

for $r, t \in [0, \ell_X]$ and $s \in [0, \ell_0'']$. Then, if we make $\eta := \eta_{\ell_X, \ell_X}^{\ell_0''}$, we have commutative squares of canonical maps

$$\begin{array}{ccc}
[X^*]_r \otimes_B \underline{W}_0^s \otimes_B [X]_t & \xrightarrow{\sigma} & X^* \otimes_B \underline{W}_0 \otimes_B X \\
\downarrow{\scriptstyle \eta_{r,t}^s} & & \downarrow{\scriptstyle \eta} \\
[X^*]_r \otimes_{R_0} [\widetilde{W}_0'']^s \otimes_{R_0} [X]_t & \xrightarrow{\widetilde{\sigma}} & X^* \otimes_{R_0} \widetilde{W}_0'' \otimes_{R_0} X.
\end{array}$$

Since R_0 is semisimple and $\mathcal{F}(\widetilde{W}''_0)$ is an additive filtration, then $\widetilde{\sigma}$ is injective. Then σ is injective too. Now it is clear that the isomorphism of S-S-bimodules

$$W_0^X = X^* \otimes_B \underline{W}_0 \otimes_B X \xrightarrow{\eta} X^* \otimes_{R_0} \widetilde{W}''_0 \otimes_{R_0} X =: \dddot{W}_0^X$$

translates the filtration $\mathcal{F}(W_0^X)$ onto the filtration $\mathcal{F}(\dddot{W}_0^X)$ defined by

$$[\dddot{W}_0^X]^m := \sum_{r+2\ell_X s+t\leq m} [X^*]_r \otimes_{R_0} [\widetilde{W}''_0]^s \otimes_{R_0} X_t, \quad \text{for } m \in [0, 2\ell_X(\ell''_0+1)].$$

Then, it will be enough to show that the filtration $\mathcal{F}(\dddot{W}_0^X)$ is S_0-S_0-free. By assumption, we also have an R_0-S_0-bimodule filtration

$$\mathcal{F}(\widetilde{X}){:}\ 0 = \widetilde{X}_0 \subseteq \widetilde{X}_1 \subseteq \cdots \subseteq \widetilde{X}_{\ell_X-1} \subseteq \widetilde{X}_{\ell_X} = \widetilde{X}$$

of some R_0-S_0-subbimodule $\widetilde{X}$ of X such that $\mathcal{F}(\widetilde{X}) \otimes_{S_0} S \cong \mathcal{F}'(X)$. Recall that the filtration $\mathcal{F}'(X)$ has the same terms of the filtration $\mathcal{F}(X)$, but considered as R_0-S-bimodules. Using that S_0 is semisimple and (15.4), we obtain the following chain of isomorphisms

$$\begin{aligned}[X^*]_r \otimes_{R_0} [\widetilde{W}''_0]^s \otimes_{R_0} X_t &\cong \left[\left(\widetilde{X} \otimes_{S_0} S\right)^*\right]_r \otimes_{R_0} [\widetilde{W}''_0]^s \otimes_{R_0} \left(\widetilde{X}_t \otimes_{S_0} S\right)\\ &\cong S \otimes_{S_0} [\widetilde{X}^*]_r \otimes_{R_0} [\widetilde{W}''_0]^s \otimes_{R_0} \widetilde{X}_t \otimes_{S_0} S.\end{aligned}$$

Then, we can consider the S_0-S_0-subbimodule $\widetilde{W}_0^X = \widetilde{X}^* \otimes_{R_0} \widetilde{W}''_0 \otimes_{R_0} \widetilde{X}$ of $\dddot{W}_0^X$, and its filtration $\mathcal{F}(\widetilde{W}_0^X)$ of S_0-S_0-bimodules defined by

$$[\widetilde{W}_0^X]^m := \sum_{r+2\ell_X s+t\leq m} [\widetilde{X}^*]_r \otimes_{R_0} [\widetilde{W}''_0]^s \otimes_{R_0} \widetilde{X}_t, \quad \text{for } m \in [0, 2\ell_X(\ell''_0+1)].$$

Then, we know that $S \otimes_{S_0} \mathcal{F}(\widetilde{W}_0^X) \otimes_{S_0} S \cong \mathcal{F}(\dddot{W}_0^X)$. The filtration $\mathcal{F}(\widetilde{W}_0^X)$ is additive because, by assumption, $S_0 \otimes_k S_0^{op}$ is semisimple. □

The following lemma clarifies the structure of $\mathcal{A}^X$: it provides a simpler description of W^X. However, it seems natural to start with the definition of W^X given in (12.7).

Lemma 15.17. *Assume that X is an admissible $\mathcal{A}'$-module, where $\mathcal{A}'$ is a proper subditalgebra of the layered ditalgebra $\mathcal{A}$. Consider the splitting (S, P) of $\Gamma = \mathrm{End}_{\mathcal{A}'}(X)^{op}$, the layer (R, W) of $\mathcal{A}$, and $B = [T']_0$ as before. Then*

$$\begin{array}{ccc}(X^* \otimes_B \underline{W}'_1 \otimes_B X) \oplus P^* & \xrightarrow{\Psi} & P^*\\ v \otimes w \otimes x + \gamma & \mapsto & \gamma^w_{v,x} + \gamma\end{array}$$

is a surjective morphism of S-S-bimodules with kernel $\mathcal{L}$. Therefore, we have an induced isomorphism $\widehat{\Psi} : [(X^* \otimes_B \underline{W}'_1 \otimes_B X) \oplus P^*]/\mathcal{L} \longrightarrow P^*$. Its inverse satisfies $\widehat{\Psi}^{-1}(\gamma) = \gamma$, for $\gamma \in P^*$.

Proof. We show first that Ψ is well defined. For $\nu \in X^*$, $w \in \underline{W}'_1$, $x \in X$, $b, b' \in B$ and $p \in P$, we have

$$\begin{aligned}\gamma^w_{\nu b, b'x}(p) &= \nu(bp^1(w)[b'x]) \\ &= \nu((bp^1(w)b')[x]) \\ &= \nu(p^1(bwb')[x]) \\ &= \gamma^{bwb'}_{\nu,x}(p).\end{aligned}$$

Ψ is a morphism of S-S-bimodules because, for $s, s' \in S$ and $p \in P$, we have

$$\begin{aligned}\gamma^w_{s'\nu,xs}(p) &= (s'\nu)[p^1(w)[xs]] \\ &= s'\left[\nu(p^1(w)[xs])\right] \\ &= s'\left[\nu(p^1(w)s^0[x])\right] \\ &= s'\left[\nu((p \circ s)^1(w)[x])\right] \\ &= s'\left[\nu((sp)^1(w)[x])\right] \\ &= s'\left[\gamma^w_{\nu,x}(sp)\right] \\ &= \left[s'(\gamma^w_{\nu,x})s\right](p).\end{aligned}$$

It is clear that $\mathcal{L} = \operatorname{Ker}(\Psi)$. □

Proposition 15.18. *Let $\mathcal{A}$ be a ditalgebra with a completely R_0-free triangular layer (R, W), where R_0 is a semisimple subalgebra of R such that $R_0 \otimes_k R_0^{op}$ is a semisimple algebra. Consider an initial subditalgebra $\mathcal{A}'$ of $\mathcal{A}$ and an admissible triangular $\mathcal{A}'$-module X; denote by (S, P) the associated splitting of the algebra $\Gamma = \operatorname{End}_{\mathcal{A}'}(X)^{op}$. Furthermore, assume that S_0 is a semisimple subalgebra of S such that $S_0 \otimes_k S_0^{op}$ is a semisimple algebra, and that X is a triangular completely $[R_0, S_0]$-free $\mathcal{A}'$-module. Then, the ditalgebra $\mathcal{A}^X = (T^X, \delta^X)$ admits a completely S_0-free triangular layer (S, W^X).*

Proof. Suppose that the initial subditalgebra $\mathcal{A}'$ of $\mathcal{A}$ is associated to the R-R-bimodule decompositions $W_0 = W'_0 \oplus W''_0$ and $W_1 = W'_1 \oplus W''_1$. Since $\mathcal{A}$ is completely R_0-free, by (15.10) and (15.13), $\mathcal{F}(W_0)^{W''_0}$ and $\mathcal{F}(W_1)^{W''_1}$ are R_0-R_0-free filtrations, where $\mathcal{F}(W_0)$ and $\mathcal{F}(W_1)$ are the filtrations realizing the triangularity of the layer of $\mathcal{A}$ (see (5.1)).

Let us keep the notation of the proof of (15.16).

Now we also have an additive R_0-R_0-bimodule filtration $\mathcal{F}(\widetilde{W}''_1)$ such that the product map induces an isomorphism of R-R-bimodule filtrations

$$R \otimes_{R_0} \mathcal{F}(\widetilde{W}''_1) \otimes_{R_0} R \xrightarrow{\cong} \mathcal{F}(W''_1) := \mathcal{F}(W_1)^{W''_1}.$$

Make $\widetilde{U}'' := X^* \otimes_{R_0} \widetilde{W}_1'' \otimes_{R_0} X$ and $\underline{U}'' := X^* \otimes_B \underline{W}_1'' \otimes_B X$. Recall that $\underline{W}_1^s = B[W_1'']^s B$, for $s \in [0, \ell_1'']$, and consider the R-R-bimodule filtrations $\mathcal{F}(\widetilde{U}'')$ and $\mathcal{F}(\underline{U}'')$ defined, respectively, by

$$[\widetilde{U}'']^m := \sum_{r+2\ell_X s+t \leq m} [X^*]_r \otimes_{R_0} [\widetilde{W}_1'']^s \otimes_{R_0} X_t, \quad \text{for } m \in [0, 2\ell_X(\ell_1''+1)],$$

and

$$[\underline{U}'']^m := \sum_{r+2\ell_X s+t \leq m} [X^*]_r \otimes_B \underline{W}_1^s \otimes_B X_t, \quad \text{for } m \in [0, 2\ell_X(\ell_1''+1)].$$

As in the proof of (15.16), we can verify that the canonical isomorphism $\widetilde{U}'' \xrightarrow{\cong} \underline{U}''$ determines an isomorphism of filtrations $\mathcal{F}(\widetilde{U}'') \xrightarrow{\cong} \mathcal{F}(\underline{U}'')$. Notice that the isomorphism described in (15.17) determines an isomorphism $\underline{U}'' \oplus P^* \longrightarrow W_1^X$, which maps the filtration $\mathcal{F}(P)^* \oplus \mathcal{F}(\underline{U}'')$ onto the filtration $\mathcal{F}(W_1^X)$. Then, from (15.11), we are reduced to show that the S-S-bimodule filtrations $\mathcal{F}(\underline{U}'')$ and $\mathcal{F}(P)^*$ are S_0-S_0-free. The same argument given in the proof of (15.16), proves that $\mathcal{F}(\widetilde{U}'')$ is S_0-S_0-free, and hence so is the filtration $\mathcal{F}(\underline{U}'')$.

Consider the S_0-S_0-bimodule $\widetilde{P}$ obtained from the S-S-bimodule P by restriction and, similarly, consider the S_0-S_0-bimodule filtration $\mathcal{F}(\widetilde{P})$ obtained from the S-S-bimodule filtration $\mathcal{F}(P)$ by restriction. Since $S_0 \otimes_k S_0^{op}$ is semisimple, $\mathcal{F}(\widetilde{P})$ is an additive filtration. The filtration $\mathcal{F}(P)$ is S_0-S_0-free because $\mathcal{F}(P) \cong (S \otimes_{S_0} \mathcal{F}(\widetilde{P}) \otimes_{S_0} S)$. Therefore

$$\begin{aligned}\mathcal{F}(P)^* &\cong (S \otimes_{S_0} \mathcal{F}(\widetilde{P}) \otimes_{S_0} S)^* \cong S \otimes_{S_0} (S \otimes_{S_0} \mathcal{F}(\widetilde{P}))^* \\ &\cong S \otimes_{S_0} \mathcal{F}(\widetilde{P})^* \otimes_{S_0} S.\end{aligned}$$

Thus, $\mathcal{F}(P)^*$ is an S_0-S_0-free filtration, and then $\mathcal{F}(W_1^X)$ is an S_0-S_0-free filtration too. □

Exercise 15.19. *Assume that R_0 is a subalgebra of the k-algebra R and assume that $\mathcal{F}(W) = \mathcal{F}(W') \oplus \mathcal{F}(W'')$ is an equality of R-R-bimodule filtrations. Moreover, suppose that $\mathcal{F}(W)$ is an R_0-R_0-free bimodule filtration. Then:*

(1) The filtration $\mathcal{F}(W') = \mathcal{F}(W)_{W'}$ is R_0-R_0-free.
(2) $\mathcal{F}(W)$ admits a decomposition $\mathcal{F}(W) = \mathcal{F}(W') \oplus \mathcal{F}(V'')$, where $\mathcal{F}(V'')$ is an R_0-R_0-free filtration of some R-R-subbimodule V'' of W.
(3) The R-R-subbimodule W'' of W admits an R_0-R_0-free filtration $\mathcal{F}'(W'')$.

References. The notion of an R_0-free ditalgebra given here is very similar to the one considered in [6]. In some sense, it generalizes the idea of "semifree bocs" (resp. "layered bocs") used by Drozd (resp. Crawley) in the proof of Drozd's Tame and Wild Theorem. In Section 23 we introduce the notions of "nested" or "seminested ditalgebra", which are more closely related to these ideas. In the application of ditalgebras (or bocses) to problems over non-algebraically closed ground fields, the free R_0-ditalgebras will very probably play an important role.

16

$\mathcal{A}^X$ is a Roiter ditalgebra, for suitable X

In this section, we show that, given a complete triangular admissible $\mathcal{A}'$-module X, where $\mathcal{A}'$ is an initial subditalgebra of the Roiter ditalgebra $\mathcal{A}$, we always obtain a Roiter ditalgebra $\mathcal{A}^X$.

Lemma 16.1. *Assume that we have an admissible $\mathcal{A}'$-module X, where $\mathcal{A}'$ is a proper subditalgebra of the layered ditalgebra $\mathcal{A}$. Consider the layered ditalgebra $\mathcal{A}^X$ and the associated functor $F^X : \mathcal{A}^X\text{-Mod}\longrightarrow\mathcal{A}\text{-Mod}$. Assume that M is an S-module such that there is an object $\overline{X \otimes_S M} \in \mathcal{A}\text{-Mod}$ having as underlying B-module structure the canonical structure on $X \otimes_S M$: where $b(x \otimes m) = bx \otimes m$, for $b \in B$, $x \in X$ and $m \in M$. Then, there exists a unique object $\overline{M} \in \mathcal{A}^X\text{-Mod}$ with underlying S-module M such that $F^X(\overline{M}) = \overline{X \otimes_S M}$. If $X^* \otimes_B X \xrightarrow{\varepsilon} S$ is the evaluation map and $S \otimes_S M \xrightarrow{\sigma} M$ is the product map, then the action $*$ of A^X on $\overline{M}$ is determined by the action $\circ$ of A on $\overline{X \otimes_S M}$ by the formula*

$$(\nu \otimes w \otimes x) * m = \sigma(\varepsilon \otimes 1)[\nu \otimes w \circ (x \otimes m)],$$

for $\nu \in X^*$, $w \in W_0''$, $x \in X$ and $m \in M$.

Proof. The evaluation map $X^* \otimes_B X \xrightarrow{\varepsilon} S$, given by $\varepsilon(\nu \otimes x) = \nu(x)$, for $\nu \in X^*$ and $x \in X$, is clearly a morphism of S-S-bimodules. Then, the composition

$$\begin{aligned} W_0^X \otimes_S M = X^* \otimes_B BW_0''B \otimes_B X \otimes_S M &\xrightarrow{1\otimes\circ} \\ X^* \otimes_B X \otimes_S M &\xrightarrow{\varepsilon\otimes 1} S \otimes_S M \xrightarrow{\sigma} M \end{aligned}$$

determines an A^X-module structure $*$ on the S-module M. The corresponding module is denoted by $\overline{M}$. We keep the notation $\cdot$ for the A-module structure of the object $F^X(\overline{M})$. To show that $F^X(\overline{M}) = \overline{X \otimes_S M}$, we have to show that $\cdot = \circ$. By assumption, we already know that, for $w \in W_0'$, $w \cdot (x \otimes m) = wx \otimes m = w \circ (x \otimes m)$. Write, for $w \in W_0''$, $w \circ (x \otimes m) = \sum_t x^t \otimes m^t$, with

$x^t \in X$ and $m^t \in M$. Then

$$\begin{aligned} w \cdot (x \otimes m) &= \textstyle\sum_i x_i \otimes (\nu_i \otimes w \otimes x) * m \\ &= \textstyle\sum_i x_i \otimes \sigma(\varepsilon \otimes 1)[\nu_i \otimes w \circ (x \otimes m)] \\ &= \textstyle\sum_{i,t} x_i \otimes \nu_i(x^t)m^t \\ &= \textstyle\sum_{i,t} x_i \nu_i(x^t) \otimes m^t \\ &= \textstyle\sum_t x^t \otimes m^t. \end{aligned}$$

Now assume that $\widehat{M}$ is another object of $\mathcal{A}^X$-Mod with underlying S-module M such that $F^X(\widehat{M}) = \overline{X \otimes_S M}$, and denote by $\circledast$ its A^X-module structure. Then, for $\nu \in X^*$, $w \in W_0''$, $x \in X$ and $m \in M$, we have

$$\begin{aligned} (\nu \otimes w \otimes x) * m &= \sigma(\varepsilon \otimes 1)[\nu \otimes w \circ (x \otimes m)] \\ &= \sigma(\varepsilon \otimes 1)[\nu \otimes w \cdot (x \otimes m)] \\ &= \sigma(\varepsilon \otimes 1)[\nu \otimes \left(\textstyle\sum_i x_i \otimes (\nu_i \otimes w \otimes x) \circledast m\right)] \\ &= \textstyle\sum_i (\nu(x_i)\nu_i \otimes w \otimes x) \circledast m \\ &= (\nu \otimes w \otimes x) \circledast m. \end{aligned}$$

Therefore, $\overline{M} = \widehat{M}$. □

Lemma 16.2. *Assume that we have an admissible $\mathcal{A}'$-module X, where $\mathcal{A}'$ is a proper subditalgebra of the layered ditalgebra $\mathcal{A}$. Consider the layered ditalgebra $\mathcal{A}^X$ and the associated functor $F^X : \mathcal{A}^X$-Mod⟶$\mathcal{A}$-Mod. Denote by $F_\circ$ the tensor product functor $X \otimes_S - : S$-Mod⟶R-Mod, and define*

$$\xi_X : \mathrm{Phom}_{S\text{-}W^X}(M, N) \longrightarrow \mathrm{Phom}_{R\text{-}W}(F_\circ(M), F_\circ(N)),$$

for $f = (f^0, f^1) \in \mathrm{Phom}_{S\text{-}W^X}(M, N)$, $w \in W_1$, $x \in X$ and $m \in M$, by the formulas

$$\begin{cases} \xi_X(f)^0(x \otimes m) = x \otimes f^0(m) + \sum_j xp_j \otimes f^1(\gamma_j)[m] \text{ and} \\ \xi_X(f)^1(w)[x \otimes m] = \sum_i x_i \otimes f^1(\nu_i \otimes w \otimes x)[m], \end{cases}$$

where $\xi_X(f) = (\xi_X(f)^0, \xi_X(f)^1)$. Then, the pair $(F_\circ, \xi_X)$ is a frame for the functor F^X (see definition (9.1)). For each $M, N \in S$-Mod, we have the commutative diagram

$$\begin{array}{ccc} \mathrm{Phom}_{S\text{-}W^X}(M, N) & \xrightarrow{\xi_X} & \mathrm{Phom}_{R\text{-}W}(F_\circ(M), F_\circ(N)) \\ \downarrow \widehat{\Phi} & & \downarrow \Phi \\ \mathrm{Phom}_{S\text{-}W'^X}(M, N) & \xrightarrow{\xi'_X} & \mathrm{Phom}_{R\text{-}W'}(F_\circ(M), F_\circ(N)), \end{array}$$

where $\widehat{\Phi}$ and Φ are the restriction maps defined by $\widehat{\Phi}(f^0, f^1) = (f^0, f^1_{|W_1'^X})$ and $\Phi(g^0, g^1) = (g^0, g^1_{|W_1'})$; and ξ'_X denotes the map corresponding to the functor $F'^X : \mathcal{A}'^X$-Mod⟶$\mathcal{A}'$-Mod.

If X is a complete triangular admissible $\mathcal{A}'$-module, where $\mathcal{A}'$ is an initial subditalgebra of the triangular ditalgebra $\mathcal{A}$, then ξ_X is injective. Moreover, if $f \in \mathrm{Phom}_{S\text{-}W^X}(M,N)$ with f^0 bijective, then $\Phi\xi_X(f) : F'^X(M) \longrightarrow F'^X(N)$ is an isomorphism of $\mathcal{A}'$-Mod with inverse $\Phi\xi_X(h) : F'^X(N) \longrightarrow F'^X(M)$, for some $h \in \mathrm{Phom}_{S\text{-}W^X}(N,M)$.

Proof. It is not hard to see that indeed $(F_\circ, \xi_X)$ is a frame for F^X and that the square in the statement of our lemma is commutative.

So assume that X is a complete triangular $\mathcal{A}'$-module, where $\mathcal{A}'$ is an initial subditalgebra of $\mathcal{A}$. From (14.9) and (14.10), $\mathcal{A}'$ as well as $\mathcal{A}'^X$ are triangular ditalgebras. Notice that, since $W_0'^X = 0$, then $\mathrm{Phom}_{S\text{-}W'^X}(M,N) \subseteq \mathrm{Hom}_{\mathcal{A}'^X}(M,N)$. Thus, $\mathcal{A}'^X$ is a Roiter ditalgebra.

Take $f = (f^0, f^1) \in \mathrm{Phom}_{S\text{-}W^X}(M,N)$, where f^0 is an isomorphism. Then, $\widehat{\Phi}(f) : M \longrightarrow N$ is an isomorphism in $\mathcal{A}'^X$-Mod and we can consider its inverse $h_|$. Notice that $h_|$ can be extended to $h = (h_|^0, h^1) \in \mathrm{Phom}_{S\text{-}W^X}(N,M)$, making $h^1(X^* \otimes_B \underline{W_1''} \otimes_B X) = 0$. Thus, $h_| = \widehat{\Phi}(h)$. Since F'^X is a functor, it follows that $\Phi\xi_X(f) = \xi'_X\widehat{\Phi}(f) = F'^X\widehat{\Phi}(f)$ is an isomorphism in $\mathcal{A}'$-Mod, with inverse $\Phi\xi_X(h) = \xi'_X\widehat{\Phi}(h) = F'^X\widehat{\Phi}(h)$.

Finally, if the assumption on f is that $\xi_X(f) = 0$, then $0 = \Phi\xi_X(f) = \xi'_X\widehat{\Phi}(f) = F'^X(\widehat{\Phi}(f))$. Since F'^X is faithful, we obtain that $\widehat{\Phi}(f) = 0$. Thus, $f^0 = 0$ and $f^1(W_1'^X) = 0$. But we also know, for $\nu \in X^*$, $w \in W_1''$, $x \in X$ and $m \in M$, that $\sum_i x_i \otimes f^1(\nu_i \otimes w \otimes x)[m] = \xi_X(f)^1(w)[x \otimes m] = 0$, and then, applying $\sigma(\nu \otimes 1)$ to this, we obtain $f^1(\nu \otimes w \otimes x)[m] = 0$. Hence $f = 0$. □

Proposition 16.3. *Assume that we have a complete triangular admissible $\mathcal{A}'$-module X, where $\mathcal{A}'$ is an initial subditalgebra of the Roiter ditalgebra $\mathcal{A}$. Then, $\mathcal{A}^X$ is a Roiter ditalgebra.*

Proof. Consider the frame $(F_\circ, \xi_X)$ of the functor $F^X : \mathcal{A}^X\text{-Mod} \longrightarrow \mathcal{A}\text{-Mod}$, constructed in last lemma, and let us show that (9.2) can be applied here. We already know that the statements (1), (2) and (3) of (9.2) hold. Now, we concentrate on (9.2)(4). Let $M \in S$-Mod and $\overline{X \otimes_S M} \in \mathcal{A}$-Mod be an object with underlying R-module $X \otimes_S M$. Take $N \in \mathcal{A}^X$-Mod and suppose that $\xi_X(f^0, f^1) \in \mathrm{Phom}_{R\text{-}W}(F_\circ(M), F_\circ(N))$ is an isomorphism $\overline{X \otimes_S M} \longrightarrow F^X(N)$ in $\mathcal{A}$-Mod. We must show that there exists an object $\overline{M} \in \mathcal{A}^X$-Mod with underlying S-module M such that $F^X(\overline{M}) = \overline{X \otimes_S M}$. From (16.1), it will be enough to show that the action of B on $\overline{X \otimes_S M}$, which we denote by $\cdot$, is the canonical action of B on $X \otimes_S M$. We want to show that $b \cdot (x \otimes m) = bx \otimes m$, for $b \in B$, $x \in X$ and $m \in M$. Denote by $(g^0, g^1) = (\xi_X(f)^0, \xi_X(f)^1) \in \mathrm{Hom}_{\mathcal{A}}(\overline{X \otimes_S M}, F^X(N))$. We know that g^0 :

$\overline{X \otimes_S M} \longrightarrow X \otimes_S N$ is an isomorphism of R-modules, then it will be enough to show that $g^0(b \cdot (x \otimes m)) = g^0(bx \otimes m)$. First notice that, for $\underline{w} \in \underline{W}_1$, we still have

$$g^1(\underline{w})[x \otimes m] = \sum_i x_i \otimes f^1(\nu_i \otimes \underline{w} \otimes x)[m],$$

where $g^1 \in \mathrm{Hom}_{B\text{-}B}(\underline{W}_1, \mathrm{Hom}_k(X \otimes_S M, X \otimes_S N))$. See, for instance, Case 1 in the proof of the claim in the proof of (12.10). Then

$$\begin{aligned}
&g^0(b \cdot (x \otimes m)) - g^0(bx \otimes m)\\
&\quad = bg^0(x \otimes m) - g^1(\delta(b))[x \otimes m] - g^0(bx \otimes m)\\
&\quad = bx \otimes f^0(m) + \textstyle\sum_j b(xp_j) \otimes f^1(\gamma_j)[m]\\
&\qquad - \textstyle\sum_i x_i \otimes f^1(\nu_i \otimes \delta(b) \otimes x)[m]\\
&\qquad - bx \otimes f^0(m) - \textstyle\sum_j (bx)p_j \otimes f^1(\gamma_j)[m]\\
&\quad = \textstyle\sum_j \left[b(xp_j) - (bx)p_j\right] \otimes f^1(\gamma_j)[m]\\
&\qquad - \textstyle\sum_i x_i \otimes f^1(\nu_i \otimes \delta(b) \otimes x)[m]\\
&\quad = \textstyle\sum_{j,i} x_i \nu_i \left[b(xp_j) - (bx)p_j\right] \otimes f^1(\gamma_j)[m]\\
&\qquad - \textstyle\sum_i x_i \otimes f^1(\nu_i \otimes \delta(b) \otimes x)[m]\\
&\quad = \textstyle\sum_i x_i \otimes f^1\left(\sum_j \nu_i \left[b(xp_j) - (bx)p_j\right] \gamma_j\right)[m]\\
&\qquad - \textstyle\sum_i x_i \otimes f^1(\gamma^b_{\nu_i,x})[m]\\
&\quad = 0.
\end{aligned}$$

The statement $(9.2)(4)^{op}$ is shown dually, using the last part of (16.2). Indeed, take $M \in S$-Mod and $\overline{X \otimes_S M} \in \mathcal{A}$-Mod, with underlying R-module $X \otimes_S M$. Then, take $N \in \mathcal{A}^X$-Mod and $f \in \mathrm{Phom}_{S\text{-}W^X}(N, M)$ such that $t = \xi_X(f) : F^X(N) \longrightarrow \overline{X \otimes_S M}$ is an isomorphism in $\mathcal{A}$-Mod. Then, as before, we only have to verify that B acts on $\overline{X \otimes_S M}$ by the rule $b \cdot (x \otimes m) = bx \otimes m$. Since t is invertible, so is $F_r(t) = F_r(\xi_X(f)) = \Phi\xi_X(f) = \xi'_X\widehat{\Phi}(f) = F'^X(\widehat{\Phi}(f))$. But, F'^X is full and faithful, hence $\widehat{\Phi}(f)$ is an isomorphism, which implies that f^0 is an isomorphism. From (16.2), we obtain that $\Phi\xi_X(f) : F'^X(N) \longrightarrow F'^X(M)$ is an isomorphism of $\mathcal{A}'$-Mod with inverse $\Phi\xi_X(h) : F'^X(M) \longrightarrow F'^X(N)$, for some $h \in \mathrm{Phom}_{S\text{-}W^X}(M, N)$. Consider the inverse $g : \overline{X \otimes_S M} \longrightarrow F^X(N)$ of t in $\mathcal{A}$-Mod. Thus, $F_r(g) = \Phi\xi_X(h)$. Thus, $g^0 = \xi_X(h)^0$ and $g^1_{|\underline{W}'_1} = \xi_X(h)^1$. Then, we can proceed as we did for case (4), replacing f by h in the last chain of equalities (where $\delta(b) \in \underline{W}'_1$). □

Exercise 16.4. *Assume that $\mathcal{A} = (T, \delta)$ is a Roiter ditalgebra with triangular layer (R, W). Suppose that $\mathcal{A}' = (T', \delta')$ is a proper subditalgebra of $\mathcal{A}$, associated to the decompositions $W_0 = W'_0 \oplus W''_0$ and $W_1 = W'_1 \oplus W''_1$. By construction, T' is the subalgebra of T generated by R and $W' = W'_0 \oplus W'_1$. Assume that the layer (R, W') of T' is triangular. From (12.3), we know that*

$\mathcal{A}'$ *is a Roiter ditalgebra. Consider the inclusion morphism of ditalgebras* $r : \mathcal{A}' \longrightarrow \mathcal{A}$ *and the associated restriction functor* $F_r : \mathcal{A}\text{-Mod} \longrightarrow \mathcal{A}'\text{-Mod}$. *Show that* F_r *is an exact functor, which induces an epimorphism of bifunctors*

$$F_r^* : \mathrm{Ext}_{\mathcal{A}}(-, ?) \longrightarrow \mathrm{Ext}_{\mathcal{A}'}(F_r(-), F_r(?)).$$

Discuss the correlation with exercise (9.7).

Hint: Consider the following commutative diagram

$$\begin{array}{ccc}
\mathrm{Phom}_{R\text{-}W}(M, N) & \xrightarrow{i_1^*} & \mathrm{Phom}_{R\text{-}W'}(F_r(M), F_r(N)) \\
\downarrow{\scriptstyle \partial_{\mathcal{A}}} & & \downarrow{\scriptstyle \partial_{\mathcal{A}'}} \\
\mathrm{Hom}_R(W_0 \otimes_R M, N) & \xrightarrow{i_0^*} & \mathrm{Hom}_R(W_0' \otimes_R F_r(M), F_r(N)) \\
\downarrow{\scriptstyle \psi_{\mathcal{A}}} & & \downarrow{\scriptstyle \psi_{\mathcal{A}'}} \\
\mathrm{Ext}_{\mathcal{A}}(M, N) & \xrightarrow{F_r^*} & \mathrm{Ext}_{\mathcal{A}'}(F_r(M), F_r(N)),
\end{array}$$

where the exact columns are given by (6.13); i_1^* *and* i_0^* *are induced by the* R-R*-bimodule sections* $i_1 : W_1' \longrightarrow W_1$ *and* $i_0 : W_0' \longrightarrow W_0$*, respectively.*

Exercise 16.5. *Under the same assumptions of exercise (16.4), show that whenever* $M \in \mathcal{A}\text{-Mod}$ *and* $N \in \mathcal{A}'\text{-Mod}$ *satisfy that* $F_r(M) \cong N$ *in* $\mathcal{A}'\text{-Mod}$, *there is* $\overline{N} \in \mathcal{A}\text{-Mod}$ *with* $\overline{N} \cong M$ *in* $\mathcal{A}\text{-Mod}$ *and* $F_r(\overline{N}) = N$.

Exercise 16.6. *Assume that we have a complete triangular* $\mathcal{A}'$*-module* X*, where* $\mathcal{A}'$ *is an initial subditalgebra of the Roiter ditalgebra* $\mathcal{A}$*. Then, the associated functor* $F_X : \mathcal{A}^X\text{-Mod} \longrightarrow \mathcal{A}\text{-Mod}$ *is exact and induces the following exact sequence of bifunctors*

$$\begin{aligned} 0 \rightarrow \mathrm{Ext}_{\mathcal{A}^X}(-, ?) &\xrightarrow{F_X^*} \mathrm{Ext}_{\mathcal{A}}(F_X(-), F_X(?)) \\ &\xrightarrow{F_r^*} \mathrm{Ext}_{\mathcal{A}'}(F_r F_X(-), F_r F_X(?)) \rightarrow 0. \end{aligned}$$

Hints: Use (5.14) to show that F_X *is exact. Use (6.21), (16.5) and (16.1) to show that any element in* $\mathrm{Ext}_{\mathcal{A}}(F_X(M), F_X(N))$ *mapped to zero by* F_r^* *is equivalent to a conflation of the form*

$$F_X(N) \xrightarrow{F_X(f)} F_X(E) \xrightarrow{F_X(g)} F_X(N).$$

Now, having in mind the commutative square of (13.4), consider the composable pair $\zeta : N \xrightarrow{f} E \xrightarrow{g} M$ *in* $\mathcal{A}^X\text{-Mod}$*, and apply (6.24) to* $G := F'^X$ *and the split conflation* $GF_{\hat{r}}(\zeta) = F_r F^X(\zeta)$ *to get that* $F_{\hat{r}}(\zeta)$ *is a split conflation in* $\mathcal{E}_{\mathcal{A}'^X}$. *Derive that* ζ *is a conflation in* $\mathcal{E}_{\mathcal{A}^X}$.

References. (16.6) generalizes (10.5) of [6].

17
Examples and applications

In this section, we describe some general settings where admissible modules appear. We will see also that the basic ditalgebra constructions presented before in Section 8 can be realized as particular cases of the construction $\mathcal{A} \mapsto \mathcal{A}^X$.

Lemma 17.1. *Assume that $\mathcal{A}$ is a Roiter ditalgebra with layer (R, W) such that W_1 is finitely generated R-R-bimodule. Suppose that $X = X_1 \oplus \cdots \oplus X_n$ is a finite direct sum of finite-dimensional non-isomorphic indecomposable modules $X_i \in \mathcal{A}$-Mod. Assume, furthermore, that each algebra $\Gamma_i = \mathrm{End}_{\mathcal{A}}(X_i)^{op}$ decomposes as $\Gamma_i = kId_{X_i} \oplus P_i$, where P_i is the Jacobson radical of Γ_i. Then, $\Gamma = \mathrm{End}_{\mathcal{A}}(X)^{op}$ admits a splitting (S, P), where P is the Jacobson radical of Γ, and X is an admissible $\mathcal{A}$-module.*

Proof. By (5.13), $\mathcal{A}$-mod is a Krull–Schmidt category. Thus, idempotents split and $\mathrm{End}_{\mathcal{A}}(M)$ is local, for every indecomposable $M \in \mathcal{A}$-mod. Hence, given a morphism $f : M \longrightarrow N$ between indecomposables in $\mathcal{A}$-mod, either f is an isomorphism or, for any morphism $g : N \longrightarrow M$, we have $gf \in \mathrm{radEnd}_{\mathcal{A}}(M)$ and $fg \in \mathrm{radEnd}_{\mathcal{A}}(N)$. Then, our assumptions guarantee that

$$\Gamma = \begin{pmatrix} \Gamma_1 & \cdots & \mathrm{Hom}_{\mathcal{A}}(X_n, X_1) \\ & \ddots & \\ \mathrm{Hom}_{\mathcal{A}}(X_1, X_n) & \cdots & \Gamma_n \end{pmatrix}$$

and

$$P = \begin{pmatrix} P_1 & \cdots & \mathrm{Hom}_{\mathcal{A}}(X_n, X_1) \\ & \ddots & \\ \mathrm{Hom}_{\mathcal{A}}(X_1, X_n) & \cdots & P_n \end{pmatrix}.$$

Consider the idempotents $e_i = (X \xrightarrow{\pi_i} X_i \xrightarrow{\sigma_i} X)$ associated to the direct sum decomposition $X = \oplus_i X_i$, and the subalgebra $S = ke_1 + \cdots + ke_n$ of Γ. Thus,

$\Gamma = S \oplus P$ is a splitting of Γ, where each element $s \in S$ has the form $s = (s^0, 0)$, since it is a linear combination of morphisms having zero second component. Then, X is an admissible $\mathcal{A}$-module. □

Remark 17.2. *In last lemma, if the layer (R, W) of $\mathcal{A}$ satisfies $W_1 = 0$, then we can replace kId_{X_i} by any division subalgebra S_i of Γ_i and we obtain the same conclusion.*

Corollary 17.3. *Assume that $\mathcal{A}$ is a Roiter ditalgebra with layer (R, W) such that W_1 is finitely generated R-R-bimodule. Suppose that the ground field k is algebraically closed. Assume that $X = X_1 \oplus \cdots \oplus X_n$ is a finite direct sum of finite-dimensional non-isomorphic indecomposable objects $X_i \in \mathcal{A}$-Mod. Then, $\Gamma = \mathrm{End}_{\mathcal{A}}(X)^{op}$ admits a splitting (S, P), where P is the Jacobson radical of Γ and X is an admissible $\mathcal{A}$-module.*

Proof. Since $\mathcal{A}$-mod is a Krull–Schmidt category and each $X_i \in \mathcal{A}$-mod is indecomposable, the algebra $\Gamma_i = \mathrm{End}_{\mathcal{A}}(X_i)^{op}$ is local and the quotient $D_i = \Gamma_i / P_i$, where P_i is the Jacobson radical of Γ_i, is a finite-dimensional division k-algebra. Since k is algebraically closed, $D_i = k$ and $\Gamma_i = kId_{X_i} \oplus P_i$. Now, we can apply (17.1). □

Lemma 17.4. *Assume that $\mathcal{A}$ a Roiter ditalgebra with layer (R, W), where W_1 is finitely generated as an R-R-bimodule. Assume that X is an admissible $\mathcal{A}$-module which is a direct sum of finite-dimensional non-isomorphic indecomposable objects in $\mathcal{A}$-Mod. Suppose that in the splitting (S, P) of the finite-dimensional k-algebra $\Gamma = \mathrm{End}_{\mathcal{A}}(X)^{op}$, P is the Jacobson radical of Γ. Then, X is a complete triangular $\mathcal{A}$-module. If R_0 is any subalgebra of R, then X is $[R_0, S]$-free. If, furthermore, $S \otimes_k S^{op}$ is semisimple, then X is completely $[R_0, S]$-free (see (15.12)).*

Proof. We have the canonical S-S-bimodule filtration $\mathcal{F}(P)$ of P given by the powers of the radical $0 = P^{(n)} \subseteq P^{(n-1)} \subseteq \cdots \subseteq P^{(1)} = P$, which satisfies $P^{(i)} P^{(j)} \subseteq P^{(i+j)}$, for all i, j. We also have the B-S-bimodule filtration $\mathcal{F}(X) : 0 = XP^{(n)} \subseteq XP^{(n-1)} \subseteq \cdots \subseteq XP^{(1)} = XP \subseteq X$, of length $\ell_X = n + 1$, where $B = [T]_0$. Since $\mathcal{A}$-mod is a Krull–Schmidt category and X is a finite direct sum of non-isomorphic indecomposable objects, then $S \cong \Gamma / P$ is a basic finite-dimensional semisimple k-algebra, thus both filtrations are right additive. Then, X is a triangular module, as in (14.6). X is complete by (13.3). The filtration ${}_{R_0}\mathcal{F}(X)$ is a right S-free filtration because ${}_{R_0}\mathcal{F}(X) \otimes_S S \cong {}_{R_0}\mathcal{F}(X)$.

If we now assume that $S \otimes_k S^{op}$ is semisimple, the filtration $\mathcal{F}(P)$ is additive. Thus, the filtration $\mathcal{F}(P)$ is an S-S-free bimodule filtration because, clearly, the product map determines an isomorphism of filtrations $S \otimes_S \mathcal{F}(P) \otimes_S S \cong \mathcal{F}(P)$. □

Lemma 17.5. *Assume that $\mathcal{A}$ is a ditalgebra with layer (R, W). Assume furthermore that $W_1 = 0$, and, therefore, $\delta(W_0) = 0$. Let $\varphi : A \longrightarrow S$ be an epimorphism in the category of k-algebras. Then, define X as the A-module obtained from the regular left S-module by restriction of scalars using φ. Then, X is a complete triangular $\mathcal{A}$-module. Moreover, if R_0 is a subalgebra of R and $S_0 = \varphi(R_0)$, then X is completely $[R_0, S_0]$-free.*

Proof. Since $\Gamma = \mathrm{End}_{\mathcal{A}}(X)^{op} = \mathrm{End}_A(X)^{op} = \mathrm{End}_S(S)^{op} = S \oplus P$, with $P = 0$, then X is an admissible $\mathcal{A}$-module (every $\mathcal{A}$-morphism has null second component because $W_1 = 0$). Moreover, $\mathcal{A}$-Mod can be identified with A-Mod and $\mathcal{A}^X$-Mod can be identified with S-Mod. The $\mathcal{A}$-module X is complete because the functor $F^X : \mathcal{A}^X\text{-Mod} \longrightarrow \mathcal{A}\text{-Mod}$ can be identified with the restriction functor $F_\varphi : S\text{-Mod} \longrightarrow A\text{-Mod}$. Since F_φ is full and faithful, so is F^X.

The $\mathcal{A}$-module X is trivially triangular with filtrations $\mathcal{F}(X) : 0 \subseteq S = X$ and $\mathcal{F}(P) : 0 = P$. Take $\widetilde{X} = S_0$ and consider the R_0-S_0-filtration $\mathcal{F}(\widetilde{X}) : 0 \subseteq S_0 = \widetilde{X}$. Then, $\mathcal{F}(\widetilde{X}) \otimes_{S_0} S \cong {}_{R_0}\mathcal{F}(X)$ (see (15.15)). □

Definition 17.6. *Given a functor $F_z : \mathcal{A}^z\text{-Mod} \longrightarrow \mathcal{A}\text{-Mod}$, between categories of modules over layered ditalgebras, we say that $(\mathcal{A}^z, F_z)$ is (completely) realizable by an admissible module iff there exists a (complete) admissible $\mathcal{A}'$-module X, over some proper subditalgebra $\mathcal{A}'$ of $\mathcal{A}$, and an isomorphism $\theta : \mathcal{A}^z \longrightarrow \mathcal{A}^X$ of layered ditalgebras such that the functors $F^X, F_z F_\theta : \mathcal{A}^X\text{-Mod} \longrightarrow \mathcal{A}\text{-Mod}$ are isomorphic.*

Corollary 17.7. *Given a ditalgebra $\mathcal{A}$ with layer (R, W) and an epimorphism in the category of k-algebras $\varphi : R \longrightarrow S$. Consider the S-S-bimodule $W_i^S := S \otimes_R W_i \otimes_R S$, for $i \in \{0, 1\}$, and form $W^S = W_0^S \oplus W_1^S$. Then, φ and the canonical morphism $W \to W^S$, which maps each $w \in W$ onto $1 \otimes w \otimes 1$, induce a morphism of t-algebras $\overline{\varphi} : T \to T^S = T_S(W^S)$. The formula $\delta^S(s' \otimes w \otimes s) = \sigma_{s',s}(\delta(w))$, for $s, s' \in S$ and $w \in \langle W \rangle$, where $\sigma_{s',s} : \langle W \rangle \to T^S$ is the insertion map given by $\sigma_{s',s}(w_1 \cdots w_n) = s' \otimes w_1 \otimes 1 \otimes 1 \otimes w_2 \otimes \cdots \otimes 1 \otimes 1 \otimes w_n \otimes s$, for $w_1, \ldots, w_n \in W$, extends to a differential δ^S on T^S in such a way that*

$$\mathcal{A} = (T, \delta) \xrightarrow{\overline{\varphi}} (T^S, \delta^S) = \mathcal{A}^S$$

is a morphism of layered ditalgebras whose associated restriction functor $F_{\overline{\varphi}} : \mathcal{A}^S\text{-Mod} \longrightarrow \mathcal{A}\text{-Mod}$ is full and faithful. Moreover, $(\mathcal{A}^S, F_{\overline{\varphi}})$ is completely realizable by an admissible module.

Proof. We can consider the algebra R as a proper subditalgebra of $\mathcal{A}$. Namely, consider the proper subditalgebra $\mathcal{A}'$ of $\mathcal{A}$ associated to the trivial R-R-bimodule decompositions $W_r = W_r' \oplus W_r''$, where $W_r' = 0$, for $r \in \{0, 1\}$. Thus $\mathcal{A}'$ is a ditalgebra with layer $(R, 0)$ and we can identify R-Mod with $\mathcal{A}'$-Mod.

Now we can look at the object $X := S \in \mathcal{A}'\text{-Mod} = R\text{-Mod}$, obtained from the regular S-module by restriction of scalars using $\varphi : B = R \longrightarrow S$. From the last lemma, we know that X is a complete admissible $\mathcal{A}'$-module. Then, the functor $F^X : \mathcal{A}^X\text{-Mod} \longrightarrow \mathcal{A}\text{-Mod}$ will be full and faithful.

Under our assumptions above, we now show that the ditalgebra $\mathcal{A}^X$ admits the natural realization described in the statement of this corollary. Notice that we have canonical isomorphisms $S \otimes_R W_i \otimes_R S \longrightarrow X^* \otimes_R W_i \otimes_R X$, which induce an isomorphism $\theta : T^S \longrightarrow T^X$ of layered t-algebras. It is convenient to use the dual basis $(1, Id_S)$ for the projective right S-module X. Here, the differential of $\mathcal{A}^X$ has a simple expression because $P = 0$ and, therefore, λ and ρ are trivial. Moreover, we can check that $\delta^X \theta = \theta \delta^S$, by evaluating directly at generators of T^S in W^S. Then, $\theta : \mathcal{A}^S \longrightarrow \mathcal{A}^X$ is an isomorphism of layered ditalgebras.

Finally, we have to show an isomorphism of functors $F^X \cong F_{\theta\overline{\varphi}}$. For $M \in \mathcal{A}^X\text{-Mod}$, consider the product map $\sigma_M^0 : F^X(M) = S \otimes_S M \longrightarrow M = F_{\theta\overline{\varphi}}(M)$, which is clearly an isomorphism of R-modules. It is an isomorphism of A-modules because, for $w \in W_0$, $x \in X$ and $m \in M$, any $\sigma_M^0(w \cdot (x \otimes m)) = \sigma_M^0(1 \otimes (Id_S \otimes w \otimes x) * m) = (Id_S \otimes w \otimes x) * m = (Id_S \otimes w \otimes 1) * (xm) = (Id_S \otimes w \otimes 1) * \sigma_M^0(x \otimes m) \;\; = w\sigma_M^0(x \otimes m)$. We claim that $\sigma_M = (\sigma_M^0, 0) : F^X(M) \longrightarrow F_{\theta\overline{\varphi}}(M)$ determines an isomorphism of functors $\sigma : F^X \longrightarrow F_{\theta\overline{\varphi}}$. Take $f = (f^0, f^1) \in \mathrm{Hom}_{\mathcal{A}^X}(M, N)$ and let us see that $\sigma_N F^X(f) = F_{\theta\overline{\varphi}}(f)\sigma_M$. This is equivalent to show that $\sigma_N^0 F^X(f)^0 = F_{\theta\overline{\varphi}}(f)^0 \sigma_M^0$ and $\sigma_N^0 F^X(f)^1 = F_{\theta\overline{\varphi}}(f)^1 \sigma_M^0$. The first equality is clear, for the second, take $w \in W_1$, $x \in X$ and $m \in M$, then

$$\begin{aligned}
\sigma_N^0 F^X(f)^1(w)[x \otimes m] &= \sigma_N^0(1 \otimes f^1(Id_S \otimes w \otimes x)[m]) \\
&= f^1(Id_S \otimes w \otimes x)[m] \\
&= f^1(Id_S \otimes w \otimes 1)[xm] \\
&= f^1\theta(1 \otimes w \otimes 1)[xm] \\
&= f^1\theta\overline{\varphi}(w)[xm] \\
&= F_{\theta\overline{\varphi}}(f)^1(w)[xm] \\
&= F_{\theta\overline{\varphi}}(f)^1(w)\sigma_M^0[x \otimes m].
\end{aligned}$$

□

Remark 17.8. *Given a ditalgebra $\mathcal{A}$ with layer (R, W) and an ideal I_R of R, we can consider the canonical projection $\varphi : R \longrightarrow R/I_R$ and the corresponding ditalgebra $\mathcal{A}^{R/I_R}$ constructed above. The isomorphism of R/I_R-R/I_R-bimodules $R/I_R \otimes_R W \otimes_R R/I_R \longrightarrow W/(I_R W + W I_R)$ yields an isomorphism of layered ditalgebras $\mathcal{A}^{R/I_R} \longrightarrow \mathcal{A}^s$, where $\mathcal{A}^s$ is the layered ditalgebra obtained from $\mathcal{A}$ by factoring out the ideal I_R. Moreover, the pair $(\mathcal{A}^s, F_s)$ defined in (8.16) is completely realizable by an admissible module.*

Corollary 17.9. *Given a ditalgebra $\mathcal{A} = (T, \delta)$ with layer (R, W) and an R-R-bimodule decomposition $W_0 = W_0' \oplus W_0''$ with $\delta(W_0') = 0$, we can consider the ditalgebra $\mathcal{A}^a$, obtained from $\mathcal{A}$ by absorption of the subbimodule W_0', and the associated functor $F_a : \mathcal{A}^a\text{-Mod} \longrightarrow \mathcal{A}\text{-Mod}$, as in (8.20). Then, the pair $(\mathcal{A}^a, F_a)$ is completely realizable by an admissible module.*

Proof. Consider the R-R-bimodule decomposition $W_1 = W_1' \oplus W_1''$, where $W_1' = 0$. Then, we can consider the proper subditalgebra $\mathcal{A}'$ associated to this decomposition. Having in mind the notation of the definition of $\mathcal{A}^a$ and $\mathcal{A}^X$, notice that $B = R^a$ is a subalgebra of T freely generated by R and W_0'. Now take X as the regular B-module, which is a complete admissible $\mathcal{A}'$-module by (17.5). We have the associated splitting, $(S, P) = (B, 0)$, for $\mathrm{End}_{\mathcal{A}'}(X)^{op}$. Take the dual basis $(1, Id_B)$ of the projective right S-module X. Then, consider the isomorphisms $W_0^X = X^* \otimes_B \underline{W}_0 \otimes_B X \cong X^* \otimes_R W_0'' \otimes_R X \cong B \otimes_R W_0'' \otimes_R B \cong BW_0''B = W_0^a$ and $W_1^X = X^* \otimes_B \underline{W}_1 \otimes_B X \cong X^* \otimes_R W_1 \otimes_R X \cong B \otimes_R W_1 \otimes_R B \cong BW_1B = W_1^a$, which give us the isomorphism $\theta : W^X \longrightarrow W^a$, see (12.2). It extends to a morphism of algebras $\overline{\theta} : T^X \longrightarrow T^a$. It is easy to see that $\overline{\theta}\delta^X = \delta\overline{\theta}$, and, therefore, we have an isomorphism of layered ditalgebras $\overline{\theta} : \mathcal{A}^X \longrightarrow \mathcal{A}^a$. Now, for $M \in \mathcal{A}^X\text{-Mod}$, consider the isomorphism of B-modules $\sigma^0 : F^X(M) = B \otimes_B M \longrightarrow M = F_a F_{\overline{\theta}^{-1}}(M)$. It can be seen that $\sigma = (\sigma^0, 0) : F^X \longrightarrow F_a F_{\overline{\theta}^{-1}}$ is an isomorphism of functors. □

Remark 17.10. *Assume that $\mathcal{A}$ is a layered ditalgebra and X is an admissible $\mathcal{A}$-module. Denote by (S, P) the splitting of the algebra $\Gamma = \mathrm{End}_{\mathcal{A}}(X)^{op}$. In the following, $\mathcal{I}_X$ will denote the class of A-modules of the form $X \otimes_S N$, for some $N \in S\text{-Mod}$. From (16.1), we know that for a complete admissible $\mathcal{A}$-module X, the full subcategory of $\mathcal{A}\text{-Mod}$ with class of objects $\mathcal{I}_X$ coincides with the image of F^X.*

Lemma 17.11. *Assume that $\mathcal{A}$ is a ditalgebra with layer (R, W). Suppose that X_1 and X_2 are admissible $\mathcal{A}$-modules. Assume that* $\mathrm{Hom}_{\mathcal{A}}(\mathcal{I}_{X_1}, \mathcal{I}_{X_2}) = 0$

and $\operatorname{Hom}_{\mathcal{A}}(\mathcal{I}_{X_2}, \mathcal{I}_{X_1}) = 0$. *Then,* $X = X_1 \oplus X_2$ *is an admissible* $\mathcal{A}$*-module. If* X_1 *and* X_2 *are complete (or triangular), so is* X. *If we have the splittings* $\operatorname{End}_{\mathcal{A}}(X_i)^{op} = S_i \oplus P_i$, *for* $i \in \{1, 2\}$, *then* $\operatorname{End}_{\mathcal{A}}(X)^{op}$ *has the splitting* (S, P), *where* S *and* P *can be identified with* $S_1 \times S_2$ *and* $P_1 \times P_2$, *respectively. Moreover, if* R_0 *and* $(S_i)_0$ *are subalgebras of* R *and* S_i, *respectively, and* X_i *is (completely)* $[R_0, (S_i)_0]$*-free, for* $i \in \{1, 2\}$, *then* X *is (completely)* $[R_0, S_0]$*-free, where* $S_0 = (S_1)_0 \times (S_2)_0$.

Proof. Assume that we have the splittings $\operatorname{End}_{\mathcal{A}}(X_i)^{op} = S_i \oplus P_i$, for $i \in \{1, 2\}$. Our assumptions guarantee that

$$\Gamma = \operatorname{End}_{\mathcal{A}}(X)^{op} \cong \begin{pmatrix} \operatorname{End}_{\mathcal{A}}(X_1)^{op} & 0 \\ 0 & \operatorname{End}_{\mathcal{A}}(X_2)^{op} \end{pmatrix} \cong (S_1 \times S_2) \oplus (P_1 \times P_2),$$

and then X is an admissible $\mathcal{A}$-module. We can identify $S_1 \times S_2$ and $P_1 \times P_2$ with their images S and P, respectively, in Γ under the last isomorphism.

Now, we want to show that $F^X : \mathcal{A}^X\text{-Mod} \longrightarrow \mathcal{A}\text{-Mod}$ is full and faithful, knowing that $F^{X_i} : \mathcal{A}^{X_i}\text{-Mod} \longrightarrow \mathcal{A}\text{-Mod}$ are so for $i \in \{1, 2\}$.

We claim that $\mathcal{A}^X \cong \mathcal{A}^{X_1} \times \mathcal{A}^{X_2}$. Indeed, $\mathcal{A}^X$ has layer $(S_1 \times S_2, W^X)$, where $W_0^X = 0$ and $W_1^X = \left[\left(X^* \otimes_B \underline{W}_1 \otimes_B X\right) \oplus P^*\right] / \mathcal{L}$. Similarly, $\mathcal{A}^{X_i}$ has layer (S_i, W^{X_i}), where $W_0^{X_i} = 0$ and $W_1^{X_i} = \left[\left(X_i^* \otimes_B \underline{W}_1 \otimes_B X_i\right) \oplus P_i^*\right] / \mathcal{L}_i$. The inclusions $\left(X_i^* \otimes_B \underline{W}_1 \otimes_B X_i\right) \oplus P_i^* \longrightarrow \left(X^* \otimes_B \underline{W}_1 \otimes_B X\right) \oplus P^*$ induce morphisms $W_1^{X_i} \longrightarrow W_1^X$, which determine the morphism θ in the following commutative square of isomorphisms

$$\begin{array}{ccc} W_1^{X_1} \times W_1^{X_2} & \xrightarrow{\theta} & W_1^X \\ \Big\downarrow {\scriptstyle \widehat{\Psi}_1 \times \widehat{\Psi}_2} & & \Big\downarrow {\scriptstyle \widehat{\Psi}} \\ P_1^* \times P_2^* & \xrightarrow{\cong} & P^*, \end{array}$$

where the isomorphisms $\widehat{\Psi}_1$, $\widehat{\Psi}_2$ and $\widehat{\Psi}$ are given in (15.17).

Since $T^{X_1} \times T^{X_2}$ is freely generated by $(S_1 \times S_2, W_1^{X_1} \times W_1^{X_2})$, see (10.1), the morphism θ extends to an isomorphism of algebras $\overline{\theta} : T^{X_1} \times T^{X_2} \longrightarrow T^X$. Let us identify $\operatorname{Hom}_S(P, S)$ with $\operatorname{Hom}_{S_1}(P_1, S_1) \oplus \operatorname{Hom}_{S_2}(P_2, S_2)$. Then, if $\{(p_{j_r}, \gamma_{j_r}) \mid j_r \in J_r\}$ is a dual basis for P_r over S_r, for $r \in \{1, 2\}$, we have that $\{(p_{j_r}, \gamma_{j_r}) \mid j_r \in J_r, r \in \{1, 2\}\}$ is a dual basis for $P_1 \times P_2$ over S. Similarly, we can construct a dual basis $\{(x_{i_r}, \nu_{i_r}) \mid i_r \in I_r, r \in \{1, 2\}\}$, for X over S. We claim that the exterior rectangle of the following diagram commutes

$$\begin{array}{ccccc} P^* & \xrightarrow{\widehat{\Psi}^{-1}} & W_1^X & \xrightarrow{\delta^X} & T^X \\ \| & & \uparrow{\scriptstyle \theta} & & \uparrow{\scriptstyle \overline{\theta}} \\ P_1^* \times P_2^* & \xrightarrow{\widehat{\Psi}_1^{-1}\times\widehat{\Psi}_2^{-1}} & W_1^{X_1} \times W_1^{X_2} & \xrightarrow{\delta^{X_1}\times\delta^{X_2}} & T^{X_1} \times T^{X_2}. \end{array}$$

Indeed, by (11.8), if $\gamma \in P^*$ and $\gamma = \gamma^1 + \gamma^2$ with $\gamma^r \in P_r^*$, for $r \in \{1, 2\}$, then

$$\begin{aligned} \delta^X \widehat{\Psi}^{-1}(\gamma) &= \mu(\gamma) \\ &= \textstyle\sum_{i_r, j_s, r, s} \gamma(p_{i_r} p_{j_s}) \gamma_{j_s} \otimes \gamma_{i_r} \\ &= \textstyle\sum_{i_r, j_r, r} \gamma(p_{i_r} p_{j_r}) \gamma_{j_r} \otimes \gamma_{i_r} \\ &= \overline{\theta}\left[\textstyle\sum_{i_1, j_1} \gamma(p_{i_1} p_{j_1}) \gamma_{j_1} \otimes \gamma_{i_1} + \sum_{i_2, j_2} \gamma(p_{i_2} p_{j_2}) \gamma_{j_2} \otimes \gamma_{i_2}\right] \\ &= \overline{\theta}\left[\mu_1(\gamma^1) + \mu_2(\gamma^2)\right] \\ &= \overline{\theta}(\delta^{X_1} \times \delta^{X_2})(\widehat{\Psi}_1^{-1} \times \widehat{\Psi}_2^{-1})[\gamma]. \end{aligned}$$

Then, the right-hand square of last diagram commutes, and $\mathcal{A}^{X_1} \times \mathcal{A}^{X_2} \xrightarrow{\overline{\theta}} \mathcal{A}^X$ is an isomorphism of layered ditalgebras.

Now, the functor $\mathcal{A}^{X_1}\text{-Mod} \times \mathcal{A}^{X_2}\text{-Mod} \xrightarrow{F^{X_1} \oplus F^{X_2}} \mathcal{A}^X\text{-Mod}$ is faithful because F^{X_1} and F^{X_2} are so, see (10.3). The hypothesis on $\mathcal{I}_{X_1}$ and $\mathcal{I}_{X_2}$ guarantee that, given $M_i, N_i \in \mathcal{A}^{X_i}\text{-Mod}$, $\mathrm{Hom}_{\mathcal{A}}(X_i \otimes_{S_i} M_i, X_j \otimes_{S_j} N_j) = 0$, for $i \neq j$, and this, together with the fullness of F^{X_1} and F^{X_2}, imply that $F^{X_1} \oplus F^{X_2}$ is full too. We claim that there is an isomorphism of functors between $F^{X_1} \oplus F^{X_2}$ and the composition

$$\mathcal{A}^{X_1}\text{-Mod} \times \mathcal{A}^{X_2}\text{-Mod} \xrightarrow{F_1 \oplus F_2} \left(\mathcal{A}^{X_1} \times \mathcal{A}^{X_2}\right)\text{-Mod} \xrightarrow{F_{\overline{\theta}^{-1}}} \mathcal{A}^X\text{-Mod} \xrightarrow{F^X} A\text{-Mod},$$

where $F_i : \mathcal{A}^{X_i}\text{-Mod} \longrightarrow \left(\mathcal{A}^{X_1} \times \mathcal{A}^{X_2}\right)\text{-Mod}$ is the functor induced by the projection map $\pi_i : \mathcal{A}^{X_1} \times \mathcal{A}^{X_2} \longrightarrow \mathcal{A}^{X_i}$. Denote the composition $F_{\overline{\theta}^{-1}}(F_1 \oplus F_2)$ by G. For $M_1 \in \mathcal{A}^{X_1}\text{-Mod}$ and $M_2 \in \mathcal{A}^{X_2}\text{-Mod}$, put $M = F_1(M_1) \oplus F_2(M_2)$ and consider the canonical isomorphism of A-modules

$$\begin{aligned} F^X G(M_1, M_2) = X \otimes_S M &\xrightarrow{\sigma^0} (X_1 \otimes_{S_1} M_1) \oplus (X_2 \otimes_{S_2} M_2) \\ &= (F^{X_1} \oplus F^{X_2})(M_1, M_2). \end{aligned}$$

We shall denote for $x \in X = X_1 \oplus X_2$ and $m \in M = F_1(M_1) \oplus F_2(M_2)$, by x_i and m_i the i-component of x and m, respectively. Thus, $\sigma^0(x \otimes m) = x_1 \otimes m_1 + x_2 \otimes m_2$. We claim that $\sigma = (\sigma^0, 0) : F^X G \longrightarrow F^{X_1} \oplus F^{X_2}$ is an isomorphism of functors. In order to verify this, take $f_1 \in \mathrm{Hom}_{\mathcal{A}_1}(M_1, N_1)$ and

$f_2 \in \mathrm{Hom}_{\mathcal{A}_2}(M_2, N_2)$. Thus

$$G(f_1, f_2)^0 = \begin{pmatrix} F_1(f_1)^0 & 0 \\ 0 & F_2(f_2)^0 \end{pmatrix} = \begin{pmatrix} f_1^0 & 0 \\ 0 & f_2^0 \end{pmatrix},$$

and, for $v \in W_1^X$

$$G(f_1, f_2)^1[v] = \begin{pmatrix} F_1(f_1)^1[\overline{\theta}^{-1}(v)] & 0 \\ 0 & F_2(f_2)^1[\overline{\theta}^{-1}(v)] \end{pmatrix}$$

$$= \begin{pmatrix} f_1^1[\pi_1\overline{\theta}^{-1}(v)] & 0 \\ 0 & f_2^1[\pi_2\overline{\theta}^{-1}(v)] \end{pmatrix}.$$

Then

$$\begin{aligned}
\sigma^0\left(F^X G(f_1, f_2)\right)^0[x \otimes m] &= \sigma^0\left(x \otimes G(f_1, f_2)^0[m]\right) \\
&\quad + \sigma^0\left(\textstyle\sum_{j_r, r} x p_{j_r} \otimes G(f_1, f_2)^1(\gamma_{j_r})[m]\right) \\
&= \sigma^0\left(x \otimes \left[f_1^0(m_1) + f_2^0(m_2)\right]\right) \\
&\quad + \sigma^0\left(\textstyle\sum_{j_r, r} x p_{j_r} \otimes f_1^1(\pi_1\overline{\theta}^{-1}(\gamma_{j_r}))[m_1]\right) \\
&\quad + \sigma^0\left(\textstyle\sum_{j_r, r} x p_{j_r} \otimes f_2^1(\pi_2\overline{\theta}^{-1}(\gamma_{j_r}))[m_2]\right) \\
&= \sigma^0\left(x \otimes f_1^0(m_1) + \textstyle\sum_{j_1} x p_{j_1} \otimes f_1^1(\gamma_{j_1})[m_1]\right) \\
&\quad + \sigma^0\left(x \otimes f_2^0(m_2) + \textstyle\sum_{j_2} x p_{j_2} \otimes f_2^1(\gamma_{j_2})[m_2]\right) \\
&= x_1 \otimes f_1^0(m_1) + \textstyle\sum_{j_1} x_1 p_{j_1} \otimes f_1^1(\gamma_{j_1})[m_1] \\
&\quad + x_2 \otimes f_2^0(m_2) + \textstyle\sum_{j_2} x_2 p_{j_2} \otimes f_2^1(\gamma_{j_2})[m_2] \\
&= \left[F^{X_1}(f_1) \oplus F^X(f_2)\right]^0 \sigma^0[x \otimes m].
\end{aligned}$$

Moreover, notice that, for $w \in W_1, x \in X, r, s \in \{1, 2\}$ and $i_r \in I_r$, we have that $\gamma^w_{v_{i_r}, x_s} = 0$ whenever $r \neq s$. Indeed, recall first that, by assumption, any $p \in P$ splits as $p = p_1 \oplus p_2$, for some $p_1 \in P_1$ and $p_2 \in P_2$. Hence, for $w \in W_1$ and $x \in X$

$$\begin{aligned}
p^1(w)[x] = (p_1 \oplus p_2)^1(w)[x] &= \begin{pmatrix} p_1^1(w) & 0 \\ 0 & p_2^1(w) \end{pmatrix} \begin{pmatrix} x_1 \\ x_2 \end{pmatrix} \\
&= p_1^1(w)[x_1] + p_2^1(w)[x_2].
\end{aligned}$$

Then, if $r \neq s$, we have $\gamma^w_{\nu_{i_r},x_s}(p) = \nu_{i_r}(p^1(w)[x_s]) = \nu_{i_r}(p^1_s(w)[x_s]) = 0$. Then

$$
\begin{aligned}
&\sigma^0\left[F^X G(f_1, f_2)\right]^1(w)[x \otimes m]\\
&\quad = \sigma^0\left(\textstyle\sum_{i_r,r} x_{i_r} \otimes G(f_1, f_2)^1(\nu_{i_r} \otimes w \otimes x)[m]\right)\\
&\quad = \sigma^0\left(\textstyle\sum_{i_r,r,s} x_{i_r} \otimes G(f_1, f_2)^1(\nu_{i_r} \otimes w \otimes x_s)[m]\right)\\
&\quad = \sigma^0\left(\textstyle\sum_{i_r,r,s} x_{i_r} \otimes G(f_1, f_2)^1(\gamma^w_{\nu_{i_r},x_s})[m]\right)\\
&\quad = \sigma^0\left(\textstyle\sum_{i_r,r} x_{i_r} \otimes G(f_1, f_2)^1(\gamma^w_{\nu_{i_r},x_r})[m]\right)\\
&\quad = \sigma^0\left(\textstyle\sum_{i_r,r} x_{i_r} \otimes f_1^1(\pi_1\overline{\theta}^{-1}(\gamma^w_{\nu_{i_r},x_r}))[m_1]\right)\\
&\qquad + \sigma^0\left(\textstyle\sum_{i_r,r} x_{i_r} \otimes f_2^1(\pi_2\overline{\theta}^{-1}(\gamma^w_{\nu_{i_r},x_r}))[m_2]\right)\\
&\quad = \textstyle\sum_{i_1} x_{i_1} \otimes f_1^1(\gamma^w_{\nu_{i_1},x_1})[m_1]\\
&\qquad + \textstyle\sum_{i_2} x_{i_2} \otimes f_2^1(\gamma^w_{\nu_{i_2},x_2})[m_2]\\
&\quad = \textstyle\sum_{i_1} x_{i_1} \otimes f_1^1(\nu_{i_1} \otimes w \otimes x_1)[m_1]\\
&\qquad + \textstyle\sum_{i_2} x_{i_2} \otimes f_2^1(\nu_{i_2} \otimes w \otimes x_2)[m_2]\\
&\quad = F^{X_1}(f_1)^1(w)[x_1 \otimes m_1]\\
&\qquad + F^{X_2}(f_2)^1(w)[x_2 \otimes m_2]\\
&\quad = \left[F^{X_1}(f_1) \oplus F^{X_2}(f_2)\right]^1(w)\sigma^0[x \otimes m].
\end{aligned}
$$

We have shown the existence of an isomorphism of functors $F^X G \cong F^{X_1} \oplus F^{X_2}$. From (10.3) we know that G is an equivalence of categories. It follows that F^X is full and faithful, which means that X is a complete admissible $\mathcal{A}$-module.

Now assume that X_i is triangular and let $\mathcal{F}(X_i)$ and $\mathcal{F}(P_i)$ be the filtrations realizing its triangularity, for $i \in \{1, 2\}$. Then, it is clear that the filtrations $\mathcal{F}(X) = \mathcal{F}(X_1) \oplus \mathcal{F}(X_2)$ and $\mathcal{F}(P) = \mathcal{F}(P_1) \oplus \mathcal{F}(P_2)$, see (15.9), where the bimodules of $\mathcal{F}(X_i)$ are considered as B-S-bimodules and the bimodules of $\mathcal{F}(P_i)$ are considered as S-S-bimodules by restriction through the canonical projections $S \longrightarrow S_i$, realize the triangularity of $X = X_1 \oplus X_2$.

Moreover, if X_i is (completely) $[R_0, (S_i)_0]$-free, for $i \in \{1, 2\}$, make $S_0 = (S_1)_0 \times (S_2)_0$ and consider the filtrations $\mathcal{F}(\widetilde{X}_i)$ and $\mathcal{F}(\widetilde{P}_i)$ such that $\mathcal{F}(\widetilde{X}_i) \otimes_{(S_i)_0} S_i \cong_{R_0} \mathcal{F}(X_i)$ and $S_i \otimes_{(S_i)_0} \mathcal{F}(\widetilde{P}_i) \otimes_{(S_i)_0} S_i \cong \mathcal{F}(P_i)$. Then, consider the filtration of R_0-S_0-bimodules $\mathcal{F}(\widetilde{X}) := \mathcal{F}(\widetilde{X}_1) \oplus \mathcal{F}(\widetilde{X}_2)$, where all the bimodules in the sequence $\mathcal{F}(\widetilde{X}_i)$ are considered as R_0-S_0-

bimodules by restriction of scalars through the canonical projections $S_0 \longrightarrow (S_i)_0$. Then, we have $\mathcal{F}(\widetilde{X}) \otimes_{S_0} S = [\mathcal{F}(\widetilde{X}_1) \oplus \mathcal{F}(\widetilde{X}_2)] \otimes_{S_0} S \cong [\mathcal{F}(\widetilde{X}_1) \otimes_{(S_1)_0} S_1] \oplus [\mathcal{F}(\widetilde{X}_2) \otimes_{(S_2)_0} S_2] \cong \mathcal{F}(X_1) \oplus \mathcal{F}(X_2) = \mathcal{F}(X)$. Then, X is $[R_0, S_0]$-free.

Similarly, defining $\mathcal{F}(\widetilde{P}) = \mathcal{F}(\widetilde{P}_1) \oplus \mathcal{F}(\widetilde{P}_2)$, we obtain that $S \otimes_{S_0} \mathcal{F}(\widetilde{P}) \otimes_{S_0} S \cong \mathcal{F}(P)$, and then X is completely $[R_0, S_0]$-free. □

Corollary 17.12. *Assume that $\mathcal{A}'$ is a proper subditalgebra of the ditalgebra $\mathcal{A}$ with layer (R, W). Assume furthermore that $W_1' = 0$, and, therefore, $\delta(W_0') = 0$. Suppose that $X_1 = Y_1 \oplus \cdots \oplus Y_n$ is a finite direct sum of non-isomorphic finite-dimensional indecomposable objects $Y_i \in \mathcal{A}'$-Mod. Assume, furthermore, that each algebra $\mathrm{End}_{\mathcal{A}'}(Y_i)^{op}$ decomposes as $T_i \oplus J_i$, where J_i is its Jacobson radical. Then, $\Gamma_1 = \mathrm{End}_{\mathcal{A}'}(X_1)^{op}$ admits the splitting (S_1, P_1), where P_1 is the Jacobson radical of Γ_1. Moreover, suppose that $R \xrightarrow{\varphi} S_2$ is an epimorphism in the category of k-algebras. Consider the object X_2 in $\mathcal{A}'$-Mod obtained from the regular S_2-module by restriction through the composition $B \longrightarrow R \xrightarrow{\varphi} S_2$. Finally, assume that $\mathrm{Hom}_{\mathcal{A}'}(\mathcal{I}_{X_i}, \mathcal{I}_{X_j}) = 0$ for $i \neq j$. Then, $X = X_1 \oplus X_2$ is a complete triangular admissible $\mathcal{A}'$-module, and, therefore, we have a full and faithful functor $F^X : \mathcal{A}^X$-Mod$\longrightarrow \mathcal{A}$-Mod. Moreover, if R_0 is a k-subalgebra of R and $(S_2)_0 = \varphi(R_0)$, then X is $[R_0, S_1 \times (S_2)_0]$-free.*

Proof. From (17.2) and (17.4), we know that X_1 is a complete, triangular and an $[R_0, S_1]$-free admissible $\mathcal{A}'$-module. From (17.5), X_2 is complete, triangular and an $[R_0, (S_2)_0]$-free admissible $\mathcal{A}'$-module. The corollary follows then from (17.11). □

Lemma 17.13. *Assume that $\mathcal{A}$ is a ditalgebra with layer (R, W) such that $\delta(W_0) = W_1$. Then, define X as the A-module obtained from the regular left R-module by restriction of scalars using the projection $\pi : A \longrightarrow R$. Then, X is a complete triangular $\mathcal{A}$-module. Moreover, if R_0 is any subalgebra of R, then X is completely $[R_0, R_0]$-free.*

Proof. Consider $f = (f^0, f^1) \in \mathrm{Phom}_{R\text{-}W}(X, X)$. Then, $f \in \Gamma = \mathrm{End}_{\mathcal{A}}(X)^{op}$ iff, for any $a \in A$ and $x \in X$

$$af^0(x) = f^0(ax) + f^1(\delta(a))[x].$$

Recall that since $\delta(R) = 0$, f^0 is a morphism of left R-modules, thus using that W_0 acts trivially on X, the last equality is equivalent to $f^1(\delta(a)) = 0$, for all $a \in A$. Since $\delta(W_0) = W_1$, this is equivalent to $f^1 = 0$. Then $\Gamma = \mathrm{End}_{\mathcal{A}}(X)^{op} = \mathrm{End}_R(R)^{op} = S \oplus P$, with $S = R$ and $P = 0$. Then, X is an admissible $\mathcal{A}$-module.

Now let us look at the functor $F^X : \mathcal{A}^X\text{-Mod} \longrightarrow \mathcal{A}\text{-Mod}$. Here, $\mathcal{A}^X$ has layer (S, W^X), where $W_0^X = 0$ and $W_1^X = 0$. Thus, $\mathcal{A}^X$-Mod can be identified with R-Mod. Let us fix the dual basis $(1, Id_R)$, for the right R-module X. Then, for $M \in \mathcal{A}^X$-Mod, the A-module $F^X(M) = R \otimes_R M$ is the canonical A-module with action determined by the left action on X. Fix $M, N \in \mathcal{A}^X$-Mod. Then, using that W_0 acts trivially on $F^X(M)$ and $F^X(N)$, the argument used in the last paragraph shows that $\mathrm{Hom}_{\mathcal{A}}(F^X(M), F^X(N)) = \mathrm{Hom}_A(F^X(M), F^X(N)) = \mathrm{Hom}_R(F^X(M), F^X(N)) = \mathrm{Hom}_R(X \otimes_R M, X \otimes_R N)$. But the composition $\mathrm{Hom}_R(M, N) \xrightarrow{F^X} \mathrm{Hom}_R(X \otimes_R M, X \otimes_R N) \cong \mathrm{Hom}_R(M, N)$ is the identity, thus F^X is full and faithful. Then X is a complete admissible $\mathcal{A}$-module.

The final part of the proof is completed easily as in the proof of (17.5). □

Lemma 17.14. *Assume that $\mathcal{A}$ is a ditalgebra with layer (R, W). Assume that we have a decomposition of R-R-bimodules $W_0 = W_0' \oplus W_0''$ and $W_1 = W_1' \oplus W_1''$, with $\delta(W_0') = W_1'$. Consider the layered ditalgebra $\mathcal{A}^r$, obtained from $\mathcal{A}$ by regularization of the bimodule W_0', and the associated functor $F_r : \mathcal{A}^r\text{-Mod} \longrightarrow \mathcal{A}\text{-Mod}$, as in (8.19). Then, the pair $(\mathcal{A}^r, F_r)$ is completely realizable by an admissible module.*

Proof. Consider the proper subditalgebra $\mathcal{A}'$ of $\mathcal{A}$ associated to the given R-R-bimodule decompositions. Then, (17.13) certifies that the object $X \in \mathcal{A}'$-Mod, whose B-module structure is determined from the regular R-module by restriction of scalars using the projection $\pi : B \longrightarrow R$, is a complete admissible $\mathcal{A}'$-module. Then, the functor $F^X : \mathcal{A}^X\text{-Mod} \longrightarrow \mathcal{A}\text{-Mod}$ is full and faithful. We have the isomorphisms of R-R-bimodules $W_0^X = X^* \otimes_B BW_0''B \otimes_B X \cong X^* \otimes_R W_0'' \otimes_R X \cong W_0''$ and $W_1^X \cong X^* \otimes_B BW_1''B \otimes_B X \cong X^* \otimes_R W_1'' \otimes_R X \cong W_1''$, which give us the isomorphism $\theta : W^X \longrightarrow W^r$. It extends to a morphism of algebras $\overline{\theta} : T^X \longrightarrow T^r$. It is easy to see that $\overline{\theta}\delta^X = \delta\overline{\theta}$ and, therefore, we have an isomorphism of layered ditalgebras $\overline{\theta} : \mathcal{A}^X \longrightarrow \mathcal{A}^r$. Now, consider the canonical morphism $\eta : \mathcal{A} \longrightarrow \mathcal{A}^r$, as in (8.19), thus $F_r = F_\eta$, and, for $M \in \mathcal{A}^X$-Mod, consider the isomorphism of B-modules $\sigma^0 : F^X(M) = R \otimes_R M \longrightarrow M = F_{\overline{\theta}^{-1}\eta}(M)$. It can be seen that $\sigma = (\sigma^0, 0) : F^X \longrightarrow F_{\overline{\theta}^{-1}\eta}$ is an isomorphism of functors. □

Exercise 17.15. *Discuss the relation of exercise (16.6) with exercises (9.8), (9.9) and (9.10).*

18
The exact categories $\mathcal{P}(\Lambda)$, $\mathcal{P}^1(\Lambda)$ and Λ-Mod

Throughout this section, Λ denotes a fixed artin algebra, over the commutative artin ring k, and J denotes its radical. We introduce the exact categories $\mathcal{P}(\Lambda)$ and $\mathcal{P}^1(\Lambda)$ and we recollect some of their basic properties. In particular, we study the cokernel functor $\mathrm{Cok} : \mathcal{P}(\Lambda) \longrightarrow \Lambda\text{-Mod}$ and its action on almost split conflations.

Definition 18.1. *Consider the category $\mathcal{M}(\Lambda)$ of morphisms of Λ-Mod. Thus, an object X of $\mathcal{M}(\Lambda)$ is a triple (X_1, X_2, φ_X), where X_1 and X_2 are left Λ-modules and $\varphi_X \in \mathrm{Hom}_\Lambda(X_1, X_2)$. A morphism $f : X \longrightarrow Y$ in $\mathcal{M}(\Lambda)$ is a pair (f_1, f_2), where $f_i \in \mathrm{Hom}_\Lambda(X_i, Y_i)$, for $i \in \{1, 2\}$, such that $\varphi_Y f_1 = f_2 \varphi_X$. The composition in $\mathcal{M}(\Lambda)$ is performed componentwise. By definition, $\mathcal{P}(\Lambda)$ (resp. $p(\Lambda)$) is the full subcategory of $\mathcal{M}(\Lambda)$ whose objects are all the morphisms between projective (resp. finitely generated projective) Λ-modules.*

Lemma 18.2. *Let $\widetilde{\Lambda}$ denote the triangular matrix algebra $\begin{pmatrix} \Lambda & 0 \\ \Lambda & \Lambda \end{pmatrix}$. Then, $\widetilde{\Lambda}$ is an artin algebra and there is an equivalence of categories*

$$\Psi : \widetilde{\Lambda}\text{-Mod} \longrightarrow \mathcal{M}(\Lambda).$$

The functor Ψ maps each $\widetilde{\Lambda}$-module M onto the triple (M_1, M_2, φ), where $M_1 = e_1 M$, $M_2 = e_2 M$ and $\varphi : M_1 \cong \Lambda \otimes_\Lambda M_1 \longrightarrow M_2$ is the morphism of Λ-modules induced by the action of the southwestern corner of $\widetilde{\Lambda}$ on M_1. Here e_1 and e_2 denote the canonical idempotents of the algebra $\widetilde{\Lambda}$. Each morphism $f : M \longrightarrow N$ in $\widetilde{\Lambda}$-Mod is mapped by the functor Ψ to the pair (f_1, f_2), where $f_i : M_i \longrightarrow N_i$ is the corresponding restriction of f, for $i \in \{1, 2\}$. The canonical exact structure of $\widetilde{\Lambda}$-Mod is transfered by Ψ to the natural exact structure in $\mathcal{M}(\Lambda)$ formed by the composable pairs of morphisms

$X \xrightarrow{f} E \xrightarrow{g} Y$ *such that*

$$0 \longrightarrow X_i \xrightarrow{f_i} E_i \xrightarrow{g_i} Y_i \longrightarrow 0$$

are exact sequences in Λ-Mod.

Lemma 18.3. *The category* $\mathcal{P}(\Lambda)$ *is closed under conflations in* $\mathcal{M}(\Lambda)$ *and, therefore, inherits from* $\mathcal{M}(\Lambda)$ *an exact structure* $\mathcal{E}$*. Consider, for a projective* Λ*-module* P*, the following objects of* $\mathcal{P}(\Lambda)$

$$S(P) := (P, 0, 0), \quad U(P) := (P, P, 1_P) \text{ and } T(P) := (0, P, 0).$$

Then, the indecomposable $\mathcal{E}$*-projective objects of* $\mathcal{P}(\Lambda)$ *are the objects isomorphic to* $U(P)$ *or* $T(P)$*, where* P *is an indecomposable projective* Λ*-module. The indecomposable* $\mathcal{E}$*-injective objects of* $\mathcal{P}(\Lambda)$ *are the objects isomorphic to* $U(P)$ *or* $S(P)$*, where* P *is an indecomposable projective* Λ*-module. Every* $\mathcal{E}$*-projective (resp.* $\mathcal{E}$*-injective) object in* $\mathcal{P}(\Lambda)$ *is isomorphic to a direct summand of* $U(P) \oplus T(P)$ *(resp. of* $U(P) \oplus S(P)$*), for some projective* Λ*-module* P.

Proof. For $i \in \{1, 2\}$, denote by $\mathcal{F}_i : \mathcal{P}(\Lambda) \longrightarrow \Lambda$-Mod the projection defined by $\mathcal{F}_i(X) = X_i$, for any object $X \in \mathcal{P}(\Lambda)$, and $\mathcal{F}_i(f_1, f_2) = f_i$, for any morphism (f_1, f_2) of $\mathcal{P}(\Lambda)$. Then, notice that, for any projective Λ-module P, there are natural isomorphisms of functors from $\mathcal{P}(\Lambda)$ to k-Mod

$$\begin{aligned}
\mathrm{Hom}_{\mathcal{P}(\Lambda)}(U(P), -) &\cong \mathrm{Hom}_\Lambda(P, \mathcal{F}_1(-)) \\
\mathrm{Hom}_{\mathcal{P}(\Lambda)}(-, U(P)) &\cong \mathrm{Hom}_\Lambda(\mathcal{F}_2(-), P) \\
\mathrm{Hom}_{\mathcal{P}(\Lambda)}(T(P), -) &\cong \mathrm{Hom}_\Lambda(P, \mathcal{F}_2(-)) \\
\mathrm{Hom}_{\mathcal{P}(\Lambda)}(-, S(P)) &\cong \mathrm{Hom}_\Lambda(\mathcal{F}_1(-), P).
\end{aligned}$$

It follows that the objects $U(P)$ and $T(P)$ are $\mathcal{E}$-projectives (and the objects $U(P)$ and $S(P)$ are $\mathcal{E}$-injectives) in $\mathcal{P}(\Lambda)$, for any projective Λ-module P. Now notice that, for any $X = (X_1, X_2, \varphi_X) \in \mathcal{P}(\Lambda)$, there are conflations in $\mathcal{E}$

$$T(X_1) \xrightarrow{\begin{pmatrix} f \\ g \end{pmatrix}} T(X_2) \oplus U(X_1) \xrightarrow{(f', g')} X, \text{ and } X \xrightarrow{\begin{pmatrix} r \\ s \end{pmatrix}} U(X_2) \oplus S(X_1) \xrightarrow{(r', s')} S(X_2),$$

where

$$\begin{aligned}
f &= (0, \varphi_X) : T(X_1) \longrightarrow T(X_2) \\
g &= (0, -1_{X_1}) : T(X_1) \longrightarrow U(X_1) \\
f' &= (0, 1_{X_2}) : T(X_2) \longrightarrow X \\
g' &= (1_{X_1}, \varphi_X) : U(X_1) \longrightarrow X,
\end{aligned}$$

and, similarly

$$\begin{aligned} r &= (\varphi_X, 1_{X_2}) : X \longrightarrow U(X_2) \\ s &= (-1_{X_1}, 0) : X \longrightarrow S(X_1) \\ r' &= (1_{X_2}, 0) : U(X_2) \longrightarrow S(X_2) \\ s' &= (\varphi_X, 0) : S(X_1) \longrightarrow S(X_2). \end{aligned}$$

It follows that the indecomposable $\mathcal{E}$-injective or $\mathcal{E}$-projective objects have the form announced in the statement of the lemma. □

Lemma 18.4. *Consider the cokernel functor* Cok : $\mathcal{P}(\Lambda) \longrightarrow \Lambda$-Mod, *which, by definition, maps each* $X \in \mathcal{P}(\Lambda)$ *onto* $\text{Cok}X = X_2/\text{Im}\,\varphi_X$, *and each morphism* $f : X \longrightarrow Y$ *in* $\mathcal{P}(\Lambda)$ *onto the induced morphism* $\text{Cok}f : \text{Cok}X \longrightarrow \text{Cok}Y$ *in* Λ-Mod. *Then, the ideal* $\mathcal{I}$ *of the category* $\mathcal{P}(\Lambda)$, *formed by the morphisms which factor through the* $\mathcal{E}$*-injective objects of* $\mathcal{P}(\Lambda)$, *coincides with the kernel of the functor* Cok, *which is full and dense. Therefore, the functor* Cok *induces an equivalence of categories* $\mathcal{P}(\Lambda)/\mathcal{I} \simeq \Lambda$-Mod.

Proof. Assume that the morphism $f : X \longrightarrow Y$ in $\mathcal{P}(\Lambda)$ is such that $\text{Cok}(f) = 0$. Then, $f_2 = \varphi_Y \rho$, for some $\rho \in \text{Hom}_\Lambda(X_2, Y_1)$. Then, the morphism f equals the composition

$$X \xrightarrow{\begin{pmatrix} g \\ h \end{pmatrix}} U(Y_1) \oplus S(X_1) \xrightarrow{(r,s)} Y,$$

where

$$\begin{aligned} g &= (\rho\varphi_X, \rho) : X \longrightarrow U(Y_1) \\ h &= (1_{X_1}, 0) : X \longrightarrow S(X_1) \\ r &= (1_{Y_1}, \varphi_Y) : U(Y_1) \longrightarrow Y \\ s &= (f_1 - \rho\varphi_X, 0) : S(X_1) \longrightarrow Y. \end{aligned}$$

Then, the kernel of the functor Cok is contained in the ideal $\mathcal{I}$. The other inclusion is clear because the functor Cok maps the objects $U(P)$ and $S(P)$ to zero. □

Lemma 18.5. *The category* $p(\Lambda)$ *inherits from* $\mathcal{M}(\Lambda)$ *an exact structure, which we denote with the same symbol* $\mathcal{E}$. *The Krull–Schmidt exact category* $(p(\Lambda), \mathcal{E})$ *admits almost split conflations.*

Proof. It will be enough to show that the full subcategory $\mathcal{C}$ of $\widetilde{\Lambda}$-Mod, corresponding to $p(\Lambda)$ under the equivalence Ψ of (18.2), inherits an exact structure which admits almost split sequences. According to the Auslander–Smalø theorem, it will be enough to show that $\mathcal{C}$ is a full subcategory closed under direct sums and summands, closed under extensions and contravariantly finite (see (7.16) and [4]). Indeed, it is clear that $\mathcal{P}(\Lambda)$ is a full subcategory of $\mathcal{M}(\Lambda)$ closed under direct sums and summands, and closed under extensions. Let us show that each $X = (X_1, X_2, \varphi_X) \in \mathcal{M}(\Lambda)$ admits a right $\mathcal{P}(\Lambda)$-approximation. Consider any projective resolutions

$$P_2^1 \xrightarrow{d_2^1} P_2^0 \xrightarrow{d_2^0} X_2 \longrightarrow 0 \quad \text{and} \quad P_1^0 \xrightarrow{d_1^0} X_1 \longrightarrow 0$$

of X_2 and X_1 in Λ-Mod. Consider also any lifting $\widehat{\varphi}_X : P_1^0 \longrightarrow P_2^0$ with $d_2^0 \widehat{\varphi}_X = \varphi_X d_1^0$. Then, put $P := (P_2^1 \oplus P_1^0, P_2^0, \varphi_P)$, where $\varphi_P = (d_2^1, \widehat{\varphi}_X) : P_2^1 \oplus P_1^0 \longrightarrow P_2^0$, and consider the morphism $t := ((0, d_1^0), d_2^0) : P \longrightarrow X$ in $\mathcal{M}(\Lambda)$. Now, take any morphism $s = (s_1, s_2) : Q \longrightarrow X$ in $\mathcal{M}(\Lambda)$, with $Q \in \mathcal{P}(\Lambda)$. Then, there are morphisms

$$\begin{pmatrix} s_1' \\ r_1'' \end{pmatrix} : Q_1 \longrightarrow P_2^1 \oplus P_1^0 \quad \text{and} \quad r_2 : Q_2 \longrightarrow P_2^0$$

such that $d_1^0 r_1'' = s_1$ and $d_2^0 r_2 = s_2$. Since s is a morphism in $\mathcal{M}(\Lambda)$, we have

$$d_2^0 \widehat{\varphi}_X r_1'' = \varphi_X d_1^0 r_1'' = \varphi_X s_1 = s_2 \varphi_Q = d_2^0 r_2 \varphi_Q.$$

Then, the morphism $\delta := r_2 \varphi_Q - \widehat{\varphi}_X r_1'' : Q_1 \longrightarrow \operatorname{Ker}(d_2^0) \subseteq P_2^0$ factors through d_2^1. Take a morphism $r_1' : Q_1 \longrightarrow P_2^1$ such that $d_2^1 r_1' = \delta$. Then, the pair $r = \left(\begin{pmatrix} r_1' \\ r_1'' \end{pmatrix}, r_2 \right)$ is a morphism from Q to P in $\mathcal{M}(\Lambda)$ such that $tr = s$. Indeed, $r : Q \longrightarrow P$ is a morphism because

$$(d_2^1, \widehat{\varphi}_X) \begin{pmatrix} r_1' \\ r_1'' \end{pmatrix} = d_2^1 r_1' + \widehat{\varphi}_X r_1'' = r_2 \varphi_Q - \widehat{\varphi}_X r_1'' + \widehat{\varphi}_X r_1'' = r_2 \varphi_Q.$$

Since $d_2^0 r_2 = s_2$ and $(0, d_1^0) \begin{pmatrix} r_1' \\ r_1'' \end{pmatrix} = d_1^0 r_1'' = s_1$, we have that $tr = s$. Thus, $t : P \longrightarrow X$ is a right $\mathcal{P}(\Lambda)$-approximation of X. It follows that $\mathcal{C}$ is contravariantly finite in $\widetilde{\Lambda}$-mod, and, therefore, $\mathcal{C}$ has almost split sequences. Thus, $(p(\Lambda), \mathcal{E})$ has almost split conflations. □

Lemma 18.6. *Consider the exact category $(\mathcal{P}(\Lambda), \mathcal{E})$ as in (18.3) and its associated k-bifunctor* $\operatorname{Ext}_{\mathcal{P}(\Lambda)}(-, ?)$. *Consider, for $i \in \{1, 2\}$, the projection functor $\mathcal{F}_i : \mathcal{P}(\Lambda) \longrightarrow \Lambda$-Mod such that $\mathcal{F}_i(X_1, X_2, \varphi_X) = X_i$, for any object $(X_1, X_2, \varphi_X) \in \mathcal{P}(\Lambda)$, and $\mathcal{F}_i(f_1, f_2) = f_i$, for any morphism (f_1, f_2) in $\mathcal{P}(\Lambda)$. Then, there is an exact sequence of k-bifunctors over $\mathcal{P}(\Lambda)$*

$$0 \longrightarrow \operatorname{Hom}_{\mathcal{P}(\Lambda)}(-, ?) \xrightarrow{\sigma} \operatorname{Hom}_\Lambda(\mathcal{F}_1(-), \mathcal{F}_1(?)) \times \operatorname{Hom}_\Lambda(\mathcal{F}_2(-), \mathcal{F}_2(?)) \xrightarrow{\partial}$$
$$\operatorname{Hom}_\Lambda(\mathcal{F}_1(-), \mathcal{F}_2(?)) \xrightarrow{\eta} \operatorname{Ext}_{\mathcal{P}(\Lambda)}(-, ?) \longrightarrow 0.$$

Proof. Given $X, Y \in \mathcal{P}(\Lambda)$, the morphism $\sigma = \sigma_{X,Y}$ is the inclusion map. Given $(f_1, f_2) \in \mathrm{Hom}_\Lambda(X_1, Y_1) \times \mathrm{Hom}_\Lambda(X_2, Y_2)$, by definition, take $\partial_{X,Y}(f_1, f_2) = \varphi_Y f_1 - f_2\varphi_X$. Finally, given any morphism $h \in \mathrm{Hom}_\Lambda(X_1, Y_2)$, consider the triple $Z = Z(h) = (Y_1 \oplus X_1, Y_2 \oplus X_2, \varphi_Z)$ in $\mathcal{P}(\Lambda)$, where

$$\varphi_Z = \begin{pmatrix} \varphi_Y & h \\ 0 & \varphi_X \end{pmatrix} : Y_1 \oplus X_1 \longrightarrow Y_2 \oplus X_2.$$

Then, define $\eta_{X,Y}(h) \in \mathrm{Ext}_{\mathcal{P}(\Lambda)}(X, Y)$ as the equivalence class of the conflation $Y \xrightarrow{s} Z \xrightarrow{t} X$, where $s_i : Y_i \longrightarrow Z_i$ is the canonical injection and $t_i : Z_i \longrightarrow X_i$ is the canonical projection in Λ-Mod. The exactness of the sequence evaluated at each pair $X, Y \in \mathcal{P}(\Lambda)$ can be verified easily. This is also the case for the naturality of σ and ∂. Let us show that $\eta_{X,Y}$ is natural at the variable X. For this, take any morphism $g : X' \longrightarrow X$ in $\mathcal{P}(\Lambda)$ and $h \in \mathrm{Hom}_\Lambda(X_1, Y_2)$. We want to show that $\mathrm{Ext}_{\mathcal{P}(\Lambda)}(g, Y)\eta_{X,Y}(h) = \eta_{X',Y}\mathrm{Hom}_\Lambda(\mathcal{F}_1(g), Y_2)(h)$. By definition, $\eta_{X,Y}(h)$ is the equivalence class $[\zeta] \in \mathrm{Ext}_{\mathcal{P}(\Lambda)}(X, Y)$ of the conflation $\zeta : Y \longrightarrow Z \longrightarrow X$ described by the diagram

$$\begin{array}{ccccc} Y_1 & \longrightarrow & Y_1 \oplus X_1 & \longrightarrow & X_1 \\ \downarrow{\scriptstyle \varphi_Y} & & \downarrow \begin{pmatrix} \varphi_Y & h \\ 0 & \varphi_X \end{pmatrix} & & \downarrow{\scriptstyle \varphi_X} \\ Y_2 & \longrightarrow & Y_2 \oplus X_2 & \longrightarrow & X_2. \end{array}$$

Similarly, $\eta_{X',Y}(hg_1)$ is the equivalence class $[\zeta'] \in \mathrm{Ext}_{\mathcal{P}(\Lambda)}(X', Y)$ of the conflation $\zeta' : Y \longrightarrow Z' \longrightarrow X'$ described by the diagram

$$\begin{array}{ccccc} Y_1 & \longrightarrow & Y_1 \oplus X'_1 & \longrightarrow & X'_1 \\ \downarrow{\scriptstyle \varphi_Y} & & \downarrow \begin{pmatrix} \varphi_Y & hg_1 \\ 0 & \varphi_{X'} \end{pmatrix} & & \downarrow{\scriptstyle \varphi_{X'}} \\ Y_2 & \longrightarrow & Y_2 \oplus X'_2 & \longrightarrow & X'_2. \end{array}$$

Notice that the following diagram in $\mathcal{P}(\Lambda)$ is commutative

$$\begin{array}{ccccc} \zeta' : Y & \longrightarrow & Z' & \longrightarrow & X' \\ \| & & \downarrow{\scriptstyle \psi} & & \downarrow{\scriptstyle g} \\ \zeta : Y & \longrightarrow & Z & \longrightarrow & X, \end{array}$$

where $\psi : Z' \longrightarrow Z$ is the morphism defined by

$$\begin{array}{ccccc} Z' & : & Y_1 \oplus X'_1 & \longrightarrow & Y_2 \oplus X'_2 \\ \downarrow{\scriptstyle \psi} & & {\scriptstyle \psi_1 = \begin{pmatrix} 1 & 0 \\ 0 & g_1 \end{pmatrix}} \downarrow & & \downarrow {\scriptstyle \psi_2 = \begin{pmatrix} 1 & 0 \\ 0 & g_2 \end{pmatrix}} \\ Z & : & Y_1 \oplus X_1 & \longrightarrow & Y_2 \oplus X_2. \end{array}$$

Therefore

$$\begin{aligned}\mathrm{Ext}_{\mathcal{P}(\Lambda)}(g, Y)\eta_{X,Y}(h) = & \ [\zeta g] = [\zeta'] = \eta_{X',Y}(hg_1) \\ & = \eta_{X',Y}\mathrm{Hom}_\Lambda(\mathcal{F}_1(g), Y_2)(h).\end{aligned}$$

The proof of the naturality of $\eta_{X,Y}$ in the variable Y is similar. □

Lemma 18.7. *Consider the exact category $(\mathcal{P}(\Lambda), \mathcal{E})$ as in (18.3) and its associated k-bifunctor* $\mathrm{Ext}_{\mathcal{P}(\Lambda)}(-, ?)$. *Consider the triple $S(\Lambda) := (\Lambda, 0, 0)$ in $\mathcal{P}(\Lambda)$. Since* $\mathrm{End}_{\mathcal{P}(\Lambda)}(S(\Lambda))^{op} \cong \mathrm{End}_\Lambda(\Lambda)^{op} \cong \Lambda$, *we can consider* $\mathrm{Hom}_{\mathcal{P}(\Lambda)}(S(\Lambda), -)$ *and* $\mathrm{Ext}_{\mathcal{P}(\Lambda)}(S(\Lambda), -)$ *as k-functors from $\mathcal{P}(\Lambda)$ to* Λ-Mod. *Consider also the natural kernel and cokernel functors* Ker , Cok : $\mathcal{P}(\Lambda) \longrightarrow \Lambda$-Mod. *Then, there are isomorphisms of functors*

$$\mathrm{Ker}\,(-) \cong \mathrm{Hom}_{\mathcal{P}(\Lambda)}(S(\Lambda), -) \quad \textit{and} \quad \mathrm{Cok}(-) \cong \mathrm{Ext}_{\mathcal{P}(\Lambda)}(S(\Lambda), -).$$

Proof. Take $X = (X_1, X_2, \varphi_X) \in \mathcal{P}(\Lambda)$ and consider the map

$$\begin{aligned}\theta_X : \mathrm{Hom}_{\mathcal{P}(\Lambda)}(S(\Lambda), X) &\longrightarrow \mathrm{Ker}\,\varphi_X = \mathrm{Ker}\,X \\ (f_1, f_2) &\mapsto f_1(1).\end{aligned}$$

Then it is easy to see that $\theta = \{\theta_X\}_X$ is an isomorphism of functors. For the second part, we have the existence of an isomorphism ν_X in the following commutative diagram with exact rows

$$\begin{array}{ccccccc}\mathrm{Hom}_\Lambda(\Lambda, X_1) & \xrightarrow{\partial} & \mathrm{Hom}_\Lambda(\Lambda, X_2) & \xrightarrow{\eta} & \mathrm{Ext}_{\mathcal{P}(\Lambda)}(S(\Lambda), X) & \longrightarrow & 0 \\ \downarrow \cong & & \downarrow \cong & & \downarrow \nu_X & & \\ X_1 & \xrightarrow{\varphi_X} & X_2 & \longrightarrow & \mathrm{Cok}\,\varphi_X & \longrightarrow & 0,\end{array}$$

where the first row is taken from (18.6). The family $\nu = \{\nu_X\}_X$ is a natural isomorphism because $\eta = \eta_{S(\Lambda),-}$ is a natural epimorphism. □

Definition 18.8. *We denote by $\mathcal{P}^1(\Lambda)$ the full subcategory of $\mathcal{P}(\Lambda)$ whose objects are the triples (P, Q, φ), where $\varphi : P \to Q$ is a morphism of Λ-modules with image contained in JQ. The full subcategory of $\mathcal{P}^1(\Lambda)$ with objects (P, Q, φ) such that* $\mathrm{Ker}\,(\varphi) \subseteq JP$ *is denoted by $\mathcal{P}^2(\Lambda)$.*

Lemma 18.9. *Any object $X \in \mathcal{P}(\Lambda)$ is of the form $X \cong S(P) \oplus U(Q) \oplus \overline{X}$, where P, Q are projective Λ-modules and $\overline{X} \in \mathcal{P}^2(\Lambda)$. In particular, $X \in \mathcal{P}^1(\Lambda)$ lies in $\mathcal{P}^2(\Lambda)$ iff it has no direct summand of the form $S(P)$.*

Proof. Since Λ is an artin algebra, Λ/J is semisimple and J is nilpotent. By Bass theorem, every left Λ-module $M \in \Lambda$-Mod admits a projective cover (see [1]). Let $X = (X_1, X_2, \varphi_X) \in \mathcal{P}(\Lambda)$, consider the Λ-module $M := \mathrm{Cok}(\varphi_X)$

and the canonical projection $\zeta : X_2 \longrightarrow M$. Then, if $\nu : \overline{X}_2 \longrightarrow M$ is a projective cover of M, there is a projective Λ-module Q and a commutative diagram

$$\begin{array}{ccccc} \mathrm{Ker}\,(\zeta) & \xrightarrow{\sigma} & X_2 & \xrightarrow{\zeta} & M \\ \downarrow \cong & & \downarrow \cong & & \downarrow \cong \\ \mathrm{Ker}\,(\nu) \oplus Q & \xrightarrow{\overline{\sigma}\oplus 1_Q} & \overline{X}_2 \oplus Q & \xrightarrow{\nu \oplus 0} & M \oplus 0, \end{array}$$

where σ and $\overline{\sigma}$ are the inclusion maps. Now, assume that $\nu' : \overline{X}_1 \longrightarrow \mathrm{Ker}\,(\nu)$ is a projective cover, then so is the direct sum map $\nu' \oplus 1_Q : \overline{X}_1 \oplus Q \longrightarrow \mathrm{Ker}\,(\nu) \oplus Q$. Then, as before, there is a projective Λ-module P and a commutative diagram

$$\begin{array}{ccccc} X_1 & \xrightarrow{\varphi_X|} & \mathrm{Ker}\,(\zeta) & \xrightarrow{\sigma} & X_2 \\ \downarrow \cong & & \downarrow \cong & & \downarrow \cong \\ \overline{X}_1 \oplus Q \oplus P & \xrightarrow{\nu'\oplus 1_Q \oplus 0} & \mathrm{Ker}\,(\nu) \oplus Q \oplus 0 & \xrightarrow{\overline{\sigma}\oplus 1_Q \oplus 0} & \overline{X}_2 \oplus Q \oplus 0. \end{array}$$

This means that $X \cong S(P) \oplus U(Q) \oplus \overline{X}$, where the map $\varphi_{\overline{X}}$ of $\overline{X} = (\overline{X}_1, \overline{X}_2, \varphi_{\overline{X}})$ is the composition $\varphi_{\overline{X}} = \overline{\sigma}\nu'$. Since $\mathrm{Im}\,(\nu') = \mathrm{Ker}\,(\nu) \subseteq J\overline{X}_2$ and $\mathrm{Ker}\,(\nu') \subseteq J\overline{X}_1$, $\overline{X} \in \mathcal{P}^2(\Lambda)$. □

Lemma 18.10. *The cokernel functor* Cok : $\mathcal{P}(\Lambda) \longrightarrow \Lambda$-Mod *satisfies:*

(1) Cok *is a full and dense functor.*
(2) If $X \in \mathcal{P}^2(\Lambda)$, then $f \in \mathrm{End}_{\mathcal{P}(\Lambda)}(X)$ is nilpotent, whenever $\mathrm{Cok}(f) = 0$.
(3) Assume that $f : X \longrightarrow Y$ is a morphism in $\mathcal{P}(\Lambda)$ with $Y \in \mathcal{P}^2(\Lambda)$ (resp. $X \in \mathcal{P}^2(\Lambda)$). Then, f is a retraction (resp. a section) of $\mathcal{P}(\Lambda)$ iff $\mathrm{Cok}(f)$ *is a retraction (resp. a section) of* Λ-Mod.
(4) The restriction of Cok *to $\mathcal{P}^2(\Lambda)$, denoted by* $\mathrm{Cok}^2 : \mathcal{P}^2(\Lambda) \longrightarrow \Lambda$-Mod, *is dense, full and reflects isomorphisms.*
(5) $X \in \mathcal{P}^2(\Lambda)$ is indecomposable iff $\mathrm{Cok}(X)$ *is so.*

Proof.

(1) Since any Λ-module admits a projective cover, any Λ-module M admits a minimal projective presentation $P \xrightarrow{f} Q \xrightarrow{g} M \longrightarrow 0$. Then, the restriction $P \xrightarrow{f} \mathrm{Im}\, f$ and $Q \xrightarrow{g} M$ are projective covers. Thus, $\mathrm{Ker}\, f \subseteq JP$ and $\mathrm{Ker}\, g = \mathrm{Im}\, f \subseteq JQ$, which means that Cok^2 is dense. Since any morphism $M \longrightarrow M'$ lifts to a morphism of the projective presentations, Cok^2 is full.

(2) Given $X \in \mathcal{P}^2(\Lambda)$ and $f \in \mathrm{End}_{\mathcal{P}(\Lambda)}(X)$ with $\mathrm{Cok}(f) = 0$, we know that $f_2 = \varphi_X \rho$, for some $\rho \in \mathrm{Hom}_\Lambda(X_2, X_1)$. Since $\mathrm{Im}\,(\varphi_X) \subseteq JX_2$, we have that $\mathrm{Im}\,(f_2) \subseteq JX_2$. Thus, if n is the nilpotence index of the radical J, we

obtain that $f^n = (f_1^n, f_2^n) = (f_1^n, 0)$ is an endomorphism of X in $\mathcal{P}(\Lambda)$. Then, $\varphi_X f_1^n = 0\varphi_X = 0$, which implies that $\operatorname{Im} f_1^n \subseteq \operatorname{Ker} \varphi_X \subseteq JX_1$. Using again that $J^n = 0$, we obtain $f^{n\times n} = 0$.

(3) Assume that $\operatorname{Cok}(f)$ is a retraction, say with right inverse h. Since Cok is full, $h = \operatorname{Cok}(g)$. Then, $\operatorname{Cok}(fg) = \operatorname{Cok}(f)\operatorname{Cok}(g) = 1_{\operatorname{Cok}(Y)}$ or, equivalently, $\operatorname{Cok}(1_Y - fg) = 0$. From (2), we obtain that $1_Y - fg$ is nilpotent and, therefore, $fg = 1_Y - (1_Y - fg)$ is invertible. Thus, f is a retraction. The other statement is dual.

(4) It follows from (1) and (3).

(5) It follows from (4). Indeed, if $Z \in \mathcal{P}^2(\Lambda)$ and $\operatorname{Cok}Z = M \oplus N$, consider $X, Y \in \mathcal{P}^2(\Lambda)$ with $\operatorname{Cok}X \cong M$ and $\operatorname{Cok}Y \cong N$. Then, $\operatorname{Cok}(X \oplus Y) \cong \operatorname{Cok}X \oplus \operatorname{Cok}Y \cong \operatorname{Cok}Z$ and, therefore, $Z \cong X \oplus Y$. The other implication is clear because $\operatorname{Cok}Z \neq 0$ whenever $0 \neq Z \in \mathcal{P}^2(\Lambda)$. □

Lemma 18.11. *Let $\mathcal{E}_1$ be the class of conflations $Y \xrightarrow{g} E \xrightarrow{f} X$ of $(\mathcal{P}(\Lambda), \mathcal{E})$ such that $Y, E, X \in \mathcal{P}^1(\Lambda)$. Then, $\mathcal{E}_1$ is an exact structure for $\mathcal{P}^1(\Lambda)$.*

Proof. Clearly, $\mathcal{E}_1$ is a class of exact pairs of $\mathcal{P}^1(\Lambda)$ closed under isomorphisms.

Claim. Let $f : X \longrightarrow Y$ be a morphism in $\mathcal{P}^1(\Lambda)$. Then, f is a deflation (resp. inflation) in $\mathcal{P}(\Lambda)$ iff f is a deflation (resp. inflation) in $\mathcal{P}^1(\Lambda)$.

Proof of the claim: Assume there is a conflation $Z \xrightarrow{g} X \xrightarrow{f} Y$ in $\mathcal{P}(\Lambda)$. We will show that $Z \in \mathcal{P}^1(\Lambda)$. By assumption, we have a commutative diagram with split exact rows

$$\begin{array}{ccccccccc} 0 & \longrightarrow & Z_1 & \xrightarrow{g_1} & X_1 & \xrightarrow{f_1} & Y_1 & \longrightarrow & 0 \\ & & \downarrow{\scriptstyle \varphi_Z} & & \downarrow{\scriptstyle \varphi_X} & & \downarrow{\scriptstyle \varphi_Y} & & \\ 0 & \longrightarrow & Z_2 & \xrightarrow{g_2} & X_2 & \xrightarrow{f_2} & Y_2 & \longrightarrow & 0. \end{array}$$

Thus, $X_2 = g_2(Z_2) \oplus X_2'$, for some submodule X_2' of X_2, and $\varphi_X g_1(z_1) \in JX_2 = Jg_2(Z_2) \oplus JX_2'$, for $z_1 \in Z_1$, where J is the radical of Λ. Then, $g_2\varphi_Z(z_1) = \varphi_X g_1(z_1) = \sum_j p_j g_2(z_2^j) + x_2'$, for some $p_j \in J$, $z_2^j \in Z_2$ and $x_2' \in JX_2'$. Thus

$$g_2[\varphi_Z(z_1) - \sum_j p_j z_2^j] = x_2' \in g_2(Z_2) \cap X_2' = 0,$$

and $\varphi_Z(z_1) = \sum_j p_j z_2^j \in JZ_2$. That is $Z \in \mathcal{P}^1(\Lambda)$, as claimed.

Our claim implies that the conditions (1), (3) and $(3)^{op}$ of the definition of exact structure for $\mathcal{E}_1$ in $\mathcal{P}^1(\Lambda)$ follow from the corresponding properties for

$\mathcal{E}$ in $\mathcal{P}(\Lambda)$ (see (6.3)). Now, assume that $d : Y \longrightarrow Z$ is a deflation of $\mathcal{P}^1(\Lambda)$ and $f : Z' \longrightarrow Z$ is a morphism in $\mathcal{P}^1(\Lambda)$. Then, $(f, d) : Z' \oplus Y \longrightarrow Z$ is a deflation in $\mathcal{P}(\Lambda)$ and its kernel has the form $\begin{pmatrix} d' \\ -f' \end{pmatrix} : Y' \longrightarrow Z' \oplus Y$. By our claim, $Y' \in \mathcal{P}^1(\Lambda)$. On the other hand, we have exact sequences

$$0 \longrightarrow Y_i' \xrightarrow{\begin{pmatrix} d_i' \\ -f_i' \end{pmatrix}} Z_i' \oplus Y_i \xrightarrow{(f_i, d_i)} Z_i \longrightarrow 0,$$

thus, we have pullback diagrams in Λ-Mod

$$\begin{array}{ccc} Y_i' & \xrightarrow{d_i'} & Z_i' \\ {\scriptstyle f_i'}\downarrow & & \downarrow{\scriptstyle f_i} \\ Y_i & \xrightarrow{d_i} & Z_i . \end{array}$$

Since d_i is surjective, so is d_i'. Then, $d' : Y' \longrightarrow Z'$ is a deflation. We have verified axiom (2) of (6.3) for $\mathcal{E}_1$, and we are done. □

Lemma 18.12. *Although $\mathcal{P}^1(\Lambda)$ is not a subcategory of $\mathcal{P}(\Lambda)$ closed under conflations, the inclusion functor $(\mathcal{P}^1(\Lambda), \mathcal{E}_1) \longrightarrow (\mathcal{P}(\Lambda), \mathcal{E})$ is exact. Moreover, its restriction $(p^1(\Lambda), \mathcal{E}_1) \longrightarrow (p(\Lambda), \mathcal{E})$, where $p^1(\Lambda) = p(\Lambda) \cap \mathcal{P}^1(\Lambda)$, preserves almost split conflations.*

Proof. Assume that $Y \xrightarrow{f} E \xrightarrow{g} X$ is an almost split conflation in $p^1(\Lambda)$. Let Z be an indecomposable object of $p(\Lambda)$ and $h : Z \longrightarrow X$ a non-isomorphism. If $Z \in p^1(\Lambda)$, we already know that h factors through g. So assume that $Z \notin p^1(\Lambda)$, thus, from (18.9), $Z \cong (P, P, 1_P)$, for some indecomposable projective Λ-module P. We may assume that in fact $Z = (P, P, 1_P)$. Since the exact sequence

$$0 \longrightarrow Y_1 \xrightarrow{f_1} E_1 \xrightarrow{g_1} X_1 \longrightarrow 0$$

splits, there is a morphism of Λ-modules $g_1' : X_1 \longrightarrow E_1$ with $g_1 g_1' = 1_{X_1}$. Make $\overline{h}_1 := g_1' h_1$ and $\overline{h}_2 := \varphi_E g_1' h_1$. Then, we have a morphism $\overline{h} := (\overline{h}_1, \overline{h}_2) : Z \longrightarrow E$ in $p(\Lambda)$ satisfying that $g_1 \overline{h}_1 = g_1 g_1' h_1 = h_1$ and $g_2 \overline{h}_2 = g_2 \varphi_E g_1' h_1 = \varphi_X g_1 g_1' h_1 = \varphi_X h_1 = h_2$. Thus, $g\overline{h} = h$. □

Proposition 18.13. *Every conflation $\zeta : Y \xrightarrow{f} E \xrightarrow{g} X$ in $(\mathcal{P}(\Lambda), \mathcal{E})$, induces an exact sequence in Λ-Mod*

$$0 \longrightarrow \operatorname{Ker} Y \xrightarrow{f'} \operatorname{Ker} E \xrightarrow{g'} \operatorname{Ker} X \xrightarrow{\delta} \operatorname{Cok} Y \xrightarrow{\overline{f}} \operatorname{Cok} E \xrightarrow{\overline{g}} \operatorname{Cok} X \longrightarrow 0.$$

If, moreover, ζ *is an almost split conflation in* $(p(\Lambda), \mathcal{E})$ *with* Cok$X \neq 0$*, then*

$$0 \longrightarrow \text{Cok}Y \xrightarrow{\overline{f}} \text{Cok}E \xrightarrow{\overline{g}} \text{Cok}X \longrightarrow 0$$

is an almost split sequence in Λ-mod.

Proof. Recall, from the general theory of categories with exact structures that, for any $Z \in \mathcal{P}(\Lambda)$, there is an exact sequence

$$0 \longrightarrow \text{Hom}_{\mathcal{P}(\Lambda)}(Z, Y) \xrightarrow{f^*} \text{Hom}_{\mathcal{P}(\Lambda)}(Z, E) \xrightarrow{g^*} \text{Hom}_{\mathcal{P}(\Lambda)}(Z, X) \xrightarrow{\partial}$$

$$\text{Ext}_{\mathcal{P}(\Lambda)}(Z, Y) \xrightarrow{f^0} \text{Ext}_{\mathcal{P}(\Lambda)}(Z, E) \xrightarrow{g^0} \text{Ext}_{\mathcal{P}(\Lambda)}(Z, X).$$

Since Cok clearly preserves deflations, from (18.7) applied to $Z = S(\Lambda)$, we get the first exact sequence of our proposition.

Now, we assume that ζ is an almost split conflation satisfying that Cok$X \neq 0$. We want to show that g' is surjective, or equivalently that the map $g^* : \text{Hom}_{\mathcal{P}(\Lambda)}(S(\Lambda), E) \to \text{Hom}_{\mathcal{P}(\Lambda)}(S(\Lambda), X)$ is surjective. If $h \in \text{Hom}_{\mathcal{P}(\Lambda)}(S(\Lambda), X)$, then h factors through g, because ζ is an almost split conflation, and Cok$X \neq 0$ implies that $X_2 \neq 0$ and thus h cannot be a retraction. Thus g^* is surjective.

Now we show that the exact sequence

$$0 \longrightarrow \text{Cok}Y \xrightarrow{\overline{f}} \text{Cok}E \xrightarrow{\overline{g}} \text{Cok}X \longrightarrow 0$$

is almost split. Since X is indecomposable and Cok$X \neq 0$, from (18.9), we know that $X \in \mathcal{P}^2(\Lambda)$. Then, since g is not a retraction, by (18.10)(3), $\overline{g} = \text{Cok}(g)$ is not a retraction. In particular, Cok$Y \neq 0$ and, as before, $Y \in \mathcal{P}^2(\Lambda)$. From (18.10)(5), Cok$Y$ is indecomposable. Then, we only have to verify that the morphism $\overline{g}$ is right almost split. Take $h : M \longrightarrow \text{Cok}X$ a non-retraction. Since Cok : $\mathcal{P}(\Lambda) \longrightarrow \Lambda$-Mod is full and dense, we can assume that $M = \text{Cok}Z$ and that $h = \text{Cok}(t)$, for some morphism $t : Z \longrightarrow X$ in $\mathcal{P}(\Lambda)$. Again, (18.10)(3) guarantees that t is not a retraction. Then, it factors through g and, therefore, h factors through $\overline{g}$. □

References. This section retakes some of the study of the canonical exact structure of $\mathcal{P}(\Lambda)$ from [5], but the material has been remastered and there are some new remarks. The study of the cokernel functor Cok : $\mathcal{P}(\Lambda) \longrightarrow \Lambda$-Mod can be traced back to work of M. Auslander (see [2]).

19

Passage from ditalgebras to finite-dimensional algebras

We will construct, for each finite-dimensional algebra Λ, which admits a splitting pair (S, J) with $J = \mathrm{rad}\Lambda$, a finite-dimensional Roiter ditalgebra $\mathcal{D}^\Lambda$ and an equivalence of categories $\Xi_\Lambda : \mathcal{D}^\Lambda\text{-Mod} \longrightarrow \mathcal{P}^1(\Lambda)$. The use of Ξ_Λ together with the cokernel functor $\mathcal{P}^1(\Lambda) \longrightarrow \Lambda\text{-Mod}$ studied in the last section constitutes the main device to derive results for the given algebra from properties known for some class of ditalgebras containing $\mathcal{D}^\Lambda$.

Definition 19.1. *Assume that Λ is a finite-dimensional k-algebra and denote by J its radical. Assume furthermore that there is a k-subalgebra S of Λ with $\Lambda = S \oplus J$, as S-S-bimodules. In this case we say that Λ splits over its radical. The Drozd's ditalgebra $\mathcal{D}^\Lambda$, associated to Λ, is the ditalgebra (T, δ) with layer (R, W) defined by*

$$R = R^\Lambda = \begin{pmatrix} S & 0 \\ 0 & S \end{pmatrix}, \; W_0 = W_0^\Lambda = \begin{pmatrix} 0 & 0 \\ J^* & 0 \end{pmatrix} \; \textit{and} \; W_1 = W_1^\Lambda = \begin{pmatrix} J^* & 0 \\ 0 & J^* \end{pmatrix}.$$

By definition, $T = T_R(W)$ and, if the comultiplication associated to the product map $J \otimes_S J \longrightarrow J$ is denoted by μ, then the differential $\delta = \delta^\Lambda$ of $\mathcal{D}^\Lambda$ is extended, using additivity and Leibniz rule (see (4.4)), from the morphisms of R-R-bimodules appearing at the top line of the following diagrams

$$\begin{array}{ccccc} W_1 & \xrightarrow{\delta_1} & W_1 \otimes_R W_1 & \subseteq T_R(W) \\ \| & & \downarrow \cong & \\ \begin{pmatrix} J^* & 0 \\ 0 & J^* \end{pmatrix} & \xrightarrow{\begin{pmatrix} \mu & 0 \\ 0 & \mu \end{pmatrix}} & \begin{pmatrix} J^* \otimes_S J^* & 0 \\ 0 & J^* \otimes_S J^* \end{pmatrix}, & \end{array}$$

$$\begin{array}{ccccc}
W_0 & \xrightarrow{\delta_0} & (W_0 \otimes_R W_1) \oplus (W_1 \otimes_R W_0) & \subseteq T_R(W) \\
\| & & \downarrow \cong & \\
\begin{pmatrix} 0 & 0 \\ J^* & 0 \end{pmatrix} & \xrightarrow{\begin{pmatrix} \begin{pmatrix} 0 & 0 \\ -\mu & 0 \end{pmatrix} \\ \begin{pmatrix} 0 & 0 \\ \mu & 0 \end{pmatrix} \end{pmatrix}} & \begin{pmatrix} 0 & 0 \\ J^* \otimes_S J^* & 0 \end{pmatrix} \oplus \begin{pmatrix} 0 & 0 \\ J^* \otimes_S J^* & 0 \end{pmatrix}, &
\end{array}$$

where the vertical isomorphisms are induced by matrix multiplication.

Remark 19.2. *Consider the algebra* $R = R^\Lambda = \begin{pmatrix} S & 0 \\ 0 & S \end{pmatrix}$, *with the matrix product, as in last definition. Given S-S-bimodules* N_{ij}, *with* $i, j \in \{1, 2\}$, *we can form the R-R-bimodule* $N = \begin{pmatrix} N_{11} & N_{12} \\ N_{21} & N_{22} \end{pmatrix}$, *where the action of R on N is given again by the matrix product.*

Consider the canonical decomposition $1 = e_1 + e_2$ *of the unit of R as a sum of orthogonal idempotents and the canonical morphism of algebras* $\xi : S \longrightarrow R$ *such that* $s \mapsto s(e_1 + e_2)$. *Then, any R-R-bimodule M is, by restriction through* ξ, *an S-S-bimodule. Moreover, each* $e_i M e_j$ *is an S-S-subbimodule of such M and there is an isomorphism of R-R-bimodules*

$$M \cong \begin{pmatrix} e_1 M e_1 & e_1 M e_2 \\ e_2 M e_1 & e_2 M e_2 \end{pmatrix}.$$

Given two such matrix bimodules, there is an isomorphism of R-R-bimodules

$$\begin{pmatrix} M_{11} & M_{12} \\ M_{21} & M_{22} \end{pmatrix} \otimes_R \begin{pmatrix} N_{11} & N_{12} \\ N_{21} & N_{22} \end{pmatrix}$$
$$\cong \begin{pmatrix} M_{11} \otimes_S N_{11} \oplus M_{12} \otimes_S N_{21} & M_{11} \otimes_S N_{12} \oplus M_{12} \otimes_S N_{22} \\ M_{21} \otimes_S N_{11} \oplus M_{22} \otimes_S N_{21} & M_{21} \otimes_S N_{12} \oplus M_{22} \otimes_S N_{22} \end{pmatrix}$$

given by the matrix multiplication map

$$\left[\begin{pmatrix} m_{11} & m_{12} \\ m_{21} & m_{22} \end{pmatrix} \otimes \begin{pmatrix} n_{11} & n_{12} \\ n_{21} & n_{22} \end{pmatrix} \right]$$
$$\mapsto \begin{pmatrix} m_{11} \otimes n_{11} + m_{12} \otimes n_{21} & m_{11} \otimes n_{12} + m_{12} \otimes n_{22} \\ m_{21} \otimes n_{11} + m_{22} \otimes n_{21} & m_{21} \otimes n_{12} + m_{22} \otimes n_{22} \end{pmatrix}.$$

This is the isomorphism used in last definition.

Definition 19.3. *A k-algebra R is called trivial iff $R \cong k \times \cdots \times k$. A finite-dimensional k-algebra Λ is said to trivially split over its radical iff Λ admits a trivial k-subalgebra S and an S-S-bimodule decomposition $\Lambda = S \oplus J$, where J is the radical of Λ.*

We say that a field extension $k \leq K$ preserves the radical of the finite-dimensional k-algebra Λ iff the radical of Λ^K coincides with J^K, where J is the radical of Λ.

Remark 19.4. *Assume that Λ is a finite-dimensional k-algebra and let K be any field extension of k. Then:*

(1) If Λ trivially splits over its radical J, then Λ^K is a K-algebra which trivially splits over its radical J^K.
(2) If Λ is basic and k is an algebraically closed field, then Λ trivially splits over its radical.
(3) If $k \leq K$ is a MacLane separable field extension (see [35]), then it preserves the radical Λ (see [36]). If k is algebraically closed, then any field extension $k \leq K$ is MacLane separable.

Proposition 19.5. *Suppose that Λ is a finite-dimensional k-algebra, which splits over its radical J. Let S be a subalgebra of Λ with $\Lambda = S \oplus J$. Then $\mathcal{D}^\Lambda$ is a Roiter ditalgebra with triangular layer (R^Λ, W^Λ).*

The differential δ^Λ of $\mathcal{D}^\Lambda$ admits the following explicit description on generators: Take $\gamma \in J^*$ and assume that $\mu(\gamma) = \sum_r \gamma_r^1 \otimes \gamma_r^2$, then

$$\delta^\Lambda\left(\begin{pmatrix} 0 & 0 \\ \gamma & 0 \end{pmatrix}\right) = \sum_r \left(\begin{pmatrix} 0 & 0 \\ 0 & \gamma_r^1 \end{pmatrix} \otimes \begin{pmatrix} 0 & 0 \\ \gamma_r^2 & 0 \end{pmatrix}\right) - \sum_r \left(\begin{pmatrix} 0 & 0 \\ \gamma_r^1 & 0 \end{pmatrix} \otimes \begin{pmatrix} \gamma_r^2 & 0 \\ 0 & 0 \end{pmatrix}\right),$$

$$\delta^\Lambda\left(\begin{pmatrix} \gamma & 0 \\ 0 & 0 \end{pmatrix}\right) = \sum_r \left(\begin{pmatrix} \gamma_r^1 & 0 \\ 0 & 0 \end{pmatrix} \otimes \begin{pmatrix} \gamma_r^2 & 0 \\ 0 & 0 \end{pmatrix}\right), \text{ and}$$

$$\delta^\Lambda\left(\begin{pmatrix} 0 & 0 \\ 0 & \gamma \end{pmatrix}\right) = \sum_r \left(\begin{pmatrix} 0 & 0 \\ 0 & \gamma_r^1 \end{pmatrix} \otimes \begin{pmatrix} 0 & 0 \\ 0 & \gamma_r^2 \end{pmatrix}\right).$$

If, moreover, Λ trivially splits over its radical (or if k is a perfect field), then the layer (R^Λ, W^Λ) of $\mathcal{D}^\Lambda$ is completely R^Λ-free.

Proof. Notice first that the explicit description of the map $\delta = \delta^\Lambda$ on generators follows from the definition of δ and (19.2). As before, write $R = R^\Lambda$ and $W = W^\Lambda$.

We show first that $\mathcal{D}^\Lambda$ is indeed a ditalgebra. For this it only remains to verify that $\delta^2 = 0$. The fact that $\delta^2(W_1) = 0$ follows from the coassociativity of μ, see (11.7), and the commutativity of the following diagram

$$\begin{array}{ccc}
W_1 & = & \begin{pmatrix} J^* & 0 \\ 0 & J^* \end{pmatrix} \\
\Big\downarrow{\scriptstyle \delta_1} & & \Big\downarrow{\begin{pmatrix} \mu & 0 \\ 0 & \mu \end{pmatrix}} \\
W_1 \otimes_R W_1 & \xrightarrow{\cong} & \begin{pmatrix} J^* \otimes_S J^* & 0 \\ 0 & J^* \otimes_S J^* \end{pmatrix} \\
\Big\downarrow{\scriptstyle \delta_1 \otimes 1 - 1 \otimes \delta_1} & & \Big\downarrow{\begin{pmatrix} \mu \otimes 1 - 1 \otimes \mu & 0 \\ 0 & \mu \otimes 1 - 1 \otimes \mu \end{pmatrix}} \\
W_1 \otimes_R W_1 \otimes_R W_1 & \xrightarrow{\cong} & \begin{pmatrix} J^* \otimes_S J^* \otimes_S J^* & 0 \\ 0 & J^* \otimes_S J^* \otimes_S J^* \end{pmatrix}.
\end{array}$$

Now we show that $\delta^2(W_0) = 0$. Take $\gamma \in J^*$ and assume that $\mu(\gamma) = \sum_r \gamma_r^1 \otimes \gamma_r^2$. Moreover, assume that $\mu(\gamma_r^1) = \sum_{s_r} \gamma_{s_r}^{1,1} \otimes \gamma_{s_r}^{1,2}$ and $\mu(\gamma_r^2) = \sum_{t_r} \gamma_{t_r}^{2,1} \otimes \gamma_{t_r}^{2,2}$. Then, using the explicit description of δ

$$\delta^2\left(\begin{pmatrix} 0 & 0 \\ \gamma & 0 \end{pmatrix}\right)$$

$$= \delta\left[\sum_r \left(\begin{pmatrix} 0 & 0 \\ 0 & \gamma_r^1 \end{pmatrix} \otimes \begin{pmatrix} 0 & 0 \\ \gamma_r^2 & 0 \end{pmatrix}\right) - \sum_r \left(\begin{pmatrix} 0 & 0 \\ \gamma_r^1 & 0 \end{pmatrix} \otimes \begin{pmatrix} \gamma_r^2 & 0 \\ 0 & 0 \end{pmatrix}\right)\right].$$

Using again this explicit description of δ and Leibniz rule, we can expand the last term into the following sum in $W \otimes_R W \otimes_R W \subseteq T$

$$\begin{aligned}
&\textstyle\sum_{r,s_r} \begin{pmatrix} 0 & 0 \\ 0 & \gamma_{s_r}^{1,1} \end{pmatrix} \otimes \begin{pmatrix} 0 & 0 \\ 0 & \gamma_{s_r}^{1,2} \end{pmatrix} \otimes \begin{pmatrix} 0 & 0 \\ \gamma_r^2 & 0 \end{pmatrix} \\
&-\textstyle\sum_{r,t_r} \begin{pmatrix} 0 & 0 \\ 0 & \gamma_r^1 \end{pmatrix} \otimes \begin{pmatrix} 0 & 0 \\ 0 & \gamma_{t_r}^{2,1} \end{pmatrix} \otimes \begin{pmatrix} 0 & 0 \\ \gamma_{t_r}^{2,2} & 0 \end{pmatrix} \\
&+\textstyle\sum_{r,t_r} \begin{pmatrix} 0 & 0 \\ 0 & \gamma_r^1 \end{pmatrix} \otimes \begin{pmatrix} 0 & 0 \\ \gamma_{t_r}^{2,1} & 0 \end{pmatrix} \otimes \begin{pmatrix} \gamma_{t_r}^{2,2} & 0 \\ 0 & 0 \end{pmatrix} \\
&-\textstyle\sum_{r,s_r} \begin{pmatrix} 0 & 0 \\ 0 & \gamma_{s_r}^{1,1} \end{pmatrix} \otimes \begin{pmatrix} 0 & 0 \\ \gamma_{s_r}^{1,2} & 0 \end{pmatrix} \otimes \begin{pmatrix} \gamma_r^2 & 0 \\ 0 & 0 \end{pmatrix} \\
&+\textstyle\sum_{r,s_r} \begin{pmatrix} 0 & 0 \\ \gamma_{s_r}^{1,1} & 0 \end{pmatrix} \otimes \begin{pmatrix} \gamma_{s_r}^{1,2} & 0 \\ 0 & 0 \end{pmatrix} \otimes \begin{pmatrix} \gamma_r^2 & 0 \\ 0 & 0 \end{pmatrix} \\
&-\textstyle\sum_{r,t_r} \begin{pmatrix} 0 & 0 \\ \gamma_r^1 & 0 \end{pmatrix} \otimes \begin{pmatrix} \gamma_{t_r}^{2,1} & 0 \\ 0 & 0 \end{pmatrix} \otimes \begin{pmatrix} \gamma_{t_r}^{2,2} & 0 \\ 0 & 0 \end{pmatrix}.
\end{aligned}$$

From the coassociativity of μ and the isomorphisms described in the last remark, we obtain that this sum is equal to zero.

Now we show that (R, W) is a triangular layer for $\mathcal{D}^\Lambda$. We can consider the ditalgebra $\mathcal{A}$ with layer $(\Lambda, 0)$, then the object $X = \Lambda$ is an admissible $\mathcal{A}$-module, see (12.4). We can identify Λ with $\mathrm{End}_{\mathcal{A}}(X)^{op}$, which have the splitting (S, P) where $P = J$ and S is semisimple. The powers of the radical define right additive filtrations for P and X, which make X a triangular admissible $\mathcal{A}$-module, as in (14.6). Then, we can construct the S-S-bimodule filtration $\mathcal{F}(P)$, of length ℓ_P, and consider the dual filtration $\mathcal{F}(P)^*$, as in (14.7). Define filtrations of R-R-bimodules of the same length ℓ_P for W_0 and W_1 making

$$W_0^i := \begin{pmatrix} 0 & 0 \\ [P^*]_i & 0 \end{pmatrix} \text{ and } W_1^i := \begin{pmatrix} [P^*]_i & 0 \\ 0 & [P^*]_i \end{pmatrix}.$$

To verify the conditions of triangularity for these filtrations, we fix a dual basis $\{(p_j, \gamma_j)\}_j$ for P compatible with the filtration $\mathcal{F}(P)$, as in (14.7). From (11.8), we know that, for $\gamma \in P^*$, $\mu(\gamma) = \sum_{i,j} \gamma(p_i p_j)\gamma_j \otimes \gamma_i$. Then, from the last item of (14.7), we know that whenever $\gamma \in [P^*]_{t+1}$ satisfies $\gamma(p_i p_j) \neq 0$, we must have that $\gamma_i, \gamma_j \in [P^*]_t$. Then, having in mind the explicit description of δ, it is clear that $\delta(W_0^{t+1}) \subseteq (W_1 \otimes_R W_0^t) \oplus (W_0^t \otimes_R W_1) \subseteq A_t W_1 A_t$, where A_t is the subalgebra of $A = [T_R(W)]_0$ generated by R and W_0^t. Similarly, using again the explicit description of δ, if $\gamma \in [P^*]_{t+1}$, we have that $\delta\left(\begin{pmatrix} \gamma & 0 \\ 0 & 0 \end{pmatrix}\right), \delta\left(\begin{pmatrix} 0 & 0 \\ 0 & \gamma \end{pmatrix}\right) \in W_1^t \otimes_R W_1^t$, and hence $\delta(W_1^{t+1}) \subseteq AW_1^t AW_1^t$. Thus, (R, W) is a triangular layer for $\mathcal{D}^\Lambda$. By (5.6), $\mathcal{D}^\Lambda$ is a Roiter ditalgebra.

Our last claim is that both filtrations of W_0 and W_1 are R-R-free, when Λ trivially splits over its radical (or when the field k is perfect). This is true because $R \otimes_k R^{op}$ is a finite-dimensional semisimple k-algebra. □

In the following, we describe the algebraic connection between $\mathcal{D}^\Lambda$-Mod and Λ-Mod. This connection is not direct, so we first recall some notation and some required properties of Λ-modules.

Lemma 19.6. *Suppose that Λ is a finite-dimensional k-algebra which splits over its radical J. Let S be a subalgebra of Λ such that $\Lambda = S \oplus J$. For any Λ-module M, there is a projective cover*

$$\Lambda \otimes_S (M/JM) \longrightarrow M.$$

In particular, every projective Λ-module is induced from an S-module.

Proof. Since S is semisimple and $\Lambda \otimes -$ preserves direct sums, $P = \Lambda \otimes_S (M/JM)$ is a projective Λ-module. Then, there exists a morphism $q : P \longrightarrow M$ such that $\nu q = \pi$, where $\nu : M \longrightarrow M/JM$ is the canonical projection and $\pi :$

$P \longrightarrow M/JM$ is the product map. Since ν is a projective cover, q is surjective. Since S is semisimple, the sequence

$$0 \longrightarrow J \otimes_S (M/JM) \longrightarrow \Lambda \otimes_S (M/JM) \xrightarrow{\pi} M/JM \longrightarrow 0$$

is exact, and $JP = \text{Ker}\,\pi$. Thus, Ker $q \subseteq JP$ and q is a projective cover. □

Remark 19.7. *Suppose that Λ is a finite-dimensional k-algebra which splits over its radical J, say $\Lambda = S \oplus J$. Consider its Drozd's ditalgebra $\mathcal{D} = \mathcal{D}^\Lambda = (T, \delta)$. Notice that $A_\mathcal{D} = [T]_0$ is the matrix algebra*

$$A = A_\mathcal{D} = \begin{pmatrix} S & 0 \\ J^* & S \end{pmatrix}.$$

Then, each object M in $\mathcal{D}^\Lambda$-Mod determines a triple (M_1, M_2, ψ_M), where M_1, M_2 are S-modules and $\psi_M \in \text{Hom}_{S\text{-}S}(J^*, \text{Hom}_k(M_1, M_2))$. Indeed, if we consider the canonical decomposition of the unit $1 = e_1 + e_2$ of A, into a sum of orthogonal idempotents, then $M_1 = e_1 M$, $M_2 = e_2 M$, and, for $\gamma \in J^*$ and $m_1 \in M_1$, $\psi_M(\gamma)[m_1] = \begin{pmatrix} 0 & 0 \\ \gamma & 0 \end{pmatrix} m_1$.

Conversely, for any triple $\tau_M = (M_1, M_2, \psi_M)$, where M_1, M_2 are left S-modules and $\psi_M \in \text{Hom}_{S\text{-}S}(J^*, \text{Hom}_k(M_1, M_2))$, we can construct the matrix A-module $\overline{\tau}_M = \begin{pmatrix} M_1 \\ M_2 \end{pmatrix}$, where the action of A is given by

$$\begin{pmatrix} s & 0 \\ \gamma & s' \end{pmatrix} \begin{pmatrix} m_1 \\ m_2 \end{pmatrix} = \begin{pmatrix} sm_1 \\ s'm_2 + \psi_M(\gamma)[m_1] \end{pmatrix},$$

for $m_1 \in M_1$, $m_2 \in M_2$, $s, s' \in S$ and $\gamma \in J^*$.

If $M \in A$-Mod and $\tau_M = (M_1, M_2, \psi_M)$ is its corresponding triple, write $\widehat{M} := \overline{\tau}_M$, for the corresponding matrix A-module. There is a natural isomorphism of A-modules $\theta_M^0 : M \longrightarrow \widehat{M}$, which maps m onto $\begin{pmatrix} e_1 m \\ e_2 m \end{pmatrix}$.

Denote by $[\mathcal{D}\text{-Mod}]_0$ the full subcategory of $\mathcal{D}$-Mod formed by the matrix $A_\mathcal{D}$-modules defined by triples as above. We exhibit explicitely a quasi inverse $\mathcal{D}\text{-Mod} \longrightarrow [\mathcal{D}\text{-Mod}]_0$ for the corresponding inclusion functor:

The natural isomorphism of A-modules $\theta_M^0 : M \longrightarrow \widehat{M}$, determines the isomorphism $\theta_M := (\theta_M^0, 0) : M \longrightarrow \widehat{M}$ in $\mathcal{D}$-Mod.

Consider the functor $\widehat{\cdot} : \mathcal{D}\text{-Mod} \longrightarrow [\mathcal{D}\text{-Mod}]_0$, which maps each object M onto $\widehat{M}$ and each morphism $f \in \text{Hom}_\mathcal{D}(M, N)$ onto $\widehat{f} := \theta_N f \theta_M^{-1}$. The functor $\widehat{\cdot}$ is clearly an equivalence (a quasi inverse for the inclusion of the given subcategory $[\mathcal{D}\text{-Mod}]_0$). The explicit description of $\widehat{f} = (\widehat{f}^0, \widehat{f}^1)$ is *given, for* $\gamma_1, \gamma_2 \in J^*$, $m_1 \in M_1$ *and* $m_2 \in M_2$, *by the formulas*

$$\widehat{f^0}\begin{pmatrix} m_1 \\ m_2 \end{pmatrix} = \begin{pmatrix} f^0(m_1) \\ f^0(m_2) \end{pmatrix} \text{ and}$$

$$\left[\widehat{f^1}\begin{pmatrix} \gamma_1 & 0 \\ 0 & \gamma_2 \end{pmatrix}\right]\begin{pmatrix} m_1 \\ m_2 \end{pmatrix} = \begin{pmatrix} \left[f^1\begin{pmatrix} \gamma_1 & 0 \\ 0 & 0 \end{pmatrix}\right][m_1] \\ \left[f^1\begin{pmatrix} 0 & 0 \\ 0 & \gamma_2 \end{pmatrix}\right][m_2] \end{pmatrix}.$$

Proposition 19.8. *Suppose that* Λ *is a finite-dimensional k-algebra which splits over its radical* J, *say* $\Lambda = S \oplus J$. *Then, there is an equivalence functor* $\Xi_\Lambda : \mathcal{D}^\Lambda\text{-Mod} \longrightarrow \mathcal{P}^1(\Lambda)$. *It can be described as follows.*

Consider any finite right S-dual basis $(p_j, \gamma_j)_j$ *for* J, *with* $p_j \in J$ *and* $\gamma_j \in J^*$. *If* $M \in \mathcal{D}^\Lambda\text{-Mod}$ *with corresponding triple* (M_1, M_2, ψ_M), *as in (19.7), then the receipt of* $\Xi_\Lambda(M) : \Lambda \otimes_S M_1 \longrightarrow \Lambda \otimes_S M_2$ *can be described, for* $\lambda \in \Lambda$ *and* $m_1 \in M_1$, *by*

$$\Xi_\Lambda(M)[\lambda \otimes m_1] = \sum_j \lambda p_j \otimes \psi_M(\gamma_j)[m_1].$$

Suppose $M, N \in \mathcal{D}^\Lambda\text{-Mod}$ *and* $f = (f^0, f^1) \in \mathrm{Hom}_{\mathcal{D}^\Lambda}(M, N)$, *then* $\Xi_\Lambda(f) = (\Xi_\Lambda(f)_1, \Xi_\Lambda(f)_2)$ *is the morphism in* $\mathcal{P}^1(\Lambda)$ *from* $\Xi_\Lambda(M) : \Lambda \otimes_S M_1 \longrightarrow \Lambda \otimes_S M_2$ *to* $\Xi_\Lambda(N) : \Lambda \otimes_S N_1 \longrightarrow \Lambda \otimes_S N_2$, *such that for any* $\lambda \in \Lambda$, $m_1 \in M_1$ *and* $m_2 \in M_2$

$$\Xi_\Lambda(f)_1(\lambda \otimes m_1) = \lambda \otimes f^0[m_1] + \textstyle\sum_j \lambda p_j \otimes f^1\begin{pmatrix} \gamma_j & 0 \\ 0 & 0 \end{pmatrix}[m_1], \text{ and}$$

$$\Xi_\Lambda(f)_2(\lambda \otimes m_2) = \lambda \otimes f^0[m_2] + \textstyle\sum_j \lambda p_j \otimes f^1\begin{pmatrix} 0 & 0 \\ 0 & \gamma_j \end{pmatrix}[m_2].$$

Proof. Step 1: We exhibit an equivalence $\mathcal{M} \xrightarrow{F} \mathcal{P}^1(\Lambda)$, where $\mathcal{M}$ is some very natural category of morphisms.

Consider the ditalgebra $\mathcal{A} = (\Lambda, 0)$ with layer $(\Lambda, 0)$. Then, we can identify $\mathcal{A}$-Mod with Λ-Mod, and $X = \Lambda$ is a complete admissible $\mathcal{A}$-module, by (13.3). Also, identify Λ with $\mathrm{End}_{\mathcal{A}}(X)^{op}$ and its splitting (S, P) with (S, J). Now, consider the layered ditalgebra $\mathcal{B} := \mathcal{A}^X$ and the corresponding full and faithful functor $F^X : \mathcal{B}\text{-Mod} \longrightarrow \mathcal{A}\text{-Mod} = \Lambda\text{-Mod}$. From (19.6), we know that every projective Λ-module is induced from S, and, therefore, (from (16.1)), F^X induces an equivalence $F^X : \mathcal{B}\text{-Mod} \longrightarrow \Lambda\text{-Proj}$, where the last category is the full subcategory of Λ-Mod formed by the projective Λ-modules. Having in mind the receipt of F^X, notice that the image $F^X(\psi) : \Lambda \otimes_S M_1 \longrightarrow \Lambda \otimes_S M_2$ of a morphism $\psi = (\psi^0, \psi^1) \in \mathrm{Hom}_{\mathcal{B}}(M_1, M_2)$ factors through $J \otimes_S M_2$ iff $\psi^0 = 0$.

Denote by $\mathcal{M}$ the following category: its objects are the morphisms of $\mathcal{B}$-Mod with zero first component; a morphism (f_1, f_2) from $(0, \psi_M^1) : M_1 \longrightarrow M_2$ to $(0, \psi_N^1) : N_1 \longrightarrow N_2$ in $\mathcal{M}$ is, by definition, a pair $(f_1, f_2) \in \mathrm{Hom}_{\mathcal{B}}(M_1, N_1) \times \mathrm{Hom}_{\mathcal{B}}(M_2, N_2)$ such that $f_2(0, \psi_M^1) = (0, \psi_N^1) f_1$ in $\mathcal{B}$-Mod; the morphisms in $\mathcal{M}$ are composed componentwise. Our previous paragraph implies that F^X induces an equivalence of categories $F : \mathcal{M} \longrightarrow \mathcal{P}^1(\Lambda)$.

Step 2: We exhibit an isomorphism of categories $H : \mathcal{M} \longrightarrow [\mathcal{D}^\Lambda\text{-Mod}]_0$.

Recall from (19.7), that for any triple $\tau_M = (M_1, M_2, \psi_M)$, where M_1, M_2 are left S-modules and $\psi_M \in \mathrm{Hom}_{S\text{-}S}(J^*, \mathrm{Hom}_k(M_1, M_2))$, we can construct the A-module $\overline{\tau}_M$. Notice that, if we have another such triple $\tau_N = (N_1, N_2, \psi_N)$, then we have natural isomorphisms

$$[\mathrm{Hom}_S(M_1, N_1) \times \mathrm{Hom}_S(M_2, N_2)] \xrightarrow{\zeta_0} \mathrm{Hom}_R(\overline{\tau}_M, \overline{\tau}_N),$$

defined by

$$[\zeta_0(f_1^0, f_2^0)] \begin{pmatrix} m_1 \\ m_2 \end{pmatrix} = \begin{pmatrix} f_1^0(m_1) \\ f_2^0(m_2) \end{pmatrix}, \quad \text{and}$$

$$\begin{array}{c} [\mathrm{Hom}_{S\text{-}S}(J^*, \mathrm{Hom}_k(M_1, N_1)) \times \mathrm{Hom}_{S\text{-}S}(J^*, \mathrm{Hom}_k(M_2, N_2))] \\ \Big\downarrow \zeta_1 \\ \mathrm{Hom}_{R\text{-}R}(W_1, \mathrm{Hom}_k(\overline{\tau}_M, \overline{\tau}_N)), \end{array}$$

defined by

$$\left[\zeta_1(f_1^1, f_2^1) \begin{pmatrix} \gamma_1 & 0 \\ 0 & \gamma_2 \end{pmatrix}\right] \begin{pmatrix} m_1 \\ m_2 \end{pmatrix} = \begin{pmatrix} f_1^1(\gamma_1)[m_1] \\ f_2^1(\gamma_2)[m_2] \end{pmatrix},$$

for all $m_1 \in M_1$, $m_2 \in M_2$ and $\gamma_1, \gamma_2 \in J^*$.

Now, we proceed to the definition of $H : \mathcal{M} \longrightarrow [\mathcal{D}^\Lambda\text{-Mod}]_0$. Every object $(0, \psi_M) : M_1 \longrightarrow M_2$ in $\mathcal{M}$ determines a triple $\tau_M := (M_1, M_2, \psi_M)$, define $H(0, \psi_M) := \overline{\tau}_M$. Now, assume we have a morphism (f_1, f_2) from $(0, \psi_M) : M_1 \longrightarrow M_2$ to $(0, \psi_N) : N_1 \longrightarrow N_2$ in $\mathcal{M}$. Consider their components $f_1 = (f_1^0, f_1^1)$ and $f_2 = (f_2^0, f_2^1)$, and the triples τ_M and τ_N associated to the objects, then define $H(f_1, f_2) \in \mathrm{Phom}_{R\text{-}W}(\overline{\tau}_M, \overline{\tau}_N)$ by the rules $H(f_1, f_2)^0 = \zeta_0(f_1^0, f_2^0)$ and $H(f_1, f_2)^1 = \zeta_1(f_1^1, f_2^1)$.

Once we have shown that H is a well-defined functor, it will be clear that it is full and faithful. With the notation above, (f_1, f_2) is a morphism in $\mathcal{M}$ which, therefore, satisfies $f_2(0, \psi_M) = (0, \psi_N) f_1$ in $\mathcal{B}$-Mod. This means that, for $\gamma \in J^*$ with $\mu(\gamma) = \sum_t \gamma_t^1 \otimes \gamma_t^2$, we have the following equation (*):

$$f_2^0 \psi_M(\gamma) + \sum_t f_2^1(\gamma_t^1)\psi_M(\gamma_t^2) = \psi_N(\gamma) f_1^0 + \sum_t \psi_N(\gamma_t^1) f_1^1(\gamma_t^2).$$

Moreover, $(f^0, f^1) := H(f_1, f_2) \in \mathrm{Hom}_{\mathcal{D}}(\overline{\tau}_M, \overline{\tau}_N)$ iff the equation $wf^0(m) = f^0(wm) + f^1(\delta(w))(m)$ is satisfied, for any $w \in W_0$ and $m \in \overline{\tau}_M$. This means that

$$\begin{pmatrix} 0 & 0 \\ \gamma & 0 \end{pmatrix} f^0\left[\begin{pmatrix} m_1 \\ m_2 \end{pmatrix}\right] = f^0\left[\begin{pmatrix} 0 & 0 \\ \gamma & 0 \end{pmatrix}\begin{pmatrix} m_1 \\ m_2 \end{pmatrix}\right] + f^1\left(\delta\left[\begin{pmatrix} 0 & 0 \\ \gamma & 0 \end{pmatrix}\right]\right)\begin{pmatrix} m_1 \\ m_2 \end{pmatrix},$$

for any $\gamma \in J^*$. Using the definition of $(f^0, f^1) = H(f_1, f_2)$ and the explicit description for δ given in (19.5), we see that the equation (*) is equivalent to the former one. Then, it remains to show that H is a functor. So, take another object $(0, \psi_L) : L_1 \longrightarrow L_2$ in $\mathcal{M}$ and a morphism (g_1, g_2) from $(0, \psi_N)$ to $(0, \psi_L)$ in $\mathcal{M}$. Let us show that $H[(g_1, g_2)(f_1, f_2)] = H(g_1, g_2)H(f_1, f_2)$. First notice that

$$\zeta_0(g_1^0, g_2^0)\zeta_0(f_1^0, f_2^0)\begin{pmatrix} m_1 \\ m_2 \end{pmatrix} = \begin{pmatrix} g_1^0 f_1^0(m_1) \\ g_2^0 f_2^0(m_2) \end{pmatrix} = \zeta_0(g_1^0 f_1^0, g_2^0 f_2^0)\begin{pmatrix} m_1 \\ m_2 \end{pmatrix}.$$

Then

$$\begin{aligned}
[H(g_1, g_2)H(f_1, f_2)]^0 &= H(g_1, g_2)^0 H(f_1, f_2)^0 \\
&= \zeta_0(g_1^0, g_2^0)\zeta_0(f_1^0, f_2^0) \\
&= \zeta_0(g_1^0 f_1^0, g_2^0 f_2^0) \\
&= (H(g_1 f_1, g_2 f_2))^0 \\
&= (H\,[(g_1, g_2)(f_1, f_2)])^0 .
\end{aligned}$$

Now, write $m = \begin{pmatrix} m_1 \\ m_2 \end{pmatrix}$ and $w = \begin{pmatrix} \gamma_1 & 0 \\ 0 & \gamma_2 \end{pmatrix}$, assume that $\mu(\gamma_i) = \sum_{t_i} \gamma_{t_i}^1 \otimes \gamma_{t_i}^2$, for $i \in \{1, 2\}$. We have $w = w_1 + w_2$, with $w_1 = e_1 w e_1$ and $w_2 = e_2 w e_2$. Write, for $i \in \{1, 2\}$, $\delta(w_i) = \sum_{t_i} w_{t_i}^1 \otimes w_{t_i}^2$, where $w_{t_i}^j$ are the matrices described in (19.5). That means

$$w_{t_1}^1 = \begin{pmatrix} \gamma_{t_1}^1 & 0 \\ 0 & 0 \end{pmatrix}, w_{t_1}^2 = \begin{pmatrix} \gamma_{t_1}^2 & 0 \\ 0 & 0 \end{pmatrix}, w_{t_2}^1 = \begin{pmatrix} 0 & 0 \\ 0 & \gamma_{t_2}^1 \end{pmatrix} \text{ and } w_{t_2}^2 = \begin{pmatrix} 0 & 0 \\ 0 & \gamma_{t_2}^2 \end{pmatrix}.$$

Then, $\delta(w) = \sum_{i,t_i} w^1_{t_i} \otimes w^2_{t_i}$. Therefore, if $\xi := [H(g_1, g_2)H(f_1, f_2)]^1\,(w)[m]$, we have

$$\begin{aligned}
\xi &= H(g_1, g_2)^0 H(f_1, f_2)^1(w)[m] + H(g_1, g_2)^1(w)H(f_1, f_2)^0[m] \\
&\quad + \textstyle\sum_{i,t_i} H(g_1, g_2)^1(w^1_{t_i})H(f_1, f_2)^1(w^2_{t_i})[m] \\
&= \zeta_0(g^0_1, g^0_2)\zeta_1(f^1_1, f^1_2)(w)[m] + \zeta_1(g^1_1, g^1_2)(w)\zeta_0(f^0_1, f^0_2)[m] \\
&\quad + \textstyle\sum_{i,t_i} \zeta_1(g^1_1, g^1_2)(w^1_{t_i})\zeta_1(f^1_1, f^1_2)(w^2_{t_i})[m] \\
&= \zeta_0(g^0_1, g^0_2)\begin{pmatrix} f^1_1(\gamma_1)[m_1] \\ f^1_2(\gamma_2)[m_2] \end{pmatrix} + \zeta_1(g^1_1, g^1_2)(w)\begin{pmatrix} f^0_1(m_1) \\ f^0_2(m_2) \end{pmatrix} \\
&\quad + \textstyle\sum_{i,t_i} \zeta_1(g^1_1, g^1_2)(w^1_{t_i})\zeta_1(f^1_1, f^1_2)(w^2_{t_i})[m] \\
&= \begin{pmatrix} g^0_1 f^1_1(\gamma_1)[m_1] \\ g^0_2 f^1_2(\gamma_2)[m_2] \end{pmatrix} + \begin{pmatrix} g^1_1(\gamma_1)[f^0_1(m_1)] \\ g^1_2(\gamma_2)[f^0_2(m_2)] \end{pmatrix} \\
&\quad + \textstyle\sum_{t_1} \zeta_1(g^1_1, g^1_2)(w^1_{t_1})\zeta_1(f^1_1, f^1_2)(w^2_{t_1})[m] \\
&\quad + \textstyle\sum_{t_2} \zeta_1(g^1_1, g^1_2)(w^1_{t_2})\zeta_1(f^1_1, f^1_2)(w^2_{t_2})[m].
\end{aligned}$$

Let us compute separately the last two sums

$$\begin{aligned}
&\textstyle\sum_{t_1} \zeta_1(g^1_1, g^1_2)(w^1_{t_1})\zeta_1(f^1_1, f^1_2)(w^2_{t_1})[m] \\
&\quad = \textstyle\sum_{t_1} \zeta_1(g^1_1, g^1_2)(w^1_{t_1})\begin{pmatrix} f^1_1(\gamma^2_{t_1})[m_1] \\ 0 \end{pmatrix} \\
&\quad = \textstyle\sum_{t_1} \begin{pmatrix} g^1_1(\gamma^1_{t_1}) f^1_1(\gamma^2_{t_1})[m_1] \\ 0 \end{pmatrix}.
\end{aligned}$$

Similarly

$$\begin{aligned}
&\textstyle\sum_{t_2} \zeta_1(g^1_1, g^1_2)(w^1_{t_2})\zeta_1(f^1_1, f^1_2)(w^2_{t_2})[m] \\
&\quad = \textstyle\sum_{t_2} \zeta_1(g^1_1, g^1_2)(w^1_{t_2})\begin{pmatrix} 0 \\ f^1_2(\gamma^2_{t_2})[m_2] \end{pmatrix} \\
&\quad = \textstyle\sum_{t_2} \begin{pmatrix} 0 \\ g^1_2(\gamma^1_{t_2}) f^1_2(\gamma^2_{t_2})[m_2] \end{pmatrix}.
\end{aligned}$$

Moreover, if $\xi' := H\,[(g_1, g_2)(f_1, f_2)]^1\,(w)[m]$, then

$$\begin{aligned}
\xi' &= [H(g_1 f_1, g_2 f_2)]^1\,(w)[m] \\
&= \zeta_1\left[(g_1 f_1)^1, (g_2 f_2)^1)\right](w)[m] \\
&= \begin{pmatrix} (g_1 f_1)^1(\gamma_1)[m_1] \\ (g_2 f_2)^1(\gamma_2)[m_2] \end{pmatrix} \\
&= \begin{pmatrix} g^0_1 f^1_1(\gamma_1)[m_1] + g^1_1(\gamma_1) f^0_1[m_1] + \sum_{t_1} g^1_1(\gamma^1_{t_1}) f^1_1(\gamma^2_{t_1})[m_1] \\ g^0_2 f^1_2(\gamma_2)[m_2] + g^1_2(\gamma_2) f^0_2[m_2] + \sum_{t_2} g^1_2(\gamma^1_{t_2}) f^1_2(\gamma^2_{t_2})[m_2] \end{pmatrix}.
\end{aligned}$$

Then, $H[(g_1, g_2)(f_1, f_2)]^1 = [H(g_1, g_2)H(f_1, f_2)]^1$. Now, it is clear that H preserves the composition. Clearly it preserves identities and induces a bijection on the classes of objects. Thus, we have constructed the isomorphism $H : \mathcal{M} \longrightarrow [\mathcal{D}\text{-Mod}]_0$.

Step 3: Define the functor Ξ_Λ as the following composition of equivalences

$$\mathcal{D}\text{-Mod} \xrightarrow{\widehat{\cdot}} [\mathcal{D}\text{-Mod}]_0 \xrightarrow{H^{-1}} \mathcal{M} \xrightarrow{F} \mathcal{P}^1(\Lambda),$$

where $\widehat{\cdot}$ was defined in the last remark. Thus, Ξ_Λ is an equivalence too. Let us describe its receipt explicitely.

Given $M \in \mathcal{D}$-Mod, consider its associated triple $\tau_M = (e_1M, e_2M, \psi_M)$. Then, $\Xi_\Lambda(M) = FH^{-1}(\widehat{M}) = F^X(0, \psi_M) \in \mathrm{Hom}_\Lambda(\Lambda \otimes_S e_1M, \Lambda \otimes_S e_2M)$, where

$$\Xi_\Lambda(M)(\lambda \otimes m_1) = \left(F^X(0, \psi_M)\right)^0 (\lambda \otimes m_1) = \sum_j \lambda p_j \otimes \psi_M(\gamma_j)[m_1].$$

Each $f \in \mathrm{Hom}_{\mathcal{D}}(M, N)$ determines $\widehat{f} = (\widehat{f}^{\,0}, \widehat{f}^{\,1}) \in \mathrm{Phom}_{R\text{-}W}\ (\overline{\tau}_M, \overline{\tau}_N)$. Then, $H^{-1}(\widehat{f}) = (\widehat{f}_1, \widehat{f}_2) \in \mathrm{Hom}_{\mathcal{M}}((0, \psi_M), (0, \psi_N))$. Thus, from the last formula of (19.7) and the definition of H, we obtain

$$\begin{pmatrix} \widehat{f}^{\,0}_1(m_1) \\ \widehat{f}^{\,0}_2(m_2) \end{pmatrix} = \widehat{f}^{\,0} \begin{pmatrix} m_1 \\ m_2 \end{pmatrix} = \begin{pmatrix} f^0(m_1) \\ f^0(m_2) \end{pmatrix}$$

and

$$\begin{pmatrix} \left[\widehat{f}^{\,1}_1(\gamma_1)\right][m_1] \\ \left[\widehat{f}^{\,1}_2(\gamma_2)\right][m_2] \end{pmatrix} = \left[\widehat{f}^{\,1} \begin{pmatrix} \gamma_1 & 0 \\ 0 & \gamma_2 \end{pmatrix}\right] \begin{pmatrix} m_1 \\ m_2 \end{pmatrix} = \begin{pmatrix} \left[f^1 \begin{pmatrix} \gamma_1 & 0 \\ 0 & 0 \end{pmatrix}\right][m_1] \\ \left[f^1 \begin{pmatrix} 0 & 0 \\ 0 & \gamma_2 \end{pmatrix}\right][m_2] \end{pmatrix}.$$

Since, $\Xi_\Lambda(f) = FH^{-1}(\widehat{f}) = F[(\widehat{f}_1, \widehat{f}_2)] = (F^X(\widehat{f}_1), F^X(\widehat{f}_2))$, then

$$\begin{aligned} \Xi_\Lambda(f)_1[\lambda \otimes m_1] &= F^X(\widehat{f}_1)^0[\lambda \otimes m_1] \\ &= \lambda \otimes \widehat{f}^{\,0}_1(m_1) + \textstyle\sum_j \lambda p_j \otimes \widehat{f}^{\,1}_1(\gamma_j)[m_1] \\ &= \lambda \otimes f^0(m_1) + \textstyle\sum_j \lambda p_j \otimes f^1 \begin{pmatrix} \gamma_j & 0 \\ 0 & 0 \end{pmatrix} [m_1], \end{aligned}$$

and, similarly

$$\Xi_\Lambda(f)_2[\lambda \otimes m_2] = \lambda \otimes f^0(m_2) + \textstyle\sum_j \lambda p_j \otimes f^1 \begin{pmatrix} 0 & 0 \\ 0 & \gamma_j \end{pmatrix} [m_2].$$

□

Lemma 19.9. *Suppose that Λ is a finite-dimensional k-algebra which splits over its radical J, say $\Lambda = S \oplus J$. Consider the associated Drozd's ditalgebra $\mathcal{D} = \mathcal{D}^\Lambda$ and the equivalence functor $\Xi_\Lambda : \mathcal{D}^\Lambda\text{-Mod} \longrightarrow \mathcal{P}^1(\Lambda)$. Given $M \in$*

$\mathcal{D}^{\Lambda}$-Mod, *if* (M_1, M_2, ψ_M) *is its corresponding triple, then* ψ_M *can be recovered from* $\Xi_{\Lambda}(M) = (\Lambda \otimes_S M_1 \xrightarrow{\phi_M} \Lambda \otimes_S M_2)$ *by the following*

$$\psi_M(\gamma)[m_1] = \sum_t \gamma(\sigma(1 \otimes \eta_{2,t})\phi_M(1 \otimes m_1))m_{2,t},$$

where $\gamma \in J^*$, $m_1 \in M_1$, $\sigma : \Lambda \otimes_S S \longrightarrow \Lambda$ *is the canonical isomorphism, and* $(m_{2,t}, \eta_{2,t})_t$ *is a dual basis of the left* S*-module* M_2 *(*$m_{2,t} \in M_2$ *and* $\eta_{2,t} \in \mathrm{Hom}_S(M_2, S)$*).*

Consider another object $N \in \mathcal{D}^{\Lambda}$-Mod, *with associated triple* (N_1, N_2, ψ_N) *and write* $\Xi_{\Lambda}(N) = (\Lambda \otimes_S N_1 \xrightarrow{\phi_N} \Lambda \otimes_S N_2)$*. Let* $(h_1, h_2) : \Xi_{\Lambda}(M) \longrightarrow \Xi_{\Lambda}(N)$ *be any morphism in* $\mathcal{P}^1(\Lambda)$*, then the morphism* $h \in \mathrm{Hom}_{\mathcal{D}}(M, N)$ *such that* $\Xi_{\Lambda}(h) = (h_1, h_2)$ *is determined by* $\widehat{h} \in \mathrm{Hom}_{\mathcal{D}}(\widehat{M}, \widehat{N})$*, where, for* $m_1 \in M_1$, $m_2 \in M_2$ *and* $\gamma_1, \gamma_2 \in J^*$*, we have*

$$\begin{aligned}
\widehat{h}^0 \begin{pmatrix} m_1 \\ m_2 \end{pmatrix} &= \begin{pmatrix} \sum_u \pi_S(\sigma(1 \otimes \nu_{1,u})h_1(1 \otimes m_1))n_{1,u} \\ \sum_v \pi_S(\sigma(1 \otimes \nu_{2,v})h_2(1 \otimes m_2))n_{2,v} \end{pmatrix} \quad \textit{and} \\
\left[\widehat{h}^1 \begin{pmatrix} \gamma_1 & 0 \\ 0 & \gamma_2 \end{pmatrix}\right] \begin{pmatrix} m_1 \\ m_2 \end{pmatrix} &= \begin{pmatrix} \sum_u \gamma_1 \left(\pi_J[\sigma(1 \otimes \nu_{1,u})h_1(1 \otimes m_1)]\right) n_{1,u} \\ \sum_v \gamma_2 \left(\pi_J[\sigma(1 \otimes \nu_{2,v})h_2(1 \otimes m_2)]\right) n_{2,v} \end{pmatrix},
\end{aligned}$$

where, $\Lambda \xrightarrow{\pi_S} S$ *and* $\Lambda \xrightarrow{\pi_J} J$ *are the natural projections,* $\sigma : \Lambda \otimes_S S \rightarrow \Lambda$ *is the canonical isomorphism, and* $(n_{1,u}, \nu_{1,u})_u$ *and* $(n_{2,v}, \nu_{2,v})_v$ *denote dual basis for the left* S*-modules* N_1 *and* N_2*, respectively.*

Proof. We will use freely the notation of the proof of last proposition. Recall that, in the proof of (13.3), we described the inverse G of the map

$$\mathrm{Hom}_{\mathcal{B}}(M_1, M_2) \xrightarrow{F^X} \mathrm{Hom}_{\mathcal{A}}(X \otimes_S M_1, X \otimes_S M_2).$$

Having in mind the identifications of $\mathrm{Hom}_{\Lambda}(\Lambda \otimes_S M_1, \Lambda \otimes_S M_2)$ with $\mathrm{Hom}_{\mathcal{A}}(X \otimes_S M_1, X \otimes_S M_2)$, which maps ϕ_M onto $(\phi_M, 0)$, and of $\Gamma = \mathrm{End}_{\mathcal{A}}(X)^{op}$ with Λ, which maps each $\mathcal{A}$-endomorphism $(g, 0)$ of X onto $g(1)$, we can translate the receipt of G, for $f := (\phi_M, 0)$, $G(f) = (0, \psi_M)$, $\gamma \in J^*$ and $m_1 \in M_1$, as follows

$$\begin{aligned}
\psi_M(\gamma)[m_1] &= G(f)^1(\gamma)[m_1] \\
&= \textstyle\sum_t \gamma(\pi_P(f_{t,m_1}))m_{2,t} \\
&= \textstyle\sum_t \gamma(\pi_J(\sigma(1 \otimes \eta_{2,t})\phi_M(1 \otimes m_1)))m_{2,t} \\
&= \textstyle\sum_t \gamma(\sigma(1 \otimes \eta_{2,t})\phi_M(1 \otimes m_1))m_{2,t},
\end{aligned}$$

as in the statement of our lemma.

For the final part of the proof, notice that the equality $\Xi_{\Lambda}(h) = (h_1, h_2)$ is equivalent to $H(G(h_1, 0), G(h_2, 0)) = \widehat{h}$, where $G : \mathrm{Hom}_{\mathcal{A}}(\Lambda \otimes_S M_i, \Lambda \otimes N_i) \longrightarrow \mathrm{Hom}_{\mathcal{B}}(M_i, N_i)$ is the inverse of the restriction of F^X, for $i \in \{1, 2\}$, as above. Thus, for $m_1 \in M_1$

$$\begin{aligned} G(h_1,0)^0[m_1] &= \textstyle\sum_u \pi_S((h_1)_{u,m_1})n_{1,u} \\ &= \textstyle\sum_u \pi_S(\sigma(1\otimes \nu_{1,u})h_1(1\otimes m_1))n_{1,u}. \end{aligned}$$

Similarly, $G(h_2,0)^0[m_2] = \sum_v \pi_S(\sigma(1\otimes \nu_{2,v})h_2(1\otimes m_2))n_{2,v}$, for $m_2 \in M_2$. Moreover, for $\gamma \in J^*$ and $m_1 \in M_1$

$$\begin{aligned} G(h_1,0)^1(\gamma)[m_1] &= \textstyle\sum_u \gamma(\pi_P[(h_1)_{t,m_1}])n_{1,u} \\ &= \textstyle\sum_u \gamma(\pi_J[\sigma(1\otimes \nu_{1,u})h_1(1\otimes m_1)])n_{1,u}. \end{aligned}$$

Similarly, $G(h_2,0)^1(\gamma)[m_2] = \sum_v \gamma(\pi_J[\sigma(1\otimes \nu_{2,v})h_2(1\otimes m_2)])n_{2,v}$, for $m_2 \in M_2$. Then, $\widehat{h} = H(G(h_1,0), G(h_2,0))$ is given as in the statement of the lemma. □

Proposition 19.10. *Let Λ be a finite-dimensional algebra over the field k, which splits over its radical. Let $\mathcal{D} = \mathcal{D}^\Lambda$ be its associated Drozd's ditalgebra. Consider the exact structures $\mathcal{E}_\mathcal{D}$ of $\mathcal{D}$-Mod and $\mathcal{E}_1$ of $\mathcal{P}^1(\Lambda)$, as described in (6.8) and (18.11), respectively. Then, the equivalence $\Xi_\Lambda : \mathcal{D}\text{-Mod} \longrightarrow \mathcal{P}^1(\Lambda)$ maps $\mathcal{E}_\mathcal{D}$ into $\mathcal{E}_1$. It induces an isomorphism of bifunctors*

$$\mathrm{Ext}_\mathcal{D}(-,?) \longrightarrow \mathrm{Ext}_{\mathcal{P}^1(\Lambda)}(\Xi_\Lambda(-), \Xi_\Lambda(?)).$$

Proof. If $\zeta : Y \xrightarrow{f} E \xrightarrow{g} X$ is a conflation of $\mathcal{E}_\mathcal{D}$, then, from (5.14), there is an equivalence of conflations in $\mathcal{D}$-Mod of the form

$$\begin{array}{ccccc} Y & \xrightarrow{f} & E & \xrightarrow{g} & X \\ \| & & \downarrow\cong & & \| \\ Y & \xrightarrow{(r^0,0)} & E' & \xrightarrow{(s^0,0)} & X, \end{array}$$

where $0 \longrightarrow Y \xrightarrow{r^0} E' \xrightarrow{s^0} X \longrightarrow 0$ is a split exact sequence of R-modules. Then, $\Xi_\Lambda(\zeta)$ is equivalent to

$$\Xi_\Lambda(Y) \xrightarrow{\Xi(r^0,0)} \Xi_\Lambda(E') \xrightarrow{\Xi(s^0,0)} \Xi_\Lambda(X),$$

which satisfies that each of its "components"

$$0 \longrightarrow \Lambda \otimes_S Y_i \xrightarrow{1\otimes r_i^0} \Lambda \otimes_S E'_i \xrightarrow{1\otimes s_i^0} \Lambda \otimes_S X_i \longrightarrow 0$$

is a split exact sequence (see (19.8)). Here, $r_i^0 : Y_i \longrightarrow E'_i$ and $s_i^0 : E'_i \longrightarrow X_i$ are the restrictions of r^0 and s^0 to the corresponding i-components. Thus, $\Xi_\Lambda(\zeta) \in \mathcal{E}_1$. The remaining part will follow from the commutativity of the following diagram

$$\begin{array}{ccccc}
\mathrm{Hom}_{R^\Lambda}(W_0^\Lambda \otimes_{R^\Lambda} X, Y) & \xrightarrow{\psi_{\mathcal{D}}} & \mathrm{Ext}_{\mathcal{D}}(X, Y) & \longrightarrow & 0 \\
\downarrow{\scriptstyle \theta} & & \downarrow{\scriptstyle \xi} & & \\
\mathrm{Hom}_\Lambda(\Xi_\Lambda(X)_1, J\Xi_\Lambda(Y)_2) & \xrightarrow{\eta} & \mathrm{Ext}_{\mathcal{P}^1(\Lambda)}(\Xi_\Lambda(X), \Xi_\Lambda(Y)) & \longrightarrow & 0,
\end{array}$$

where J is the radical of Λ, $\psi_{\mathcal{D}}$ is the epimorphism defined in (6.13), η is the epimorphism constructed in a similar way as η in (18.6), ξ is induced by Ξ_Λ, and θ is the following canonical isomorphism. First denote by ρ the following composition of natural isomorphisms

$$\mathrm{Hom}_S(J^*, Y_2) \cong J \otimes_S Y_2 \cong \mathrm{Hom}_\Lambda(\Lambda, J \otimes_S Y_2).$$

Thus, if $(p_j, \gamma_j)_j$ is a dual basis for the projective right S-module J, $\varphi \in \mathrm{Hom}_S(J^*, Y_2)$ and $\lambda \in \Lambda$, then

$$\rho(\varphi)[\lambda] = \sum_j \lambda p_j \otimes \varphi(\gamma_j).$$

Then, by definition, θ is the composition

$$\begin{aligned}
\mathrm{Hom}_{R^\Lambda}(W_0^\Lambda \otimes_{R^\Lambda} X, Y) &\cong \mathrm{Hom}_S(J^* \otimes_S X_1, Y_2) \cong \mathrm{Hom}_S(X_1, \mathrm{Hom}_S(J^*, Y_2)) \\
&\cong \mathrm{Hom}_S(X_1, \mathrm{Hom}_\Lambda(\Lambda, J \otimes_S Y_2)) \cong \mathrm{Hom}_\Lambda(\Lambda \otimes_S X_1, J \otimes_S Y_2).
\end{aligned}$$

Thus, given $u \in \mathrm{Hom}_{R^\Lambda}(W_0^\Lambda \otimes_{R^\Lambda} X, Y)$, $\lambda \in \Lambda$ and $x_1 \in X_1$

$$\theta(u)[\lambda \otimes x_1] = \sum_j \lambda p_j \otimes u\left[\begin{pmatrix} 0 & 0 \\ \gamma_j & 0 \end{pmatrix} \otimes x_1\right].$$

Now notice that there are canonical isomorphisms h_1 and h_2 making the following diagram commutative in Λ-Mod

$$\begin{array}{ccc}
\Lambda \otimes_S (Y \oplus_u X)_1 & \xrightarrow{h_1} & (\Lambda \otimes_S Y_1) \oplus (\Lambda \otimes_S X_1) \\
{\scriptstyle \Xi_\Lambda(Y \oplus_u X)}\downarrow & & \downarrow \begin{pmatrix} \Xi_\Lambda(Y) & \theta(u) \\ 0 & \Xi_\Lambda(X) \end{pmatrix} \\
\Lambda \otimes_S (Y \oplus_u X)_2 & \xrightarrow{h_2} & (\Lambda \otimes_S Y_2) \oplus (\Lambda \otimes_S X_2).
\end{array}$$

If we denote by $h = (h_1, h_2)$ this isomorphism in $\mathcal{P}(\Lambda)$, then we obtain an equivalence of conflations

$$\begin{array}{ccccc}
\Xi_\Lambda(Y) & \longrightarrow & \Xi_\Lambda(Y \oplus_u X) & \longrightarrow & \Xi_\Lambda(X) \\
\| & & \downarrow{\scriptstyle h} & & \| \\
\Xi_\Lambda(Y) & \longrightarrow & Z(\theta(u)) & \longrightarrow & \Xi_\Lambda(X).
\end{array}$$

Here $Z(\theta(u))$ is the object of $\mathcal{P}^1(\Lambda)$ associated to the morphism $\theta(u)$ which is to appear as the middle term of the canonical representative of the class $\eta(\theta(u))$ (see the construction of η in the proof of (18.6)). Thus, $\xi\psi_{\mathcal{D}}(u) = \eta\theta(u)$, as claimed. □

References. The construction of a bocs corresponding to a finite-dimensional algebra Λ was given by Drozd in [28]. The description of $\mathcal{D}^\Lambda$ and the functor $\Xi_\Lambda : \mathcal{D}^\Lambda\text{-Mod} \longrightarrow \mathcal{P}^1(\Lambda)$ adapts Crawley's one in [19] to ditalgebra language. The proof given here of the fact that Ξ_Λ is an equivalence follows the proposal of [6].

20

Scalar extension and ditalgebras

Here we will consider field extensions K of our fixed ground field k, define scalar extension of ditalgebras and modules, and study the compatibility of basic ditalgebra constructions with scalar extension. As usual, given a k-vector space (resp. a k-algebra) B, we denote by B^K the induced K-vector space (resp. K-algebra) $B \otimes_k K$.

Definition 20.1. *Given any field extension $k \leq K$ and a k-t-algebra T, we can form the extended K-t-algebra T^K, with $[T^K]_i = [T]_i^K$, for all i. Indeed, for any n, we have a commutative square of isomorphisms*

$$\begin{array}{ccc} \left([T]_1 \otimes_{[T]_0} \cdots \otimes_{[T]_0} [T]_1\right)^K & \rightarrow & [T]_n^K \\ \cong \uparrow & & \uparrow \cong \\ [T^K]_1 \otimes_{[T^K]_0} \cdots \otimes_{[T^K]_0} [T^K]_1 & \rightarrow & [T^K]_n. \end{array}$$

Moreover, given a k-ditalgebra $\mathcal{A} = (T, \delta)$, we can form the extended K-ditalgebra $\mathcal{A}^K = (T^K, \delta^K)$, where $\delta^K = \delta \otimes 1_K$. Clearly, if $\mathcal{A}$ is layered by (R, W), with $W = W_0 \oplus W_1$, then $\mathcal{A}^K$ is layered by (R^K, W^K), with $W^K = W_0^K \oplus W_1^K$. For any morphism $\phi : \mathcal{A} \longrightarrow \mathcal{B}$ of k-ditalgebras, we have the corresponding morphism $\phi^K = \phi \otimes 1 : \mathcal{A}^K \longrightarrow \mathcal{B}^K$ of K-ditalgebras. If ϕ is a morphism of layered ditalgebras, so is ϕ^K.

Lemma 20.2. *Given any field extension $k \leq K$ and a k-ditalgebra $\mathcal{A}$, consider the extended K-ditalgebra $\mathcal{A}^K$. Then, there is a scalar extension functor $(-)^K$: $\mathcal{A}$-Mod$\longrightarrow \mathcal{A}^K$-Mod defined as follows: If $M \in \mathcal{A}$-Mod, then the space M^K is equipped with its natural A_A^K-module structure. Let us describe the receipt on morphisms. Consider $M, N \in \mathcal{A}$-Mod and the canonical morphisms*

$$\mathrm{Hom}_k(M, N) \xrightarrow{\sigma_0} \mathrm{Hom}_K(M^K, N^K) \text{ and}$$
$$\mathrm{Hom}_{A\text{-}A}(V, \mathrm{Hom}_k(M, N)) \xrightarrow{\sigma_1} \mathrm{Hom}_{A^K\text{-}A^K}(V^K, \mathrm{Hom}_K(M^K, N^K)),$$

given by $\sigma_0(f^0) = f^0 \otimes 1_K$ *and* $\sigma_1(f^1)[v \otimes t] = f^1(v) \otimes t1_K$, *for* $v \in V$ *and* $t \in K$. *Then, for* $f = (f^0, f^1) \in \mathrm{Hom}_{\mathcal{A}}(M, N)$, *its image under* $(-)^K$ *is defined by* $f^K = (\sigma_0(f^0), \sigma_1(f^1))$.

Moreover, the trivial scalar extension $(-)^k : \mathcal{A}\text{-Mod} \longrightarrow \mathcal{A}^k\text{-Mod}$ *is an equivalence of categories.*

Proof. It is straightforward to verify that $f^K \in \mathrm{Hom}_{\mathcal{A}^K}(M^K, N^K)$. Assume that $f \in \mathrm{Hom}_{\mathcal{A}}(M, N)$ and $g \in \mathrm{Hom}_{\mathcal{A}}(N, L)$. We have to see that $(gf)^K = g^K f^K$. Clearly, $(gf)^{K0} = g^{K0} f^{K0}$. In order to see that the other component coincides too, take $v \in V$, $m \in M$ and $t, t' \in K$, assume that $\delta(v) = \sum_r v_r^1 \otimes v_r^2$, with $v_r^1, v_r^2 \in V$, then our identification $V^K \otimes V^K = [T^K]_1 \otimes_{[T^K]_0} [T^K]_1 \xrightarrow{\cong} [T^K]_2$ justifies the formula $\delta^K(v \otimes t) = \sum_r [v_r^1 \otimes 1] \otimes [v_r^2 \otimes t]$. Hence

$$
\begin{aligned}
((gf)^K)^1(v \otimes t)[m \otimes t'] &= [(gf)^1(v) \otimes t1_K][m \otimes t'] \\
&= g^0 f^1(v)[m] \otimes tt' + g^1(v) f^0[m] \otimes tt' \\
&\quad + \textstyle\sum_r g^1(v_r^1) f^1(v_r^2)[m] \otimes tt' \\
&= g^{K0} f^{K1}(v \otimes t)[m \otimes t'] + g^{K1}(v \otimes t) f^{K0}[m \otimes t'] \\
&\quad + \textstyle\sum_r g^{K1}[v_r^1 \otimes 1] f^{K1}[v_r^2 \otimes t][m \otimes t'] \\
&= (g^K f^K)^1(v \otimes t)[m \otimes t'].
\end{aligned}
$$

Then, $(-)^K$ preserves the composition of morphisms of $\mathcal{A}$-Mod. It clearly preserves identities.

For the last statement of our lemma, where $K = k$, notice first that, for $M \in A$-Mod, the product map $\pi_M : M^k \longrightarrow M$ is an isomorphism of A-modules, where M^k is an A-module by restriction using $A \longrightarrow A^k$ given by $a \mapsto a \otimes 1$. In fact, π is a natural isomorphism from $(-)^k : A\text{-Mod} \longrightarrow A\text{-Mod}$ to the identity functor on A-Mod. Then, notice that $\sigma_0(f^0) = \pi_N^{-1} f^0 \pi_M$ and $\sigma_1(f^1) = \sigma_0 f^1 \pi_V$. It follows that σ_0 and σ_1 are bijective, and $(-)^k : \mathcal{A}\text{-Mod} \longrightarrow \mathcal{A}^k\text{-Mod}$ is full and faithful. The former functor is dense because $(\pi_M, 0) : M^k \longrightarrow M$ is an isomorphism in $\mathcal{A}^k$-Mod. □

Lemma 20.3. *Assume we have a field extension* $k \leq K$ *and a morphism of ditalgebras* $\phi : \mathcal{A} \longrightarrow \mathcal{B}$. *Then, the following square of functors commutes*

$$
\begin{array}{ccc}
\mathcal{B}^K\text{-Mod} & \xrightarrow{F_{\phi^K}} & \mathcal{A}^K\text{-Mod} \\
\uparrow{\scriptstyle (-)^K} & & \uparrow{\scriptstyle (-)^K} \\
\mathcal{B}\text{-Mod} & \xrightarrow{F_\phi} & \mathcal{A}\text{-Mod}.
\end{array}
$$

In the following series of lemmas, we show that scalar extension commutes with basic ditalgebra constructions and with the functors associated to them. We look at the more concrete realizations of these operations, which appear in practice.

Lemma 20.4. *Let $\mathcal{A}$ be a ditalgebra with layer (R, W) and assume that there is an idempotent $e \in R$ satisfying that $eR(1-e) = 0 = (1-e)Re$. Let $\mathcal{A}^d$ be the ditalgebra obtained from $\mathcal{A}$ by deletion of the idempotent $1-e$, as in (8.17). Thus, $\mathcal{A}^d$ has layer $(R^d, W^d) = (eRe, eWe)$ and there is a canonical morphism $\phi_d : \mathcal{A} \longrightarrow \mathcal{A}^d$. Given any field extension $k \leq K$, we can first consider the extension $\mathcal{A}^K$, with layer (R^K, W^K), where R^K has the idempotent $\hat{e} = e \otimes 1$ satisfying $\hat{e}R^K(1-\hat{e}) = 0 = (1-\hat{e})R^K\hat{e}$. Then, we can form the ditalgebra $\mathcal{A}^{Kd}$ obtained from $\mathcal{A}^K$ by deletion of the idempotent $1-\hat{e}$, and we have the corresponding morphism of ditalgebras $\phi_{d^K} : \mathcal{A}^K \to \mathcal{A}^{Kd}$. Then, there is an isomorphism of layered ditalgebras $\xi_d : \mathcal{A}^{Kd} \longrightarrow \mathcal{A}^{dK}$, such that $(\phi_d)^K = \xi_d \phi_{d^K}$. It induces an isomorphism of categories in the following commutative diagram*

$$\begin{array}{ccccc}
\mathcal{A}^{dK}\text{-Mod} & \xrightarrow{F_{\xi_d}} & \mathcal{A}^{Kd}\text{-Mod} & \xrightarrow{F_{\phi_{d^K}}} & \mathcal{A}^{K}\text{-Mod} \\
\uparrow{\scriptstyle (-)^K} & & & & \uparrow{\scriptstyle (-)^K} \\
\mathcal{A}^{d}\text{-Mod} & & \xrightarrow{\qquad F_{\phi_d} \qquad} & & \mathcal{A}\text{-Mod}.
\end{array}$$

Proof. Write $\mathcal{A} = (T, \delta)$, $\mathcal{A}^d = (T^d, \delta^d)$ and $\mathcal{A}^K = (T^K, \delta^K)$. Since T^{Kd} is freely generated by $(\hat{e}R^K\hat{e}, \hat{e}W^K\hat{e})$ the morphism of algebras $R^{Kd} = \hat{e}R^K\hat{e} \cong [eRe]^K = R^{dK}$ and the morphism of R^{Kd}-bimodules $W^{Kd} = \hat{e}W^K\hat{e} \cong [eWe]^K = W^{dK}$, induce an isomorphism of t-algebras $\xi_d : T^{Kd} \longrightarrow T^{dK}$. Let us verify that $\xi_d\delta^{Kd} = \delta^{dK}\xi_d$. Both maps are zero when evaluated at elements of R^{Kd}. Since T^{Kd} is generated by R^{Kd} and W^{Kd}, we have to compute the value of these maps on elements of W^{Kd}. Take $w \in W$ and $t \in K$, assume that $\delta(w) = \sum_r w_1^r w_2^r \cdots w_{t_r}^r$, with $w_j^r \in W$. Then, $\delta^K(w \otimes t) = \sum_r w_1^r \cdots w_{t_r}^r \otimes t = \sum_r (w_1^r \otimes 1)(w_2^r \otimes 1) \cdots (w_{t_r}^r \otimes t)$. Since $\delta^d\phi_d = \phi_d\delta$, $\delta^d(ewe) = \sum_r ew_1^r ew_2^r e \cdots ew_{t_r}^r e$. Hence

$$\begin{aligned}
\xi_d\delta^{Kd}(\hat{e}(w \otimes t)\hat{e}) &= \xi_d \left[\textstyle\sum_r \hat{e}(w_1^r \otimes 1)\hat{e}(w_2^r \otimes 1)\hat{e} \cdots \hat{e}(w_{t_r}^r \otimes t)\hat{e} \right] \\
&= \textstyle\sum_r ew_1^r ew_2^r e \cdots ew_{t_r}^r e \otimes t \\
&= \delta^d(ewe) \otimes t \\
&= \delta^{dK}(ewe \otimes t) \\
&= \delta^{dK}\xi_d(\hat{e}(w \otimes t)\hat{e}).
\end{aligned}$$

The equality $(\phi_d)^K = \xi_d \phi_{d^K}$ is immediately verified on generators, and the commutativity of the diagram now follows from (20.3). □

Lemma 20.5. *Let $\mathcal{A}$ be a ditalgebra with layer (R, W) and assume that there are decompositions of R-R-bimodules $W_0 = W_0' \oplus W_0''$ and $W_1 = \delta(W_0') \oplus W_1''$. Let $\mathcal{A}^r$ be the ditalgebra obtained from $\mathcal{A}$ by regularization of the bimodule W_0', as in (8.19). Thus, $\mathcal{A}^r$ has layer $(R^r, W^r) = (R, W_0'' \oplus W_1'')$ and there is a canonical morphism $\phi_r : \mathcal{A} \longrightarrow \mathcal{A}^r$. Given any field extension $k \leq K$, we can first consider the extension $\mathcal{A}^K$, with layer (R^K, W^K) and the decomposition of R^K-bimodules $W_0^K = (W_0')^K \oplus (W_0'')^K$ and $W_1^K = \delta^K[(W_0')^K] \oplus (W_1'')^K$. Then, we can form the ditalgebra $\mathcal{A}^{Kr}$ obtained from $\mathcal{A}^K$ by regularization of the bimodule $(W_0')^K$ and we have the corresponding morphism of ditalgebras $\phi_{r^K} : \mathcal{A}^K \to \mathcal{A}^{Kr}$. Then, there is an isomorphism of layered ditalgebras $\xi_r : \mathcal{A}^{Kr} \longrightarrow \mathcal{A}^{rK}$, such that $(\phi_r)^K = \xi_r \phi_{r^K}$. It induces an isomorphism of categories in the following commutative diagram*

$$\begin{array}{ccccc} \mathcal{A}^{rK}\text{-Mod} & \xrightarrow{F_{\xi_r}} & \mathcal{A}^{Kr}\text{-Mod} & \xrightarrow{F_{\phi_{r^K}}} & \mathcal{A}^K\text{-Mod} \\ \uparrow (-)^K & & & & \uparrow (-)^K \\ \mathcal{A}^r\text{-Mod} & & \xrightarrow{\quad F_{\phi_r} \quad} & & \mathcal{A}\text{-Mod}. \end{array}$$

Proof. We proceed as for the proof of last lemma. Write $\mathcal{A} = (T, \delta)$, $\mathcal{A}^r = (T^r, \delta^r)$ and $\mathcal{A}^K = (T^K, \delta^K)$. Since T^{Kr} is freely generated by (R^{Kr}, W^{Kr}), the identity morphism $R^{Kr} = R^K = R^{rK}$ and the morphism of R^{Kr}-bimodules $W^{Kr} = (W_0'')^K \oplus (W_1'')^K \cong W^{rK}$ induce an isomorphism of t-algebras $\xi_r : T^{Kr} \longrightarrow T^{rK}$. The verification of $\xi_r \delta^{Kr} = \delta^{rK} \xi_r$ is similar to the corresponding statement in last lemma. Now we use that, for $w \in W$ with $\delta(w) = \sum_r w_1^r w_2^r \cdots w_{t_r}^r$, where $w_j^r \in W$, we have $\delta^r(\overline{w}) = \sum_r \overline{w}_1^r \overline{w}_2^r \cdots \overline{w}_{t_r}^r$, where the overline denotes the canonical projection $W \longrightarrow W^r$. The equality $(\phi_r)^K = \xi_r \phi_{r^K}$ is immediately verified on generators, and the commutativity of the diagram follows from (20.3). □

Lemma 20.6. *Let $\mathcal{A} = (T, \delta)$ be a ditalgebra with layer (R, W) and assume that there is a decomposition of R-bimodules $W_0 = W_0' \oplus W_0''$ and $\delta(W_0') = 0$. Let $\mathcal{A}^a$ be the ditalgebra obtained from $\mathcal{A}$ by absorption of the bimodule W_0', as in (8.20). Thus, $\mathcal{A}^a$ has layer (R^a, W^a), where R^a is the subalgebra of T generated by R and W_0', $W_0^a = R^a W_0'' R^a$ and $W_1^a = R^a W_1 R^a$. We have a canonical morphism $\phi_a : \mathcal{A} \longrightarrow \mathcal{A}^a$. Given any field extension $k \leq K$, we can first consider the extension $\mathcal{A}^K$, with layer (R^K, W^K) and the bimodule decomposition of R^K-bimodules $W_0^K = (W_0')^K \oplus (W_0'')^K$, where $\delta^K[(W_0')^K] = 0$. Then, we can form the ditalgebra $\mathcal{A}^{Ka}$ obtained from $\mathcal{A}^K$ by* *absorption of the bimodule* $(W_0')^K$ *and we have the corresponding morphism of ditalgebras* $\phi_{a^K} : \mathcal{A}^K \to \mathcal{A}^{Ka}$. *Then, there is an isomorphism of layered ditalgebras* $\xi_a : \mathcal{A}^{Ka} \longrightarrow \mathcal{A}^{aK}$, *such that* $(\phi_a)^K = \xi_a \phi_{a^K}$. *It induces an isomorphism of categories in the following commutative diagram*

$$\begin{array}{ccccc} \mathcal{A}^{aK}\text{-Mod} & \xrightarrow{F_{\xi_a}} & \mathcal{A}^{Ka}\text{-Mod} & \xrightarrow{F_{\phi_{a^K}}} & \mathcal{A}^K\text{-Mod} \\ \uparrow{\scriptstyle (-)^K} & & & & \uparrow{\scriptstyle (-)^K} \\ \mathcal{A}^a\text{-Mod} & & \xrightarrow{F_{\phi_a}} & & \mathcal{A}\text{-Mod}. \end{array}$$

Proof. Here, of course, $\xi_a : T^{Ka} \longrightarrow T^{aK}$ is the identity map, which is an isomorphism of layered t-algebras because $R^{Ka} = R^{aK}$ and $W_i^{Ka} = W_i^{aK}$, for $i \in \{0, 1\}$. □

Lemma 20.7. *Let* $\mathcal{A}$ *be a ditalgebra with layer* (R, W) *and assume that* W_0' *is an* R-R-*subbimodule of* W_0 *such that* $\delta(W_0') \subseteq W_1$. *Let* $\mathcal{A}^p$ *be the ditalgebra obtained from* $\mathcal{A}$ *by factoring out the subbimodule* W_0', *as in (8.18). Thus,* $\mathcal{A}^p$ *has layer* (R, W^p), *where* $W_0^p = W_0/W_0'$ *and* $W_1^p = W_1/\delta(W_0')$, *and there is a canonical morphism of layered ditalgebras* $\phi_p : \mathcal{A} \longrightarrow \mathcal{A}^p$. *Given any field extension* $k \leq K$, *we can first consider the extension* $\mathcal{A}^K$, *with layer* (R^K, W^K) *and the* R^K-*subbimodule* $(W_0')^K$ *of* W_0^K, *which satisfies* $\delta^K((W_0')^K) \subseteq W_1^K$. *Then, we can form the ditalgebra* $\mathcal{A}^{Kp}$ *obtained from* $\mathcal{A}^K$ *by factoring out the subbimodule* $(W_0')^K$, *and we have the corresponding morphism of ditalgebras* $\phi_{p^K} : \mathcal{A}^K \to \mathcal{A}^{Kp}$. *Then, there is an isomorphism of layered ditalgebras* $\xi_p : \mathcal{A}^{Kp} \longrightarrow \mathcal{A}^{pK}$, *such that* $(\phi_p)^K = \xi_p \phi_{p^K}$. *It induces an isomorphism of categories in the following commutative diagram*

$$\begin{array}{ccccc} \mathcal{A}^{pK}\text{-Mod} & \xrightarrow{F_{\xi_p}} & \mathcal{A}^{Kp}\text{-Mod} & \xrightarrow{F_{\phi_{p^K}}} & \mathcal{A}^K\text{-Mod} \\ \uparrow{\scriptstyle (-)^K} & & & & \uparrow{\scriptstyle (-)^K} \\ \mathcal{A}^p\text{-Mod} & & \xrightarrow{F_{\phi_p}} & & \mathcal{A}\text{-Mod}. \end{array}$$

Proof. We proceed as before. Write $\mathcal{A} = (T, \delta)$, $\mathcal{A}^p = (T^p, \delta^p)$ and $\mathcal{A}^K = (T^K, \delta^K)$. Since T^{Kp} is freely generated by (R^{Kp}, W^{Kp}), the identity morphism $R^{Kp} = R^K = R^{pK}$ and the morphism of R^{Kp}-bimodules

$$\begin{aligned} W^{Kp} = (W_0)^K/(W_0')^K \oplus (W_1)^K/\delta^K((W_0')^K) \to [W_0/W_0']^K \oplus [W_1/\delta(W_0')]^K \\ = W^{pK}, \end{aligned}$$

induce an isomorphism of t-algebras $\xi_p : T^{Kp} \longrightarrow T^{pK}$. Then, it is not hard to see that $\xi_p : \mathcal{A}^{Kp} \longrightarrow \mathcal{A}^{pK}$ is the required isomorphism of layered ditalgebras. □

Definition 20.8. *Given a k-ditalgebra $\mathcal{A}$, $X \in \mathcal{A}$-Mod and a field extension $k \leq K$, we say that X extends to K iff the canonical morphism of K-algebras*

$$\theta = \theta_X : \mathrm{End}_{\mathcal{A}}(X) \otimes_k K \longrightarrow \mathrm{End}_{\mathcal{A}^K}(X^K),$$

given by $\theta(f \otimes t)^0 = f^0 \otimes t1_K$, and $\theta(f \otimes t)^1[v \otimes t'] = f^1(v) \otimes tt'1_K$, for $f = (f^0, f^1) \in \mathrm{End}_{\mathcal{A}}(X)$, $v \in V$ and $t, t' \in K$, is an isomorphism.

Lemma 20.9. *Assume we have a field extension $k \leq K$ and a k-ditalgebra $\mathcal{A}$ with layer (R, W). Then, $X \in \mathcal{A}$-Mod extends to K in each one of the following cases:*

(1) $W_1 = 0$ and X is a finitely presented A-module;
(2) $\delta(W_0) = W_1$, there is an epimorphism of k-algebras $\varphi : R \longrightarrow S$ and X is the A-module obtained from the regular left S-module by restriction using the composition $A \xrightarrow{\pi} R \xrightarrow{\varphi} S$, where π is the canonical projection;
(3) $X = X_1 \oplus X_2$ in $\mathcal{A}$-Mod, where X_1 and X_2 are $\mathcal{A}$-modules which extend to K and satisfy $\mathrm{Hom}_{\mathcal{A}}(X_i, X_j) = 0$ and $\mathrm{Hom}_{\mathcal{A}^K}(X_i^K, X_j^K) = 0$, for $i \neq j$.

Proof. In the first two cases, the endomorphisms of X (and X^K) have zero second component. In each one of these cases, let us give an isomorphism of algebras σ such that θ equals the following composition of isomorphisms

$$\mathrm{End}_{\mathcal{A}}(X)^K \cong \mathrm{End}_A(X)^K \xrightarrow{\sigma} \mathrm{End}_{A^K}(X^K) \cong \mathrm{End}_{\mathcal{A}^K}(X^K).$$

In case (1), σ is by the canonical map, which is an isomorphism because the A-module X is finitely presented (Change of Rings Theorem). In case (2), notice that X^K coincides with the A^K-module obtained from the regular S^K-module by restriction, using the composition of epimorphisms of K-algebras $A^K \xrightarrow{\pi^K} R^K \xrightarrow{\varphi^K} S^K$. Then, we can take σ as the following composition

$$\mathrm{End}_A(X)^K = \mathrm{End}_S(X)^K \cong S^{opK} = S^{Kop} \cong \mathrm{End}_{S^K}(X^K) = \mathrm{End}_{A^K}(X^K).$$

For case (3), we just have to notice that θ coincides with the composition: $\mathrm{End}_{\mathcal{A}}(X)^K \cong \mathrm{End}_{\mathcal{A}}(X_1)^K \times \mathrm{End}_{\mathcal{A}}(X_2)^K \cong \mathrm{End}_{\mathcal{A}^K}(X_1^K) \times \mathrm{End}_{\mathcal{A}^K}(X_2^K) \cong \mathrm{End}_{\mathcal{A}^K}(X^K)$. □

Lemma 20.10. *Assume $\mathcal{A}$ is a ditalgebra with layer (R, W), and that X is an admissible $\mathcal{A}'$-module, where $\mathcal{A}'$ is the proper subditalgebra of $\mathcal{A}$, associated to the R-R-bimodule decompositions $W_0 = W_0' \oplus W_0''$ and $W_1 = W_1' \oplus W_1''$.*

Assume K is a field extension of k and that X extends to K. Then, X^K is an admissible $\mathcal{A}'^K$-module, where $\mathcal{A}'^K$ is the the proper subditalgebra of $\mathcal{A}^K$, associated to the R^K-R^K-bimodule decompositions $W_0^K = (W_0')^K \oplus (W_0'')^K$ and $W_1^K = (W_1')^K \oplus (W_1'')^K$. If we have the associated splitting $\Gamma = \mathrm{End}_{\mathcal{A}'}(X)^{op} = S \oplus P$, then $\overline{\Gamma} := \mathrm{End}_{\mathcal{A}'^K}(X^K)^{op} = \overline{S} \oplus \overline{P}$, with $\overline{S} := \theta(S^K)$ and $\overline{P} := \theta(P^K)$, where θ is the canonical isomorphism of (20.8).

Proof. It is straightforward. We only have to notice that $\theta(f^0, 0)$ has zero second component too. □

Lemma 20.11. *Assume $\mathcal{A} = (T, \delta)$ is a ditalgebra with layer (R, W), and that X is an admissible $\mathcal{A}'$-module, where $\mathcal{A}'$ is a proper subditalgebra of $\mathcal{A}$. Assume that K is a field extension of k and that X extends to K. Thus, X^K is an admissible $\mathcal{A}'^K$-module over the proper subditalgebra $\mathcal{A}'^K$ of $\mathcal{A}^K$. Consider the layered ditalgebras $\mathcal{A}^X$ and $\mathcal{A}^{KX^K}$ obtained from $\mathcal{A}$ and $\mathcal{A}^K$ using the admissible modules X and X^K, respectively, as in (12.9). Then, there is an isomorphism of layered ditalgebras $\xi_X : \mathcal{A}^{KX^K} \longrightarrow \mathcal{A}^{XK}$. It induces an isomorphism of categories in the following (up to isomorphism of functors) commutative diagram*

$$\begin{array}{ccccc}
\mathcal{A}^{XK}\text{-Mod} & \xrightarrow{F_{\xi_X}} & \mathcal{A}^{KX^K}\text{-Mod} & \xrightarrow{F_{X^K}} & \mathcal{A}^K\text{-Mod} \\
{\scriptstyle (-)^K}\uparrow & & & & \uparrow{\scriptstyle (-)^K} \\
\mathcal{A}^X\text{-Mod} & & \xrightarrow{\quad F_X \quad} & & \mathcal{A}\text{-Mod}.
\end{array}$$

Proof. With the notation of (12.9) and (20.10), the extension $\mathcal{A}^{XK}$ has layer (S^K, W^{XK}), and $\mathcal{A}^{KX^K}$ has layer $(\overline{S}, W^{KX^K})$.

In order to define an isomorphism $\tau : T^{XK} \longrightarrow T^{KX^K}$, we need to fix some notation. We have $\Gamma = \mathrm{End}_{\mathcal{A}'}(X)^{op} = S \oplus P$, its extension $\Gamma^K = [\mathrm{End}_{\mathcal{A}'}(X)^{op}]^K = S^K \oplus P^K$, $\overline{\Gamma} = \mathrm{End}_{\mathcal{A}'^K}(X^K)^{op} = \overline{S} \oplus \overline{P}$, and the K-algebra isomorphism $\theta_X : \Gamma^K \longrightarrow \overline{\Gamma}$ of (20.8). We shall denote by $\theta_S : S^K \longrightarrow \overline{S}$ and $\theta_P : P^K \longrightarrow \overline{P}$, the restrictions of θ_X. Then, for $x \in X$, $f \in \Gamma$ and $t, t' \in K$, we have $(x \otimes t)\theta_X(f \otimes t') = [\theta_X(f \otimes t')]^0[x \otimes t] = [f^0 \otimes t'1_K][x \otimes t] = f^0(x) \otimes tt' = xf \otimes tt'$. Thus, the right action of Γ^K on X^K, obtained by restriction through θ_X from the right $\overline{\Gamma}$-module X^K associated to the splitting $(\overline{S}, \overline{P})$ of $\overline{\Gamma}$, coincides with the canonical right action of Γ^K on X^K, determined by the right action of Γ on X. We freely use the formulas $(x \otimes t)\theta_P(p \otimes t') = xp \otimes tt'$, for $p \in P$, and $(x \otimes t)\theta_S(s \otimes t') = xs \otimes tt'$, for $s \in S$.

Consider the canonical maps $\alpha_X : X^{*K} \longrightarrow X^{K*} = \mathrm{Hom}_{S^K}(X^K, S^K)$, an isomorphism of S^K-B^K-bimodules, and $\alpha_P : P^{*K} \longrightarrow P^{K*} = \mathrm{Hom}_{S^K}$

(P^K, S^K), an isomorphism of S^K-bimodules. Consider, also, the maps

$$\beta_X : X^{K*} = \mathrm{Hom}_{S^K}(X^K, S^K) \longrightarrow \mathrm{Hom}_{\overline{S}}(X^K, \overline{S}) =: X^{K\circledast}$$
$$\beta_P : P^{K*} = \mathrm{Hom}_{S^K}(P^K, S^K) \longrightarrow \mathrm{Hom}_{\overline{S}}(\overline{P}, \overline{S}) =: \overline{P}^{\circledast},$$

given by $\beta_X(\nu) = \theta_S \nu$ and $\beta_P(\gamma) = \theta_S \gamma \theta_P^{-1}$. Then, β_X is an isomorphism of S^K-B^K-bimodules, and so is the composition $\overline{\beta}_X := \beta_X \alpha_X : X^{*K} \longrightarrow X^{K\circledast}$. Similarly, β_P is an isomorphism of S^K-bimodules, and so is the composition $\overline{\beta}_P := \beta_P \alpha_P : P^{*K} \longrightarrow \overline{P}^{\circledast}$.

Now, define $\tau_0 := \theta_S$ and τ_1 as the following vertical composition of isomorphisms of S^K-S^K-bimodules

$$\begin{array}{rcl} \widehat{W}^{XK} & = & \left([X^* \otimes_B \underline{W} \otimes_B X] \oplus P^*\right)^K \\ & & \downarrow \zeta \\ & & [X^{*K} \otimes_{B^K} \underline{W}^K \otimes_{B^K} X^K] \oplus P^{*K} \\ & & \downarrow (\overline{\beta}_X \otimes 1 \otimes 1) \oplus \overline{\beta}_P \\ \widehat{W}^{KX^K} & = & [X^{K\circledast} \otimes_{B^K} \underline{W}^K \otimes_{B^K} X^K] \oplus \overline{P}^{\circledast}. \end{array}$$

Recall from (12.7), that we have to consider the S-S-subbimodule $\mathcal{L} = \mathcal{L}_X$ of

$$\widehat{W}^X = \left([X^* \otimes_B \underline{W}_0 \otimes_B X] \oplus [X^* \otimes_B \underline{W}_1 \otimes_B X] \oplus P^*\right),$$

generated by the elements $\gamma^w_{\nu,x} - \nu \otimes w \otimes x$, where $\nu \in X^*$, $w \in BW_1'B$ and $x \in X$. Thus $\mathcal{L}_X^K$ is an S^K-S^K-subbimodule of $\widehat{W}^{XK}$, and we can consider the corresponding $\overline{S}$-$\overline{S}$-subbimodule $\mathcal{L}_{X^K}$ of $\widehat{W}^{KX^K}$. We claim that $\tau_1(\mathcal{L}_X^K) = \mathcal{L}_{X^K}$ and, therefore, τ_0 and τ_1 induce an isomorphism of layered t-algebras $\tau : T^{XK} \longrightarrow T^{KX^K}$. Fix $\nu \in X^*$, $w \in BW_1'B$, $x \in X$ and $t \in K$. Then, clearly, $\tau_1([\nu \otimes w \otimes x] \otimes t) = \overline{\beta}_X(\nu \otimes 1) \otimes (w \otimes 1) \otimes (x \otimes t)$. We claim that $\tau_1(\gamma^w_{\nu,x} \otimes t) = \gamma^{w\otimes 1}_{\overline{\beta}_X(\nu\otimes 1), x\otimes t}$. Indeed, for $p \in P$ and $t' \in K$

$$\begin{aligned} \tau_1(\gamma^w_{\nu,x} \otimes t)[\theta_P(p \otimes t')] &= \beta_P \alpha_P(\gamma^w_{\nu,x} \otimes t)[\theta_P(p \otimes t')] \\ &= \beta_P(\gamma^w_{\nu,x} \otimes t1_K)[\theta_P(p \otimes t')] \\ &= \theta_S(\gamma^w_{\nu,x} \otimes t1_K)[p \otimes t'] \\ &= \theta_S[\gamma^w_{\nu,x}(p) \otimes tt'] \\ &= \theta_S[\nu(p^1(w)[x]) \otimes tt'] \\ &= \overline{\beta}_X(\nu \otimes 1)[p^1(w)[x] \otimes tt'] \\ &= \overline{\beta}_X(\nu \otimes 1)\left[(p^1(w) \otimes t'1_K)[x \otimes t]\right] \\ &= \overline{\beta}_X(\nu \otimes 1)\left[[\theta_P(p \otimes t')^1(w \otimes 1)](x \otimes t)\right] \\ &= \gamma^{w\otimes 1}_{\overline{\beta}_X(\nu\otimes 1), x\otimes t}\left[\theta_P(p \otimes t')\right]. \end{aligned}$$

It follows that $\tau_1(\mathcal{L}_X^K) = \mathcal{L}_{X^K}$ (see the formulas in the proof of (15.17)).

Now we have to verify that $\tau(\delta^X \otimes 1) = \delta^{X^K}\tau : T^{XK} \longrightarrow T^{KX^K}$. For this we need to recall the definition of δ^X. We will first gradually examine its ingredients and study their compatibility with scalar extension.

Consider the comultiplication $P^* \xrightarrow{\mu} P^* \otimes_S P^*$, associated to the splitting (S, P) of Γ, see (11.7), and the comultiplication $\overline{\mu}$, associated to the splitting $(\overline{S}, \overline{P})$ of $\overline{\Gamma}$. Then, the following diagram commutes

$$\begin{array}{ccc} P^{*K} & \xrightarrow{\mu\otimes 1} & (P^* \otimes_S P^*)^K \\ \Big\downarrow{\scriptstyle \overline{\beta}_P} & & \Big\downarrow{\scriptstyle (\overline{\beta}_P\otimes\overline{\beta}_P)\zeta_1} \\ \overline{P}^{\circledast} & \xrightarrow{\overline{\mu}} & \overline{P}^{\circledast} \otimes_{\overline{S}} \overline{P}^{\circledast}, \end{array}$$

where ζ_1 is the canonical isomorphism.

Indeed, consider a dual basis $(p_j, \gamma_j)_j$ for the right S-module P, then the family $(\theta_P(p_j \otimes 1), \overline{\beta}_P(\gamma_j \otimes 1))_j$ is a dual basis for the right $\overline{S}$-module $\overline{P}$. Then, if $\Delta := (\overline{\beta}_P \otimes \overline{\beta}_P)\zeta_1(\mu \otimes 1)[\gamma \otimes t]$, using the formulas of (11.8), we obtain

$$\begin{aligned}
\Delta &= (\overline{\beta}_P \otimes \overline{\beta}_P)\zeta_1[\textstyle\sum_{i,j} \gamma(p_i p_j)\gamma_j \otimes \gamma_i \otimes t] \\
&= (\overline{\beta}_P \otimes \overline{\beta}_P)\left[\textstyle\sum_{i,j}(\gamma(p_i p_j) \otimes 1)(\gamma_j \otimes 1) \otimes (\gamma_i \otimes t)\right] \\
&= \textstyle\sum_{i,j} \theta_S[\gamma(p_i p_j) \otimes 1]\overline{\beta}_P(\gamma_j \otimes 1) \otimes \overline{\beta}_P(\gamma_i \otimes t) \\
&= \textstyle\sum_{i,j} \theta_S\left[\gamma(p_i p_j) \otimes t\right]\overline{\beta}_P(\gamma_j \otimes 1) \otimes \overline{\beta}_P(\gamma_i \otimes 1) \\
&= \textstyle\sum_{i,j} \theta_S(\gamma \otimes t1_K)\theta_P^{-1}\left[\theta_P(p_i \otimes 1)\theta_P(p_j \otimes 1)\right]\overline{\beta}_P(\gamma_j \otimes 1) \otimes \overline{\beta}_P(\gamma_i \otimes 1) \\
&= \overline{\mu}[\theta_S(\gamma \otimes t1_K)\theta_P^{-1}] \\
&= \overline{\mu}\overline{\beta}_P(\gamma \otimes t).
\end{aligned}$$

Consider the morphisms $\lambda : X^* \longrightarrow P^* \otimes_S X^*$ and $\rho : X \longrightarrow X \otimes_S P^*$, associated to the admissible $\mathcal{A}'$-module X, see (11.10), and consider, alternatively the morphisms $\overline{\lambda} : X^{K\circledast} \longrightarrow \overline{P}^{\circledast} \otimes_{\overline{S}} X^{K\circledast}$ and $\overline{\rho} : X^K \longrightarrow X^K \otimes_{\overline{S}} \overline{P}^{\circledast}$, associated to the admissible $\mathcal{A}'^K$-module X^K. Then, the following squares commute

$$\begin{array}{ccc} X^{*K} & \xrightarrow{\lambda\otimes 1} & (P^* \otimes_S X^*)^K \\ \Big\downarrow{\scriptstyle \overline{\beta}_X} & & \Big\downarrow{\scriptstyle (\overline{\beta}_P\otimes\overline{\beta}_X)\zeta_2} \\ X^{K\circledast} & \xrightarrow{\overline{\lambda}} & \overline{P}^{\circledast} \otimes_{\overline{S}} X^{K\circledast} \end{array} \qquad \begin{array}{ccc} X^{K} & \xrightarrow{\rho\otimes 1} & (X \otimes_S P^*)^K \\ \Big\| & & \Big\downarrow{\scriptstyle (1\otimes\overline{\beta}_P)\zeta_3} \\ X^{K} & \xrightarrow{\overline{\rho}} & X^K \otimes_{\overline{S}} \overline{P}^{\circledast}, \end{array}$$

where ζ_2 and ζ_3 are the canonical isomorphisms. To verify this, choose a finite dual basis $(x_i, \nu_i)_i$ for the right S-module X, and notice that $(x_i \otimes 1, \overline{\beta}_X[\nu_i \otimes 1])_i$ is a dual basis for the right $\overline{S}$-module X^K. Then, for $x \in X$, $\nu \in X^*$ and

$t \in K$, if $\Delta' := (\overline{\beta}_P \otimes \overline{\beta}_X)\zeta_2(\lambda \otimes 1)[\nu \otimes t]$, using (11.11), we obtain

$$\begin{aligned}
\Delta' &= (\overline{\beta}_P \otimes \overline{\beta}_X)\zeta_2[\textstyle\sum_{i,j} \nu(x_i p_j)\gamma_j \otimes \nu_i \otimes t] \\
&= (\overline{\beta}_P \otimes \overline{\beta}_X)[\textstyle\sum_{i,j}(\nu(x_i p_j) \otimes 1)(\gamma_j \otimes 1) \otimes (\nu_i \otimes t)] \\
&= \textstyle\sum_{i,j} \theta_S(\nu(x_i p_j) \otimes 1)\overline{\beta}_P(\gamma_j \otimes 1) \otimes \overline{\beta}_X(\nu_i \otimes t) \\
&= \textstyle\sum_{i,j} \theta_S(\nu \otimes t1_K)[(x_i p_j \otimes 1)]\overline{\beta}_P(\gamma_j \otimes 1) \otimes \overline{\beta}_X(\nu_i \otimes 1) \\
&= \textstyle\sum_{i,j} \theta_S(\nu \otimes t1_K)[(x_i \otimes 1)\theta_P(p_j \otimes 1)]\overline{\beta}_P(\gamma_j \otimes 1) \otimes \overline{\beta}_X(\nu_i \otimes 1) \\
&= \overline{\lambda}[\theta_S(\nu \otimes t1_K)] \\
&= \overline{\lambda\beta_X}[\nu \otimes t],
\end{aligned}$$

and also

$$\begin{aligned}
(1 \otimes \overline{\beta}_P)\zeta_3(\rho \otimes 1)[x \otimes t] &= (1 \otimes \overline{\beta}_P)\zeta_3[\textstyle\sum_j xp_j \otimes \gamma_j \otimes t] \\
&= (1 \otimes \overline{\beta}_P)[\textstyle\sum_j (xp_j \otimes 1) \otimes (\gamma_j \otimes t)] \\
&= \textstyle\sum_j (xp_j \otimes 1) \otimes \overline{\beta}_P(\gamma_j \otimes t) \\
&= \textstyle\sum_j (x \otimes t)\theta_P(p_j \otimes 1) \otimes \overline{\beta}_P(\gamma_j \otimes 1) \\
&= \overline{\rho}[x \otimes t].
\end{aligned}$$

We also need to study the compatibility of the insertion maps defined in (12.8) with scalar extension. We have the following commutative diagram

$$\begin{array}{ccc}
\oplus_{n\geq 1}(\underline{W}^{\otimes n})^K \cong \langle \underline{W} \rangle^K & \xrightarrow{\sigma_{\nu,x}\otimes 1} & T^{XK} \\
\downarrow{\scriptstyle \zeta_4} & & \downarrow{\scriptstyle \tau} \\
\oplus_{n\geq 1}(\underline{W}^K)^{\otimes n} \cong \langle \underline{W}^K \rangle & \xrightarrow{\sigma_{\overline{\beta}_X(\nu\otimes 1),(x\otimes 1)}} & T^{KX^K},
\end{array}$$

where ζ_4 is the identity map. To verify this, it is enough to compute both maps on a typical generator $w_1 w_2 \cdots w_n \otimes t \in \langle \underline{W} \rangle^K$, with $w_i \in \underline{W}$ and $t \in K$. There, if $\chi = \tau(\sigma_{\nu,x} \otimes 1)[w_1 w_2 \cdots w_n \otimes t]$, then

$$\begin{aligned}
\chi &= \tau\left[\textstyle\sum \nu \otimes w_1 \otimes x_{i_1} \otimes \nu_{i_1} \otimes w_2 \otimes \cdots \otimes \nu_{i_{n-1}} \otimes w_n \otimes x \otimes t\right] \\
&= \textstyle\sum \overline{\beta}_X(\nu \otimes 1) \otimes (w_1 \otimes 1) \otimes (x_{i_1} \otimes 1) \otimes \overline{\beta}_X(\nu_{i_1} \otimes 1) \otimes (w_2 \otimes 1) \otimes \cdots \\
&\qquad \cdots \otimes \overline{\beta}_X(\nu_{i_{n-1}} \otimes 1) \otimes (w_n \otimes 1) \otimes (x \otimes t) \\
&= \sigma_{\overline{\beta}_X(\nu\otimes 1),(x\otimes 1)}[(w_1 \otimes 1)(w_2 \otimes 1) \cdots (w_n \otimes t)] \\
&= \sigma_{\overline{\beta}_X(\nu\otimes 1),(x\otimes 1)}\zeta_4[w_1 w_2 \cdots w_n \otimes t].
\end{aligned}$$

Now, for $\nu \in X^*$, $w \in \underline{W}_0 \cup \underline{W}_1$, $x \in X$ and $t \in K$, we have

$$\begin{aligned}
\delta^{X^K}\tau[(\nu \otimes w \otimes x) \otimes t] &= \delta^{X^K}\left[\overline{\beta}_X(\nu \otimes 1) \otimes (w \otimes 1) \otimes (x \otimes t)\right] \\
&= \overline{\lambda}[\overline{\beta}_X(\nu \otimes 1)] \otimes (w \otimes 1) \otimes (x \otimes t) \\
&\quad + \sigma_{\overline{\beta}_X(\nu\otimes 1),(x\otimes 1)}(\delta \otimes 1)[w \otimes t] \\
&\quad + (-1)^{\deg(w\otimes 1)+1}\overline{\beta}_X(\nu \otimes 1) \otimes (w \otimes 1) \otimes \overline{\rho}(x \otimes t).
\end{aligned}$$

Moreover, we have

$$\begin{aligned}\tau(\delta^X \otimes 1)[(\nu \otimes w \otimes x) \otimes t] = \tau[\lambda(\nu) \otimes w \otimes x \otimes t] \\ + \tau[\sigma_{\nu,x}(\delta(w)) \otimes t] \\ + (-1)^{\deg(w)+1}\tau[\nu \otimes w \otimes \rho(x) \otimes t].\end{aligned}$$

We shall compare each of the three terms in the last equalities, using the previous commutative diagrams

$$\begin{aligned}\tau[\nu \otimes w \otimes \rho(x) \otimes t] &= \overline{\beta}_X(\nu \otimes 1) \otimes (w \otimes 1) \otimes \left(\textstyle\sum_j xp_j \otimes 1\right) \otimes \overline{\beta}_P(\gamma_j \otimes t) \\ &= \overline{\beta}_X(\nu \otimes 1) \otimes (w \otimes 1) \\ &\quad \otimes (1 \otimes \overline{\beta}_P)\zeta_3[\textstyle\sum_j xp_j \otimes \gamma_j \otimes t] \\ &= \overline{\beta}_X(\nu \otimes 1) \otimes (w \otimes 1) \otimes (1 \otimes \overline{\beta}_P)\zeta_3(\rho \otimes 1)[x \otimes t] \\ &= \overline{\beta}_X(\nu \otimes 1) \otimes (w \otimes 1) \otimes \overline{\rho}(x \otimes t);\end{aligned}$$

$$\begin{aligned}\tau[\lambda(\nu) \otimes w \otimes x \otimes t] &= \tau[\textstyle\sum_{i,j} \nu(x_i p_j)\gamma_j \otimes \nu_i \otimes w \otimes x \otimes t] \\ &= \overline{\beta}_P(\textstyle\sum_{i,j} \nu(x_i p_j)\gamma_j \otimes 1) \otimes \overline{\beta}_X(\nu_i \otimes 1) \otimes (w \otimes 1) \\ &\quad \otimes (x \otimes t) \\ &= (\overline{\beta}_P \otimes \overline{\beta}_X)\zeta_2\left[\textstyle\sum_{i,j} \nu(x_i p_j)\gamma_j \otimes \nu_i \otimes 1\right] \otimes (w \otimes 1) \\ &\quad \otimes (x \otimes t) \\ &= (\overline{\beta_P} \otimes \overline{\beta}_X)\zeta_2(\lambda \otimes 1)[\nu \otimes 1] \otimes (w \otimes 1) \otimes (x \otimes t) \\ &= \overline{\lambda}(\overline{\beta}_X(\nu \otimes 1)) \otimes (w \otimes 1) \otimes (x \otimes t).\end{aligned}$$

Finally, if $w \in \underline{W}$ and $\delta(w) = \sum_r w_1^r w_2^r \cdots w_{t_r}^r$, with $w_j^r \in \underline{W}$, then

$$\begin{aligned}\tau[\sigma_{\nu,x}(\delta(w)) \otimes t] &= \tau(\sigma_{\nu,x} \otimes 1)[\delta(w) \otimes t] \\ &= \sigma_{\overline{\beta}_X(\nu\otimes 1),(x\otimes 1)}\zeta_4[\delta(w) \otimes t] \\ &= \sigma_{\overline{\beta}_X(\nu\otimes 1),(x\otimes 1)}[\textstyle\sum_r (w_1^r \otimes 1)(w_2^r \otimes 1) \cdots (w_{t_r}^r \otimes t)] \\ &= \sigma_{\overline{\beta}_X(\nu\otimes 1),(x\otimes 1)}(\delta \otimes 1)[w \otimes t].\end{aligned}$$

Then, $\tau : \mathcal{A}^{XK} \longrightarrow \mathcal{A}^{KX^K}$ is an isomorphism of layered K-ditalgebras. We claim that the isomorphism $\xi_X := \tau^{-1}$ satisfies the statement of the lemma.

In order to verify the commutativity, up to isomorphism of functors, in the statement of the lemma, consider, for $M \in \mathcal{A}^X$-Mod, the morphism

$$\eta^0 : F_X(M)^K \longrightarrow F_{X^K} F_{\tau^{-1}}(M^K)$$

given at the level of underlying vector spaces by the canonical isomorphism of vector spaces $\eta^0 : (X \otimes_S M)^K \longrightarrow X^K \otimes_{\overline{S}} M^K$. In fact, η^0 preserves the structure of A^K-modules. In order to verify this, denote by $*$ the action of A^X on M and the action of A^{XK} on M^K; denote by $\cdot$ the action of A on $F^X(M) = X \otimes_S M$ and the action of A^K on $F^X(M)^K = (X \otimes_S M)^K$; finally, denote by $\circledast$ the action of A^{KX^K} on $F_{\tau^{-1}}(M^K)$ and by $\odot$ the action of A^K on

$F_{X^K}F_{\tau^{-1}}(M^K) = X^K \otimes_{\overline{S}} M^K$. Then, we have, for $b \in B$ and $t \in K$

$$\begin{aligned}\eta^0[(b\otimes t)\cdot(x\otimes m\otimes t')] &= \eta^0[bx\otimes m\otimes tt']\\ &= (bx\otimes 1)\otimes(m\otimes tt')\\ &= (bx\otimes t)\otimes(m\otimes t')\\ &= (b\otimes t)\odot[(x\otimes 1)\otimes(m\otimes t')]\\ &= (b\otimes t)\odot\eta^0[x\otimes m\otimes t'],\end{aligned}$$

and, for $w \in \underline{W}_0$, $x \in X$, $m \in M$ and $t, t' \in K$

$$\begin{aligned}\eta^0[(w\otimes t)\cdot(x\otimes m\otimes t')] &= \eta^0[(w\cdot[x\otimes m])\otimes tt')]\\ &= \eta^0[\textstyle\sum_i x_i\otimes[(\nu_i\otimes w\otimes x)*m]\otimes tt']\\ &= \textstyle\sum_i(x_i\otimes 1)\otimes[(\nu_i\otimes w\otimes x)*m]\otimes tt'\\ &= \textstyle\sum_i(x_i\otimes 1)\otimes[(\nu_i\otimes w\otimes x)\otimes t]*(m\otimes t')\\ &= \textstyle\sum_i(x_i\otimes 1)\otimes\tau[(\nu_i\otimes w\otimes x)\otimes t)]\circledast(m\otimes t')\\ &= \textstyle\sum_i(x_i\otimes 1)\otimes[\overline{\beta}_X(\nu_i\otimes 1)\otimes(w\otimes t)\otimes(x\otimes 1)]\\ &\quad\circledast(m\otimes t')\\ &= (w\otimes t)\odot[(x\otimes 1)\otimes(m\otimes t')]\\ &= (w\otimes t)\odot\eta^0[x\otimes m\otimes t'].\end{aligned}$$

Since A^K is freely generated by B^K and $\underline{W}_0^K$, the map η^0 is an isomorphism of A^K-modules.

Finally, given $f = (f^0, f^1) \in \mathrm{Hom}_{\mathcal{A}^X}(M, N)$, we have to see that the following square commutes in $\mathcal{A}^K$-Mod

$$\begin{array}{ccc} F_X(M)^K & \xrightarrow{(\eta^0,0)} & F_{X^K}F_{\tau^{-1}}(M^K)\\ {\scriptstyle F_X(f)^K}\downarrow & & \downarrow{\scriptstyle F_{X^K}F_{\tau^{-1}}(f^K)}\\ F_X(N)^K & \xrightarrow{(\eta^0,0)} & F_{X^K}F_{\tau^{-1}}(N^K).\end{array}$$

Equivalently, we have to show that, for all $v \in V$ and $t \in K$, the equalities of k-linear transformations $\eta^0 F_X(f)^{K0} = [F_{X^K}F_{\tau^{-1}}(f^K)]^0\eta^0$ and $\eta^0 F_X(f)^{K1}(v\otimes t) = [F_{X^K}F_{\tau^{-1}}(f^K)]^1(v\otimes t)\eta^0$ hold. If $x \in X$, $m \in M$ and $t \in K$, then, if we make $\Delta_1 := \eta^0 F_X(f)^{K0}[x\otimes m\otimes t]$, we have

$$\begin{aligned}\Delta_1 &= \eta^0(F_X(f)^0\otimes 1_K)[x\otimes m\otimes t]\\ &= \eta^0[x\otimes f^0(m)\otimes t+\textstyle\sum_j xp_j\otimes f^1(\gamma_j)[m]\otimes t]\\ &= (x\otimes 1)\otimes(f^0(m)\otimes t)+\textstyle\sum_j(xp_j\otimes 1)\otimes(f^1(\gamma_j)[m]\otimes t)\\ &= (x\otimes 1)\otimes(f^0(m)\otimes t)+\textstyle\sum_j(xp_j\otimes 1)\otimes(f^K)^1(\gamma_j\otimes 1)[m\otimes t]\\ &= (x\otimes 1)\otimes F_{\tau^{-1}}(f^K)^0[m\otimes t]\\ &\quad+\textstyle\sum_j(x\otimes 1)\theta_P(p_j\otimes 1)\otimes F_{\tau^{-1}}(f^K)^1(\overline{\beta}_P(\gamma_j\otimes 1))[m\otimes t]\\ &= \left[F_{X^K}(F_{\tau^{-1}}(f^K))\right]^0[(x\otimes 1)\otimes(m\otimes t)]\\ &= \left[F_{X^K}F_{\tau^{-1}}(f^K)\right]^0\eta^0[x\otimes m\otimes t].\end{aligned}$$

Moreover, if $v \in V$, $x \in X$, $m \in M$ and $t, t' \in K$, and we make $\Delta_2 := \eta^0 F_X(f)^{K1}(v \otimes t)[x \otimes m \otimes t']$, we have

$$\begin{aligned}
\Delta_2 &= \eta^0 \left[(F_X(f)^1(v) \otimes t 1_K)[x \otimes m \otimes t'] \right] \\
&= \eta^0 \left[F_X(f)^1(v)[x \otimes m] \otimes tt' \right] \\
&= \textstyle\sum_i (x_i \otimes 1) \otimes (f^1(\sigma_{\nu_i,x}(v))[m] \otimes tt') \\
&= \textstyle\sum_i (x_i \otimes 1) \otimes (f^{K1}(\sigma_{\nu_i,x}(v) \otimes t)[m \otimes t']) \\
&= \textstyle\sum_i (x_i \otimes 1) \otimes (f^{K1}(\sigma_{\nu_i,x} \otimes 1_K)(v \otimes t)[m \otimes t']) \\
&= \textstyle\sum_i (x_i \otimes 1) \otimes f^{K1} \left[\tau^{-1} \sigma_{\overline{\beta}_X(\nu_i \otimes 1), x \otimes 1} \zeta_4(v \otimes t) \right] [m \otimes t'] \\
&= \textstyle\sum_i (x_i \otimes 1) \otimes F_{\tau^{-1}}(f^K)^1 \left[\sigma_{\overline{\beta}_X(\nu_i \otimes 1), x \otimes 1} \zeta_4(v \otimes t) \right] [m \otimes t'] \\
&= F_{X^K}(F_{\tau^{-1}}(f^K))^1 \zeta_4(v \otimes t)[(x \otimes 1) \otimes (m \otimes t')] \\
&= \left[F_{X^K} F_{\tau^{-1}}(f^K) \right]^1 (v \otimes t) \eta^0 [x \otimes m \otimes t'].
\end{aligned}$$

□

Lemma 20.12. *If $k \leq K$ is a field extension which preserves the radical of the finite-dimensional k-algebra Λ, as in (19.3), we have the following well-defined and (up to isomorphism) commutative diagrams of functors*

$$\begin{array}{ccc}
\mathcal{P}^1(\Lambda^K) & \xrightarrow{\text{Cok}^1} & \Lambda^K\text{-Mod} \\
{\scriptstyle (-)^K}\uparrow & & \uparrow{\scriptstyle (-)^K} \\
\mathcal{P}^1(\Lambda) & \xrightarrow{\text{Cok}^1} & \Lambda\text{-Mod}
\end{array}
\qquad
\begin{array}{ccc}
\mathcal{P}^2(\Lambda^K) & \xrightarrow{\text{Cok}^2} & \Lambda^K\text{-Mod} \\
{\scriptstyle (-)^K}\uparrow & & \uparrow{\scriptstyle (-)^K} \\
\mathcal{P}^2(\Lambda) & \xrightarrow{\text{Cok}^2} & \Lambda\text{-Mod},
\end{array}$$

where Cok^1, Cok^2 *are the restrictions of the cokernel functor of (18.10).*

Proof. It is straightforward. □

Lemma 20.13. *Suppose that Λ is a finite-dimensional k-algebra which splits over its radical J, say $\Lambda = S \oplus J$, and consider the associated Drozd's ditalgebra $\mathcal{D}^\Lambda$, as in (19.1). Assume that the field extension $k \leq K$ preserves the radical of Λ. Consider the extended ditalgebra $\mathcal{D}^{\Lambda K}$ of $\mathcal{D}^\Lambda$ and the extended algebra Λ^K. Then, Λ^K admits the splitting (S^K, J^K), where S^K is a semisimple K-subalgebra of Λ^K. Consider the associated Drozd's ditalgebra $\mathcal{D}^{\Lambda^K}$. Then, there is an isomorphism of layered ditalgebras $\xi_\Lambda : \mathcal{D}^{\Lambda^K} \rightarrow \mathcal{D}^{\Lambda K}$, which induces an isomorphism of categories in the following well-defined and (up to isomorphism)-commutative diagram of functors*

$$\begin{array}{ccccc}
\mathcal{D}^{\Lambda K}\text{-Mod} & \xrightarrow{F_{\xi_\Lambda}} & \mathcal{D}^{\Lambda^K}\text{-Mod} & \xrightarrow{\Xi_{\Lambda^K}} & \mathcal{P}^1(\Lambda^K) \\
{\scriptstyle (-)^K}\uparrow & & & & \uparrow{\scriptstyle (-)^K} \\
\mathcal{D}^\Lambda\text{-Mod} & & \xrightarrow{\qquad\Xi_\Lambda\qquad} & & \mathcal{P}^1(\Lambda).
\end{array}$$

Proof. Write $\mathcal{D}^\Lambda = (T^\Lambda, \delta^\Lambda)$, thus $\mathcal{D}^{\Lambda K} = (T^{\Lambda K}, \delta^\Lambda \otimes 1)$, and $\mathcal{D}^{\Lambda^K} = (T^{\Lambda^K}, \delta^{\Lambda^K})$. The canonical isomorphism $\beta : J^{*K} \longrightarrow J^{K*}$, determines isomorphisms

$$R^{\Lambda K} = \begin{pmatrix} S & 0 \\ 0 & S \end{pmatrix}^K \xrightarrow{\theta'} \begin{pmatrix} S^K & 0 \\ 0 & S^K \end{pmatrix} = R^{\Lambda^K},$$

$$(W_0^\Lambda)^K = \begin{pmatrix} 0 & 0 \\ J^* & 0 \end{pmatrix}^K \xrightarrow{\theta_0} \begin{pmatrix} 0 & 0 \\ J^{K*} & 0 \end{pmatrix} = W_0^{\Lambda^K} \text{ and}$$

$$(W_1^\Lambda)^K = \begin{pmatrix} J^* & 0 \\ 0 & J^* \end{pmatrix}^K \xrightarrow{\theta_1} \begin{pmatrix} J^{K*} & 0 \\ 0 & J^{K*} \end{pmatrix} = W_1^{\Lambda^K}.$$

They induce an isomorphism θ of layered t-algebras between the underlying t-algebras of $\mathcal{D}^{\Lambda K}$ and $\mathcal{D}^{\Lambda^K}$. In order to prove that θ is an isomorphism of ditalgebras, we have to verify the commutativity of the diagram

$$\begin{array}{ccc} T^{\Lambda K} = T_{R^\Lambda}(W^\Lambda)^K & \xrightarrow{\theta} & T_{R^{\Lambda^K}}(W^{K\Lambda^K}) = T^{\Lambda^K} \\ \Big\downarrow {\scriptstyle \delta^\Lambda \otimes 1} & & \Big\downarrow {\scriptstyle \delta^{\Lambda^K}} \\ T^{\Lambda K} = T_{R^\Lambda}(W^\Lambda)^K & \xrightarrow{\theta} & T_{R^{\Lambda^K}}(W^{K\Lambda^K}) = T^{\Lambda^K}. \end{array}$$

For this, denote by $\overline{\mu}$, $\overline{\delta}_0$ and $\overline{\delta}_1$ the ingredients to define δ^{Λ^K} corresponding to the K-algebra Λ^K, as explained in (19.1) for μ, δ_0, δ_1 and Λ. Then, tensor by K the two diagrams defining δ_0 and δ_1, and compare them with the corresponding diagrams defining $\overline{\delta}_0$ and $\overline{\delta}_1$ to obtain the commutative diagrams

$$\begin{array}{ccc} W_0^{\Lambda K} & \xrightarrow{\theta_0} & W_0^{\Lambda^K} \\ \Big\downarrow {\scriptstyle \delta_0 \otimes 1} & & \Big\downarrow {\scriptstyle \overline{\delta}_0} \\ [(W_0^\Lambda \otimes_{R^\Lambda} W_1^\Lambda) \oplus (W_1^\Lambda \otimes_{R^\Lambda} W_0^\Lambda)]^K & \xrightarrow{\zeta_0} & \left(W_0^{\Lambda^K} \otimes_{R^{\Lambda^K}} W_1^{\Lambda^K}\right) \oplus \left(W_1^{\Lambda^K} \otimes_{R^{\Lambda^K}} W_0^{\Lambda^K}\right) \end{array}$$

and

$$\begin{array}{ccc} W_1^{\Lambda K} & \xrightarrow{\theta_1} & W_1^{\Lambda^K} \\ \Big\downarrow {\scriptstyle \delta_1 \otimes 1} & & \Big\downarrow {\scriptstyle \overline{\delta}_1} \\ \left(W_1^\Lambda \otimes_{R^\Lambda} W_1^\Lambda\right)^K & \xrightarrow{\zeta_1} & W_1^{\Lambda^K} \otimes_{R^{\Lambda^K}} W_1^{\Lambda^K}, \end{array}$$

where ζ_0 and ζ_1 are the canonical isomorphisms. For the last statement, we need to have in mind the first square diagram in the proof of (20.11), where μ and $\overline{\mu}$ are compared (that is, the formula: $\overline{\mu}\beta = (\beta \otimes \beta)\zeta(\mu \otimes 1)$). Then, since the morphisms $\theta(\delta^\Lambda \otimes 1)$ and $\delta^{\Lambda^K}\theta$ coincide on generators of the form $w_i \otimes t$, with $w_i \in W_i^\Lambda$ and $t \in K$, they coincide everywhere.

Now we proceed to check the commutativity of the diagram in the statement of the lemma with $\xi_\Lambda = \theta^{-1} : \mathcal{D}^{\Lambda^K} \to \mathcal{D}^{\Lambda K}$. Assume that $M \in \mathcal{D}^\Lambda$-Mod is given by the triple (M_1, M_2, ψ_M). Consider the extension M^K and the $A_{\mathcal{D}^{\Lambda K}}$-module $\underline{M}^K = F_{\xi_\Lambda}(M^K) \in \mathcal{D}^{\Lambda^K}$-Mod. If we denote by τ the morphism

$$\mathrm{Hom}_{S\text{-}S}(J^*, \mathrm{Hom}_k(M_1, M_2))^K \longrightarrow \mathrm{Hom}_{S^K\text{-}S^K}(J^{K*}, \mathrm{Hom}_K(M_1^K, M_2^K)),$$

then it is easy to see that the triple associated to $\underline{M}^K$ is $(M_1^K, M_2^K, \psi_{\underline{M}^K})$, with $\psi_{\underline{M}^K} = \tau(\psi_M \otimes 1)$. From the formulas in (19.8), it follows that the next diagram, where η_1 and η_2 are the canonical isomorphisms, commutes

$$\begin{array}{ccc}
\Lambda^K \otimes_{S^K} M_1^K & \xrightarrow{\Xi_{\Lambda^K}(\underline{M}^K)} & \Lambda^K \otimes_{S^K} M_2^K \\
\uparrow \eta_1 & & \uparrow \eta_2 \\
(\Lambda \otimes_S M_1)^K & \xrightarrow{\Xi_\Lambda(M)\otimes 1} & (\Lambda \otimes_S M_2)^K .
\end{array}$$

It is not hard to see that $\eta = (\eta_1, \eta_2) : \Xi_\Lambda(M)^K \to \Xi_{\Lambda^K} F_{\xi_\Lambda}(M^K)$ is in fact an isomorphism of functors. □

Exercise 20.14. *Assume we have a field extension $k \leq K$ and a finite-dimensional Roiter k-ditalgebra $\mathcal{A}$ with semisimple layer. Show that any $X \in \mathcal{A}$-mod extends to K.*

Hint: Use (6.16) and consider a presentation $A^m \to A^n \to X \to 0$ in $\mathcal{A}$-mod.

21

Bimodules

Here we consider $\mathcal{A}$-E-bimodules over a k-ditalgebra $\mathcal{A}$ and a k-algebra E, and begin the study of the functors F^E induced on bimodule categories by functors F between ditalgebra module categories.

Recall that given two k-algebras A and E, we required that the ground field k acts centrally on any A-E-bimodule M. This means that the structure of the A-E-bimodule M is determined by its left A-module structure and a morphism of k-algebras $\alpha_M : E \longrightarrow \mathrm{End}_A(M)^{op}$, where the k-algebra structure of $\mathrm{End}_A(M)$ is naturally given by $(\lambda f)[m] = \lambda f[m]$, for $\lambda \in k$, $f \in \mathrm{End}_A(M)$ and $m \in M$. Then, the notion of a bimodule can be extended as follows.

Definition 21.1. *Given a ditalgebra $\mathcal{A}$ and a k-algebra E, an $\mathcal{A}$-E-bimodule is an object $M \in \mathcal{A}$-*Mod*, together with a k-algebra homomorphism $\alpha_M : E \rightarrow \mathrm{End}_{\mathcal{A}}(M)^{op}$. If N is another $\mathcal{A}$-E-bimodule, then*

$$\mathrm{Hom}_{\mathcal{A}\text{-}E}(M,N) = \{f \in \mathrm{Hom}_{\mathcal{A}}(M,N) \mid f\alpha_M(e) = \alpha_N(e)f, \textit{ for all } e \in E\}.$$

*The category of $\mathcal{A}$-E-bimodules, where the rule of composition is the same as for $\mathcal{A}$-*Mod*, is denoted by $\mathcal{A}$-E-*Mod.

An $\mathcal{A}$-E-bimodule M is proper iff α_M factors through the canonical embedding map

$$\mathrm{End}_A(M)^{op} \longrightarrow \mathrm{End}_{\mathcal{A}}(M)^{op} \textit{ such that } f \mapsto (f, 0).$$

*The full subcategory of $\mathcal{A}$-E-*Mod *formed by the proper $\mathcal{A}$-E-bimodules will be denoted by $\mathcal{A}$-E-*Mod$_p$*. The canonical embedding functor $L_{\mathcal{A}}$:* A-Mod $\rightarrow$ $\mathcal{A}$-Mod*, which maps each A-morphism f to $(f, 0)$, induces a functor $L_{\mathcal{A}}^E$:* A-E-Mod $\rightarrow$ $\mathcal{A}$-E-Mod*. Clearly, the proper $\mathcal{A}$-E-bimodules coincide with the images of the A-E-bimodules under the faithful functor $L_{\mathcal{A}}^E$; we will identify this class of objects with the A-E-bimodules. The notice previous to this definition guarantees that A-E-*Mod *can be identified with the category of A-E-bimodules in the usual sense.*

Lemma 21.2. *Assume that $\mathcal{A}$ is a ditalgebra over the field k. Then:*

(1) *If E is any k-algebra and M, N are proper $\mathcal{A}$-E-bimodules, their hom space in $\mathcal{A}$-E-Mod_p is given by*

$$\mathrm{Hom}_{\mathcal{A}\text{-}E}(M,N) = \mathrm{Hom}_{\mathcal{A}}(M,N) \cap [\mathrm{Hom}_E(M,N) \times \mathrm{Hom}_{A\text{-}A}(V, \mathrm{Hom}_E(M,N))].$$

(2) *Suppose K is a field extension of k. Then, there is an isomorphism of K-categories $\Psi : \mathcal{A}\text{-}K\text{-}\mathrm{Mod}_p \rightarrow \mathcal{A}^K\text{-}\mathrm{Mod}$. It can be described as follows. Denote by $A = A_{\mathcal{A}}$, then any proper $\mathcal{A}$-K-bimodule M is an A-K-bimodule and hence an A^K-module, so an object of $\mathcal{A}^K$-Mod. Given the proper $\mathcal{A}$-K-bimodules M, N, and a morphism $f = (f^0, f^1) \in \mathrm{Hom}_{\mathcal{A}\text{-}K}(M,N)$, we know from the last remark that $f^0 \in \mathrm{Hom}_K(M,N)$ and $f^1 \in \mathrm{Hom}_{A\text{-}A}(V, \mathrm{Hom}_K(M,N))$. In order to define $\Psi(f)$, first recall that the K-vector space $\mathrm{Hom}_K(M,N)$ admits a structure of an A-A-bimodule, but also of an A^K-A^K-bimodule (using the formulas $[g(a \otimes c)](m) = g(amc)$ and $[(a \otimes c)g](m) = ag(m)c$, for $a \in A$, $m \in M$, $c \in K$ and $g \in \mathrm{Hom}_K(M,N)$). With these structures in mind, there is an isomorphism (natural in the variables V, M and N)*

$$\mathrm{Hom}_{A\text{-}A}(V, \mathrm{Hom}_K(M,N)) \xrightarrow{\psi} \mathrm{Hom}_{A^K\text{-}A^K}(V^K, \mathrm{Hom}_K(M,N)),$$

which maps each $f^1 \in \mathrm{Hom}_{A\text{-}A}(V, \mathrm{Hom}_K(M,N))$ onto the map $\psi(f^1)$ defined by the receipt $\psi(f^1)(v \otimes c) = f^1(v)c$, for $v \in V$ and $c \in K$. Then, the receipt $\Psi(f^0, f^1) = (f^0, \psi(f^1))$ defines the isomorphism functor Ψ.

Proof.

(1) Since M and N are proper bimodules, we have $\alpha_M(e) = (\alpha_M(e)^0, 0)$ and $\alpha_N(e) = (\alpha_N(e)^0, 0)$, for all $e \in E$. Assume that $h = (h^0, h^1) \in \mathrm{Hom}_{\mathcal{A}}(M,N)$. Then, $h \in \mathrm{Hom}_{\mathcal{A}\text{-}E}(M,N)$ *iff*, for all $e \in E$, we have $(h^0, h^1)(\alpha_M(e)^0, 0) = (\alpha_N(e)^0, 0)(h^0, h^1)$. Equivalently, *iff* $(h^0\alpha_M(e)^0, h^1(v)\alpha_M(e)^0) = (\alpha_N(e)^0 h^0, \alpha_N(e)^0 h^1(v))$, for any $v \in V$ and $e \in E$; that is, $h^0 \in \mathrm{Hom}_E(M,N)$ and $h^1 \in \mathrm{Hom}_{A\text{-}A}(V, \mathrm{Hom}_E(M,N))$.

(2) This part admits a straightforward verification. Just remark that the inverse of ψ is given by $\psi^{-1}(g)[v] = g[v \otimes 1]$, for any morphism $g \in \mathrm{Hom}_{A^K\text{-}A^K}(V^K, \mathrm{Hom}_K(M,N))$.

□

Lemma 21.3. *Assume that E is a k-algebra and $\mathcal{A}$, $\mathcal{A}'$ are k-ditalgebras. Then any functor $F : \mathcal{A}'\text{-}\mathrm{Mod} \longrightarrow \mathcal{A}\text{-}\mathrm{Mod}$ induces a functor*

$$F^E : \mathcal{A}'\text{-}E\text{-}\mathrm{Mod} \longrightarrow \mathcal{A}\text{-}E\text{-}\mathrm{Mod},$$

which maps any object $M \in \mathcal{A}'\text{-}E\text{-}\mathrm{Mod}$ onto $F(M) = F^E(M)$, where the structure of $\mathcal{A}\text{-}E$-bimodule on $F(M)$ is given by the morphism

$$\alpha_{F(M)} : E \xrightarrow{\alpha_M} \mathrm{End}_{\mathcal{A}'}(M)^{op} \xrightarrow{F} \mathrm{End}_{\mathcal{A}}(F(M))^{op}.$$

Moreover, if F is full and faithful, then F^E is so. In this case, if an $\mathcal{A}\text{-}E$-bimodule M is isomorphic as an $\mathcal{A}$-module to $F(N)$, for some $\mathcal{A}'$-module N, then N admits a natural structure of $\mathcal{A}'\text{-}E$-bimodule such that $M \cong F^E(N)$ in $\mathcal{A}\text{-}E\text{-}\mathrm{Mod}$.

Proof. Take $f \in \mathrm{Hom}_{\mathcal{A}'\text{-}E}(M, N)$. Thus, $f\alpha_M(e) = \alpha_N(e)f$, for all $e \in E$. Applying F, we obtain

$$F(f)\alpha_{F(M)}(e) = F(f)F(\alpha_M(e)) = F(\alpha_N(e))F(f) = \alpha_{F(N)}(e)F(f).$$

Hence, the restriction $F^E : \mathrm{Hom}_{\mathcal{A}'\text{-}E}(M, N) \longrightarrow \mathrm{Hom}_{\mathcal{A}\text{-}E}(F(M), F(N))$ of F is well defined. Now assume that $F(f) \in \mathrm{Hom}_{\mathcal{A}\text{-}E}(F(M), F(N))$, for some $f \in \mathrm{Hom}_{\mathcal{A}'}(M, N)$. Then, as before, $F(f)F(\alpha_M(e)) = F(\alpha_N(e))F(f)$. Hence, F maps $f\alpha_M(e) - \alpha_N(e)f$ to zero, and, if F is faithful, $f \in \mathrm{Hom}_{\mathcal{A}'\text{-}E}(M, N)$; that is, F^E is a full and faithful functor, whenever F is so. Finally, assume this is the case, and $F(N) \cong M$ in $\mathcal{A}$-Mod, where $M \in \mathcal{A}\text{-}E\text{-}\mathrm{Mod}$. Then, given an isomorphism $f \in \mathrm{Hom}_{\mathcal{A}}(F(N), M)$, we can give an $\mathcal{A}\text{-}E$-bimodule structure to $F(N)$ defining $\alpha_{F(N)}$ as the composition

$$E \xrightarrow{\alpha_M} \mathrm{End}_{\mathcal{A}}(M)^{op} \xrightarrow{c_f} \mathrm{End}_{\mathcal{A}}(F(N))^{op},$$

where c_f denotes conjugation by f: $g \mapsto f^{-1}gf$. Then, the k-algebra morphism α_N given by the composition

$$E \xrightarrow{\alpha_{F(N)}} \mathrm{End}_{\mathcal{A}}(F(N))^{op} \xrightarrow{F^{-1}} \mathrm{End}_{\mathcal{A}'}(N)^{op}$$

clearly provides the wanted $\mathcal{A}'\text{-}E$-bimodule structure for N. □

Definition 21.4. *If E is a k-algebra and $\mathcal{A}$ is a k-ditalgebra, the forgetful functor $U_{\mathcal{A}} : \mathcal{A}\text{-}\mathrm{Mod} \to k\text{-}\mathrm{Mod}$, which maps each morphism (f^0, f^1) to f^0, induces a functor $U_{\mathcal{A}}^E : \mathcal{A}\text{-}E\text{-}\mathrm{Mod} \to k\text{-}E\text{-}\mathrm{Mod} = \mathrm{Mod}\text{-}E$. For each $M \in \mathcal{A}\text{-}E\text{-}\mathrm{Mod}$, we will consider the length of the right module $U_{\mathcal{A}}^E(M)$ over the algebra E and we denote this number by $\ell_E(M)$. The full subcategory of $\mathcal{A}\text{-}E\text{-}\mathrm{Mod}$ formed by the finite E-length bimodules is denoted by $\mathcal{A}\text{-}E\text{-}\mathrm{mod}$ and its intersection with $\mathcal{A}\text{-}E\text{-}\mathrm{Mod}_p$ by $\mathcal{A}\text{-}E\text{-}\mathrm{mod}_p$. Notice that if $\mathcal{A}$ is a layered ditalgebra with layer (R, W), then we can consider that $U_{\mathcal{A}} : \mathcal{A}\text{-}\mathrm{Mod} \to R\text{-}\mathrm{Mod}$ and $U_{\mathcal{A}}^E : \mathcal{A}\text{-}E\text{-}\mathrm{Mod} \to R\text{-}E\text{-}\mathrm{Mod}$.*

Suppose that $\mathcal{A}$ and $\mathcal{A}'$ are ditalgebras and E is a k-algebra. Thus, a length map ℓ_E has been attached to both $\mathcal{A}\text{-}E\text{-}\mathrm{Mod}$ and $\mathcal{A}'\text{-}E\text{-}\mathrm{Mod}$. Then we say that a functor $F : \mathcal{A}\text{-}E\text{-}\mathrm{Mod} \longrightarrow \mathcal{A}'\text{-}E\text{-}\mathrm{Mod}$ is length controlling iff $\ell_E(M)$ finite *implies* $\ell_E(F(M))$ *finite, and, furthermore,* $\ell_E(M) \leq \ell_E(F(M))$*, for all objects in* $M \in \mathcal{A}$-E-Mod. *If* $\ell_E(M) = \ell_E(F(M))$*, for all* M*, then* F *is called length preserving.*

Remark 21.5. *If* $\mathcal{A}$ *is any ditalgebra and* E *is any k-algebra, the length map* ℓ_E *attached to* $\mathcal{A}$-E-Mod *is an isomorphism invariant.*

Lemma 21.6. *Assume that* $\phi : \mathcal{A} \longrightarrow \mathcal{B}$ *is a morphism of ditalgebras. Then:*

(1) *for any k-algebra* E*, the induced functor* $F_\phi^E : \mathcal{B}$-E-Mod$\longrightarrow \mathcal{A}$-$E$-Mod *is length preserving and it sends proper bimodules to proper bimodules;*
(2) *if* $E = K$ *is any field extension of* k*,* $\phi^K : \mathcal{A}^K \to \mathcal{B}^K$ *is the extension morphism defined by* ϕ *(as in (20.1)), and* Ψ *denotes the isomorphism functor given in (21.2), then the following square commutes*

$$\begin{array}{ccc} \mathcal{B}\text{-}K\text{-Mod}_p & \xrightarrow{F_\phi^K} & \mathcal{A}\text{-}K\text{-Mod}_p \\ \Psi \downarrow & & \downarrow \Psi \\ \mathcal{B}^K\text{-Mod} & \xrightarrow{F_{\phi^K}} & \mathcal{A}^K\text{-Mod}. \end{array}$$

Proof. For the first part, let $M \in \mathcal{B}$-E-Mod and consider the right E-module $U_{\mathcal{B}}^E(M)$, that is the space M with E-action $\circ$ on the right such that $e \in E$ acts on $m \in M$, by the rule $m \circ e = e^0(m)$, where $\alpha_M(e) = (e^0, e^1) \in \mathrm{End}_{\mathcal{B}}(M)^{op}$.

Consider the other E-action $\diamond$ on the space M given by the E-module $U_{\mathcal{A}}^E(F_\phi^E(M))$, that is where each $e \in E$ acts on the element m by the rule $m \diamond e = \underline{e}^0(m)$, where $\alpha_{F_\phi(M)}(e) = (\underline{e}^0, \underline{e}^1) \in \mathrm{End}_{\mathcal{A}}(F_\phi(M))^{op}$.

Now recall that the receipt of the induced functor is $F_\phi(e^0, e^1) = (e^0, e^1\phi_1)$, and hence both E-actions coincide and, therefore, F_ϕ^E is a length preserving functor. It clearly sends proper bimodules to proper bimodules.

Part (2) admits a straightforward verification. □

Lemma 21.7. *Assume that* X *is an admissible* $\mathcal{A}'$*-module, where* $\mathcal{A}'$ *is a proper subditalgebra of the layered ditalgebra* $\mathcal{A}$*. Fix any field extension* K *of* k*, assume that* X *extends to* K*, and consider the associated functors* $F^X : \mathcal{A}^X$-Mod $\to \mathcal{A}$-Mod *and* $F^{X^K} : \mathcal{A}^{K X^K}$-Mod $\to \mathcal{A}^K$-Mod*, as in (20.11). Notice that the induced functor* $(F^X)^K : \mathcal{A}^X$-K-Mod $\to \mathcal{A}$-K-Mod *restricts to a functor* $(F^X)^K : \mathcal{A}^X$-K-Mod$_p \to \mathcal{A}$-K-Mod$_p$*. Then, with the notation of (20.11), the following diagram of functors commutes (up to isomorphism of functors)*

$$\begin{array}{ccccc} \mathcal{A}^X\text{-}K\text{-Mod}_p & & \xrightarrow{(F^X)^K} & & \mathcal{A}\text{-}K\text{-Mod}_p \\ \Psi \downarrow & & & & \downarrow \Psi \\ \mathcal{A}^{XK}\text{-Mod} & \xrightarrow{F_{\xi_X}} & \mathcal{A}^{K X^K}\text{-Mod} & \xrightarrow{F^{X^K}} & \mathcal{A}^K\text{-Mod}. \end{array}$$

Proof. We shall verify that defining

$$\begin{array}{rcl}\eta_M^0 : (F^X)^K\Psi^{-1}F_{\xi_X^{-1}}(M) & \longrightarrow & \Psi^{-1}F^{X^K}(M)\\ x\otimes m & \mapsto & (x\otimes 1)\otimes m,\end{array}$$

for $M \in \mathcal{A}^{KX^K}$-Mod, $\eta_M = (\eta_M^0, 0)$ gives a natural isomorphism of functors.

First recall that each $N \in \mathcal{A}^X$-K-Mod$_p$ can be seen as an A^X-K-bimodule with right action of K defined by $nt := \alpha_N^0(t)[n]$, where $\alpha_N = (\alpha_N^0, 0)$, $n \in N$ and $t \in K$. Thus, the structure of A-K-bimodule of $F^X(N) = X \otimes_S N$ is given on the right by $\alpha^0_{F^X(N)}(t)[x\otimes n] = (1\otimes\alpha_N^0(t))[x\otimes n] = x\otimes nt$, for $x \in X$ and $n \in N$. Now recall the notation of (12.1) and (12.2), where B is the subalgebra of $A = A_{\mathcal{A}}$ freely generated by (R, W_0'), and A is freely generated by $(B, \underline{W})$. Thus, to verify that η_M^0 is a morphism of A-K-bimodules, it is enough to see that it is a morphism of B-K-bimodules (which is clear) and that $\eta_M^0(w(x\otimes m)) = w\eta_M^0(x\otimes m)$, for all $w \in \underline{W}_0$. Denote by $*$ the action of $(A^K)^{X^K}$ on M, by $\widehat{*}$ the action of A^X on M, as well as the action of A^{XK} on M. Adopt the notation of the proof of (20.11). Then

$$\begin{aligned}\eta_M^0(w(x\otimes m)) &= \eta_M^0\left(\textstyle\sum_i x_i\otimes\sigma_{v_i,x}(w)\,\widehat{*}m\right)\\ &= \eta_M^0\left(\textstyle\sum_i x_i\otimes\left(\sigma_{v_i,x}(w)\otimes 1\right)\widehat{*}m\right)\\ &= \eta_M^0\left(\textstyle\sum_i x_i\otimes\left(\xi_X\left[\sigma_{\overline{\beta}_X(v_i\otimes 1),x\otimes 1}(w\otimes 1)\right]\right)\widehat{*}m\right)\\ &= \textstyle\sum_i (x_i\otimes 1)\otimes\left(\xi_X\left[\sigma_{\overline{\beta}_X(v_i\otimes 1),x\otimes 1}(w\otimes 1)\right]\right)\widehat{*}m\\ &= \textstyle\sum_i (x_i\otimes 1)\otimes\left[\sigma_{\overline{\beta}_X(v_i\otimes 1),x\otimes 1}(w\otimes 1)\right] * m\\ &= (w\otimes 1)[(x\otimes 1)\otimes m]\\ &= w(x\otimes 1)\otimes m\\ &= w\eta_M^0(x\otimes m).\end{aligned}$$

Now, take $f \in \mathrm{Hom}_{\mathcal{A}^{KX^K}}(M, N)$ and let us verify that the following square commutes

$$\begin{array}{ccc}(F^X)^K\Psi^{-1}F_{\xi_X^{-1}}(M) & \xrightarrow{\eta_M} & \Psi^{-1}F^{X^K}(M)\\ \Big\downarrow{\scriptstyle (F^X)^K\Psi^{-1}F_{\xi_X^{-1}}(f)} & & \Big\downarrow{\scriptstyle \Psi^{-1}F^{X^K}(f)}\\ (F^X)^K\Psi^{-1}F_{\xi_X^{-1}}(N) & \xrightarrow{\eta_N} & \Psi^{-1}F^{X^K}(N).\end{array}$$

Equivalently, for any $v \in V = V_{\mathcal{A}}$

$$\begin{aligned}[\Psi^{-1}F^{X^K}(f)]^0\eta_M^0 &= \eta_N^0[(F^X)^K\Psi^{-1}F_{\xi_X^{-1}}(f)]^0\\ [\Psi^{-1}F^{X^K}(f)]^1(v)\eta_M^0 &= \eta_N^0[(F^X)^K\Psi^{-1}F_{\xi_X^{-1}}(f)]^1(v).\end{aligned}$$

In order to prove the second equality, we may assume that $v = w \in \underline{W}_1$. Take $x \in X$ and $m \in M$, then

$$\begin{aligned}
[\Psi^{-1}F^{X^K}(f)]^0\eta^0_M(x \otimes m) &= [\Psi^{-1}F^{X^K}(f)]^0[(x \otimes 1) \otimes m] \\
&= [F^{X^K}(f)]^0[(x \otimes 1) \otimes m] \\
&= (x \otimes 1) \otimes f^0(m) \\
&\quad + \textstyle\sum_j (x \otimes 1)\theta_P(p_j \otimes 1) \otimes f^1(\overline{\beta}_P(\gamma_j \otimes 1))[m] \\
&= (x \otimes 1) \otimes f^0(m) \\
&\quad + \textstyle\sum_j (xp_j \otimes 1) \otimes \left(f^1\xi_X^{-1}\right)(\gamma_j \otimes 1)[m] \\
&= \eta^0_N \Big[x \otimes f^0(m) \\
&\quad + \textstyle\sum_j xp_j \otimes \psi^{-1}\left(f^1\xi_X^{-1}\right)(\gamma_j)[m]\Big] \\
&= \eta^0_N[(F^X)^K\Psi^{-1}F_{\xi_X^{-1}}(f)]^0(x \otimes m).
\end{aligned}$$

Moreover, for $w \in \underline{W}_1$, $x \in X$ and $m \in M$, we have

$$\begin{aligned}
[\Psi^{-1}F^{X^K}(f)]^1(w)\eta^0_M[x \otimes m] &= [\Psi^{-1}F^{X^K}(f)]^1(w)[(x \otimes 1) \otimes m] \\
&= [F^{X^K}(f)]^1(w \otimes 1)[(x \otimes 1) \otimes m] \\
&= \textstyle\sum_i (x_i \otimes 1) \otimes f^1(\sigma_{\overline{\beta}_X(v_i \otimes 1), x \otimes 1}(w \otimes 1))[m] \\
&= \textstyle\sum_i (x_i \otimes 1) \otimes f^1(\xi_X^{-1}[\sigma_{v_i,x}(w) \otimes 1])[m] \\
&= \eta^0_N \left(\textstyle\sum_i x_i \otimes \psi^{-1}\left(f^1\xi_X^{-1}\right)(\sigma_{v_i,x}(w))[m]\right) \\
&= \eta^0_N \left(\left[(F^X)^K\Psi^{-1}F_{\xi_X^{-1}}(f)\right]^1(w)[x \otimes m]\right).
\end{aligned}$$

□

Definition 21.8. *If we fix a k-algebra E, notice that the notion of bimodule, at the begining of (21.1), makes sense for any k-category $\mathcal{C}$ (not only for $\mathcal{A}$-Mod). Thus, we can consider the category of $\mathcal{C}$-E-bimodules, which we denote by $\mathcal{C}^E$. Then, for instance, if Λ is a finite-dimensional algebra and $\mathcal{C} = \mathcal{P}^1(\Lambda)$, an object in $\mathcal{P}^1(\Lambda)^E$ is an object $X \in \mathcal{P}^1(\Lambda)$, together with a k-algebra homomorphism $\alpha_X : E \to \mathrm{End}_{\mathcal{P}^1(\Lambda)}(X)^{op}$. If Y is another $\mathcal{P}^1(\Lambda)$-E-bimodule, then*

$$\mathrm{Hom}_{\mathcal{P}^1(\Lambda)^E}(X, Y) = \{\beta \in \mathrm{Hom}_{\mathcal{P}^1(\Lambda)}(X, Y) \mid \beta\alpha_X(e) = \alpha_Y(e)\beta, \text{ for all } e \in E\}.$$

Remark 21.9. *In the context of last definition, the argument in the proof of (21.3) shows that: any k-functor $F : \mathcal{B} \to \mathcal{C}$ induces a functor $F^E : \mathcal{B}^E \to \mathcal{C}^E$; moreover, if F is full and faithful, then F^E is so; in this case, if a $\mathcal{C}$-E-bimodule X is isomorphic as an object in $\mathcal{C}$ to $F(M)$, for some $\mathcal{B}$-module M, then M admits a natural structure of $\mathcal{B}$-E-bimodule such that $X \cong F^E(M)$ in $\mathcal{C}^E$.*

All this is applicable, in particular, to the functors $\Xi_\Lambda : \mathcal{D}^\Lambda$-Mod$\longrightarrow\mathcal{P}^1(\Lambda)$ and $\Xi^E_\Lambda : \mathcal{D}^\Lambda$-E-Mod$\longrightarrow\mathcal{P}^1(\Lambda)^E$, when Λ is a finite-dimensional algebra which splits over its radical.

Lemma 21.10. *Suppose that Λ is a finite-dimensional k-algebra which splits over its radical J, say $\Lambda = S \oplus J$, and consider the equivalence functor $\Xi_\Lambda : \mathcal{D}^\Lambda$-Mod$\longrightarrow \mathcal{P}^1(\Lambda)$. Consider any k-algebra E and any object $X = (P, Q, f) \in \mathcal{P}^1(\Lambda)^E$. Then, P and Q admit a natural structure of Λ-E-bimodules, and, therefore, P/JP and Q/JQ are S-E-bimodules. Moreover, if $M \in \mathcal{D}^\Lambda$-$E$-Mod satisfies that $\Xi^E_\Lambda(M) \cong (P, Q, f)$ in $\mathcal{P}^1(\Lambda)^E$, then*

$$\ell_E(M) = \ell_E(P/JP) + \ell_E(Q/JQ).$$

Proof. We proceed in several steps.

(a) Whenever $X = (P_1, P_2, f) \in \mathcal{P}^1(\Lambda)^E$, P_1 and P_2 admit a natural structure of Λ-E-bimodules. It is given on P_j by the composition with the corresponding projection

$$E \xrightarrow{\alpha_X} \mathrm{End}_{\mathcal{P}^1(\Lambda)}(X)^{op} \xrightarrow{\pi_j} \mathrm{End}_\Lambda(P_j)^{op}.$$

Composing with $\mathrm{End}_\Lambda(P_j)^{op} \longrightarrow \mathrm{End}_\Lambda(P_j/JP_j)^{op} = \mathrm{End}_S(P_j/JP_j)^{op}$, we get naturally an S-E-bimodule structure for P_j/JP_j.

(b) If $(P_1, P_2, f) \cong (P'_1, P'_2, f')$ in $\mathcal{P}^1(\Lambda)^E$, then we clearly obtain isomorphisms of Λ-E-bimodules $P_j \cong P'_j$, and isomorphisms of S-E-bimodules $P_j/JP_j \cong P'_j/JP'_j$.

(c) Given any $M \in \mathcal{D}^\Lambda$-$E$-Mod, and the corresponding triple (M_1, M_2, ψ_M), there are in principle two natural structures of S-E-bimodules which can be considered on each S-module M_j. They correspond to the upper and lower paths in the next diagram

$$\begin{array}{ccccccc} E & \xrightarrow{\alpha_M} & \mathrm{End}_{\mathcal{D}^\Lambda}(M)^{op} & \xrightarrow{\Xi_\Lambda} & \mathrm{End}_{\mathcal{P}^1(\Lambda)}(\Xi_\Lambda(M))^{op} & \xrightarrow{\pi_j} & \mathrm{End}_\Lambda(\Lambda \otimes_S M_j)^{op} \\ \| & & & & & & \downarrow q_j \\ E & \xrightarrow{\alpha_M} & \mathrm{End}_{\mathcal{D}^\Lambda}(M)^{op} & \xrightarrow{U_{\mathcal{D}^\Lambda}} & \mathrm{End}_{R^\Lambda}(M)^{op} & \xrightarrow{p_j} & \mathrm{End}_S(M_j)^{op}, \end{array}$$

where q_j is obtained as the composition

$$\mathrm{End}_\Lambda(\Lambda \otimes_S M_j)^{op} \longrightarrow \mathrm{End}_S([\Lambda \otimes_S M_j]/[J \otimes_S M_j])^{op} \xrightarrow{c_{\sigma_j}} \mathrm{End}_S(M_j)^{op},$$

where c_{σ_j} is the conjugation with the composition σ_j of isomorphisms of S-modules $[\Lambda \otimes_S M_j]/[J \otimes_S M_j] \longrightarrow [\Lambda/J] \otimes_S M_j \longrightarrow M_j$. Let us verify that they produce the same action of E on M_j. For $e \in E$ and $m_j \in M_j$, we denote by $m_j \star e$ the action of e on m_j given by the lower path of the diagram, and by $m_j \circ e$ the action of e on m_j given by the upper path of the diagram. The corresponding S-E-bimodules are denoted by $M^\star_j$ and M°_j, respectively. For $e \in E$, write $\alpha_M(e) = (\alpha_M(e)^0, \alpha_M(e)^1)$, and $\Xi_\Lambda(\alpha_M(e)) = (\Xi_\Lambda(\alpha_M(e))_1, \Xi_\Lambda(\alpha_M(e))_2)$. Then, for $m_j \in M_j$, we have

$m_j \star e = \alpha_M(e)_j^0[m_j]$. Moreover, if we look first to the right E-module structure in $\Lambda \otimes_S M_j$, which we also denote by $\circ$, having in mind the description of the action of Ξ_Λ on morphisms given in (19.8), we have

$$\begin{aligned}(1 \otimes m_j) \circ e &= \Xi_\Lambda(\alpha_M(e))_j(1 \otimes m_j)\\ &= 1 \otimes \alpha_M(e)_j^0[m_j] + \sum_i p_i \otimes \alpha_M(e)_j^1(w_{i,j})[m_j],\end{aligned}$$

for appropriate elements $w_{i,j} \in W_1^\Lambda$. Then, taking the class of this element modulo $J \otimes_S M_j$, we get: $\overline{(1 \otimes m_j) \circ e} = \overline{1 \otimes m_j \star e}$ in $[\Lambda \otimes_S M_j]/[J \otimes_S M_j]$. Then, their images under the isomorphism σ_j coincide: $m_j \star e = m_j \circ e$. And we obtain $M_j^\star = M_j^\circ$, as claimed.

(d) Assume that we have objects $(P_1, P_2, f) \in \mathcal{P}^1(\Lambda)^E$ and $M \in \mathcal{D}^\Lambda$-$E$-Mod, satisfying $\Xi_\Lambda^E(M) \cong (P_1, P_2, f)$ in $\mathcal{P}^1(\Lambda)^E$. Consider the structures of S-E-bimodules for P_j/JP_j described in (a). Then, if M corresponds to the triple (M_1, M_2, ψ_M), from (b) and (c) we know the existence of isomorphisms of S-E-bimodules

$$M_j^\star = M_j^\circ \cong [\Lambda \otimes_S M_j]/[J \otimes_S M_j] \cong P_j/JP_j.$$

(e) It is clear that $\ell_E(M) = \ell_E(M_1^\star) + \ell_E(M_2^\star)$. From (d), $M_j^\star \cong P_j/JP_j$, as S-E-bimodules, and the lemma is proved.

□

Lemma 21.11. *Suppose that Λ is a finite-dimensional k-algebra which splits over its radical J, say $\Lambda = S \oplus J$, and let E be any k-algebra. Denote by $\mathcal{P}^1(\Lambda)_p^E$ the full subcategory of $\mathcal{P}^1(\Lambda)^E$ formed by the triples in $\mathcal{P}^1(\Lambda)$ of the form $M = (\Lambda \otimes_S M_1, \Lambda \otimes_S M_2, \phi)$, where M_1 and M_2 are S-E-bimodules and ϕ is a morphism of Λ-E-bimodules. Such a triple is considered an object of $\mathcal{P}^1(\Lambda)^E$ equipped with the algebra morphism $\beta_M : E \longrightarrow \mathrm{End}_{\mathcal{P}^1(\Lambda)}(M)^{op}$ such that $\beta_M(e) = (\beta_M(e)_1, \beta_M(e)_2)$, and $\beta_M(e)_i(\lambda \otimes m_i) = \lambda \otimes m_i e$, for $m_i \in M_i, \lambda \in \Lambda, e \in E$. Thus, if $M, N \in \mathcal{P}^1(\Lambda)_p^E$ and $u = (u_1, u_2) \in \mathrm{Hom}_{\mathcal{P}^1(\Lambda)}(M, N)$, then $u \in \mathrm{Hom}_{\mathcal{P}^1(\Lambda)^E}(M, N)$ iff $u_i \in \mathrm{Hom}_{\Lambda\text{-}E}(\Lambda \otimes_S M_i, \Lambda \otimes_S N_i)$, for $i \in \{1, 2\}$. Then, the equivalence functor $\Xi_\Lambda^E : \mathcal{D}^\Lambda$-$E$-Mod$\longrightarrow \mathcal{P}^1(\Lambda)^E$, restricts to an equivalence*

$$\mathcal{D}^\Lambda\text{-}E\text{-Mod}_p \xrightarrow{\Xi_\Lambda^E} \mathcal{P}^1(\Lambda)_p^E.$$

If $E = K$ is a field extension of k preserving the radical of Λ, then $\mathcal{P}^1(\Lambda)_p^K$ is naturally identified with a full and dense subcategory of $\mathcal{P}^1(\Lambda^K)$. The composition with the inclusion yields an equivalence

$$\mathcal{D}^\Lambda\text{-}K\text{-Mod}_p \xrightarrow{\Xi_\Lambda^K} \mathcal{P}^1(\Lambda^K).$$

Proof. Assume that the splitting of Λ over its radical J has the form $\Lambda = S \oplus J$, with S a subalgebra of Λ. By definition, $\mathcal{D}^\Lambda$ has layer (R^Λ, W^Λ), where

$$R = R^\Lambda = \begin{pmatrix} S & 0 \\ 0 & S \end{pmatrix}, W_0^\Lambda = \begin{pmatrix} 0 & 0 \\ J^* & 0 \end{pmatrix} \text{ and } W_1^\Lambda = \begin{pmatrix} J^* & 0 \\ 0 & J^* \end{pmatrix}.$$

Recall from (19.7) that $A = A_{\mathcal{D}^\Lambda} = \begin{pmatrix} S & 0 \\ J^* & S \end{pmatrix}$ and, therefore, the objects $M \in \mathcal{D}^\Lambda$-Mod determine triples (M_1, M_2, ψ_M), where M_1, M_2 are S-modules and $\psi_M \in \mathrm{Hom}_{S\text{-}S}(J^*, \mathrm{Hom}_k(M_1, M_2))$. Let us show first the following:

Claim 1. $M \in \mathcal{D}^\Lambda$-$E$-$\mathrm{Mod}_p$ iff M_1 and M_2 admit a structure of S-E-bimodules and $\psi_M \in \mathrm{Hom}_{S\text{-}S}(J^*, \mathrm{Hom}_E(M_1, M_2))$.

Proof of claim 1: If $M \in \mathcal{D}^\Lambda$-$E$-$\mathrm{Mod}_p$ has associated morphism of algebras $\alpha_M : E \longrightarrow \mathrm{End}_{\mathcal{D}}(M)^{op}$, the formula $me := \alpha_M(e)^0[m]$, for $m \in M$ and $e \in E$, defines a right E-module structure on M such that M is an A-E-bimodule. Hence, M is an S-E-bimodule (by restriction through $S \to A$), and each M_j is an S-E-subbimodule of M. Moreover, for $m_1 \in M_1$, $\gamma \in J^*$ and $e \in E$

$$\psi_M(\gamma)[m_1 e] = \begin{pmatrix} 0 & 0 \\ \gamma & 0 \end{pmatrix} (m_1 e) = \left[\begin{pmatrix} 0 & 0 \\ \gamma & 0 \end{pmatrix} m_1 \right] e = (\psi_M(\gamma)[m_1]) e.$$

Thus, $\psi_M \in \mathrm{Hom}_{S\text{-}S}(J^*, \mathrm{Hom}_E(M_1, M_2))$. Conversely, assume that M_1 and M_2 admit a structure of S-E-bimodules and $\psi_M \in \mathrm{Hom}_{S\text{-}S}(J^*, \mathrm{Hom}_E(M_1, M_2))$. Then, $M = M_1 \oplus M_2$ is an S-E-bimodule. It is in fact an A-E-bimodule because, for $\gamma \in J^*$, $m_1 \in M_1$, $m_2 \in M_2$ and $e \in E$

$$\begin{aligned} \begin{pmatrix} s & 0 \\ \gamma & s' \end{pmatrix} [(m_1 + m_2)e] &= \begin{pmatrix} s & 0 \\ \gamma & s' \end{pmatrix} [m_1 e + m_2 e] \\ &= s(m_1 e) + s'(m_2 e) + \psi_M(\gamma)[m_1 e] \\ &= (s m_1)e + (s' m_2)e + (\psi_M(\gamma)[m_1]) e \\ &= \left[\begin{pmatrix} s & 0 \\ \gamma & s' \end{pmatrix} (m_1 + m_2) \right] e. \end{aligned}$$

Claim 1 is proved. Now, recall from (19.8) the receipt of the functor $\Xi_\Lambda : \mathcal{D}^\Lambda\text{-Mod} \longrightarrow \mathcal{P}^1(\Lambda)$. Consider any finite right S-dual basis $\{(p_j, \gamma_j)\}_j$ for J, with $p_j \in J$ and $\gamma_j \in J^*$. Then, the receipt of $\Xi_\Lambda(M) : \Lambda \otimes_S M_1 \longrightarrow \Lambda \otimes_S M_2$ can be described, for $\lambda \in \Lambda$ and $m_1 \in M_1$, by

$$\Xi_\Lambda(M)[\lambda \otimes m_1] = \sum_j \lambda p_j \otimes \psi_M(\gamma_j)[m_1].$$

From our claim 1, we know that, for $M \in \mathcal{D}^\Lambda$-$E$-$\mathrm{Mod}_p$, each map $\psi_M(\gamma)$ is E-linear. Hence, we obtain that $\Xi_\Lambda(M)$ is E-linear, thus the object $\Xi_\Lambda(M)$ equipped with $\beta_{\Xi_\Lambda(M)}$ belongs to $\mathcal{P}^1(\Lambda)_p^E$. It remains to show that $\Xi_\Lambda^E(M)$ is the bimodule described above, that is $\beta_{\Xi_\Lambda(M)}$ coincides with the algebra morphism $\alpha_{\Xi_\Lambda(M)} : E \longrightarrow \mathrm{End}_{\mathcal{P}^1(\Lambda)}(\Xi_\Lambda(M))^{op}$ associated to the bimodule $\Xi^E_\Lambda(M)$. For $e \in E, \lambda \in \Lambda, m_i \in M_i, i \in \{1,2\}$, we have to see that

$$\left[\alpha_{\Xi_\Lambda(M)}(e)\right]_i (\lambda \otimes m_i) = \left[\beta_{\Xi_\Lambda(M)}(e)\right]_i (\lambda \otimes m_i).$$

We check this for $i = 1$, the other case is similar

$$\begin{aligned}\alpha_{\Xi_\Lambda(M)}(e)_1(\lambda \otimes m_1) &= \Xi_\Lambda(\alpha_M(e))_1(\lambda \otimes m_1)\\ &= \lambda \otimes \alpha_M(e)^0(m_1) + \textstyle\sum_j \lambda p_j \otimes \alpha_M(e)^1 \left[\begin{pmatrix}\gamma_j & 0\\ 0 & 0\end{pmatrix}\right][m_1]\\ &= \lambda \otimes m_1 e\\ &= \beta_{\Xi_\Lambda(M)}(e)_1(\lambda \otimes m_1),\end{aligned}$$

because M is a proper bimodule (thus $\alpha_M(e)^1 = 0$). Then, $\alpha_{\Xi_\Lambda(M)}(e) = \beta_{\Xi_\Lambda(M)}(e)$, and $\Xi^E_\Lambda(M) \in \mathcal{P}^1(\Lambda)^E_p$.

Since $\mathcal{P}^1(\Lambda)^E_p$ is a full subcategory of $\mathcal{P}^1(\Lambda)^E$, we do not have to check, for $M, N \in \mathcal{D}^\Lambda$-Mod and $h \in \mathrm{Hom}_{\mathcal{D}^\Lambda}(M,N)$, that the components of $\Xi_\Lambda(h) = (\Xi_\Lambda(h)_1, \Xi_\Lambda(h)_2)$ are E-linear.

All this shows that the functor $\Xi^E_\Lambda : \mathcal{D}^\Lambda\text{-}E\text{-Mod}_p \longrightarrow \mathcal{P}^1(\Lambda)^E_p$ is well defined (and clearly full and faithful, since Ξ^E_Λ was so: Remark (21.9)). From the following claim 2, we get that the functor $\Xi^E_\Lambda : \mathcal{D}^\Lambda\text{-}E\text{-Mod}_p \longrightarrow \mathcal{P}^1(\Lambda)^E_p$ is dense.

Claim 2. Let $\phi_M : \Lambda \otimes_S M_1 \longrightarrow \Lambda \otimes_S M_2$ be an object of $\mathcal{P}^1(\Lambda)^E_p$. Then, there exist $N \in \mathcal{D}^\Lambda\text{-}E\text{-Mod}_p$ and isomorphisms of S-E-bimodules $\theta_i : N_i \longrightarrow M_i$, for $i \in \{1,2\}$, such that the following diagram commutes

$$\begin{array}{ccc}\Lambda \otimes_S N_1 & \xrightarrow{\Xi_\Lambda(N)} & \Lambda \otimes_S N_2\\ {\scriptstyle 1\otimes\theta_1}\big\downarrow & & \big\downarrow{\scriptstyle 1\otimes\theta_2}\\ \Lambda \otimes_S M_1 & \xrightarrow{\phi_M} & \Lambda \otimes_S M_2.\end{array}$$

Proof of claim 2: Adopt the notation of the proof of (19.8), where the functor Ξ_Λ is defined as the following composition of equivalences

$$\mathcal{D}\text{-Mod} \xrightarrow{\widehat{\ }} [\mathcal{D}\text{-Mod}]_0 \xrightarrow{H^{-1}} \mathcal{M} \xrightarrow{F} \mathcal{P}^1(\Lambda).$$

Thus, F is induced by an equivalence $F^X : \mathcal{B}\text{-Mod} \longrightarrow \mathcal{A}\text{-Mod}$, with $X = \Lambda$. Recall that, in the proof of (13.3), we described the inverse G of the map

$$\mathrm{Hom}_{\mathcal{B}}(M_1, M_2) \xrightarrow{F^X} \mathrm{Hom}_{\mathcal{A}}(X \otimes_S M_1, X \otimes_S M_2).$$

Having in mind the identifications of $\mathrm{Hom}_\Lambda(\Lambda \otimes_S M_1, \Lambda \otimes_S M_2)$ with $\mathrm{Hom}_{\mathcal{A}}(X \otimes_S M_1, X \otimes_S M_2)$, which maps ϕ_M onto $(\phi_M, 0)$, and of $\Gamma = \mathrm{End}_{\mathcal{A}}(X)^{op}$ with Λ, which maps each $\mathcal{A}$-endomorphism $(g, 0)$ of X onto

$g(1)$, we can translate the receipt of G, for $f := (\phi_M, 0)$, $G(f) = (0, \psi_M)$, $\gamma \in J^*$ and $m_1 \in M_1$, as $\psi_M(\gamma)[m_1] = \sum_t \gamma(\sigma(1 \otimes \eta_t)\phi_M(1 \otimes m_1))m_t$, as we did in the proof of (19.9), where $\sigma : \Lambda \otimes_S S \longrightarrow \Lambda$ is the canonical isomorphism, and $(m_t, \eta_t)_t$ is a dual basis of the left S-module M_2 ($m_t \in M_2$ and $\eta_t \in \mathrm{Hom}_S(M_2, S)$).

Take $m_1 \in M_1$ and $e \in E$, we want to show that $\psi_M(\gamma)[m_1 e] = \psi_M(\gamma)[m_1]e$, equivalently

$$\sum_t \gamma(\sigma(1 \otimes \eta_t)\phi_M(1 \otimes m_1 e))m_t = \sum_t \gamma(\sigma(1 \otimes \eta_t)\phi_M(1 \otimes m_1))m_t e.$$

Assume that $\phi_M(1 \otimes m_1) = \sum_z \lambda_z \otimes n_z$. Hence, $\phi_M(1 \otimes m_1 e) = \phi_M(1 \otimes m_1)e = \sum_z \lambda_z \otimes n_z e$. For each z, from the definition of dual basis, we know that $\sum_t \eta_t(n_z e)m_t = n_z e = \left[\sum_t \eta_t(n_z)m_t\right] e$. Then

$$\begin{aligned}\sum_{z,t} \gamma[\lambda_z \eta_t(n_z e)]m_t &= \sum_z \gamma[\lambda_z] \sum_t \eta_t(n_z e)m_t \\ &= \sum_z \gamma[\lambda_z] \sum_t \eta_t(n_z)m_t e \\ &= \sum_{z,t} \gamma\left[\lambda_z \eta_t(n_z)\right] m_t e.\end{aligned}$$

Thus, we have verified that $\psi_M \in \mathrm{Hom}_{S\text{-}S}(J^*, \mathrm{Hom}_E(M_1, M_2))$. Define $N := \overline{\tau}_M$, the matrix A-module associated to the triple $\tau_M = (M_1, M_2, \psi_M)$ as defined in (19.7). The object N determines a triple (N_1, N_2, ψ_N), as in (19.7). Thus

$$N_1 = \begin{pmatrix} M_1 \\ 0 \end{pmatrix}, N_2 = \begin{pmatrix} 0 \\ M_2 \end{pmatrix}, \text{ and for } n_1 = \begin{pmatrix} m_1 \\ 0 \end{pmatrix} \in N_1,$$

we have, by definition of the action of A on N and the definition of ψ_N, that

$$\psi_N(\gamma)[n_1] = \begin{pmatrix} 0 & 0 \\ \gamma & 0 \end{pmatrix} n_1 = \begin{pmatrix} 0 & 0 \\ \gamma & 0 \end{pmatrix} \begin{pmatrix} m_1 \\ 0 \end{pmatrix} = \begin{pmatrix} 0 \\ \psi_M(\gamma)[m_1] \end{pmatrix},$$

for any $\gamma \in J^*$. From claim 1, we obtain that $N \in \mathcal{D}^\Lambda\text{-}E\text{-}\mathrm{Mod}_p$. Therefore, if we denote by $\theta_i : N_i \longrightarrow M_i$ the canonical isomorphism of S-E-bimodules, for $i \in \{1, 2\}$, we obtain

$$\begin{aligned}(1 \otimes \theta_2)\Xi_\Lambda(N)[\lambda \otimes n_1] &= (1 \otimes \theta_2)\left[\sum_j \lambda p_j \otimes \psi_N(\gamma_j)[n_1]\right] \\ &= \sum_j \lambda p_j \otimes \psi_M(\gamma_j)[m_1] \\ &= \left[F^X(0, \psi_M)\right]^0 [\lambda \otimes m_1] \\ &= \left[F^X G(\phi_M, 0)\right]^0 [\lambda \otimes m_1] \\ &= \phi_M[\lambda \otimes m_1] \\ &= \phi_M(1 \otimes \theta_1)[\lambda \otimes n_1].\end{aligned}$$

Our claim 2 is proved.

To verify the last statement of the lemma, for $E = K$, recall that $\Lambda^K = S^K \oplus J^K$ splits over its radical J^K. Notice that we can regard $\mathcal{P}^1(\Lambda)_p^K$ as a full subcategory of $\mathcal{P}^1(\Lambda^K)$. Indeed, notice first that, for any $N \in S$-K-Mod, we have the isomorphism of Λ-K-bimodules $\Lambda \otimes_S N \longrightarrow \Lambda^K \otimes_{S^K} N$ which maps $\lambda \otimes n$ onto $(\lambda \otimes 1) \otimes n$. Now, if P is any projective Λ^K-module, we have the isomorphism $\sigma_P : P \longrightarrow \Lambda^K \otimes_{S^K} P/J^K P$ which is obtained applying (19.6) to the K-algebra Λ^K. Then, we can consider the following composition $\widehat{\sigma}_P$ of isomorphisms of Λ-K-bimodules

$$P \xrightarrow{\sigma_P} \Lambda^K \otimes_{S^K} P/J^K P = \Lambda^K \otimes_{S^K} P/JP \cong \Lambda \otimes_S P/JP.$$

Thus, any object (P, Q, f) in $\mathcal{P}^1(\Lambda^K)$ is isomorphic in $\mathcal{P}^1(\Lambda^K)$ to one having the special form $(\Lambda \otimes_S P/JP, \Lambda \otimes_S Q/JQ, \widehat{\sigma}_Q f \widehat{\sigma}_P^{-1})$. Thus, $\mathcal{P}^1(\Lambda)_p^K$ is naturally identified with a full and dense subcategory of $\mathcal{P}^1(\Lambda^K)$. □

Lemma 21.12. *Assume Λ is a finite-dimensional k-algebra which splits over its radical. Assume that $k \leq K$ is a field extension preserving the radical of Λ. Consider the Drozd's ditalgebras $\mathcal{D}^\Lambda$ and $\mathcal{D}^{\Lambda^K}$, the equivalences F_{ξ_Λ}, Ξ_Λ and Ξ_{Λ^K} of (20.13), the isomorphism Ψ of (21.2) and the equivalence Ξ_Λ^K of (21.11). Then, the following diagram of equivalences commutes (up to isomorphism)*

$$\begin{array}{ccccc}
\mathcal{D}^\Lambda\text{-}K\text{-Mod}_p & & \xrightarrow{\quad\Xi_\Lambda^K\quad} & & \mathcal{P}^1(\Lambda^K) \\
\downarrow{\scriptstyle \Psi} & & & & \| \\
\mathcal{D}^{\Lambda K}\text{-Mod} & \xrightarrow{F_{\xi_\Lambda}} & \mathcal{D}^{\Lambda^K}\text{-Mod} & \xrightarrow{\Xi_{\Lambda^K}} & \mathcal{P}^1(\Lambda^K).
\end{array}$$

Proof. Let $M \in \mathcal{D}^\Lambda$-$K$-Mod$_p$ with associated triple (M_1, M_2, ψ_M). Then, consider the following square

$$\begin{array}{ccc}
\Lambda^K \otimes_{S^K} M_1 & \xrightarrow{\Xi_{\Lambda^K} F_{\xi_\Lambda} \Psi(M)} & \Lambda^K \otimes_{S^K} M_2 \\
\uparrow{\scriptstyle \sigma_1} & & \uparrow{\scriptstyle \sigma_2} \\
\Lambda \otimes_S M_1 & \xrightarrow{\Xi_\Lambda^K(M)} & \Lambda \otimes_S M_2,
\end{array}$$

where $\sigma_i(\lambda \otimes m_i) = (\lambda \otimes 1) \otimes m_i$, for $i \in \{1, 2\}$, $m_i \in M_i$ and $\lambda \in \Lambda$. Clearly, each σ_i is an isomorphism of Λ-K-bimodules. Let us see that the square commutes

$$\begin{aligned}
\Xi_{\Lambda^K} F_{\xi_\Lambda} \Psi(M)\sigma_1[\lambda \otimes m_1] &= \Xi_{\Lambda^K} F_{\xi_\Lambda} \Psi(M)[(\lambda \otimes 1) \otimes m_1] \\
&= \textstyle\sum_j [(\lambda \otimes 1)(p_j \otimes 1)] \otimes \begin{pmatrix} 0 & 0 \\ \beta(\gamma_j \otimes 1) & 0 \end{pmatrix} m_1 \\
&= \textstyle\sum_j [\lambda p_j \otimes 1] \otimes \begin{pmatrix} 0 & 0 \\ \gamma_j & 0 \end{pmatrix} m_1 \\
&= \sigma_2 \left[\textstyle\sum_j \lambda p_j \otimes \begin{pmatrix} 0 & 0 \\ \gamma_j & 0 \end{pmatrix} m_1 \right] \\
&= \sigma_2 \Xi_\Lambda^K(M)[\lambda \otimes m_1],
\end{aligned}$$

where $\beta : J^{*K} \longrightarrow J^{K*}$ is the canonical isomorphism and $(p_j, \gamma_j)_j$ is a dual basis for the right S-module J. Thus, $\sigma_M := (\sigma_1, \sigma_2) : \Xi_\Lambda^K(M) \longrightarrow \Xi_{\Lambda^K} F_{\xi_\Lambda} \Psi(M)$ is an isomorphism in $\mathcal{P}^1(\Lambda^K)$.

Now, take $h = (h^0, h^1) \in \mathrm{Hom}_{\mathcal{D}^{\Lambda}\text{-}K}(M, N)$ and let us verify that the following diagram commutes

$$\begin{array}{ccc} \Xi_\Lambda^K(M) & \xrightarrow{\sigma_M} & \Xi_{\Lambda^K} F_{\xi_\Lambda} \Psi(M) \\ \Big\downarrow{\scriptstyle \Xi_\Lambda^K(h)} & & \Big\downarrow{\scriptstyle \Xi_{\Lambda^K} F_{\xi_\Lambda} \Psi(h)} \\ \Xi_\Lambda^K(N) & \xrightarrow{\sigma_N} & \Xi_{\Lambda^K} F_{\xi_\Lambda} \Psi(N). \end{array}$$

We have to show that $(\sigma_N)_i \Xi_\Lambda^K(h)_i(\lambda \otimes m_i) = [\Xi_{\Lambda^K} F_{\xi_\Lambda} \Psi(h)]_i(\sigma_M)_i(\lambda \otimes m_i)$, for all $i \in \{1, 2\}$, $\lambda \in \Lambda$ and $m_i \in M_i$. We consider only the case $i = 1$, since the other one is similar. If we make $\Delta := [\Xi_{\Lambda^K} F_{\xi_\Lambda} \Psi(h)]_1(\sigma_M)_1[\lambda \otimes m_1]$, then

$$\begin{aligned}
\Delta &= [\Xi_{\Lambda^K} F_{\xi_\Lambda}(h^0, \psi(h^1))]_1[(\lambda \otimes 1) \otimes m_1] \\
&= [\Xi_{\Lambda^K}(h^0, \psi(h^1)\xi_\Lambda)]_1[(\lambda \otimes 1) \otimes m_1] \\
&= (\lambda \otimes 1) \otimes h^0[m_1] \\
&\quad + \textstyle\sum_j (\lambda \otimes 1)(p_j \otimes 1) \otimes \psi(h^1)\xi_\Lambda \begin{pmatrix} \beta(\gamma_j \otimes 1) & 0 \\ 0 & 0 \end{pmatrix} [m_1] \\
&= (\sigma_N)_1 \left[\lambda \otimes h^0[m_1] + \textstyle\sum_j \lambda p_j \otimes h^1 \begin{pmatrix} \gamma_j & 0 \\ 0 & 0 \end{pmatrix} [m_1] \right] \\
&= (\sigma_N)_1 \Xi_\Lambda^K(h)_1[\lambda \otimes m_1].
\end{aligned}$$

And, $\sigma : \Xi_\Lambda^K \longrightarrow \Xi_{\Lambda^K} F_{\xi_\Lambda} \Psi$ is an isomorphism of functors. □

22

Parametrizing bimodules and wildness

In this section, we introduce the notion of wildness; we give some examples and some useful criteria for wildness. We will also exhibit sufficient conditions for wildness and a tensor-like behavior of some important functors.

Definition 22.1. *A functor $F : \mathcal{C} \longrightarrow \mathcal{C}'$ between k-categories preserves indecomposability iff, for any indecomposable object M in $\mathcal{C}$, the object $F(M)$ of $\mathcal{C}'$ is indecomposable. The functor F preserves isomorphism classes iff $F(M) \cong F(N)$ implies $M \cong N$, for any objects M, N of $\mathcal{C}$. Finally, we say that F preserves isomorphism classes of indecomposables iff F preserves indecomposability and $F(M) \cong F(N)$ implies $M \cong N$, for any indecomposable objects M, N of $\mathcal{C}$.*

Lemma 22.2. *Any full and faithful functor $F : \mathcal{A}$-Mod$\longrightarrow \mathcal{A}'$-Mod, where $\mathcal{A}$ and $\mathcal{A}'$ are Roiter ditalgebras, preserves indecomposability and isomorphism classes.*

Proof. By (5.12), any $\mathcal{A}$-module (as well as any $\mathcal{A}'$-module) is indecomposable iff it does not admit non-trivial idempotent endomorphisms. Hence, F preserves indecomposability. □

Definition 22.3. *A ditalgebra $\mathcal{A}$ over the field k is wild iff there is an $A_{\mathcal{A}}$-$k\langle x, y\rangle$-bimodule Z, free of finite rank as a right $k\langle x, y\rangle$-module, such that the composition functor*

$$k\langle x, y\rangle\text{-Mod} \xrightarrow{Z\otimes_{k\langle x,y\rangle}-} A_{\mathcal{A}}\text{-Mod} \xrightarrow{L_{\mathcal{A}}} \mathcal{A}\text{-Mod}$$

preserves isomorphism classes of indecomposables. In this case, we say that the bimodule Z produces the wildness of $\mathcal{A}$.

The k-algebra Λ is wild iff the regular ditalgebra $\mathcal{R}_\Lambda = (\Lambda, 0)$, with layer $(\Lambda, 0)$, is wild. Equivalently, iff there is a Λ-$k\langle x, y\rangle$-bimodule Z, free of finite *rank when considered as a right $k\langle x, y\rangle$-module, such that the tensor functor $Z \otimes_{k\langle x,y\rangle} - : k\langle x, y\rangle\text{-Mod} \longrightarrow \Lambda\text{-Mod}$ preserves isomorphism classes of indecomposables.*

Using Drozd's *Tame and Wild Theorem* (to be proved in Section 27) and some geometric facts on varieties of modules, we will show in Section 33 the following equivalence of possible characterizations of wildness.

Proposition 22.4. *Let Λ be a finite-dimensional algebra over the algebraically closed field k. Then, the following statements are equivalent:*

(1) Λ is wild.
(2) There is a Λ-$k\langle x, y\rangle$-bimodule Z, free of finite rank as a right $k\langle x, y\rangle$-module, such that the functor $Z \otimes_{k\langle x,y\rangle} - : k\langle x, y\rangle\text{-mod} \longrightarrow \Lambda\text{-mod}$ preserves isomorphism classes of indecomposables.
(3) There is a Λ-$k\langle x, y\rangle$-bimodule Z, free of finite rank as a right $k\langle x, y\rangle$-module, such that the functor $Z \otimes_{k\langle x,y\rangle} - : k\langle x, y\rangle\text{-mod} \longrightarrow \Lambda\text{-mod}$ preserves indecomposability and isomorphism classes.
(4) There is a Λ-$k\langle x, y\rangle$-bimodule Z, free of finite rank as a right $k\langle x, y\rangle$-module, such that the functor $Z \otimes_{k\langle x,y\rangle} - : k\langle x, y\rangle\text{-mod} \longrightarrow \Lambda\text{-mod}$ preserves isomorphism classes.

For the moment, we only show the following:

Lemma 22.5. *Let $\mathcal{A}$ be a wild Roiter ditalgebra over the field k, with layer (R, W) such that W_1 is a finitely generated R-R-bimodule. Then, there is an $A_{\mathcal{A}}$-$k\langle x, y\rangle$-bimodule Z, free of finite rank as a right $k\langle x, y\rangle$-module, such that the composition functor*

$$k\langle x, y\rangle\text{-mod} \xrightarrow{Z\otimes_{k\langle x,y\rangle}-} A_{\mathcal{A}}\text{-mod} \xrightarrow{L_A} \mathcal{A}\text{-mod}$$

preserves indecomposability and isomorphism classes. Recall that $\mathcal{A}$-mod means the full subcategory of $\mathcal{A}$-Mod formed by the finite-dimensional objects. In particular, if an algebra Λ is wild, then there is a Λ-$k\langle x, y\rangle$-bimodule Z, free of finite rank as a right $k\langle x, y\rangle$-module, such that the functor $Z \otimes_{k\langle x,y\rangle} - : k\langle x, y\rangle\text{-mod} \longrightarrow \Lambda\text{-mod}$ preserves indecomposability and isomorphism classes.

Proof. By assumption, there is an $A_{\mathcal{A}}$-$k\langle x, y\rangle$-bimodule Z, free of finite rank as a right $k\langle x, y\rangle$-module, such that the composition functor F

$$k\langle x, y\rangle\text{-mod} \xrightarrow{Z\otimes_{k\langle x,y\rangle}-} A_{\mathcal{A}}\text{-mod} \xrightarrow{L_A} \mathcal{A}\text{-mod}$$

preserves isomorphism classes of indecomposables. We claim that this composition functor F preserves isomorphism classes. Take $M, N \in k\langle x, y\rangle$-mod and assume that $F(N) \cong F(M)$ in $\mathcal{A}$-mod. Consider the decompositions $M = \oplus_{i=1}^m M_i$ and $N = \oplus_{j=1}^n N_j$ of M and N as direct sums of indecomposables. Then, $\oplus_{i=1}^m F(M_i) = F(M) \cong F(N) = \oplus_{j=1}^n F(N_j)$. From (5.13), we know that $\mathcal{A}$-mod is a Krull–Schmidt category and, therefore, $m = n$ and, for some reindexation $\sigma \in S_n$, $F(M_{\sigma(i)}) \cong F(N_i)$, for all $i \in [1, n]$. Since F preserves isomorphism classes of indecomposables, then $M_{\sigma(i)} \cong N_i$, for all i. Thus, $M \cong N$. □

Remark 22.6. *In last definition the phrase "there is a Λ-$k\langle x, y\rangle$-bimodule Z, free of finite rank as a right $k\langle x, y\rangle$-module" can be substituted by "there is a Λ-$k\langle x, y\rangle$-bimodule Z, finitely generated and projective as a right $k\langle x, y\rangle$-module". Indeed, the class of projective right $k\langle x, y\rangle$-modules coincides with the class of free right $k\langle x, y\rangle$-modules. This follows from a well-known theorem (see [17](2.4)) stating that any right ideal of a free associative k-algebra R is a free right R-module. Then, by a theorem of Kaplanski (see [47](4.17)), every submodule of a free right R-module is free.*

Lemma 22.7. *Assume that $H : \mathcal{A}'\text{-Mod} \longrightarrow \mathcal{A}\text{-Mod}$ is a functor obtained as a finite composition of functors of the form F^X, for some admissible module X, or F_ϕ, for some morphism $\phi : \mathcal{A} \to \mathcal{A}'$ of ditalgebras. Let $A := A_{\mathcal{A}}$ and $A' := A_{\mathcal{A}'}$. Then, H induces by restriction a functor $\underline{H}$ which makes the right square of the following diagram commutative. If Δ is any k-algebra and Z is an A'-Δ-bimodule, then $H(Z)$ is an A-Δ-bimodule and the first square in the following diagram commutes up to isomorphism*

$$\begin{array}{ccccc} \Delta\text{-Mod} & \xrightarrow{Z\otimes_\Delta -} & A'\text{-Mod} & \xrightarrow{L_{\mathcal{A}'}} & \mathcal{A}'\text{-Mod} \\ \| & & \downarrow{\scriptstyle \underline{H}} & & \downarrow{\scriptstyle H} \\ \Delta\text{-Mod} & \xrightarrow{H(Z)\otimes_\Delta -} & A\text{-Mod} & \xrightarrow{L_{\mathcal{A}}} & \mathcal{A}\text{-Mod}. \end{array}$$

If Z is a projective right Δ-module, so is $H(Z)$. In particular, making $\Delta := A'$, we obtain that $\underline{H} \cong H(A') \otimes_{A'} -$ is exact and preserves direct sums.

Proof. It is enough to show it for functors of the form $H = F^X$ or F_ϕ. The case $H = F_\phi$ is clear: here, $\underline{H} = F_{\phi_0} : A'\text{-Mod} \longrightarrow A\text{-Mod}$ is the functor induced by the restriction $\phi_0 : A \to A'$ of ϕ. For the case $H = F^X$, notice first that $F^X(f^0, 0) = (1_X \otimes f^0, 0)$, for any morphism in $\mathcal{A}'$-Mod with zero second component, therefore the formula $\underline{F}^X(f^0) = 1_X \otimes f^0$ defines a functor making the right square of the diagram in the statement of the lemma commutative. Now, consider for each $M \in \Delta$-Mod the isomorphism

$$\theta_M : \underline{F}^X(Z \otimes_\Delta M) \longrightarrow F^X(Z) \otimes_\Delta M,$$

given at the level of vector spaces by the canonical isomorphism $X \otimes_S (Z \otimes_\Delta M) \longrightarrow (X \otimes_S Z) \otimes_\Delta M$ mapping $x \otimes [z \otimes m]$ onto $[x \otimes z] \otimes m$. Then, with our usual notations (see (12.10)), we have that θ_M is an isomorphism of B-modules. Moreover, for $w \in \underline{W}_0$, we have

$$\begin{aligned}\theta_M(w \cdot (x \otimes [z \otimes m])) &= \theta_M\left(\sum_i x_i \otimes \sigma_{v_i,x}(w) * [z \otimes m]\right)\\ &= \theta_M\left(\sum_i x_i \otimes \left[(\sigma_{v_i,x}(w) * z) \otimes m\right]\right)\\ &= (w \cdot [x \otimes z]) \otimes m\\ &= w \cdot ([x \otimes z] \otimes m)\\ &= w \cdot \theta_M(x \otimes [z \otimes m]).\end{aligned}$$

It is straightforward to verify that $\theta : \underline{F}^X(Z \otimes_\Delta -) \longrightarrow F^X(Z) \otimes_\Delta -$ is a natural transformation. Finally, if Z_Δ is projective, since X_S is projective too, $F^X(Z)_\Delta = X \otimes_S Z_\Delta$ is projective. □

Lemma 22.8. *Let $\phi : \mathcal{A} \longrightarrow \mathcal{A}^z$ be a morphism of Roiter ditalgebras such that the induced functor $F_\phi : \mathcal{A}^z$-Mod$\longrightarrow \mathcal{A}$-Mod preserves isomorphism classes of indecomposables. This is the case, for instance, if $\mathcal{A}$ is a Roiter ditalgebra and $\phi : \mathcal{A} \longrightarrow \mathcal{A}^z$ is the canonical morphism for $z \in \{a, r, d, p\}$ as defined in (8.18), (8.20), (8.19) and (8.17). Then, if $\mathcal{A}^z$ is wild, so is $\mathcal{A}$. Moreover, if Z produces the wildness of $\mathcal{A}^z$, then $F_\phi(Z)$ produces the wildness of $\mathcal{A}$.*

Proof. By the results recalled above, for $z \in \{a, r, d, p\}$, the canonical morphism of ditalgebras $\phi : \mathcal{A} \longrightarrow \mathcal{A}^z$ induces a full and faithful functor F_ϕ. By (9.3) and (22.2), F_ϕ preserves indecomposability and isomorphism classes. In the general case, by assumption, $\mathcal{A}^z$ is wild, and we know the existence of an A^z-$k\langle x, y\rangle$-bimodule Z, where $A^z := A_{\mathcal{A}^z}$, Z is free of finite rank as a right $k\langle x, y\rangle$-module, such that the composition functor

$$k\langle x, y\rangle\text{-Mod} \xrightarrow{Z \otimes_{k\langle x,y\rangle} -} A^z\text{-Mod} \xrightarrow{L_{\mathcal{A}^z}} \mathcal{A}^z\text{-Mod}$$

preserves isomorphism classes of indecomposables. From (22.7), we have an up to isomorphism commutative diagram

$$\begin{array}{ccccc} k\langle x, y\rangle\text{-Mod} & \xrightarrow{Z \otimes_{k\langle x,y\rangle} -} & A^z\text{-Mod} & \xrightarrow{L_{\mathcal{A}^z}} & \mathcal{A}^z\text{-Mod} \\ \| & & \downarrow & & \downarrow F_\phi \\ k\langle x, y\rangle\text{-Mod} & \xrightarrow{F_\phi(Z) \otimes_{k\langle x,y\rangle} -} & A\text{-Mod} & \xrightarrow{L_\mathcal{A}} & \mathcal{A}\text{-Mod}. \end{array}$$

Hence, the lower composition preserves isomorphism classes of indecomposables. Finally, we just have to notice that the A-$k\langle x, y\rangle$-bimodule $F_\phi(Z)$ is clearly free of finite rank as a right $k\langle x, y\rangle$-module. □

Corollary 22.9. *If A^z is a wild algebra and there is an epimorphism of algebras $\phi : A \longrightarrow A^z$, then the algebra A is wild too.*

Proof. Consider the Roiter ditalgebras $\mathcal{A} := (A, 0)$ and $\mathcal{A}^z := (A^z, 0)$, with layers $(A, 0)$ and $(A^z, 0)$. Thus, $\mathcal{A}$-Mod and $\mathcal{A}^z$-Mod can be identified with A-Mod and A^z-Mod, respectively. Moreover, the given ϕ determines a morphism $\phi : \mathcal{A} \longrightarrow \mathcal{A}^z$. It induces a functor $F_\phi : \mathcal{A}^z\text{-Mod} \longrightarrow \mathcal{A}\text{-Mod}$ which is, by Silver's theorem (see (2.10)), full and faithful. Then, we can apply (22.2) and the last lemma. □

Lemma 22.10. *Assume that X is a complete triangular admissible module over an initial subditalgebra of the Roiter ditalgebra $\mathcal{A}$. Consider the associated ditalgebra $\mathcal{A}^X$, see (12.9). Then, if $\mathcal{A}^X$ is wild, so is $\mathcal{A}$.*

Proof. From (16.3), we know that $\mathcal{A}^X$ is a Roiter ditalgebra. By (13.5) the functor F^X is full and faithful. Then, (22.2) guarantees that F^X preserves indecomposablity and isomorphism classes. Then, we can proceed as in the proof (22.8), using the functor F^X instead of F_ϕ, and notice that $F^X(Z) = X \otimes_S Z$ is finitely generated projective as a right $k\langle x, y\rangle$-module. Therefore, from (22.6) we obtain that $\mathcal{A}$ is wild. □

Lemma 22.11. *Assume that $\mathcal{A}$ is a Roiter ditalgebra with layer (R, W). Then, the functor $U_\mathcal{A} : \mathcal{A}\text{-Mod} \longrightarrow R\text{-Mod}$ preserves isomorphism classes and reflects indecomposability. Moreover, if R is a wild algebra, then $\mathcal{A}$ is a wild ditalgebra.*

Proof. Consider the Roiter ditalgebra $\mathcal{A}^z = (R, 0)$, with layer $(R, 0)$. Then, the projection map $\phi : T_\mathcal{A} \longrightarrow R$ is a retraction, with right inverse the inclusion $\sigma : R \longrightarrow T_\mathcal{A}$. They determine layered ditalgebra morphisms $\phi : \mathcal{A} \longrightarrow \mathcal{A}^z$ and $\sigma : \mathcal{A}^z \longrightarrow \mathcal{A}$ with $\phi\sigma = Id$. The functor $F_\sigma : \mathcal{A}\text{-Mod} \longrightarrow \mathcal{A}^z\text{-Mod} \cong R\text{-Mod}$, can be identified with the forgetful functor $U_\mathcal{A}$. Clearly, the functor F_ϕ preserves isomorphism classes. If, for some $M \in \mathcal{A}$-Mod, the R-module $U_\mathcal{A}(M)$ is indecomposable and $e \in \text{End}_\mathcal{A}(M)$ is an idempotent element, then $e^0 \in \text{End}_R(U_\mathcal{A}(M))$ is a trivial idempotent. Thus, either $e^0 = 0$ or $e^0 = I_M$. In the first case, by (5.4), e should be a nilpotent idempotent, hence $e = 0$. In the second case, by (5.8), e is an idempotent isomorphism, thus $e = 1_M$. Then, the functor $U_\mathcal{A}$ reflects indecomposability. Since $U_\mathcal{A} F_\phi = F_\sigma F_\phi = 1_{R\text{-Mod}}$, we obtain that F_ϕ preserves indecomposability and isomorphism classes. Finally, we can apply (22.8) to the functor F_ϕ to obtain that $\mathcal{A}$ is wild if $\mathcal{A}^z$ is so, that is if R is so. □

More generally, we have the following:

Lemma 22.12. *Assume that we have a morphism of Roiter ditalgebras $r : \mathcal{A}' \longrightarrow \mathcal{A}$ and a morphism of algebras $\pi : A \longrightarrow A'$ such that $\pi r_{|A'} = Id_{A'}$. Then, if $\mathcal{A}'$ is wild, so is $\mathcal{A}$.*

Proof. If $\mathcal{A}'$ is wild, there is an A'-$k\langle x, y\rangle$-bimodule Z, free of finite rank as a right $k\langle x, y\rangle$-module, such that the composition functor

$$k\langle x, y\rangle\text{-Mod} \xrightarrow{Z\otimes_{k\langle x,y\rangle}-} A'\text{-Mod} \xrightarrow{L_{\mathcal{A}'}} \mathcal{A}'\text{-Mod}$$

preserves isomorphism classes of indecomposables. By assumption, we have the commutative square

$$\begin{array}{ccc} A'\text{-Mod} & \xrightarrow{F_\pi} & A\text{-Mod} \\ {\scriptstyle L_{\mathcal{A}'}}\downarrow & & \downarrow{\scriptstyle L_{\mathcal{A}}} \\ \mathcal{A}'\text{-Mod} & \xleftarrow{F_r} & \mathcal{A}\text{-Mod}, \end{array}$$

where F_r and F_π are the restriction functors induced, respectively, by r and π. Consider the A-$k\langle x, y\rangle$-bimodule $F_\pi(Z)$, which is free of finite rank as a right $k\langle x, y\rangle$-module. The functors $F_\pi(Z \otimes -)$, $F_\pi(Z) \otimes -$: $k\langle x, y\rangle$-Mod$\longrightarrow$$A$-Mod can be identified. If $L_{\mathcal{A}}(F_\pi(Z) \otimes M) \cong L_{\mathcal{A}}(F_\pi(Z) \otimes N)$, for some indecomposables $N, M \in k\langle x, y\rangle$-Mod, then

$$L_{\mathcal{A}'}(Z \otimes M) = F_r L_{\mathcal{A}}(F_\pi(Z) \otimes M) \cong F_r L_{\mathcal{A}}(F_\pi(Z) \otimes N) = L_{\mathcal{A}'}(Z \otimes N),$$

and, therefore, $M \cong N$.

Assume that $M \in k\langle x, y\rangle$-Mod is indecomposable and take an idempotent $e \in \mathrm{End}_{\mathcal{A}}(L_{\mathcal{A}}(F_\pi(Z) \otimes M))$, then $F_r(e) \in \mathrm{End}_{\mathcal{A}'}(L_{\mathcal{A}'}(Z \otimes M))$ is an idempotent. Hence, $F_r(e) = (e^0, F_r(e)^1) \in \{(0, 0), (1, 0)\}$. In particular, $e^0 = 0$ or $e^0 = I_{Z\otimes M}$. In the first case, by (5.4), e should be a nilpotent idempotent, and $e = 0$. In the second case, e is an idempotent isomorphism, thus e is the identity. Then, $L_{\mathcal{A}}(F_\pi(Z) \otimes -)$ preserves indecomposables too, and $\mathcal{A}$ is wild. □

Corollary 22.13. *If $\mathcal{A}'$ is a triangular proper subditalgebra of the Roiter ditalgebra $\mathcal{A}$ and $\mathcal{A}'$ is wild, then $\mathcal{A}$ is wild. If we denote by $\pi : A \longrightarrow A'$ the canonical projection of algebras, as in (12.3), and the A'-$k\langle x, y\rangle$-bimodule Z produces the wildness of $\mathcal{A}'$, then the A-$k\langle x, y\rangle$-bimodule $F_\pi(Z)$ produces the wildness of $\mathcal{A}$.*

Proof. From (12.3), we know that $\mathcal{A}'$ is a Roiter ditalgebra. Moreover, if $r : \mathcal{A}' \longrightarrow \mathcal{A}$ is the inclusion morphism, we have $\pi r_{|A'} = Id_{A'}$, and we can apply (22.12). □

Lemma 22.14. *Consider the Roiter ditalgebras $\mathcal{A}$ and $\mathcal{A}'$, and write $A := A_{\mathcal{A}}$ and $A' := A_{\mathcal{A}'}$. Let $F : \mathcal{A}$-Mod$\longrightarrow$$\mathcal{A}'$-Mod be an equivalence of categories* *which restricts to another equivalence $\underline{F} : A\text{-Mod} \longrightarrow A'\text{-Mod}$. Then, if $\mathcal{A}$ is wild, so is $\mathcal{A}'$.*

Proof. From (22.2), F preserves indecomposability and isomorphism classes. Let B be an A-$k\langle x, y\rangle$-bimodule, free of finite rank and a right $k\langle x, y\rangle$-module, which realizes the wildness of $\mathcal{A}$. Thus, the functor $L_{\mathcal{A}}(B \otimes_{k\langle x,y\rangle} -)$ preserves isomorphism classes of indecomposables. Then, so does the composition $FL_{\mathcal{A}}(B \otimes_{k\langle x,y\rangle} -)$.

We have the equivalence $\underline{F} : A\text{-Mod} \to A'\text{-Mod}$. By Morita's theorem, there is an A'-A-bimodule P, projective and finitely generated as a right A-module, such that $\underline{F} \cong P \otimes_A -$. Consider the A'-$k\langle x, y\rangle$-bimodule $B' := P \otimes_A B$. Since P_A is a direct summand of A^r, for some r, then $B'_{k\langle x,y\rangle}$ is a direct summand of $B^r_{k\langle x,y\rangle}$ and is, therefore, finitely generated and projective.

Finally, $FL_{\mathcal{A}}(B \otimes_{k\langle x,y\rangle} -) = L_{\mathcal{A}'}\underline{F}(B \otimes_{k\langle x,y\rangle} -) \cong L_{\mathcal{A}'}(B' \otimes_{k\langle x,y\rangle} -)$ preserves isomorphism classes of indecomposables; therefore, from (22.6), we obtain that $\mathcal{A}'$ is wild. □

Corollary 22.15. *If A and A' are Morita equivalent algebras over any field k, then A is wild iff A' is so.*

Proof. Suppose A is wild. By assumption, there is an equivalence of categories $\underline{F} : A\text{-Mod} \longrightarrow A'\text{-Mod}$. Consider the Roiter ditalgebras $\mathcal{A}$ and $\mathcal{A}'$, with corresponding layers $(A, 0)$ and $(A', 0)$, associated respectively to A and A'. Then, $\underline{F}$ determines trivially an equivalence F as needed in (22.14). □

Lemma 22.16. *Let $h \in k[x, y]$ be a non-zero polynomial and $\Lambda := k[x, y]_h$ the localization of $k[x, y]$ at the powers of h. Assume furthermore that the field k is infinite. Then, Λ is wild.*

Proof. We will use the following.

Claim. If $h \in k[x, y]$ is a non-zero polynomial, then there is $g \in k[x, y]$ with $g(0, 0) \neq 0$, such that $k[x, y]_h \cong k[x, y]_g$ as k-algebras.

Proof of the claim: Suppose that $h(0, 0) = 0$. By assumption, $0 \neq h(x, y) = \sum_{i=0}^{n} a_i(y) \cdot x^i \in k[y][x] = k[x, y]$, with $a_i(y) \in k[y]$, not all zero.

Case 1: $a_0(y) \neq 0$. Since k is infinite, we can choose $\lambda \in k$ with $a_0(\lambda) \neq 0$. Let $\varphi : k[x, y] \to k[x, y]$ be the isomorphism such that $\varphi(x) := x$ and $\varphi(y) := y + \lambda$. Make $g := \varphi(h)$; then, φ induces an isomorphism $k[x, y]_h \to k[x, y]_g$ and $g(0, 0) = h(0, \lambda) = a_0(\lambda) \neq 0$.

Case 2: $a_0(y) = 0$. Consider the coefficients $a_i(y)$ of h as rational functions, thus $h \in k(y)[x]$. Then, there are at most a finite number of elements λ of $k(y)$ (hence of k) such that $h(\lambda, y) = 0$. Thus, we can choose $\lambda \in k$ with $h(\lambda, y) \neq 0$. Let $\varphi : k[x, y] \to k[x, y]$ be the isomorphism such that $\varphi(x) := x + \lambda$ and $\varphi(y) := y$, and make $h' := \varphi(h) \neq 0$. Again, $k[x, y]_h \cong k[x, y]_{h'}$, but now, when we write $h'(x, y) = \sum_{i=0}^{n} a'_i(y) \cdot x^i \in k[y][x]$, we have $a'_0(y) \neq 0$. Indeed, $h'(x, y) = h(x + \lambda, y) = \sum_{i=0}^{n} a_i(y)(x + \lambda)^i$, and $a'_0(y) = \sum_{i=0}^{n} a_i(y) \cdot \lambda^i = h(\lambda, y) \neq 0$. Case 2 has been reduced to case 1, and the claim is proved.

From (22.9) and our claim, we may assume that $h(0, 0) \neq 0$.

Let B be the Λ-$k\langle x, y\rangle$-bimodule which as a right $k\langle x, y\rangle$-module is free of rank $r(B) = 4$, and the action of the generators x and y of $\Lambda = k[x, y]_h$ is given by the following matrices in $k\langle x, y\rangle^{4\times 4}$

$$A_1 := \begin{pmatrix} 0 & 0 & 0 & 0 \\ 0 & 0 & 0 & 0 \\ 1 & 0 & 0 & 0 \\ 0 & x & y & 0 \end{pmatrix} \qquad A_2 := \begin{pmatrix} 0 & 0 & 0 & 0 \\ 1 & 0 & 0 & 0 \\ 0 & 0 & 0 & 0 \\ 0 & 1 & x & 0 \end{pmatrix}.$$

Since $A_1 A_2 = A_2 A_1$, to see that these matrices define a left Λ-module structure on $k\langle x, y\rangle^4$, we only need to show that $h(A_1, A_2)$ is invertible in $k\langle x, y\rangle^{4\times 4}$. Notice that A_1 as well as A_2 are lower triangular with zeros on the principal diagonal, thus every monomial $A_1^s A_2^t$ with $s + t > 0$ is a nilpotent matrix. If $\alpha := h(0, 0) \in k$, $h(A_1, A_2)$ is the sum of αI and a nilpotent matrix, since $\alpha \neq 0$, $h(A_1, A_2)$ is invertible.

Consider now the functor $F := B \otimes_{k\langle x,y\rangle} - : k\langle x, y\rangle\text{-Mod} \longrightarrow \Lambda\text{-Mod}$. If $M \in k\langle x, y\rangle$-Mod is given by the k-linear actions $T_x, T_y : M \to M$ of the generators x and y of $k\langle x, y\rangle$, on the Λ-module $F(M) = M^4$ the generators x and y of Λ act through the matrices

$$C_1 := \begin{pmatrix} 0 & 0 & 0 & 0 \\ 0 & 0 & 0 & 0 \\ I & 0 & 0 & 0 \\ 0 & T_x & T_y & 0 \end{pmatrix} \qquad C_2 := \begin{pmatrix} 0 & 0 & 0 & 0 \\ I & 0 & 0 & 0 \\ 0 & 0 & 0 & 0 \\ 0 & I & T_x & 0 \end{pmatrix},$$

and if M' is an other $k\langle x, y\rangle$-module with actions $T'_x, T'_y : M' \longrightarrow M'$, for $F(M')$, we will have the matrices C'_1 and C'_2 analogous to the preceeding ones but with primes.

Let φ be any morphism of Λ-modules from $F(M)$ to $F(M')$. Using the vector space decompositions $F(M) = M^4$ and $F(M') = M'^4$, φ takes the form

$$\varphi = \begin{pmatrix} A & C & D & E \\ G & H & J & K \\ L & N & P & Q \\ R & S & T & U \end{pmatrix},$$

for some k-linear maps $A, C, \ldots, U : M \to M'$. Since φ is Λ-linear, $\varphi \cdot C_1 = C'_1 \cdot \varphi$ and $\varphi \cdot C_2 = C'_2 \cdot \varphi$. Comparing the entries of these matrices, we obtain that φ has the form

$$\varphi = \begin{pmatrix} A & 0 & 0 & 0 \\ G & A & 0 & 0 \\ L & 0 & A & 0 \\ R & S & T & A \end{pmatrix}.$$

Comparing the entries of indexes (4,2) and (4,3) of $\varphi \cdot C_1$ and $C_1' \cdot \varphi$ we obtain that $A \cdot T_x = T_x' \cdot A$ and $A \cdot T_y = T_y' \cdot A$, so $A \in \mathrm{Hom}_{k\langle x,y\rangle}(M, M')$.

If φ is an isomorphism from $F(M)$ onto $F(M')$, A is a $k\langle x, y\rangle$-linear isomorphism from M to M' and, therefore, F preserves isomorphism classes.

If $M = M'$ is indecomposable and φ is an idempotent endomorphism of $F(M)$, it turns out that: $A^2 = A$, $G = G \cdot A + A \cdot G$, $L = L \cdot A + A \cdot L$, $T = T \cdot A + A \cdot T$, $S = S \cdot A + A \cdot S$ and $R = R \cdot A + S \cdot G + T \cdot L + A \cdot R$. Since M is indecomposable and $A \in \mathrm{End}_{k\langle x,y\rangle}(M)$ is idempotent, we have $A = 0$ or $A = I$, and in both cases we see that $G = L = T = S = R = 0$. Then, φ is zero or the identity, and $F(M)$ is indecomposable. This shows that B realizes the wildness of Λ. □

For the infinite ground field case, in the last definitions of wild ditalgebra and wild algebra, we can replace the requirement that $Z_{k\langle x,y\rangle}$ is free of finite rank by the (apparently weaker) requirement that $Z_{k\langle x,y\rangle}$ is finitely generated.

Proposition 22.17. *Let $\mathcal{A}$ be a ditalgebra over the infinite field k. Write $A := A_{\mathcal{A}}$. Then, the following statements are equivalent:*

(i) $\mathcal{A}$ is wild.

(ii) There is an A-$k\langle x, y\rangle$-bimodule C, which is finitely generated as a right $k\langle x, y\rangle$-module, and such that the functor

$$k\langle x, y\rangle\text{-Mod} \xrightarrow{C \otimes_{k\langle x,y\rangle} -} A\text{-Mod} \xrightarrow{L_{\mathcal{A}}} \mathcal{A}\text{-Mod}$$

preserves isomorphism classes of indecomposables.

(iii) There is an A-$k[x, y]$-bimodule D, which is finitely generated as a right $k[x, y]$-module, and such that the functor

$$k[x, y]\text{-Mod} \xrightarrow{D \otimes_{k[x,y]} -} A\text{-Mod} \xrightarrow{L_{\mathcal{A}}} \mathcal{A}\text{-Mod}$$

preserves isomorphism classes of indecomposables.

(iv) There are an $h \in k[x, y]$ and an A-$k[x, y]_h$-bimodule E, which is finitely generated free as a right $k[x, y]_h$-module, and such that the functor

$$k[x, y]_h\text{-Mod} \xrightarrow{E \otimes_{k[x,y]_h} -} A\text{-Mod} \xrightarrow{L_{\mathcal{A}}} \mathcal{A}\text{-Mod}$$

preserves isomorphism classes of indecomposables.

(v) *There is a wild k-algebra Γ and an A-Γ-bimodule F, which is finitely generated free as a right Γ-module, and such that the functor*

$$\Gamma\text{-Mod} \xrightarrow{F\otimes_\Gamma -} A\text{-Mod} \xrightarrow{L_A} \mathcal{A}\text{-Mod}$$

preserves isomorphism classes of indecomposables.

Proof.

(i) ⇒ (ii) is clear.

(ii) ⇒ (iii) Consider the $k\langle x, y\rangle$-$k[x, y]$-bimodule Z obtained from the regular $k[x, y]$-bimodule $k[x, y]$ by left restriction of scalars through the quotient map $k\langle x, y\rangle \to k[x, y]$. Then, $D := C \otimes_{k\langle x,y\rangle} Z$ is an A-$k[x, y]$-bimodule. Since $C_{k\langle x,y\rangle}$ is finitely generated so is $D_{k[x,y]}$.

Since $Z \otimes_{k[x,y]} - : k[x, y]\text{-Mod} \to k\langle x, y\rangle\text{-Mod}$ is full and faithful, it preserves indecomposability and isomorphism classes. By assumption, the functor $L_{\mathcal{A}}(C \otimes_{k\langle x,y\rangle} -) : k\langle x, y\rangle\text{-Mod} \to \mathcal{A}\text{-Mod}$, preserves isomorphism classes of indecomposables, then so does the composition $L_{\mathcal{A}}(D \otimes_{k[x,y]} -) : k[x, y]\text{-Mod} \to \mathcal{A}\text{-Mod}$.

(iii) ⇒ (iv) Since $k[x, y]$ is noetherian and $D_{k[x,y]}$ is finitely generated, we have an exact sequence of right $k[x, y]$-modules

$$k[x, y]^s \xrightarrow{H} k[x, y]^r \xrightarrow{H'} D \longrightarrow 0.$$

The morphism H is the multiplication by a matrix which we also denote by $H \in k[x, y]^{r\times s} \leq k(x, y)^{r\times s}$. Multiplying H by appropriate invertible matrices $P \in \mathbb{G}l_r(k(x, y))$ and $Q \in \mathbb{G}l_s(k(x, y))$, we obtain a matrix of the form

$$PHQ = \begin{pmatrix} I & 0 \\ 0 & 0 \end{pmatrix} =: G,$$

where I is the identity $(d \times d)$-matrix and d is the rank of H. The entries of P and Q are rational functions in x and y with coefficients in k, then, the denominators of these fractions have a least common multiple $h \in k[x, y]$. Then, P and Q have entries in the k-algebra $\Gamma := k[x, y]_h$. By left restriction of scalars, we have that Γ is an $k[x, y]$-Γ-bimodule which is a finitely generated free right Γ-module.

Consider the A-Γ-bimodule $E := D \otimes_{k[x,y]} \Gamma$. The matrix $H \in k[x, y]^{r\times s} \leq \Gamma^{r\times s}$ defines a morphism from Γ^s to Γ^r, which we also denote by H and whose cokernel we denote by C. Consider the following commutative diagram of right Γ-modules

$$\begin{array}{ccccccc}
k[x,y]^s \otimes_{k[x,y]} \Gamma & \xrightarrow{H\otimes I} & k[x,y]^r \otimes_{k[x,y]} \Gamma & \xrightarrow{H'\otimes I} & E & \longrightarrow & 0 \\
\downarrow \cong & & \downarrow \cong & & \downarrow \cong & & \\
\Gamma^s & \xrightarrow{H} & \Gamma^r & \longrightarrow & C & \longrightarrow & 0 \\
\downarrow Q^{-1} & & \downarrow P & & \downarrow \cong & & \\
\Gamma^s & \xrightarrow{G} & \Gamma^r & \xrightarrow{\pi} & \Gamma^{r-d} & \longrightarrow & 0,
\end{array}$$

where G denotes the morphism defined by the matrix G given before, and π is its cokernel. The rows of the diagram are exact and the morphisms of the first two columns are isomorphisms. Those of the third column are given by the universal property of cokernels. As a consequence, E_Γ is free of finite rank.

Considering Γ as a $k[x,y]$-Γ-bimodule, we have the functor

$$\Gamma \otimes_\Gamma - : \Gamma\text{-Mod} \to k[x,y]\text{-Mod},$$

which is full and faithful (it is the embedding identifying each Γ-module with the $k[x,y]$-module on which h acts through a k-linear isomorphism), and, by assumption, we know that the functor

$$L_{\mathcal{A}}(D \otimes_{k[x,y]} -) : k[x,y]\text{-Mod} \to \mathcal{A}\text{-Mod}$$

preserves isomorphism classes of indecomposables. Then, so does the composition $L_{\mathcal{A}}(E \otimes_\Gamma -) : \Gamma\text{-Mod} \to \mathcal{A}\text{-Mod}$.

(iv) $\Rightarrow$ (v) By (22.16), $\Gamma := k[x,y]_h$ is wild. We can take $F := E$.

(v) $\Rightarrow$ (i) Let B' be a Γ-$k\langle x,y\rangle$-bimodule, free of finite rank as a right $k\langle x,y\rangle$-module, that realizes the wildness of Γ. Define $B := F \otimes_\Gamma B'$, which is an $\mathcal{A}$-$k\langle x,y\rangle$-bimodule isomorphic, as a right $k\langle x,y\rangle$-module, to B'^r, where r is the rank of F as a free right Γ-module, and, therefore, is free of finite rank by the right. Since $B' \otimes_{k\langle x,y\rangle} - : k\langle x,y\rangle\text{-Mod} \to \Gamma\text{-Mod}$ and $L_{\mathcal{A}}(F \otimes_\Gamma -) : \Gamma\text{-Mod} \to \mathcal{A}\text{-Mod}$ preserve isomorphism classes of indecomposables, so does their composition $L_{\mathcal{A}}(B \otimes_{k\langle x,y\rangle} -) : k\langle x,y\rangle\text{-Mod} \to \mathcal{A}\text{-Mod}$, that is B realizes the wildness of $\mathcal{A}$.

□

Lemma 22.18. *Assume that Λ is a finite-dimensional k-algebra which splits over its radical. Consider the Drozd's ditalgebra $\mathcal{D}^\Lambda$ and write $A = A_{\mathcal{D}^\Lambda}$. Consider also the equivalence functor $\Xi_\Lambda : \mathcal{D}^\Lambda\text{-Mod} \longrightarrow \mathcal{P}^1(\Lambda)$ and the cokernel functor $\mathrm{Cok}^1 : \mathcal{P}^1(\Lambda) \longrightarrow \Lambda\text{-Mod}$.*

(1) If Δ is a k-algebra and Y is any A-Δ-bimodule, then $Z := \mathrm{Cok}^1(\Xi_\Lambda(Y))$ has a canonical structure of a Λ-Δ-bimodule. Furthermore, the composition functor

$$\Delta\text{-Mod} \xrightarrow{Y\otimes_\Delta -} A\text{-Mod} \xrightarrow{L_{\mathcal{D}^\Lambda}} \mathcal{D}^\Lambda\text{-Mod} \xrightarrow{\Xi_\Lambda} \mathcal{P}^1(\Lambda) \xrightarrow{\mathrm{Cok}^1} \Lambda\text{-Mod}$$

is naturally isomorphic to $Z \otimes_\Delta -$.

(2) In particular, taking $\Delta = A$ and the regular bimodule $Y = A$, we obtain a finite-dimensional Λ-A-bimodule $Z = Z(\Lambda)$ such that

$$A\text{-Mod} \xrightarrow{L_{\mathcal{D}^\Lambda}} \mathcal{D}^\Lambda\text{-Mod} \xrightarrow{\Xi_\Lambda} \mathcal{P}^1(\Lambda) \xrightarrow{\mathrm{Cok}^1} \Lambda\text{-Mod}$$

is naturally isomorphic to $Z \otimes_A -$. We call this bimodule $Z(\Lambda)$ the transition bimodule corresponding to Λ.

Proof. (1): Each A-Δ-bimodule Y has an associated triple (Y_1, Y_2, ψ_Y), where Y_1 and Y_2 are S-Δ-bimodules and $\psi_Y \in \mathrm{Hom}_{S\text{-}S}(J^*, \mathrm{Hom}_\Delta(Y_1, Y_2))$, with

$$\psi_Y(\gamma)[y_1] = \begin{pmatrix} 0 & 0 \\ \gamma & 0 \end{pmatrix} y_1,$$

for $\gamma \in J^*$ and $y_1 \in Y_1$. Then, $\phi_Y := \Xi_\Lambda(Y) : \Lambda \otimes_S Y_1 \longrightarrow \Lambda \otimes_S Y_2$, satisfies $\phi_Y(\lambda \otimes y_1) = \sum_j \lambda p_j \otimes \psi_Y(\gamma_j)[y_1]$.

Moreover, if $M \in \Delta\text{-Mod}$, the object $Y \otimes_\Delta M \in \mathcal{D}^\Lambda\text{-Mod}$ defines the triple $((Y \otimes_\Delta M)_1, (Y \otimes_\Delta M)_2, \psi_{Y\otimes_\Delta M})$, where $(Y \otimes_\Delta M)_i = Y_i \otimes_\Delta M$ and $\psi_{Y\otimes_\Delta M} \in \mathrm{Hom}_{S\text{-}S}(J^*, \mathrm{Hom}_k(Y_1 \otimes_\Delta M, Y_2 \otimes_\Delta M))$ satisfies that $\psi_{Y\otimes_\Delta M}(\gamma) = \psi_Y(\gamma) \otimes 1_M$, for any $\gamma \in J^*$. Write $\phi_{Y\otimes_\Delta M} := \Xi_\Lambda(Y \otimes_\Delta M)$. Then, we have a commutative diagram

$$\begin{array}{ccccc}
[\Lambda \otimes_S Y_1] \otimes_\Delta M & \xrightarrow{\phi_Y \otimes 1_M} & [\Lambda \otimes_S Y_2] \otimes_\Delta M & \xrightarrow{\pi_0 \otimes 1_M} & \mathrm{Cok}(\phi_Y) \otimes_\Delta M \\
\downarrow{\scriptstyle \theta^1_M} & & \downarrow{\scriptstyle \theta^2_M} & & \downarrow{\scriptstyle \theta_M} \\
\Lambda \otimes_S [Y_1 \otimes_\Delta M] & \xrightarrow{\phi_{Y\otimes_\Delta M}} & \Lambda \otimes_S [Y_2 \otimes_\Delta M] & \xrightarrow{\pi_M} & \mathrm{Cok}(\phi_{Y\otimes_\Delta M}),
\end{array}$$

where θ^1_M and θ^2_M are the canonical isomorphisms and induce the isomorphism of Λ-modules θ_M. Now, take $f \in \mathrm{Hom}_\Delta(M, N)$ and let us verify that the following square commutes

$$\begin{array}{ccc} Z\otimes_\Delta M = \mathrm{Cok}(\phi_Y)\otimes_\Delta M & \xrightarrow{\theta_M} & \mathrm{Cok}\,\Xi_\Lambda L(Y\otimes_\Delta M) \\ \big\downarrow{\scriptstyle 1_Z\otimes f} & & \big\downarrow{\scriptstyle \mathrm{Cok}^1\Xi_\Lambda L(Y\otimes f)} \\ Z\otimes_\Delta N = \mathrm{Cok}(\phi_Y)\otimes_\Delta N & \xrightarrow{\theta_N} & \mathrm{Cok}\,\Xi_\Lambda L(Y\otimes_\Delta N). \end{array}$$

First recall that, by definition

$$\begin{array}{ccccc} \Lambda\otimes_S[Y_1\otimes_\Delta M] & \xrightarrow{\phi_{Y\otimes_\Delta M}} & \Lambda\otimes_S[Y_2\otimes_\Delta M] & \xrightarrow{\pi_M} & \mathrm{Cok}(\phi_{Y\otimes_\Delta M}) \\ \big\downarrow{\scriptstyle \Xi_\Lambda L(Y\otimes f)_1} & & \big\downarrow{\scriptstyle \Xi_\Lambda L(Y\otimes f)_2} & & \big\downarrow{\scriptstyle \mathrm{Cok}\,\Xi_\Lambda L(Y\otimes f)} \\ \Lambda\otimes_S[Y_1\otimes_\Delta N] & \xrightarrow{\phi_{Y\otimes_\Delta N}} & \Lambda\otimes_S[Y_2\otimes_\Delta N] & \xrightarrow{\pi_N} & \mathrm{Cok}(\phi_{Y\otimes_\Delta N}), \end{array}$$

is a commutative diagram and notice that the following diagram commutes also

$$\begin{array}{ccc} [\Lambda\otimes_S Y_2]\otimes_\Delta M & \xrightarrow{\theta^2_M} & \Lambda\otimes_S[Y_2\otimes_\Delta M] \\ \big\downarrow{\scriptstyle 1_{[\Lambda\otimes Y_2]}\otimes f} & & \big\downarrow{\scriptstyle \Xi_\Lambda L(Y\otimes f)_2} \\ [\Lambda\otimes_S Y_2]\otimes_\Delta N & \xrightarrow{\theta^2_N} & \Lambda\otimes_S[Y_2\otimes_\Delta N]. \end{array}$$

Then

$$\begin{aligned} \theta_N(1_Z\otimes f)(\pi_0\otimes 1_M) &= \theta_N(\pi_0\otimes 1_N)(1_{\Lambda\otimes Y_2}\otimes f) \\ &= \pi_N\theta^2_N(1_{\Lambda\otimes Y_2}\otimes f) \\ &= \pi_N(\Xi_\Lambda L(Y\otimes f)_2)\theta^2_M \\ &= (\mathrm{Cok}\,\Xi_\Lambda L(Y\otimes f))\pi_M\theta^2_M \\ &= (\mathrm{Cok}\,\Xi_\Lambda L(Y\otimes f))\theta_M(\pi_0\otimes 1_M). \end{aligned}$$

Since $\pi_0\otimes 1_M$ is an epimorphism, we have $\theta_N(1_Z\otimes f) = \mathrm{Cok}\,\Xi_\Lambda L(Y\otimes f)\theta_M$. Hence, $\theta : Z\otimes_\Delta - \longrightarrow \mathrm{Cok}\,\Xi_\Lambda L(Y\otimes_\Delta -)$ is an isomorphism of functors.

Part (2) is clear. □

Lemma 22.19. *Assume that the finite-dimensional k-algebra Λ splits over its radical, say $\Lambda = S\oplus J$. Consider the functor $\Xi_\Lambda : \mathcal{D}^\Lambda\text{-Mod}\longrightarrow\mathcal{P}^1(\Lambda)$. Assume that $\Xi_\Lambda(M)\cong(P_1, P_2, \alpha)$, for some $M\in\mathcal{D}^\Lambda$-Mod and $(P_1, P_2, \alpha)\in\mathcal{P}^1(\Lambda)$. Then, if (M_1, M_2, ψ_M) is the triple associated to M, we have isomorphisms of left S-modules $M_i\cong P_i/JP_i$, for $i\in\{1,2\}$. In particular*

$$\ell_S(M) = \ell_S(P_1/JP_1)+\ell_S(P_2/JP_2) = \ell_\Lambda(P_1/JP_1)+\ell_\Lambda(P_2/JP_2).$$

Thus, if every simple Λ-module is one dimensional

$$\dim M = \ell_S(M) = \ell_\Lambda(P_1/JP_1)+\ell_\Lambda(P_2/JP_2).$$

Proof. If $(u_1, u_2) : \Xi_\Lambda(M) \to (P_1, P_2, \alpha)$ is an isomorphism in $\mathcal{P}^1(\Lambda)$, then we have a commutative diagram with exact rows

$$\begin{array}{ccccccccc} 0 \to & J \otimes_S M_i & \to & \Lambda \otimes_S M_i & \to & M_i & \to 0 \\ & \cong\downarrow & & u_i\downarrow & & \cong\downarrow & \\ 0 \to & JP_i & \to & P_i & \to & P_i/JP_i & \to 0, \end{array}$$

where the right isomorphism is induced by the left square of the diagram. □

Lemma 22.20. *Assume that Δ is a k-algebra such that every projective right Δ-module is free. Assume Λ is a finite-dimensional k-algebra which splits over its radical, consider its Drozd's ditalgebra $\mathcal{D} = \mathcal{D}^\Lambda$, the equivalence Ξ_Λ : $\mathcal{D}$-Mod$\longrightarrow \mathcal{P}^1(\Lambda)$ and make $A = A_\mathcal{D}$. Assume that Z is an A-Δ-bimodule, free of finite rank when considered as a right Δ-module, and consider the composition functor*

$$\Delta\text{-Mod} \xrightarrow{Z\otimes_\Delta -} A\text{-Mod} \xrightarrow{L_\mathcal{D}} \mathcal{D}\text{-Mod}.$$

Assume, furthermore, that we are in one of the following two cases:

(1) There are infinitely many non-isomorphic indecomposable finite-dimensional Δ-modules and $L_\mathcal{D}(Z \otimes_\Delta -)$ preserves isomorphism classes of indecomposables.
(2) There are infinitely many non-isomorphic Δ-modules with bounded dimension and $L_\mathcal{D}(Z \otimes_\Delta -)$ preserves isomorphism classes.

Then, no non-zero object of the form $(P, 0, 0)$ is isomorphic in $\mathcal{P}^1(\Lambda)$ to one of the form $\Xi_\Lambda(L_\mathcal{D}(Z \otimes_\Delta M))$, with $M \in \Delta$-Mod.

Assumption (1) holds, for suitable bimodules Z, for the algebra $\Delta = k\langle x, y\rangle$ and for $\Delta = k[x]_f$. If the ground field k is infinite, assumption (2) holds for the same type of algebras.

Proof. An algebra of the form $k[x]_f$ is a principal ideal domain, and hence a possible Δ as above. For the algebra, $k\langle x, y\rangle$, recall (22.6).

We have a splitting $\Lambda = S \oplus J$ of Λ over its radical J. Denote by e_1 and e_2 the canonical idempotents associated to the matrix algebra $A = A_\mathcal{D} = \begin{pmatrix} S & 0 \\ J^* & S \end{pmatrix}$. With the notation of the lemma, assume that $0 \neq (P, 0, 0) \cong \Xi_\Lambda(L_\mathcal{D}(Z \otimes_\Delta M))$ in $\mathcal{P}^1(\Lambda)$, for some $M \in \Delta$-Mod, and make $H := L_\mathcal{D}(Z \otimes_\Delta M)$. Thus, $M \neq 0$. By (22.19), $e_2 H = 0$. Since every direct summand of a free right Δ-module is free, we have free right Δ-modules $e_1 Z$ and $e_2 Z$ with ranks r_1 and r_2, respectively. Then, $0 = e_2 H = e_2 Z \otimes_\Delta M \cong M^{r_2}$, thus $r_2 = 0$ and $e_2 Z = 0$.

Case 1: Consider an infinite family of non-isomorphic indecomposables $\{M_\lambda\}_\lambda$ in Δ-mod and assume that $L_{\mathcal{D}}(Z \otimes_\Delta -)$ preserves isomorphism classes of indecomposables. Then, $\{L_{\mathcal{D}}(Z \otimes_\Delta M_\lambda)\}_\lambda$ is an infinite family of non-isomorphic indecomposable objects in $\mathcal{D}$-mod. They all satisfy that $e_2(Z \otimes_\Delta M_\lambda) = 0$, because $e_2 Z = 0$. However, the finite-dimensional indecomposable $\mathcal{D}$-modules N satisfying $e_2 N = 0$, are mapped by Ξ_Λ onto only finitely possible isomorphism classes of indecomposable objects of the form $(Q, 0, 0)$ in $\mathcal{P}^1(\Lambda)$, a contradiction.

Case 2: Consider an infinite family $\{M_\lambda\}_\lambda$ of non-isomorphic objects in Δ-mod with $\dim M_\lambda < t$, for all λ, and assume that $L_{\mathcal{D}}(Z \otimes_\Delta -)$ preserves isomorphism classes. Then, as before, $\{L_{\mathcal{D}}(Z \otimes_\Delta M_\lambda)\}_\lambda$ is an infinite family of non-isomorphic objects in $\mathcal{D}$-mod. They all satisfy that $e_2(Z \otimes_\Delta M_\lambda) = 0$, because $e_2 Z = 0$. By (22.19), each $H_\lambda := L_{\mathcal{D}}(Z \otimes_\Delta M_\lambda)$ is mapped by the equivalence Ξ_Λ onto an object of the form $(P_\lambda, 0, 0)$, with $\dim_k(P_\lambda / J P_\lambda) = \dim_k e_1 H_\lambda = \dim_k M_\lambda^{r_1} < r_1 t$. Then, the number of indecomposable direct summands of all the P_λ is bounded. Then, the family $\{(P_\lambda, 0, 0)\}_\lambda$ admits isomorphic members: a contradiction. □

Proposition 22.21. *Assume that k is an infinite field and that Λ is a finite-dimensional k-algebra, which splits over its radical. If Λ is not wild, then the Drozd's ditalgebra $\mathcal{D}^\Lambda$ is not wild.*

Proof. Make $A = A_{\mathcal{D}^\Lambda}$ and $L = L_{\mathcal{D}^\Lambda}$. Assume that Z_0 is an A-$k\langle x, y\rangle$-bimodule, free of finite rank as a right $k\langle x, y\rangle$-module, such that

$$k\langle x, y\rangle\text{-Mod} \xrightarrow{Z_0 \otimes_{k\langle x,y\rangle} -} A\text{-Mod} \xrightarrow{L} \mathcal{D}^\Lambda\text{-Mod}$$

preserves isomorphism classes of indecomposables. Recall, from (18.9) and (18.10), that $\mathrm{Cok}^1(\phi)$ is indecomposable for all indecomposable $\phi \in \mathcal{P}^1(\Lambda)$ with the exception of those of the form $\phi = (P, 0, 0)$, where P is an indecomposable Λ-module.

We claim that the functor $G := \mathrm{Cok}\,\Xi_\Lambda L(Z_0 \otimes -)$ preserves isomorphism classes of indecomposables. Indeed, if $M \in k\langle x, y\rangle$-Mod is indecomposable, by (22.20), we know that $\Xi_\Lambda L(Z_0 \otimes M)$ is an indecomposable of $\mathcal{P}^1(\Lambda)$ not of the form $(P, 0, 0)$, with P a projective Λ-module. Hence, $\mathrm{Cok}\,\Xi_\Lambda L(Z_0 \otimes M)$ is an indecomposable Λ-module. The functor

$$F := \Xi_\Lambda L(Z_0 \otimes -) : k\langle x, y\rangle\text{-Mod} \longrightarrow \mathcal{P}^1(\Lambda)$$

maps every indecomposable $k\langle x, y\rangle$-module to an object in $\mathcal{P}^2(\Lambda)$. Thus, if we have indecomposables $M, N \in k\langle x, y\rangle$-Mod such that $G(M) \cong G(N)$, since $\text{Cok}^2 : \mathcal{P}^2(\Lambda) \longrightarrow \Lambda\text{-Mod}$ is a full functor that reflects isomorphisms, we obtain that $F(N) \cong F(M)$. Since F is a composition of functors that preserve isomorphism classes of indecomposables, we obtain $M \cong N$. Thus, G preserves isomorphism classes of indecomposables, as claimed.

Then, $Z \otimes_A Z_0$ is a Λ-$k\langle x, y\rangle$-bimodule that realizes the wildness of Λ, if $Z = Z(\Lambda)$ is the transition bimodule of Λ of (22.18). Here, we know that the right A-module Z is finitely generated, and hence the right $k\langle x, y\rangle$-module $Z \otimes_A Z_0$ is finitely generated. Then, since our ground field is infinite, we can use (22.17) and we are done. □

Exercise 22.22. *Show that, for some ditalgebras $\mathcal{A}$, it is possible that the functor $L_{\mathcal{A}} : A\text{-Mod} \longrightarrow \mathcal{A}\text{-Mod}$ does not preserve indecomposability.*

Exercise 22.23. *Given any natural number t, construct a $k\langle x, y\rangle$-bimodule Z_t, free of finite rank as a right $k\langle x, y\rangle$-module, such that the tensor functor $Z_t \otimes_{k\langle x,y\rangle} - : k\langle x, y\rangle\text{-Mod} \longrightarrow k\langle x, y\rangle\text{-Mod}$ preserves indecomposability and isomorphism classes and, moreover, it satisfies that $\dim Z_t \otimes M \geq t$, for any non-zero $M \in k\langle x, y\rangle\text{-Mod}$.*

Hint: Consider the $k\langle x, y\rangle$-bimodule Z_t such that $(Z_t)_{k\langle x,y\rangle} = k\langle x, y\rangle^n$ ($n = t + 3$) where x and y act on the left over Z_t, respectively, through the matrices

$$\begin{pmatrix} 0 & 1 & \cdots & 0 & 0 & 0 & 0 \\ 0 & 0 & \ddots & \vdots & \vdots & \vdots & \vdots \\ 0 & 0 & \ddots & 1 & 0 & 0 & 0 \\ 0 & 0 & \ddots & 0 & 1 & 0 & 0 \\ 0 & 0 & \ddots & 0 & 0 & 1 & 0 \\ \vdots & \vdots & \ddots & 0 & 0 & 0 & 1 \\ 0 & 0 & \cdots & 0 & 0 & 0 & 0 \end{pmatrix} \quad \text{and} \quad \begin{pmatrix} 0 & 0 & \cdots & 0 & 0 & 0 & 0 \\ 1 & 0 & \ddots & \vdots & \vdots & \vdots & \vdots \\ x & 1 & \ddots & 0 & 0 & 0 & 0 \\ y & x & \ddots & 0 & 0 & 0 & 0 \\ 0 & y & \ddots & 1 & 0 & 0 & 0 \\ \vdots & \vdots & \ddots & x & 1 & 0 & 0 \\ 0 & 0 & \cdots & y & x & 1 & 0 \end{pmatrix}.$$

Exercise 22.24. *Let $\psi : \mathcal{A}' \longrightarrow \mathcal{A}$ be a morphism of Roiter ditalgebras such that $\psi_| : A' \longrightarrow A$ is an isomorphism of algebras. Show that $\mathcal{A}$ is wild whenever $\mathcal{A}'$ is so.*

Exercise 22.25. *Let $\mathcal{A} = (T, \delta)$ be a Roiter ditalgebra with layer (R, W). Assume that $S \subseteq W_1$ is such that $\delta(S) \subseteq ASV + VSA$, and consider the ideal I of the algebra T generated by S. Thus, I is an ideal of the ditalgebra $\mathcal{A}$. Assuming that $\mathcal{A}$ is wild, show that $\mathcal{A}/I$ is wild too. Derive the fact: $\mathcal{A}$ is a wild ditalgebra implies that A is a wild algebra.*

Hint: Apply (22.24), with the help of (9.5).

Exercise 22.26. *Let Λ be a finite-dimensional algebra which splits over its radical. If $k \leq K$ is a field extension preserving the radical of Λ, then Λ^K admits a Drozd's ditalgebra too, and we have an isomorphism of Λ-K-bimodules*

$$Z\left(\Lambda^K\right) \cong Z\left(\Lambda\right)^K,$$

where $Z(\Lambda)$ and $Z(\Lambda^K)$ denote the transition bimodules of Λ and Λ^K, respectively. Hint: use (20.12) and (20.13).

Exercise 22.27. *Consider the quotient $\Gamma := k[u, v, w]/\langle u^2, v^2, w^2, uv, vw, uw\rangle$ of the polynomial k-algebra $k[u, v, w]$ in three commutative variables. Show that the algebra Γ is wild.*

Hint: Consider the functor $G : k\langle x, y\rangle$-Mod$\longrightarrow \Gamma$-Mod given by the tensor product by the bimodule B constructed as follows: $B = k\langle x, y\rangle^2$ as a right module, and the action of the generators u, v and w of Γ on B by the left are given, respectively, by the matrices

$$\begin{pmatrix} 0 & 1 \\ 0 & 0 \end{pmatrix}, \begin{pmatrix} 0 & x \\ 0 & 0 \end{pmatrix} \text{ and } \begin{pmatrix} 0 & y \\ 0 & 0 \end{pmatrix}.$$

Show that G preserves isomorphism classes and indecomposables.

Exercise 22.28. *Assume that $\mathcal{A}$ is a ditalgebra and Δ is any algebra. Let Z be an $A_{\mathcal{A}}$-Δ-bimodule, which is finitely generated by the right, and denote by G the composition*

$$\Delta\text{-mod} \xrightarrow{Z\otimes_\Delta -} A_{\mathcal{A}}\text{-mod} \xrightarrow{L_{\mathcal{A}}} \mathcal{A}\text{-mod}.$$

Show that if G preserves isomorphism classes, then G is faithful and reflects isomorphisms.

Exercise 22.29. *Assume $\mathcal{A}$ is a Roiter k-ditalgebra with layer (R, W) such that the R-R-bimodule W_1 is finitely generated. Suppose that, for some constant c, almost every finite-dimensional indecomposable $\mathcal{A}$-module M, in a set of representatives of the isomorphism classes in $\mathcal{A}$-mod, satisfies that*

$$\dim_k \text{End}_{\mathcal{A}}(M) \leq c \times \dim_k M.$$

Show that $\mathcal{A}$ is not wild.

Hint: Assume that $\mathcal{A}$ is wild. Consider the wild algebra Γ of (22.27) and the functor G defined there. Notice that, for any $M \in k\langle x, y\rangle$-mod, $(\dim_k M)^2 \leq \dim_k \text{End}_\Gamma(G(M))$. From (22.17), we know the existence of an A-Γ-bimodule C, finitely generated free by the right, such that

$$\Gamma\text{-Mod} \xrightarrow{C\otimes -} A_{\mathcal{A}}\text{-Mod} \xrightarrow{L_{\mathcal{A}}} \mathcal{A}\text{-Mod}$$

preserves isomorphism classes of indecomposables. With the same argument given in the proof of (22.5), we obtain that

$$F := [\Gamma\text{-mod} \xrightarrow{C\otimes -} A_{\mathcal{A}}\text{-mod} \xrightarrow{L_{\mathcal{A}}} \mathcal{A}\text{-mod}]$$

preserves isomorphism classes and indecomposables. By (22.28), F is a faithful functor. Then, show that, for $M \in k\langle x, y\rangle$-mod, $\frac{1}{4}(\dim_k GM)^2 = (\dim_k M)^2 \leq \dim_k \mathrm{End}_\Gamma(GM) \leq \dim_k \mathrm{End}_{\mathcal{A}}(FGM) \leq cr \dim_k GM$, *where r is the rank of the right free* Γ*-module C. Derive a contradiction.*

23

Nested and seminested ditalgebras

In this section, we introduce the notions of nested and seminested ditalgebras and their associated marked bigraphs. Given a seminested ditalgebra $\mathcal{A}$, with marked bigraph $\mathbb{B}$, we examine important operations $\mathcal{A} \mapsto \mathcal{A}^z$ such as: reduction of an edge, deletion of idempotents, regularization and unravelling of a loop. In all these cases, the marked bigraph $\mathbb{B}^z$ associated to $\mathcal{A}^z$ is described in terms of the marked bigraph $\mathbb{B}$ of $\mathcal{A}$. In the case where Λ is a finite-dimensional k-algebra, which trivially splits over its radical, the associated Drozd's ditalgebra $\mathcal{D}^\Lambda$ is nested, and we shall describe its marked bigraph.

Definition 23.1. *Assume $1 = \sum_{j=1}^{s} e_j$ and $1 = \sum_{i=1}^{t} f_i$ are decompositions of the unit of the algebras R and S as a sum of central primitive orthogonal idempotents. Then, we say that a non-zero element u, of a given an R-S-bimodule U, is directed iff, for some i, j, $u \in e_jUf_i$. Since $U = \oplus_{i,j}e_jUf_i$, f_i and e_j are uniquely determined by a directed element u and we write $s(u) := i$ and $t(u) := j$.*

Definition 23.2. *Let R be a k-algebra. Recall that R is called trivial iff $R \cong k \times \cdots \times k$. R is called minimal iff it has the form*

$$R \cong k \times \cdots \times k \times k[x]_{f_1} \times \cdots \times k[x]_{f_t}$$

for some non-zero polynomials $f_1, \ldots, f_t \in k[x]$.

Assume that the k-algebra R has a decomposition as a finite product of indecomposable k-algebras and let $1 = \sum_{i=1}^{s} e_i$ be the unique decomposition of the unit of R as a sum of primitive central orthogonal idempotents.

The R-R-bimodule W is called freely generated by its subset $\mathbb{B}$ iff $\mathbb{B}$ consists of directed elements and there is an isomorphism of R-R-bimodules

$$W \longrightarrow \bigoplus_{w \in \mathbb{B}} Re_{t(w)} \otimes_k e_{s(w)}R,$$

sending each $w \in \mathbb{B}$ to $e_{t(w)} \otimes e_{s(w)}$.

The R-R-bimodule filtration $\mathcal{F}(W): 0 = W^0 \subseteq \cdots \subseteq W^\ell = W$ is called freely generated by the set filtration $\mathcal{F}(\mathbb{B}): \emptyset = \mathbb{B}^0 \subseteq \cdots \subseteq \mathbb{B}^\ell = \mathbb{B}$ of the subset $\mathbb{B}$ of W iff W is freely generated by its subset $\mathbb{B}$ and, moreover

$$W^j = \bigoplus_{w \in \mathbb{B}^j} RwR, \text{ for all } i \in [1, \ell].$$

The height associated to the set filtration $\mathcal{F}(\mathbb{B})$ is the map $\mathbb{h}: \mathbb{B} \rightarrow [1, \ell]$ such that $\mathbb{h}(w) = q$ iff $w \in \mathbb{B}^q \setminus \mathbb{B}^{q-1}$.

Remark 23.3. *Notice that, in the last definition, when the set filtration $\mathcal{F}(\mathbb{B})$ freely generates $\mathcal{F}(W)$, we have that each set $\mathbb{B}^j$ freely generates W^j.*

Lemma 23.4. *Assume that the k-algebra R has a decomposition as a finite product of indecomposable k-algebras. Then R admits a canonical trivial subalgebra R_0, which we call its trivial part. Namely, if $1 = \sum_{i=1}^s e_i$ is the unique decomposition of the unit of R as a sum of primitive central orthogonal idempotents, then $R_0 = \sum_{i=1}^s ke_i$ is the trivial part of R. Consider an R-R-bimodule W. Then:*

(1) W is freely generated by some (resp. finite) subset $\mathbb{B}$ iff W is freely generated by some (resp. finitely generated) R_0-R_0-subbimodule $\widetilde{W}$ of W.
(2) Suppose $\mathbb{B}$ is a subset of W formed by directed elements. Then, W is freely generated by $\mathbb{B}$ iff, for every R-R-bimodule U and every family $\{u_w\}_{w \in \mathbb{B}}$ of elements of U with $u_w \in e_{t(w)} U e_{s(w)}$, there is a unique morphism of R-R-bimodules $\phi: W \longrightarrow U$ satisfying $\phi(w) = u_w$, for all $w \in \mathbb{B}$.
(3) An R-R-bimodule filtration $\mathcal{F}(W)$ is freely generated by a set filtration $\mathcal{F}(\mathbb{B})$ for some subset $\mathbb{B}$ of W iff $\mathcal{F}(W)$ is an R_0-R_0-free filtration.

Proof.

(1) Assume W is freely generated by some (resp. finite) subset $\mathbb{B}$ and adopt the notation of (23.2). Consider $\widetilde{W} := \oplus_{w \in \mathbb{B}} R_0 e_{t(w)} \otimes_k e_{s(w)} R_0$, a (resp. finitely generated) R_0-bimodule. Then

$$\begin{aligned} R \otimes_{R_0} \widetilde{W} \otimes_{R_0} R &\cong \bigoplus_{w \in \mathbb{B}} R \otimes_{R_0} R_0 e_{t(w)} \otimes_k e_{s(w)} R_0 \otimes_{R_0} R \\ &\cong \bigoplus_{w \in \mathbb{B}} Re_{t(w)} \otimes_k e_{s(w)} R \cong W, \end{aligned}$$

and we can apply (15.8).

Conversely, if W is freely generated by some (resp. finitely generated) R_0-bimodule $\widetilde{W}$, from (15.7), we know that there is an isomorphism of R-R-bimodules $\phi: \oplus_{\alpha \in A} Re_{t(\alpha)} \otimes_k e_{s(\alpha)} R \longrightarrow W$, for some (resp. finite) set A and some maps $s, t: A \longrightarrow [1, s]$. Take $w_\alpha := \phi(e_{t(\alpha)} \otimes e_{s(\alpha)}) \in$

$e_{t(\alpha)}We_{s(\alpha)}$. Thus, the set $\mathbb{B}$ formed by the elements w_α, for $\alpha \in A$, freely generates W.

(2) Given the family $\{u_w\}_{w\in\mathbb{B}}$, consider the linear map $\oplus_w ke_{t(w)} \otimes_k ke_{s(w)} \to U$ mapping each $e_{t(w)} \otimes e_{s(w)}$ onto u_w. This is a morphism $\oplus_w R_0 e_{t(w)} \otimes_k e_{s(w)} R_0 \to U$ of R_0-R_0-bimodules, which extends to the required morphism of R-R-bimodules $W \cong \oplus_w Re_{t(w)} \otimes_k e_{s(w)} R \to U$. The converse is also easy to see.

(3) Assume that $\mathcal{F}(W)$ is an R-R-bimodule filtration, of length ℓ, R_0-R_0-free. Then, there is an additive R_0-R_0-bimodule filtration $\mathcal{F}(\widetilde{W})$ and an isomorphism of filtrations $R \otimes_{R_0} \mathcal{F}(\widetilde{W}) \otimes_{R_0} R \cong \mathcal{F}(W)$. We have, $\widetilde{W}^q = T_q \oplus \cdots \oplus T_1$, where $\widetilde{W}^q = \widetilde{W}^{q-1} \oplus T_q$. Since T_q is an $R_0 \otimes_k R_0^{op}$-module, there is an isomorphism $\phi_q : T_q \longrightarrow \oplus_{\alpha\in A_q} R_0 e_{t(\alpha)} \otimes_k e_{s(\alpha)} R_0$. Put $A := \cup_q A_q$, $\widehat{W} := \oplus_{\alpha\in A} R_0 e_{t(\alpha)} \otimes_k e_{s(\alpha)} R_0$, and $U := \oplus_{\alpha\in A} Re_{t(\alpha)} \otimes_k e_{s(\alpha)} R$. Consider the filtrations $\mathcal{F}(\widehat{W})$ and $\mathcal{F}(U)$ defined by $\widehat{W}^i := \oplus_{\alpha\in A_q, q\leq i} R_0 e_{t(\alpha)} \otimes_k e_{s(\alpha)} R_0$ and $U^i := \oplus_{\alpha\in A_q, q\leq i} Re_{t(\alpha)} \otimes_k e_{s(\alpha)} R$. Then, we can consider the isomorphism $\oplus_q \phi_q : \oplus_q T_q \longrightarrow \oplus_q \oplus_{\alpha\in A_q} R_0 e_{t(\alpha)} \otimes_k e_{s(\alpha)} R_0$, which constitutes an isomorphism of filtrations, of the same length ℓ, $\mathcal{F}(\widetilde{W}) \cong \mathcal{F}(\widehat{W})$. Then, we have the isomorphism of filtrations ϕ given by the composition of isomorphisms

$$\mathcal{F}(U) \cong R \otimes_{R_0} \mathcal{F}(\widehat{W}) \otimes_{R_0} R \cong R \otimes_{R_0} \mathcal{F}(\widetilde{W}) \otimes_{R_0} R \cong \mathcal{F}(W).$$

Then, $\phi : U \longrightarrow W$, and we can take $\mathbb{B}^i := \{\phi(e_{t(\alpha)} \otimes e_{s(\alpha)}) \mid \alpha \in A_q \text{ and } q \leq i\}$, for $i \in [1, \ell]$. □

Definition 23.5. *Let $\mathcal{A}$ be a layered k-ditalgebra. Then:*

(1) The layer (R, W) for $\mathcal{A}$ is called nested iff R is a minimal k-algebra, the layer (R, W) is triangular and, for $r \in \{0, 1\}$, the corresponding R-R-bimodule filtration $\mathcal{F}(W_r) = \{W_r^j\}_{j=0}^{\ell_r}$ is freely generated by a set filtration $\mathcal{F}(\mathbb{B}_r) = \{\mathbb{B}_r^j\}_{j=0}^{\ell_r}$ of some finite subset $\mathbb{B}_r$ of W_r.

(2) The layer (R, W) for $\mathcal{A}$ is called seminested iff R is a minimal k-algebra, the layer (R, W) is triangular, the R-R-bimodule W_1 is freely generated by a finite directed subset $\mathbb{B}_1$ of W_1 and the corresponding R-R-bimodule filtration $\mathcal{F}(W_0) = \{W_0^j\}_{j=0}^{\ell_0}$ of W_0 is freely generated by a set filtration $\mathcal{F}(\mathbb{B}_0) = \{\mathbb{B}_0^j\}_{j=0}^{\ell_0}$ of some finite subset $\mathbb{B}_0$ of W_0.

(3) The ditalgebra $\mathcal{A}$ is called nested (resp. seminested) iff it admits a nested (resp. seminested) layer. In both cases, the pair $(\mathbb{B}_0, \mathbb{B}_1)$ is called a basis of the layer (R, W).

(4) $\mathcal{A}$ is called minimal iff it admits a seminested layer (R, W), with $W_0 = 0$.

(5) *$\mathcal{A}$ is called trivial iff it admits a seminested layer (R, W), where R is a trivial k-algebra and $W_0 = 0$.*

Remark 23.6. *Every seminested ditalgebra $\mathcal{A}$ is additive triangular and, hence, a Roiter ditalgebra. In fact, if (R, W) is its seminested layer and R_0 is the trivial part of R, $\mathcal{A}$ is a R_0-free ditalgebra, as defined in (15.14). If (R, W) is nested, then $\mathcal{A}$ a completely R_0-free ditalgebra. From (23.4), we know that the corresponding filtrations are R_0-R_0-free.*

Lemma 23.7. *Assume $\mathcal{A}$ is a k-ditalgebra with nested (resp. seminested) layer (R, W) admitting the basis $(\mathbb{B}_0, \mathbb{B}_1)$. Assume that K is a field extension of k. Then, its extension $\mathcal{A}^K$ is a nested (resp. seminested) K-ditalgebra whose layer (R^K, W^K) admits the basis $(\hat{\mathbb{B}}_0, \hat{\mathbb{B}}_1)$, where $\hat{\mathbb{B}}_r$ consists, for $r \in \{0, 1\}$ (resp. for $r = 0$), of the elements of the form $\hat{w} = w \otimes 1$ with $w \in \mathbb{B}_r$. Each term $\hat{\mathbb{B}}_r^j$ of the filtration $\mathcal{F}(\hat{\mathbb{B}}_r)$ consists of the elements of the form $\hat{w} = w \otimes 1$ with $w \in \mathbb{B}_r^j$, for $r \in \{0, 1\}$ (resp. for $r = 0$).*

Proof. First notice that R^K is minimal whenever R is so, because $k^K \cong K$ and $k[x]_f^K \cong K[x]_f$. We also know that $\mathcal{A}^K$ is a triangular layered ditalgebra, whose triangular filtrations are obtained by extending those of $\mathcal{A}$.

The decomposition of the unit in the trivial k-algebra R_0 as a sum of central orthogonal primitive idempotents $1 = \sum_i e_i$ determines the corresponding decomposition $\hat{1} = \sum_i \hat{e}_i$ of the trivial K-algebra R_0^K. It is also clear that a freely generated R-R-bimodule W is extended to a freely generated R^K-R^K-bimodule. In fact, if an R-R-isomorphism

$$W \xrightarrow{\alpha} \bigoplus_{w \in \mathbb{B}} Re_{t(w)} \otimes_k e_{s(w)} R,$$

maps w to $e_{t(w)} \otimes e_{s(w)}$, then the R^K-R^K-isomorphism

$$W^K \xrightarrow{\alpha \otimes 1} \Big(\bigoplus_{w \in \mathbb{B}} Re_{t(w)} \otimes_k e_{s(w)} R\Big)^K \cong \bigoplus_{\hat{w} \in \hat{\mathbb{B}}} R^K \hat{e}_{t(\hat{w})} \otimes_K \hat{e}_{s(\hat{w})} R^K$$

maps $\hat{w}$ to $\hat{e}_{t(\hat{w})} \otimes \hat{e}_{s(\hat{w})}$. Here $t(\hat{w}) = t(w)$ and $s(\hat{w}) = s(w)$, by definition. □

Lemma 23.8. *Let $\mathcal{B}$ be a minimal k-ditalgebra with layer (R, W), and let $1 = \sum_{i=1}^s e_i$ be the corresponding decomposition of the unit element of R as a sum of primitive central orthogonal idempotents. For any $i \in [1, s]$ such that $Re_i = k$, define $S(i) = Re_i$; and for any $j \in [1, s]$ such that $Re_j \neq k$, any $n \geq 1$, and any irreducible element $\pi \in Re_j = k[x]_{f_j(x)}$, define $J(j, n, \pi) = Re_j/(\pi^n)$. Then, the $\mathcal{B}$-modules $S(i)$ and $J(j, n, \pi)$ form a complete set of finite-dimensional indecomposable objects of $\mathcal{B}$-*Mod.

Proof. Since $\mathcal{B}$ is a Roiter ditalgebra, $M \in \mathcal{B}$-Mod is indecomposable iff its endomorphism algebra $\mathrm{End}_{\mathcal{B}}(M)$ admits no non-trivial idempotent. From the equality $A_{\mathcal{B}} = R$, we obtain that $M \in \mathcal{B}$-Mod indecomposable implies that $M \in R$-Mod is indecomposable. From (22.11), we know that $U_{\mathcal{B}}(M)$ indecomposable implies that M is indecomposable as a $\mathcal{B}$-module. Then, apply the classification of finitely generated modules over a principal ideal domain to describe the finite-dimensional indecomposable R-modules. □

Definition 23.9. *A (marked directed) bigraph B is a family*

$$\mathbb{B} := (\mathcal{P}, \mathcal{P}_{\mathrm{mk}}, \mathbb{B}_0, \mathbb{B}_1, \mathbb{B}_{\mathrm{mk}}, \mathrm{mk}),$$

where $\mathcal{P}$ is a set of points and $\mathbb{B}_{\mathrm{mk}} \cup \mathbb{B}_0 \cup \mathbb{B}_1$ is a set of arrows, each element α of this set has associated its source point $s(\alpha)$ and its target point $t(\alpha)$; as usual, we write $x \xrightarrow{\alpha} y$ to indicate that the arrow α satisfies $s(\alpha) = x$ and $t(\alpha) = y$; $\mathcal{P}_{\mathrm{mk}}$ is a subset of $\mathcal{P}$, each of its elements is called a marked point; the elements of the set $\mathbb{B}_0$ are called the solid arrows of $\mathbb{B}$ and the elements of $\mathbb{B}_1$ are called the dotted arrows of $\mathbb{B}$; the set $\mathbb{B}_{\mathrm{mk}}$ is a set of loops, that is arrows ρ such that $s(\rho) = t(\rho)$, its elements are called the marked loops, there is exactly one $\rho_i \in \mathbb{B}_{\mathrm{mk}}$, for each $i \in \mathcal{P}_{\mathrm{mk}}$; finally, $\mathrm{mk} : \mathcal{P}_{\mathrm{mk}} \longrightarrow k[x]$ is a map called the mark map, which maps every marked point i onto a non-zero polynomial $f_i(x) \in k[x]$.

Assume that $\mathcal{A}$ is a seminested ditalgebra with layer (R, W) and basis $(\mathbb{B}_0, \mathbb{B}_1)$. With these ingredients, we can construct its associated bigraph

$$\mathbb{B} := (\mathcal{P}, \mathcal{P}_{\mathrm{mk}}, \mathbb{B}_0, \mathbb{B}_1, \mathbb{B}_{\mathrm{mk}}, \mathrm{mk}),$$

where $\mathcal{P}$ contains one point i for each idempotent e_i in the decomposition $1 = \sum_i e_i$ of the unit of R as a sum of central primitive orthogonal idempotents; $\mathcal{P}_{\mathrm{mk}}$ is the subset of $\mathcal{P}$ consisting of all $i \in \mathcal{P}$ such that $Re_i \ncong k$; $\mathrm{mk} : \mathcal{P}_{\mathrm{mk}} \longrightarrow k[x]$ maps each i onto $f_i(x) \in k[x]$, where $Re_i \cong k[x]_{f_i}$; $\mathbb{B}_{\mathrm{mk}}$ consists of the image ρ_i of the variable x under chosen isomorphisms $Re_i \cong k[x]_{f_i}$, for each $i \in \mathcal{P}_{\mathrm{mk}}$. The idempotent e_i is the idempotent associated to the point i of $\mathcal{A}$.

These bigraphs permit the following useful description of the objects and morphisms of $\mathcal{A}$-Mod.

Remark 23.10. *Assume that $\mathcal{A}$ is a seminested ditalgebra with layer (R, W), basis $(\mathbb{B}_0, \mathbb{B}_1)$ and associated bigraph $\mathbb{B}$. By definition a representation of $\mathbb{B}$ is a family $M = ((M_i)_{i \in \mathcal{P}}, (M_\alpha)_{\alpha \in \mathbb{B}_0 \cup \mathbb{B}_{\mathrm{mk}}})$, where M_i is a k-vectorspace and, for each arrow $i \xrightarrow{\alpha} j$ in $\mathbb{B}_0 \cup \mathbb{B}_{\mathrm{mk}}$, $M_\alpha : M_i \longrightarrow M_j$ is a k-linear transformation; moreover, we ask that $f_i(M_{\rho_i})$ is invertible, for each $\rho_i \in \mathbb{B}_{\mathrm{mk}}$, where $\mathrm{mk}(\rho_i) = f_i$. Thus, every $\mathcal{A}$-module M gives rise to the corresponding representation*

$((M_i)_{i\in\mathcal{P}}, (M_\alpha)_{\alpha\in\mathbb{B}_0\cup\mathbb{B}_{\mathrm{mk}}})$, where $M_i = e_iM$, for every arrow $\alpha : i \to j$, $M_\alpha : M_i \to N_j$ is determined by the multiplication by α. The $\mathcal{A}$-module M can be recovered from its representation as the direct sum $M = \oplus_i M_i$ and by extending the multiplication f_α of the generators $\alpha \in \mathbb{B}_0 \cup \mathbb{B}_{\mathrm{mk}}$ on M to an action of A on M.

Moreover, a morphism $f : M \longrightarrow N$ of representations of $\mathbb{B}$ is a family $f = ((f_i)_{i\in\mathcal{P}}, (f_v)_{v\in\mathbb{B}_1})$ satisfying, for each marked loop $\rho_i \in \mathbb{B}_{\mathrm{mk}}$, that $N_{\rho_i} f_i = f_i M_{\rho_i}$, and, for each solid arrow $\alpha \in \mathbb{B}_0$, that $N_\alpha f_i - f_j M_\alpha = f^1(\delta(\alpha))$, where f^1 is defined as follows. Consider the R-modules $M = \oplus_i M_i$ and $N = \oplus_i N_i$, where the generator ρ_i of $Re_i \cong k[x]_{f_i}$, for each marked point i, acts through M_{ρ_i}, and then extend the action of R over M and N to actions of A on them using the linear maps M_α, with $\alpha \in \mathbb{B}_0$. Then, each $f_v \in \mathrm{Hom}_k(M_i, M_j) \cong e_j\mathrm{Hom}_k(M, N)e_i$ defines a directed element f_v^1 of the R-R-bimodule $\mathrm{Hom}_k(M, N)$; then, $f^1 : W_1 \to \mathrm{Hom}_k(M, N)$ is, by definition, the morphism of R-R-bimodules obtained from the family of directed elements $\{f_v^1\}_{v\in\mathbb{B}_1}$ using (2) of (23.4), knowing that the R-R-bimodule W_1 is freely generated by the set $\mathbb{B}_1$. Recall that with f^1 we also denote its extension to an A-A-bimodule morphism $f^1 \in \mathrm{Hom}_{A\text{-}A}(V, \mathrm{Hom}_k(M, N))$.

Every morphism $f = (f^0, f^1) : M \to N$ in $\mathcal{A}$-Mod determines a morphism of representations $((f_i)_{i\in\mathcal{P}}, (f_v)_{v\in\mathbb{B}_1})$, taking as f_i the restriction of f^0 to a linear map $e_iM \to e_iN$, for each i; and $f_v := f^1(v)$, for each $v \in \mathbb{B}_1$. The morphism f will be an isomorphism iff every linear map f_i is invertible.

If E is any k-algebra and M is a proper bimodule $M \in \mathcal{A}$-E-Mod_p, then we obtain naturally representations where the M_i are right E-modules and each map M_α is a morphism of right E-modules; similarly, if $f : M \longrightarrow N$ is a morphism of proper $\mathcal{A}$-E-bimodules, each f_i or f_v are morphisms of right E-modules.

Lemma 23.11. *Assume that we have algebras R and S, and that $1 = \sum_{j=1}^{s} e_j$ and $1 = \sum_{i=1}^{t} f_i$ are decompositions of the unit of the algebras R and S as a sum of central primitive orthogonal idempotents, respectively. Suppose we have a right additive R-S-bimodule filtration $\mathcal{F}(U)$ of length ℓ of the R-S-bimodule U, $0 = U_0 \subseteq U_1 \subseteq \cdots \subseteq U_{\ell-1} \subseteq U_\ell = U$. Assume that the right S-module U is finitely generated projective, then we already know from (14.4) that U_S admits a finite dual basis compatible with $\mathcal{F}(U)$. In fact we have that U_S admits a finite directed dual basis $(x_d, \nu_d)_{d\in D}$ compatible with $\mathcal{F}(U)$ (with directed dual basis we mean that x_d and ν_d are directed in U and U^*, respectively, and, moreover, $s(x_d) = t(\nu_d)$ and $t(x_d) = s(\nu_d)$, for all $d \in D$). If S is a minimal algebra, we can actually choose such a dual basis with the following additional property. Take $D(i, j, u) := \{d \in$*

$D \mid t(x_d) = j, s(x_d) = i$ and $\mathbb{h}(x_d) \leq u\}$ and $D'(j,i,u) := \{d \in D \mid t(\nu_d) = i, s(\nu_d) = j$ and $\mathbb{h}(\nu_d) \leq u\}$. Then, $(x_d)_{d \in D(i,j,u)}$ is a basis for the free right Sf_i-module $e_jU_uf_i$ and $(\nu_{d'})_{d' \in D'(j,i,u)}$ is a basis for the free left Sf_i-module $f_i[U^*]_ue_j$.

Proof. It is constructed as follows: Consider, for each e_j and f_i, the filtration of R-S-bimodules, of the same length ℓ

$$\mathcal{F}(e_jUf_i) \ : \ 0 = e_jU_0f_i \subseteq e_jU_1f_i \subseteq \cdots \subseteq e_jU_{\ell-1}f_i \subseteq e_jU_\ell f_i = e_jUf_i.$$

Since $\mathcal{F}(U)$ is right additive, $\mathcal{F}(e_jUf_i)$ is right additive, and, as right S-modules, $e_jU_qf_i = T^q_{i,j} \oplus \cdots \oplus T^1_{i,j}$, where $e_jU_qf_i = e_jU_{q-1}f_i \oplus T^q_{i,j}$. Consider, for each q, the canonical projection $\pi_q : e_jUf_i \longrightarrow T^q_{i,j}$ and consider a finite dual basis $\{(x_d, \lambda_d) \mid d \in D(i,j)^q\}$ of the projective S-module $T^q_{i,j}$. Denote by $\widehat{\lambda}_d$ the composition $e_jUf_i \xrightarrow{\pi_q} T^q_{i,j} \xrightarrow{\lambda_d} S$, for all $d \in D(i,j)^q$. Thus, $x_d \in e_jUf_i$ and $\widehat{\lambda}_d \in (e_jUf_i)^*$. Then, if we make $D(i,j) := \bigcup_{q \in [1,\ell]} D(i,j)^q$, we obtain that $\{(x_d, \widehat{\lambda}_d) \mid d \in D(i,j)\}$ is a finite dual basis for $(e_jUf_i)_S$ compatible with $\mathcal{F}(e_jUf_i)$, according to (14.4).

Now, for each i, j, denote by $\pi_{i,j} : U \longrightarrow e_jUf_i$ the canonical projection, and denote by ν_d the composition $U \xrightarrow{\pi_{i,j}} e_jUf_i \xrightarrow{\widehat{\lambda}_d} S$, for all $d \in D(i,j)$. Then, using that $U = \oplus_{i,j} e_jUf_i$ is a decomposition of U as an R-S-bimodule, we obtain that $\{(x_d, \nu_d) \mid d \in D\}$, where $D := \bigcup_{i,j} D(i,j)$, is a finite dual basis for U_S. Since, for $d \in D(i,j)$, we clearly have that $x_d \in e_jUf_i$ and $\nu_d \in f_iU^*e_j$, the components of the dual basis consist of directed elements. Now, denote by $\mathbb{h}_{i,j}$ the height associated to the filtration $\mathcal{F}(e_jUf_i)$ (or its dual), and denote by $\mathbb{h}$ the height associated to the filtration $\mathcal{F}(U)$ (or its dual). Then, for any $d \in D(i,j)$, we have that $\mathbb{h}_{i,j}(x_d) = \mathbb{h}(x_d)$; we also have that $\widehat{\lambda}_d(e_jU_qf_i) = 0$ iff $\nu_d(U_q) = 0$ (for any index q), and, therefore, $\mathbb{h}_{i,j}(\widehat{\lambda}_d) = \mathbb{h}(\nu_d)$. Since $(x_d, \widehat{\lambda}_d)_{d \in D(i,j)}$ is compatible with the filtration $\mathcal{F}(e_iUf_j)$ of length ℓ, we get $\mathbb{h}(x_d) + \mathbb{h}(\nu_d) = \ell + 1$. Then $(x_d, \nu_d)_{d \in D}$ is the finite dual basis we were looking for.

Finally, if S is minimal, choose a basis $\{x_d\}_{d \in D(i,j)^q}$ of the free right Sf_i-module $T^q_{i,j}$ and consider the maps $\lambda_d \in (T^q_{i,j})^* = \mathrm{Hom}_S(T^q_{i,j}, S)$ defined by $\lambda_d(x_{d'}) = \delta_{d,d'}f_i$, where $\delta_{d,d'}$ is the Kronecker delta. Thus, $(x_d, \lambda_d)_{d \in D(i,j)^q}$ is a dual basis for the right S-module $T^q_{i,j}$. Applying the construction above to this choice of dual basis, we get a suitable dual basis $(x_d, \nu_d)_{d \in D}$ for U such that $\{x_d \mid t(x_d) = j, s(x_d) = i$ and $\mathbb{h}(x_d) \leq u\}$ is a basis for the free right Sf_i-module $e_jU_uf_i$ and $\{\nu_d \mid t(\nu_d) = i, s(\nu_d) = j$ and $\mathbb{h}(\nu_d) \leq u\}$ is a basis for the free left Sf_i-module $f_i[U^*]_ue_j$. □

Remark 23.12. *Assume $\mathcal{A}$ is a nested (resp. seminested) k-ditalgebra, with layer (R, W) and basis $(\mathbb{B}_0, \mathbb{B}_1)$. Assume that $\mathcal{A}'$ is an initial subditalgebra* *of $\mathcal{A}$ associated to R-R-bimodule decompositions $W_0 = W_0' \oplus W_0''$ and $W_1 = W_1' \oplus W_1''$. Say, with $W_0' = W_0^{\ell_0'}$ and $W_1' = W_1^{\ell_1'}$. See (14.9). Then, $\mathcal{A}'$ is a nested (resp. seminested) ditalgebra with layer (R, W') and the filtration $\mathcal{F}(W_r'')$ defined by $[W_r'']^j := W_r'' \cap W_r^{\ell_r'+j}$, for $j \in [0, \ell_r'']$, is freely generated by the set filtration $\mathcal{F}(\mathbb{B}_r'') : \emptyset = [\mathbb{B}_r'']^0 \subseteq \cdots \subseteq [\mathbb{B}_r'']^{\ell_r''} = \mathbb{B}_r''$, where $[\mathbb{B}_r'']^j$ consists of the components α'' of the basic elements α in $\mathbb{B}_r^{\ell_r'+j} \backslash \mathbb{B}_r^{\ell_r'}$, for $r \in \{0, 1\}$ (resp. for $r = 0$). For simplicity, we will usually assume that the decomposition is compatible with the given basis, that is $\mathbb{B}_r \backslash \mathbb{B}_r^{\ell_r'} \subseteq W_0''$. Then, we have $[\mathbb{B}_r'']^j = \mathbb{B}_r^{\ell_r'+j} \backslash \mathbb{B}_r^{\ell_r'}$.*

Lemma 23.13. *Assume $\mathcal{A}$ is a nested (resp. seminested) k-ditalgebra, with layer (R, W) admiting basis $(\mathbb{B}_0, \mathbb{B}_1)$. Consider the associated triangular filtrations $\{W_r^j = \oplus_{w \in \mathbb{B}_r^j} RwR\}_{j=1}^{\ell_r}$, for $r \in \{0, 1\}$ (resp. for $r = 0$). Assume that $\mathcal{A}'$ is an initial subditalgebra of $\mathcal{A}$ associated to R-R-bimodule decompositions of the form $W_0 = W_0' \oplus W_0''$ and $W_1 = W_1' \oplus W_1''$, where $W_0' = W_0^{\ell_0'}$ and $W_1' = 0$. We suppose that the decomposition is compatible with the given basis of the layer. Assume that $X_1 = Y_1 \oplus \cdots \oplus Y_t$ is a direct sum of a finite number of non-isomorphic finite-dimensional indecomposable $\mathcal{A}'$-modules $Y_1, \ldots, Y_t$. Assume that each algebra $\Gamma_i'' = \mathrm{End}_{\mathcal{A}'}(Y_i)^{op}$ decomposes as $\Gamma_i'' = k1_{Y_i} \oplus P_i''$, where P_i'' is the radical of Γ_i''. Then, $\Gamma_1' = \mathrm{End}_{\mathcal{A}'}(X_1)^{op}$ admits the splitting (S_1, P_1'), where P_1' is the radical of Γ_1'. Moreover, suppose that $\varphi : R \to S_2$ is an epimorphism of algebras, where S_2 is a minimal algebra. Consider the object X_2 in $\mathcal{A}'$-*Mod *obtained from the regular S_2-module by restriction through the composition $B \longrightarrow R \xrightarrow{\varphi} S_2$. Then, we have a splitting $\Gamma_2' := \mathrm{End}_{\mathcal{A}'}(X_2)^{op} = S_2 \oplus P_2'$, with $P_2' = 0$. Finally, assume that* $\mathrm{Hom}_{\mathcal{A}'}(\mathcal{I}_{X_i}, \mathcal{I}_{X_j}) = 0$, *for $i \neq j$. Then, from (17.12), we know that $X = X_1 \oplus X_2$ is a complete triangular admissible $\mathcal{A}'$-module. Moreover, $\mathcal{A}^X$ is a nested (resp. seminested) ditalgebra with layer (S, W^X). In order to describe a basis for its layer, assume $\Gamma := \mathrm{End}_{\mathcal{A}'}(X)^{op}$ has splitting $\Gamma = S \oplus P$, denote by $1 = \sum_{j=1}^{s} e_j$ the decomposition of the unit of R as a sum of central primitive orthogonal idempotents, by $1 = \sum_{i=1}^{h} f_i$ the canonical corresponding decomposition of the unit of S_1, that is where each f_i is obtained as the composition of the projection $X \to Y_i$ with the injection $Y_i \to X$, and denote by $1 = \sum_{j=h+1}^{t} f_j$ the canonical decomposition of the unit of the minimal algebra S_2 (into a sum of primitive central orthogonal idempotents). Thus, $1 = \sum_{j=1}^{t} f_j$ is the canonical decomposition of the unit of S. Consider the associated filtrations*

$$\mathcal{F}(P) \ : \ 0 = P_0 \subseteq \cdots \subseteq P_{\ell_P} = P,$$

given by the powers of the radical P, and

$$\mathcal{F}(X) \ : \ 0 = XP_0 \subseteq \cdots \subseteq XP_{\ell_P} \subseteq X_{\ell_X} = X.$$

Then, consider dual basis $(x_i, \nu_i)_i$ *and* $(p_j, \gamma_j)_j$ *for the right S-modules X and P, which are compatible with the given filtrations and such that each of the elements* $x_i, \nu_i, p_j, \gamma_j$ *is directed, as constructed in (23.11). Then, the nested (resp. seminested) layer* (S, W^X) *admits the basis* $(\mathbb{B}_0^X, \mathbb{B}_1^X)$ *with associated filtrations* $\mathcal{F}(\mathbb{B}_0^X)$ *and* $\mathcal{F}(\mathbb{B}_1^X)$ *given as follows*

$$[\mathbb{B}_0^X]^m = \{\nu_\beta \otimes w \otimes x_\alpha \mid w \in \mathbb{B}_0'', x_\alpha \in e_{s(w)}X, \nu_\beta \in X^* e_{t(w)} \\ \text{with } \mathbb{h}(\nu_\beta) + 2\ell_X \mathbb{h}(w) + \mathbb{h}(x_\alpha) \leq m\},$$

for $m \in [0, 2\ell_X(\ell_0'' + 1)]$; $[\mathbb{B}_1^X]^m = \{\gamma_j \mid \mathbb{h}(\gamma_j) \leq m\}$, *for* $m \in [0, \ell_P]$; *and*

$$[\mathbb{B}_1^X]^{\ell_P + m} = [\mathbb{B}_1^X]^{\ell_P} \cup \{\nu_\beta \otimes w \otimes x_\alpha \mid w \in \mathbb{B}_1, x_\alpha \in e_{s(w)}X, \nu_\beta \in X^* e_{t(w)} \\ \text{with } \mathbb{h}(\nu_\beta) + 2\ell_X \mathbb{h}(w) + \mathbb{h}(x_\alpha) \leq m\},$$

for $m \in [0, 2\ell_X(\ell_1'' + 1)]$. *The height map applied to elements of* $\mathbb{B}_0''$ *corresponds to the set filtration of* $\mathbb{B}_0''$ *described in (23.12). In the case of a seminested ditalgebra* $\mathcal{A}$, *we do not have to specify any set filtration of* $\mathbb{B}_1^X$.

Proof. We already know that X is a complete triangular admissible $\mathcal{A}'$-module. In (14.10) we described the triangular filtrations $\mathcal{F}(W_r^X) = \{[W_r^X]^m\}_m$ of $\mathcal{A}^X$. Clearly, S is a minimal algebra. In order to show that (S, W^X) is nested (resp. seminested), we shall exhibit a basis for the layer. Having in mind (23.11), consider the following chain of isomorphisms

$$\begin{aligned}
[X^*]_r \otimes_B [\underline{W}_0]^s \otimes_B X_t &\cong [X^*]_r \otimes_B (B \otimes_R [W_0'']^s \otimes_R B) \otimes_B X_t \\
&\cong \oplus_w [X^*]_r \otimes_R RwR \otimes_R X_t \\
&\cong \oplus_w [X^*]_r \otimes_R Re_{t(w)} \otimes_k e_{s(w)}R \otimes_R X_t \\
&\cong \oplus_w [X^*]_r e_{t(w)} \otimes_k e_{s(w)} X_t \\
&\cong \oplus_{w,i,j} f_j [X^*]_r e_{t(w)} \otimes_k e_{s(w)} X_t f_i \\
&\cong \oplus_{w,i,j} \left(\oplus_{d' \in D'(t(w),j,r)} Sf_j \nu_{d'}\right) \otimes_k \left(\oplus_{d \in D(i,s(w),t)} x_d Sf_i\right) \\
&\cong \oplus_{w,i,j} \oplus_{d' \in D'(t(w),j,r)} \oplus_{d \in D(i,s(w),t)} (S\nu_{d'} \otimes_k x_d S),
\end{aligned}$$

where, w runs in $[\mathbb{B}_0'']^s$. But, for $d' \in D'(t(w), j, r)$ and $d \in D(i, s(w), t)$, we have an isomorphism of S-bimodules

$$S\nu_{d'} \otimes_k x_d S \cong Sf_j \otimes_k f_i S.$$

Then, tracing back the image of the canonical generators, we obtain the first equality of our lemma.

We claim that the S-S-bimodule P^* is freely generated by $\{\gamma_j\}_j$. From (23.11) we know that $\{\gamma_j\}_j$ is a basis of P^* as a vector space, therefore we have an isomorphism $P^* = \oplus_j k\gamma_j \cong \oplus_j Se_{t(\gamma_j)} \otimes_k e_{s(\gamma_j)} S$ mapping $e_{t(\gamma_j)} \otimes_k e_{s(\gamma_j)}$ onto γ_j, and the S-S-bimodule P^* is freely generated by this basis.

The verification of the last equality of our lemma, in the nested case, is similar to that of the first one, having in mind the second one. □

In the following, we introduce the usual specific operations $\mathcal{A} \mapsto \mathcal{A}^z$ and associated reduction functors needed to prove Drozd's *Tame and Wild Theorem*. Starting from a nested (resp. seminested) ditalgebra $\mathcal{A}$, the corresponding basic operation produces a new nested (resp. seminested) ditalgebra $\mathcal{A}^z$.

Lemma 23.14. *Deletion of idempotents: Assume that $\mathcal{A}$ is a nested (resp. seminested) ditalgebra with layer (R, W) admitting basis $(\mathbb{B}_0, \mathbb{B}_1)$. Consider a collection $e_{i_1}, \ldots, e_{i_t}$ of some central primitive orthogonal idempotents of R and write $e = 1 - \sum_{j=1}^{t} e_{i_j}$. Then, the ditalgebra $\mathcal{A}^d$ obtained by deletion of the idempotents $e_{i_1}, \ldots, e_{i_t}$ (i.e. by deletion of the idempotent $1 - e$ with the terminology of (8.17)) is a nested (resp. seminested) ditalgebra. Its layer $(R^d, W^d) = (eRe, eWe)$ admits the basis $(e\mathbb{B}_0 e, e\mathbb{B}_1 e)$, with associated filtrations $\mathcal{F}(e\mathbb{B}_r e) = \{e\mathbb{B}_r^j e\}$, for $r \in \{0, 1\}$ (resp. for $r = 0$). Thus, the bigraph $\mathbb{B}^d$ of $\mathcal{A}^d$ is obtained from the bigraph $\mathbb{B}$ of $\mathcal{A}$ by deleting the points $i_1, \ldots, i_t$ and all the arrows starting or ending at them.*

Lemma 23.15. *Regularization: Assume that $\mathcal{A}$ is a nested (resp. seminested) ditalgebra with layer (R, W), admitting basis $(\mathbb{B}_0, \mathbb{B}_1)$. Assume that $\alpha \in \mathbb{B}_0$ and $v \in \mathbb{B}_1$ satisfy $\delta(\alpha) = v$. Consider the R-R-bimodule decompositions $W_0 = W_0' \oplus W_0''$, with $W_0' = R\alpha R$ and $W_0'' = \oplus_{\alpha \neq \beta \in \mathbb{B}_0} R\beta R$, and $W_1 = W_1' \oplus W_1''$, with $W_1' = RvR$ and $W_1'' = \oplus_{v \neq w \in \mathbb{B}_1} RwR$. Then, the ditalgebra $\mathcal{A}^r$ obtained by regularization of the bimodule W_0', as in (8.19), is a nested (resp. seminested) ditalgebra. We say that $\mathcal{A}^r$ is obtained from $\mathcal{A}$ by regularization (of α). Its layer $(R^r, W^r) = (R, W'')$ admits the basis $(\mathbb{B}_0 \backslash \{\alpha\}, \mathbb{B}_1 \backslash \{v\})$.*

Its bigraph $\mathbb{B}^r$ can be described in terms of the bigraph $\mathbb{B}$ of $\mathcal{A}$ as follows: $\mathcal{P}^r = \mathcal{P}$; $\mathcal{P}^r_{\mathrm{mk}} = \mathcal{P}_{\mathrm{mk}}$; $\mathbb{B}_0^r = \mathbb{B}_0 \backslash \{\alpha\}$; $\mathbb{B}_1^r = \mathbb{B}_1 \backslash \{v\}$; $\mathbb{B}^r_{\mathrm{mk}} = \mathbb{B}_{\mathrm{mk}}$; and $\mathrm{mk}^r = \mathrm{mk}$.

Proof. It follows from the description of the layer of $\mathcal{A}^r$ given in (8.19). In the nested case, the filtrations $\mathcal{F}(W_s^r)$, for $s \in \{0, 1\}$, are freely generated by the set filtrations $\mathcal{F}(\mathbb{B}_s^r)$, where $[\mathbb{B}_0^r]_i = [\mathbb{B}_0]_i \backslash \{\alpha\}$ and $[\mathbb{B}_1^r] = [\mathbb{B}_1]_i \backslash \{v\}$. □

Lemma 23.16. *Absorption of a loop: Assume that $\mathcal{A}$ is a nested (resp. seminested) ditalgebra with layer (R, W), admitting basis $(\mathbb{B}_0, \mathbb{B}_1)$. Assume that $\alpha \in e_{i_0}\mathbb{B}_0 e_{i_0}$, with $Re_{i_0} = k$, satisfies $\delta(\alpha) = 0$. Consider the R-R-bimodule decomposition $W_0 = W_0' \oplus W_0''$, where $W_0' = R\alpha R$ and $W_0'' = \oplus_{\beta \neq \alpha} R\beta R$. Then, the ditalgebra $\mathcal{A}^a$ obtained from $\mathcal{A}$ by absorption of the loop α (i.e.*

of the subbimodule W_0' *with the terminology of (8.20)) is a nested (resp. seminested) ditalgebra. Its layer* (R^a, W^a) *satisfies:* R^a *has the same trivial part of* R, *in fact* $e_{i_0}R^a \cong k[x]$ *and* $e_iR^a = e_iR$, *for* $i \neq i_0$*;* $W_0^a = R^aW_0''R^a$ *and* $W_1^a = R^aW_1R^a$ *are freely generated by* $\mathbb{B}_0^a = \mathbb{B}_0 \setminus \{\alpha\}$ *and* $\mathbb{B}_1^a = \mathbb{B}_1$, *respectively; the heights of the arrows remain unchanged. Thus, the bigraph* $\mathbb{B}^a$ *of* $\mathcal{A}^a$ *is obtained from the bigraph* $\mathbb{B}$ *of* $\mathcal{A}$ *by transfering the loop* α *from* $\mathbb{B}_0$ *to* $\mathbb{B}_{\rm mk}$*. More precisely:* $\mathcal{P}^a = \mathcal{P}$, $\mathcal{P}^a_{\rm mk} = \mathcal{P}_{\rm mk} \cup \{i_0\}$, $\mathbb{B}_0^a = \mathbb{B}_0 \setminus \{\alpha\}$, $\mathbb{B}^a_{\rm mk} = \mathbb{B}_{\rm mk} \cup \{\alpha\}$, $\mathbb{B}_1^a = \mathbb{B}_1$, *and* $\mathrm{mk}^a(i_0) = 1 \in k[x]$ *and* $\mathrm{mk}^a(j) = \mathrm{mk}(j)$, *for* $j \neq i_0$.

Before introducing the operation *reduction of an edge*, it is convenient to make some preliminary observations.

Lemma 23.17. *Consider the path algebra* $H := kE$, *where* E *is the quiver* $E : i_0 \xrightarrow{\alpha} j_0$. *Then,* H *admits only three isomorphism classes of indecomposable* H*-modules represented by: the simple injective* $H_{i_0} = [k \longrightarrow 0]$, *the simple projective* $H_{j_0} = [0 \longrightarrow k]$ *and the projective injective* $H_z = [k \xrightarrow{I_k} k]$. *Consider the endomorphism algebra*

$$\Gamma = \mathrm{End}_H(H_{i_0} \oplus H_z \oplus H_{j_0}).$$

Then, there is a natural isomorphism $\overline{\varphi} : kQ/I \longrightarrow \Gamma$, *where* kQ *is the path algebra of the quiver* $Q : i_* \xleftarrow{\alpha_1} z \xleftarrow{\alpha_2} j_*$ *and* I *is the ideal of* kQ *generated by the path* $\alpha_1\alpha_2$. *The* H*-module* $X = H_{i_0} \oplus H_z \oplus H_{j_0}$ *has a natural structure of an* H*-*Γ^{op}*-bimodule. Denote by* f_{i_*}, f_z *and* f_{j_*} *the idempotents of* Γ *associated to the direct summands* H_{i_0}, H_z *and* H_{j_0}, *respectively. Then,* $\Gamma^{op} = S \oplus P$, *where* $S = kf_{i_*} \oplus kf_z \oplus kf_{j_*}$ *is a trivial* k*-algebra, and* P *is the radical of* Γ^{op}. *The nilpotency index of* P *is* $\ell_P + 1 = 2$. *Consider the filtration of* P *given by its powers* $0 = P^{(\ell_P+1)} \subseteq P^{(1)} = P$, *make* $P_j := P^{(\ell_P - j+1)}$, *for all* $j \in [0, \ell_P]$, *to obtain the* S*-*S*-bimodule filtration* $\mathcal{F}(P) : 0 = P_0 \subseteq P_{\ell_P} = P$, *and the* H*-*S*-bimodule filtration* $\mathcal{F}(X) : 0 = [X]_0 \subseteq [X]_1 \subseteq [X]_{\ell_X} = X$, *where* $\ell_X = \ell_P + 1$ *and* $[X]_i = XP_i$, *for* $i \in [0, \ell_P]$.

Denote: $x_{i_*}^{i_0} := 1 \in e_{i_0}H_{i_0}$*;* $x_{j_*}^{j_0} := 1 \in e_{j_0}H_{j_0}$*;* $x_z^{i_0} := 1 \in e_{i_0}H_z$*; and* $x_z^{j_0} := \alpha x_z^{i_0} = 1 \in e_{j_0}H_z$. *Then,* $\{x_{i_*}^{i_0}, x_{j_*}^{j_0}, x_z^{i_0}, x_z^{j_0}\}$ *is a directed basis of the* H*-*S*-bimodule* X. *Moreover, the height of each of these basis vectors is given by* $\mathbb{h}(x_{i_*}^{i_0}) = 1 = \mathbb{h}(x_z^{j_0})$ *and* $\mathbb{h}(x_z^{i_0}) = 2 = \mathbb{h}(x_{j_*}^{j_0})$. *In the next array, we list in each column* i *the basic elements of* H_i, *and at the end of each row* j *the common height of the elements of the row* j

H_{i_0}	H_z	H_{j_0}	$\mathbb{h}$
	$x_z^{i_0}$	$x_{j_*}^{j_0}$	2
$x_{i_*}^{i_0}$	$x_z^{j_0}$		1.

Denote by p_ξ the element of Γ corresponding to each non-trivial path class ξ of kQ/I. Then, $\{p_\xi\}_\xi$ is a directed vector space basis for the S-S-bimodule P. From (23.11) we know that there is a directed dual basis $(p_\xi, \gamma_\xi)_\xi$ of the right S-module P, which is compatible with the filtration $\mathcal{F}(P)$. Moreover, $\mathbb{h}(\gamma_\xi) = \ell(\xi)$, where $\ell(\xi)$ denotes the length of the paths in ξ. Also, $s(\gamma_\xi) = s(\xi)$ and $t(\gamma_\xi) = t(\xi)$, where $s(\xi)$ and $t(\xi)$ denote the source and target points of the path class ξ.

Proof. Denote by $\sigma : H_{j_0} \longrightarrow H_z$ and $\pi : H_z \longrightarrow H_{i_0}$ the canonical injection and projection respectively. Then notice that $P = kp_1 \oplus kp_2$, where

$$p_1 = \begin{pmatrix} 0 & \pi & 0 \\ 0 & 0 & 0 \\ 0 & 0 & 0 \end{pmatrix} \text{ and } p_2 = \begin{pmatrix} 0 & 0 & 0 \\ 0 & 0 & \sigma \\ 0 & 0 & 0 \end{pmatrix}.$$

It is clear that the morphism of algebras $\varphi : kQ \longrightarrow \Gamma$ determined by $\varphi(\tau_{i_*}) = f_{i_*}$, $\varphi(\tau_{j_*}) = f_{j_*}$, $\varphi(\tau_z) = f_z$, $\varphi(\alpha_1) = p_1$ and $\varphi(\alpha_2) = p_2$, admits the ideal I as its kernel. Thus, we have an induced isomorphism $\overline{\varphi} : kQ/I \longrightarrow \Gamma$.

We have $x_z^{i_0} p_1 = x_{i_*}^{i_0}$ and $x_{j^*}^{j_0} p_2 = x_z^{j_0}$. Moreover: $e_{i_0} X f_{i_*} = k x_{i_*}^{i_0}$; $e_{i_0} X f_z = k x_z^{i_0}$; $e_{i_0} X f_{j_*} = 0$; $e_{j_0} X f_{i_*} = 0$; $e_{j_0} X f_z = k x_z^{j_0}$; and $e_{j_0} X f_{j_*} = k x_{j_*}^{j_0}$. Thus, $X = k x_{i_*}^{i_0} \oplus k x_z^{i_0} \oplus k x_z^{j_0} \oplus k x_{j_*}^{j_0}$; $X p_1 = k x_{i_*}^{i_0}$ and $X p_2 = k x_z^{j_0}$. Therefore, $XP = k x_{i_*}^{i_0} \oplus k x_z^{j_0}$. From here, we get the height values specified in the statement of the lemma.

Finally, if $\xi : i \to j$ in kQ/I, then $p_\xi : j \longrightarrow i$ in $P \subseteq \Gamma^{op}$ and, therefore, since the dual basis is directed, $\gamma_\xi : i \longrightarrow j$. □

Lemma 23.18. *Reduction of an edge: Assume $\mathcal{A}$ is a nested (resp. seminested) k-ditalgebra, with layer (R, W) admitting basis $(\mathbb{B}_0, \mathbb{B}_1)$. Assume that $\alpha \in e_{j_0} \mathbb{B}_0 e_{i_0}$, with $i_0 \neq j_0$, $Re_{i_0} = k$ and $Re_{j_0} = k$, satisfies $\delta(\alpha) = 0$. Such an arrow will be called an edge of $\mathcal{A}$. Make $e := e_{i_0} + e_{j_0}$. We can refine the triangular filtration of W_0 in such a way that $R\alpha R$ is the first non-zero term of this filtration. Then, we can consider the initial subditalgebra $\mathcal{A}'$ of $\mathcal{A}$ associated to the R-R-bimodule decompositions $W_0 = W_0' \oplus W_0''$ and $W_1 = W_1' \oplus W_1''$, where $W_0' = R\alpha R$ and $W_0'' = \oplus_{\alpha \neq \beta \in \mathbb{B}_0} R\beta R$, and $W_1' = 0$. Then, $B := [A_{\mathcal{A}'}]_0 = eB \times (1-e)R$, where $H := eB$ can be identified with the path algebra kE of the quiver $E : i_0 \xrightarrow{\alpha} j_0$. Thus, H admits only three isomorphism classes of indecomposable modules represented by H_{i_0}, H_z and H_{j_0}, as described in (23.17). Consider these three objects as B-modules by restriction via the projection $B \longrightarrow H$. Consider the B-modules $X_1 := H_{i_0} \oplus H_z \oplus H_{j_0}$ and*

$X_2 := (1-e)R$. Then, $X := X_1 \oplus X_2$ is a complete triangular admissible $\mathcal{A}'$-module. The associated ditalgebra $\mathcal{A}^e := \mathcal{A}^X$ is nested (resp. seminested). We say that $\mathcal{A}^e$ is obtained from $\mathcal{A}$ by reduction of the edge α.

Its bigraph $\mathbb{B}^e$ can be described in terms of the bigraph $\mathbb{B}$ of $\mathcal{A}$ as follows. The set $\mathcal{P}^e$ of points has "old" points $\mathcal{P}\setminus\{i_0, j_0\}$ and three "new" points i_*, j_* and z; the marked points are the same $\mathcal{P}^e_{\rm mk} = \mathcal{P}_{\rm mk}$, as well as the marked loops $\mathbb{B}^e_{\rm mk} = \mathbb{B}_{\rm mk}$; the mark maps $\mathrm{mk} = \mathrm{mk}^e$ are the same too; the solid arrows of $\mathbb{B}^e_0$ are produced by solid arrows of $\mathbb{B}_0$ by the following rules: Each arrow $\beta : i \longrightarrow j$ of $\mathbb{B}_0$, with $i, j \notin \{i_0, j_0\}$, induces in $\mathbb{B}^e_0$ the arrow $v^j_j \otimes \beta \otimes x^i_i : i \longrightarrow j$; each arrow $\beta : i \longrightarrow s_0$ of $\mathbb{B}_0$, with $i \notin \{i_0, j_0\}$ and $s_0 \in \{i_0, j_0\}$, induces in $\mathbb{B}^e_0$ two arrows $v^{s_*}_{s_0} \otimes \beta \otimes x^i_i : i \longrightarrow s_*$ and $v^z_{s_0} \otimes \beta \otimes x^i_i : i \longrightarrow z$; dually, the arrows $\beta : r_0 \longrightarrow j$ of $\mathbb{B}_0$, with $j \notin \{i_0, j_0\}$ and $r_0 \in \{i_0, j_0\}$, induce in $\mathbb{B}^e_0$ two arrows $v^j_j \otimes \beta \otimes x^{r_0}_{r_*} : r_* \longrightarrow j$ and $v^j_j \otimes \beta \otimes x^{r_0}_z : z \longrightarrow j$; finally, each arrow $\beta : r_0 \longrightarrow s_0$, different from α and such that $r_0, s_0 \in \{i_0, j_0\}$ induces four arrows $v^{s_*}_{s_0} \otimes \beta \otimes x^{r_0}_{r_*} : r_* \longrightarrow s_*$, $v^z_{s_0} \otimes \beta \otimes x^{r_0}_{r_*} : r_* \longrightarrow z$, $v^{s_*}_{s_0} \otimes \beta \otimes x^{r_0}_z : z \longrightarrow s_*$ and $v^z_{s_0} \otimes \beta \otimes x^{r_0}_z : z \longrightarrow z$. For each of the arrows $v_y \otimes \beta \otimes x$, described above, where $v_y \in X^*$ denotes the element corresponding to $y \in X$ in the fixed directed dual basis of X_S, we have the formula $\mathbb{h}(v_y \otimes \beta \otimes x) = 2\ell_X \mathbb{h}(\beta) + \mathbb{h}(v_y) + \mathbb{h}(x) = 2\ell_X \mathbb{h}(\beta) + \ell_X + 1 - \mathbb{h}(y) + \mathbb{h}(x) = 4\mathbb{h}(\beta) + \mathbb{h}(x) - \mathbb{h}(y) + 3$.

There are two types of dotted arrows in $\mathbb{B}^e_1$. The first type, of the form $v_y \otimes v \otimes x$ where $v \in \mathbb{B}_1$, are constructed from the dotted arrows of $\mathbb{B}_1$ in the same way described for the solid arrows. The other type is formed by two new dotted arrows $\gamma_{\overline{\alpha}_2} : j_* \longrightarrow z$ and $\gamma_{\overline{\alpha}_1} : z \longrightarrow i_*$, associated to the path classes $\overline{\alpha}_1$ and $\overline{\alpha}_2$ of kQ/I. In the nested case, their heights are $\mathbb{h}(\gamma_{\overline{\alpha}_1}) = 1 = \mathbb{h}(\gamma_{\overline{\alpha}_2})$ and the value of the height of each new dotted arrow of the first type can be computed as $\mathbb{h}(v_y \otimes v \otimes x) = 4\mathbb{h}(v) + \mathbb{h}(x) - \mathbb{h}(y) + 4$.

Proof. We have that $\mathrm{Hom}_{\mathcal{A}'}(\mathcal{I}_{X_s}, \mathcal{I}_{X_t}) \cong \mathrm{Hom}_{\mathcal{B}}(\mathcal{I}_{X_s}, \mathcal{I}_{X_t}) = 0$, for $s \neq t$, because e and $1-e$ are orthogonal. By (17.12), X is a complete triangular admissible $\mathcal{A}'$-module. We have the splitting $\Gamma_1 = \mathrm{End}_{\mathcal{A}'}(H_{i_0} \oplus H_z \oplus H_{j_0})^{op} = S_1 \oplus P_1$, with $S_1 = kf_{i_*} \oplus kf_z \oplus kf_{j_*}$ where f_{i_*}, f_{j_*} and f_z are the canonical idempotents associated to the given direct sum decomposition of X_1, and P_1 is the radical of Γ_1. Moreover

$$\Gamma_2 = \mathrm{End}_{\mathcal{A}'}(X_2)^{op} = \mathrm{End}_{(1-e)R}((1-e)R)^{op} \cong (1-e)R,$$

and we have the splitting $\Gamma_2 = S_2 \oplus P_2$, with $P_2 = 0$. We identify Γ_2 with $(1-e)R$, thus we have that $S_2 = \prod_{i \neq i_0, j_0} Re_i$.

Then, we have the splitting $\Gamma = \mathrm{End}_{\mathcal{A}'}(X)^{op} = S \oplus P$, where we can identify, as in (17.11), S with $S_1 \times S_2$ and P with $P_1 \times P_2$.

Thus, $S = S_1 \times S_2 \cong (kf_{i_*} \oplus kf_z \oplus kf_{j_*}) \times (\prod_{i \neq i_0, j_0} Re_i)$ is a minimal k-algebra. From now on write $f_i := e_i$, for $i \notin \{i_0, j_0\}$.

In order to describe a natural basis for the layer of $\mathcal{A}^X$, we look at the filtrations associated to X and P. Since $P_1^2 = 0$, $P^2 = 0$ and we have the filtration of S-S-bimodules $\mathcal{F}(P): 0 = [P]_0 \subseteq [P]_1 = P$. We also have the filtration of B-S-bimodules $\mathcal{F}(X): 0 = X[P]_0 \subseteq X[P]_1 \subseteq X$. In (23.17), we described directed basis for the bimodules X_1 and P_1: let us adopt the notations introduced there. For $i \notin \{i_0, j_0\}$, write $x_i^i := e_i \in Re_i = e_i X f_i$. Clearly, $\mathbb{lh}(x_i^i) = 2$, for all $i \notin \{i_0, j_0\}$. Then, as in (23.11), we can complete the given family $(x_i^j)_{i,j}$ to a directed dual basis $(x_i^j, v_j^i)_{i,j}$ compatible with the filtration $\mathcal{F}(X)$. Thus, for each $x_i^j \in e_j X f_i$, we have $v_j^i \in f_i X^* e_j$, with $\mathbb{lh}(x_i^j) + \mathbb{lh}(v_j^i) = \ell_X + 1 = 3$. It follows that $\mathbb{lh}(v_{i_0}^{i_*}) = 2 = \mathbb{lh}(v_{j_0}^z)$, $\mathbb{lh}(v_{i_0}^z) = 1 = \mathbb{lh}(v_{j_0}^{j_*})$, and $\mathbb{lh}(v_j^i) = 1$, for all $i \notin \{i_0, j_0\}$.

From the description of the filtrations $\mathcal{F}(\mathbb{B}_r^e) = \mathcal{F}(\mathbb{B}_r^X)$, given in (23.13), and (23.17), we obtain the arrows of $\mathbb{B}_r^e$ described in the statement of the lemma. $\qquad\square$

Lemma 23.19. *Assume $\mathcal{A}^e$ is the ditalgebra obtained from $\mathcal{A}$ by reduction of the edge α, and consider the associated admissible module $X = X_1 \oplus X_2$, with $X_1 = H_{i_0} \oplus H_z \oplus H_{j_0}$, as in the last lemma. Consider the quiver $\widetilde{Q}$, with points set $\{x_{i_*}^{i_0}, x_z^{i_0}, x_z^{j_0}, x_{j_*}^{j_0}\} \subseteq X_1$, and arrows $\widetilde{\alpha}_1 : x_z^{i_0} \longrightarrow x_{i_*}^{i_0}$ and $\widetilde{\alpha}_2 : x_{j_*}^{j_0} \longrightarrow x_z^{j_0}$. Then we have a quiver map $\pi : \widetilde{Q} \longrightarrow Q$, defined by $\pi(\widetilde{\alpha}_i) = \alpha_i$. We can visualize this map with the following figure*

$$\widetilde{Q}: \qquad \begin{array}{ccc} & x_z^{i_0} & x_{j_*}^{j_0} \\ \widetilde{\alpha}_1 \swarrow & & \widetilde{\alpha}_2 \swarrow \\ x_{i_*}^{i_0} & x_z^{j_0} & \end{array}$$

$$\Big\downarrow \pi$$

$$Q: \quad i_* \xleftarrow{\ \alpha_1\ } z \xleftarrow{\ \alpha_2\ } j_*$$

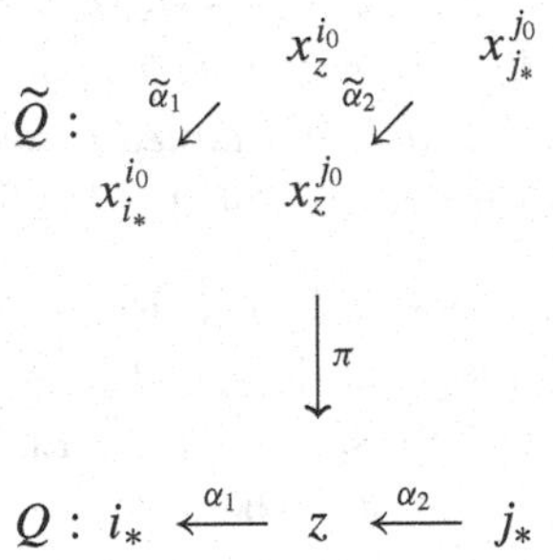

We can define a relation $<$ on the points of $\widetilde{Q}$: By definition, $x_a^b < x_{a'}^{b'}$ iff there is a non-trivial path φ of $\widetilde{Q}$ with source point x_a^b and target point $x_{a'}^{b'}$. In this case, we will denote this path by $\varphi(x_a^b, x_{a'}^{b'})$. We shall denote by $\xi(x_a^b, x_{a'}^{b'})$ the class of $\pi(\varphi(x_a^b, x_{a'}^{b'}))$ in kQ/I. Then, $z < z'$ iff $z' = z\xi$, for some non-zero non-trivial path class ξ of kQ/I, in the right $(kQ/I)^{op}$-module X_1; moreover, in this case, $\xi = \xi(z, z')$.

Furthermore, writing $\nu_{x_a^b} := \nu_b^a$, *we have the following, for* $z \in \widetilde{Q}$:

(1) $\rho(z) = \sum_{z<z'\in\widetilde{Q}} z' \otimes \gamma_{\xi(z,z')}$;
(2) $\lambda(\nu_z) = \sum_{z>z'\in\widetilde{Q}} \gamma_{\xi(z',z)} \otimes \nu_{z'}$.

In particular, $\mathbb{h}(z) = 1$ *implies* $\rho(z) = 0$, *and* $\mathbb{h}(\nu_z) = 1$ *implies* $\lambda(\nu_z) = 0$.

Proof. It follows from (11.11), applied to the directed dual basis

$$\{(x_{i_*}^{i_0}, \nu_{i_0}^{i_*}), (x_z^{i_0}, \nu_{i_0}^z), (x_z^{j_0}, \nu_{j_0}^z), (x_{j_*}^{j_0}, \nu_{j_0}^{j_*})\} \cup \{(x_i^i, \nu_i^i)_{i\neq i_0, j_0}\}$$

of X_S constructed in the proof of (23.18). □

Remark 23.20. *The formulas in last lemma are reduced to:* $\rho(x_{i_*}^{i_0}) = 0 = \rho(x_z^{j_0})$; $\rho(x_z^{i_0}) = x_{i_*}^{i_0} \otimes \gamma_{\overline{\alpha}_1}$; $\rho(x_{j_*}^{j_0}) = x_z^{j_0} \otimes \gamma_{\overline{\alpha}_2}$; $\lambda(\nu_{i_0}^z) = 0 = \lambda(\nu_{j_0}^{j_*})$; $\lambda(\nu_{i_0}^{i_*}) = \gamma_{\overline{\alpha}_1} \otimes \nu_{i_0}^z$; *and* $\lambda(\nu_{j_0}^z) = \gamma_{\overline{\alpha}_2} \otimes \nu_{j_0}^{j_*}$. *Moreover, we clearly have that* $\rho(x_i^i) = 0$ *and* $\lambda(\nu_i^i) = 0$, *for all vertex* $i \notin \{i_0, j_0\}$ *of* $\mathbb{B}$. *All these values are necessary for the explicit computation of the differential* δ^e *of* $\mathcal{A}^e$. *For example, if* $\beta : i_0 \longrightarrow j_0$ *is an arrow of* $\mathcal{A}$ *with trivial differential, then* $\delta^e(\nu_{j_0}^{j_*} \otimes \beta \otimes x_{i_*}^{i_0}) = 0$.

The description of ρ *and* λ *using the quiver* $\widetilde{Q}$ *provides a helpful geometric way to recall their values. This type of description can be applied to other more complex reduction operations* $\mathcal{A} \mapsto \mathcal{A}^X$, *where the pictorial holder is really helpful. This will be the case for the operation called unravelling of a loop described below.*

As for the edge reduction, we need some preliminary results before introducing the operation *unravelling of a loop*.

Lemma 23.21. *Consider the path algebra* kQ/I, *defined by the following quiver*

$$Q : 1 \underset{\beta_1}{\overset{\alpha_1}{\rightleftarrows}} 2 \underset{\beta_2}{\overset{\alpha_2}{\rightleftarrows}} 3 \underset{\beta_3}{\overset{\alpha_3}{\rightleftarrows}} 4 \cdots n-2 \underset{\beta_{n-2}}{\overset{\alpha_{n-2}}{\rightleftarrows}} n-1 \underset{\beta_{n-1}}{\overset{\alpha_{n-1}}{\rightleftarrows}} n,$$

and the ideal I *generated by the relations*

$$\alpha_1\beta_1 = 0 \text{ and } \beta_i\alpha_i = \alpha_{i+1}\beta_{i+1}, \text{ for } i \in [1, n-1].$$

This makes sense for $n \geq 3$. *For* $n = 2$, *we mean that* I *is generated only by* $\alpha_1\beta_1 = 0$; *and, for* $n = 1$, *we mean* $I = 0$. *Then, the algebra* kQ/I *admits the non-zero classes of paths as a vector space basis, it has dimension* $d = \sum_{i=1}^n [(2n+1)i - i^2]/2$.

Proof. In order to describe all the classes of paths of the quiver Q modulo the ideal I, consider the following picture

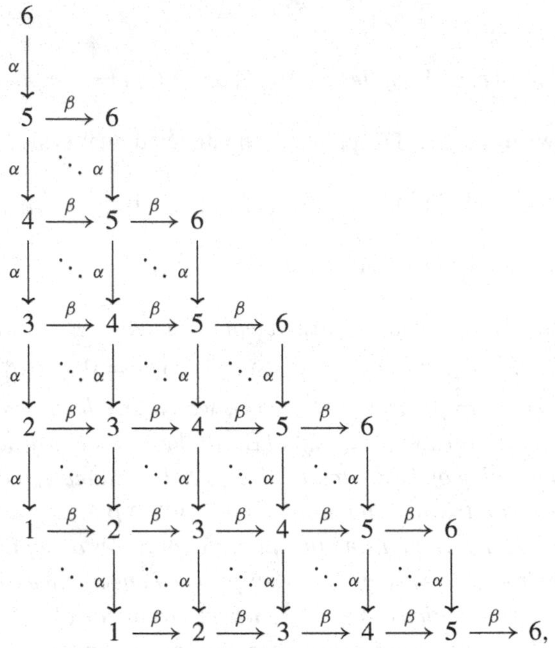

where we made the picture for $n = 6$; we have suppressed the subindexes of the arrows for simplicity; the dotted diagonals denote commutativity relations, with exception of the left bottom corner, which corresponds to the zero relation. Then, it is clear that the non-zero classes of paths (modulo I) which start at the vertex i are represented by

$$\begin{array}{l} e_i \\ \beta, \beta^2, \ldots, \beta^{n-i}, \\ \alpha, \beta\alpha, \beta^2\alpha, \ldots, \beta^{n-i+1}\alpha, \\ \alpha^2, \beta\alpha^2, \beta^2\alpha^2, \ldots, \beta^{n-i+2}\alpha^2, \\ \cdots \\ \alpha^{i-1}, \beta\alpha^{i-1}, \beta^2\alpha^{i-1}, \ldots, \beta^{n-1}\alpha^{i-1}. \end{array}$$

Then, they are exactly $[(2n+1)i - i^2]/2$. Since each power J^t of the radical J of kQ/I is the vector space generated by the classes of paths of length t, it is clear that the paths listed above are all linearly independent. Then, we can form a k-basis for kQ/I by considering all these classes of paths for all possible vertices i. Then, we get the formula given in the statement of the lemma for the dimension of this algebra. □

Lemma 23.22. *Consider the algebra $H := k[x]_{f(x)}$, where $f(x)$ is any nonzero polynomial of $k[x]$. Then, for $\lambda \in k$ with $f(\lambda) \neq 0$, we have the irreducible $\pi = x - \lambda \in H$. Now, for $n \geq 1$, consider the endomorphism algebra*

$$\Gamma = \Gamma(\lambda, n) = \mathrm{End}_H(H/H\pi^1 \oplus H/H\pi^2 \oplus \cdots \oplus H/H\pi^n).$$

Then, $\Gamma \cong kQ/I$, where Q and I are defined in (23.21).

The H-module $X = X_\lambda = H/H\pi^1 \oplus H/H\pi^2 \oplus \cdots \oplus H/H\pi^n$ has a natural structure of an H-Γ^{op}-bimodule. Denote by f_a the idempotent of Γ associated to the direct summand $H/H\pi^a$. Then, $\Gamma^{op} = S \oplus P$, where $S = \oplus_{a=1}^n kf_a$ is a trivial k-algebra and P is the radical of Γ^{op}. The nilpotency index of P is $\ell_P + 1$, where $\ell_P = 2(n-1)$. Consider the filtration of P given by its powers

$$0 = P^{(\ell_P+1)} \subseteq P^{(\ell_P)} \subseteq \cdots \subseteq P^{(1)} = P,$$

make $P_j := P^{(\ell_P - j+1)}$, for all $j \in [0, \ell_P]$, to obtain the S-S-bimodule filtration

$$\mathcal{F}(P): \qquad 0 = P_0 \subseteq P_1 \subseteq \cdots \subseteq P_{\ell_P - 1} \subseteq P_{\ell_P} = P,$$

and the H-S-bimodule filtration

$$\mathcal{F}(X): \qquad 0 = [X]_0 \subseteq [X]_1 \subseteq \cdots \subseteq [X]_{\ell_X} = X,$$

where $\ell_X = \ell_P + 1$ and $[X]_i = XP_i$, for $i \in [0, \ell_P]$.

Denote by $x_a^t = \underline{\pi}^t$ the image of π^t in $H/H\pi^a = \oplus_{j=0}^{a-1} k\underline{\pi}^j$ under the canonical projection $H \longrightarrow H/H\pi^a$. Then, $\{x_a^t \mid 1 \leq a \leq n, 0 \leq t \leq a-1\}$ is a directed basis of the vector space X. Moreover, the height of each of these basis vectors is given, for each $a \in [1, n]$ and $t \in [0, a-1]$, by

$$\mathbb{h}(x_a^t) = n + a - 2t - 1.$$

In the next array, we list in each column i the basic elements of $H_i = H/H\pi^i$, and at the end of each row j the common height of the elements of the *row j*

H_1	H_2	H_3	H_4	$\cdots$	H_{n-3}	H_{n-2}	H_{n-1}	H_n	$\mathbb{h}$
								x_n^0	$2n-1$
							x_{n-1}^0		$2n-2$
						x_{n-2}^0		x_n^1	$2n-3$
					x_{n-3}^0		x_{n-1}^1		$2n-4$
				$\iddots$	$\vdots$		$\vdots$		$\vdots$
			x_4^0						$n+3$
		x_3^0		$\cdots$					$n+2$
	x_2^0		x_4^1						$n+1$
x_1^0		x_3^1		$\cdots$					n
	x_2^1		x_4^2						$n-1$
		x_3^2		$\cdots$					$n-2$
			x_4^3						$n-3$
				$\ddots$	$\vdots$		$\vdots$		$\vdots$
					x_{n-3}^{n-4}		x_{n-1}^{n-3}		4
						x_{n-2}^{n-3}		x_n^{n-2}	3
							x_{n-1}^{n-2}		2
								x_n^{n-1}	1

Denote by p_ξ the element of Γ corresponding to each non-trivial path class ξ of kQ/I, as in Lemma (23.21). Then, $\{p_\xi\}_\xi$ is a directed vector space basis for the S-S-bimodule P. From (23.11) we know that there is a directed dual basis $(p_\xi, \gamma_\xi)_\xi$ of the right S-module P, which is compatible with the filtration $\mathcal{F}(P)$. Moreover, $\mathbb{h}(\gamma_\xi) = \ell(\xi)$, where $\ell(\xi)$ denotes the common length of all the paths in ξ. Also, $s(\gamma_\xi) = s(\xi)$ and $t(\gamma_\xi) = t(\xi)$, where $s(\xi)$ and $t(\xi)$ denote the source and target points of the path class ξ.

Proof. Write $H_i := H/H\pi^i$, for $i \in [1, n]$. Whenever we refer to the element $H_r \xrightarrow{\tau} H_s$ of Γ, we mean the element $\oplus_i H_i \longrightarrow H_r \xrightarrow{\tau} H_s \longrightarrow \oplus_i H_i$. Consider the morphism of algebras $\varphi : kQ \longrightarrow \Gamma$ defined by mapping: for $i \in [1, n]$, $e_i \mapsto 1_{H_i} : H_i \longrightarrow H_i$; for $i \in [1, n-1]$, $\alpha_i \mapsto \mu_1 : H_{i+1} \longrightarrow H_i$, induced by the identity map 1_H, and, $\beta_i \mapsto \mu_\pi : H_i \longrightarrow H_{i+1}$ induced by multiplication by π. Since the relations

$$[H_1 \xrightarrow{\mu_\pi} H_2 \xrightarrow{\mu_1} H_1] = 0 \text{ and } [H_i \xrightarrow{\mu_\pi} H_{i+1} \xrightarrow{\mu_1} H_i] = [H_i \xrightarrow{\mu_1} H_{i-1} \xrightarrow{\mu_\pi} H_i],$$

are satisfied, there is a morphism of algebras $\overline{\varphi} : kQ/I \longrightarrow \Gamma$. We shall see that $\overline{\varphi}$ is an isomorphism verifying that it is a surjective map between finite-dimensional k-algebras with the same dimension.

Let us first explore the vector space structure of H_i and of Γ. For $i \in [1, n]$, we have a canonical embedding $\theta : k[x]/k[x]\pi^i \longrightarrow H/H\pi^i$. Since $f(x)$ and π^i are relatively prime in $k[x]$, the class of $f(x)$ modulo $k[x]\pi^i$ is invertible in $k[x]/k[x]\pi^i$. Hence, θ is in fact an isomorphism of k-algebras and $H_i = \oplus_{s=0}^{i-1} k\underline{\pi}^s$.

Moreover, for $i, j \in [1, n]$, the map sending each morphism $g : H_i \longrightarrow H_j$ onto $g(1)$ induces isomorphisms of H-modules

$$\mathrm{Hom}_H(H_i, H_j) \cong \begin{cases} H_j & \text{if } j \leq i; \\ \pi^{j-i} H_j & \text{if } j > i. \end{cases}$$

This implies that we have vector space decompositions

$$\mathrm{Hom}_H(H_i, H_j) = \begin{cases} \oplus_{s=0}^{j-1} k\mu_{\pi^s} & \text{if } j \leq i; \\ \oplus_{s=j-i}^{j-1} k\mu_{\pi^s} & \text{if } j > i, \end{cases}$$

where $\mu_{\pi^s} : H_i \longrightarrow H_j$ denotes the morphism induced by multiplication by π^s. In particular, $\dim_k \Gamma = \sum_{i=1}^n \sum_{j=1}^n \dim_k \mathrm{Hom}_H(H_i, H_j) = \sum_{i=1}^n \sum_{j=1}^n \min\{i, j\}$. Then, adding the content of each column i of the $n \times n$ matrix $(\min\{i, j\})$, we obtain

$$\dim_k \Gamma = \sum_{i=1}^{n} [(1 + 2 + \cdots + i) + i(n - i)] = \sum_{i=1}^{n} [(2n + 1)i - i^2]/2 = d.$$

In order to show that $\overline{\varphi}$ is surjective, we exhibit a natural family of generators of the k-algebra Γ which lies in the image of φ. If $j < i$ and $0 \leq s \leq j - 1$, we have the factorization

$$[H_i \xrightarrow{\mu_{\pi^s}} H_j] = [H_i \xrightarrow{\mu_{\pi^s}} H_i \xrightarrow{\mu_1} H_j];$$

if $j > i$ and $j - i \leq s \leq j - 1$, we have the factorization

$$[H_i \xrightarrow{\mu_{\pi^s}} H_j] = [H_i \xrightarrow{\mu_{\pi^{j-i}}} H_j \xrightarrow{\mu_{\pi^{s-(j-i)}}} H_j];$$

finally, if $j = i$ and $1 \leq s \leq i - 1$, we have the factorization

$$[H_i \xrightarrow{\mu_{\pi^s}} H_i] = [H_i \xrightarrow{\mu_\pi} H_i]^s.$$

Therefore, the following family generates the k-algebra Γ

$$\{H_i \xrightarrow{\mu_1} H_j\}_{j<i} \cup \{H_i \xrightarrow{\mu_{\pi^{j-i}}} H_j\}_{j>i} \cup \{H_i \xrightarrow{\mu_\pi} H_i\}_{i>1} \cup \{H_i \xrightarrow{\mu_1} H_i\}_i.$$

Clearly, this family coincides with

$$\{\varphi(\alpha_j \cdots \alpha_{i-2}\alpha_{i-1})\}_{j<i} \cup \{\varphi(\beta_{j-1} \cdots \beta_{i+1}\beta_i)\}_{j>i} \cup \{\varphi(\beta_{i-1}\alpha_{i-1})\}_{i>1} \cup \{\varphi(e_i)\}_i.$$

It follows that $\overline{\varphi} : kQ/I \longrightarrow \Gamma$ is an isomorphism of algebras.

Given a non-trivial path class ξ of kQ/I, make $p_\xi := \overline{\varphi}(\xi) \in P \subseteq \Gamma^{op}$. Then, if $\xi = \xi(\gamma)$ is the path class of any non-trivial path γ of Q and $x \in X$, we have

$$xp_\xi = \overline{\varphi}(\xi(\gamma))[x] = \varphi(\gamma)[x].$$

Define the *depth* of a non-zero element $x \in X$ as $\mathbb{d}(x) = i$ iff $x \in XP^{(i)} \setminus XP^{(i+1)}$, thus we have the map $\mathbb{d} : X \longrightarrow [1, \ell_P]$. Since $[X]_t = XP_t = XP^{(\ell_P - t+1)}$, for any $t \in [1, \ell_P]$, we have that $\mathbb{d}(x) = t$ iff $\mathbb{h}(x) = \ell_P - t + 1$.

We know that $\{x_b^t\}_{b,t}$ is a basis of the k-vector space X and that the set of elements $p_\xi \in P$, where ξ runs in the set of path classes of length $\geq r$, constitutes a basis for the vector space $P^{(r)}$. Thus, whenever $x \in XP^{(r)}$, then $x = \sum_{b,t,\xi} c_b^t(\xi) x_b^t p_\xi$, where $c_b^t(\xi) \in k$ and the indexes run as follows: $b \in [1, n]$, $t \in [0, b-1]$ and ξ is a path class in kQ/I with length $\geq r$. Given any path γ of Q, we shall denote by $\beta(\gamma)$ the number of times arrows of type β_i appear in γ. For instance, $\beta(\beta_2\beta_1\alpha_1\alpha_2\beta_2) = 3$. The map β induces a map on the path classes of kQ/I by the rule $\beta(\xi) = \beta(\gamma)$, whenever $\gamma = \xi(\gamma)$.

If $a < n$, then

$$\begin{aligned} x_a^0 &= \varphi(\alpha_a \alpha_{a+1} \cdots \alpha_{n-1})[x_n^0] \\ &= x_n^0 p_{\xi(\alpha_a\alpha_{a+1}\cdots\alpha_{n-1})} \in XP^{(n-a)}. \end{aligned}$$

If $x_a^0 \in XP^{(n-a+1)}$, then $x_a^0 = \sum_{b,t,\xi} c_b^t(\xi) x_b^t p_\xi$, where ξ runs in the set of path classes of kQ/I with length $\geq n - a + 1$. Since every path γ of Q that ends at a of length $\geq n - a + 1$ satisfies that $\beta(\gamma) \neq 0$, then each $x_b^t p_{\xi(\gamma)}$ is either zero or of the form x_a^t with $t > 0$. A contradiction. It follows that $x_a^0 \in XP^{(n-a)} \setminus XP^{(n-a+1)}$, that is $\mathbb{d}(x_a^0) = n - a$. Therefore, $\mathbb{h}(x_a^0) = \ell_P - n + a + 1 = n + a - 1$. Similarly, $x_n^0 \notin XP$. Hence, $\mathbb{h}(x_n^0) = \ell_X = 2(n-1) + 1 = 2n - 1$.

Moreover, notice that for $1 \leq a \leq n$ and $1 \leq t \leq a - 1$

$$\begin{aligned} x_a^t &= \varphi(\beta_{a-1} \cdots \beta_{a-t}\alpha_{a-t} \cdots \alpha_{n-1})[x_n^0] \\ &= x_n^0 p_{\xi(\beta_{a-1}\cdots\beta_{a-t}\alpha_{a-t}\cdots\alpha_{n-1})} \in XP^{(n+2t-a)}; \end{aligned}$$

the path involved in this equation has length $n + 2t - a$ and it can be seen in the following subquiver of Q

$$a - t \underset{\beta_{a-t}}{\overset{\alpha_{a-t}}{\leftrightarrows}} \cdots a - 1 \underset{\beta_{a-1}}{\overset{\alpha_{a-1}}{\leftrightarrows}} a \underset{\beta_a}{\overset{\alpha_a}{\leftrightarrows}} \cdots n - 1 \underset{\beta_{n-1}}{\overset{\alpha_{n-1}}{\leftrightarrows}} n.$$

If $x_a^t \in XP^{(n+2t-a+1)}$, then $x_a^t = \sum_{b,t',\xi} c_b^{t'}(\xi) x_b^{t'} p_\xi$, where ξ runs on the set of path classes of kQ/I with length $\geq n+2t-a+1$. Since every path γ of Q that ends at a of length $\geq n+2t-a+1$ satisfies that $\beta(\gamma) > t$, then each $x_b^{t'} p_{\xi(\gamma)}$ is either zero or of the form $x_a^{t''}$ with $t'' > t$. A contradiction. It follows that $x_a^t \in XP^{(n+2t-a)} \setminus XP^{(n+2t-a+1)}$, that is $\mathbb{d}(x_a^t) = n+2t-a$ and, therefore, $\mathbb{h}(x_a^t) = \ell_P - \mathbb{d}(x_a^t) + 1 = n+a-2t-1$, as claimed.

For the last part, notice that for any non-zero and non-trivial path class ξ of kQ/I, $\mathbb{h}(p_\xi) = \ell_P - \ell(\xi) + 1$, because $p_\xi \in P^{(\ell(\xi))} \setminus P^{(\ell(\xi)+1)}$. Thus, $\mathbb{h}(\gamma_\xi) = \ell_P + 1 - \mathbb{h}(p_\xi) = \ell(\xi)$, as claimed. If $\xi : i \to j$ in kQ/I, then $p_\xi : j \longrightarrow i$ in $P \subseteq \Gamma^{op}$. Therefore, since the dual basis is directed, $\gamma_\xi : i \longrightarrow j$. □

Lemma 23.23. *Unravelling: Assume that $\mathcal{A}$ is a nested (resp. seminested) ditalgebra with layer (R, W), admitting basis $(\mathbb{B}_0, \mathbb{B}_1)$. Assume that $Re_{i_0} = k[x]_{f(x)}$. Assume that $n \in \mathbb{N}$ and $\lambda_1, \ldots, \lambda_q$ are distinct elements in k with $f(\lambda_j) \neq 0$, for all j. Define $g(x) = (x-\lambda_1)(x-\lambda_2)\cdots(x-\lambda_q) \in k[x]$. Make $H := k[x]_{f(x)}$ and $\pi_s := x - \lambda_s$, for $s \in [1, q]$. Consider the Re_{i_0}-module*

$$X_1 = \oplus_{s=1}^q \left[H/H\pi_s^1 \oplus H/H\pi_s^2 \oplus \cdots \oplus H/H\pi_s^n\right],$$

$X_2 := (e_{i_0}R)_{g(x)} \cong H_{g(x)}$, $X_3 := (1-e_{i_0})R$, and make $X = X_1 \oplus X_2 \oplus X_3$. Consider R as an initial subditalgebra of $\mathcal{A}$, that is the subditalgebra $\mathcal{A}'$ of $\mathcal{A}$ corresponding to the trivial R-R-bimodule decompositions $W_0 = 0 \oplus W_0''$ and $W_1 = 0 \oplus W_1''$. Thus, $B = R$ and X has a natural structure of R-module. Then, X is an admissible $\mathcal{A}'$-module. The ditalgebra obtained from $\mathcal{A}$ by unravelling at i_0, using $\lambda_1, \ldots, \lambda_q$ and n, is $\mathcal{A}^u = \mathcal{A}^X$. It is a nested (resp. seminested) ditalgebra.

Its bigraph $\mathbb{B}^u$ can be described in terms of the bigraph $\mathbb{B}$ of $\mathcal{A}$ as follows. The set $\mathcal{P}^u$ of points has "old" points $\mathcal{P} \setminus \{i_0\}$ and the "new" points i_* and $\{i_{s,a} \mid 1 \leq s \leq q, 1 \leq a \leq n\}$; the marked points are given by $\mathcal{P}^u_{\mathrm{mk}} = [\mathcal{P}_{\mathrm{mk}} \setminus \{i_0\}] \cup \{i_*\}$; the marked loops $\mathbb{B}^u_{\mathrm{mk}} = [\mathbb{B}_{\mathrm{mk}} \setminus \{\rho_{i_0}\}] \cup \{\rho_{i_*}\}$; the mark map mk^u satisfies $\mathrm{mk}^u(i_*) = f(x)g(x)$ and coincides with mk on the old marked points. The solid arrows of $\mathbb{B}_0^u$ are produced by solid arrows of $\mathbb{B}_0$ by the following rules. If $\beta : i \to j$ in $\mathbb{B}_0$, with $i, j \neq i_0$, then in $\mathbb{B}_0^u$ we have the arrow $v_j^j \otimes \beta \otimes x_i^i : i \longrightarrow j$. If $\beta : i \to i_0$ in $\mathbb{B}_0$, with $i \neq i_0$, then in $\mathbb{B}_0^u$ we have the arrow $v_{i_0}^{i_*} \otimes \beta \otimes x_i^i : i \longrightarrow i_*$, and the $q \times n(n+1)/2$ arrows $v_t^{s,a} \otimes \beta \otimes x_i^i : i \longrightarrow i_{s,a}$, for $0 \leq t \leq a-1$. If $\beta : i_0 \to j$ in $\mathbb{B}_0$, with $j \neq i_0$, then in $\mathbb{B}_0^u$ we have the arrow $v_j^j \otimes \beta \otimes x_{i_*}^{i_0} : i_* \longrightarrow j$, and the $q \times n(n+1)/2$ arrows $v_j^j \otimes \beta \otimes x_{s,a}^t : i_{s,a} \longrightarrow j$, for $0 \leq t \leq a-1$. If $\beta : i_0 \to i_0$ in $\mathbb{B}_0$, then in $\mathbb{B}_0^u$ we have the arrow $v_{i_0}^{i_*} \otimes \beta \otimes x_{i_*}^{i_0} : i_* \longrightarrow i_*$, and the $q^2 \times [n(n+1)/2]^2$ arrows $v_{t'}^{s',a'} \otimes \beta \otimes x_{s,a}^t : i_{s,a} \longrightarrow i_{s',a'}$, for $0 \leq t \leq a-1$ and $0 \leq t' \leq a'-1$; the $q \times n(n+1)/2$ arrows $v_t^{s,a} \otimes \beta \otimes x_{i_*}^{i_0} : i_* \longrightarrow i_{s,a}$, for $0 \leq t \leq a-1$; and the $q \times n(n+1)/2$

arrows $v_{i_0}^{i_*} \otimes \beta \otimes x_{s,a}^t : i_{s,a} \longrightarrow i_*$, *for* $0 \leq t \leq a-1$. *The height of each one of these arrows* $v_y \otimes \beta \otimes x$, *described above, where* $v_y \in X^*$ *denotes the element corresponding to* $y \in X$ *in the fixed directed dual basis of* X_S, *can be computed with the formula* $\mathbb{h}(v_y \otimes \beta \otimes x) = 2\ell_X \mathbb{h}(\beta) + \mathbb{h}(v_y) + \mathbb{h}(x) = 2\ell_X \mathbb{h}(\beta) + \ell_X + 1 - \mathbb{h}(y) + \mathbb{h}(x) = (4n-2)\mathbb{h}(\beta) - \mathbb{h}(y) + \mathbb{h}(x) + 2n - 1$.

There are two types of dotted arrows in $\mathbb{B}_1^u$: *the first type consists of arrows*

$$\gamma_{s,\xi} : i_{s,s(\xi)} \longrightarrow i_{s,t(\xi)} \text{ with } \mathbb{h}(\gamma_{s,\xi}) = \ell(\xi),$$

one for each $s \in [1, q]$ *and each non-zero and non-trivial path class* ξ *of the algebra* kQ/I *(as in (23.21)); the other type of dotted arrows are constructed from the dotted arrows of* $\mathbb{B}_1$ *in the same way described for the solid arrows. The height of this type of dotted arrows is computed as in the case of solid arrows, but we have to add* $\ell_P = 2n-2$ *to the right of the height formula given in the description of the solid case.*

Proof. In the following, we naturally identify $\mathcal{A}'$-Mod with R-Mod. From the proof of (23.22) we know that, for $E_s := \mathrm{End}(H/H\pi^s)^{op} \cong H/H\pi^s$, we have $E_s = k1 \oplus \mathrm{rad}E_s$. Then, we have a splitting $\Gamma_1 := \mathrm{End}_R(X_1)^{op} = S_1 \oplus P_1$, where P_1 is the radical of Γ_1 and X_1 is an admissible $\mathcal{A}'$-module; we also have that $\Gamma_2 := \mathrm{End}_R(X_2)^{op} = S_2 \oplus P_2$ and $\Gamma_3 := \mathrm{End}_R(X_3)^{op} = S_3 \oplus P_3$, with $S_2 \cong (Re_{i_0})_{g(x)}$, $S_3 \cong (1-e_{i_0})R$, $P_2 = 0$ and $P_3 = 0$. We claim that $\mathrm{Hom}_R(\mathcal{I}_{X_i}, \mathcal{I}_{X_j}) = 0$, for all ordered pairs (i, j) with $i \neq j$. Since $e_{i_0}X = X_1 \oplus X_2$ and $(1-e_{i_0})X = X_3$, this is clear for the pairs $(1, 3)$, $(3, 1)$, $(2, 3)$, $(3, 2)$. If $M, N \in S$-Mod, a morphism of R-modules

$$H/H\pi_s^i \otimes_S M \xrightarrow{f} X_2 \otimes_S N \cong H_{g(x)} \otimes_S N$$

satisfies $\pi_s^i f(1 \otimes m) = f[\pi_s^i(1 \otimes m)] = f(\pi_s^i \otimes m) = f(0) = 0$, but, since π_s is invertible in $H_{g(x)}$, we have that $f(1 \otimes m) = 0$, hence $f = 0$. Thus, the claim holds for the pair $(1, 2)$ too. For the pair $(2, 1)$, the argument is similar. Thus, we have the splitting $\Gamma := \mathrm{End}_R(X)^{op} = S \oplus P$, where S and P can be identified with $S_1 \times S_2 \times S_3 = S_1 \times (e_{i_0}R)_{g(x)} \times (1-e_{i_0})R$ and $P = P_1 \times 0 \times 0$, as in (17.11).

Since $\mathrm{Hom}_H(H/H\pi_s^t, H/H\pi_{s'}^{t'}) = 0$ whenever $s \neq s'$, for $1 \leq t, t' \leq n$ and $1 \leq s, s' \leq q$, then $\mathrm{End}_R(X_1) \cong \prod_{s=1}^q \mathrm{End}_H(X_{\lambda_s})$, where $X_{\lambda_s} = H/H\pi_s^1 \oplus \cdots \oplus H/H\pi_s^n$, for each s. Hence, by (23.22), we know that $\mathrm{End}_R(X_1)$ is the path algebra of q disjoint quivers as described in (23.21). Thus the filtration of P given by the powers of the radical has the form

$$0 = P^{(\ell_P+1)} \subseteq P^{(\ell_P)} \subseteq \cdots \subseteq P^{(1)} = P,$$

where $\ell_P := 2(n-1)$. Thus, making $P_j := P^{(\ell_P - j+1)}$, for all $j \in [0, \ell_P]$, we obtain a right additive S-S-bimodule filtration

$$\mathcal{F}(P): \qquad 0 = P_0 \subseteq P_1 \subseteq \cdots \subseteq P_{\ell_P - 1} \subseteq P_{\ell_P} = P,$$

and, from this, we get the filtration

$$\mathcal{F}(X): 0 = [X]_0 \subseteq [X]_1 \subseteq \cdots \subseteq [X]_{\ell_P} \subseteq [X]_{\ell_X} = X,$$

where $[X]_i = XP_i$, for $i \in [0, \ell_P]$, and $\ell_X = \ell_P + 1$.

Write f_i, f_{i_*}, $f_{i_{s,a}}$ for the idempotents of Γ corresponding to the direct summands $e_i R \subseteq (1 - e_{i_0})R = X_2$, $e_{i_0} R_{g(x)} = X_3$ and $H/H\pi_s^a \subseteq X_{\lambda_s} \subseteq X_1$ of X.

From (23.22), for each $s \in [1, q]$, we can choose a directed basis $\{x_{s,a}^t \mid 1 \leq a \leq n, 1 \leq t \leq a-1\}$ of X_{λ_s}, compatible with the corresponding filtration $\mathcal{F}(X_{\lambda_s})$ given there. Since $[X_1]_i = [X_{\lambda_1}]_i \oplus \cdots \oplus [X_{\lambda_q}]_i$, for each $0 \leq i \leq \ell_{X_1} = \ell_{X_{\lambda_s}}$, then $\{x_{s,a}^t \mid 1 \leq a \leq n, 1 \leq t \leq a-1, 1 \leq s \leq q\}$ is a directed basis of X_1, compatible with $\mathcal{F}(X_1)$. Here, $x_{s,a}^t \in X_{\lambda_s} f_{i_{s,a}} \subseteq X f_{i_{s,a}}$.

Now, for each $i \in \mathcal{P}\backslash\{i_0\}$, make $x_i^i := e_i \in (1 - e_{i_0})Re_i = X_3 f_i$, and complete to a directed dual basis (x_i^i, ν_i^i) of $X_3 f_i$ over $(1 - e_{i_0})R$. Then, define $x_{i_*}^{i_0} := e_{i_0} \in (e_{i_0} R)_{g(x)} = X_2 f_{i_*}$, and complete to a directed dual basis $(x_{i_*}^{i_0}, \nu_{i_0}^{i_*})$ of $X_2 f_{i_*}$ over $(e_{i_0} R)_{g(x)} \cong k[x]_{f(x)g(x)}$.

Then, there is a directed dual basis $(x_{s,a}^t, \nu_t^{s,a})_{s,a,t} \cup (x_i^i, \nu_i^i)_i \cup \{(x_{i_*}^{i_0}, \nu_{i_0}^{i_*})\}$ of X compatible with the filtration $\mathcal{F}(X)$. For the sake of simplicity, we have written with the same symbol some elements of the dual spaces and their extensions to elements in X^*. Here, $\mathbb{h}(x_i^i) = \ell_X$, for all i, as well as $\mathbb{h}(x_{i_*}^{i_0}) = \ell_X$. Thus, $\mathbb{h}(x_i^i) = 2n - 1 = \mathbb{h}(x_{i_*}^{i_0})$, while $\mathbb{h}(\nu_i^i) = 1 = \mathbb{h}(\nu_{i_0}^{i_*})$. The height of the other basic elements is given, for any s, as in (23.22), by $\mathbb{h}(x_{s,a}^t) = n + a - 2t - 1$; therefore, $\mathbb{h}(\nu_t^{s,a}) = (\ell_X + 1) - \mathbb{h}(x_{s,a}^t) = n + 2t - a + 1$.

Then, (23.13) gives us the description of $\mathcal{F}(\mathbb{B}_0^u) = \mathcal{F}(\mathbb{B}_0^X)$. Let us compute explicitely the height of each arrow: If $\beta : i \to j$ in $\mathbb{B}_0$, with $i, j \neq i_0$, then in $\mathbb{B}_0^u$ we have the arrow

$$\hat{\beta} := \nu_j^j \otimes \beta \otimes x_i^i : i \longrightarrow j,$$

with $\mathbb{h}(\hat{\beta}) = 2\ell_X \mathbb{h}(\beta) + \mathbb{h}(x_i^i) + \mathbb{h}(\nu_j^j) = (4n-2)\mathbb{h}(\beta) + (2n-1) + 1 = (4n-2)\mathbb{h}(\beta) + 2n$. If $\beta : i \to i_0$ in $\mathbb{B}_0$, with $i \neq i_0$, then in $\mathbb{B}_0^u$ we have the arrow

$$\beta^{i_*} := \nu_{i_0}^{i_*} \otimes \beta \otimes x_i^i : i \longrightarrow i_*,$$

with $\mathbb{h}(\beta^{i_*}) = 2\ell_X\mathbb{h}(\beta) + \mathbb{h}(x_i^i) + \mathbb{h}(v_{i_0}^{i_*}) = (4n-2)\mathbb{h}(\beta) + (2n-1) + 1 = (4n-2)\mathbb{h}(\beta) + 2n$; and the $q \times n(n+1)/2$ arrows

$$\beta_t^{s,a} := v_t^{s,a} \otimes \beta \otimes x_i^i : i \longrightarrow i_{s,a},$$

with $\mathbb{h}(\beta_t^{s,a}) = 2\ell_X\mathbb{h}(\beta) + \mathbb{h}(x_i^i) + \mathbb{h}(v_t^{s,a}) = (4n-2)\mathbb{h}(\beta) + (2n-1) + n + 2t - a + 1 = (4n-2)\mathbb{h}(\beta) + 3n + 2t - a$, for $0 \leq t \leq a-1$. If $\beta : i_0 \to j$ in $\mathbb{B}_0$, with $j \neq i_0$, then in $\mathbb{B}_0^u$ we have the arrow

$$\beta_{i_*} := v_j^j \otimes \beta \otimes x_{i_*}^{i_0} : i_* \longrightarrow j,$$

with $\mathbb{h}(\beta_{i_*}) = 2\ell_X\mathbb{h}(\beta) + \mathbb{h}(x_{i_*}^{i_0}) + \mathbb{h}(v_j^j) = (4n-2)\mathbb{h}(\beta) + (2n-1) + 1 = (4n-2)\mathbb{h}(\beta) + 2n$; and the $q \times n(n+1)/2$ arrows

$$\beta_{s,a}^t := v_j^j \otimes \beta \otimes x_{s,a}^t : i_{s,a} \longrightarrow j,$$

with $\mathbb{h}(\beta_{s,a}^t) = 2\ell_X\mathbb{h}(\beta) + \mathbb{h}(x_{s,a}^t) + \mathbb{h}(v_j^j) = (4n-2)\mathbb{h}(\beta) + (n + a - 2t - 1) + 1$, for $0 \leq t \leq a-1$. If $\beta : i_0 \to i_0$ in $\mathbb{B}_0$, then in $\mathbb{B}_0^u$ we have the arrow

$$\beta_{i_*}^{i_*} := v_{i_0}^{i_*} \otimes \beta \otimes x_{i_*}^{i_0} : i_* \longrightarrow i_*,$$

with $\mathbb{h}(\beta_{i_*}^{i_*}) = 2\ell_X\mathbb{h}(\beta) + \mathbb{h}(x_{i_*}^{i_0}) + \mathbb{h}(v_{i_0}^{i_*}) = (4n-2)\mathbb{h}(\beta) + (2n-1) + 1 = (4n-2)\mathbb{h}(\beta) + 2n$; the $q^2 \times [n(n+1)/2]^2$ arrows

$$\beta_{s,a,s',a'}^{t,t'} := v_{t'}^{s',a'} \otimes \beta \otimes x_{s,a}^t : i_{s,a} \longrightarrow i_{s',a'},$$

with $\mathbb{h}(\beta_{s,a,s',a'}^{t,t'}) = 2\ell_X\mathbb{h}(\beta) + \mathbb{h}(x_{s,a}^t) + \mathbb{h}(v_{t'}^{s',a'}) = (4n-2)\mathbb{h}(\beta) + n + a - 2t - 1 + n - a' + 2t' + 1 = (4n-2)\mathbb{h}(\beta) + 2n + (a - a') + 2(t' - t)$, for $0 \leq t \leq a-1$ and $0 \leq t' \leq a'-1$; the $q \times n(n+1)/2$ arrows

$$\beta_{i_*,s,a}^t := v_t^{s,a} \otimes \beta \otimes x_{i_*}^{i_0} : i_* \longrightarrow i_{s,a}$$

with $\mathbb{h}(\beta_{i_*,s,a}^t) = 2\ell_X\mathbb{h}(\beta) + \mathbb{h}(x_{i_*}^{i_0}) + \mathbb{h}(v_t^{s,a}) = (4n-2)\mathbb{h}(\beta) + 3n + 2t - a$; and the $q \times n(n+1)/2$ arrows

$$\beta_{s,a,i_*}^t := v_{i_0}^{i_*} \otimes \beta \otimes x_{s,a}^t : i_{s,a} \longrightarrow i_*$$

with $\mathbb{h}(\beta_{s,a,i_*}^t) = 2\ell_X\mathbb{h}(\beta) + \mathbb{h}(x_{s,a}^t) + \mathbb{h}(v_{i_0}^{i_*}) = (4n-2)\mathbb{h}(\beta) + n + a - 2t$.

We have seen that $\mathrm{End}_R(X_1) \cong \prod_{s=1}^q \mathrm{End}_H(X_{\lambda_s})$, therefore P can be identified with $P_{\lambda_1} \times \cdots \times P_{\lambda_s}$, and S_1 with $S_{\lambda_1} \times \cdots \times S_{\lambda_q}$, where $\mathrm{End}_H(X_{\lambda_s})^{op} = S_{\lambda_s} \oplus P_{\lambda_s}$ is a splitting over the radical for each $s \in [1, q]$. Lemma (23.22) can be applied to each one of these q algebras to construct directed dual basis $(p_{s,\xi}, \gamma_{s,\xi})_\xi$ of each right S_{λ_s}-module P_{λ_s} compatible with the canonical filtration $\mathcal{F}(P_{\lambda_s})$. All these basis can be extended to a directed dual basis $(p_{s,\xi}, \gamma_{s,\xi})_{s,\xi}$ of the right S-module P, compatible with the canonical filtration $\mathcal{F}(P)$, as wanted. Again, we are using the same symbol to denote an element

$p_{s,\xi}$ of P_{λ_s} and the element it determines canonically in P, and, dually, we use the same symbol to denote the element $\gamma_{s,\xi}$ in $P^*_{\lambda_s}$ and the element in P^* it determines. The formulas for the height follow immediately from (23.22). From (23.13) we know that P^* is freely generated by $\{\gamma_{s,\xi}\}_{s,\xi}$. □

Lemma 23.24. *Assume $\mathcal{A}^u$ is the ditalgebra obtained from $\mathcal{A}$ by unravelling at i_0, and consider the associated bimodule $X = X_1 \oplus X_2 \oplus X_3$, with $X_1 = X_{\lambda_1} \oplus \cdots \oplus X_{\lambda_q}$, as in the last lemma. In the proof of the last lemma, the dual basis $(p_{s,\xi}, \gamma_{s,\xi})$ of the radical P of $\Gamma = \mathrm{End}_{\mathcal{A}'}(X)^{op}$ is obtained from the family of isomorphisms $\mathrm{End}_H(X_{\lambda_s}) \cong kQ/I$ described in (23.22). Thus, each X_{λ_s} is a right $(kQ/I)^{op}$-module, and so is $X_1 = \oplus_{s\in[1,q]} X_{\lambda_s}$. In particular, for each non-zero non-trivial path class ξ of kQ/I, $s \in [1,q]$ and $x \in X_{\lambda_s}$, we have the element $p_{s,\xi} \in P$ and the formula $x\xi = xp_{s,\xi}$. Consider the quiver $\widetilde{Q}$, with points set $\cup_{s=1}^{q} Z_{\lambda_s}$, where $Z_{\lambda_s} := \{x^t_{s,a} \mid a \in [1,n] \text{ and } t \in [0, a-1]\} \subseteq X_{\lambda_s}$, and arrows $\alpha^t_{s,a} : x^t_{s,a+1} \longrightarrow x^t_{s,a}$ and $\beta^{t-1}_{s,a} : x^{t-1}_{s,a} \longrightarrow x^t_{s,a+1}$, for $s \in [1,q]$, $a \in [1, n-1]$, and $t \in [1, a-1]$. Moreover, consider the ideal $\widetilde{I}$ of the path algebra $k\widetilde{Q}$ generated by the relations $\beta^{t-1}_{s,a}\alpha^{t-1}_{s,a} = \alpha^t_{s,a+1}\beta^{t-1}_{s,a+1}$, for all $s \in [1,q]$, $a \in [1, n-1]$ and $t \in [1, a-1]$, that is all possible commutativity relations. Then we have a quiver map $\pi : \widetilde{Q} \longrightarrow Q$, defined by $\pi(x^t_{s,a}) = a$, $\pi(\alpha^t_{s,a}) = \alpha_a$ and $\pi(\beta^t_{s,a}) = \beta_a$, which preserves the relations. We can visualize this map with the following figure, where the general typical connected component $\widetilde{Q}_{\lambda_s}$ (there is one for each $s \in [1,q]$), for the case $n = 4$, is depicted*

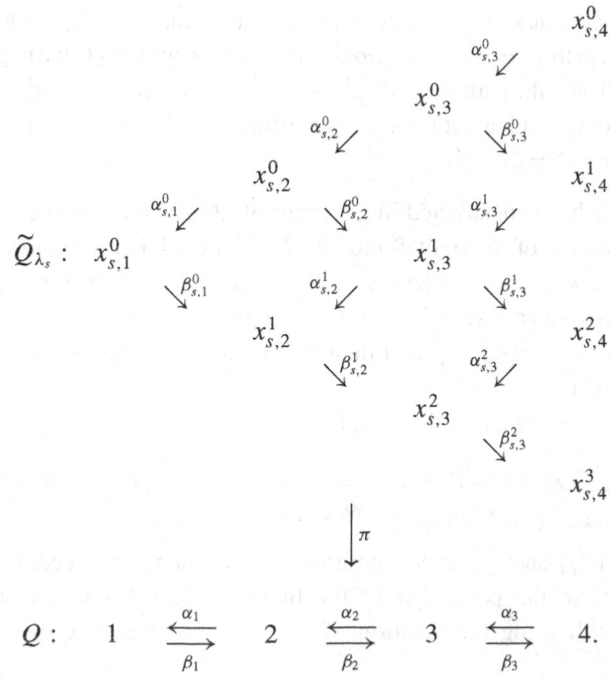

We consider a relation $<$ on the points of $\widetilde{Q}$: By definition, $x^t_{s,a} < x^{t'}_{s,a'}$ iff there is a non-zero non-trivial path class φ of $k\widetilde{Q}/\widetilde{I}$ with source point $x^t_{s,a}$ and target point $x^{t'}_{s,a'}$. Notice that in this situation the path class φ is unique: we will denote this class by $\varphi(x^t_{s,a}, x^{t'}_{s,a'})$. We shall write $\xi(x^t_{s,a}, x^{t'}_{s,a'}) := \pi(\varphi(x^t_{s,a}, x^{t'}_{s,a'}))$. Then, $z < z'$ iff $z' = z\xi$, for some non-zero non-trivial path class ξ of kQ/I, in the right $(kQ/I)^{op}$-module X_1; moreover, in this case, $\xi = \xi(z, z')$.

Furthermore, writing $\nu_{x^t_{s,a}} := \nu^{s,a}_t$, we have the following, for $s \in [1, q]$ and $z \in Z_{\lambda_s}$:

(1) $\rho(z) = \sum_{z<z'\in Z_{\lambda_s}} z' \otimes \gamma_{s,\xi(z,z')}$;
(2) $\lambda(\nu_z) = \sum_{z>z'\in Z_{\lambda_s}} \gamma_{s,\xi(z',z)} \otimes \nu_{z'}$.

In particular, $\mathbb{h}(z) = 1$ implies $\rho(z) = 0$, and $\mathbb{h}(\nu_z) = 1$ implies $\lambda(\nu_z) = 0$.

Proof. If $z < z'$, then $z' = z\xi(z, z')$ follows from the argument in the proof of (23.22). Indeed, assume $z = x^t_{s,a}$ and $z' = x^{t'}_{s',a'}$, and let us denote by $\varphi(\gamma)$ the path class in $k\widetilde{Q}/\widetilde{I}$ of the path γ of the quiver $\widetilde{Q}$. Then, $z < z'$ implies that $s = s'$ and $\varphi(z, z') = \varphi(\beta^{t'-1}_{s,a'-1} \cdots \beta^{t+1}_{s,i+1}\beta^t_{s,i}\alpha^t_{s,i}\alpha^t_{s,i+1} \cdots \alpha^t_{s,a-1})$, for some $i \in [1, n]$. Then, $\xi(z, z') = \xi(\beta_{a'-1} \cdots \beta_{i+1}\beta_i\alpha_i\alpha_{i+1} \cdots \alpha_{a-1})$. It follows that

$$\begin{aligned} z' = x^{t'}_{s,a'} &= \phi(\beta_{a'-1} \cdots \beta_{i+1}\beta_i\alpha_i\alpha_{i+1} \cdots \alpha_{a-1})[x^t_{s,a}] \\ &= x^t_{s,a}\xi(\beta_{a'-1} \cdots \beta_{i+1}\beta_i\alpha_i\alpha_{i+1} \cdots \alpha_{a-1}) = z\xi(z, z'). \end{aligned}$$

Conversely, assume that $z' = z\xi$ in X_1. If we have $z \in Z_{\lambda_s}$ and $z' \in Z_{\lambda_{s'}}$ we certainly must have that $s = s'$. Moreover, if $z' = x^{t'}_{s,a'}$ and $z = x^t_{s,a}$, then ξ is the path class of a path $\gamma := \beta_{a'-1} \cdots \beta_i\alpha_i \cdots \alpha_{a-1}$ of the quiver Q starting at a and ending at a'. Then, the path $\widetilde{\gamma} := \beta^{t'-1}_{s,a'-1} \cdots \beta^{t+1}_{s,i+1}\beta^t_{s,i}\alpha^t_{s,i}\alpha^t_{s,i+1} \cdots \alpha^t_{s,a-1}$ of $\widetilde{Q}$, with source point z and target point z', is a lifting of γ. Finally, $\varphi(z, z') = \widetilde{\gamma}$ and, hence, $\xi = \pi(\widetilde{\gamma}) = \xi(z, z')$.

(1) Recall that we have constructed in the proof of (23.23) the canonical dual basis $(p_{s,\xi}, \gamma_{s,\xi})_{s,\xi}$ of the right S-module P. From (11.11), we know that $\rho(z) = \sum_{s',\xi} zp_{s',\xi} \otimes \gamma_{s',\xi}$, where $s' \in [1, q]$ and ξ runs in the set of non-trivial non-zero path classes of kQ/I. Given $z \in Z_{\lambda_s}$, since $zp_{s',\xi} = z\xi \neq 0$ iff $s = s'$ and $z' := z\xi \in Z_{\lambda_s}$, and then $\xi = \xi(z, z\xi) = \xi(z, z')$, we obtain the first formula.

(2) Put $Z := Z_{\lambda_1} \cup \cdots \cup Z_{\lambda_q}$ and recall, from (11.11), that

$$\begin{aligned} \lambda(\nu_z) = &\textstyle\sum_{z'\in Z,s',\xi} \nu_z(z'p_{s',\xi})\gamma_{s',\xi} \otimes \nu_{z'} + \sum_{i,s',\xi} \nu_z(x^i_i p_{s',\xi})\gamma_{s',\xi} \otimes \nu^i_i \\ &+ \textstyle\sum_{s',\xi} \nu_z(x^{i_0}_{i_*} p_{s',\xi})\gamma_{s',\xi} \otimes \nu^{i_*}_{i_0}, \end{aligned}$$

where $s' \in [1, q]$ and ξ are the same indexes as for the preceeding formulas, i runs in the points set of the bigraph of $\mathcal{A}$ different from i_0 and we are still using the notation of last lemma. Clearly, $x^i_i p_{s',\xi} = 0$,

$x_{i_*}^{i_0} p_{s',\xi} = 0$, and $z' p_{s',\xi} = 0$ whenever $z' \notin Z_{\lambda_{s'}}$. Thus, we are reduced to $\lambda(\nu_z) = \sum_{z' \in Z_{\lambda_{s'}}, s', \xi} \nu_z(z' p_{s',\xi}) \gamma_{s',\xi} \otimes \nu_{z'}$. Besides, $\nu_z(z'') = 0$, for all $z'' \neq z$, and $z = z' p_{s',\xi} = z'\xi$, for some ξ iff $z' < z$ and $s = s'$. Moreover, in this case, $\xi = \xi(z', z)$. We know that $\nu_z(z) \in S$ is the idempotent of $\mathrm{End}(X)$ corresponding to the direct summand $H/H\pi_s^a$ of X containing z. This last idempotent is the target point of $\gamma_{s,\xi}$ (which is the same than the target point of ξ), thus $\nu_z(z)\gamma_{s,\xi} = \gamma_{s,\xi}$. From this, the formula (2) follows.

□

Lemma 23.25. *Assume that Λ is finite-dimensional k-algebra, which trivially splits over its radical J, see (19.3) and (19.4). Then, the Drozd's ditalgebra $\mathcal{D} = \mathcal{D}^\Lambda$ is a nested ditalgebra. The corresponding bigraph and differential can be described as follows. By assumption, we have a splitting $\Lambda = S \oplus J$, with S a trivial subalgebra of Λ. Let $1 = \sum_{i=1}^n e_i$ be the decomposition of the unit of S as a sum of central primitive orthogonal idempotents. We consider a finite directed dual basis $(p_j, \gamma_{p_j})_j$ of the right S-module J, which is compatible with the filtration of the radical J by its powers, as constructed in (23.11). Thus, $P := \{p_j\}_j$ and $P^* := \{\gamma_{p_j}\}_j$ are vector space basis for J and J^*, respectively. Consider also the structural constants $c_{i,j}^t \in k$ of the product of Λ restricted to J. Hence, $p_j p_i = \sum_t c_{j,i}^t p_t$, for each basic elements p_i and p_j of J. Then, clearly $R = R^\Lambda$ is a trivial algebra, with canonical decomposition of unit $1 = (\sum_{i=1}^n ke_i') + (\sum_{i=1}^n ke_i'')$, where $e_i' = \begin{pmatrix} e_i & 0 \\ 0 & 0 \end{pmatrix}$ and $e_i'' = \begin{pmatrix} 0 & 0 \\ 0 & e_i \end{pmatrix}$. Thus, the bigraph of $\mathcal{D}$ has the $2n$ points associated to these idempotents, which we denote with the same symbols. For each basic element $p \in e_j J e_i$, we have the basic element $\gamma_p \in e_i J^* e_j$ such that $\gamma_p(q) = \delta_{p,q} e_i$ (the Kronecker delta of the elements $p, q \in P$). Then every such basic element p determines: a solid arrow $\begin{pmatrix} 0 & 0 \\ \gamma_p & 0 \end{pmatrix}$ of $\mathcal{D}$ from e_j' to e_i''; a dotted arrow $\begin{pmatrix} \gamma_p & 0 \\ 0 & 0 \end{pmatrix}$ of $\mathcal{D}$ from e_j' to e_i'; and a dotted arrow $\begin{pmatrix} 0 & 0 \\ 0 & \gamma_p \end{pmatrix}$ of $\mathcal{D}$ from e_j'' to e_i''. These are all the arrows of $\mathcal{D}$. The value of the differential δ^Λ of $\mathcal{D}$ is given on these arrows by the formulas*

$$\begin{aligned}
\delta^\Lambda\left(\begin{pmatrix} 0 & 0 \\ \gamma_p & 0 \end{pmatrix}\right) &= \sum_{r,s,t} c_{s,r}^t \delta_{p,p_t} \left(\begin{pmatrix} 0 & 0 \\ 0 & \gamma_{p_r} \end{pmatrix} \otimes \begin{pmatrix} 0 & 0 \\ \gamma_{p_s} & 0 \end{pmatrix}\right) \\
&\quad - \sum_{r,s,t} c_{s,r}^t \delta_{p,p_t} \left(\begin{pmatrix} 0 & 0 \\ \gamma_{p_r} & 0 \end{pmatrix} \otimes \begin{pmatrix} \gamma_{p_s} & 0 \\ 0 & 0 \end{pmatrix}\right), \\
\delta^\Lambda\left(\begin{pmatrix} \gamma_p & 0 \\ 0 & 0 \end{pmatrix}\right) &= \sum_{r,s,t} c_{s,r}^t \delta_{p,p_t} \left(\begin{pmatrix} \gamma_{p_r} & 0 \\ 0 & 0 \end{pmatrix} \otimes \begin{pmatrix} \gamma_{p_s} & 0 \\ 0 & 0 \end{pmatrix}\right), \text{ and} \\
\delta^\Lambda\left(\begin{pmatrix} 0 & 0 \\ 0 & \gamma_p \end{pmatrix}\right) &= \sum_{r,s,t} c_{s,r}^t \delta_{p,p_t} \left(\begin{pmatrix} 0 & 0 \\ 0 & \gamma_{p_r} \end{pmatrix} \otimes \begin{pmatrix} 0 & 0 \\ 0 & \gamma_{p_s} \end{pmatrix}\right).
\end{aligned}$$

where the subscripts satisfy $s(p) = s(p_r)$.

Proof. It is clear that, for $p \in e_j P e_i$, we have that $\begin{pmatrix} 0 & 0 \\ \gamma_p & 0 \end{pmatrix} \in e_i'' W_0^\wedge e_j'$, $\begin{pmatrix} \gamma_p & 0 \\ 0 & 0 \end{pmatrix} \in e_i' W_1^\wedge e_j'$ and $\begin{pmatrix} 0 & 0 \\ 0 & \gamma_p \end{pmatrix} \in e_i'' W_1^\wedge e_j''$. Moreover

$$W_0^\wedge = \oplus_p R \begin{pmatrix} 0 & 0 \\ \gamma_p & 0 \end{pmatrix} R \text{ and}$$

$$W_1^\wedge = \left[\oplus_p R \begin{pmatrix} \gamma_p & 0 \\ 0 & 0 \end{pmatrix} R \right] \oplus \left[\oplus_p R \begin{pmatrix} 0 & 0 \\ 0 & \gamma_p \end{pmatrix} R \right].$$

The fact that all the described arrows are free generators for the appropriate R-R-bimodules is clear because R is trivial. Finally, the description of the differential of $\mathcal{D}$ follows from the description given in (19.5) and from (11.8). We have

$$\begin{aligned} \mu(\gamma_p) &= \textstyle\sum_{r,s} \gamma_p(p_s p_r)\gamma_{p_r} \otimes \gamma_{p_s} \\ &= \textstyle\sum_{r,s} \gamma_p(\sum_t c_{s,r}^t p_t)\gamma_{p_r} \otimes \gamma_{p_s} \\ &= \textstyle\sum_{r,s,t} c_{s,r}^t \delta_{p,p_t} e_{s(p)}\gamma_{p_r} \otimes \gamma_{p_s}; \end{aligned}$$

but $e_{s(p)}\gamma_{p_r} \neq 0$ iff $s(p) = t(\gamma_{p_r}) = s(p_r)$, and, in this case, $e_{s(p)}\gamma_{p_r} = \gamma_{p_r}$. □

Exercise 23.26. *Let R be a k-algebra with a decomposition as a finite product of indecomposable algebras and let $1 = \sum_{i=1}^n e_i$ be the unique decomposition of the unit of R as a sum of primitive central orthogonal idempotents. Show that an R-R-bimodule W freely generated by some subset $\mathbb{B}$ is a left projective $R \otimes_k R$-module, but W is not a left free $R \otimes_k R$-module in general. If the algebra R is trivial, then every R-R-bimodule is freely generated by some subset $\mathbb{B}$.*

Exercise 23.27. *Assume that $\mathcal{A}$ is a seminested k-ditalgebra and that $\mathcal{A}'$ is an initial subditalgebra of $\mathcal{A}$ as in (23.13). Assume that $X = Y_1 \oplus \cdots \oplus Y_t$ is a direct sum of a finite number of non-isomorphic finite-dimensional indecomposable $\mathcal{A}'$-modules $Y_1, \ldots, Y_t$. Assume that each algebra $\Gamma_i = \mathrm{End}_{\mathcal{A}}(Y_i)^{op}$ decomposes as $\Gamma_i = k1_{Y_i} \oplus P_i$, where P_i is the radical of Γ_i. Then, from (17.4), we know that X is a complete and triangular admissible object in* $\mathcal{A}'$-Mod*; from (23.13), $\mathcal{A}^X$ is a seminested ditalgebra. Consider the associated functor $F^X : \mathcal{A}^X\text{-Mod} \longrightarrow \mathcal{A}\text{-Mod}$ and take any morphism $f : M \longrightarrow N$ in* $\mathcal{A}^X$-Mod*. Then, the map $F^X(f)^0 : F^X(M) \longrightarrow F^X(N)$ is surjective (resp. injective) whenever $f^0 : M \longrightarrow N$ is so.*

Exercise 23.28. *Let $\mathcal{A}$ be a nested (resp. seminested) ditalgebra with layer (R, W) and marked point i_0. Assume that $Re_{i_0} = k[x]_{f(x)}$, where $f(0) \neq 0$. Consider the ideal I_R of R generated by x. Show that the following holds:*

(1) *The ditalgebra $\mathcal{A}^s$ obtained from $\mathcal{A}$ by factoring out the ideal I_R, as in (8.16), is a nested (resp. seminested) ditalgebra. We call it the ditalgebra obtained from $\mathcal{A}$ by factoring out the marked loop x. $\mathcal{A}^s$ has the same points as $\mathcal{A}$ and it has one marked point less than $\mathcal{A}$, namely i_0. If $(\mathbb{B}_0, \mathbb{B}_1)$ is the basis of the layer (R, W), then the layer (R^s, W^s) of $\mathcal{A}^s$ has basis $(\overline{\mathbb{B}}_0, \overline{\mathbb{B}}_1)$ such that for each $i \in \{0, 1\}$, each arrow $\alpha \in \mathbb{B}_i$ of $\mathcal{A}$ determines an arrow $\overline{\alpha} \in \overline{\mathbb{B}}_i$ of $\mathcal{A}^s$ (obtained as the class of α in W_i modulo the subbimodule $I_R W_i + W_i I_R$); and the mark map of $\mathcal{A}^s$ is the restriction of the mark map of $\mathcal{A}$.*

(2) *The canonical projection $\eta : \mathcal{A} \longrightarrow \mathcal{A}^s$ determines a full and faithful functor $F_s := F_\eta : \mathcal{A}^s\text{-Mod} \longrightarrow \mathcal{A}\text{-Mod}$, which induces an equivalence over the full subcategory of $\mathcal{A}$-Mod formed by the objects M with $M(x) = 0$.*

Hint: Use (8.12) and (23.4)(2).

Exercise 23.29. *Let $\mathcal{A}$ be a ditalgebra with layer (R, W), where R admits a decomposition of the unit $1 = \sum_{i=1}^n e_i$ as a sum of central orthogonal idempotents, W_0 and W_1 are freely generated by the directed sets $\mathbb{B}_0$ and $\mathbb{B}_1$, respectively. Assume $\emptyset \neq S \subseteq \mathbb{B}_0$. Then, construct the ditalgebra $\widehat{\mathcal{A}}$ with layer $(R, \widehat{W})$ as follows. First construct an R-R-bimodule W_0^t freely generated by the set of symbols $S_0^t := \{\alpha^t \mid \alpha \in S\}$, where by definition $s(\alpha^t) = t(\alpha)$ and $t(\alpha^t) = s(\alpha)$, for each $\alpha \in S$; then, construct an R-R-bimodule W_1^t freely generated by the set of symbols $S_1^t := \{v_\alpha^t \mid \alpha \in S\}$, where by definition $s(v_\alpha^t) = s(\alpha^t)$ and $t(v_\alpha^t) = t(\alpha^t)$, for each $\alpha \in S$. Next, define $\widehat{W}_0 := W_0 \oplus W_0^t$, $\widehat{W}_1 := W_1 \oplus W_1^t$, $\widehat{W} := \widehat{W}_0 \oplus \widehat{W}_1$, and $\widehat{T} := T_R(\widehat{W})$, and, finally, consider the differential $\widehat{\delta} : \widehat{T} \longrightarrow \widehat{T}$ defined by the recipe $\widehat{\delta}(w) = \delta(w)$, for $w \in W$, $\widehat{\delta}(\alpha^t) = v_\alpha^t$, for $\alpha \in S$, and $\widehat{\delta}(v_\alpha^t) := 0$, for $\alpha \in S$ (see (4.4)). Then, we have that $\widehat{\mathcal{A}} = (\widehat{T}, \widehat{\delta})$ is a ditalgebra with layer $(R, \widehat{W})$, and we can consider the morphism of ditalgebras $\sigma : \mathcal{A} \longrightarrow \widehat{\mathcal{A}}$ given by the inclusion map. As usual, we write $\widehat{A} = [\widehat{T}]_0$ and $\widehat{V} = [\widehat{T}]_1$. Now, consider the quotient ditalgebra $\mathcal{A}' := \widehat{\mathcal{A}}/I$, as in (8.8), where I is the ideal of $\widehat{\mathcal{A}}$ generated by the differences $\alpha\alpha^t - e_{t(\alpha)}$ and $\alpha^t\alpha - e_{s(\alpha)}$, where α runs in the set S. Consider also the canonical projection $\pi : \widehat{\mathcal{A}} \longrightarrow \mathcal{A}'$. Finally, consider the morphism of ditalgebras $\eta : \mathcal{A} \longrightarrow \mathcal{A}'$ defined by the composition $\eta = \pi\sigma$. Show that the restriction functor $F_\eta : \mathcal{A}'\text{-Mod} \longrightarrow \mathcal{A}\text{-Mod}$ is a full and faithful functor. For any $M \in \mathcal{A}$-Mod such that $M_\alpha : M_{s(\alpha)} \longrightarrow M_{t(\alpha)}$, the morphism multiplication by α, is invertible, for all $\alpha \in S$, there is an $N \in \mathcal{A}'$-Mod with $F_\eta(N) \cong M$.*

Hint: In order to apply (2.10), show that $\eta_0 : A \longrightarrow A' = [T']_0$ is an epimorphism of algebras and that $\eta_1 : V \longrightarrow V' := [T']_1$ induces an isomorphism

$$\mathrm{Hom}_{A'\text{-}A'}(V', U) \xrightarrow{\eta_1^*} \mathrm{Hom}_{A\text{-}A}(V, U),$$

for any A'-A'-bimodule U. We may use (8.9).

Exercise 23.30. *Given a seminested ditalgebra $\mathcal{A}$, with layer (R, W) and bigraph $(\mathcal{P}, \mathcal{P}_{\rm mk}, \mathbb{B}_0, \mathbb{B}_1, \mathbb{B}_{\rm mk}, {\rm mk})$, consider its bilinear form $(-, ?)_{\mathcal{A}}$ on $\mathbb{Z}^{\mathcal{P}}$, defined for $x, y \in \mathbb{Z}^{\mathcal{P}}$ by the rule*

$$(x, y)_{\mathcal{A}} := \sum_{i \in \mathcal{P} \setminus \mathcal{P}_{\rm mk}} y_i x_i - \sum_{\alpha \in \mathbb{B}_0} y_{t(\alpha)} x_{s(\alpha)} + \sum_{v \in \mathbb{B}_1} y_{t(v)} x_{s(v)}.$$

If $\mathcal{A}$ has no marked point, show that, for $M, N \in \mathcal{A}$-mod, we have

$$(\underline{\dim} M, \underline{\dim} N)_{\mathcal{A}} = \dim {\rm Hom}_{\mathcal{A}}(M, N) - \dim {\rm Ext}_{\mathcal{A}}(M, N).$$

Hint: use (6.13) and (23.4).

Exercise 23.31. *Let $\mathcal{A} = (T, \delta)$ be a ditalgebra with nested (resp. seminested) layer (R, W). Consider the height map determined by the filtration $\mathcal{F}_0$ of R-R-bimodules of W_0, given by the triangularity requirement in (5.1). Suppose that we have a solid arrow $\alpha \in \mathbb{B}_1$ from the vertex i_0 to the vertex j_0 of $\mathcal{A}$. Assume that*

$$\alpha' = r\alpha + \sum_t r^t_{\lambda_t+1} \beta^t_{\lambda_t} r^t_{\lambda_t} \cdots r^t_2 \beta^t_2 r^t_1 \beta^t_1 r^t_0,$$

where: $r \in Re_{j_0} \otimes_k Re_{i_0}$ is invertible; all $\lambda_t \geq 0$; all $r^t_i \in R$, $r^t_0 \in Re_{i_0}$, $r^t_{\lambda_t+1} \in Re_{j_0}$; and all $\beta^t_i \in \mathbb{B}_0$ are solid arrows with $\mathbb{h}(\beta^t_i) < \mathbb{h}(\alpha)$. Then, we can construct a new nested (resp. seminested) layer (R, W') for $\mathcal{A}$, where $W'_1 = W_1$ and W'_0 is the R-R-subbimodule of A generated by $\mathbb{B}'_0 := (\mathbb{B}_0 \setminus \{\alpha\}) \cup \{\alpha'\}$.

Exercise 23.32. *Let $\mathcal{A} = (T, \delta)$ be a ditalgebra with nested layer (R, W). Consider the filtrations of R-R-bimodules $\mathcal{F}_0 : 0 = W^0_0 \subseteq W^1_0 \subseteq \cdots \subseteq W^{\ell_0}_0 = W_0$ and $\mathcal{F}_1 : 0 = W^0_1 \subseteq W^1_1 \subseteq \cdots \subseteq W^{\ell_1}_1 = W_1$, of W_0 and W_1, respectively, given by the triangularity conditions in (5.1). We can consider the A-A-bimodule filtration $\widetilde{\mathcal{F}}_1 : 0 = V^0 \subseteq V^1 \subseteq \cdots \subseteq V^{\ell_1} = V$ of V defined by $V^s := AW^s_1 A$, for all $s \in [0, \ell_1]$. Notice that, given any $w \in W_1$, its height with respect to $\mathcal{F}_1$ coincides with its height with respect to $\widetilde{\mathcal{F}}_1$: thus, we can use the same symbol $\mathbb{h}(w)$ to denote this number. Suppose that we have a fixed dotted arrow $v \in \mathbb{B}_1$, from the vertex i_0 to the vertex j_0 of $\mathcal{A}$ and assume that*

$$v' = rv + \sum_t \left(r^t_{\lambda_t+1} \beta^t_{\lambda_t} r^t_{\lambda_t} \cdots r^t_2 \beta^t_2 r^t_1 \beta^t_1 r^t_0\right) v_t \left(s^t_{\rho_t+1} \alpha^t_{\rho_t} s^t_{\rho_t} \cdots s^t_2 \alpha^t_2 s^t_1 \alpha^t_1 s^t_0\right),$$

where: $r \in Re_{j_0} \otimes_k Re_{i_0}$ is invertible, all $r^t_i, s^t_j \in R$, $s^t_0 \in Re_{i_0}$, $r^t_{\lambda_t+1} \in Re_{j_0}$, all $\beta^t_i, \alpha^t_j \in \mathbb{B}_0$ are solid arrows with $\mathbb{h}(\beta^t_i), \mathbb{h}(\alpha^t_j), \mathbb{h}(\delta(\beta^t_i)), \mathbb{h}(\delta(\alpha^t_j)) < \mathbb{h}(v)$, and each v_t is a dotted arrow satisfying $\mathbb{h}(v_t) < \mathbb{h}(v)$. Put $\mu := \max\{\mathbb{h}(\beta^t_i), \mathbb{h}(\alpha^t_j) \mid i, j, t\}$ and make $\overline{W}_1 := \oplus_{w \in \mathbb{B}_1 \setminus \{v\}} RwR$. Finally, assume that $\delta(W^{s+1}_0) \subseteq A_s \overline{W}_1 A_s$, for all $s < \mu$. Then, we can construct a new nested *layer* (R, W') *for* $\mathcal{A}$, *where* $W'_0 = W_0$ *and* W'_1 *is the* R-R-*subbimodule of* V *generated by* $\mathbb{B}'_1 := (\mathbb{B}_1 \setminus \{v\}) \cup \{v'\}$.

References. Nested ditalgebras are very close to Drozd's *almost free bocses* in [28], and seminested ditalgebras are very close to Crawley's *layered bocses* in [19] and [20]. The operations *regularization* and *edge reduction* were introduced by Roiter, see [45]. The original idea for the *unravelling* was proposed by Drozd in [28], then refined in [19]. As remarked before, the construction of the *edge reduction* and the *unravelling* as reductions using appropriate admissible modules X was developed in [6]. The description of the differential of the unravelling through a graphical description using a translation quiver first appeared in [13].

24

Critical ditalgebras

In this section, we define critical ditalgebras and prove that they are wild. We present here an alternative to the usual proof, which involves many calculations since it provides explicitly very large bimodules producing wildness.

We start with a couple of general statements which play an important role in our argument.

Definition 24.1. *Let $\mathcal{A} = (T, \delta)$ be a ditalgebra with layer (R, W). A subset N of W_0 is called triangular iff it admits a filtration of subsets*

$$\mathcal{F}(N) : \emptyset = N_0 \subseteq N_1 \subseteq \cdots \subseteq N_\ell = N$$

such that, for any $i \in [1, \ell]$, $\delta(N_i) \subseteq \langle N_{i-1}\rangle W_1 + W_1 + W_1\langle N_{i-1}\rangle$, where $\langle N_i\rangle$ denotes the ideal of $[T]_0$ generated by N_i.

Lemma 24.2. *Assume that $\mathcal{A} = (T, \delta)$ is a ditalgebra with layer (R, W). Assume that we have a triangular subset N of W_0. Consider the ideal I of $\mathcal{A}$ generated by N (see (8.8)) and the canonical projection $\eta : \mathcal{A} \longrightarrow \mathcal{A}/I$. Then, the quotient ditalgebra $\mathcal{A}/I = (\overline{T}, \overline{\delta})$ admits the layer $(R, \eta(W))$, where $\eta(W) = \eta(W_0) \oplus \eta(W_1)$. Moreover, $\mathcal{A}/I$ is a Roiter ditalgebra whenever $\mathcal{A}$ is so, and if $\mathcal{A}/I$ is wild, so is $\mathcal{A}$. In case we have an R-R-bimodule decomposition $W_0 = \widehat{N} \oplus W_0^c$, where $\widehat{N}$ denotes the R-R-subbimodule of W_0 generated by N, then $[\overline{T}]_0$ can be identified with the subalgebra of $[T]_0$ generated by R and W_0^c.*

Proof. Consider the filtration $\mathcal{F}(N) = \{N_t\}_{t=0}^{\ell}$ of the triangular subset N and denote, for each t, by $\widehat{N}_t$ the R-R-subbimodule of W_0 generated by N_t. Consider the ditalgebra $\mathcal{A}^1$ obtained from $\mathcal{A}^0 := \mathcal{A}$ factoring out the subbimodule $\widehat{N}_1$ of W_0, as in (8.18), and the associated projection $\eta_1 : \mathcal{A} \longrightarrow \mathcal{A}^1$. Make $\zeta_0 := 1_{\mathcal{A}}$ and $\zeta_1 := \eta_1$, thus: $0 = \eta_1\zeta_0(\widehat{N}_1) = \zeta_1(\widehat{N}_1)$. Assume we have constructed the ditalgebras $\mathcal{A}^0, \ldots, \mathcal{A}^i$, where each $\mathcal{A}^j = (T^j, \delta^j)$ is obtained from $\mathcal{A}^{j-1}$

by factoring out the subbimodule $\zeta_{j-1}(\widehat{N}_j)$, where each $\zeta_j : \mathcal{A} \longrightarrow \mathcal{A}^j$ is the composition of ζ_{j-1} with the canonical projection $\eta_j : \mathcal{A}^{j-1} \longrightarrow \mathcal{A}^j$. Let us see that we can construct the ditalgebra $\mathcal{A}^{i+1}$, obtained from $\mathcal{A}^i$ by factoring out the subbimodule $\zeta_i(\widehat{N}_{i+1})$. Notice that $0 = \eta_i \zeta_{i-1}(\widehat{N}_i) = \zeta_i(\widehat{N}_i)$. Using that

$$\delta(N_{i+1}) \subseteq \langle N_i \rangle W_1 + W_1 + W_1 \langle N_i \rangle,$$

we obtain

$$\begin{aligned} \delta^i(\zeta_i(\widehat{N}_{i+1})) &= \zeta_i(\delta(\widehat{N}_{i+1})) \\ &\subseteq \langle \zeta_i(\widehat{N}_i) \rangle \zeta_i(W_1) + \zeta_i(W_1) + \zeta_i(W_1) \langle \zeta_i(\widehat{N}_i) \rangle \subseteq \zeta_i(W_1). \end{aligned}$$

Then, we can really apply the procedure of (8.18) and factor out $\zeta_i(\widehat{N}_{i+1})$.

Following this procedure, we construct $\mathcal{A}^\ell$, where ℓ is the length of the filtration $\mathcal{F}(N)$. We claim that $\mathcal{A}^\ell \cong \mathcal{A}/I$.

For $i \in [0, \ell]$, denote by I_i be the ideal of $\mathcal{A}$ generated by N_i (see (8.8)). We show by induction that there is an isomorphism $\theta_i : \mathcal{A}/I_i \longrightarrow \mathcal{A}^i$ such that $\theta_i p_i = \zeta_i$, where $p_i : \mathcal{A} \longrightarrow \mathcal{A}/I_i$ is the canonical projection. This is clear for $i = 1$, where we can take as θ_1 the isomorphism described in (8.18). Assume θ_i has been defined and denote by θ_{i+1} the composition of the isomorphisms at the bottom of the following commutative diagram

$$\begin{array}{ccccccc} \mathcal{A} & \xrightarrow{p_i} & \mathcal{A}/I_i & \xrightarrow{\theta_i} & \mathcal{A}^i & \xleftarrow{\zeta_i} & \mathcal{A} \\ \downarrow{\scriptstyle p_{i+1}} & & \downarrow{\scriptstyle \pi_i} & & \downarrow{\scriptstyle \xi_i} & & \downarrow{\scriptstyle \zeta_{i+1}} \\ \mathcal{A}/I_{i+1} & \xrightarrow{p'_i} & (\mathcal{A}/I_i)/(I_{i+1}/I_i) & \xrightarrow{\theta'_i} & \mathcal{A}^i/J_i & \xrightarrow{\zeta'_i} & \mathcal{A}^{i+1}, \end{array}$$

where J_i denotes the ideal of $\mathcal{A}^i$ generated by the subset $\zeta_i(N_{i+1})$ of $[T_{\mathcal{A}^i}]_0$, π_i and ξ_i are projections, the isomorphism p'_i is given by (8.7); the isomorphism θ'_i is induced by θ_i, since $\theta_i p_i(N_{i+1}) = \zeta_i(N_{i+1})$; and the isomorphism ζ'_i is given by (8.18). The isomorphism $\theta_\ell^{-1} : \mathcal{A}^\ell \longrightarrow \mathcal{A}/I_\ell = \mathcal{A}/I$ maps the layer $(R, \zeta_\ell(W))$ of $\mathcal{A}^\ell$ onto $(\eta(R), \eta(W))$ and we may identify $\eta(R)$ with R. If $\mathcal{A}$ is a Roiter ditalgebra, applying (9.3) successively to each $\mathcal{A}^i$, we get that $\mathcal{A}^\ell$ is a Roiter ditalgebra. Hence, by (9.4), $\mathcal{A}/I$ is a Roiter ditalgebra. The statement on wildness follows from (22.8). □

Lemma 24.3. *Let $\mathcal{A}$ be a Roiter ditalgebra with layer (R, W). Assume that W'_0 is a direct summand of W_0 with $\delta(W'_0) = 0$ and suppose that e is a central idempotent of R with $eW'_0e = W'_0$. Then, if the algebra $T_{eR}(W'_0)$ is wild, so is the ditalgebra $\mathcal{A}$.*

Proof. Consider the ditalgebra $\mathcal{A}^a$ obtained from $\mathcal{A}$ by absorption of the bimodule W'_0, with layer (R^a, W^a), as in (8.20). Then, consider the ditalgebra $\mathcal{A}^{ad}$

obtained from $\mathcal{A}^a$ deleting the idempotent $1-e$, as in (8.17). From (9.3), we know that $\mathcal{A}^{ad}$ is a Roiter ditalgebra. By assumption $T_{eR}(W_0') \cong R^{ad}$ is wild. From (22.11), we get that $\mathcal{A}^{ad}$ is wild. Then, by (22.8), $\mathcal{A}$ is wild too. □

Assume that $\mathcal{A}$ is a seminested ditalgebra with layer (R, W) and bigraph $\mathbb{B} = (\mathcal{P}, \mathcal{P}_{\text{mk}}, \mathbb{B}_0, \mathbb{B}_1, \mathbb{B}_{\text{mk}}, \text{mk})$. For any $i, j \in \mathcal{P}_{\text{mk}}$, we know that $e_j W e_i$ is an Re_j-Re_i-bimodule. Since $Re_i \cong k[x]_{f_i(x)}$ and $Re_j \cong k[y]_{f_j(y)}$, $e_j W e_i$ is a $k[x]_{f_i(x)} \otimes_k k[y]_{f_j(y)}$-module or, equivalently, a $k[x, y]_{f_i(x)f_j(y)}$-module. Thus, a typical element of $e_j W_1 e_i$, for instance, can be written as $\sum_\ell r_\ell(x, y) v_\ell$, where v_ℓ are the dotted arrows from i to j, and $r_\ell(x, y) \in k[x, y]_{f_i(x)f_j(y)}$.

Lemma 24.4. *We say that an element $r(x, y) \in k[x, y]_{f(x)g(y)}$ admits an $f(x)g(y)$-zero iff there exist $\lambda, \mu \in k$ with $f(\lambda) \neq 0$, $g(\mu) \neq 0$ and $r(\lambda, \mu) = 0$. Any element $r(x, y) \in k[x, y]_{f(x)g(y)}$ can be written as $r(x, y) = ur_0(x, y)$, where $r_0(x, y) \in k[x, y]$ has no irreducible factor in common with $f(x)g(y)$ (or $r_0(x, y) = 0$) and u is an invertible element in $k[x, y]_{f(x)g(y)}$.*

In general, any $r(x, y) \in k[x, y]_{f(x)g(y)}$ which admits an $f(x)g(y)$-zero must be a non-invertible element in $k[x, y]_{f(x)g(y)}$. If the field k is algebraically closed, the converse holds.

Proof. Write first $r(x, y) = (f(x)g(y))^{-m} r_1(x, y)$, where $r_1(x, y) \in k[x, y]$ and $m \geq 0$. Then, write $r_1(x, y) = wr_0(x, y)$, where $r_0(x, y) \in k[x, y]$ and w is the product of all irreducible factors of $r_1(x, y)$ in $k[x, y]$ that divide $f(x)g(y)$. Thus, $r_0(x, y) \in k[x, y]$ shares no irreducible factors with $f(x)g(y)$.

If $r(\lambda, \mu) = 0$ with $f(\lambda) \neq 0$ and $g(\mu) \neq 0$, then $w(\lambda, \mu) \neq 0$, because $w(x, y)$ divides some power of $f(x)g(y)$. Then, $r_0(\lambda, \mu) = 0$ and, $r_0(x, y)$ cannot be a divisor of any power of $f(x)g(y)$. Thus, $r_0(x, y)$ (and, therefore, $r(x, y)$) is not invertible in $k[x, y]_{f(x)g(y)}$.

Finally, assume that the field k is algebraically closed. Given an ideal I of $k[x, y]$, denote by $Z(I) \subseteq k^2$ the set of zeros of I; given a subset $Z \subseteq k^2$, denote by $I(Z)$ the ideal of $k[x, y]$ consisting of the polynomials which are zero on every point of Z. Assume that for every $(\lambda, \mu) \in k^2$ with $r_0(\lambda, \mu) = 0$, we have that $f(\lambda) = 0$ or $g(\mu) = 0$. Then, $Z(\langle r_0(x, y)\rangle) \subseteq Z(\langle f(x)g(y)\rangle)$, where $\langle r_0(x, y)\rangle$ is the ideal of $k[x, y]$ generated by $r_0(x, y)$. Hence, by Hilbert's theorem, we have

$$\sqrt{\langle f(x)g(y)\rangle} = I(Z(\langle f(x)g(y)\rangle)) \subseteq I(Z(\langle r_0(x, y)\rangle)) = \sqrt{\langle r_0(x, y)\rangle}.$$

In particular, $f(x)g(y) \in \sqrt{\langle r_0(x, y)\rangle}$. Therefore, there exist $s, t \in \mathbb{N}$ such that $f^s(x)g^t(y) = r_0'(x, y)r_0(x, y)$: a contradiction. Then, there exist $\lambda, \mu \in k$ with $f(\lambda) \neq 0$, $g(\mu) \neq 0$ and $r_0(\lambda, \mu) = 0$, as claimed. □

Definition 24.5. *Assume that $\mathcal{A}$ is a seminested ditalgebra with layer (R, W), differential δ and bigraph $\mathbb{B} = (\mathcal{P}, \mathcal{P}_{\rm mk}, \mathbb{B}_0, \mathbb{B}_1, \mathbb{B}_{\rm mk}, {\rm mk})$. Then, we say that $\mathcal{A}$ is critical iff we are in one of the following cases:*

(1) $R = k[x]_{f(x)}$, there is at least one solid arrow $\alpha \in \mathbb{B}_0$ and one dotted arrow $v \in \mathbb{B}_1$. Moreover, we assume that $\delta(\alpha) = r(x, y)v$, for some $r(x, y) \in k[x, y]_{f(x)f(y)}$ which admits an $f(x)f(y)$-zero.

(2) $R = k[x]_{f(x)} \times k[y]_{g(y)}$, there is at least one solid arrow $\alpha \in \mathbb{B}_0$ and one dotted arrow $v \in \mathbb{B}_1$. Moreover, we assume that α is not a loop and $\delta(\alpha) = r(x, y)v$, for some $r(x, y) \in k[x, y]_{f(x)g(y)}$, which admits an $f(x)g(y)$-zero.

(3) $R = k[x]_{f(x)}$, where the polynomial $f(x)$ admits an element $\mu \in k$ such that $f(\mu) \neq 0$. Assume there is at least one solid arrow α in $\mathcal{A}$. Moreover, we assume that $\delta(\alpha) = 0$.

(4) $R = k[x]_{f(x)} \times k$ or $R = k[x]_{f(x)} \times k[y]_{g(y)}$, where the polynomial $f(x)$ admits an element $\mu \in k$ such that $f(\mu) \neq 0$. Assume there is at least one solid arrow α in $\mathcal{A}$. Moreover, we assume that α is not a loop and $\delta(\alpha) = 0$.

Remark 24.6. *Assume that $\mathcal{A}$ is a seminested ditalgebra and let $\mathcal{A}^u$ be obtained from $\mathcal{A}$ by unravelling at i_0 with n and $\lambda_1, \ldots, \lambda_q$. In order to have simple expressions for the value of the differential δ^u on some special arrows, let us agree on the following notation. By definition, $\tau x^t_{s,a} := x^{t+1}_{s,a}$ (and $x^t_{s,a} =: \tau^{-1} x^{t+1}_{s,a}$) for all $t \in [0, a-2]$, $s \in [1, q]$ and $a \in [1, n]$. Hence, $\tau^i x^t_{s,a} = x^{t+i}_{s,a}$, whenever $t + i \leq a - 1$; if $t + i > a - 1$, we make $\tau^i x^t_{s,a} = 0$. The corresponding convention is made for the powers of τ^{-1}.*

Now, suppose that α is a solid loop of $\mathcal{A}$ based at i_0 and let $v_1, \ldots, v_p$ be the collection of dotted loops of $\mathcal{A}$ based at the vertex i_0. Assume, furthermore, that $\delta(\alpha) = \sum_\ell r_\ell(x, y)v_\ell$, for some $r_\ell(x, y) \in k[x, y]$. Then, given any $s_1, s_2 \in [1, q]$, from the definition of δ^u and (23.24), we have:

(1) There exists scalars $\{c^\ell_{i,j} \mid \ell \in [1, p], i \in [0, n_\ell], j \in [0, m_\ell]\}$ such that, for any $z_1 \in \widetilde{Q}_{\lambda_{s_1}}$ and $z_2 \in \widetilde{Q}_{\lambda_{s_2}}$

$$\begin{aligned}\delta^u(v_{z_2} \otimes \alpha \otimes z_1) = &\textstyle\sum_{z_2 > z'_2 \in Z_{\lambda_{s_2}}} \gamma_{s_2,\xi(z'_2,z_2)} \otimes v_{z'_2} \otimes \alpha \otimes z_1 \\ &+ \textstyle\sum_{\ell,i,j} c^\ell_{i,j} v_{\tau^{-i} z_2} \otimes v_\ell \otimes \tau^j z_1 \\ &- \textstyle\sum_{z_1 < z'_1 \in Z_{\lambda_{s_1}}} v_{z_2} \otimes \alpha \otimes z'_1 \otimes \gamma_{s_1,\xi(z_1,z'_1)}.\end{aligned}$$

These coefficients $c^\ell_{i,j}$ are given by any choice of expression for $\delta(\alpha)$ of the form

$$\delta(\alpha) = \sum_{\ell,i,j} c^\ell_{i,j}(x - \lambda_{s_2})^i v_\ell (x - \lambda_{s_1})^j.$$

This follows from the fact that for any $s \in [1, q]$ *and* $z_s \in Z_{\lambda_s}$, $\pi_s z_s = \tau z_s$ *and* $v_{z_s}\pi_s = v_{\tau^{-1}z_s}$, *where* $\pi_s = (x - \lambda_s)$.

(2) *Assume* z_1 *is a projective of* $\widetilde{Q}_{\lambda_{s_1}}$ *(this means that* $z_1 = x^{a-1}_{s_1,a}$, *for some* $a \in [1, n]$*) and* z_2 *is an injective of* $\widetilde{Q}_{\lambda_{s_2}}$ *(this means that* $z_2 = x^0_{s_2,a}$, *for some* $a \in [1, n]$*). Then*

$$\begin{aligned}\delta^u(v_{z_2} \otimes \alpha \otimes z_1) = & \textstyle\sum_{z_2 > z_2' \in Z_{\lambda_{s_2}}} \gamma_{s_2,\xi(z_2',z_2)} \otimes v_{z_2'} \otimes \alpha \otimes z_1 \\ & + \textstyle\sum_\ell r_\ell(\lambda_{s_1}, \lambda_{s_2}) v_{z_2} \otimes v_\ell \otimes z_1 \\ & - \textstyle\sum_{z_1 < z_1' \in Z_{\lambda_{s_1}}} v_{z_2} \otimes \alpha \otimes z_1' \otimes \gamma_{s_1,\xi(z_1,z_1')}.\end{aligned}$$

This holds because $0 = \pi_s x^{a-1}_{s,a} = (x - \lambda_s)x^{a-1}_{s,a}$ *implies that* $x x^{a-1}_{s,a} = \lambda_s x^{a-1}_{s,a}$. *Similarly,* $0 = v_{x^0_{s,a}}\pi_s = v_{x^0_{s,a}}(y - \lambda_s)$ *implies that* $v_{x^0_{s,a}} y = \lambda_s v_{x^0_{s,a}}$, *for any* $s \in [1, q]$.

(3) *If* z_1 *is maximal projective of* $\widetilde{Q}_{\lambda_{s_1}}$, *the last group of summands in the last formula is zero; if* z_2 *is minimal injective of* $\widetilde{Q}_{\lambda_{s_2}}$, *the fist group of summands in last formula is zero. Under both assumptions, we obtain*

$$\delta^u(v_{z_2} \otimes \alpha \otimes z_1) = \sum_\ell r_\ell(\lambda_{s_2}, \lambda_{s_1}) v_{z_2} \otimes v_\ell \otimes z_1.$$

Lemma 24.7. *Assume that* $\mathcal{A}$ *is a seminested ditalgebra with layer* (R, W) *and differential* δ. *Assume that* $\mathcal{A}$ *has only one point and there are at least one solid arrow* α *and one dotted arrow* v. *Assume* $R = k[x]_{f(x)}$ *and* $\delta(\alpha) = r(x, y)v$, *for some* $r(x, y) \in k[x, y]_{f(x)f(y)}$, *which admits an* $f(x)f(y)$*-zero. Then,* $\mathcal{A}$ *is a wild ditalgebra.*

Proof. From (24.4), $r(x, y)$ can be written as $r(x, y) = ur_0(x, y)$, where $r_0(x, y) \in k[x, y]$ has no irreducible factor in common with $f(x)f(y)$ and u is an invertible element in $k[x, y]_{f(x)f(y)}$. By an appropriate change of basis of the form $v' = f(x)^{-a} f(y)^{-b} v$, if necessary, we can assume that $r(x, y) \in k[x, y]$ has an $f(x)f(y)$-zero $(\lambda_1, \lambda_2) \in k^2$. We can write

$$r(x, y) = \sum_{i,j} c_{i,j}(y - \lambda_2)^i (x - \lambda_1)^j,$$

for some appropriate scalars $c_{i,j} \in k$. Clearly, $c_{0,0} = r(\lambda_1, \lambda_2) = 0$.

Consider the ditalgebra $\mathcal{A}^u$ obtained from $\mathcal{A}$ by unravelling at the vertex $i_0 = 1$ using $n := 10$ and λ_1, in case $\lambda_1 = \lambda_2$, and $n := 9$, together with λ_1 and λ_2, in case $\lambda_1 \neq \lambda_2$.

Recall that, for $H := k[x]_{f(x)}$, we have for $s \in \{1, 2\}$, $X_{\lambda_s} = H/H\pi_s \oplus \cdots \oplus H/H\pi_s^n$, where $\pi_s = x - \lambda_s$. Thus, with the notation of (23.24), $X_1 = X_{\lambda_1}$ if $\lambda_1 = \lambda_2$ and $X_1 = X_{\lambda_1} \oplus X_{\lambda_2}$ if $\lambda_1 \neq \lambda_2$. Make also $q := 1$ if $\lambda_1 = \lambda_2$ and $q := 2$, if $\lambda_1 \neq \lambda_2$. We have described in (23.24) a quiver $\widetilde{Q}$, with one connected component $\widetilde{Q}_{\lambda_1} = \widetilde{Q}_{\lambda_q}$ if $\lambda_1 = \lambda_2$, and with two connected components $\widetilde{Q}_{\lambda_1}, \widetilde{Q}_{\lambda_q}$ if $\lambda_1 \neq \lambda_2$.

We shall write $z_a := x_{1,a}^{a-1}$, for $a \in [1, n]$, and $z := x_{q,n}^0$. Moreover, we adopt the the τ-notation of (24.6). We simplify further the notation by writing $w \otimes \alpha \otimes w'$ instead of $v_w \otimes \alpha \otimes w'$, whenever $w, w' \in \widetilde{Q}$. According to (24.6), if we write

$$\delta(\alpha) = \sum_{i,j} c_{i,j}(y - \lambda_q)^i v(x - \lambda_1)^j,$$

then we have, for $w_1 \in \widetilde{Q}_{\lambda_1}$ and $w_2 \in \widetilde{Q}_{\lambda_q}$

$$\begin{aligned} \delta^u(w_2 \otimes \alpha \otimes w_1) = & \textstyle\sum_{w_2 > w_2' \in \widetilde{Q}_{\lambda_q}} \gamma_{w_2'} w_2' \otimes \alpha \otimes w_1 \\ & + \textstyle\sum_{i,j} c_{i,j} \tau^{-i} w_2 \otimes v \otimes \tau^j w_1 \\ & - \textstyle\sum_{w_1 < w_1' \in \widetilde{Q}_{\lambda_1}} w_2 \otimes \alpha \otimes w_1' \gamma_{w_1'}, \end{aligned}$$

for some $\gamma_{w_2'}, \gamma_{w_1'} \in P^*$. Now, consider the "central axis" of the quiver $\widetilde{Q}_{\lambda_1}$, that is the set of points $\tau^{-s} z_i$ in $\widetilde{Q}_{\lambda_1}$ determined by the equation $i = 2s + 1$ (that is the points $w \in \widetilde{Q}_{\lambda_1}$ with height $\mathbb{h}(w) = n$).

Consider the following subset N of arrows of W_0^X

$$N := \{w \otimes \alpha \otimes w' \mid w \in \widetilde{Q}_{\lambda_q}, w' \in \widetilde{Q}_{\lambda_1} \text{ and } \mathbb{h}(w') < n\},$$

and consider the filtration

$$\mathcal{F}(N) : \emptyset = N_0 \subseteq N_1 \subseteq \cdots \subseteq N_\ell = N,$$

where $\ell = 3n - 2$ and each N_t is, by definition, the subset of N formed by the elements $w \otimes \alpha \otimes w'$ such that $2n - \mathbb{h}(w) + \mathbb{h}(w') \leq t$. Recall that $\mathbb{h}(v_w) + \mathbb{h}(w) = \ell_X + 1 = 2n$, thus the former inequality is equivalent to $\mathbb{h}(v_w) + \mathbb{h}(w') \leq t$ and, indeed, $N_0 = \emptyset$. Moreover, the maximal possible value of $\mathbb{h}(v_w) + \mathbb{h}(w')$ is $(2n - 1) + (n - 1) = 3n - 2$, and $N_\ell = N$.

Then, the formula for the differential δ^u above shows that N is triangular with filtration $\mathcal{F}(N)$, as in (24.1). From (24.2), we obtain that the quotient ditalgebra $\mathcal{A}^X/I$, where I is the ideal of $\mathcal{A}^X$ generated by N, is a Roiter ditalgebra. If $\eta : \mathcal{A}^X \longrightarrow \mathcal{A}^X/I$ the canonical projection, the layer of the quotient is $(S, \eta(W^X))$. Denote by $\widehat{N}$ the S-S-subbimodule of W_0^X generated by the set of arrows N. Then, $W_0^X = \widehat{N} \oplus W_0^c$, where $W_0^c = \oplus_\beta S\beta S$, where β runs in the set of arrows of W_0^X which are not in N. Recall that $\eta(W_0^X)$ can be identified with W_0^c.

Now, consider the five arrows $\alpha_j := z \otimes \alpha \otimes \tau^{-j} z_{2j+1} : i_{1,2j+1} \longrightarrow i_{q,n}$, $j \in [0, 4]$. For simplicity, we denote by $\widehat{\delta}$ the differential of the quotient $\mathcal{A}^X/I$ and we write with the same symbol an arrow of $\mathcal{A}^X$ and its class modulo I.

Notice that

$$\begin{aligned}
\widehat{\delta}(\alpha_0) &= \quad \widehat{\delta}(z \otimes \alpha \otimes z_1) \quad = 0\\
\widehat{\delta}(\alpha_1) &= \widehat{\delta}(z \otimes \alpha \otimes \tau^{-1} z_3) = c_{0,1} z \otimes v \otimes z_3\\
\widehat{\delta}(\alpha_2) &= \widehat{\delta}(z \otimes \alpha \otimes \tau^{-2} z_5) = c_{0,1} z \otimes v \otimes \tau^{-1} z_5 + c_{0,2} z \otimes v \otimes z_5\\
\widehat{\delta}(\alpha_3) &= \widehat{\delta}(z \otimes \alpha \otimes \tau^{-3} z_7) = c_{0,1} z \otimes v \otimes \tau^{-2} z_7 + c_{0,2} z \otimes v \otimes \tau^{-1} z_7\\
&\qquad\qquad\qquad\qquad + c_{0,3} z \otimes v \otimes z_7\\
\widehat{\delta}(\alpha_4) &= \widehat{\delta}(z \otimes \alpha \otimes \tau^{-4} z_9) = c_{0,1} z \otimes v \otimes \tau^{-3} z_9 + c_{0,2} z \otimes v \otimes \tau^{-2} z_9\\
&\qquad\qquad\qquad\qquad + c_{0,3} z \otimes v \otimes \tau^{-1} z_9 + c_{0,4} z \otimes v \otimes z_9.
\end{aligned}$$

We shall prove that the expressions for $\widehat{\delta}(\alpha_0), \ldots, \widehat{\delta}(\alpha_4)$ are all zero. In this case, we can consider the direct summand $W_0' := \oplus_{j=0}^4 S\alpha_j S$ of W_0^c and the central idempotent of S defined by $e := f_{i_{q,n}} + \sum_{j=0}^4 f_{i_{1,2j+1}}$, and then notice that the tensor algebra $T_{eS}(W_0')$ is isomorphic to the path algebra of the quiver

$$\begin{array}{ccccc}
i_{1,1} & & & & i_{1,3}\\
 & \searrow^{\alpha_0} & & {}^{\alpha_1}\swarrow & \\
 & & i_{q,n} & & \\
 & {}^{\alpha_2}\nearrow & \uparrow \alpha_3 & \nwarrow^{\alpha_4} & \\
i_{1,5} & & i_{1,7} & & i_{1,9}
\end{array}$$

And then, using (24.3), we obtain that $\mathcal{A}^X/I$ is wild. The star algebra mentioned above is wild; we will come back to review this fact later in (30.2). From (24.2), we get that $\mathcal{A}^X$ is wild, and from (22.10), $\mathcal{A}$ is wild.

We divide the proof into the following cases:

Case 1: $c_{0,1} = c_{0,2} = c_{0,3} = c_{0,4} = 0$. This case is trivial, given the stated expressions of each $\widehat{\delta}(\alpha_j)$.

Case 1′: $c_{1,0} = c_{2,0} = c_{3,0} = c_{4,0} = 0$. This case is similar to the previous one: we have to consider the same unravelling $\mathcal{A}^u$, then make $z_a' := x_{q,a}^0$, for $a \in [1, n]$, and $z' := x_{1,n}^{n-1}$; the five arrows $\alpha_j' := \tau^j z_{2j+1}' \otimes \alpha \otimes z'$, for $j \in [0, 4]$; and the triangular subset $N' := \{w \otimes \alpha \otimes w' \mid w \in \widetilde{Q}_{\lambda_q}, w' \in \widetilde{Q}_{\lambda_1}$ and $\mathbb{h}(w) > n\}$; then, if I' denotes the ideal of $\mathcal{A}^u$ generated by N', the quotient ditalgebra $\mathcal{A}^u/I'$ "contains" the wild path algebra with quiver

$$\begin{array}{ccccc}
i_{q,1} & & & & i_{q,3}\\
 & \nwarrow^{\alpha_0'} & & {}^{\alpha_1'}\nearrow & \\
 & & i_{1,n} & & \\
 & {}^{\alpha_2'}\swarrow & \downarrow \alpha_3' & \searrow^{\alpha_4'} & \\
i_{q,5} & & i_{q,7} & & i_{q,9}
\end{array}$$

Case 2: $c_{1,0} \neq 0$. We claim that, for any $i \in [3, n]$, we have that $\tau^t z \otimes v \otimes \tau^{-s} z_i = 0$, for all $t \in [0, n-2]$ and $s \in [0, (i-1)/2 - 1]$. In the following argument, which justifies our claim, we need that $t \leq n-2$ in order that $\tau^{t+1} z$ makes sense, and we need that $s \in [0, (i-1)/2 - 1]$ in order that $\tau^{-s} z_i$ is a point in the graph $\widetilde{Q}_{\lambda_1}$ below the central axis (recall that a point $\tau^{-s} z_i$ in the central axis satisfies the equation $i = 2s + 1$) and, therefore, $w \otimes \alpha \otimes \tau^{-s} z_i \in N$, for any $w \in \widetilde{Q}_{\lambda_q}$.

Fix the point z_i. We prove our claim making a double induction: first over s and then over t. Assume first that $s = 0$ and $t = 0$. Then, $z \otimes v \otimes z_i = 0$, because

$$0 = \widehat{\delta}(\tau z \otimes \alpha \otimes z_i) = c_{1,0} z \otimes v \otimes z_i.$$

Now, take $t \in [1, n-2]$ and assume that $\tau^{t'} z \otimes v \otimes z_i = 0$, for all $t' \in [0, t-1]$. Then

$$\begin{aligned} 0 = \widehat{\delta}(\tau^{t+1} z \otimes \alpha \otimes z_i) &= c_{1,0} \tau^t z \otimes v \otimes z_i \\ &\quad + \sum_{j=2}^{t+1} c_{j,0} \tau^{t+1-j} z \otimes v \otimes z_i, \end{aligned}$$

and, therefore, $\tau^t z \otimes v \otimes z_i = 0$. We have completed the base of the induction: $s = 0$. Now, take $s \in [1, (i-1)/2 - 1]$ and assume that $\tau^t z \otimes v \otimes \tau^{-s'} z_i = 0$, for all $t \in [0, n-2]$ and $s' \in [0, s-1]$. Then

$$\begin{aligned} 0 = \widehat{\delta}(\tau z \otimes \alpha \otimes \tau^{-s} z_i) = \textstyle\sum_{j=1}^{s} c_{0,j} \tau z \otimes v \otimes \tau^{j-s} z_i + c_{1,0} z \otimes v \\ \otimes \tau^{-s} z_i + \textstyle\sum_{j=1}^{s} c_{1,j} z \otimes v \otimes \tau^{j-s} z_i, \end{aligned}$$

and, therefore, $z \otimes v \otimes \tau^{-s} z_i = 0$. Finally, assume that $t \in [1, n-2]$ and $\tau^{t'} z \otimes v \otimes \tau^{-s} z_i = 0$, for all $t' \in [0, t-1]$. Then, we have

$$\begin{aligned} 0 = \widehat{\delta}(\tau^{t+1} z \otimes \alpha \otimes \tau^{-s} z_i) &= \textstyle\sum_{j=1}^{s} c_{0,j} \tau^{t+1} z \otimes v \otimes \tau^{j-s} z_i \\ &\quad + c_{1,0} \tau^t z \otimes v \otimes \tau^{-s} z_i \\ &\quad + \textstyle\sum_{j=1}^{s} c_{1,j} \tau^t z \otimes v \otimes \tau^{j-s} z_i \\ &\quad + \textstyle\sum_{j=0}^{s} c_{2,j} \tau^{t-1} z \otimes v \otimes \tau^{j-s} z_i \\ &\quad + \cdots \\ &\quad + \textstyle\sum_{j=0}^{s} c_{t+1,j} z \otimes v \otimes \tau^{j-s} z_i, \end{aligned}$$

and, therefore, $\tau^t z \otimes v \otimes \tau^{-s} z_i = 0$. Our claim is proved, and, as a consequence, we obtain that $\widehat{\delta}(z \otimes \alpha \otimes \tau^{-j} z_{2j+1}) = 0$, for $j \in [0, 4]$.

Case 3: There is $m \in [2, 4]$ with $c_{m,0} \neq 0$ and $c_{1,0} = c_{2,0} = \cdots = c_{m-1,0} = 0$. We claim that, for any $i \in [3, n]$, we have that $\tau^t z \otimes v \otimes \tau^{-s} z_i = 0$, for all $t \in [0, n-m-1]$ and $s \in [0, (i-1)/2 - 1]$. In the following argument, which justifies our claim, we need that $t \leq n-m-1$ in order that $\tau^{t+m}z$ makes sense, and we need that $s \in [0, (i-1)/2 - 1]$ in order that $\tau^{-s}z_i$ is a point in the graph $\widetilde{Q}_{\lambda_1}$ below the central axis, and, as before, we derive that $w \otimes \alpha \otimes \tau^{-s}z_i \in N$, for any $w \in \widetilde{Q}_{\lambda_q}$.

Fix the point z_i. Again, we prove our claim making a double induction: first over s and then over t. Assume first that $s = 0$ and $t = 0$. Then, $z \otimes v \otimes z_i = 0$, because

$$0 = \widehat{\delta}(\tau^m z \otimes \alpha \otimes z_i) = c_{m,0} z \otimes v \otimes z_i.$$

Now, take $t \in [1, n-m-1]$ and assume that $\tau^{t'}z \otimes v \otimes z_i = 0$, for all $t' \in [0, t-1]$. Then

$$0 = \widehat{\delta}(\tau^{m+t}z \otimes \alpha \otimes z_i) = c_{m,0}\tau^t z \otimes v \otimes z_i + \sum_{j=m+1}^{m+t} c_{j,0}\tau^{m+t-j}z \otimes v \otimes z_i,$$

and, therefore, $\tau^t z \otimes v \otimes z_i = 0$. Now, take $s \in [1, (i-1)/2 - 1]$ and assume that $\tau^t z \otimes v \otimes \tau^{-s'}z_i = 0$, for all $t \in [0, n-m-1]$ and $s' \in [0, s-1]$. Then, we obtain

$$\begin{aligned} 0 = \widehat{\delta}(\tau^m z \otimes \alpha \otimes \tau^{-s}z_i) = & \textstyle\sum_{p=0}^{m-1} c_{p,0}\tau^{m-p}z \otimes v \otimes \tau^{-s}z_i \\ & + \textstyle\sum_{p=0}^{m-1}\sum_{j=1}^{s} c_{p,j}\tau^{m-p}z \otimes v \otimes \tau^{j-s}z_i \\ & + c_{m,0}z \otimes v \otimes \tau^{-s}z_i \\ & + \textstyle\sum_{j=1}^{s} c_{m,j}z \otimes v \otimes \tau^{j-s}z_i, \end{aligned}$$

and, therefore, $z \otimes v \otimes \tau^{-s}z_i = 0$.

Now, take $t \in [1, n-m-1]$ and assume that $\tau^{t'}z \otimes v \otimes \tau^{-s}z_i = 0$, for any $t' \in [0, t-1]$. Then, we obtain

$$\begin{aligned} 0 = \widehat{\delta}(\tau^{m+t}z \otimes \alpha \otimes \tau^{-s}z_i) = & \textstyle\sum_{p=0}^{m-1} c_{p,0}\tau^{m+t-p}z \otimes v \otimes \tau^{-s}z_i \\ & + \textstyle\sum_{p=0}^{m-1}\sum_{j=1}^{s} c_{p,j}\tau^{m+t-p}z \otimes v \otimes \tau^{j-s}z_i \\ & + c_{m,0}\tau^t z \otimes v \otimes \tau^{-s}z_i \\ & + \textstyle\sum_{j=1}^{s} c_{m,j}\tau^t z \otimes v \otimes \tau^{j-s}z_i \\ & + \textstyle\sum_{p=m+1}^{m+t}\sum_{j=1}^{s} c_{p,j}\tau^{m+t-p}z \otimes v \otimes \tau^{j-s}z_i \\ & + \textstyle\sum_{p=m+1}^{m+t} c_{p,0}\tau^{m+t-p}z \otimes v \otimes \tau^{-s}z_i, \end{aligned}$$

and, therefore, $\tau^t z \otimes v \otimes \tau^{-s}z_i = 0$. Our claim is proved, and, as a consequence, we obtain that $\widehat{\delta}(z \otimes \alpha \otimes \tau^{-j}z_{2j+1}) = 0$, for $j \in [0, 4]$. □

Remark 24.8. *Simultaneous unwinding: assume that $\mathcal{A}$ is a nested (resp. seminested) ditalgebra with layer (R, W) and differential δ. Assume that $Re_{i_0} = k[x]_{f(x)}$ and $Re_{j_0} = k[y]_{g(y)}$, with $i_0 \neq j_0$. Suppose that $m \in \mathbb{N}$ and $\lambda_1, \ldots, \lambda_q \in k$ are distinct elements with $f(\lambda_s) \neq 0$, for all $s \in [1, q]$; and $n \in \mathbb{N}$ and $\mu_1, \ldots, \mu_p \in k$ are distinct elements with $g(\mu_r) \neq 0$, for all*

$r \in [1, p]$. *Make* $H^x := k[x]_{f(x)}$ *and* $H^y := k[y]_{g(y)}$; $\pi_{x,s} := x - \lambda_s \in H^x$, *for* $s \in [1, q]$, *and* $\pi_{y,r} := y - \mu_r \in H^y$, *for* $r \in [1, p]$. *Consider, for* $s \in [1, q]$, *the* Re_{i_0}*-module* $X^x_{\lambda_s} := H^x/H^x\pi^1_{x,s} \oplus \cdots \oplus H^x/H^x\pi^m_{x,s}$, *and make* $X^x := \oplus_{s=1}^q X^x_{\lambda_s}$. *Similarly, consider, for* $r \in [1, p]$, *the* Re_{j_0}*-module* $X^y_{\mu_r} := H^y/H^y\pi^1_{y,r} \oplus \cdots \oplus H^y/H^y\pi^n_{y,r}$, *and make* $X^y := \oplus_{r=1}^p X^y_{\mu_r}$. *Finally, consider* $X' := Re$, *where* $e = 1 - (e_{i_0} + e_{j_0})$. *Then, make* $X := X^x \oplus X^y \oplus X'$ *and consider* R *as an initial subditalgebra of* $\mathcal{A}$. *Then,* X *is an admissible* R*-module. The ditalgebra obtained from* $\mathcal{A}$ *by unwinding at* i_0 *and* j_0, *using* $(m; \lambda_1, \ldots, \lambda_q)$ *and* $(n; \mu_1, \ldots, \mu_p)$ *is* $\mathcal{A}^X$. *It is a nested (resp. seminested) ditalgebra. In fact,* $\mathcal{A}^X$ *can be obtained from* $\mathcal{A}$ *by first unravelling at* i_0 *with* $(m; \lambda_1, \ldots, \lambda_q)$, *then unravelling at* j_0 *with* $(n; \mu_1, \ldots, \mu_p)$, *and then deleting the idempotents corresponding to the points* i_* *and* j_*. *It is convenient, to associate to it a special name and a special simple notation. The same construction can be performed for an arbitrary number of marked points, but we shall only use it for two points.*

So, X *is an admissible* R*-module and an argument similar to the one used in the proof of (23.23) shows that* $\mathcal{A}^X$ *is a nested (resp. seminested) ditalgebra. We have a splitting* $\mathrm{End}_R(X)^{op} = S \oplus P$, *with* $S = (\oplus_{s=1}^q \oplus_{a=1}^m kf^x_{s,a}) \oplus (\oplus_{r=1}^p \oplus_{a=1}^n kf^y_{r,a}) \oplus Re$ *minimal. We can describe the bigraph of* $\mathcal{A}^X$ *as we did for* $\mathcal{A}^u$ *in (23.23). We can choose basic elements* $x^t_{s,a} \in X^x_{\lambda_s}$, *with* $s \in [1, q]$, $a \in [1, m]$ *and* $t \in [1, a-1]$, *and basic elements* $y^t_{r,a} \in X^y_{\mu_r}$, *with* $r \in [1, p]$, $a \in [1, n]$ *and* $t \in [1, a-1]$, *and make* $x^i_i := e_i \in Re_i$, *for all vertex* $i \notin \{i_0, j_0\}$, *which can be extended to a dual basis*

$$(x^t_{s,a}, \nu_{x^t_{s,a}})_{s,a,t} \cup (y^t_{r,a}, \nu_{y^t_{r,a}})_{r,a,t} \cup (x^i_i, \nu^i_i)_{i \notin \{i_0, j_0\}}$$

of the right S*-module* X. *We also have connected components* $\widetilde{Q}^x_{\lambda_s}$, *for* $s \in [1, q]$, *and* $\widetilde{Q}^y_{\mu_r}$, *for* $r \in [1, p]$, *of the quiver* $\widetilde{Q}$, *associated to the quiver* Q *of the endomorphism algebra* $\mathrm{End}_R(X)$. *Now,* Q *admits* $q + p$ *components (each one a copy of the quiver described in (23.21) corresponding to each factor* $\mathrm{End}_{H^x}(X^x_{\lambda_s})$ *or* $\mathrm{End}_{H^y}(X^y_{\mu_r})$ *of* $\mathrm{End}_R(X^x \oplus X^y)$*).*

Now, assume that $\alpha : i_0 \to j_0$ *is a solid arrow of* $\mathcal{A}$, *such that, for* $s \in [1, q]$ *and* $r \in [1, p]$, *we have*

$$\delta(\alpha) = \sum_{i,j,d} c_{s,r,i,j,d}(y - \mu_r)^i v_d (x - \lambda_s)^j,$$

for some coefficients $c_{s,r,i,j,d} \in k$ *and some dotted arrows* $v_1, \ldots, v_d : i_0 \to j_0$ *of* $\mathcal{A}$. *Then, if we adopt the* τ*-notation and the other notational conventions used for the unravelling in (24.6) and in the proof of (24.7), we have an analogous formula for the differential of a solid arrow* $z \otimes \alpha \otimes w$ *in* $\mathcal{A}^X$, *with* $w \in \widetilde{Q}^x_{\lambda_s}$

and $z \in \widetilde{Q}^y_{\mu_r}$. *That is*

$$\begin{aligned}\delta^X(z \otimes \alpha \otimes w) = & \textstyle\sum_{z>z' \in \widetilde{Q}^y_{\mu_r}} \gamma_{z'} z' \otimes \alpha \otimes w \\ & + \textstyle\sum_{s,r,i,j,d} c_{s,r,i,j,d} \tau^{-i} z \otimes v_d \otimes \tau^j w \\ & - \textstyle\sum_{w<w' \in \widetilde{Q}^x_{\lambda_s}} z \otimes \alpha \otimes w' \gamma_{w'},\end{aligned}$$

for some $\gamma_{z'}, \gamma_{w'} \in P^*$.

Recall from (11.8) and the definition of δ^X, *that for the dotted arrows of* $\mathcal{A}^X$ *of the form* $\gamma_\xi \in P^*$, *where* ξ *is a path class of* kQ/I, *see (23.22) and (23.24), we have*

$$\delta^X(\gamma_\xi) = \mu(\gamma_\xi) = \sum_{\xi_1, \xi_2} \gamma_\xi(p_{\xi_1} p_{\xi_2}) \gamma_{\xi_2} \otimes \gamma_{\xi_1},$$

where ξ_1, ξ_2 *run in the set of path classes of* kQ/I *and* p_{ξ_1}, p_{ξ_2} *denote the corresponding elements in* $P = \mathrm{radEnd}_R(X)^{op}$. *We have that* $p_{\xi_1} p_{\xi_2} = p_{\xi_2 \xi_1}$ *if* ξ_1 *and* ξ_2 *are composable path classes and zero otherwise. Moreover, if* ξ *is a path class determined by the projection of a path class in* $\widetilde{Q}^x_{\lambda_s}$, *then* $\gamma_\xi(p_{\xi'}) = f^x_{s,t(\xi)}$ *if* $\xi = \xi'$ *and zero otherwise. Then*

$$\delta^X(\gamma_\xi) = \sum_{\xi_1, \xi_2 \text{ with } \xi = \xi_2 \xi_1} \gamma_{\xi_2} \otimes \gamma_{\xi_1}.$$

Lemma 24.9. *Assume that* $\mathcal{A}$ *is a seminested ditalgebra with layer* (R, W) *and differential* δ, *that* $\mathcal{A}$ *has only two points* $1, 2$, *and that there are at least one solid arrow* α *and one dotted arrow* v. *Suppose* $Re_1 = k[x]_{f(x)}$, $Re_2 = k[y]_{g(y)}$, $\alpha : 1 \to 2$ *and* $v : 1 \to 2$. *Moreover, assume that* $\delta(\alpha) = r(x, y)v$, *for some element* $r(x, y) \in k[x, y]_{f(x)g(y)}$ *which admits an* $f(x)g(y)$*-zero. Then,* $\mathcal{A}$ *is a wild ditalgebra.*

Proof. We can give a very similar proof to the one of (24.7), using (24.8). As before, we can assume that $r(x, y) \in k[x, y]$. We know the existence of scalars λ_1 and μ_1 such that $f(\lambda_1) \neq 0$, $g(\mu_1) \neq 0$ and $r(\lambda_1, \mu_1) = 0$. We shall make a simultaneous unwinding at the point 1 (using $m = 9$ and λ_1) and at the point 2 (using $n = 9$ and μ_1). Consider the connected components $\widetilde{Q}_1 := \widetilde{Q}^x_{\lambda_1}$ and $\widetilde{Q}_2 := \widetilde{Q}^y_{\mu_1}$ of the quiver $\widetilde{Q}$, described in (24.8), associated to the quiver Q of the endomorphism algebra $\mathrm{End}_R(X^x \oplus X^y)$. Now, Q admits two components (each one a copy of the quiver described in (23.21) corresponding to each factor $\mathrm{End}_{H^x}(X^x_{\lambda_1})$ and $\mathrm{End}_{H^y}(X^y_{\mu_1})$ of $\mathrm{End}_R(X)$). We have an analogous formula for the differential of arrows $w \otimes \alpha \otimes w'$ in $\mathcal{A}^X$ with $w \in \widetilde{Q}_2$ and $w' \in \widetilde{Q}_1$. That is, if we write

$$\delta(\alpha) = \sum_{i,j} c_{i,j}(y - \mu_1)^i v (x - \lambda_1)^j,$$

then we have, for $w_2 \in \widetilde{Q}_2$ and $w_1 \in \widetilde{Q}_1$

$$\begin{aligned}\delta^X(w_2 \otimes \alpha \otimes w_1) = & \textstyle\sum_{w_2 > w_2' \in \widetilde{Q}_2} \gamma_{w_2'} w_2' \otimes \alpha \otimes w_1 \\ & + \textstyle\sum_{i,j} c_{i,j} \tau^{-i} w_2 \otimes v \otimes \tau^j w_1 \\ & - \textstyle\sum_{w_1 < w_1' \in \widetilde{Q}_1} w_2 \otimes \alpha \otimes w_1' \gamma_{w_1'},\end{aligned}$$

for some $\gamma_{w_2'}, \gamma_{w_1'} \in P^*$. Then, if we make $q = 2$, we can complete this proof with the same formal argument given in the proof of (24.7). □

Lemma 24.10. *Assume that $\mathcal{A}$ is a seminested ditalgebra with layer (R, W) and differential δ. Assume that $\mathcal{A}$ has at least one solid arrow α. Assume that either α is a loop based at a unique vertex 1 of $\mathcal{A}$, with $R = Re_1 = k[x]_{f(x)}$, or α is not a loop and $R = Re_1 \times Re_2$, with $Re_1 = k[x]_{f(x)}$. Thus, vertex 2 may be marked or not. In all cases, we suppose that there is an element $\mu \in k$ such that $f(\mu) \neq 0$ and $\delta(\alpha) = 0$. Then, $\mathcal{A}$ is a wild ditalgebra.*

Proof. We provide a proof in the same spirit as the one given for the other critical ditalgebras. We first suppose that, in case α is not a loop, we have $\alpha : 1 \to 2$. Consider the ditalgebra $\mathcal{A}^u$ obtained from $\mathcal{A}$ by unravelling at the vertex $i_0 = 1$ with $n = 10$ and μ. Consider the associated quiver $\widetilde{Q}$ and consider its central axis. Again, we mean the points $\tau^{-s} z_i$ with $i = 2s + 1$, where $z_i := x_i^{i-1}$. Define $z := x_n^0$ if α is a loop and $z := v_2^2$ otherwise. Then, consider the following subset N of arrows of W_0^u

$$N := \{z \otimes \alpha \otimes w \mid w \in \widetilde{Q} \text{ with } \mathbb{h}(w) < n\},$$

and consider the filtration

$$\mathcal{F}(N) : \emptyset = N_0 \subseteq N_1 \subseteq \cdots \subseteq N_\ell = N,$$

where $\ell = n - 1$ and each N_t is, by definition, the subset of N formed by the elements $z \otimes \alpha \otimes w$ with $\mathbb{h}(w) \leq t$.

In case α is not a loop recall that, with the notation of (23.24), we have $\rho(w) = \sum_{w < w' \in \widetilde{Q}} w' \gamma_{w'}$, for some $\gamma_{w'} \in P^*$. We also know that $\lambda(v_2^2) = 0$, from the general formula for λ given in (11.11), because $x_2^2 P = 0$. Then, since $\delta(\alpha) = 0$, for $w \in \widetilde{Q}$, we have

$$\delta^u(v_2^2 \otimes \alpha \otimes w) = - \sum_{w < w' \in \widetilde{Q}} v_2^2 \otimes \alpha \otimes w' \gamma_{w'},$$

for some $\gamma_{w'} \in P^*$. Then, in case α is a loop, from (24.6), we also have the formula

$$\delta^u(z \otimes \alpha \otimes w) = - \sum_{w < w' \in \widetilde{Q}} z \otimes \alpha \otimes w' \gamma_{w'},$$

for some $\gamma_{w'} \in P^*$. Thus, in any case, N is a triangular subset of W_0^u and we can consider the ditalgebra $\mathcal{A}^u/I$, where I is the ideal of $\mathcal{A}^u$ generated by N. Then, if we define $q := i_n$ in case α is a loop and $q := 2$ otherwise, we can consider in $\mathcal{A}^u/I$ the five solid arrows $\alpha_j := z \otimes \alpha \otimes \tau^{-j} z_{2j+1} : i_{2j+1} \rightarrow q$, for $j \in [0, 4]$. It is clear that the differential of $\alpha_0, \ldots, \alpha_4$ is zero in $\mathcal{A}^u/I$, which therefore "contains" the wild algebra with quiver

$$\begin{array}{ccccc} i_1 & & & & i_3 \\ & \overset{\alpha_0}{\searrow} & & \overset{\alpha_1}{\swarrow} & \\ & & q & & \\ & \overset{\alpha_2}{\nearrow} & \uparrow \alpha_3 & \overset{\alpha_4}{\nwarrow} & \\ i_5 & & i_7 & & i_9 \end{array}$$

Then, $\mathcal{A}$ is wild. The case $\alpha : 2 \rightarrow 1$ is similar. □

Corollary 24.11. *Assume that k is an algebraically closed field. Assume that $\mathcal{A}$ is a seminested ditalgebra, with differential δ and layer (R, W). Then, $\mathcal{A}$ is wild in the following cases:*

(1) $\delta(\alpha) = 0$, for some solid arrow α with either $Re_{s(\alpha)} \not\cong k$ or $Re_{t(\alpha)} \not\cong k$.
(2) $\delta(\alpha) = cv$, for some solid arrow α, some dotted arrow v and some non-invertible element $c \in C := Re_{t(\alpha)} \otimes_k Re_{s(\alpha)}$, where $Re_{t(\alpha)} \not\cong k$ and $Re_{s(\alpha)} \not\cong k$.

Proof. After deleting all the vertices of $\mathcal{A}$ different from $t(\alpha)$ and $s(\alpha)$, if necessary, and using (22.8), we can assume that $\mathcal{A}$ has only the points $t(\alpha)$ and $s(\alpha)$ (which may coincide). Then, by (24.4), $\mathcal{A}$ is a critical ditalgebra of one of the types listed in (24.5). Critical ditalgebras are wild by (24.7), (24.9) and (24.10). □

Exercise 24.12. *Consider the algebra $R = k[x]_{f(x)} \times k[y]_{g(y)}$, $e_1 := (1, 0)$ and $e_2 := (0, 1)$. Fix elements $g_1(x), \ldots, g_s(x) \in k[x]_{f(x)}$ and $h_1(y), \ldots, h_s(y) \in k[y]_{g(y)}$. Then, for each pair (p, q) of families $p_1(x), \ldots, p_s(x) \in k[x]_{f(x)}$ and $q_1(y), \ldots, q_s(y) \in k[y]_{g(y)}$, we can consider the nested ditalgebra $\mathcal{A}_{p,q}$ with differential δ and layer (R, W), defined as follows: $W_0 := Re_2 \otimes_k Re_1 = R\alpha R$, where $\alpha = e_2 \otimes e_1$ is the unique solid arrow $\alpha : 1 \longrightarrow 2$; clearly, $Re_1 = k[x]_{f(x)}$ and $Re_2 = k[y]_{g(y)}$; consider any R-R-bimodule W_1 freely generated by s dotted arrows $v_1, \ldots, v_s$ from 1 to 2; make*

$$\delta(\alpha) = \sum_{t=1}^{s} h_t(y) q_t(y) v_t p_t(x) g_t(x);$$

and $\delta(v_t) = 0$, *for all* t. *It is clear that we have indeed a nested ditalgebra. For the simplest case where all the elements* $p_t(x)$ *and* $q_t(y)$ *are equal to 1, we simply write* $\mathcal{A}_0 := \mathcal{A}_{p,q}$. *Show that* $\mathcal{A}_{p,q}$ *is wild whenever* $\mathcal{A}_0$ *is so.*

References. The list and the proof that *critical semifree bocses* are wild was given by Drozd in [28], and then refined in [19] for *layered bocses*. They both provide big bimodules producing their wildness. In [49] Z. Yingbo and Xu Yunge provide smaller bimodules producing this wildness. A primitive version of the proof that critical ditalgebras are wild, without exhibiting explicit bimodules producing wildness, was presented in [51].

25
Reduction functors

Here we explore the properties of the functors $F_z : \mathcal{A}^z\text{-Mod} \longrightarrow \mathcal{A}\text{-Mod}$ associated to each one of the basic operations $z \in \{a, d, r, e, u\}$, described in previous sections. The name *reduction functor* applied to them makes reference to the behavior of these functors towards the following notions.

Definition 25.1. *Assume that $\mathcal{A}$ is a seminested k-ditalgebra with layer (R, W). Suppose E is some k-algebra and consider the decomposition of the unit element of R as a sum of central orthogonal primitive idempotents $1 = \sum_{i=1}^{s} e_i$. If $M \in \mathcal{A}\text{-}E\text{-Mod}$, see (21.4), then each $M_i = e_i M$ is an $Re_i\text{-}E$-bimodule, and we can look at its length $\ell_E(M_i)$. The length vector of M over E is*

$$\underline{\ell}_E(M) = (\ell_E(M_1), \ldots, \ell_E(M_s)).$$

Denote by $(\mathbb{B}_0, \mathbb{B}_1)$ a basis for the seminested layer (R, W); by $\mathcal{P}_{\text{mk}}$ and $\mathcal{P} \backslash \mathcal{P}_{\text{mk}}$ the sets of marked and unmarked points, respectively, of the bigraph $\mathbb{B}$ of $\mathcal{A}$. Then, the norm $\|M\|$ of the $\mathcal{A}\text{-}E$-bimodule M is given by

$$\|M\| = \sum_{\alpha \in \mathbb{B}_0} \ell_E(M_{t(\alpha)}) \ell_E(M_{s(\alpha)}) + \sum_{i \in \mathcal{P}_{\text{mk}}} \ell_E(M_i)^2.$$

Recall that we have a forgetful functor $U_{\mathcal{A}} : \mathcal{A}\text{-}E\text{-Mod} \longrightarrow R\text{-}E\text{-Mod}$, which preserves isomorphisms. Hence, the length vector and the norm are isomorphism invariants in $\mathcal{A}\text{-}E\text{-Mod}$.

Notice that, when $E = k$, we have $\mathcal{A}\text{-}k\text{-Mod} = \mathcal{A}\text{-}k\text{-Mod}_p \cong \mathcal{A}^k\text{-Mod} \cong \mathcal{A}\text{-Mod}$. Thus, given a k-ditalgebra $\mathcal{A}$ and $M \in \mathcal{A}\text{-Mod}$, the number $\|M\|$ is computed with the given formula with $E = k$.

Lemma 25.2. *Assume that $\mathcal{A}^a$ is obtained by absorption of a loop from the seminested ditalgebra $\mathcal{A}$, as in (23.16) and let $F_a : \mathcal{A}^a\text{-Mod} \longrightarrow \mathcal{A}\text{-Mod}$ be the associated functor, as in (8.20). Given a k-algebra E, we have:*

(1) *For any* $N \in \mathcal{A}^a$-E-Mod, *we have* $\underline{\ell}_E(F_a^E(N)) = \underline{\ell}_E(N)$.
(2) $F_a^E : \mathcal{A}^a$-E-Mod$\longrightarrow\mathcal{A}$-$E$-Mod *is a length preserving equivalence.*
(3) F_a^E *restricts to an equivalence* $\mathcal{A}^a$-E-Mod$_p\longrightarrow\mathcal{A}$-$E$-Mod$_p$.
(4) *For any* $N \in \mathcal{A}^a$-E-mod, *we have* $\|N\| = \|F_a^E(N)\|$.

Proof. (2) and (3) are clear by (8.20), (21.3) and (21.6)(1). (1) and (4) follow from the definition of F_a, the fact that the trivial part of the layer of $\mathcal{A}^a$ is the trivial part of the layer of $\mathcal{A}$ and (23.16). □

Lemma 25.3. *Assume that* $\mathcal{A}^r$ *is obtained by regularization of an arrow* α *from the seminested ditalgebra* $\mathcal{A}$*, as in (23.15), and let* $F_r : \mathcal{A}^r$-Mod$\longrightarrow\mathcal{A}$-Mod *be the associated functor, as in (8.19). Given a k-algebra E, we have:*

(1) *For any* $N \in \mathcal{A}^r$-E-Mod, *we have* $\underline{\ell}_E(F_r^E(N)) = \underline{\ell}_E(N)$.
(2) $F_r^E : \mathcal{A}^r$-E-Mod$\longrightarrow\mathcal{A}$-$E$-Mod *is a length preserving equivalence.*
(3) F_r^E *restricts to an equivalence* $\mathcal{A}^r$-E-Mod$_p\longrightarrow\mathcal{A}$-$E$-Mod$_p$.
(4) *If* $N \in \mathcal{A}^r$-E-mod, *then* $\|N\| \leq \|F_r^E(N)\|$. *This inequality is strict if* $F_r^E(N)_{t(\alpha)} \neq 0$ *and* $F_r^E(N)_{s(\alpha)} \neq 0$.

Proof. (2) and (3) follow from (8.19), (21.3) and (21.6)(1). The fact that the equivalence F_r is induced by restriction of an epimorphism of ditalgebras $\phi_r : \mathcal{A} \to \mathcal{A}^r$, which maps R identically onto $R^r = R$, implies (1). The last part follows from the inspection of the layer described in (23.15), because we have $\|F_r^E(N)\| - \|N\| = \ell_E(F_r^E(N)_{t(\alpha)})\ell_E(F_r^E(N)_{s(\alpha)})$. □

Lemma 25.4. *Assume that* $\mathcal{A}^d$ *is obtained by deletion of a finite collection* $\{e_i \mid i \in \Omega\}$ *of idempotents where* Ω *is a set of points of the seminested ditalgebra* $\mathcal{A}$*, as in (23.14), and let* $F_d : \mathcal{A}^d$-Mod$\longrightarrow\mathcal{A}$-Mod *be the associated functor, as in (8.17). Given a k-algebra E, we have:*

(1) *For* $N \in \mathcal{A}^d$-E-Mod, *the length vector* $\underline{\ell}_E(F_d^E(N))$ *satisfies* $\underline{\ell}_E(F_d^E(N))_i = 0$*, for all* $i \in \Omega$*, and* $\underline{\ell}_E(F_d^E(N))_i = \underline{\ell}_E(N)_i$*, otherwise.*
(2) $F_d^E : \mathcal{A}^d$-E-Mod$\longrightarrow\mathcal{A}$-$E$-Mod *is a length preserving functor inducing an equivalence from* $\mathcal{A}^d$-E-Mod *to the full subcategory of* $\mathcal{A}$-E-Mod *formed by the bimodules* M *with* $M_i = 0$*, for all* $i \in \Omega$.
(3) F_d^E *restricts to an equivalence from* $\mathcal{A}^d$-E-Mod$_p$ *to the full subcategory of* $\mathcal{A}$-E-Mod$_p$ *formed by the proper bimodules* M *satisfying* $M_i = 0$*, for all* $i \in \Omega$.
(4) *If* $N \in \mathcal{A}^d$-E-mod, *then* $\|N\| = \|F_d^E(N)\|$.

Proof. (2) and (3) follow from (8.17), (21.3) and (21.6)(1). The rest is straightforward, having in mind the description of the layer given in (23.14). □

Lemma 25.5. *Assume that X is an admissible module over the proper subditalgebra $\mathcal{A}'$ of the layered ditalgebra $\mathcal{A}$. Consider the layered ditalgebra $\mathcal{A}^X$ and the associated functor $F_X : \mathcal{A}^X\text{-Mod} \longrightarrow \mathcal{A}\text{-Mod}$. Assume that for $M \in \mathcal{A}\text{-Mod}$ there is an S-module N such that $X \otimes_S N \cong M$ in B-Mod. Then, there is $\overline{N} \in \mathcal{A}^X\text{-Mod}$ with $F_X(\overline{N}) \cong M$ in $\mathcal{A}$-Mod. Moreover, given a k-algebra E, if $M \in \mathcal{A}\text{-}E\text{-Mod}_p$, $N \in S\text{-}E\text{-Mod}$ and $X \otimes_S N \cong M$ in $B\text{-}E$-Mod, then $\overline{N}$ admits a natural structure of proper $\mathcal{A}^X\text{-}E$-bimodule satisfying that $U^E_{\mathcal{A}^X}(\overline{N}) = N$ and $F^E_X(\overline{N}) \cong M$ in $\mathcal{A}\text{-}E\text{-Mod}_p$.*

Proof. Take $M \in \mathcal{A}\text{-}E\text{-Mod}_p$ and $N \in S\text{-}E\text{-Mod}$. Let $f^0 : X \otimes_S N \longrightarrow M$ be an isomorphism of $B\text{-}E$-bimodules. Transfer the structure of $A\text{-}E$-bimodule of M over $X \otimes_S N$ using f^0, to construct the $A\text{-}E$-bimodule $\overline{X \otimes_S N}$. Thus, the pair $f = (f^0, 0) : \overline{X \otimes_S N} \longrightarrow M$ is an isomorphism in $\mathcal{A}\text{-}E\text{-Mod}_p$. By definition, we have, for $x \in X$, $n \in N$, $b \in B$ and $e \in E$, that $b(x \otimes n)e = (f^0)^{-1}[b(f^0(x \otimes n))e] = (f^0)^{-1}[f^0(bx \otimes ne)] = (bx) \otimes (ne)$. Then, we can apply (16.1) to obtain $\overline{N} \in \mathcal{A}^X$-Mod, with the same underlying S-module structure as N, such that $F_X(\overline{N}) = \overline{X \otimes_S N}$ in $\mathcal{A}$-Mod. We have that $\overline{N}$ is an $S\text{-}E$-bimodule. We claim that it is in fact an $A^X\text{-}E$-bimodule. Indeed, having in mind the notation and the formula defining the left A^X-module structure for $\overline{N}$ given in (16.1), we have, for $\nu \in X^*$, $w \in W''_0$, $x \in X$, $n \in N$ and $e \in E$

$$\begin{aligned}
[(\nu \otimes w \otimes x) * n]e &= \sigma(\varepsilon \otimes 1)[\nu \otimes w \circ (x \otimes n)]e \\
&= \sigma(\varepsilon \otimes 1)[\nu \otimes [w \circ (x \otimes n)]e] \\
&= \sigma(\varepsilon \otimes 1)[\nu \otimes w \circ [(x \otimes n)e]] \\
&= \sigma(\varepsilon \otimes 1)[\nu \otimes [w \circ (x \otimes ne)]] \\
&= (\nu \otimes w \otimes x) * (ne).
\end{aligned}$$

Then, $\overline{N} \in \mathcal{A}^X\text{-}E\text{-Mod}_p$ and $F^E_X(\overline{N}) = \overline{X \otimes_S N} \cong M$ in $\mathcal{A}\text{-}E$-Mod. □

Remark 25.6. *Under the general assumptions of the last lemma, assume furthermore that the layer (S, W^X) of the ditalgebra $\mathcal{A}^X$ has in the first component an algebra S such that the unit decomposes as a sum $1 = \sum_{i=1}^t f_i$ of central orthogonal primitive idempotents. Take $M \in \mathcal{A}\text{-}E\text{-Mod}_p$. Then, the condition $M \cong X \otimes_S N$ in $B\text{-}E$-Mod, for some $S\text{-}E$-bimodule N, can be replaced by $M \cong \oplus_{i=1}^t Xf_t \otimes_{Sf_t} f_t N$ in $B\text{-}E$-Mod, for some $S\text{-}E$-bimodule N.*

Lemma 25.7. *Let $\mathcal{A}$ be a seminested ditalgebra with layer (R, W). Assume that $\mathcal{A}^X$ is a obtained from $\mathcal{A}$ by reduction, using the triangular complete admissible $\mathcal{A}'$-module X, where $\mathcal{A}'$ is an initial subditalgebra of $\mathcal{A}$. Thus, $\mathcal{A}^X$ has layer (S, W^X), where $\Gamma = \mathrm{End}_{\mathcal{A}'}(X)^{op}$ admits the splitting $\Gamma = S \oplus P$. Assume that this layer is seminested. We shall denote by e_i (resp. f_j) the canonical primitive central idempotent associated to the point i of R (resp. to the point j of S). Let*

$F_X : \mathcal{A}^X\text{-Mod} \longrightarrow \mathcal{A}\text{-Mod}$ *be the associated functor and take any k-algebra* E. *Then, we have:*

(1) For any $N \in \mathcal{A}^X\text{-}E\text{-Mod}$, *we have* $\underline{\ell}_E(F_X^E(N))^t = [X]\underline{\ell}_E(N)^t$, *where* $[X]$ *is the ranks matrix of* X, *that is the matrix with non-negative integral entries* $[X]_{i,j} = \text{rank}_{Sf_j}(e_i X f_j)$.

(2) $F_X^E : \mathcal{A}^X\text{-}E\text{-Mod} \longrightarrow \mathcal{A}\text{-}E\text{-Mod}$ *is a length controlling, full and faithful functor which restricts to proper bimodules:* $\mathcal{A}^X\text{-}E\text{-Mod}_p \longrightarrow \mathcal{A}\text{-}E\text{-Mod}_p$.

Proof. Recall that the triangular bimodule X admits a right additive B-S-bimodule filtration $\mathcal{F}(X): \quad 0 = X^0 \subseteq X^1 \subseteq \cdots \subseteq X^{\ell_X} = X$, such that $X^t P \subseteq X^{t-1}$, for all $t \in [1, \ell_X]$. We know that N is an S-E-bimodule via $ne := \alpha_N(e)^0(n)$, where $\alpha_N : E \longrightarrow \text{End}_{\mathcal{A}^X}(N)^{op}$ is the given $\mathcal{A}$-E-bimodule structure of $N \in \mathcal{A}^X\text{-}E\text{-Mod}$, $n \in N$ and $e \in E$. Thus, each $e_i X^t \otimes_S N$ inherits a natural structure of an Re_i-E-bimodule. Namely, $(x \otimes n) \star e := x \otimes (ne)$, for $x \in e_i X^t$ and $n \in N$. We denote the length of submodules or quotients of these modules with the symbol $\ell_E^\star$.

Recall also, from (12.10) and (21.3), that $\alpha_{F_X(N)} : E \longrightarrow \text{End}_{\mathcal{A}}(F_X(N))^{op}$ satisfies, for $e \in E$, $n \in N$ and $x \in X$

$$\begin{aligned}\left(\alpha_{F_X(N)}(e)\right)^0 [x \otimes n] &= F_X(\alpha_N(e))^0[x \otimes n] \\ &= x \otimes \alpha_N(e)^0(n) + \sum_\xi x p_\xi \otimes \alpha_N(e)^1(\gamma_\xi)[n],\end{aligned}$$

where $(p_\xi, \gamma_\xi)_\xi$ is a dual basis for the projective right S-module P. Consider the structure of Re_i-E-bimodule on $e_i F_X(N) = e_i X \otimes_S N$ determined by the $\mathcal{A}$-E-bimodule $F_X^E(N)$, that is $(x \otimes n) \cdot e = \alpha_{F_X(N)}(e)^0[x \otimes n]$, for $x \in e_i X$ and $n \in N$. From the previous formula for $\alpha_{F_X(N)}(e)^0$, we immediately obtain that each $e_i X^t \otimes_S N$ is an Re_i-E-subbimodule of $e_i F_X^E(N)$. We write the length of submodules or quotients of these modules with the symbol ℓ_E. We claim that $\ell_E^\star(e_i X^t \otimes_S N) = \ell_E(e_i X^t \otimes_S N)$, for any $t \in [0, \ell_X]$. If we write $r(x \otimes n) := \sum_\xi x p_\xi \otimes \alpha_N(e)^1(\gamma_\xi)[n]$, for $x \in e_i X^t$ and $n \in N$, then we have that $r(e_i X^{t+1} \otimes_S N) \subseteq e_i X^t \otimes_S N$ and, therefore

$$\ell_E^\star\left((e_i X^{t+1} \otimes_S N)/(e_i X^t \otimes_S N)\right) = \ell_E\left((e_i X^{t+1} \otimes_S N)/(e_i X^t \otimes_S N)\right).$$

Moreover, the sequences

$$0 \rightarrow e_i X^t \otimes_S N \rightarrow e_i X^{t+1} \otimes_S N \rightarrow (e_i X^{t+1} \otimes_S N)/(e_i X^t \otimes_S N) \rightarrow 0$$

are exact with the $\cdot$ structure as well as with the $\star$ structure. Then, we can show our claim by induction on t. The base of the induction $t = 1$ is clear. So assume the claim holds for $t \in [1, \ell_X - 1]$ and let us show it for $t + 1$. From the given exact sequences, we obtain

$$\begin{aligned}\ell_E(e_iX^{t+1}\otimes_S N) &= \ell_E(e_iX^t\otimes_S N)+\ell_E\left((e_iX^{t+1}\otimes_S N)/(e_iX^t\otimes_S N)\right)\\ &= \ell^\star_E(e_iX^t\otimes_S N)+\ell^\star_E\left((e_iX^{t+1}\otimes_S N)/(e_iX^t\otimes_S N)\right)\\ &= \ell^\star_E(e_iX^{t+1}\otimes_S N),\end{aligned}$$

and our claim is proved. Then, for $t=\ell_X$, $X=X^t$, and we have

$$\begin{aligned}\ell_E(e_iX\otimes_S N) &= \ell^\star_E(e_iX\otimes_S N)\\ &= \ell^\star_E\left(\oplus_j e_iXf_j\otimes_S f_jN\right)\\ &= \textstyle\sum_j \ell^\star_E(e_iXf_j\otimes_{Sf_j} f_jN)\\ &= \textstyle\sum_j \operatorname{rank}[e_iXf_j]\ell_E(f_jN).\end{aligned}$$

We have an induced functor $F^E_X:\mathcal{A}^X\text{-}E\text{-Mod}\longrightarrow\mathcal{A}\text{-}E\text{-Mod}$, which is full and faithful by (13.5) and (21.3). From the definition of F_X, it is clear that this functor restricts to proper bimodules $F^E_X:\mathcal{A}^X\text{-}E\text{-Mod}_p\longrightarrow\mathcal{A}\text{-}E\text{-Mod}_p$.

By definition (see (12.4)), each $f_j=(f^0_j,0)\in \operatorname{End}_{\mathcal{A}'}(X)^{op}$ is a non-zero idempotent, thus $Xf_j\neq 0$, and $e_iXf_j\neq 0$, for some i. Then, $\ell_E(F^E_X(N))=\sum_i \ell_E(e_iX\otimes_S N)=\sum_{i,j}\operatorname{rank}[e_iXf_j]\,\ell_E(f_jN)\geq\sum_j\ell_E(f_jN)=\ell_E(N)$, and the functor $F^E_X:\mathcal{A}^X\text{-}E\text{-Mod}\longrightarrow\mathcal{A}\text{-}E\text{-Mod}$ is length controlling. □

Lemma 25.8. *Assume that $\mathcal{A}^e$ is obtained by reduction of an edge $\alpha: i_0\rightarrow j_0$ from the seminested ditalgebra $\mathcal{A}$, as in (23.18). Consider the associated functor $F_e:\mathcal{A}^e\text{-Mod}\longrightarrow\mathcal{A}\text{-Mod}$ and any k-algebra E. Then:*

(1) $F^E_e:\mathcal{A}^e\text{-}E\text{-Mod}\longrightarrow\mathcal{A}\text{-}E\text{-Mod}$ is a length controlling equivalence.
(2) If E is a semisimple k-algebra, the functor F^E_e restricts to an equivalence from $\mathcal{A}^e\text{-}E\text{-Mod}_p$ onto $\mathcal{A}\text{-}E\text{-Mod}_p$.
(3) If $N\in\mathcal{A}^e\text{-}E\text{-mod}$, then $\|N\|\leq\|F^E_e(N)\|$. If $F^E_e(N)_{i_0}\neq 0$ and $F^E_e(N)_{j_0}\neq 0$, this inequality is strict.

Proof. We adopt the notation of (23.18) and its proof. Thus, $\mathcal{A}^e=\mathcal{A}^X$ is seminested and $F_e=F^X:\mathcal{A}^e\text{-Mod}\longrightarrow\mathcal{A}\text{-Mod}$, where $X=H_{i_0}\oplus H_z\oplus H_{j_0}\oplus(1-e)R$, with $e=e_{i_0}+e_{j_0}$. Recall that $\mathcal{P}^e=\{i_*,z,j_*\}\cup\mathcal{P}\backslash\{i_0,j_0\}$.

From (25.7), we already know that the functor $F^E_e:\mathcal{A}^e\text{-}E\text{-Mod}\longrightarrow\mathcal{A}\text{-}E\text{-Mod}$ is length controlling, full and faithful. Then, its restriction to proper bimodules $F^E_e:\mathcal{A}^e\text{-}E\text{-Mod}_p\longrightarrow\mathcal{A}\text{-}E\text{-Mod}_p$ is full and faithful too. Assume that E is semisimple and let us show that the restricted one F^E_e is dense.

Recall that the algebra eB can be identified with the path algebra of the quiver $i_0\stackrel{\alpha}{\longrightarrow}j_0$. Thus, given any B-E-bimodule M, we can consider the multiplication by α map $M_\alpha: e_{i_0}M\longrightarrow e_{j_0}M$. Since E is semisimple, there is a right E-module direct sum decomposition $e_{i_0}M=M'_{i_*}\oplus M'_z$, where $M'_{i_*}:=$

$\operatorname{Ker} M_\alpha$. Again, semisimplicity of E entails a right E-module decomposition $e_{j_0}M = \alpha M'_z \oplus M'_{j_*}$. Then, $eM = e_{i_0}M \oplus e_{j_0}M = M'_{i_*} \oplus (M'_z \oplus \alpha M'_z) \oplus M'_{j_*} \cong [H_{i_0} \otimes_k M'_{i_*}] \oplus [H_z \otimes_k M'_z] \oplus [H_{j_0} \otimes_k M'_{j_*}]$, where the last isomorphism is an eB-E-bimodule isomorphism. Now, consider the S-E-bimodule N defined by $f_{i_*}N := M'_{i_*}$, $f_zN := M'_z$, $f_{j_*}N := M'_{j_*}$ and, for $i \notin \{i_0, j_0\}$, we take $f_iN := M_i$ (recall that we are identifying f_iS with e_iR through the isomorphism of algebras $f_iS \cong \operatorname{End}_R(Re_i) \cong Re_i$). Notice that we have the isomorphisms of B-E-bimodules: $Xf_{i_*} \otimes_{Sf_{i_*}} f_{i_*}N \cong H_{i_0} \otimes_k M'_{i_*}$, $Xf_z \otimes_{Sf_z} f_zN \cong H_z \otimes_k M'_z$, $Xf_{j_*} \otimes_{Sf_{j_*}} f_{j_*}N \cong H_{j_0} \otimes_k M'_{j_*}$ and $Xf_i \otimes_{Sf_i} f_iN \cong Re_i \otimes_{Re_i} M_i \cong M_i$, for $i \notin \{i_0, j_0\}$. Hence, we have an isomorphism $M \cong \oplus_{t\in\mathcal{P}^e} Xf_t \otimes_{Sf_t} f_tN$ of B-E-bimodules, and we can apply (25.6) and (25.5) to obtain $F_e^E(\overline{N}) \cong M$ in $\mathcal{A}$-E-Mod_p, for some $\overline{N} \in \mathcal{A}^e$-$E$-$\operatorname{Mod}_p$ with underlying S-E-bimodule N. We have completed the proof of (2).

To order to show (1), it is enough to verify that $F_e : \mathcal{A}^e\text{-Mod} \longrightarrow \mathcal{A}\text{-Mod}$ is dense, and then we can apply (21.3). The density of F_e can be obtained with the same argument as before, taking in $E = k$.

Now, given any $N \in \mathcal{A}^e$-E-mod, we examine the behavior of the norm. Notice that $e_{i_0}F_e^E(N) = e_{i_0}X \otimes_S N = \oplus_{t\in\mathcal{P}^e} e_{i_0}Xf_t \otimes_{Sf_t} f_tN \cong f_{i_*}N \oplus f_zN$ and similarly for $e_{j_0}F_e^E(N)$. Thus, we know that $\ell_E(F_e^E(N)_i) = \ell_E(N_i)$, for all $i \in \mathcal{P}\setminus\{i_0, j_0\}$; $\ell_E(F_e^E(N)_{i_0}) = \ell_E(N_{i_*}) + \ell_E(N_z)$; and $\ell_E(F_e^E(N)_{j_0}) = \ell_E(N_{j_*}) + \ell_E(N_z)$. Then, having in mind the description of the arrows of $\mathcal{A}^e$ given in (23.18) and the fact that $\mathcal{P}^e_{\text{mk}} = \mathcal{P}_{\text{mk}}$ does not contain i_0 and j_0, we have

$$\|F_e^E(N)\| - \|N\| = \ell_E(F_e^E(N)_{j_0})\ell_E(F_e^E(N)_{i_0}).$$

This difference corresponds to the contribution of α to the norm $\|F_e^E(N)\|$, because the arrow α disappeared in the passage from $\mathcal{A}$ to $\mathcal{A}^e$. Hence, $\|F_e^E(N)\| \geq \|N\|$, and the inequality is strict iff $\ell_E(F_e^E(N)_{i_0}) \neq 0$ and $\ell_E(F_e^E(N)_{j_0}) \neq 0$. □

Lemma 25.9. *Suppose that $\mathcal{A}^u$ is obtained by unravelling at i_0, using n and $\lambda_1, \ldots, \lambda_q \in k$, from the seminested ditalgebra $\mathcal{A}$, as in (23.23). Let $F_u : \mathcal{A}^u\text{-Mod} \longrightarrow \mathcal{A}\text{-Mod}$ be the associated functor and let E be any k-algebra. Then:*

(1) F_u^E induces a length controlling equivalence from $\mathcal{A}^u$-E-Mod to a full subcategory of $\mathcal{A}$-E-Mod which contains all $\mathcal{A}$-E-bimodules of norm $\leq n$.

(2) If E is a division k-algebra, F_u^E induces an equivalence from $\mathcal{A}^u$-E-Mod_p to a full subcategory of $\mathcal{A}$-E-Mod_p which contains all proper bimodules of norm $\leq n$.

(3) If $N \in \mathcal{A}^u$-E-mod, then $\|N\| \leq \|F_u^E(N)\|$. If multiplication by $g(x) = (x - \lambda_1)(x - \lambda_2)\cdots(x - \lambda_q)$ does not induce an invertible morphism on $F_u^E(N)_{i_0}$, then this inequality is strict.

Proof. We adopt the notation of (23.23) and its proof. Thus, $\mathcal{A}^u = \mathcal{A}^X$ is seminested and $F_u = F^X : \mathcal{A}^u\text{-Mod} \longrightarrow \mathcal{A}\text{-Mod}$, where $X = X_1 \oplus X_2 \oplus X_3$, with $Re_{i_0} = k[x]_{f(x)} =: H$, $\pi_s = x - \lambda_s$, $X_1 = \oplus_{s=1}^q(\oplus_{a=1}^n H/H\pi_s^a)$, $X_2 = H_{g(x)}$ and $X_3 = (1 - e_{i_0})R$. Recall that $\mathcal{P}^u = \{i_*\} \cup \mathcal{P}\setminus\{i_0\}$.

From (25.7), we already know that $F_u^E : \mathcal{A}^u\text{-}E\text{-Mod} \longrightarrow \mathcal{A}\text{-}E\text{-Mod}$ is a length controlling, full and faithful functor. Then, its restriction to proper bimodules $F_u^E : \mathcal{A}^u\text{-}E\text{-Mod}_p \longrightarrow \mathcal{A}\text{-}E\text{-Mod}_p$ is full and faithful too. Assume that E is a division k-algebra and let us show that the restricted one F_u^E is dense on a full subcategory of $\mathcal{A}\text{-}E\text{-Mod}_p$, which contains all proper bimodules of norm $\leq n$.

Assume that M is an H-E-bimodule with $\ell_E(M) \leq n$. Consider the (noncommutative) principal ideal domain $D := E[x] \cong k[x] \otimes_k E$. Thus, M can be considered as a right D-module of finite length. Then, by Fitting's Lemma, there is a right D-module decomposition $M = M_{i_*} \oplus M_c$, where $g(x)$ acts invertibly on M_{i_*} and $g(x)^n M_c = 0$. Thus, M_c is a right finitely generated torsion module over the principal ideal domain D and, therefore, it has the form $M_c \cong D/d_1D \oplus \cdots \oplus D/d_wD$, for some non-zero non-units $d_1, \ldots, d_w \in D$ (see [17](8.2)). Notice that $k[x]$ is contained in the center of D and recall that $g(x)^n$ annihilates M_c and, hence, each direct summand D/d_iD. Fix $i \in [1, w]$ and consider a prime decomposition $d_i = p_1 \cdots p_v$ of d_i in the (noncommutative) unique factorization domain D (see [18](2) and [34](3)). Thus, $(x - \lambda_1)^n \cdots (x - \lambda_q)^n = g(x)^n = p_1 \cdots p_v d$, for some $d \in D$. Then, from the uniqueness of prime factorizations, for each $j \in [1, v]$, we have $D/p_jD \cong D/(x - \lambda_{s(j)})D$, for some $s(j) \in [1, q]$. Again, since $(x - \lambda_{s(j)})$ annihilates the right D-module D/p_jD, we obtain that $(x - \lambda_{s(j)}) = p_jd'$, for some $d' \in D = E[x]$. Since p_j is not invertible, d' is invertible. It follows that $d_i = (x - \lambda_1)^{a_1} \cdots (x - \lambda_q)^{a_q} u$, for some $a_1, \ldots, a_q \in [0, n]$ and some unit $u \in D$. Then, $M_c \cong \oplus_{s=1}^q \oplus_{a=1}^n n(s, a)[D/(x - \lambda_s)^a D]$, a D-module decomposition with each multiplicity $n(s, a) \geq 0$. Now, recall that we have an isomorphism of k-algebras $(H/H\pi_s^a) \cong k[x]/\pi_s^a k[x]$, and $H/H\pi_s^a \otimes_k E \cong (k[x]/\pi_s^a k[x]) \otimes_k E \cong D/\pi_s^a D$. Thus, $M_c \cong \oplus_{s=1}^q \oplus_{a=1}^n n(s, a)[H/H\pi_s^a \otimes_k E]$.

Then, given $M \in \mathcal{A}\text{-}E\text{-Mod}$, with norm $\leq n$, we know that the $e_{i_0}R$-E-bimodule $e_{i_0}M$ has length $\ell_E(e_{i_0}M) \leq n$ and, from the previous argument, admits a decomposition as an H-E-bimodule of the form

$$e_{i_0}M = M_{i_*} \oplus [\oplus_{s=1}^q \oplus_{a=1}^n M(s, a)],$$

where $M(s, a) \cong n(s, a)[H/H\pi_s^a \otimes_k E]$ and $g(x)$ acts invertibly on M_{i_*}.

Given $M \in \mathcal{A}$-E-Mod_p, we can consider the S-E-bimodule N defined by: $f_{i_*}N := M_{i_*}$; $f_{i_{s,a}}N := n(s,a)E$, that is a right E-vector space with E-dimension $n(s,a)$, for $s \in [1,q]$ and $a \in [1,n]$; and $f_iN := M_i$, for any $i \in \mathcal{P}\backslash\{i_0\}$ (recall, again, that we are identifying f_iS with Re_i through the isomorphism of algebras $f_iS \cong \mathrm{End}_R(Re_i) \cong Re_i$). Notice that we have isomorphisms of R-E-bimodules $Xf_{i_*} \otimes_{Sf_{i_*}} f_{i_*}N \cong H_{g(x)} \otimes_{H_{g(x)}} M_{i_*} \cong M_{i_*}$, $Xf_i \otimes_{Sf_i} f_iM \cong Re_i \otimes_{Re_i} M_i \cong M_i$, for $i \in \mathcal{P}\backslash\{i_0\}$; and, finally, $Xf_{i_{s,a}} \otimes_{Sf_{i_{s,a}}} f_{i_{s,a}}N \cong H/\pi_s^a H \otimes_k n(s,a)E \cong n(s,a)[H/\pi_s^a H \otimes_k E] \cong M(s,a)$. Hence, we have an isomorphism of B-E-bimodules $M \cong \oplus_{t\in\mathcal{P}^u} Xf_t \otimes_{Sf_t} f_tN$, and we can apply (25.6) and (25.5) to obtain $F_u^E(\overline{N}) \cong M$ in $\mathcal{A}$-E-Mod_p, for some $\overline{N} \in \mathcal{A}^u$-$E$-$\mathrm{Mod}_p$ with underlying S-E-bimodule N. We have completed the proof of (2).

To show (1) it is enough to see that any $M \in \mathcal{A}$-E-Mod with norm $\leq n$ is isomorphic in $\mathcal{A}$-Mod to $F_u(N)$, for some $N \in \mathcal{A}^u$-Mod, and then we can apply (21.3). For this, decompose, as before, $e_{i_0}M = M_{i_*} \oplus M_c$, where $g(x)$ acts invertibly on M_{i_*} and $g(x)^n M_c = 0$. Then, M_c is a $k[x]/g(x)^n k[x]$-module and, hence, it has a decomposition of the form $M_c = \oplus_{s=1}^q \oplus_{a=1}^n M(s,a)$, where each submodule $M(s,a)$ is isomorphic to a possibly infinite direct sum of copies of the module $k[x]/\pi_s^a k[x]$. Thus, $M(s,a) \cong k[x]/\pi_s^a k[x] \otimes_k V(s,a)$, for an appropriate k-vectorspace $V(s,a)$. Then, as before, we can consider the S-module N defined by: $f_{i_*}N := M_{i_*}$; $f_{i_{s,a}}N := V(s,a)$, for $s \in [1,q]$ and $a \in [1,n]$; and $f_iN := M_i$, for any $i \in \mathcal{P}\backslash\{i_0\}$. Then, we have isomorphisms of R-modules $Xf_{i_*} \otimes_{Sf_{i_*}} f_{i_*}N \cong H_{g(x)} \otimes_{H_{g(x)}} M_{i_*} \cong M_{i_*}$, $Xf_i \otimes_{Sf_i} f_iM \cong Re_i \otimes_{Re_i} M_i \cong M_i$, for $i \in \mathcal{P}\backslash\{i_0\}$; and, finally, $Xf_{i_{s,a}} \otimes_{Sf_{i_{s,a}}} f_{i_{s,a}}N \cong H/\pi_s^a H \otimes_k V(s,a) \cong M(s,a)$. Hence, we have an isomorphism of B-modules $M \cong \oplus_{t\in\mathcal{P}^u} Xf_t \otimes_{Sf_t} f_tN$, and we can apply (25.6) and (25.5) to complete the proof of (1).

Finally, take $N \in \mathcal{A}^u$-E-mod and let us examine $\|F_u^E(N)\|$. We have, $e_{i_0}F_u^E(N) = e_{i_0}X \otimes_S N \cong [e_{i_0}Xf_{i_*} \otimes_{Sf_{i_*}} f_{i_*}N] \oplus [\oplus_{s,a} e_{i_0}Xf_{i_{s,a}} \otimes_{Sf_{i_{s,a}}} f_{i_{s,a}}N] \cong f_{i_*}N \oplus (\oplus_{s=1}^q \oplus_{a=1}^n \oplus_{t=0}^{a-1} f_{i_{s,a}}N)$ and, clearly, $e_iF_u^E(N) \cong N_i$, for $i \in \mathcal{P}\backslash\{i_0\}$. Indeed, we have $e_{i_0}Xf_{i_{s,a}} \otimes_{Sf_{i_{s,a}}} f_{i_{s,a}}N \cong H/H\pi_s^a \otimes_k f_{i_{s,a}}N \cong \oplus_{t=0}^{a-1} f_{i_{s,a}}N$. Then, we know that $\ell_E(F_u^E(N)_i) = \ell_E(N_i)$, for all $i \in \mathcal{P}\backslash\{i_0\}$, and $\ell_E(F_u^E(N)_{i_0}) = \ell_E(N_{i_*}) + \sum_{s=1}^q \sum_{a=1}^n \sum_{t=0}^{a-1} \ell_E(N_{i_{s,a}})$. Then, having in mind the description of the arrows of $\mathcal{A}^u$ given in (23.23) and the fact that $\mathcal{P}_{\mathrm{mk}}^u = (\mathcal{P}_{\mathrm{mk}}\backslash\{i_0\}) \cup \{i_*\}$, we have

$$\begin{aligned}
\|F_u^E(N)\| - \|N\| &\geq \ell_E(F_u^E(N)_{i_0})^2 - \ell_E(N_{i_*})^2 \\
&= [\ell_E(N_{i_*}) + \textstyle\sum_{s,a,t} \ell_E(N_{i_{s,a}})]^2 - \ell_E(N_{i_*})^2 \\
&= 2\ell_E(N_{i_*})\big[\textstyle\sum_{s,a,t} \ell_E(N_{i_{s,a}})\big] + \big[\textstyle\sum_{s,a,t} \ell_E(N_{i_{s,a}})\big]^2 .
\end{aligned}$$

Hence, $\|F_u^E(N)\| \geq \|N\|$, and the inequality is strict if some $N_{i_{s,a}} \neq 0$. Since $F_u^E(N)_{i_0} = e_{i_0}X \otimes_S N \cong [H_{g(x)} \otimes_{H_{g(x)}} N_{i_*}] \oplus [\oplus_{s,a} H/H\pi_s^a \otimes_k f_{i_{s,a}}N]$, we know that $g(x)$ acts invertibly in this module iff $f_{i_{s,a}}N = 0$, for all s, a, and we are done. □

Definition 25.10. *Assume that the seminested ditalgebra $\mathcal{A}^z$ is obtained from the seminested ditalgebra $\mathcal{A}$ by one of the operations $\mathcal{A} \mapsto \mathcal{A}^z$, where $z \in \{a, r, d, e, u\}$, as in (23.16), (23.15), (23.14), (23.18) or (23.23). Each such operation has an associated functor $F_z : \mathcal{A}^z\text{-Mod} \longrightarrow \mathcal{A}\text{-Mod}$. We call F_z a basic reduction functor (of type z). By definition, a reduction functor is a finite composition of basic reduction functors.*

Remark 25.11. *Assume that the seminested ditalgebra $\mathcal{A}^z$ is obtained from the seminested ditalgebra $\mathcal{A}$ by one of the operations $\mathcal{A} \mapsto \mathcal{A}^z$, where $z \in \{a, r, d, e, u\}$, as in (23.16), (23.15), (23.14), (23.18) or (23.23). Let K be any field extension of k and E any K-algebra. Consider the corresponding operations on the induced ditalgebra $\mathcal{A}^K \mapsto \mathcal{A}^{Kz}$ as described in (20.4), (20.5), (20.6) and (20.11). Notice that the admissible modules associated to the edge reduction and the unravelling extend to K by (20.9) and, therefore, (20.11) is indeed applicable. Consider also the associated functors $F_{z^K} : \mathcal{A}^{Kz}\text{-Mod} \longrightarrow \mathcal{A}^K\text{-Mod}$, and the isomorphisms $\xi_z : \mathcal{A}^{Kz} \longrightarrow \mathcal{A}^{zK}$ of layered ditalgebras introduced in these lemmas. Define the corresponding induced functor as the composition*

$$\widehat{F}_z := F_{\xi_z} F_{z^K} : \mathcal{A}^{zK}\text{-Mod} \longrightarrow \mathcal{A}^K\text{-Mod},$$

for $z \in \{a, r, d, e, u\}$.

Notice that, for any K-algebra E, $F_{\xi_z}^E : \mathcal{A}^{zK}\text{-}E\text{-Mod} \longrightarrow \mathcal{A}^K\text{-}E\text{-Mod}$ is an isomorphism of categories which preserves norm and length, in fact $\underline{\ell}_E(F_{\xi_z}^E(M)) = \underline{\ell}_E(M)$, for all $M \in \mathcal{A}^{zK}\text{-}E\text{-Mod}$. Then, Lemmas (25.2), (25.3), (25.4), (25.8) and (25.9) induce the following five corollaries.

Corollary 25.12. *Assume that $\mathcal{A}^a$ is obtained by absorption of a loop from the seminested ditalgebra $\mathcal{A}$, as in (23.16), let K be any field extension of k and consider the associated induced functor $\widehat{F}_a : \mathcal{A}^{aK}\text{-Mod} \longrightarrow \mathcal{A}^K\text{-Mod}$. Then, given any K-algebra E, we have:*

(1) For any $N \in \mathcal{A}^{aK}\text{-}E\text{-Mod}$, we have $\underline{\ell}_E(\widehat{F}_a^E(N)) = \underline{\ell}_E(N)$.
(2) $\widehat{F}_a^E : \mathcal{A}^{aK}\text{-}E\text{-Mod} \longrightarrow \mathcal{A}^K\text{-}E\text{-Mod}$ is a length preserving equivalence.
(3) $\widehat{F}_a^E$ restricts to an equivalence $\mathcal{A}^{aK}\text{-}E\text{-Mod}_p \longrightarrow \mathcal{A}^K\text{-}E\text{-Mod}_p$.
(4) For any $N \in \mathcal{A}^{aK}\text{-}E\text{-mod}$, we have $\|N\| = \|\widehat{F}_a^E(N)\|$.

Corollary 25.13. *Assume that $\mathcal{A}^r$ is obtained by regularization of an arrow α from the seminested ditalgebra $\mathcal{A}$, let K be any field extension of k and consider the associated induced functor $\widehat{F}_r : \mathcal{A}^{rK}\text{-Mod}\longrightarrow\mathcal{A}^K\text{-Mod}$. Then, given any K-algebra E, we have:*

1. *For any $N \in \mathcal{A}^{rK}\text{-}E\text{-Mod}$, we have $\underline{\ell}_E(\widehat{F}_r^E(N)) = \underline{\ell}_E(N)$.*
2. *$\widehat{F}_r^E : \mathcal{A}^{rK}\text{-}E\text{-Mod}\longrightarrow\mathcal{A}^K\text{-}E\text{-Mod}$ is a length preserving equivalence.*
3. *$\widehat{F}_r^E$ restricts to an equivalence $\mathcal{A}^{rK}\text{-}E\text{-Mod}_p\longrightarrow\mathcal{A}^K\text{-}E\text{-Mod}_p$.*
4. *If $N \in \mathcal{A}^{rK}\text{-}E\text{-mod}$, then $\|N\| \leq \|\widehat{F}_r^E(N)\|$. If $\widehat{F}_r^E(N)_{t(\alpha)} \neq 0$ and $\widehat{F}_r^E(N)_{s(\alpha)} \neq 0$, this inequality is strict.*

Corollary 25.14. *Assume that $\mathcal{A}^d$ is obtained by deletion of a finite collection of idempotents $\{e_i \mid i \in \Omega\}$, where Ω is a set of points of the seminested ditalgebra $\mathcal{A}$, let K be any field extension of k and consider the associated induced functor $\widehat{F}_d : \mathcal{A}^{dK}\text{-Mod}\longrightarrow\mathcal{A}^K\text{-Mod}$. Then, given any K-algebra E, we have:*

1. *For any $N \in \mathcal{A}^{dK}\text{-}E\text{-Mod}$, the length vector $\underline{\ell}_E(\widehat{F}_d^E(N))$ satisfies that $\underline{\ell}_E(\widehat{F}_d^E(N))_i = 0$, for all $i \in \Omega$, and $\underline{\ell}_E(\widehat{F}_d^E(N))_i = \underline{\ell}_E(N)_i$, otherwise.*
2. *$\widehat{F}_d^E : \mathcal{A}^{dK}\text{-}E\text{-Mod}\longrightarrow\mathcal{A}^K\text{-}E\text{-Mod}$ is a length preserving functor inducing an equivalence from $\mathcal{A}^{dK}\text{-}E\text{-Mod}$ to the full subcategory of $\mathcal{A}^K\text{-}E\text{-Mod}$ formed by the bimodules M with $M_i = 0$, for all $i \in \Omega$.*
3. *$\widehat{F}_d^E$ restricts to an equivalence from $\mathcal{A}^{dK}\text{-}E\text{-Mod}_p$ to the full subcategory of $\mathcal{A}^K\text{-}E\text{-Mod}_p$ formed by the proper bimodules M satisfying $M_i = 0$, for all $i \in \Omega$.*
4. *If $N \in \mathcal{A}^{dK}\text{-}E\text{-mod}$, then $\|N\| = \|\widehat{F}_d^E(N)\|$.*

Corollary 25.15. *Assume that $\mathcal{A}^e$ is obtained by reduction of an edge $\alpha : i_0\longrightarrow j_0$ from the seminested ditalgebra $\mathcal{A}$, let K be any field extension of k and consider the associated induced functor $\widehat{F}_e : \mathcal{A}^{eK}\text{-Mod}\longrightarrow\mathcal{A}^K\text{-Mod}$. Then, given any K-algebra E, we have:*

1. *$\widehat{F}_e^E : \mathcal{A}^{eK}\text{-}E\text{-Mod}\longrightarrow\mathcal{A}^K\text{-}E\text{-Mod}$ is a length controlling equivalence.*
2. *If E is a semisimple K-algebra, the functor $\widehat{F}_e^E$ restricts to an equivalence from $\mathcal{A}^{eK}\text{-}E\text{-Mod}_p$ onto $\mathcal{A}^K\text{-}E\text{-Mod}_p$.*
3. *If $N \in \mathcal{A}^{eK}\text{-}E\text{-mod}$, then $\|N\| \leq \|\widehat{F}_e^E(N)\|$. This inequality is strict when $\widehat{F}_e^E(N)_{i_0} \neq 0$ and $\widehat{F}_e^E(N)_{j_0} \neq 0$.*

Corollary 25.16. *Suppose that $\mathcal{A}^u$ is obtained by the unravelling at i_0, using n and $\lambda_1, \ldots, \lambda_q \in k$, from the seminested ditalgebra $\mathcal{A}$, let K be any field extension of k and let $\widehat{F}_u : \mathcal{A}^{uK}\text{-Mod}\longrightarrow\mathcal{A}^K\text{-Mod}$ be the corresponding induced functor. Then, given any K-algebra E, we have:*

(1) $\widehat{F}_u^E$ *induces a length controlling equivalence from* $\mathcal{A}^{uK}$-E-Mod *to a full subcategory of* $\mathcal{A}^K$-E-Mod, *which contains all* $\mathcal{A}^K$-E-*bimodules of norm* $\leq n$.

(2) *If* E *is a division* K-*algebra,* $\widehat{F}_u^E$ *induces an equivalence from* $\mathcal{A}^{uK}$-E-Mod_p *to a full subcategory of* $\mathcal{A}^K$-E-Mod_p, *which contains all proper bimodules of norm* $\leq n$.

(3) *If* $N \in \mathcal{A}^{uK}$-E-mod, *then* $\|N\| \leq \|\widehat{F}_u^E(N)\|$. *If multiplication by* $g(x) = (x - \lambda_1)(x - \lambda_2)\cdots(x - \lambda_q)$ *does not induce an invertible morphism on* $\widehat{F}_u^E(N)_{i_0}$, *then this inequality is strict.*

Exercise 25.17. *Assume that* X *is an admissible module over the proper subditalgebra* $\mathcal{A}'$ *of the layered ditalgebra* $\mathcal{A}$. *Assume that the layer* (S, W^X) *of the ditalgebra* $\mathcal{A}^X$ *has in the first component a minimal algebra* S *and consider its canonical decomposition of the unit* $1 = \sum_{i=1}^t f_i$ *as a sum of central primitive orthogonal idempotents. Consider the associated functor* $F^X : \mathcal{A}^X\text{-Mod} \longrightarrow \mathcal{A}\text{-Mod}$. *Then, for* $N \in \mathcal{A}^X$-Mod *with finitely generated free underlying* S-*module, the* B-*module* $F^X(N)$ *is isomorphic to a direct sum of left* B-*modules of the form* Xf_i. *Moreover, each* Xf_i *is a direct summand of* ${}_BF^X(N)$ *with multiplicity equal to the rank of* N_i *as a left free* Sf_i-*module. Given* $M \in \mathcal{A}$-Mod *such that,* ${}_BM \cong \oplus_i m_i Xf_i$ *in* B-Mod, *then* $M \cong F^X(N)$ *in* $\mathcal{A}$-Mod, *for some* $N \in \mathcal{A}^X$-Mod.

Exercise 25.18. *Assume that* $\mathcal{A}^s$ *is obtained from the seminested ditalgebra* $\mathcal{A}$ *by factoring out the marked loop* x *at* i_0, *as in (23.28) and consider the associated functor* $F_s : \mathcal{A}^s\text{-Mod} \longrightarrow \mathcal{A}\text{-Mod}$. *Then, given any* k-*algebra* E, *we have:*

(1) *For any* $N \in \mathcal{A}^s$-E-Mod, *we have* $\underline{\ell}_E(F_s^E(N)) = \underline{\ell}_E(N)$.

(2) $F_s^E : \mathcal{A}^s\text{-}E\text{-Mod} \longrightarrow \mathcal{A}\text{-}E\text{-Mod}$ *is a length preserving functor, inducing an equivalence from* $\mathcal{A}^s$-E-Mod *onto the full subcategory of* $\mathcal{A}$-E-Mod *formed by the bimodules* M *such that* $xM = 0$.

(3) F_s^E *restricts to an equivalence from* $\mathcal{A}^s$-E-Mod_p *to the full subcategory of* $\mathcal{A}$-E-Mod_p *formed by the proper bimodules* M *satisfying that* $xM = 0$.

(4) *For any* $N \in \mathcal{A}^s$-E-mod, *we have* $\|N\| \leq \|F_s^E(N)\|$, *and the inequality is strict when* $F_s^E(N)_{i_0} \neq 0$.

Exercise 25.19. *Assume that the ditalgebra* $\mathcal{A}^z$ *is obtained from the seminested ditalgebra by absorption, regularization, deletion of idempotents, edge reduction, unravelling, or factoring out a marked loop, as in (23.16), (23.15), (23.14), (23.18), (23.23) or (23.28). That is a basic operation of type* $z \in \{a, d, r, e, u, s\}$. *Assume that* I *is a triangular ideal of* $\mathcal{A}$, *as in (8.22). Then there* *is a triangular ideal I^z of $\mathcal{A}^z$ such that the associated reduction functor F_z : $\mathcal{A}^z$-Mod$\longrightarrow\mathcal{A}$-Mod induces a functor $\overline{F}_z : (\mathcal{A}^z, I^z)$-Mod$\longrightarrow(\mathcal{A}, I)$-Mod. Moreover, for any $N \in \mathcal{A}^z$-Mod and $M \in (\mathcal{A}, I)$-Mod with $F_z(N) \cong M$, there is $N' \in (\mathcal{A}^z, I^z)$-Mod with $\overline{F}_z(N') \cong M$.*

Hint: Since F_a, F_r, F_d, F_s are restriction functors, this follows from (8.24). For $z \in \{e, u\}$, the functor $F_z = F^X$, for a suitable admissible triangular $\mathcal{A}'$-module X over a proper subditalgebra $\mathcal{A}'$ of $\mathcal{A}$ with $\delta(W_0') = 0$. Use (14.11) and (12.10) to show that $H^X * N = 0$ iff $H \cdot F^X(N) = 0$.

Exercise 25.20. *Assume $\mathcal{A}$ is a seminested ditalgebra over the algebraically closed field k and assume $\mathcal{A}$ has some marked point i_0. Let I be a triangular ideal of $\mathcal{A}$ such that $(\mathcal{A}, I)$ is of finite representation type: that means that there are only finitely many isomorphism classes of indecomposable objects in $(\mathcal{A}, I)$-mod. Show that there is a seminested ditalgebra $\mathcal{A}'$, obtained from $\mathcal{A}$ by a finite sequence of elementary reductions, and a triangular ideal I' of $\mathcal{A}'$ such that:*

(1) $|\mathcal{P}_{\mathrm{mk}}(\mathcal{A}')| < |\mathcal{P}_{\mathrm{mk}}(\mathcal{A})|$;
(2) the associated reduction functor $F : \mathcal{A}'$-mod$\longrightarrow\mathcal{A}$-mod induces an equivalence $\overline{F} : (\mathcal{A}', I')$-mod$\longrightarrow(\mathcal{A}, I)$-mod;
(3) for any $N \in (\mathcal{A}', I')$-mod, $\|N\| \leq \|F(N)\|$. The inequality is strict whenever $F(N)(x) \neq 0$, where x is the marked loop at i_0.

Hint: First notice that if (R, W) is the layer of $\mathcal{A}$, then $I \cap Re_{i_0} \neq 0$, because $(\mathcal{A}, I)$ is of finite representation type. If $Re_{i_0} = k[x]_{f(x)}$, choose a monic polynomial $h(x) \in k[x]$ admitting no common root with $f(x)$, that generates $I \cap Re_{i_0}$. Then, take the unravelling of the loop at i_0 associated to a number r greater than the norm of any indecomposable object in $(\mathcal{A}, I)$-mod and to the roots of $h(x)$. Use the proofs of (23.23) and (25.9) to show that $h(x) \in I^u$, thus $N_{i_*} = 0$ for all $N \in (\mathcal{A}^u, I^u)$-mod. Then, delete the idempotent corresponding to the vertex i_* of $\mathcal{A}^u$ to construct $\mathcal{A}' := \mathcal{A}^{ud}$.

26

Modules over non-wild ditalgebras

In this section, we prove Theorem (26.9), which states that some part of the study of $\mathcal{A}$-E-bimodules with bounded length, of a non-wild seminested ditalgebra $\mathcal{A}$ over an algebraically closed field, can be reduced to the study of bimodules over a finite number of minimal ditalgebras.

Lemma 26.1. *Assume that R is a minimal k-algebra and let W be an R-R-bimodule such that, for given central primitive orthogonal idempotents e_{i_0} and e_{j_0}, the R-R-bimodule $e_{j_0}We_{i_0}$ is freely generated by the elements $v_1, \dots, v_n$. Suppose that we have an invertible matrix $Q = (Q_{ij}) \in M_{n\times n}(Re_{j_0} \otimes_k Re_{i_0})$. Define the elements $v'_i := \sum_{j=1}^n Q_{ij}v_j \in e_{j_0}We_{i_0}$, for $i \in [1, n]$. Then, $e_{j_0}We_{i_0}$ is freely generated by the elements $v'_1, \dots, v'_n$. In the sequel, we will often consider suitable changes of basis of this type. For instance, in some cases, we will use the next lemma, which can be derived from the Quillen–Suslin theorem confirming Serre's Conjecture.*

Proof. By (23.4)(2), we can consider the unique morphism of R-R-bimodules $\phi_Q : e_{j_0}We_{i_0} \longrightarrow e_{j_0}We_{i_0}$, which maps v_i onto v'_i, for all $i \in [1, n]$. Similarly, we have the morphism of bimodules $\phi_{Q^{-1}} : e_{j_0}We_{i_0} \longrightarrow e_{j_0}We_{i_0}$, with $\phi_{Q^{-1}}(v_j) = \sum_{t=1}^n Q^{-1}_{jt}v_t$, for all j. Thus, $\phi_{Q^{-1}}(v'_i) = v_i$, for all i, and ϕ_Q is an isomorphism of R-R-bimodules with inverse $\phi_{Q^{-1}}$. □

Lemma 26.2. *Assume that R is a commutative ring such that every finitely generated projective R-module is free. This is the case if $R = D[y]$, where D is a principal ideal domain, or if $R = k[x, y]_{f(x)g(y)}$. By definition, an element $q = (q_1, \dots, q_n) \in R^n$ is unimodular iff the ideal of R generated by $q_1, \dots, q_n$ equals R. For any unimodular $q \in R^n$, there is an invertible matrix $Q \in M_{n\times n}(R)$ with first row q.*

Proof. The ring $D[y]$ satisfies the above property by the Quillen–Suslin theorem. The fact that the ring $k[x, y]_{f(x)g(y)}$ has the above property is shown in [19]. Both non-trivial facts pertain to commutative algebra and will not be proved in detail here. Assume q is unimodular and let $p_1, \ldots, p_n \in R$ satisfy $\sum_{i=1}^n p_i q_i = 1$. Then, the morphism of R-modules $\varphi : R^n \longrightarrow R$, given by $\varphi(x_1, \ldots, x_n) = \sum_{i=1}^n x_i p_i$, admits the section $\psi : R \longrightarrow R^n$, given by $\psi(x) = xq$. Then, $R^n = K \oplus Rq$, where K is the kernel of φ. By assumption, every finitely generated projective R-module is free, and this is the case of K. Thus, R^n admits a basis of the form $B = \{q, b_1, \ldots, b_{n-1}\}$, where $\{b_1, \ldots, b_{n-1}\}$ is a basis for K. Then, the matrix $A := (q^t, b_1^t, \ldots, b_{n-1}^t)$ is an invertible matrix in $M_{n\times n}(R)$. Indeed, it is the change of coordinates matrix between B and the canonical basis of R^n. Then, $Q := A^t$ is the matrix we were looking for. □

Lemma 26.3. *Given a ditalgebra $\mathcal{A}$ with layer (R, W) and an epimorphism in the category of k-algebras $\varphi : R \longrightarrow S$, consider the induced layered ditalgebra $\mathcal{A}^S$, as defined in (17.7). Consider also the canonical morphism $\overline{\varphi} : \mathcal{A} \longrightarrow \mathcal{A}^S$ and the induced functor $F_{\overline{\varphi}} : \mathcal{A}^S$-Mod$\longrightarrow \mathcal{A}$-Mod. We have the following commutative diagram of restriction and forgetful functors*

$$\begin{array}{ccc} \mathcal{A}^S\text{-Mod} & \xrightarrow{F_{\overline{\varphi}}} & \mathcal{A}\text{-Mod} \\ \downarrow{\scriptstyle U_{\mathcal{A}^S}} & & \downarrow{\scriptstyle U_{\mathcal{A}}} \\ S\text{-Mod} & \xrightarrow{F_{\varphi}} & R\text{-Mod}. \end{array}$$

Assume that $M \in \mathcal{A}$-Mod satisfies $U_{\mathcal{A}}(M) \cong F_\varphi(N)$ in R-Mod, for some $N \in S$-Mod, then there exists $\overline{N} \in \mathcal{A}^S$-Mod with $M \cong F_{\overline{\varphi}}(\overline{N})$ and $U_{\mathcal{A}^S}(\overline{N}) = N$. Moreover, given a k-algebra E, if $M \in \mathcal{A}$-E-Mod$_p$, $N \in S$-E-Mod with $F_\varphi^E(N) \cong U_{\mathcal{A}}^E(M)$ in R-E-Mod, then N admits a natural structure of proper $\mathcal{A}^S$-E-bimodule $\overline{N}$ with $U_{\mathcal{A}^S}^E(\overline{N}) = N$ satisfying that $F_{\overline{\varphi}}^E(\overline{N}) \cong M$ in $\mathcal{A}$-E-Mod$_p$.

Proof. By definition, we have the S-S-bimodules $W_i^S = S \otimes_R W_i \otimes_R S$, for $i \in \{0, 1\}$, and $W^S = W_0^S \oplus W_1^S$. The morphism $\varphi : R \longrightarrow S$ and the canonical morphism $W \longrightarrow W^S$, which maps each $w \in W$ onto $1 \otimes w \otimes 1$, induce the morphism of t-algebras $\overline{\varphi} : T_A \longrightarrow T_{\mathcal{A}^S} = T^S = T_S(W^S)$. By assumption, we have the morphism of left R-modules $\pi : W_0 \otimes_R M \longrightarrow M$, which determines the left A-module structure of M. We also know that the S-module N, when restricted to R using φ is isomorphic to the R-module M. Call σ the corresponding isomorphism. We also have the isomorphism $q : S \otimes_S N \longrightarrow N$ and the product morphism $p : S \otimes_R N \longrightarrow N$ in S-Mod. Then, the composition

$$W_0^S \otimes_S N = S \otimes_R W_0 \otimes_R S \otimes_S N \xrightarrow{1\otimes q} S \otimes_R W_0 \otimes_R N \xrightarrow{1\otimes\sigma} S \otimes_R W_0 \otimes_R M$$
$$\xrightarrow{1\otimes\pi} S \otimes_R M \xrightarrow{1\otimes\sigma^{-1}} S \otimes_R N \xrightarrow{p} N$$

defines a structure of left A^S-module on N, which we denote by $\overline{N}$, such that $(\sigma, 0) : F_{\overline{\varphi}}(\overline{N}) \longrightarrow M$ is an isomorphism in $\mathcal{A}$-Mod. Indeed, σ provides an isomorphism of A-modules from the restriction of $\overline{N}$ to A, using $\overline{\varphi}_| : A \longrightarrow A^S$, onto M.

For the moreover part, we have an isomorphism of R-E-bimodules $\sigma : N \longrightarrow M$; then, as before, we define an A^S-module structure $\overline{N}$ over N, using σ and the A-module structure of M. We know that $U_{A^S}(\overline{N}) = N$ and, hence, $\overline{N}$ has a natural structure of a right E-module. Since N is an S-E-bimodule, so is $\overline{N}$. We claim that in fact $\overline{N}$ is a proper A^S-E-bimodule (with $U_{\mathcal{A}}^E(\overline{N}) = N$). Indeed, we have that $\sigma(ne) = \sigma(n)e$, for all $n \in N$ and $e \in E$. Thus, for $s, s' \in S$, $n \in N$, $w \in W_0$ and $e \in E$, we have

$$\begin{aligned}
(s \otimes w \otimes s')[ne] &= (s \otimes w \otimes s')[\sigma^{-1}(\sigma(n)e)] \\
&= s(\sigma^{-1}(w\sigma(s'\sigma^{-1}(\sigma(n)e)))) \\
&= s(\sigma^{-1}(w\sigma(s'(ne)))) \\
&= s(\sigma^{-1}(w\sigma(s'n)e)) \\
&= [s(\sigma^{-1}(w\sigma(s'n)))]e \\
&= [(s \otimes w \otimes s')n]e.
\end{aligned}$$

Then notice that $(\sigma, 0) : F_{\overline{\varphi}}^E(\overline{N}) \longrightarrow M$ is an isomorphism in $\mathcal{A}$-E-Mod_p. □

Lemma 26.4. *Localization at a marked point: Assume $\mathcal{A}$ is a nested (resp. seminested) k-ditalgebra, with layer (R, W) admitting basis $(\mathbb{B}_0, \mathbb{B}_1)$. Assume that i_0 is a marked point of $\mathcal{A}$; thus, $Re_{i_0} = k[x]_{f(x)}$, for some $f(x) \in k[x]$. Assume that $g(x) \in k[x]$ is any non-zero polynomial and consider the canonical epimorphism of k-algebras $\varphi : R \longrightarrow S$, where*

$$S = Re_1 \times \cdots \times Re_{i_0-1} \times k[x]_{f(x)g(x)} \times Re_{i_0+1} \times \cdots \times Re_n,$$

where $R = \prod_{i=1}^n Re_i$ is the decomposition of the minimal k-algebra R into a product of indecomposable k-algebras. Then, the ditalgebra $\mathcal{A}^l := \mathcal{A}^S$, as constructed in (17.7), is nested (resp. seminested). We call $\mathcal{A}^l$ the ditalgebra obtained from $\mathcal{A}$ by localization at the marked point i_0 with the polynomial $g(x)$.

Its bigraph $\mathbb{B}^l$ is obtained from the bigraph $\mathbb{B}$ of $\mathcal{A}$ as follows. The sets of points $\mathcal{P}^l$ and $\mathcal{P}$ coincide; the marked points are the same $\mathcal{P}^l_{\mathrm{mk}} = \mathcal{P}_{\mathrm{mk}}$, as well as the marked loops $\mathbb{B}^l_{\mathrm{mk}} = \mathbb{B}_{\mathrm{mk}}$; the mark maps mk *and* mk^l *coincide at every* *marked point with the exception of* i_0, *where* $\mathrm{mk}^l(i_0) = \mathrm{mk}(i_0)g(x)$; *each solid (resp. dotted) arrow* α *in* $\mathbb{B}_0$ *(resp. in* $\mathbb{B}_1$*), determines a solid (resp. dotted) arrow* $1 \otimes \alpha \otimes 1$ *in* $\mathbb{B}_0^l$ *(resp. in* $\mathbb{B}_1^l$*), with the same source and target points.*

Proof. It follows, for instance, from the realization of $\mathcal{A}^l$ as a ditalgebra of the form $\mathcal{A}^X$, for a suitable admissible module X (see (17.7)). □

Lemma 26.5. *Assume that* $\mathcal{A}^l$ *is obtained by localization at a marked point, using the polynomial* $g(x)$*, from the seminested ditalgebra* $\mathcal{A}$*, as in (26.4), and let* $F_l := F_{\overline{\varphi}} : \mathcal{A}^l\text{-Mod} \longrightarrow \mathcal{A}\text{-Mod}$ *be the associated functor, as in (17.7). Given a k-algebra E, we have:*

(1) For any $N \in \mathcal{A}^l\text{-}E\text{-Mod}$*, we have* $\underline{\ell}_E(F_l^E(N)) = \underline{\ell}_E(N)$.
(2) $F_l^E : \mathcal{A}^l\text{-}E\text{-Mod} \longrightarrow \mathcal{A}\text{-}E\text{-Mod}$ *is a length preserving functor inducing an equivalence from* $\mathcal{A}^l\text{-}E\text{-Mod}$ *onto the full subcategory of* $\mathcal{A}\text{-}E\text{-Mod}$ *formed by the bimodules M with* $M(g(x))$ *invertible.*
(3) F_l^E *restricts to an equivalence from* $\mathcal{A}^l\text{-}E\text{-Mod}_p$ *onto the full subcategory of* $\mathcal{A}\text{-}E\text{-Mod}_p$ *formed by the proper bimodules with* $M(g(x))$ *invertible.*
(4) If $N \in \mathcal{A}^l\text{-}E\text{-mod}$*, then* $||N|| = ||F_l^E(N)||$.

Proof. Let $\varphi : R \longrightarrow S$ be the canonical epimorphism of algebras, as in (26.4), and consider the morphism of ditalgebras $\overline{\varphi} : \mathcal{A} \to \mathcal{A}^l$ constructed in (17.7). The fact that the functor F_l is induced by restriction of the morphism $\overline{\varphi}$, which maps the set of primitive central orthogonal idempotents of R onto the corresponding set of idempotents of S implies (1), in particular F_l^E is length preserving. From this and the inspection of the layer described in (26.4), we get (4).

From (17.7), F_l^E is full and faithful. From (21.6)(1), F_l^E restricts to proper bimodules.

Consider the induced functor $F_\varphi^E : S\text{-}E\text{-Mod} \longrightarrow R\text{-}E\text{-Mod}$. If $M \in \mathcal{A}\text{-}E\text{-Mod}$ satisfies that $M(g(x))$ is invertible, then $U_{\mathcal{A}}^E(M) \in R\text{-}E\text{-Mod}$ satisfies that $M(g(x))$ is invertible. Clearly, M admits a left S-module structure satisfying $g(x)^{-1}m = M(g(x))^{-1}(m)$, for $m \in M$. Since $M \in R\text{-}E\text{-Mod}$, the map $M(g(x))$ is a right E-module morphism, then so is the multiplication by $g(x)^{-1}$. It follows that M admits the structure of an S-E-bimodule, which we denote by N. Thus, $F_\varphi^E(N) = U_{\mathcal{A}}^E(M)$. If $M \in \mathcal{A}\text{-}E\text{-Mod}_p$, we obtain from (26.3) a proper bimodule $\overline{N} \in \mathcal{A}^l\text{-}E\text{-Mod}_p$ with $F_l^E(\overline{N}) \cong M$ in $\mathcal{A}\text{-}E\text{-Mod}$. Then, (3) follows.

To prove (2), show, as before, that, for every $M \in \mathcal{A}\text{-}E\text{-Mod}$ with $M(g(x))$ invertible, there exists $\overline{N} \in \mathcal{A}^l\text{-Mod}$ with $F_l(\overline{N}) \cong M$ in $\mathcal{A}\text{-Mod}$, and then apply (21.3). □

Corollary 26.6. *Assume that $\mathcal{A}^l$ is obtained by localization at a marked point, using the polynomial $g(x)$, from the seminested ditalgebra $\mathcal{A}$, let K be any field extension of k and let $\widehat{F}_l : \mathcal{A}^{lK}$-Mod$\longrightarrow\mathcal{A}^K$-Mod be the associated induced functor. Then, given any K-algebra E, we have:*

(1) For any $N \in \mathcal{A}^{lK}$-E-Mod, we have $\underline{\ell}_E(\widehat{F}_l^E(N)) = \underline{\ell}(N)$.

(2) $\widehat{F}_l^E : \mathcal{A}^{lK}$-$E$-Mod$\longrightarrow\mathcal{A}^K$-$E$-Mod is a length preserving functor inducing an equivalence from $\mathcal{A}^{lK}$-E-Mod to the full subcategory of $\mathcal{A}^K$-E-Mod formed by the bimodules M with $M(g(x))$ invertible.

(3) $\widehat{F}_l^E$ restricts to an equivalence from $\mathcal{A}^{lK}$-E-Mod$_p$ to the full subcategory of $\mathcal{A}^K$-E-Mod$_p$ formed by the proper bimodules M satisfying that $M(g(x))$ is invertible.

(4) If $N \in \mathcal{A}^{lK}$-E-mod, then $\|N\| = \|\widehat{F}_l^E(N)\|$.

Lemma 26.7. *Let $\mathcal{A}$ be a seminested ditalgebra, let $\mathcal{A}^u$ be the ditalgebra obtained from $\mathcal{A}$ by unravelling at i_0 using n and $\lambda_1, \ldots, \lambda_q \in k$, and consider the points of the bigraph of $\mathcal{A}^u$ of the form $i_{s,a}$, $s \in [1, q]$ and $a \in [1, n]$. Consider the ditalgebra $\mathcal{A}^{ud}$ obtained from $\mathcal{A}^u$ by deleting all the idempotents corresponding to these points. Finally, consider the polynomial $h(x) := (x - \lambda_1)\cdots(x - \lambda_q)$ and the ditalgebra $\mathcal{A}^l$ obtained from $\mathcal{A}$ by localizing at i_0 using $h(x)$. Then, there is an isomorphism of layered ditalgebras $\xi : \mathcal{A}^l \longrightarrow \mathcal{A}^{ud}$. Moreover, the following diagram of functors associated to these constructions commutes up to isomorphism*

$$\begin{array}{ccc}
\mathcal{A}^l\text{-Mod} & \xrightarrow{F_l} & \mathcal{A}\text{-Mod} \\
{\scriptstyle F_\xi}\Big\uparrow & & \Big\uparrow{\scriptstyle F_u} \\
\mathcal{A}^{ud}\text{-Mod} & \xrightarrow{F_d} & \mathcal{A}^u\text{-Mod.}
\end{array}$$

Proof. From the description of the seminested layers for $\mathcal{A}^l$ and $\mathcal{A}^{ud}$, we see that both ditalgebras have essentially the same bigraph. We shall describe an isomorphism of ditalgebras $\xi : \mathcal{A}^l \longrightarrow \mathcal{A}^{ud}$. First, adopting the notation of the proof of (23.23), recall that if the layer of $\mathcal{A}$ is (R, W), then the layer (S, W^X) of $\mathcal{A}^u$ has the form $S = S_1 \times \widehat{S}$, with $\widehat{S} = (e_{i_0}R)_{h(x)} \times (1 - e_{i_0})R$. Thus, the layer $(\widehat{S}, W^{\widehat{S}})$ of $\mathcal{A}^l$ and the layer $(S^d, (W^X)^d)$ of $\mathcal{A}^{ud}$ have an isomorphic first component. We shall identify them. Then, we can define: for any arrow $i \xrightarrow{w} j$ of $\mathcal{A}$, with $i_0 \notin \{i, j\}$, $\xi(1 \otimes w \otimes 1) = \xi(e_j \otimes w \otimes e_i) = \nu_j^j \otimes w \otimes x_i^i$; for any arrow $i \xrightarrow{w} i_0$ of $\mathcal{A}$, with $i \neq i_0$, $\xi(e_{i_0} \otimes w \otimes e_i) = \nu_{i_0}^{i_*} \otimes w \otimes x_i^i$; for any arrow $i_0 \xrightarrow{w} j$ of $\mathcal{A}$, with $j \neq i_0$, $\xi(e_j \otimes w \otimes e_{i_0}) = \nu_j^j \otimes w \otimes x_{i_*}^{i_0}$; for any arrow $i_0 \xrightarrow{w} i_0$ of $\mathcal{A}$, $\xi(e_{i_0} \otimes w \otimes e_{i_0}) = \nu_{i_0}^{i_*} \otimes w \otimes x_{i_*}^{i_0}$. Then, we can extend ξ to an isomorphism of t-algebras $\xi : T^l = T_{\widehat{S}}(W^{\widehat{S}}) \longrightarrow T^{ud}$. Using the expressions for δ^u and δ^l given in (12.9) and (17.7), it is easy to see that $\delta^{ud}\xi = \xi\delta^l$.

Let us define an isomorphism $\varphi : F_u F_d \longrightarrow F_l F_\xi$. We will use here the description of F_l as the functor $F_{\widehat{S}}$ associated with the admissible module $\widehat{S}$. Assume that $M \in \mathcal{A}^{ud}$-Mod, recall that $X = X_1 \oplus \widehat{S}$, thus any $x \in X$ can be written uniquely as $x = x_1 + \widehat{x}$, with $x \in X_1$ and $\widehat{x} \in \widehat{S}$. Then, define for $x \in X$ and $m \in M$

$$\varphi^0(x \otimes m) = \widehat{x} \otimes m.$$

We claim that $\varphi^0 : F_u F_d(M) = X \otimes_S F_d(M) \longrightarrow F_l F_\xi(M) = \widehat{S} \otimes_{\widehat{S}} M$ is an isomorphism of A-modules. For $r \in R$, we have that $\varphi^0(r \cdot (x \otimes m)) = \varphi^0(rx \otimes m) = r\widehat{x} \otimes m = r \cdot (\widehat{x} \otimes m) = r \cdot \varphi^0(x \otimes m)$. Now if $w \in \mathbb{B}_0$, $x \in X$ and $m \in M$, then

$$\begin{aligned}\varphi^0(w \cdot (x \otimes m)) &= \varphi^0[\textstyle\sum_{s,a,t} x^t_{s,a} \otimes (v^{s,a}_t \otimes w \otimes x) * m \\ &\quad + \textstyle\sum_{i \neq i_0} x^i_i \otimes (v^i_i \otimes w \otimes x) * m \\ &\quad + x^{i_0}_{i_*} \otimes (v^{i_*}_{i_0} \otimes w \otimes x) * m] \\ &= \textstyle\sum_{i \neq i_0} e_i \otimes (v^i_i \otimes w \otimes x) * m \\ &\quad + e_{i_0} \otimes (v^{i_*}_{i_0} \otimes w \otimes x) * m,\end{aligned}$$

because $\widehat{x^t_{s,a}} = 0$, $\widehat{x^i_i} = x^i_i = e_i$ and $\widehat{e_{i_0}} = x^{i_0}_{i_*} = e_{i_0}$. Let $*'$ denote the action of $W^l_0 = \widehat{S} \otimes_R W_0 \otimes_R \widehat{S}$ over $M = F_\xi(M)$, thus $(e_j \otimes w \otimes \widehat{x}) *' m = \xi(e_j \otimes w \otimes \widehat{x}) * m$. Notice that, for $\widehat{x} \in \widehat{S}$, $\xi(e_j \otimes w \otimes \widehat{x}) = v^j_j \otimes w \otimes \widehat{x}$, if $j \neq i_0$, and $\xi(e_{i_0} \otimes w \otimes \widehat{x}) = v^{i_*}_{i_0} \otimes w \otimes \widehat{x}$. Moreover, for $m \in F_d(M)$, we have that $(v \otimes w \otimes x) * m = (v \otimes w \otimes \widehat{x}) * m$, for all $v \in X^*$ and $x \in X$. Therefore

$$\begin{aligned}w \cdot \varphi^0(x \otimes m) &= w \cdot (\widehat{x} \otimes m) \\ &= 1 \otimes (1 \otimes w \otimes \widehat{x}) *' m \\ &= \textstyle\sum_{j \neq i_0} e_j \otimes (e_j \otimes w \otimes \widehat{x}) *' m + e_{i_0} \otimes (e_{i_0} \otimes w \otimes \widehat{x}) *' m.\end{aligned}$$

From the previous series of equations, the last term coincides with $\varphi^0(w \cdot (x \otimes m))$, we have that φ^0 is a morphism of A-modules. Then, $(\varphi^0, 0) : F_u F_d(M) \longrightarrow F_l F_\xi(M)$ is an isomorphism in $\mathcal{A}$-Mod. It remains to show that the following diagram commutes in $\mathcal{A}$-Mod, for any morphism $f : M \longrightarrow N$ in $\mathcal{A}^{ud}$-Mod

$$\begin{array}{ccc} X \otimes_S F_d(M) & \xrightarrow{(\varphi^0,0)} & \widehat{S} \otimes_{\widehat{S}} M \\ {\scriptstyle F_u F_d(f)} \downarrow & & \downarrow {\scriptstyle F_l F_\xi(f)} \\ X \otimes_S F_d(N) & \xrightarrow{(\varphi^0,0)} & \widehat{S} \otimes_{\widehat{S}} N. \end{array}$$

In the following, we combine the standard notation (ν_i, x_i) and (γ_j, p_j) for the dual basis of X and P over S, respectively, with the previous notation

$$\begin{aligned}
[(\varphi^0, 0)F_u F_d(f)]^0(x \otimes m) &= \varphi^0(F_u(F_d(f)))^0(x \otimes m) \\
&= \varphi^0[x \otimes F_d(f)^0(m) + \textstyle\sum_j xp_j \otimes F_d(f)^1(\gamma_j)[m]] \\
&= \varphi^0[x \otimes f^0(m) + \textstyle\sum_j xp_j \otimes \overline{f}^1(\gamma_j)[m]] \\
&= \widehat{x} \otimes f^0(m) \\
&= \widehat{x} \otimes F_\xi(f)^0(m) \\
&= (F_\ell F_\xi(f))^0 \varphi^0[x \otimes m] \\
&= [(F_\ell F_\xi(f))(\varphi^0, 0)]^0[x \otimes m].
\end{aligned}$$

Moreover, for $w \in W_1$

$$\begin{aligned}
[(\varphi^0, 0)F_u F_d(f)]^1(w)(x \otimes m) &= \varphi^0(F_u(F_d(f)))^1(w)(x \otimes m) \\
&= \varphi^0(\textstyle\sum_i x_i \otimes F_d(f)^1(\nu_i \otimes w \otimes x)[m]) \\
&= \varphi^0(\textstyle\sum_{i \in \mathcal{P}} x_i \otimes f^1(\nu_i \otimes w \otimes \widehat{x})[m]) \\
&= \textstyle\sum_{i \in \mathcal{P}} e_i \otimes f^1 \xi(1 \otimes w \otimes \widehat{x})[m] \\
&= 1 \otimes F_\xi(f)^1(1 \otimes w \otimes \widehat{x})[m] \\
&= [F_l F_\xi(f)]^1(w)[\widehat{x} \otimes m] \\
&= [F_l F_\xi(f)]^1(w)\varphi^0[x \otimes m] \\
&= [F_l F_\xi(f)(\varphi^0, 0)]^1(w)[x \otimes m].
\end{aligned}$$

□

Notation 26.8. *Recall that by a reduction functor we mean a finite composition of functors of the form* $F_z : \mathcal{A}^z\text{-Mod} \longrightarrow \mathcal{A}\text{-Mod}$, *where* $\mathcal{A}$ *is some seminested ditalgebra and* $z \in \{a, r, d, e, u\}$. *If* $F : \mathcal{B}\text{-Mod} \longrightarrow \mathcal{A}\text{-Mod}$ *is a reduction functor, say* $F = F_{z_1} F_{z_2} \cdots F_{z_n}$, *with* F_{z_i} *a basic reduction functor and* K *is a field extension of* k, *then*

$$\widehat{F} := \widehat{F}_{z_1} \widehat{F}_{z_2} \cdots \widehat{F}_{z_n} : \mathcal{B}^K\text{-Mod} \longrightarrow \mathcal{A}^K\text{-Mod}$$

denotes the corresponding functor induced by F *(see (25.11)).*

The proof of the following theorem reveals why it is necessary to consider the notion of seminested ditalgebra, since some changes of basis which do not preserve nestedness of layers have to be considered. Thus, even starting with a nested ditalgebra (such as the Drozd's ditalgebra $\mathcal{D}^\Lambda$ of a basic finite-dimensional algebra Λ, over an algebraically closed field), we might obtain, at some stage of the procedure, some non-nested ditalgebra.

Theorem 26.9. *Let* $\mathcal{A}$ *be a non-wild seminested ditalgebra over the algebraically closed field* k *and take* $d \in \mathbb{N}$. *Then, there are minimal* k*-ditalgebras*

$\mathcal{B}_1, \ldots, \mathcal{B}_p$ and a family of reduction functors $F_i : \mathcal{B}_i$-Mod$\longrightarrow\mathcal{A}$-Mod such that, for any field extension K of k and every K-algebra E, we have:

(1) $\widehat{F}_i^E : \mathcal{B}_i^K$-$E$-Mod$\longrightarrow\mathcal{A}^K$-$E$-Mod is length controlling, for any $i \in [1, p]$.
(2) If $M \in \mathcal{A}^K$-E-mod has length $\leq d$, then there exist some $i \in [1, p]$ and some $N \in \mathcal{B}_i^K$-E-mod with $\widehat{F}_i^E(N) \cong M$ in $\mathcal{A}^K$-E-Mod. Moreover, if E is a division K-algebra and M is a proper bimodule, then N can be chosen to be a proper bimodule.

The functors F_i are reduction functors, so each $\mathcal{B}_i$ is obtained from $\mathcal{A}$ by a finite sequence of basic operations of the form $\mathcal{C} \mapsto \mathcal{C}^z$, for $z \in \{a, d, r, e, u\}$.

Proof. Notice that all the ditalgebras we consider in this proof appear as $\mathcal{C}^z$ for some non-wild $\mathcal{C}$, and hence, by (22.8) and (22.10), $\mathcal{C}^z$ is also non-wild.

Let us consider a new statement, which we will label (26.9′), constructed from (26.9) by replacing item 2 by the following:

(2′) *If $M \in \mathcal{A}^K$-E-mod has norm $\leq d$, then there exist some $i \in [1, p]$ and some $N \in \mathcal{B}_i^K$-E-mod with $\widehat{F}_i^E(N) \cong M$ in $\mathcal{A}^K$-E-Mod. Moreover, if E is a division K-algebra and M is a proper bimodule, then N can be chosen to be a proper bimodule.*

Then, notice that our Theorem (26.9) follows from the new statement (26.9′). Furthermore, (26.9′) is equivalent to the newer statement, labelled (26.9″), which is obtained from (26.9′) replacing (2′) by the following:

(2″) *If $M \in \mathcal{A}^K$-E-mod is a sincere module with norm $\leq d$, then there exist some $i \in [1, p]$ and some $N \in \mathcal{B}_i^K$-E-mod with $\widehat{F}_i^E(N) \cong M$ in $\mathcal{A}^K$-E-Mod. Moreover, if E is a division K-algebra and M is a proper bimodule, then N can be chosen to be a proper bimodule.*

Indeed, consider the k-ditalgebras $\mathcal{A}^{d_1}, \ldots, \mathcal{A}^{d_t}$ obtained from $\mathcal{A}$ by deletion of a finite number of idempotents of R. Then, apply (26.9″) to each one of $\mathcal{A}^{d_1}, \ldots, \mathcal{A}^{d_t}$ to obtain minimal ditalgebras $\mathcal{B}_{ij}$ and reduction functors F_{ij} : $\mathcal{B}_{ij}$-Mod$\longrightarrow\mathcal{A}^{d_i}$-Mod satisfying the corresponding statements (1) and (2″) for each $\mathcal{A}^{d_i}$. Then, the compositions

$$\mathcal{B}_{ij}^K\text{-}E\text{-Mod}\xrightarrow{\widehat{F}_{ij}^E}\mathcal{A}^{d_i K}\text{-}E\text{-Mod}\xrightarrow{\widehat{F}_{d_i}^E}\mathcal{A}^K\text{-}E\text{-Mod}$$

are induced by the compositions of reduction functors $F_{d_i}F_{ij}$. The functors $\widehat{F}_{d_i}^E\widehat{F}_{ij}^E$ are length controlling and satisfy (2′), because, given $M \in \mathcal{A}^K$-E-Mod with $||M|| \leq d$, we have $M \cong \widehat{F}_{d_i}^E(N)$, for some $i \in [1, p]$ and some sincere $N \in \mathcal{A}^{d_i K}$-$E$-Mod, and then $\widehat{F}_{ij}^E(\overline{N}) \cong N$, for some $\overline{N} \in \mathcal{B}_{ij}^K$-$E$-Mod. Moreover, if E is a division K-algebra and M is a proper bimodule, so is N by (25.14), and so will be $\overline{N}$ by our assumption (2″).

We shall prove (26.9″) by induction on d. If $d = 0$ and $M \in \mathcal{A}^K$-E-Mod is sincere with $||M|| = 0$, then $W_0 = 0$ and $\mathcal{A}$ is minimal. So assume that $d > 0$ and (26.9″) holds for any $d' < d$, or, equivalently, that (26.9′) holds for any $d' < d$. Now, we have to consider the sincere modules $M \in \mathcal{A}^K$-E-Mod with $||M|| = d$.

Since $\mathcal{A}$ is a seminested ditalgebra, we can choose a minimal solid arrow $\alpha : i_0 \longrightarrow j_0$ in $\mathbb{B}_0$, that is a solid arrow α with minimal height $\mathbb{h}(\alpha)$. Then, by triangularity, we have that $\delta(\alpha) \in W_1$.

Case 1: $\delta(\alpha) = 0$ and $i_0 = j_0$.

Since $\mathcal{A}$ is not wild, by (24.11), $Re_{i_0} \cong k$. Consider the ditalgebra $\mathcal{A}^a$ obtained from $\mathcal{A}$ by absorption of the loop α. Then, we can apply (25.12) to every sincere $M \in \mathcal{A}^K$-E-Mod with norm d to obtain a sincere $N \in \mathcal{A}^{aK}$-E-Mod with $\widehat{F}_a^E(N) \cong M$ having the same norm d. We have that $\mathcal{A}^a$ has one solid arrow less that $\mathcal{A}$. Repeating this argument, if necessary, either we end up with a seminested ditalgebra with no solid arrows (that is a minimal ditalgebra) or at some step we obtain a ditalgebra with a minimal solid arrow α in one of the following cases.

Case 2: $\delta(\alpha) = 0$ and $i_0 \neq j_0$.

Since $\mathcal{A}$ is not wild, by (24.11), $Re_{i_0} \cong k$ and $Re_{j_0} \cong k$. Then, we can consider the ditalgebra $\mathcal{A}^e$ obtained from $\mathcal{A}$ by reduction of the edge α and apply (25.15). By our induction hypothesis, we can apply (26.9′) to $\mathcal{A}^e$ and $d - 1$. This provides reduction functors $F_i : \mathcal{B}_i$-Mod$\longrightarrow \mathcal{A}^e$-Mod, which can be composed with $F_e : \mathcal{A}^e$-Mod$\longrightarrow \mathcal{A}$-Mod to get the required family of functors.

Case 3: $\delta(\alpha) \neq 0$.

Write $C := Re_{i_0} \otimes_k Re_{j_0}$. Then, $e_{j_0} W_1 e_{i_0}$ is a C-module and we can write

$$\delta(\alpha) = \sum_{i=1}^{j} c_i v_i,$$

for some $v_1, \ldots, v_j \in e_{j_0} \mathbb{B}_1 e_{i_0}$ and $0 \neq c_i \in C$.

Subcase 3.1: Some c_t is invertible in C.

This is the case, for instance, if $Re_{i_0} \cong k$ and $Re_{j_0} \cong k$. Make $v'_t := \sum_{i=1}^{j} c_i v_i$ and consider the change of basis for W_1 where $v_1, \ldots, v_t, \ldots, v_j$ is replaced by $v_1, \ldots, v'_t, \ldots, v_j$ using (26.1) (more precisely, consider the matrix Q with $Q_{ti} = c_i$ for $i \in [1, j]$; $Q_{ii} = 1$ for $i \neq t$; and $Q_{iq} = 0$ for $t \neq i \neq q$; where Q is invertible because c_t is so).

Then, we can apply regularization, that is (25.13), to obtain a length preserving equivalence $\widehat{F}_r^E : \mathcal{A}^{rK}\text{-}E\text{-Mod} \longrightarrow \mathcal{A}^K\text{-}E\text{-Mod}$. Now, we can apply our induction hypothesis to $\mathcal{A}^r$ and $d-1$. Composing the reduction functors this provides with F_r, we obtain the required family of functors. Indeed, if $M \in \mathcal{A}^K\text{-}E\text{-Mod}$ is sincere with $||M|| = d$, then there is $N \in \mathcal{A}^{rK}\text{-}E\text{-Mod}$ with $\widehat{F}_r^E(N) \cong M$ and $||N|| < d$ (we can use (25.13) and the fact that M is sincere). Then, we can apply our induction hypothesis to N and we are done.

Subcase 3.2: $Re_{j_0} \cong k$.

Since we have already considered the *Subcase 3.1*, we may assume that $Re_{i_0} \not\cong k$. Thus, $Re_{i_0} = k[x]_{f(x)}$ and $C \cong k[x]_{f(x)}$. By an appropriate change of basis of W_1 of the form $v_i' = f(x)^{-p} v_i$, for all i, using again (26.1), we may assume that $c_i \in k[x]$. Consider the ditalgebra $\mathcal{A}^u$ obtained from $\mathcal{A}$ by unravelling at i_0 using d and the different roots of $c_1(x)$. From (25.16), we know that the sincere bimodules $M \in \mathcal{A}^K\text{-}E\text{-Mod}$ with norm $||M|| \leq d$ and $M(c_1(x))$ not invertible are of the form $\widehat{F}_u^E(N) \cong M$, where $N \in \mathcal{A}^{uK}\text{-}E\text{-Mod}$ has norm $||N|| \leq d-1$. Moreover, if E is a division K-algebra and M is proper, then N can be taken to be proper. Then, we can apply our induction hypothesis to $\mathcal{A}^u$ and $d-1$, and then apply (2′) to N.

Moreover, if $\mathcal{A}^l$ is the ditalgebra obtained from $\mathcal{A}$ by localizing at the vertex i_0 using the polynomial $c_1(x)$, the sincere $\mathcal{A}^K\text{-}E$-bimodules with norm $||M|| = d$ and $M(c_1(x))$ invertible are of the form $\widehat{F}_l^E(N) \cong M$, where $N \in \mathcal{A}^{lK}\text{-}E\text{-Mod}$ is sincere and has norm $||N|| = ||M|| = d$ (see (26.6)). According to the description of the differential δ^l of $\mathcal{A}^l$ given in (17.7), we have that

$$\delta^l(1 \otimes \alpha \otimes 1) = \sum_{i=1}^{j} c_i(x)(1 \otimes v_i \otimes 1).$$

But now c_1 is invertible and we can proceed as in *Subcase 3.1*.

Subcase 3.3: $Re_{i_0} \cong k$.

This case is dual to *Subcase 3.2*.

Subcase 3.4: $Re_{i_0} \not\cong k$ and $Re_{j_0} \not\cong k$.

Assume that $Re_{i_0} = k[x]_{f(x)}$ and $Re_{j_0} = k[y]_{g(y)}$. Hence, $C = k[x, y]_{f(x)g(y)}$. After an appropriate change of basis of W_1, of the form $v_i' = f(x)^{-p} g(y)^{-q} v_i$, using (26.1), we may assume that all the c_i are polynomials in $k[x, y]$. Let $h(x, y)$ be the highest common factor of the $c_i(x, y)$ and assume that $h(x, y)q_i(x, y) = c_i(x, y)$, for all

i. Since the $q_i(x, y)$ are coprime in $k(x)[y]$, there are polynomials $s_i(x, y) \in k[x, y]$ and a non-zero polynomial $u(x) \in k[x]$ such that

$$u(x) = \sum_{i=1}^{j} s_i(x, y)q_i(x, y).$$

If the module M satisfies that $M(u(x))$ is not invertible, we can proceed as in *Subcase 3.2*, by unravelling at i_0 using d and the different roots of $u(x)$.

Otherwise, we can consider the ditalgebra $\mathcal{A}^l$ obtained from $\mathcal{A}$ by localizing at the point i_0 using $u(x)$. Then, $M \cong \widehat{F}_l^E(N)$, where $N \in \mathcal{A}^{lK}$-E-Mod is sincere and has the same norm than M. From the previous formula for $u(x)$, we obtain in the ditalgebra $\mathcal{A}^l$ that

$$1 = \sum_{i=1}^{j} [s_i(x, y)u(x)^{-1}]q_i(x, y).$$

The terms in this formula all belong to the algebra $H := k[x, y]_{u(x)}$. Here, $H \cong D[y]$, where $D = k[x]_{u(x)}$ is a principal ideal domain. From (26.2), we obtain an invertible matrix $Q \in M_{j\times j}(H)$ with first row $(q_1, \ldots, q_j)$. Consider the change of basis for $W_1^l = S \otimes_R W_1 \otimes_R S$, which replaces each $w_i := 1 \otimes v_i \otimes 1$ by w_i' defined by the formula

$$(w_1', \ldots, w_j')^t = Q(w_1, \ldots, w_j)^t.$$

Hence

$$\delta^l(1 \otimes \alpha \otimes 1) = \sum_i c_i w_i = \sum_i h q_i w_i = h w_1'.$$

Since $\mathcal{A}^l$ is not wild, by (24.11), h must be invertible. Now, we can replace the basis $w_1', w_2', \ldots$ of W_1^l by $hw_1', w_2', \ldots$ and apply regularization, that is (25.13), to finish the proof. □

The last remark in the statement of the theorem is obtained from (26.7).

Exercise 26.10. *Let $\mathcal{A}$ be a seminested ditalgebra over any field k. Consider, for $M, N \in \mathcal{A}$*-mod *the number*

$$\Delta_{\mathcal{A}}(M, N) := \dim \mathrm{Hom}_{\mathcal{A}}(M, N) - \dim \mathrm{Ext}_{\mathcal{A}}(M, N).$$

Show that, for any reduction functor $F : \mathcal{B}$-mod$\longrightarrow\mathcal{A}$-mod, *we have the inequality $\Delta_{\mathcal{B}}(M, N) \geq \Delta_{\mathcal{A}}(F(M), F(N))$. If $\mathcal{B}$ is a minimal ditalgebra, then* $\Delta_{\mathcal{B}}(M, N) = \dim \mathrm{Hom}_{\mathcal{B}}(M, N)$.

Hint: Use (6.13) and the fact that, for any basic reduction functor F_z : $\mathcal{A}^z$-Mod$\longrightarrow\mathcal{A}$-Mod, *the space* $\mathrm{Ext}_{\mathcal{A}^z}(M,N)$ *embeds in* $\mathrm{Ext}_{\mathcal{A}}(F_z(M),F_z(N))$.

Exercise 26.11. *Let* R *be a trivial* k*-algebra and* W_0 *an* R-R*-bimodule freely generated by a directed element* $\alpha\in W_0$, *with* $t(\alpha)\neq s(\alpha)$. *Let* B *be an algebra freely generated by the algebra* R *and the* R-R*-bimodule* W_0. *Assume that the modules* $M,M'\in B$-mod *satisfy* $\mathrm{Ext}_B(M,M)=0$ *and* $\mathrm{Ext}_B(M',M')=0$. *Show that* $M\cong M'$ *in* B-mod *iff* $M\cong M'$ *in* R-mod.

Hint: Examine the decompositions of M *and* M' *as direct sums of indecomposables in* B-mod *and the possible extensions between indecomposables.*

Exercise 26.12. *Let* $\mathcal{A}$ *be a seminested ditalgebra without marked points. Suppose that* $M,N\in\mathcal{A}$-mod *satisfy that* $\mathrm{Ext}_{\mathcal{A}}(M,M)=0$ *and* $\mathrm{Ext}_{\mathcal{A}}(N,N)=0$. *Then, show that* $M\cong N$ *in* $\mathcal{A}$-mod *iff* $\underline{\ell}(M)=\underline{\ell}(N)$.

Hint: We can assume that M *and* N *are sincere. Make an induction on* $||M||$. *Examine the layer* (R,W) *of* $\mathcal{A}$, *look at the possibilities for the value of the differential of a minimal solid arrow* $\alpha\in\mathbb{B}_0$. *If* $\delta(\alpha)\neq 0$, *make a change of basis of the layer and use regularization. If* $\delta(\alpha)=0$, *show that* α *cannot be a loop by examining the subditalgebra of* $\mathcal{A}$ *generated by* R *and* α, *which admits no sincere modules without self-extensions (see (16.4) and (32.2)). Then, use reduction of an edge having in mind (16.6), (25.6) and (26.11).*

Exercise 26.13. *Let* $\mathcal{A}$ *be a seminested ditalgebra with layer* (R,W), *over the algebraically closed field* k. *Suppose that* $M,N\in\mathcal{A}$-mod *satisfy* $\mathrm{Ext}_{\mathcal{A}}(M,M)=0$ *and* $\mathrm{Ext}_{\mathcal{A}}(N,N)=0$. *Show that* $M\cong N$ *in* $\mathcal{A}$-mod *iff* ${}_RM\cong{}_RN$ *in* R-mod.

Hint: Take a basic R*-module* X *such that* ${}_RM\in\mathrm{add}\,(X)$. *Consider* R *as an initial subditalgebra of* $\mathcal{A}$ *and notice that* X *is a complete triangular admissible* $\mathcal{A}'$*-module. Then, consider the seminested ditalgebra* $\mathcal{A}^X$, *which has no marked point, and the associated functor* $F^X:\mathcal{A}^X$-mod$\longrightarrow\mathcal{A}$-mod. *Then, apply (26.12), (16.6) and (25.17).*

27
Tameness and wildness

Here we prove Drozd's *Tame and Wild Theorem* (27.10) and some of its consequences, using the preparation of the previous sections.

Definition 27.1. *A ditalgebra $\mathcal{A}$ over an algebraically closed field k is called tame iff, for all $d \in \mathbb{N}$, there is a finite collection $\{(\Gamma_i, Z_i)\}_{i=1}^n$, where $\Gamma_i = k[x]_{f_i}$ and Z_i is an A-Γ_i-bimodule, which is free of finite rank as a right Γ_i-module, such that for every indecomposable $M \in \mathcal{A}$-*Mod *with* $\dim_k M = d$ *there is an $i \in [1, n]$ and a simple Γ_i-module S such that $Z_i \otimes_{\Gamma_i} S \cong M$ in* $\mathcal{A}$-Mod. *A finitely generated k-algebra Λ is called tame iff the ditalgebra $(\Lambda, 0)$ is tame.*

The definition of a tame ditalgebra can be reformulated in various forms. For instance, we have the following.

Lemma 27.2. *Let $\mathcal{A}$ be a ditalgebra over the algebraically closed field k. Then, $\mathcal{A}$ is tame iff, for all $d \in \mathbb{N}$, there is a finite collection $\{(\Gamma_i, Z_i)\}_{i=1}^n$, where $\Gamma_i = k[x]_{f_i}$ and Z_i is an A-Γ_i-bimodule, which is free of finite rank as a right Γ_i-module, such that for every indecomposable $M \in \mathcal{A}$-*Mod *with* $\dim_k M = d$ *there is an $i \in [1, n]$ and a Γ_i-module N such that $Z_i \otimes_{\Gamma_i} N \cong M$ in* $\mathcal{A}$-Mod.

Proof. Since k is algebraically closed, from the structure theorem for finite-dimensional modules over the principal ideal domain $k[x]$, we know that $k[x]_f$ is tame, for any $f \in k[x]$. Our lemma follows from this. □

Notation 27.3. *When we say that almost all modules in a class $\mathcal{C}$ of objects in* $\mathcal{A}$-Mod *satisfy some property, we mean that every $Z \in \mathcal{C}$ has this property, with the exception of those lying in a finite union of isomorphism classes in $\mathcal{C}$.*

We give another variation of the definition of tameness.

Definition 27.4. *A ditalgebra $\mathcal{A}$ over an algebraically closed field k is called strictly tame iff, for all $d \in \mathbb{N}$, there is a finite collection $\{(\Gamma_i, Z_i)\}_{i=1}^n$, where $\Gamma_i = k[x]_{f_i}$ and Z_i is an A-Γ_i-bimodule, which is free of finite rank as a right Γ_i-module, such that the functors $\Gamma_i\text{-Mod} \xrightarrow{Z_i \otimes_{\Gamma_i} -} A\text{-Mod} \xrightarrow{L_A} \mathcal{A}\text{-Mod}$ preserve isomorphism classes of indecomposables and almost every d-dimensional indecomposable $M \in \mathcal{A}\text{-Mod}$ is of the form $M \cong Z_i \otimes_{\Gamma_i} N$, for some $i \in [1, n]$ and $N \in \Gamma_i\text{-Mod}$. A finitely generated k-algebra Λ is called strictly tame iff the regular ditalgebra $(\Lambda, 0)$, with layer $(\Lambda, 0)$ is strictly tame.*

From (27.2), we know that strict tameness implies tameness. Indeed, the exceptional classes $[M]$ of d-dimensional $\mathcal{A}$-modules, which are not covered by the functors $Z_i \otimes_{\Gamma_i} -$ in the last definition, can be individually covered by the A-$k[x]$-bimodule $B_M := M \otimes_k k[x]$ (since $B_M \otimes_{k[x]} S \cong M \otimes_k k[x] \otimes_{k[x]} S \cong M \otimes_k k \cong M$, for any simple $k[x]$-module S). Again, the definition of strict tameness can be reformulated as follows.

Lemma 27.5. *A ditalgebra $\mathcal{A}$ over an algebraically closed field k is strictly tame iff, for all $d \in \mathbb{N}$, there is a finite collection $\{(\Gamma_i, Z_i)\}_{i=1}^n$, where $\Gamma_i = k[x]_{f_i}$ and Z_i is an A-Γ_i-bimodule, which is finitely generated as a right Γ_i-module, such that the functors $\Gamma_i\text{-Mod} \xrightarrow{Z_i \otimes_{\Gamma_i} -} A\text{-Mod} \xrightarrow{L_A} \mathcal{A}\text{-Mod}$ preserve isomorphism classes of indecomposables and almost every d-dimensional indecomposable $M \in \mathcal{A}\text{-Mod}$ is of the form $M \cong Z_i \otimes_{\Gamma_i} N$, for some $i \in [1, n]$ and $N \in \Gamma_i\text{-Mod}$.*

Proof. The argument is similar to the one used in the proof of (22.17). Given a fixed dimension d, we consider an exact sequence of right Γ_i-modules

$$\Gamma_i^s \xrightarrow{H} \Gamma_i^r \xrightarrow{H'} Z_i \longrightarrow 0.$$

The morphism H is the multiplication by a matrix, which we also denote by $H \in \Gamma_i^{r \times s} \leq k(x)^{r \times s}$. Multiplying H by appropriate invertible matrices $P \in \mathbb{G}l_r(k(x))$ and $Q \in \mathbb{G}l_s(k(x))$, we obtain a matrix of the form

$$PHQ = \begin{pmatrix} I & 0 \\ 0 & 0 \end{pmatrix},$$

where I is the identity $(\rho \times \rho)$-matrix and ρ is the rank of H. The entries of P and Q are rational functions in x with coefficients in k, then the denominators of these fractions have a least common multiple $h \in k[x] \subseteq k[x]_{f_i}$. Then, P and Q have entries in the k-algebra $\Gamma_i' := k[x]_{hf_i}$. By left restriction of scalars, we have that Γ_i' is a Γ_i-Γ_i'-bimodule, which is a finitely generated free right Γ_i'-module.

Consider the A-Γ_i-bimodule $Z_i' := Z_i \otimes_{\Gamma_i} \Gamma_i'$. Then, Z_i' is an A-Γ_i'-bimodule, which is free of finite rank as a right Γ_i'-module. Moreover, we have the functor

$$\Gamma_i' \otimes_{\Gamma_i'} - : \Gamma_i'\text{-Mod} \to \Gamma_i\text{-Mod},$$

which is full and faithful (it is the embedding identifying each Γ_i'-module with the Γ_i-module on which h acts through a k-linear isomorphism) and, by assumption, we know that the functor $L_{\mathcal{A}}(Z_i \otimes_{\Gamma_i} -) : \Gamma_i\text{-Mod} \to \mathcal{A}\text{-Mod}$ preserves isomorphism classes of indecomposables. Then, the composition

$$L_{\mathcal{A}}(Z_i' \otimes_{\Gamma_i'} -) : \Gamma_i'\text{-Mod} \to \mathcal{A}\text{-Mod}$$

also preserves isomorphism classes of indecomposables.

Notice that given a rational algebra $\Gamma = k[x]_{f(x)}$ its prime elements, up to a unit factor, have the form $\pi_\lambda = x - \lambda$, with $\lambda \in D(f) := \{\lambda \in k \mid f(\lambda) \neq 0\}$. Thus, the family $\{\Gamma/(\pi_\lambda^a) \mid \lambda \in D(f) \text{ and } a \in \mathbb{N}\}$ constitutes a complete set of representatives of the non-isomorphic indecomposable Γ-modules. We claim that, if Z is an A-Γ-bimodule such that $L_{\mathcal{A}}(Z \otimes_\Gamma -)$ preserves isomorphism classes of indecomposables, then, given $\lambda \in D(f)$, there is at most one $a \in \mathbb{N}$ with $\dim_k Z \otimes_\Gamma \Gamma/(\pi_\lambda^a) = d$. Indeed, if $\dim_k Z \otimes_\Gamma \Gamma/(\pi_\lambda^a) = \dim_k Z \otimes_\Gamma \Gamma/(\pi_\lambda^b)$ and we suppose $a < b$, then $(\pi_\lambda^b) \subseteq (\pi_\lambda^a)$ and there is an epimorphism of left A-modules $Z \otimes_\Gamma \Gamma/(\pi_\lambda^b) \cong Z/Z\pi_\lambda^b \longrightarrow Z/Z\pi_\lambda^a \cong Z \otimes_\Gamma \Gamma/(\pi_\lambda^a)$, which must be an isomorphism. This leads to the contradiction $\Gamma/(\pi_\lambda^b) \cong \Gamma/(\pi_\lambda^a)$. Then, there are only a finite number of isomorphism classes $[N]$ of indecomposables $N \in \Gamma\text{-Mod}$ with $\dim_k L_{\mathcal{A}}(Z \otimes_\Gamma N) = d$ and the polynomial h does not act invertibly on N.

Now, apply the last argument to the rational algebra Γ_i and the bimodule Z_i. Thus, almost for every isomorphism class $[M]$ of d-dimensional indecomposable $\mathcal{A}$-modules M, there is an indecomposable $N \in \Gamma_i\text{-Mod}$, where h acts invertibly, satisfying that $M \cong Z_i \otimes_{\Gamma_i} N$. Hence, $N \cong \Gamma_i' \otimes_{\Gamma_i'} N'$ for some $N' \in \Gamma_i'\text{-Mod}$. Thus, $M \cong Z_i' \otimes_{\Gamma_i'} N'$. □

For the case of a finite-dimensional algebra, the denominators in the definition of tameness can be eliminated (see [40](2.5)).

Lemma 27.6. *A finite-dimensional algebra Λ over an algebraically closed field k is tame iff, for all $d \in \mathbb{N}$, there are a finite number of Λ-$k[x]$-bimodules $Z_1, \ldots, Z_n$ which are free of finite rank as right $k[x]$-modules and such that every indecomposable Λ-module M with $\dim_k M = d$ is isomorphic to $Z_i \otimes_{k[x]} S$, for some $i \in [1, n]$ and some simple $k[x]$-module S.*

The fact that a finite-dimensional algebra over an algebraically closed field k cannot be simultaneously wild and tame is well known. Its proof requires a geometric argument, which can be extended to the case of semipresented ditalgebras, as defined below.

Definition 27.7. *A Roiter ditalgebra $\mathcal{A}$ with layer (R, W) is called presented iff the following four conditions are satisfied:*

(1) R is a finitely generated k-algebra, which admits a decomposition as a finite product of indecomposable k-algebras; hence, its unit decomposes as a finite sum $1 = \sum_{i=1}^{n} e_i$ of central primitive orthogonal idempotents.

(2) There is a set filtration, of some finite directed subset $\mathbb{B}_0 \subseteq W_0$

$$\emptyset = \mathbb{B}_0^0 \subseteq \mathbb{B}_0^1 \subseteq \cdots \subseteq \mathbb{B}_0^{\ell_0} = \mathbb{B}_0,$$

such that each $\mathbb{B}_r^0$ generates the R-R-bimodule W_0^j, for $j \in [0, \ell_0]$, where

$$0 = W_0^0 \subseteq W_0^1 \subseteq \cdots \subseteq W_0^{\ell_0} = W_0$$

is the the triangular filtration of W_0, as in (5.1).

(3) There is a set filtration, of some finite directed subset $\mathbb{B}_1 \subseteq W_1$

$$\emptyset = \mathbb{B}_1^0 \subseteq \mathbb{B}_1^1 \subseteq \cdots \subseteq \mathbb{B}_1^{\ell_1} = \mathbb{B}_1$$

such that each $\mathbb{B}_1^j$ generates the R-R-bimodule W_1^j, for $j \in [0, \ell_1]$, where

$$0 = W_1^0 \subseteq W_1^1 \subseteq \cdots \subseteq W_1^{\ell_1} = W_1$$

is the triangular filtration of W_1, as in (5.1).

(4) The R-R-bimodule W_1 is freely generated by $\mathbb{B}_1$.

The Roiter ditalgebra $\mathcal{A}$ is called semipresented iff conditions (1), (2) and (4) are satisfied.

Remark 27.8.

(1) Every nested (resp. seminested) ditalgebra is presented (resp. semipresented).

(2) If Λ is a finite-dimensional k-algebra, the ditalgebra $(\Lambda, 0)$ with layer $(\Lambda, 0)$ is presented.

The proof of the following result, based on an elementary geometric argument, will be given in Section 33.

Proposition 27.9. *Let $\mathcal{A}$ be a semipresented ditalgebra over the algebraically closed field k. Assume there is an $A_{\mathcal{A}}$-$k\langle x, y\rangle$-bimodule Z, which is free of finite rank as a right $k\langle x, y\rangle$-module, such that the composition functor*

$$k\langle x, y\rangle\text{-mod} \xrightarrow{Z\otimes_{k\langle x,y\rangle}-} A\text{-mod} \xrightarrow{L_A} \mathcal{A}\text{-mod}$$

preserves isomorphism classes of indecomposables. Then, $\mathcal{A}$ is not tame. Thus, semipresented ditalgebras cannot be simultaneously tame and wild, when k is algebraically closed (see (22.5)).

Theorem 27.10. *Assume that the ground field k is algebraically closed and that $\mathcal{A}$ is a seminested ditalgebra. Then:*

(1) $\mathcal{A}$ is either wild or tame, but not both. Moreover, $\mathcal{A}$ is strictly tame iff it is tame.
(2) If $\mathcal{A}$ is wild, there is a reduction functor $F : \mathcal{C}\text{-Mod} \longrightarrow \mathcal{A}\text{-Mod}$, where $\mathcal{C}$ is a critical ditalgebra.

Proof. In the proof of (26.9) an inductive procedure is performed. New ditalgebras are inductively constructed: $\mathcal{A} \mapsto \mathcal{A}^{z_1} \mapsto \mathcal{A}^{z_1 z_2} \mapsto \cdots \mapsto \mathcal{A}^{z_1 z_2 \cdots z_t}$. It is shown there that either the procedure can be continued to show the statement of that theorem or, at some stage of it, we obtain a ditalgebra $\mathcal{A}^{z_1 z_2 \cdots z_t}$ to which Corollary (24.11) can be applied. This implies the existence of a reduction functor $F : \mathcal{C}\text{-Mod} \longrightarrow \mathcal{A}\text{-Mod}$, where $\mathcal{C}$ is a critical ditalgebra. Then, Section 24, (22.8) and (22.10) imply that $\mathcal{A}$ is wild too.

Now, if $\mathcal{A}$ is not wild, by (26.9), there are minimal k-ditalgebras $\mathcal{B}_1, \ldots, \mathcal{B}_p$ and a family of reduction functors $F_i : \mathcal{B}_i\text{-Mod} \longrightarrow \mathcal{A}\text{-Mod}$ such that, whenever $M \in \mathcal{A}\text{-Mod}$ satisfies that $\dim_k M \leq d$, there is some $N \in \mathcal{B}_i\text{-Mod}$ with $F_i(N) \cong M$ in $\mathcal{A}\text{-Mod}$. We know that the indecomposable $\mathcal{B}_i$-modules are of the form $S(z)$ or $J(z, m, x - \lambda)$, for some point z of $\mathcal{B}_i$ (see (23.8)). There are only finitely many isomorphism classes of indecomposable $\mathcal{A}$-modules of the form $[F_i(S(z))]$, which we leave aside. Assume that N has the form $J(z, m, x - \lambda)$. In order to construct the bimodules of definition (27.4), write $\Gamma := \Gamma_{i,z} := B_i e_z \cong k[x]_{f_z}$ for each marked point z of $\mathcal{B}_i$, and, having in mind (22.7), consider the up to isomorphism commutative diagram

$$\begin{array}{ccccc}
\Gamma\text{-Mod} & \xrightarrow{\Gamma \otimes_\Gamma -} & B_i\text{-Mod} & \xrightarrow{L_{\mathcal{B}_i}} & \mathcal{B}_i\text{-Mod} \\
\| & & \downarrow{\scriptstyle \underline{F}_i} & & \downarrow{\scriptstyle F_i} \\
\Gamma\text{-Mod} & \xrightarrow{F_i(\Gamma) \otimes_\Gamma -} & A\text{-Mod} & \xrightarrow{L_{\mathcal{A}}} & \mathcal{A}\text{-Mod}.
\end{array}$$

Since $F_i L_{\mathcal{B}_i}(\Gamma \otimes_\Gamma -)$ preserves indecomposability and isomorphism classes, so does the lower row of the diagram $L_{\mathcal{A}}(F_i(\Gamma) \otimes_\Gamma -)$. Moreover, since F_i is a reduction functor, $F_i(\Gamma)$ is an A-Γ-bimodule which is projective and finitely generated by the right. Hence, since Γ is a principal ideal domain, $F_i(\Gamma)$ is in fact a free right Γ-module of finite rank. We have that $L_{\mathcal{A}}(F_i(\Gamma) \otimes_\Gamma J(z, m, x - \lambda)) \cong F_i(L_{\mathcal{B}_i}(J(z, m, x - \lambda))) \cong F_i(N) \cong M$. Then, $\mathcal{A}$ is strictly tame. By (27.9), the ditalgebra $\mathcal{A}$ is not wild. □

Lemma 27.11. *Assume that Λ' and Λ are finite-dimensional algebras and $F : \Lambda'\text{-Mod} \longrightarrow \Lambda\text{-Mod}$ is an equivalence of their module categories. Then, there is $s \in \mathbb{N}$ such that, for any $M' \in \Lambda'\text{-Mod}$, $\dim_k M' \leq s \times \dim_k F(M')$.*

Proof. Consider the functor $G : \Lambda\text{-Mod} \longrightarrow \Lambda'\text{-Mod}$, with $FG \cong I_{\Lambda\text{-Mod}}$ and $GF \cong I_{\Lambda'\text{-Mod}}$. By Morita's theorem, $F \cong P \otimes_{\Lambda'} -$ and $G \cong Q \otimes_{\Lambda} -$, where P is a Λ-Λ'-bimodule and Q is a Λ'-Λ-bimodule, both finitely generated and projective by the right. Choose $s \in \mathbb{N}$ such that Q is a direct summand of the right Λ-module Λ^s. Then, for any $M \in \Lambda$-mod, we have $\dim(Q \otimes_\Lambda M) \leq \dim(\Lambda^s \otimes_\Lambda M) \leq s \times \dim M$. Then, for $M' \in \Lambda'$-mod, since $M' \cong Q \otimes_\Lambda P \otimes_{\Lambda'} M'$, we obtain $\dim M' \leq s \times \dim(P \otimes_{\Lambda'} M')$. □

Lemma 27.12. *Let Λ' and Λ be finite-dimensional algebras over the algebraically closed field k. If they are Morita equivalent algebras and Λ' is tame (resp. strictly tame), then so is Λ.*

Proof. Consider an equivalence $G : \Lambda'\text{-mod} \longrightarrow \Lambda\text{-mod}$ of the form $G \cong Q \otimes_{\Lambda'} -$, where Q is a Λ-Λ'-bimodule, finitely generated and projective by the right. By (27.11), there is a natural number s such that, for any $M' \in \Lambda'$-mod, $\dim M' \leq s \times \dim(Q \otimes_{\Lambda'} M')$.

Assume that Λ' is strictly tame, fix any $d \in \mathbb{N}$ and define $d' := s \times d$. Then, there is a finite collection $\{(\Gamma_i, Z'_i)\}_{i=1}^n$, where $\Gamma_i = k[x]_{f_i}$ and Z'_i is a Λ'-Γ_i-bimodule, which is free of finite rank as a right Γ_i-module, such that the functors $Z'_i \otimes_{\Gamma_i} - : \Gamma_i\text{-Mod} \longrightarrow \Lambda'\text{-Mod}$ preserve isomorphism classes of indecomposables and, with the possible exception of only a finite number of isomorphism classes $[M']$ of indecomposables $M' \in \Lambda'$-Mod with dimension $\leq d'$, every isomorphism class $[M']$ of an indecomposable $M' \in \Lambda'$-Mod with $\dim_k M' \leq d'$ is of the form $[M'] = [Z'_i \otimes_{\Gamma_i} N]$, for some $i \in [1, n]$ and $N \in \Gamma_i$-Mod.

For each $i \in [1, n]$, define $Z_i := Q \otimes_{\Lambda'} Z'_i$, a Λ-Γ_i-bimodule, which is free of finite rank by the right. Let us denote by $\mathcal{C}_d$ (resp. $\mathcal{C}_{\leq d'}$) the set of isoclasses of d-dimensional indecomposable Λ-modules (resp. isoclasses of indecomposable Λ'-modules with dimension $\leq d'$). Our choice of d' guarantees that the functor $Q \otimes_{\Lambda'} -$ induces a bijection between a subset of $\mathcal{C}_{\leq d'}$ and $\mathcal{C}_d$. Since almost all classes $[M'] \in \mathcal{C}_{\leq d'}$ have the form $[M'] = [Z'_i \otimes_{\Gamma_i} N]$, then almost all classes $[M] \in \mathcal{C}_d$, have the form $[M] = [Q \otimes_{\Lambda'} M'] = [Q \otimes_{\Lambda'} Z'_i \otimes_{\Gamma_i} N] = [Z_i \otimes_{\Gamma_i} N]$, as we needed to show. Moreover, it is clear that $Z \otimes_{\Gamma_i} -$ coincides with the composition of $Z'_i \otimes_{\Gamma_i} -$ with $Q \otimes_{\Lambda'} -$, which preserves isomorphism classes of indecomposables. Thus, Λ is strictly tame too.

The proof of the fact that Λ is tame whenever Λ' is so is similar. □

Lemma 27.13. *Let Λ be a finite-dimensional k-algebra and denote by J its radical. Take $M \in \Lambda$-Mod and let $P \xrightarrow{f} Q \longrightarrow M \longrightarrow 0$ be its minimal projective presentation. Then, if Λ splits over its radical, the following inequalities hold*

$$\ell_\Lambda(Q/JQ) \leq \dim_k M \text{ and } \ell_\Lambda(P/JP) \leq \dim_k \Lambda \times \dim_k M.$$

Proof. Clearly, $\ell_\Lambda(Q/JQ) = \ell_\Lambda(M/JM) \leq \dim_k(M/JM) \leq \dim_k M$. From (19.6), we have that $Q \cong \Lambda \otimes_S (Q/JQ)$. Then, if $t := \ell_\Lambda(Q/JQ) = \ell_S(Q/JQ)$, we know that Q/JQ is a direct summand of S^t and, as a consequence, $\ell_\Lambda(Q) = \ell_\Lambda(\Lambda \otimes_S Q/JQ) \leq \ell_\Lambda(\Lambda \otimes_S S^t) \leq \ell_\Lambda(\Lambda^t) \leq \dim_k \Lambda \times t \leq \dim_k \Lambda \times \dim_k M$. Then, $\ell_\Lambda(P/JP) = \ell_\Lambda(\mathrm{Im}\,(f)/J\mathrm{Im}\,(f)) \leq \ell_\Lambda(\mathrm{Im}\,(f)) \leq \ell_\Lambda(Q) \leq \dim_k \Lambda \times \dim_k M$. □

Theorem 27.14. *Let Λ be a finite-dimensional algebra over the algebraically closed field k. Consider the following statements:*

(1) Λ is not wild;
(2) $\mathcal{D}^\Lambda$ is not wild;
(3) $\mathcal{D}^\Lambda$ is strictly tame;
(4) $\mathcal{D}^\Lambda$ is tame;
(5) Λ is strictly tame;
(6) Λ is tame.

Then, (1), (5) and (6) are equivalent. If Λ is basic, all the statements are equivalent.

Proof. By (22.15) and (27.12), we may assume that Λ is basic. Thus, by (23.25), $\mathcal{D}^\Lambda$ is a nested ditalgebra. By (27.10), $\mathcal{D}^\Lambda$ is tame iff it is strictly tame.

If Λ is not wild, from (22.21), we know that $\mathcal{D}^\Lambda$ is not wild. By (27.10), $\mathcal{D}^\Lambda$ is tame iff it is strictly tame.

Now, assume that $\mathcal{D} := \mathcal{D}^\Lambda$ is strictly tame and let us show that so is Λ. Fix $d \in \mathbb{N}$ and make $d' := (1 + \dim \Lambda)d$. Then, there is a finite collection $\{(\Gamma_i, Z_i)\}_{i=1}^n$, where $\Gamma_i = k[x]_{f_i}$ and Z_i is an A-Γ_i-bimodule, which is free of finite rank as a right Γ_i-module, such that the functors $\Gamma_i\text{-Mod} \xrightarrow{Z_i \otimes_{\Gamma_i} -} A\text{-Mod} \xrightarrow{L_\mathcal{D}} \mathcal{D}\text{-Mod}$ preserve isomorphism classes of indecomposables and, with the possible exception of only a finite number of isomorphism classes $[M]$ of indecomposables $M \in \mathcal{D}$-Mod with dimension $\leq d'$, every isomorphism class $[M]$ of an indecomposable $M \in \mathcal{D}$-Mod, with $\dim_k M \leq d'$, is of the form $[M] = [Z_i \otimes_{\Gamma_i} N]$, for some $i \in [1, n]$ and $N \in \Gamma_i$-Mod.

Consider the functor $\Xi_\Lambda : \mathcal{D}^\Lambda\text{-Mod} \longrightarrow \mathcal{P}^1(\Lambda)$ of (19.8). By (22.20), the objects of the form $\Xi_\Lambda(Z_i \otimes_{\Gamma_i} M)$, with $M \in \Gamma_i$-Mod indecomposable, cannot be of the form $(P, 0, 0)$. Therefore, the image of each functor $\Xi_\Lambda L_{\mathcal{D}}(Z_i \otimes_{\Gamma_i} -)$ restricted to the full subcategory of indecomposable of Γ_i-modules, lies in $\mathcal{P}^2(\Lambda)$, and its composition with the cokernel functor $\mathcal{P}^2(\Lambda) \longrightarrow \Lambda\text{-Mod}$ gives a functor which preserves isomorphism classes of indecomposables (see (18.10)). By (22.18)(1), the mentioned composition is naturally isomorphic to $Z'_i \otimes_{\Gamma_i} -$, where $Z'_i = \mathrm{Cok}\,\Xi_\Lambda(Z_i)$ is a Λ-Γ_i-bimodule which is finitely generated by the right. Now, assume that H is an indecomposable d-dimensional Λ-module and $\mathrm{Cok}\,\Xi_\Lambda(M) \cong H$. Make $\Xi_\Lambda(M) = (P \longrightarrow Q)$. Then, by (22.19) and (27.13), $\dim M = \ell_\Lambda(P/JP) + \ell_\Lambda(Q/JQ) \leq \dim H \times (1 + \dim \Lambda) = d'$. Then $H \cong Z'_i \otimes_{\Gamma_i} N$, for some i and some N. Then, we can apply (27.5) to obtain that Λ is strictly tame. As we remarked before, this implies that Λ is tame.

By (27.8), $(\Lambda, 0)$ is a presented ditalgebra. Hence, if Λ is tame, (27.9) implies that Λ is not wild. □

Exercise 27.15. *Assume Λ is a finite-dimensional algebra over the algebraically closed field k. Let $P_1, \ldots, P_d$ be a complete set of representatives of the indecomposable projective Λ-modules. Then, for each $M \in \Lambda\text{-mod}$, we can consider its minimal projective presentation $\phi_M \in \mathcal{P}^2(\Lambda)$*

$$\oplus_{i=1}^d m_i P_i \xrightarrow{\phi_M} \oplus_{i=1}^d n_i P_i \longrightarrow M \longrightarrow 0.$$

Then, we can associate the vector $\underline{p}(M) := (m_1, \ldots, m_d; n_1, \ldots, n_d)$. We will say that a Λ-module M is tense iff $\mathrm{Ext}_{\mathcal{P}^1(\Lambda)}(\phi_M, \phi_M) = 0$. Here the extension group is taken relative to the natural exact structure in $\mathcal{P}^1(\Lambda)$ described in (18.11). Show that, for any tense modules $M, N \in \Lambda\text{-mod}$, $M \cong N$ iff $\underline{p}(M) = \underline{p}(N)$.

Hint: We can assume that Λ is basic and consider the nested Drozd's ditalgebra $\mathcal{D}^\Lambda$. Then, apply (26.12) to $\mathcal{D}^\Lambda$-mod, having in mind (19.10) and (22.19).

References. Drozd's Theorem was proved in [28]; it validates a conjecture due to P. Donovan and M. R. Freislich (see [26]). Here we followed Crawley–Boevey's refined proof of this theorem given in [20], with our more detailed and ring-theoretical point of view. Statement (27.15) was first noticed in this context and later generalized for artin algebras in [8].

28
Modules over non-wild ditalgebras revisited

In this section, we shall prove that the finite number p of minimal ditalgebras necessary to study the modules of bounded length of a non-wild seminested ditalgebra $\mathcal{A}$ over an algebraically closed field k, as in Theorem (26.9), can be reduced to $p = 1$.

Definition 28.1. *Let $\mathcal{A}$ be a seminested ditalgebra. Denote by $\mathbb{L}(\mathcal{A})$ the set of all maps from the set of vertices of $\mathcal{A}$ to $\mathbb{N} \cup \{0\}$. The elements of $\mathbb{L}(\mathcal{A})$ are called the length vectors of $\mathcal{A}$. By definition, given $\mathcal{L} \subseteq \mathbb{L}(\mathcal{A})$, the set $\mathcal{Z}(\mathcal{L})$ consists of all the vertices i of $\mathcal{A}$ such that $\ell_i = 0$, for all $\ell \in \mathcal{L}$. Given two length vectors $\ell, \ell' \in \mathbb{L}(\mathcal{A})$, we write $\ell \leq \ell'$ iff $\ell_i \leq \ell'_i$, for all vertex i. Given $\mathcal{L} \subseteq \mathbb{L}(\mathcal{A})$, then $\overline{\mathcal{L}}$ denotes the subset of $\mathbb{L}(\mathcal{A})$ formed by the length vectors ℓ such that $\ell \leq \ell' + \ell''$, for some $\ell', \ell'' \in \mathcal{L}$. Notice that $\mathcal{L} \subseteq \overline{\mathcal{L}}$ and that $\mathcal{Z}(\overline{\mathcal{L}}) = \mathcal{Z}(\mathcal{L})$.*

The norm of a length vector $\ell \in \mathbb{L}(\mathcal{A})$ is defined by the rule

$$\|\ell\| := \sum_{\alpha \in \mathbb{B}_0} \ell_{t(\alpha)} \ell_{s(\alpha)} + \sum_{i \in \mathcal{P}_{\text{mk}}} \ell_i^2,$$

where $\mathbb{B}_0$ is the set of solid arrows of $\mathcal{A}$ and $\mathcal{P}_{\text{mk}}$ is the set of marked points of $\mathcal{A}$.

Lemma 28.2. *Assume that the ditalgebra $\mathcal{A}^z$ is obtained from the seminested ditalgebra $\mathcal{A}$ by absorption, regularization, deletion of idempotents, edge reduction or unravelling, as in (23.16), (23.15), (23.14), (23.18) or (23.23). Define the additive map*

$$t_z : \mathbb{L}(\mathcal{A}^z) \longrightarrow \mathbb{L}(\mathcal{A}),$$

for $z \in \{a, r, d, e, u\}$, as follows: $t_a := id$; $t_r := id$; $t_d(\ell)_i := \ell_i$, if the idempotent e_i was not deleted, and $t_d(\ell)_i := 0$, otherwise; $t_e(\ell)_i := \ell_i$, for all $i \notin \{i_0, j_0\}$, $t_e(\ell)_{i_0} := \ell_{i_*} + \ell_z$ and $t_e(\ell)_{j_0} := \ell_{j_*} + \ell_z$; $t_u(\ell)_i := \ell_i$, for all $i \neq i_0$, and $t_u(\ell)_{i_0} := \ell_{i_*} + \sum_{s=1}^{q} \sum_{a=1}^{n} \sum_{t=0}^{a-1} \ell_{i_{s,a}}$. We use the notation of the *context where each basic operation $\mathcal{A} \mapsto \mathcal{A}^z$ was defined. Then, for any k-algebra E:*

(1) if $F_z^E : \mathcal{A}^z\text{-}E\text{-Mod} \longrightarrow \mathcal{A}\text{-}E\text{-Mod}$ is induced by the reduction functor associated to $\mathcal{A}^z$, we have, for any $M \in \mathcal{A}^z\text{-}E\text{-mod}$, $\underline{\ell}_E(F_z^E(M)) = t_z(\underline{\ell}_E(M))$;

(2) for any $\ell \in \mathbb{L}(\mathcal{A}^z)$, $\|t_z(\ell)\| \geq \|\ell\|$;

(3) whenever $\ell \leq \ell'$ in $\mathbb{L}(\mathcal{A}^z)$, we have $t_z(\ell) \leq t_z(\ell')$.

Proof. (1) and (2) follow from (25.2), (25.3), (25.4), (25.8) and (25.9). For $z \in \{e, u\}$, we have to look at the analysis given in the corresponding proof. (3) is clear in each case. □

Lemma 28.3. *In the context of Lemma (28.2), for each finite subset $\mathcal{L}$ of $\mathbb{L}(\mathcal{A})$, we consider the finite subset $\mathcal{L}^z := t_z^{-1}(\mathcal{L})$ of $\mathbb{L}(\mathcal{A}^z)$. Notice that $t_z(\overline{\mathcal{L}^z}) \subseteq \overline{\mathcal{L}}$ and $\overline{(\mathcal{L}^z)} \subseteq (\overline{\mathcal{L}})^z$.*

Proof. Indeed, if $\ell \leq \ell' + \ell''$, with $\ell', \ell'' \in \mathcal{L}^z$, then by (28.2)(3), $t_z(\ell) \leq t_z(\ell') + t_z(\ell'')$. Thus, $t_z(\ell) \in \overline{\mathcal{L}}$, and $\ell \in (\overline{\mathcal{L}})^z$. □

Notation 28.4. *Given a seminested ditalgebra $\mathcal{A}$, a finite set of length vectors $\mathcal{L}$ of $\mathcal{A}$, and a non-negative integer j, we make*

$$\mathcal{L}(j) := \{\ell \in \mathcal{L} \mid \|\ell\| = j\}.$$

Lemma 28.5. *Assume that $\mathcal{A}^z$ is obtained from a seminested ditalgebra $\mathcal{A}$ by regularization or edge reduction. Fix some finite set $\mathcal{L}$ of length vectors for $\mathcal{A}$. Then:*

(1) if $\|t_z(\ell)\| = \|\ell\|$, for all $\ell \in \mathcal{L}^z$, then $|\mathcal{L}^z(j)| = |\mathcal{L}(j)|$, for all $j \geq 0$;

(2) if there exists $\ell_0 \in \mathcal{L}^z$ such that $\|t_z(\ell_0)\| > \|\ell_0\|$, consider

$$m := \max\{\|t_z(\ell)\| \mid \ell \in \mathcal{L}^z \text{ and } \|t_z(\ell)\| > \|\ell\|\}, \text{ then}$$
$$|\mathcal{L}^z(m)| < |\mathcal{L}(m)| \text{ and, for } j > m, \ |\mathcal{L}^z(j)| = |\mathcal{L}(j)|.$$

Proof. Let us first make two remarks in the edge reduction case:

(a) Given any $\ell' \in \mathbb{L}(\mathcal{A})$, we can define: $\ell_i := \ell'_i$, for all vertex i of $\mathcal{A}^e$ different from z, i_*, j_*; $\ell_z := 0$; $\ell_{i_*} := \ell'_{i_0}$ and $\ell_{j_*} := \ell'_{j_0}$. Then, $t_e(\ell) = \ell'$ and the map $t_e : \mathbb{L}(\mathcal{A}^e) \longrightarrow \mathbb{L}(\mathcal{A})$ is surjective.

(b) Given $\ell \in \mathbb{L}(\mathcal{A}^e)$, we have $\|t_e(\ell)\| - \|\ell\| = t_e(\ell)_{i_0} t_e(\ell)_{j_0}$, where $t_e(\ell)_{i_0} = \ell_{i_*} + \ell_z$ and $t_e(\ell)_{j_0} = \ell_{j_*} + \ell_z$. Then, $\|t_e(\ell)\| = \|\ell\|$ iff $t_e(\ell)_{i_0} t_e(\ell)_{j_0} = 0$, and this implies that $\ell_z = 0$. Then, whenever $\ell, \ell' \in \mathbb{L}(\mathcal{A}^e)$ satisfy $\|t_e(\ell)\| = \|\ell\|$, $\|t_e(\ell')\| = \|\ell'\|$ and $t_e(\ell) = t_e(\ell')$, we have that $\ell = \ell'$.

(1) Our assumption clearly implies that the map $t_z : \mathbb{L}(\mathcal{A}^z) \longrightarrow \mathbb{L}(\mathcal{A})$ restricts to a map $t'_z : \mathcal{L}^z(j) \longrightarrow \mathcal{L}(j)$, for any $j \geq 0$. In the regularization case, $t'_r = id$ is clearly bijective. In the edge reduction case, our assumption and the previous remark (b) imply that t'_e is injective. Since t_e is surjective, our assumption implies that t'_e is surjective too.

(2) First we show that, for $j \geq m$, the map $t_z : \mathbb{L}(\mathcal{A}^z) \longrightarrow \mathbb{L}(\mathcal{A})$ restricts to a well-defined injective map $t'_z : \mathcal{L}^z(j) \longrightarrow \mathcal{L}(j)$. Let $\ell \in \mathcal{L}^z(j)$, then, by (28.2)(2), $\|t_z(\ell)\| \geq \|\ell\| = j \geq m$. Since $\|t_z(\ell)\| > \|\ell\|$ would contradict the maximality of m, we have that $\|t_z(\ell)\| = \|\ell\| = j$, and $t_z(\ell) \in \mathcal{L}(j)$. Thus, t'_z is well defined. Let us show that t'_z is injective. In the regularization case, $t_z = id$ is already injective. In the edge reduction case, t'_e is injective by the previous remark (b).

Let us now show that t'_z is surjective for $j > m$. Take $\ell \in \mathcal{L}(j)$. Since t_z is surjective, $t_z(\ell') = \ell$, for some $\ell' \in \mathcal{L}^z$. We know that $\|\ell'\| \leq \|t_z(\ell')\| = \|\ell\| = j$. If $\|\ell'\| \neq j$, then $\|\ell'\| < j$. By the maximality of m, $j \leq m$, which is impossible. Thus, $\|\ell'\| = j$ and t'_z is surjective.

Finally, assume that $j = m$. We want to exhibit some $\ell \in \mathcal{L}(m) \setminus t_z(\mathcal{L}^z(m))$. By assumption, we can choose $\ell' \in \mathcal{L}^z$ such that $m = \|t_z(\ell')\| > \|\ell'\|$. Thus, $\ell := t_z(\ell') \in \mathcal{L}(m)$. Let us see that $\ell \notin \operatorname{Im} t'_z$. Assume that $\ell'' \in \mathcal{L}^z(m)$ satisfies $t_z(\ell'') = \ell$. If t_z is injective, then $\ell'' = \ell'$ and we get the contradiction $m = \|\ell''\| = \|\ell'\| < m$. Then, t_z is not injective and we are in the edge reduction case $z = e$. As remarked above, $\|t_e(\ell')\| > \|\ell'\|$ iff $t_e(\ell')_{i_0} t_e(\ell')_{j_0} \neq 0$. Since $t_e(\ell') = t_e(\ell'')$, we also have that $\|t_e(\ell'')\| > \|\ell''\|$, a contradiction.

□

Lemma 28.6. *Let $\mathcal{A}$ be a seminested ditalgebra and fix some finite set $\mathcal{L}$ of length vectors for $\mathcal{A}$. Assume that $\mathcal{A}^z$ is obtained from $\mathcal{A}$ by absorption or elimination of a finite set of idempotents in $\mathcal{Z}(\mathcal{L})$. Then*

$$|\mathcal{L}^z(j)| = |\mathcal{L}(j)|, \quad \textit{for all } j \geq 0.$$

Proof. In both cases, $t_z : \mathbb{L}(\mathcal{A}^z) \longrightarrow \mathbb{L}(\mathcal{A})$ is injective and $\|t_z(\ell)\| = \|\ell\|$, for all $\ell \in \mathbb{L}(\mathcal{A}^z)$. In the absorption case, $t_a = id$ is clearly bijective. In the deletion of idempotents case, $z = d$, since we are deleting idempotents in $\mathcal{Z}(\mathcal{L})$, we have that $t_z : \mathcal{L}^z \longrightarrow \mathcal{L}$ is surjective. Then, in both cases, the map t_z induces a bijection $\mathcal{L}^z(j) \longrightarrow \mathcal{L}(j)$, for all $j \geq 0$. □

Definition 28.7. *Let $\mathcal{A}$ be a seminested ditalgebra with layer (R, W), and consider the decomposition of the unit $1 = \sum_{i=1}^n e_i$ of R, as a sum of primitive central orthogonal idempotents. Then, the support of M, denoted by* supp$M$, *is the subset of vertices i of $\mathcal{A}$ such that $M_i := e_i M \neq 0$. Given a fixed vertex*

v of $\mathcal{A}$, we say that an $\mathcal{A}$-module M is concentrated at v iff $\mathrm{supp}M = \{v\}$ *and* $\alpha M = 0$, *for any solid arrow α of $\mathcal{A}$.*

Proposition 28.8. *Let $\mathcal{A}$ be a non-wild seminested ditalgebra over the algebraically closed field k. Assume $\ell \in \mathbb{N}$ and v is a marked vertex of $\mathcal{A}$, say with marked loop z. Then, there is a finite subset $\mathcal{S}(\ell, v)$ of k such that any indecomposable $M \in \mathcal{A}$-*Mod *with dimension $\ell_k(M) = \ell$ and such that $M_v \neq 0$ and* $\mathrm{spec}M(z) \not\subseteq \mathcal{S}(\ell, v)$, *there is $N \in \mathcal{A}$-*mod *concentrated at v with $N \cong M$.*

Proof. This proof follows the same strategy used in the proof of (26.9). We adapt the case-by-case arguments given there.

For any seminested k-ditalgebra $\mathcal{C}$, any length $\ell \in \mathbb{N}$ and any marked vertex v of $\mathcal{C}$, denote by $\mathcal{I}_{\mathcal{C}}(\ell, v)$ the class of indecomposables $M \in \mathcal{C}$-Mod with $\ell_k(M) \leq \ell$ and $M_v \neq 0$. By $\mathcal{I}'_{\mathcal{C}}(\ell, v)$, we denote the class of indecomposables $M \in \mathcal{I}_{\mathcal{C}}(\ell, v)$, which are not isomorphic to an object in $\mathcal{C}$-mod concentrated at v. We will say that a subset $\mathcal{S}$ of k *bounds* a class $\mathcal{I}$ of $\mathcal{C}$-modules iff $\mathrm{spec}M(z) \subseteq \mathcal{S}$, for all $M \in \mathcal{I}$, where z is the marked loop at v. With this terminology in mind, we have to prove:

(P): *For any marked vertex v of $\mathcal{A}$ and any $\ell \in \mathbb{N}$, there is a finite subset $\mathcal{S}(\ell, v)$ of k such that $\mathcal{S}(\ell, v)$ bounds $\mathcal{I}'_{\mathcal{A}}(\ell, v)$.*

Given a marked vertex v of a seminested k-ditalgebra $\mathcal{C}$ and $d \in \mathbb{N}$, we can define the classes $\mathcal{I}_{\mathcal{C}}[d, v]$ and $\mathcal{I}'_{\mathcal{C}}[d, v]$ as before, replacing ℓ by d, and $\ell_k(M) \leq \ell$ by $\|M\|_k \leq d$. Then, (P) will follow from:

(P′): *For any marked vertex v of $\mathcal{A}$ and any $d \in \mathbb{N}$, there is a finite subset $\mathcal{S}[d, v]$ of k such that $\mathcal{S}[d, v]$ bounds $\mathcal{I}'_{\mathcal{A}}[d, v]$.*

Furthermore, if we define $\mathcal{I}'_{\mathcal{C}}[\![d, v]\!]$ as the class of sincere objects in $\mathcal{I}'_{\mathcal{C}}[d, v]$, then statement (P′) is equivalent to the following:

(P″): *For any marked vertex v of $\mathcal{A}$ and any $d \in \mathbb{N}$, there is a finite subset $\mathcal{S}[\![d, v]\!]$ of k such that $\mathcal{S}[\![d, v]\!]$ bounds $\mathcal{I}'_{\mathcal{A}}[\![d, v]\!]$.*

Indeed, consider the k-ditalgebras $\mathcal{A}^{d_1}, \ldots, \mathcal{A}^{d_t}$ obtained from $\mathcal{A}$ by deletion of a finite number of idempotents different from e_v. Thus v is indeed a marked vertex of any such $\mathcal{A}^{d_i}$. Then, apply (P″) to each one of these ditalgebras to obtain finite subsets $\mathcal{S}_{\mathcal{A}^{d_i}}[\![d, v]\!]$ of k such that each $\mathcal{S}_{\mathcal{A}^{d_i}}[\![d, v]\!]$ bounds $\mathcal{I}'_{\mathcal{A}^{d_i}}[\![d, v]\!]$. Then, make

$$\mathcal{S}_{\mathcal{A}}[d, v] := \left(\cup_{i=1}^{t} \mathcal{S}_{\mathcal{A}^{d_i}}[\![d, v]\!]\right) \cup \mathcal{S}_{\mathcal{A}}[\![d, v]\!].$$

Now, if $M \in \mathcal{I}'_{\mathcal{A}}[d, v] \setminus \mathcal{I}'_{\mathcal{A}}[\![d, v]\!]$, then $M \cong F_{d_i}(N)$, for some $i \in [1, t]$ and some sincere $N \in \mathcal{A}^{d_i}$-Mod. The deletion of idempotents $\mathcal{A} \mapsto \mathcal{A}^{d_i}$ did not delete the marked vertex v of $\mathcal{A}$, and $N_v \neq 0$, because $M_v \neq 0$. Then, $N \in$

$\mathcal{I}'_{\mathcal{A}^{d_i}}[[d, v]]$, and $\mathcal{S}_{\mathcal{A}^{d_i}}[[d, v]]$ bounds N, so it also bounds M. Hence $\mathcal{S}_{\mathcal{A}}[d, v]$ bounds $\mathcal{I}'_{\mathcal{A}}[d, v]$.

We shall prove (P″) by induction on d. If $d = 0$, then $\mathcal{I}'_{\mathcal{A}}[[d, v]] = \emptyset$ and there is nothing to show. Notice also that, for any minimal k-ditalgebra $\mathcal{B}$, each indecomposable $\mathcal{B}$-module M, with $M_v \neq 0$ for some vertex v, is obviously concentrated at v.

So assume as in the proof of (26.9) that $d > 0$ and (P″) holds for any $d' < d$, or, equivalently, that (P′) holds for any $d' < d$. Now, we have to consider a fixed marked vertex v of $\mathcal{A}$ and the sincere modules $M \in \mathcal{I}'_{\mathcal{A}}[[d, v]]$. We can assume that $\mathcal{A}$ is not minimal.

Since $\mathcal{A}$ is a seminested ditalgebra we can consider a minimal solid arrow $\alpha : i_0 \longrightarrow j_0$ in $\mathcal{A}$. Then, $\delta(\alpha) \in W_1$.

Case 1: $\delta(\alpha) = 0$ and $i_0 = j_0$.

Since $\mathcal{A}$ is not wild, by (24.11), $Re_{i_0} \cong k$. Consider the ditalgebra $\mathcal{A}^a$ obtained from $\mathcal{A}$ by absorption of the loop α. Then, we can apply (25.2) to every sincere $M \in \mathcal{A}$-Mod with norm d to obtain a sincere $N \in \mathcal{A}^a$-Mod with $F_a(N) \cong M$ having the same norm d. We have that $\mathcal{A}^a$ has one solid arrow less than $\mathcal{A}$. Since $Re_{i_0} \cong k$, we have that $v \neq i_0$ and v is a marked vertex of $\mathcal{A}^a$. Since $F_a(\mathcal{I}'_{\mathcal{A}^a}[[d, v]]) = \mathcal{I}'_{\mathcal{A}}[[d, v]]$, then any finite subset $\mathcal{S}$ of k that bounds $\mathcal{I}'_{\mathcal{A}^a}[[d, v]]$ will bound $\mathcal{I}'_{\mathcal{A}}[[d, v]]$.

Repeating this argument, if necessary, either we end up with a seminested ditalgebra with no solid arrows (that is a minimal ditalgebra) or at some step we obtain a ditalgebra with a minimal solid arrow in one of the following cases.

Case 2: $\delta(\alpha) = 0$ and $i_0 \neq j_0$.

Since $\mathcal{A}$ is not wild, by (24.11), $Re_{i_0} \cong k$ and $Re_{j_0} \cong k$. Then, we can consider the ditalgebra $\mathcal{A}^e$ obtained from $\mathcal{A}$ by reduction of the edge α and apply (25.8). Moreover, since v is marked, $v \notin \{i_0, j_0\}$; thus, v is a marked vertex of $\mathcal{A}^e$. Then, we can apply our induction hypothesis to $\mathcal{A}^e$ and $d - 1$ to get a finite subset $\mathcal{S}_{\mathcal{A}^e}[d-1, v]$ of k, which bounds $\mathcal{I}'_{\mathcal{A}^e}[d-1, v]$. Since $F_e(N) \cong M$ with $M \in \mathcal{I}'_{\mathcal{A}}[[d, v]]$ implies that $N \in \mathcal{I}'_{\mathcal{A}^e}[d-1, v]$, we get that $\mathcal{S}_{\mathcal{A}^e}[d-1, v]$ bounds $M \in \mathcal{I}'_{\mathcal{A}}[[d, v]]$.

Case 3: $\delta(\alpha) \neq 0$.

Write $C := Re_{i_0} \otimes_k Re_{j_0}$. Then, $e_{j_0} W_1 e_{i_0}$ is a C-module and we can write

$$\delta(\alpha) = \sum_{i=1}^{j} c_i v_i,$$

for some $v_1, \ldots, v_j \in e_{j_0} \mathbb{B}_1 e_{i_0}$ and $0 \neq c_i \in C$.

Subcase 3.1: Some c_t is invertible in C.

This is the case, for instance, if $Re_{i_0} \cong k$ and $Re_{j_0} \cong k$. Make $v'_t := \sum_{i=1}^{j} c_i v_i$ and consider the change of basis for W_1 where $v_1, \ldots, v_t, \ldots, v_j$ is replaced by $v_1, \ldots, v'_t, \ldots, v_j$ using (26.1).

Then, we can apply regularization, that is (25.3), to obtain an equivalence $F_r : \mathcal{A}^r\text{-Mod} \longrightarrow \mathcal{A}\text{-Mod}$. We can apply our induction hypothesis to $\mathcal{A}^r$ and $d-1$. We know that v is a marked vertex of $\mathcal{A}^r$ and there is a finite subset $\mathcal{S}_{\mathcal{A}^r}[d-1, v]$ of k which bounds $\mathcal{I}'_{\mathcal{A}^r}[d-1, v]$. Since $F_r(N) \cong M$ with $M \in \mathcal{I}'_{\mathcal{A}}[[d, v]]$ implies that $N \in \mathcal{I}'_{\mathcal{A}^r}[d-1, v]$, we get that $\mathcal{S}_{\mathcal{A}^r}[d-1, v]$ bounds $\mathcal{I}'_{\mathcal{A}}[[d, v]]$ too.

Subcase 3.2: $Re_{j_0} \cong k$.

Since we have already considered the *Subcase 3.1*, we may assume that $Re_{i_0} \ncong k$. Thus, $Re_{i_0} = k[x]_{f(x)}$ and $C \cong k[x]_{f(x)}$. By an appropriate change of basis of W_1 of the form $v'_i = f(x)^{-p} v_i$, for all i, using again (26.1), we may assume that $c_i \in k[x]$. Consider the ditalgebra $\mathcal{A}^u$ obtained from $\mathcal{A}$ by unravelling at i_0 using d and the different roots $\lambda_1, \ldots, \lambda_q \in k$ of $c_1(x)$. From (25.9), we know that the sincere modules $M \in \mathcal{A}\text{-Mod}$ with norm $\|M\| \leq d$ and $M(c_1(x))$ not invertible are of the form $F_u(N) \cong M$, where $N \in \mathcal{A}^u\text{-Mod}$ has norm $\|N\| \leq d-1$.

Subcase 3.2.1: $v \neq i_0$.

Here, we can apply our induction hypothesis to $\mathcal{A}^u$, the marked vertex v of $\mathcal{A}^u$ and $d-1$, to obtain a finite subset $\mathcal{S}_{\mathcal{A}^u}[d-1, v]$ of k, which bounds $\mathcal{I}'_{\mathcal{A}^u}[d-1, v]$, and hence bounds M.

Moreover, if $\mathcal{A}^l$ is the ditalgebra obtained from $\mathcal{A}$ by localizing at the vertex i_0 using the polynomial $c_1(x)$, the sincere $\mathcal{A}$-modules with norm $\|M\| \leq d$ and $M(c_1(x))$ invertible are of the form $F_l(N) \cong M$, where $N \in \mathcal{A}^l\text{-Mod}$ has norm $\|N\| = \|M\| \leq d$. According to the description of the differential δ^l of $\mathcal{A}^l$ given in (17.7), we have that

$$\delta^l(1 \otimes \alpha \otimes 1) = \sum_{i=1}^{j} c_i(x)(1 \otimes v_i \otimes 1).$$

But now, c_1 is invertible and we can proceed as in *Subcase 3.1*. That is we can bound all these indecomposables N with some finite subset $\mathcal{S}_{\mathcal{A}^l}[[d, v]]$ of k. Then, $\mathcal{S}_{\mathcal{A}}[[d, v]] := \mathcal{S}_{\mathcal{A}^l}[[d, v]] \cup \mathcal{S}_{\mathcal{A}^u}[d-1, v]$ bounds $\mathcal{I}'_{\mathcal{A}}[[d, v]]$.

Subcase 3.2.2: $v = i_0$.

Here, i_* is a marked vertex of $\mathcal{A}^u$ and, therefore, we can apply our induction hypothesis to $\mathcal{A}^u$, i_* and $d-1$, to obtain a finite subset $\mathcal{S}_{\mathcal{A}^u}[d-1, i_*]$ of k, which bounds $\mathcal{I}'_{\mathcal{A}^u}[d-1, i_*]$. Here, $z = x$ and

$H = Re_{i_0} = k[x]_{f(x)}$. From the proof of (25.9), recall that we have an isomorphism of H-modules

$$\begin{aligned} e_{i_0}M &\cong F_u(N)_{i_0} \\ &\cong N_{i_*} \oplus \left(\oplus_{s=1}^{q} \oplus_{a=1}^{n} [H/(x-\lambda_s)^a H] \otimes_k N_{i_{s,a}}\right). \end{aligned}$$

Thus, $\operatorname{spec} M(x) \subseteq \operatorname{spec} N(x) \cup \{\lambda_1, \ldots, \lambda_q\}$. Then, we can bound all the modules $M \in \mathcal{I}'_{\mathcal{A}}[\![d, v]\!]$ which satisfy that $M(c_1(x))$ is not invertible with the finite set $\mathcal{S}_{\mathcal{A}^u}[d-1, i_*] \cup \{\lambda_1, \ldots, \lambda_q\}$. In order to bound all the $M \in \mathcal{I}'_{\mathcal{A}}[\![d, v]\!]$ which satisfy that $M(c_1(x))$ is invertible with a finite subset of k, we proceed as in Subcase 3.2.1.

Subcase 3.3: $Re_{i_0} \cong k$.

This case is dual to *Subcase 3.2*.

Subcase 3.4: $Re_{i_0} \ncong k$ and $Re_{j_0} \ncong k$.

Assume that $Re_{i_0} = k[x]_{f(x)}$ and $Re_{j_0} = k[x]_{g(x)}$. Hence, $C = k[x, y]_{f(x)g(y)}$. After an appropriate change of basis of W_1, of the form $v'_i = f(x)^{-p} g(y)^{-q} v_i$, using (26.1), we may assume that all the c_i are polynomials in $k[x, y]$. Let $h(x, y)$ be the highest common factor of the $c_i(x, y)$, and assume that $h(x, y)q_i(x, y) = c_i(x, y)$, for all i. Since the $q_i(x, y)$ are coprime in $k(x)[y]$, there are polynomials $s_i(x, y) \in k[x, y]$ and a non-zero polynomial $u(x) \in k[x]$ such that

$$u(x) = \sum_{i=1}^{j} s_i(x, y) q_i(x, y).$$

If the module M satisfies that $M(u(x))$ is not invertible, we can proceed as in *Subcase 3.2*, by unravelling at i_0 using d and the different roots of $u(x)$.

Otherwise, we can consider the ditalgebra $\mathcal{A}^l$ obtained from $\mathcal{A}$ by localizing at the point i_0 using $u(x)$. Then, $M \cong F_l(N)$, where $N \in \mathcal{A}^l$-Mod is sincere and has the same norm than M. From the previous formula for $u(x)$, we obtain in the ditalgebra $\mathcal{A}^l$ that

$$1 = \sum_{i=1}^{j} [s_i(x, y) u(x)^{-1}] q_i(x, y).$$

The terms in this formula all belong to the algebra $H := k[x, y]_{u(x)}$. Here, $H \cong D[y]$, where $D = k[x]_{u(x)}$ is a principal ideal domain. From (26.2), we obtain an invertible matrix $Q \in M_{j \times j}(H)$ with the first row $(q_1, \ldots, q_j)$. Consider the change of basis for $W_1^l = S \otimes_R$

$W_1 \otimes_R S$, which replaces each $w_i := 1 \otimes v_i \otimes 1$ by w'_i defined by the formula

$$(w'_1, \dots, w'_j)^t = Q(w_1, \dots, w_j)^t.$$

Hence

$$\delta^l(1 \otimes \alpha \otimes 1) = \sum_i c_i w_i = \sum_i hq_i w_i = hw'_1.$$

Since $\mathcal{A}^l$ is not wild, by (24.11), h must be invertible. Now, we can replace the basis $w'_1, w'_2, \dots$ of W^l_1 by $hw'_1, w'_2, \dots$ and apply regularization, that is (25.13), to finish the proof. □

Definition 28.9. *Let $\mathcal{L}$ be a finite set of length vectors of the seminested ditalgebra $\mathcal{A}$. Then, a marked vertex i of $\mathcal{A}$ is called $\mathcal{L}$-isolated iff*

(1) $i \notin \mathcal{Z}(\mathcal{L})$;

(2) for any indecomposable $M \in \mathcal{A}$-mod with $M_i \neq 0$ and $\underline{\ell}_k(M) \in \overline{\mathcal{L}}$, there exists $N \in \mathcal{A}$-mod concentrated at the vertex i with $N \cong M$ in $\mathcal{A}$-mod.

Lemma 28.10. *Assume that the ditalgebra $\mathcal{A}^z$ is obtained from the seminested ditalgebra $\mathcal{A}$ by absorption, regularization, deletion of a set $\{e_i \mid i \in \Omega\}$ of idempotents, edge reduction or unravelling, as in (23.16), (23.15), (23.14), (23.18) or (23.23). Denote by $\mathcal{P}_{\mathrm{mk}}(\mathcal{A})$ and $\mathcal{P}_{\mathrm{mk}}(\mathcal{A}^z)$ the set of marked points of $\mathcal{A}$ and $\mathcal{A}^z$, respectively; write $\mathcal{P}_{\mathrm{mk}}(\mathcal{A}, z) := \mathcal{P}_{\mathrm{mk}}(\mathcal{A})$, if $z \neq d$, and $\mathcal{P}_{\mathrm{mk}}(\mathcal{A}, d) := \mathcal{P}_{\mathrm{mk}}(\mathcal{A})\setminus\Omega$; and define the map*

$$j_z : \mathcal{P}_{\mathrm{mk}}(\mathcal{A}, z) \longrightarrow \mathcal{P}_{\mathrm{mk}}(\mathcal{A}^z),$$

for $z \in \{a, r, d, e, u\}$, as follows: $j_z(i) := i$, for all $i \in \mathcal{P}_{\mathrm{mk}}(\mathcal{A}, z)$, if $z \in \{a, r, d, e\}$; $j_u(i) := i$, for all $i \in \mathcal{P}_{\mathrm{mk}}(\mathcal{A}, u) \setminus \{i_0\}$, and $j_u(i_0) := i_*$. We use the notation of the context where each basic operation $\mathcal{A} \mapsto \mathcal{A}^z$ was defined. Then, j_z is always an injective map and, for all $z \in \{r, d, e, u\}$, it is also surjective. Consider the associated reduction functor $F_z : \mathcal{A}^z\text{-mod} \longrightarrow \mathcal{A}\text{-mod}$. Then, for any $i \in \mathcal{P}_{\mathrm{mk}}(\mathcal{A}, z)$:

(1) for each $N \in \mathcal{A}^z$-mod the vector space $N_{j_z(i)}$ is isomorphic to a subspace of $F_z(N)_i$;

(2) if $N \in \mathcal{A}^z$-mod is concentrated at $j_z(i)$, then $F_z(N)$ is concentrated at i;

(3) Assume furthermore, that the basic operation $\mathcal{A} \mapsto \mathcal{A}^z$ is not an unravelling at the point i. Then $F_z(N)_i$ and $N_{j_z(i)}$ are isomorphic vector spaces. Moreover, for any $N \in \mathcal{A}^z$-mod, such that $F_z(N) \cong M$ where $M \in \mathcal{A}$-mod is concentrated at the point i, there is $N' \in \mathcal{A}^z$-mod concentrated at the point $j_z(i)$ such that $N \cong N'$ in $\mathcal{A}^z$-mod.

Proof. Clearly, each j_z is surjective, for all $z \in \{r, d, e, u\}$. In the absorption case, a new marked point appears in $\mathcal{A}^a$. Now, take $z \in \{a, r, d, e, u\}$ and $i \in \mathcal{P}_{\text{mk}}(\mathcal{A}, z)$.

Since F_a, F_d and F_r are restriction functors it is clear that $N_{j_z(i)}$ and $F_z(N)_i$ are isomorphic spaces. From the analysis in the proof of (25.8) and (25.9), we know that $N_{j_z(i)}$ and $F_z(N)_i$, are isomorphic spaces, unless we are in the case of the unravelling at the vertex $i = i_0$, where we only have that $N_{j_z(i)} = N_{i_*}$ is isomorphic to a direct summand of the space $F_z(N)_i$. Then, (1) holds, as well as the first part of (3).

Item (2) is clear for the restriction functors F_a, F_r and F_d (notice that since i is a marked point of $\mathcal{A}$, it only makes sense to consider absorption of a loop based at a vertex $\neq i$). For the cases $z \in \{e, u\}$, it follows from the definition of F_X in (12.10) and the description of the arrows of $\mathcal{A}^z$ in (23.18) and (23.23).

In order to prove (3), assume that $N \in \mathcal{A}^z$-mod is such that $F_z(N) \cong M$, where $M \in \mathcal{A}$-mod is concentrated at i.

If $z \in \{a, r, d\}$, define the representation $N' \in \mathcal{A}^z$-mod such that $N'_j = 0$, for all $j \neq i$; $N'_i = M_i$; and $\alpha N' = 0$, for all solid arrow α of $\mathcal{A}^z$. Then, clearly N' is concentrated at $i = j_z(i)$ and $F_z(N') = M$, because F_z is a restriction functor. Since F_z is full and faithful, $F_z(N') = M \cong F_z(N)$ implies that $N' \cong N$ in $\mathcal{A}^z$-mod.

Now assume that $z \in \{e, u\}$ and let us construct $\overline{N'} \in \mathcal{A}^z$-mod concentrated at $j_z(i)$ such that $\overline{N'} \cong N$ in $\mathcal{A}^z$-mod. The object $\overline{N'}$ is constructed in the proof of (25.8) and (25.9) as follows. First construct the S-module N' defined by $N'_j = 0$, for all $j \neq i$, and $N'_i = M_i = M_{j_z(i)}$. In case $z = u$, this construction coincides with the one given in (25.9) because $\mathcal{A} \mapsto \mathcal{A}^z$ is not an unravelling at the point i. Then, there is an isomorphism of B-modules $f^0 : X \otimes_S N' \longrightarrow M$. From (25.5), we obtain $F_z(\overline{N'}) \cong M$ in $\mathcal{A}$-mod, for some $\overline{N'} \in \mathcal{A}^z$-mod with $U_{\mathcal{A}^z}(\overline{N'}) = N'$. We claim that $\overline{N'}$ is concentrated at i. We already have that $\text{supp}\overline{N'} = \{i\}$. Let us recall from the proof of (25.5) that $\overline{N'}$ is obtained using (16.1) from the object $\overline{X \otimes_S N'} \in \mathcal{A}$-mod, which is isomorphic to M through an isomorphism of the form $(f^0, 0) : X \otimes_S N' \longrightarrow M$. Thus, every solid arrow α of $\mathcal{A}$ acts trivially on $\overline{X \otimes_S N'}$. Then, by (16.1), every solid arrow of $\mathcal{A}^z$ acts trivially on $\overline{N'}$ and, therefore, $\overline{N'}$ is concentrated at the point $i = j_z(i)$. As before, since F_z is full and faithful, $F_z(\overline{N'}) \cong M \cong F_z(N)$ implies that $\overline{N'} \cong N$ in $\mathcal{A}^z$-mod. □

Proposition 28.11. *Let $\mathcal{L}$ be a finite set of length vectors of a seminested ditalgebra $\mathcal{A}$. Assume that the ditalgebra $\mathcal{A}^z$ is obtained from the seminested ditalgebra $\mathcal{A}$ by absorption, regularization, deletion of some set of idempotents in $\mathcal{Z}(\mathcal{L})$, edge reduction or unravelling, as in (23.16), (23.15), (23.14), (23.18) or (23.23). Then, the following statements hold:*

(1) If $\mathcal{A} \mapsto \mathcal{A}^z$ is not an unravelling at the $\mathcal{L}$-isolated point $i \in \mathcal{P}_{\text{mk}}(\mathcal{A}, z)$, then $j_z(i)$ is an $\mathcal{L}^z$-isolated point of $\mathcal{A}^z$.

(2) Assume that $\mathcal{A} \mapsto \mathcal{A}^z$ is a regularization or a deletion of some idempotents in $\mathcal{Z}(\mathcal{L})$. Then, if $i \in \mathcal{P}_{\text{mk}}(\mathcal{A}, z)$ and the point $j_z(i)$ is $\mathcal{L}^z$-isolated, we have that i is $\mathcal{L}$-isolated.

Proof.

(1) Assume that $\mathcal{A} \mapsto \mathcal{A}^z$ is not an unravelling at the $\mathcal{L}$-isolated point $i \in \mathcal{P}_{\text{mk}}(\mathcal{A}, z)$. Since $i \in \mathcal{P}_{\text{mk}}(\mathcal{A}, z) \setminus \mathcal{Z}(\mathcal{L})$, we have $j_z(i) \in \mathcal{P}_{\text{mk}}(\mathcal{A}^z) \setminus \mathcal{Z}(\mathcal{L}^z)$. Take an indecomposable $N \in \mathcal{A}^z$-mod with $0 \neq N_{j_z(i)}$ and $\underline{\ell}_k(N) \in \overline{(\mathcal{L}^z)}$. From (28.10)(1), $0 \neq N_{j_z(i)}$ is a subspace of $F_z(N)_i$, thus $F_z(N)_i \neq 0$. Moreover, $\underline{\ell}_k(F_z(N)) = t_z(\underline{\ell}_k(N)) \in t_z(\overline{\mathcal{L}^z}) \subseteq \overline{\mathcal{L}}$ and, since i is $\mathcal{L}$-isolated, there is $M \in \mathcal{A}$-mod concentrated at i with $F_z(N) \cong M$. From (28.10)(3), there is $N' \in \mathcal{A}^z$-mod concentrated at $j_z(i)$ with $N \cong N'$. Thus, $j_z(i)$ is $\mathcal{L}^z$-isolated, as claimed.

(2) Assume that $\mathcal{A} \mapsto \mathcal{A}^z$ is a regularization or a deletion of a set of idempotents in $\mathcal{Z}(\mathcal{L})$. In both cases, it is easy to see that $(\overline{\mathcal{L}})^z = \overline{\mathcal{L}^z}$. Take $i \in \mathcal{P}_{\text{mk}}(\mathcal{A}, z)$ and assume that the point $j_z(i)$ is $\mathcal{L}^z$-isolated. Thus, $j_z(i) \notin \mathcal{Z}(\mathcal{L}^z)$ implies that $i \notin \mathcal{Z}(\mathcal{L})$.

In order to show that i is $\mathcal{L}$-isolated, take an indecomposable object $M \in \mathcal{A}$-mod with $\underline{\ell}_k(M) \in \overline{\mathcal{L}}$ and $M_i \neq 0$. From (25.3) and (25.4), we know there is $N \in \mathcal{A}^z$-mod with $F_z(N) \cong M$. From (28.10)(3), we know that there are isomorphisms $M_i \cong F_z(N)_i \cong N_{j_z(i)}$, thus $N_{j_z(i)} \neq 0$. Moreover, $t_z(\underline{\ell}_k(N)) = \underline{\ell}_k(F_z(N)) = \underline{\ell}_k(M) \in \overline{\mathcal{L}}$, implies that $\underline{\ell}_k(N) \in (\overline{\mathcal{L}})^z = \overline{\mathcal{L}^z}$. Since $j_z(i)$ is $\mathcal{L}^z$-isolated, there is $N' \in \mathcal{A}^z$-mod concentrated at $j_z(i)$ with $N' \cong N$. Then, $F_z(N') \cong F_z(N) \cong M$, where $F_z(N') \in \mathcal{A}$-mod is concentrated at i, by (28.10)(2). □

Corollary 28.12. *Given a seminested ditalgebra $\mathcal{A}$ and a finite set $\mathcal{L}$ of length vectors of $\mathcal{A}$, we denote by $\mathbb{I}(\mathcal{A}, \mathcal{L})$ the set of all marked points of $\mathcal{A}$ which are $\mathcal{L}$-isolated. Assume that the ditalgebra $\mathcal{A}^z$ is obtained from the seminested ditalgebra $\mathcal{A}$ by absorption, regularization, deletion of some idempotents in $\mathcal{Z}(\mathcal{L})$, edge reduction or unravelling. Then:*

(1) If $\mathcal{A} \mapsto \mathcal{A}^z$ is not an unravelling at an $\mathcal{L}$-isolated point of $\mathcal{A}$, then

$$|\mathbb{I}(\mathcal{A}, \mathcal{L})| \leq |\mathbb{I}(\mathcal{A}^z, \mathcal{L}^z)|.$$

(2) If $\mathcal{A} \mapsto \mathcal{A}^z$ is a regularization or a deletion of some idempotents in $\mathcal{Z}(\mathcal{L})$, then

$$|\mathbb{I}(\mathcal{A}, \mathcal{L})| = |\mathbb{I}(\mathcal{A}^z, \mathcal{L}^z)|.$$

Proof.

(1) From Proposition (28.11)(1), the map j_z induces a well-defined map $j'_z : \mathbb{I}(\mathcal{A}, \mathcal{L}) \longrightarrow \mathbb{I}(\mathcal{A}^z, \mathcal{L}^z)$, which is injective because j_z is injective.

(2) From the previous argument and Proposition (28.11)(2), the map $j'_z : \mathbb{I}(\mathcal{A}, \mathcal{L}) \longrightarrow \mathbb{I}(\mathcal{A}^z, \mathcal{L}^z)$ is bijective. □

Lemma 28.13. *Let k be an algebraically closed field. Assume that $\mathcal{A}$ is a seminested k-ditalgebra with layer (R, W) and differential δ. Fix a solid arrow $\alpha : i \longrightarrow j$ of $\mathcal{A}$ and take any finite set $\mathcal{L}$ of length vectors of $\mathcal{A}$ with $i, j \notin \mathcal{Z}(\mathcal{L})$. Then, we have:*

(1) If i and j are different $\mathcal{L}$-isolated points of $\mathcal{A}$ and $\delta(\alpha) \in I_j W_1 + W_1 I_i$, where I_i, I_j are ideals of $e_i R$, $e_j R$, respectively, then $I_i = e_i R$ or $I_j = e_j R$.

(2) If $i = j$ is an $\mathcal{L}$-isolated point of $\mathcal{A}$ and $\delta(\alpha) \in I' W_1 + W_1 I''$, where I', I'' are ideals of $e_i R$, then $I' = e_i R$ or $I'' = e_i R$.

(3) If j is unmarked, i is an $\mathcal{L}$-isolated point of $\mathcal{A}$ and $\delta(\alpha) \in W_1 I_i$, where I_i is an ideal of $e_i R$, then $I_i = e_i R$.

(4) If i is unmarked, j is an $\mathcal{L}$-isolated point of $\mathcal{A}$ and $\delta(\alpha) \in I_j W_1$, where I_j is an ideal of $e_j R$, then $I_j = e_j R$.

Proof.

Case 1: Suppose our claim does not hold. Then, we can assume that I_i and I_j are maximal ideals of $e_i R$ and $e_j R$, respectively. Since k is algebraically closed, we obtain $e_i R/I_i \cong k$ and $e_j R/I_j \cong k$.

Since $i \neq j$, we can consider an object $M \in \mathcal{A}$-mod such that $M_i = e_i R/I_i$; $M_j = e_j R/I_j$; $M_t = 0$, for all $t \neq i, j$; $\alpha M \neq 0$; and $\beta M = 0$, for any solid arrow $\beta \neq \alpha$ of $\mathcal{A}$. Since $i, j \notin \mathcal{Z}(\mathcal{L})$, we have $\underline{\ell}_k(M) \in \overline{\mathcal{L}}$.

We claim that $M \cong N$, where $N \in \mathcal{A}$-mod satisfies $\alpha N = 0$. Indeed, if M is indecomposable, since i and j are $\mathcal{L}$-isolated we have objects $M', M'' \in \mathcal{A}$-mod with M' concentrated at i, M'' concentrated at j, and $M' \cong M \cong M''$, which is impossible. Then, M is decomposable, thus $M \cong N = N' \oplus N''$ with $N'_i \cong M_i$ and $N'_t = 0$, for all $t \neq i$; and $N''_j \cong M_j$ and $N''_t = 0$, for all $t \neq j$. Then, $\alpha N' = 0$ and $\alpha N'' = 0$. Therefore, $\alpha N = 0$.

Therefore, there is an isomorphism $f = (f^0, f^1) : M \longrightarrow N$. Thus $f^0 : M \longrightarrow N$ is an isomorphism of left R-modules. By assumption, we have $\delta(\alpha) = \sum_r h_r w_r + \sum_s w_s g_s$, where $w_r, w_s \in W_1$, $g_s \in I_i$ and $h_r \in I_j$. Therefore, for $m \in M$

$$\alpha f^0[m] - f^0[\alpha m] = \sum_r h_r f^1(w_r)[m] + \sum_s f^1(w_s)[g_s m] = 0.$$

It follows that $f^0[\alpha M] = 0$, and hence that $\alpha M = 0$, a contradiction.

Case 2: Suppose our claim does not hold. Then, we can assume that I' and I'' are maximal ideals of Re_i.

Subcase 2.1 If $I' = I''$. As before, consider $M \in \mathcal{A}$-mod such that $M_i = e_i R/I' \cong k$; $M_t = 0$, for all $t \neq i$; $\alpha M \neq 0$; and $\beta M = 0$, for any solid arrow $\beta \neq \alpha$ of $\mathcal{A}$. Since $i \notin \mathcal{Z}(\mathcal{L})$, we have $\underline{\ell}_k(M) \in \overline{\mathcal{L}}$. Since M is indecomposable and i is $\mathcal{L}$-isolated, we have $N \in \mathcal{A}$-mod concentrated at i with $M \cong N$. In particular, $\alpha N = 0$. Then, we proceed as in last case.

Subcase 2.2 If $I' \neq I''$. The algebraic closure of k implies that $Q' := e_i R/I' \cong k$ and $Q'' := e_i R/I'' \cong k$. Now consider $M \in \mathcal{A}$-mod such that $M_i := Q' \oplus Q''$, as an $e_i R$-module; $M_t = 0$, for all $t \neq i$; left multiplication by α on M is given by the linear map $\begin{pmatrix} 0 & \tau \\ 0 & 0 \end{pmatrix}$, determined by any non-zero linear map $\tau : Q'' \longrightarrow Q'$; and $\beta M = 0$, for any solid arrow $\beta \neq \alpha$ of $\mathcal{A}$. Since $i \notin \mathcal{Z}(\mathcal{L})$, we have $\underline{\ell}_k(M) \in \overline{\mathcal{L}}$.

Assume that M is indecomposable. Since $M_i \neq 0$, $\underline{\ell}_k(M) \in \overline{\mathcal{L}}$ and i is $\mathcal{L}$-isolated, we have an object $N \in \mathcal{A}$-mod with N concentrated at i and $M \cong N$. In particular, $M_i \cong N_i$, which is impossible because N_i is an indecomposable Re_i-module. Then, $M \cong L = L' \oplus L''$, for some non-zero L', $L'' \in \mathcal{A}$-mod. Therefore, L' and L'' have dimension one and, hence, are indecomposable. Again, since i is $\mathcal{L}$-isolated, L' and L'' are isomorphic to $\mathcal{A}$-modules N' and N'' respectively, concentrated at i. Thus, $\alpha N' = 0$ and $\alpha N'' = 0$, and there is an isomorphism $f : M \longrightarrow N := N' \oplus N''$, with $\alpha N = 0$. We may assume that $N'_i \cong Q'$ and $N''_i \cong Q''$, as R-modules. Notice that, since $I' \neq I''$, then $\mathrm{Hom}_R(Q', N''_i) = 0$ and $\mathrm{Hom}_R(Q'', N'_i) = 0$. Thus, the isomorphism $f^0 : M_i = Q' \oplus Q'' \longrightarrow N'_i \oplus N''_i = N_i$ splits as $f^0 = f' \oplus f''$. As before, we have $\delta(\alpha) = \sum_r h_r w_r + \sum_s w_s g_s$, where $w_r, w_s \in W_1$, $g_s \in I''$ and $h_r \in I'$. Therefore, for $m \in M$

$$\alpha f^0[m] - f^0[\alpha m] = \sum_r h_r f^1(w_r)[m] + \sum_s f^1(w_s)[g_s m].$$

We will show the contradiction: $\alpha M = 0$. By definition of the action of α on M, we already know that $\alpha Q' = 0$. So, assume $m \in Q''$ and let us show that $\alpha m = 0$.

Since $\alpha N = 0$, $\alpha f^0[m] = 0$; by definition of the action of α on M, $\alpha Q'' \subseteq Q'$; since $h_r \in I'$ annihilates N'_i, $h_r f^1(w_r)(M_i) \subseteq N''_i$; and since $g_s \in I''$ annihilates Q'', $\sum f^1(w_s)[g_s m] = 0$. Thus, $f^0[\alpha m] = -\sum h_r f^1(w_r)[m] \in N'_i \cap N''_i = 0$. It follows that $f^0[\alpha m] = 0$ and, hence, that $\alpha m = 0$. We have indeed shown the contradiction: $\alpha M = 0$.

Case 3: Suppose our claim does not hold. Then, we can assume that I_i is a maximal ideal of $e_i R$. We have $e_i R/I_i \cong k$. Now consider $M \in \mathcal{A}$-mod such that $M_i = e_i R/I_i$; $M_j = k$; $M_t = 0$, for all $t \neq i$; $\alpha M \neq 0$; and $\beta M = 0$, for any solid arrow $\beta \neq \alpha$ of $\mathcal{A}$. Again, $i, j \notin \mathcal{Z}(\mathcal{L})$ imply $\underline{\ell}_k(M) \in \overline{\mathcal{L}}$.

If M is indecomposable, since $M_i \neq 0$, $\underline{\ell}_k(M) \in \overline{\mathcal{L}}$ and i is $\mathcal{L}$-isolated, we have an object $N \in \mathcal{A}$-mod with N concentrated at i and $M \cong N$. This entails the contradiction $k = M_j \cong N_j = 0$. Then, M is decomposable, thus $M \cong N = N' \oplus N''$ with $N'_i \cong M_i$ and $N'_t = 0$, for all $t \neq i$; and $N''_j \cong M_j$ and $N''_t = 0$, for all $t \neq j$. Then, $\alpha N' = 0$ and $\alpha N'' = 0$. Therefore, $\alpha N = 0$. Then, proceeding as in the proof of case 1, we obtain a contradiction.

Case 4: The proof is similar to the previous one. □

Lemma 28.14. *Assume that $\mathcal{A}$ is a non-wild seminested ditalgebra over the algebraically closed field k. Assume $r \in \mathbb{N}$ and i_0 is a marked point of $\mathcal{A}$. Then, there is an unravelling $\mathcal{A}^u$ at i_0 such that, for any indecomposable $N \in \mathcal{A}^u$-mod with $N_{i_*} \neq 0$ and $\|F_u(N)\| \leq r$, there is $N' \in \mathcal{A}^u$-mod concentrated at i_* with $N \cong N'$. Here, $F_u : \mathcal{A}^u\text{-mod} \longrightarrow \mathcal{A}\text{-mod}$ denotes the associated reduction functor.*

Proof. Let us denote by x the marked loop at i_0 and make $H := Re_{i_0} = k[x]_{f(x)}$. From (28.8), we know there is a finite subset $\mathcal{S}[r, i_0]$ of k such that, any indecomposable $M \in \mathcal{A}$-mod with $\|M\| \leq r$, $M_{i_0} \neq 0$ and $\operatorname{spec} M(x) \not\subseteq \mathcal{S}[r, i_0]$ is isomorphic to an object $M' \in \mathcal{A}$-mod concentrated at i_0.

Assume that $\mathcal{A}^u$ is obtained from $\mathcal{A}$ by unravelling at i_0 using $\mathcal{S}[r, i_0]$ and r. Assume that $\mathcal{S}[r, i_0] = \{\lambda_1, \ldots, \lambda_q\}$. Take an indecomposable $N \in \mathcal{A}^u$-mod with $N_{i_*} \neq 0$ and $\|F_u(N)\| \leq r$. Then, $F_u(N)$ is indecomposable with $F_u(N)_{i_0} \neq 0$ and $\|F_u(N)\| \leq r$. We have an H-module isomorphism

$$F_u(N)_{i_0} \cong N_{i_*} \oplus \left(\oplus_{s=1}^{q} \oplus_{a=1}^{n} \left[H/(x - \lambda_s)^a H\right] \otimes_k N_{i_{s,a}}\right).$$

See the analysis in the proof of (25.9). Here, $\operatorname{spec} N_{i_*}(x) \cap \mathcal{S}[r, i_0] = \emptyset$, because $g(x) = (x - \lambda_1) \cdots (x - \lambda_q)$ acts invertibly on N_{i_*}. Since, $N_{i_*} \neq 0$, there is at least one proper value λ in $\operatorname{spec} N_{i_*}(x) \setminus \mathcal{S}[r, i_0]$. Then, $\operatorname{spec} F_u(N)(x) \not\subseteq \mathcal{S}[r, i_0]$ and there is $M' \in \mathcal{A}$-mod concentrated at the vertex i_0 with $F_u(N) \cong M'$.

We know that $F_u(N)_{i_0} \cong M'_{i_0}$ as $e_{i_0} R$-modules, and the second one is indecomposable. Thus, if some $N_{i_{s,a}} \neq 0$, this implies that $N_{i_*} = 0$, a contradiction.

Then, every $N_{i_{s,a}}$ is zero and supp$N = \{i_*\}$. This implies that the construction in the proof of (25.9), of an $\mathcal{A}^u$-module $N' \in \mathcal{A}^u$-mod with $F_u(N') \cong M'$, provides us with an $N' \in \mathcal{A}^u$-mod concentrated at i_* and with $F_u(N) \cong F_u(N')$. Therefore, $N \cong N'$ in $\mathcal{A}^u$-mod and we are done. □

Definition 28.15. *Assume that $\mathcal{A}$ is a seminested k-ditalgebra and $\mathcal{L}$ is a finite set of length vectors of $\mathcal{A}$. Then, a composition of basic reduction functors F : $\mathcal{B}$-mod$\longrightarrow\mathcal{A}$-mod is called properly $\mathcal{L}$-dense iff, for any field extension K of k and any K-algebra E, the induced functor $\widehat{F}^E : \mathcal{B}^K$-$E$-mod$\longrightarrow\mathcal{A}^K$-$E$-mod is length controlling and, for any $\mathcal{A}^K$-E-bimodule M, with $\underline{\ell}_E(M) \in \mathcal{L}$, there is a $\mathcal{B}^K$-E-bimodule N such that $\widehat{F}^E(N) \cong M$; moreover, if E is a division K-algebra, N can be chosen to be a proper bimodule when M is so.*

Remark 28.16. *Assume $\mathcal{L}$ is a finite set of length vectors of a seminested ditalgebra $\mathcal{A}$. Notice that if $\mathcal{A}^z$ is obtained from $\mathcal{A}$ by some basic operation of type $z \in \{a, r, e\}$, then the corresponding functor $F_z : \mathcal{A}^z$-mod$\longrightarrow\mathcal{A}$-mod is properly $\mathcal{L}$-dense. In the deletion of a subset of idempotents case, if we delete only points in $\mathcal{Z}(\mathcal{L})$, then F_d is properly $\mathcal{L}$-dense. In the unravelling case, if it is relative to a finite subset of scalars of k and a natural number $r \geq \max\{\|\ell\| \mid \ell \in \mathcal{L}\}$, then F_u is properly $\mathcal{L}$-dense. All this follows immediately from Corollaries (25.12), (25.13), (25.14), (25.15) and (25.16).*

Proposition 28.17. *Assume that $\mathcal{A}$ is a non-wild seminested ditalgebra over the algebraically closed field k. Assume $\mathcal{L}$ is a finite set of length vectors of $\mathcal{A}$ and fix any marked vertex i_0 of $\mathcal{A}$ not in $\mathcal{Z}(\mathcal{L})$. Then, there is an unravelling $\mathcal{A}^u$ at i_0 such that:*

(1) the vertex $j_u(i_0) = i_*$ of $\mathcal{A}^u$ is $\mathcal{L}^u$-isolated, and
(2) the associated functor $F_u : \mathcal{A}^u$-mod$\longrightarrow\mathcal{A}$-mod is properly $\mathcal{L}$-dense.

Proof. Define $r := \max\{\|\ell\| \mid \ell \in \overline{\mathcal{L}}\}$, let $\mathcal{A}^u$ be the unravelling at i_0, as in (28.14), and $F_u : \mathcal{A}^u$-mod$\longrightarrow\mathcal{A}$-mod its associated reduction functor.

(1) First notice that $i_0 \notin \mathcal{Z}(\mathcal{L})$ implies that $i_* \notin \mathcal{Z}(\mathcal{L}^u)$. Now, take an indecomposable $N \in \mathcal{A}^u$-mod with $N_{i_*} \neq 0$ and $\ell := \underline{\ell}_k(N) \in \overline{\mathcal{L}^u}$. Then, $\underline{\ell}_k(F_u(N)) = t_u(\ell) \in t_u(\overline{\mathcal{L}^u}) \subseteq \overline{\mathcal{L}}$. Thus, $\|F_u(N)\| \leq r$ and by (28.14) there exists $N' \in \mathcal{A}^u$-mod concentrated at i_* with $N \cong N'$.
(2) The associated reduction functor F_u is properly $\mathcal{L}$-dense, because $\mathcal{L} \subseteq \overline{\mathcal{L}}$, as we remarked in (28.16). □

Definition 28.18. *Let $\mathcal{A}$ be a ditalgebra over the algebraically closed field k. Then, a family $\mathcal{M}$ of finite-dimensional non-isomorphic indecomposable*

$\mathcal{A}$-modules is called a one-parameter family of indecomposables iff there exists a rational algebra $\Gamma = k[x]_{h(x)}$ and an A-Γ-bimodule Z which is free of finite rank as a right Γ-module such that $\mathcal{M} = \{Z \otimes_\Gamma \Gamma/(x-\lambda) \mid \lambda \in k$ and $h(\lambda) \neq 0\}$. Two one-parameter families of indecomposables $\mathcal{M}$ and $\mathcal{M}'$ are equivalent iff for almost all $M \in \mathcal{M}$ there is $M' \in \mathcal{M}'$ with $M \cong M'$.

Lemma 28.19. *Assume that $\mathcal{A}$ is a seminested ditalgebra over the algebraically closed field k. Recall from (27.4), (27.2) and (27.10)(1) that if $\mathcal{A}$ is not wild, then, for all $\ell \in \mathbb{N}$, there is a finite number of one-parameter families $\mathcal{M}_1, \ldots, \mathcal{M}_m$ such that, for almost every indecomposable $M \in \mathcal{A}$-mod with $\ell_k(M) = \ell$, there exist $i \in [1, m]$ and $M_i \in \mathcal{M}_i$ with $M \cong M_i$ (see (27.3)). The numbers $\ell \in \mathbb{N}$ and $\ell_k(M)$ in last statement can be replaced respectively by a dimension vector $\ell \in \mathbb{L}(\mathcal{A})$ and $\underline{\ell}_k(M)$. Assume that $\mathcal{A}$ is not wild. For $\ell \in \mathbb{L}(\mathcal{A})$, we shall denote by $\mu(\mathcal{A}, \ell)$ the minimal number m of one-parameter families $\mathcal{M}_1, \ldots, \mathcal{M}_m$ satisfying last property. The number $\mu(\mathcal{A}, \ell)$ coincides with the maximal number of pairwise non-equivalent one-parameter families $\mathcal{M}_1, \ldots, \mathcal{M}_m$ of indecomposable $\mathcal{A}$-modules with dimension vector ℓ. Finally, given a finite set $\mathcal{L}$ of dimension vectors of $\mathcal{A}$, we write $\mu(\mathcal{A}, \mathcal{L}) := \sum_{\ell \in \mathcal{L}} \mu(\mathcal{A}, \ell)$. Then:*

(1) For any finite set $\mathcal{L}$ of length vectors of $\mathcal{A}$, we have

$$|\mathbb{I}(\mathcal{A}, \mathcal{L})| \leq \mu(\mathcal{A}, \overline{\mathcal{L}}).$$

(2) If $\mathcal{A}^z$ is obtained from $\mathcal{A}$ by a basic reduction operation $z \in \{a, r, d, e, u\}$, then

$$\mu(\mathcal{A}^z, \overline{(\mathcal{L}^z)}) \leq \mu(\mathcal{A}, \overline{\mathcal{L}}).$$

Proof.

(1) Every sequence $i_1, \ldots, i_m$ of distinct $\mathcal{L}$-isolated vertices of $\mathcal{A}$ is, in particular, a finite sequence of distinct marked vertices of $\mathcal{A}$, not in $\mathcal{Z}(\mathcal{L})$, hence it determines a sequence $\mathcal{M}_1, \ldots, \mathcal{M}_m$ of pairwise non-equivalent one-parameter families of indecomposable $\mathcal{A}$-modules with dimension vectors in $\overline{\mathcal{L}}$.

(2) The associated full and faithful functor $F_z : \mathcal{A}^z\text{-mod} \longrightarrow \mathcal{A}\text{-mod}$ maps non-equivalent sequences of one-parameter families with dimension vectors in $\overline{(\mathcal{L}^z)}$ onto non-equivalent sequences of one-parameter families with dimension vectors in $\overline{\mathcal{L}}$. □

Definition 28.20. *Denote by $\mathbb{N}^*$ the set $\{-1, 0, 1, 2, \ldots, \infty\}$ with its natural linear order, and by $\mathbb{H}$ the set of support finite maps $h : \mathbb{N}^* \longrightarrow \{0, 1, 2, \ldots\}$, that is the set of maps h such that $h(x) = 0$, for almost all $x \in \mathbb{N}^*$. Given*

$h, h' \in \mathbb{H}$, we will write $h < h'$ iff $h(x) < h'(x)$, for some $x \in \mathbb{N}^*$, and $h(y) \leq h'(y)$, for all $y \in \mathbb{N}^*$ with $x < y$. Notice that whenever we have a decreasing sequence

$$h_1 \geq h_2 \geq \cdots \geq h_s \geq h_{s+1} \geq \cdots$$

in $\mathbb{H}$, there is a natural number n such that $h_s = h_n$, for all $s \geq n$. Thus, there are no infinite strictly decreasing sequences in $\mathbb{H}$ (although there are, for any given $h \in \mathbb{H}$ with $h(i) \neq 0$ for some $i \geq 0$, strictly decreasing finite sequences starting at h of any length).

Proposition 28.21. *Assume that $\mathcal{A}_0$ is a non-wild seminested ditalgebra over the algebraically closed field k and fix a finite set $\mathcal{L}_0$ of length vectors of $\mathcal{A}_0$. Consider the family $\mathcal{R} = \mathcal{R}(\mathcal{A}_0, \mathcal{L}_0)$ of pairs $(\mathcal{A}, \mathcal{L})$ where $\mathcal{A} = \mathcal{A}_0^{z_1 \cdots z_t}$ is obtained from $\mathcal{A}_0$ by a finite number of basic operations $z_1, \ldots, z_t \in \{a, r, d, e, u\}$ and $\mathcal{L} = \mathcal{L}_0^{z_1 \cdots z_t}$. Moreover, we require that the associated reduction functor $F : \mathcal{A}\text{-mod} \longrightarrow \mathcal{A}_0\text{-mod}$ is properly $\mathcal{L}_0$-dense. For each $(\mathcal{A}, \mathcal{L}) \in \mathcal{R}$, let us define a map $h(\mathcal{A}, \mathcal{L}) \in \mathbb{H}$ by the following*

$$\begin{cases} h(\mathcal{A}, \mathcal{L})(-1) := \text{ number of solid arrows of } \mathcal{A}; \\ h(\mathcal{A}, \mathcal{L})(n) := |\mathcal{L}(n)|, \text{ for } n \in \mathbb{N} \cup \{0\}; \\ h(\mathcal{A}, \mathcal{L})(\infty) := \mu(\mathcal{A}_0, \overline{\mathcal{L}}_0) - |\mathbb{I}(\mathcal{A}, \mathcal{L})|. \end{cases}$$

Then, if $(\mathcal{A}, \mathcal{L}) \in \mathcal{R}$ and $\mathcal{A}$ is not a minimal ditalgebra, there exists $(\mathcal{A}', \mathcal{L}') \in \mathcal{R}$ such that, $h(\mathcal{A}', \mathcal{L}') < h(\mathcal{A}, \mathcal{L})$.

Proof. Take a pair $(\mathcal{A}, \mathcal{L}) \in \mathcal{R}$, say $\mathcal{A} = \mathcal{A}_0^{z_1 \cdots z_t}$ and $\mathcal{L} = \mathcal{L}_0^{z_1 \cdots z_t}$. We know that the associated reduction functor $F : \mathcal{A}\text{-mod} \longrightarrow \mathcal{A}_0\text{-mod}$ is properly $\mathcal{L}_0$-dense. Thus, $\widehat{F}^E : \mathcal{A}^K\text{-}E\text{-mod} \longrightarrow \mathcal{A}_0^K\text{-}E\text{-mod}$ is length controlling and satisfies the density conditions specified in (28.15), for each field extension K of k and each K-algebra E

Since $\mathcal{A}$ is a non-wild seminested ditalgebra, (28.19) guarantees that each $h(\mathcal{A}, \mathcal{L})(\infty)$ is indeed a non-negative integer.

Assume that the given $\mathcal{A}$ is a non-minimal seminested ditalgebra, say with layer (R, W) and differential δ. Then, we can choose a minimal solid arrow $\alpha : i_0 \longrightarrow j_0$ of $\mathcal{A}$. Thus, $\delta(\alpha) \in W_1$.

Case 1: $\delta(\alpha) = 0$ and $i_0 = j_0$.

Since $\mathcal{A}$ is not wild, by (24.11), $Re_{i_0} \cong k$. Consider the ditalgebra $\mathcal{A}^a$ obtained from $\mathcal{A}$ by absorption of the loop α. By (28.12), $h(\mathcal{A}^a, \mathcal{L}^a)(\infty) \leq h(\mathcal{A}, \mathcal{L})(\infty)$. By (28.6), $h(\mathcal{A}^a, \mathcal{L}^a)(j) = h(\mathcal{A}, \mathcal{L})(j)$, for any $j \in \mathbb{N} \cup \{0\}$. We know that $h(\mathcal{A}^a, \mathcal{L}^a)(-1) = h(\mathcal{A}, \mathcal{L})(-1) - 1$. Thus, $h(\mathcal{A}^a, \mathcal{L}^a) < h(\mathcal{A}, \mathcal{L})$. Moreover, F_a :

$\mathcal{A}^a$-mod$\longrightarrow\mathcal{A}$-mod is properly $\mathcal{L}_0^{z_1\cdots z_t}$-dense and the functor F is properly $\mathcal{L}_0$-dense, hence $FF_a : \mathcal{A}^a$-mod$\longrightarrow\mathcal{A}_0$-mod is properly $\mathcal{L}_0$-dense.

Case 2: $\delta(\alpha) = 0$ and $i_0 \neq j_0$.

Since $\mathcal{A}$ is not wild, by (24.11), $Re_{i_0} \cong k$ and $Re_{j_0} \cong k$. Then, we can consider the ditalgebra $\mathcal{A}^e$ obtained from $\mathcal{A}$ by reduction of the edge α. From (28.12)(1), we know that $h(\mathcal{A}^e, \mathcal{L}^e)(\infty) \leq h(\mathcal{A}, \mathcal{L})(\infty)$. If we are in case (2) of Lemma (28.5), then $h(\mathcal{A}^e, \mathcal{L}^e) < h(\mathcal{A}, \mathcal{L})$, and we are done. Moreover, $F_e : \mathcal{A}^e$-mod$\longrightarrow\mathcal{A}$-mod is properly $\mathcal{L}$-dense and the functor F is properly $\mathcal{L}_0$-dense, hence FF_e : $\mathcal{A}^e$-mod$\longrightarrow\mathcal{A}_0$-mod is properly $\mathcal{L}_0$-dense.

Otherwise, we are in case (1) of the same Lemma (28.5), then by the remark (b) at the begining of its proof, we know that $z \in \mathcal{Z}(\mathcal{L}^e)$. Then, we consider the ditalgebra $\mathcal{A}^{ed}$ obtained from $\mathcal{A}^e$ by deletion of the idempotent e_z. So its corresponding reduction functor $F_d : \mathcal{A}^{ed}$-mod$\longrightarrow\mathcal{A}^e$-mod is properly $\mathcal{L}^e$-dense. Then, by (28.5)(1) and (28.6), $h(\mathcal{A}^{ed}, \mathcal{L}^{ed})(j) = h(\mathcal{A}^e, \mathcal{L}^e)(j) = h(\mathcal{A}, \mathcal{L})(j)$, for all $j \geq 0$. By (28.12)(2), $h(\mathcal{A}^{ed}, \mathcal{L}^{ed})(\infty) = h(\mathcal{A}^e, \mathcal{L}^e)(\infty) \leq h(\mathcal{A}, \mathcal{L})(\infty)$. Then, we have that $h(\mathcal{A}^{ed}, \mathcal{L}^{ed}) < h(\mathcal{A}, \mathcal{L})$, because $h(\mathcal{A}^{ed}, \mathcal{L}^{ed})(-1) \leq h(\mathcal{A}, \mathcal{L})(-1) - 1$. Finally, F_d is properly $\mathcal{L}^e$-dense, F_e is properly $\mathcal{L}$-dense and F is properly $\mathcal{L}_0$-dense imply that FF_eF_d is properly $\mathcal{L}_0$-dense.

Case 3: $\delta(\alpha) = v$ is one of the free generators of W_1.

Consider the ditalgebra $\mathcal{A}^r$ obtained from $\mathcal{A}$ by regularization of α. As in the previous case, from (28.12)(1), we know that $h(\mathcal{A}^r, \mathcal{L}^r)(\infty) \leq h(\mathcal{A}, \mathcal{L})(\infty)$. And, if we are in case (2) of Lemma (28.5), then $h(\mathcal{A}^r, \mathcal{L}^r) < h(\mathcal{A}, \mathcal{L})$, and we are done. Otherwise, we are in case (1) of the same Lemma (28.5), then $h(\mathcal{A}^r, \mathcal{L}^r)(j) = h(\mathcal{A}, \mathcal{L})(j)$, for all $j \geq 0$. By (28.12)(2), $h(\mathcal{A}^r, \mathcal{L}^r)(\infty) = h(\mathcal{A}, \mathcal{L})(\infty)$. Then, we have that $h(\mathcal{A}^r, \mathcal{L}^r) < h(\mathcal{A}, \mathcal{L})$, because $h(\mathcal{A}^r, \mathcal{L}^r)(-1) = h(\mathcal{A}, \mathcal{L})(-1) - 1$. As before, $F_r : \mathcal{A}^r$-mod$\longrightarrow\mathcal{A}$-mod is properly $\mathcal{L}$-dense and, therefore, FF_r is properly $\mathcal{L}_0$-dense.

Case 4: Either i_0 or j_0 is in $\mathcal{Z}(\mathcal{L})$.

Assume for instance that $i_0 \in \mathcal{Z}(\mathcal{L})$. Here, we consider the ditalgebra $\mathcal{A}^d$ obtained from $\mathcal{A}$ by deletion of the idempotent corresponding to i_0. Then, $h(\mathcal{A}^d, \mathcal{L}^d) < h(\mathcal{A}, \mathcal{L})$, because of (28.12), (28.6) and the fact that at least the arrow α is eliminated in this procedure. Here again, since $i_0 \in \mathcal{Z}(\mathcal{L})$, F_d is properly $\mathcal{L}$-dense and, therefore, FF_d is properly $\mathcal{L}_0$-dense.

From now on, we assume we are not in this case.

Case 5: There is a marked point $i \in \{i_0, j_0\}$ of $\mathcal{A}$ which is not $\mathcal{L}$-isolated. From (28.17), since we are not in case (4), we know the existence of an unravelling $\mathcal{A} \mapsto \mathcal{A}^u$ at the marked point i such that $j_u(i)$ is $\mathcal{L}^u$-isolated and such that the associated reduction functor $F_u : \mathcal{A}^u\text{-mod} \longrightarrow \mathcal{A}\text{-mod}$ is properly $\mathcal{L}$-dense. Then, the map j_u induces a map $j'_u : \mathbb{I}(\mathcal{A}, \mathcal{L}) \longrightarrow \mathbb{I}(\mathcal{A}^u, \mathcal{L}^u)$, which is injective because j_u is injective (see the proof of (28.12)(1)) and not surjective. Then, $h(\mathcal{A}^u, \mathcal{L}^u)(\infty) < h(\mathcal{A}, \mathcal{L})(\infty)$. Thus, $h(\mathcal{A}^u, \mathcal{L}^u) < h(\mathcal{A}, \mathcal{L})$. Since F_u is properly $\mathcal{L}$-dense, FF_u is properly $\mathcal{L}_0$-dense.

Case 6: Assume that we are not in cases (1), (2), (4) or (5). We will show that after a suitable change of basis we are reduced to case (3).

Write $C := Re_{i_0} \otimes_k Re_{j_0}$. Then, $e_{j_0} W_1 e_{i_0}$ is a C-module, and we can write

$$\delta(\alpha) = \sum_{i=1}^{j} c_i v_i,$$

for some $v_1, \ldots, v_j \in e_{j_0} \mathbb{B}_1 e_{i_0}$ and $0 \neq c_i \in C$. Consider the ideal I of C generated by the elements $c_1, \ldots, c_j$. It will be enough to show that $I = C$. Indeed, in this case, there are $c'_1, \ldots, c'_j \in C$ such that $1 = \sum_i c'_i c_i$. Then, by Lemma (26.2), there is an invertible matrix $Q \in M_{j \times j}(C)$ with the first row $(c_1, \ldots, c_j)$. Consider the change of basis of W_1, which replaces each v_i by v'_i defined by the formula $(v'_1, \ldots, v'_j)^t = Q(v_1, \ldots, v_j)^t$. Then, as remarked in (26.1), we obtain a new set of free generators for W_1, with one of its elements equal to $\delta(\alpha)$, and we are reduced to case (3).

We shall prove that $I = C$ in each one of the following cases.

Subcase 6.1: $Re_{i_0} \cong k$ and $Re_{j_0} \cong k$.

In this case, some c_t is invertible, thus $I = C$ and we are done.

Subcase 6.2: $Re_{j_0} \cong k$ and $Re_{i_0} \ncong k$.

Here, $Re_{i_0} = k[x]_{f(x)}$ and $C = k[x]_{f(x)}$. We can assume after a suitable change of basis that every c_i is a polynomial in $k[x]$ coprime to $f(x)$. Assume that $I \neq C$. Then, since k is algebraically closed, by Hilbert's theorem, there exists $\lambda \in k$ with $f(\lambda) \neq 0$ but $c_i(\lambda) = 0$, for all $i \in [1, j]$. Thus, every c_i belongs to the proper ideal $\langle x - \lambda \rangle$ of C generated by $x - \lambda$. Thus, $\delta(\alpha) \in W_1 \langle x - \lambda \rangle$. We know that $i_0, j_0 \notin \mathcal{Z}(\mathcal{L})$ because we are not in case (4) and the point i_0 is $\mathcal{L}$-isolated because we are not in case (5). Then, we have a contradiction with (28.13)(3).

Subcase 6.3: $Re_{i_0} \cong k$ and $Re_{j_0} \ncong k$.

This case is dual to *Subcase 6.2*.

Subcase 6.4: $Re_{i_0} \ncong k$ and $Re_{j_0} \ncong k$.

Assume that $Re_{i_0} = k[x]_{f(x)}$ and $Re_{j_0} = k[y]_{g(y)}$. Hence, $C = k[x, y]_{f(x)g(y)}$. After an appropriate change of basis of W_1, of the form $v'_i = f(x)^{-p}g(y)^{-q}v_i$, using (26.1), we may assume that all the c_i are polynomials in $k[x, y]$. Again, assume that $I \neq C$. Then, since k is algebraically closed, by Hilbert's theorem, there exist $\lambda, \mu \in k$ with $f(\lambda) \neq 0$ and $g(\mu) \neq 0$ but $c_i(\lambda, \mu) = 0$, for all $i \in [1, j]$ (see the argument in the proof of (24.4)). Then, we can rewrite each $c_i(x, y) = \sum_{s,t} c^i_{s,t}(x - \lambda)^s(y - \mu)^t$, with $c^i_{s,t} \in k$ and $c^i_{0,0} = 0$. Thus,

$$c_i(x, y) = (x - \lambda)q_i(x, y) + (y - \mu)q'_i(x, y),$$

for some polynomials $q_i(x, y)$ and $q'_i(x, y)$. Hence, $\delta(\alpha) \in \langle y - \mu\rangle W_1 + W_1\langle x - \lambda\rangle$. As in the previous subcase, we have that i_0 and j_0 are $\mathcal{L}$-isolated. If $i_0 \neq j_0$, this contradicts (28.13)(1). If $i_0 = j_0$, we have a contradiction with (28.13)(2). □

Theorem 28.22. *Let $\mathcal{A}$ be a non-wild seminested ditalgebra over the algebraically closed field k. Then, for each positive number r, there is a minimal ditalgebra $\mathcal{B}$ and a reduction functor $F : \mathcal{B}$-mod$\longrightarrow\mathcal{A}$-mod such that, for any field extension K of k and any K-algebra E, we have:*

(1) the induced functor $\widehat{F}^E : \mathcal{B}^K$-$E$-mod$\longrightarrow\mathcal{A}^K$-$E$-mod is length controlling;

(2) for any $\mathcal{A}^K$-E-bimodule M, with length $\leq r$, there is a $\mathcal{B}^K$-E-bimodule N such that $\widehat{F}^E(N) \cong M$; moreover, if E is a division K-algebra and M is a proper bimodule, N can be chosen to be a proper bimodule.

Proof. Consider the finite set $\mathcal{L}$ of length vectors ℓ of $\mathcal{A}$ such that $\sum_i \ell_i \leq r$. Consider the family $\mathcal{R} = \mathcal{R}(\mathcal{A}, \mathcal{L})$ defined in Proposition (28.21), as well as the maps $h(\mathcal{A}', \mathcal{L}') \in \mathbb{H}$ defined there, for each $(\mathcal{A}', \mathcal{L}') \in \mathcal{R}$.

From the notice in (28.20), we know there are no infinite strictly decreasing sequences in $\mathbb{H}$. Then, applying (28.21) to $(\mathcal{A}, \mathcal{L})$ and then successively to the new pairs (with non-wild and non-minimal first component) produced by this proposition, we obtain a finite sequence

$$(\mathcal{A}, \mathcal{L}) = (\mathcal{A}_0, \mathcal{L}_0), (\mathcal{A}_1, \mathcal{L}_1), \ldots, (\mathcal{A}_t, \mathcal{L}_t) \in \mathcal{R}$$

with a strictly decreasing sequence of associated maps in $\mathbb{H}$

$$h(\mathcal{A}, \mathcal{L}) = h(\mathcal{A}_0, \mathcal{L}_0) > h(\mathcal{A}_1, \mathcal{L}_1) > \cdots > h(\mathcal{A}_t, \mathcal{L}_t),$$

where $\mathcal{A}_t$ is a minimal ditalgebra. By definition of $\mathcal{R}$, for each field extension K of k and each K-algebra E, the associated induced reduction functor $\widehat{F}^E : \widehat{\mathcal{A}}_t^K\text{-}E\text{-mod} \longrightarrow \widehat{\mathcal{A}}^K\text{-}E\text{-mod}$ has the required properties, because $F : \mathcal{A}_t\text{-mod} \longrightarrow \mathcal{A}\text{-mod}$ is properly $\mathcal{L}$-dense. □

Exercise 28.23. *Let $\mathcal{A}_0$ be a seminested ditalgebra over the algebraically closed field k, having no marked points. Suppose that I_0 is a triangular ideal of $\mathcal{A}$ such that $(\mathcal{A}_0, I_0)$ is of finite representation type. Consider the family $\mathcal{R} = \mathcal{R}(\mathcal{A}_0, I_0)$ of pairs $(\mathcal{A}, I)$, where $\mathcal{A} = \mathcal{A}_0^{z_1 z_2 \cdots z_t}$ is a seminested ditalgebra without marked points obtained from $\mathcal{A}_0$ by a successive sequence of basic operations of type $z_1, \ldots, z_t \in \{a, d, r, e, u, s\}$, as in (25.19), and $I = I_0^{z_1 z_2 \cdots z_t}$. Moreover, we require that the associated functor $F : \mathcal{A}\text{-mod} \longrightarrow \mathcal{A}_0\text{-mod}$ induces an equivalence $\overline{F} : (\mathcal{A}, I)\text{-mod} \longrightarrow (\mathcal{A}_0, I_0)\text{-mod}$. For each pair $(\mathcal{A}, I) \in \mathcal{R}$, define $M(\mathcal{A}, I)$ as the direct sum in $(\mathcal{A}, I)$-mod of any chosen set of representatives of all the indecomposable objects in $(\mathcal{A}, I)$-mod, and consider the natural number*

$$c(\mathcal{A}, I) := |\mathcal{P}(\mathcal{A})| + \|M(\mathcal{A}, I)\|,$$

where $\mathcal{P}(\mathcal{A})$ denotes the set of points of $\mathcal{A}$ and $\|M(\mathcal{A}, I)\|$ is the norm of $M(\mathcal{A}, I)$ in $\mathcal{A}$-mod. Show that whenever $(\mathcal{A}, I) \in \mathcal{R}$ with $\mathcal{A}$ non-trivial, there exists $(\mathcal{A}', I') \in \mathcal{R}$ with $c(\mathcal{A}', I') < c(\mathcal{A}, I)$.

Hint: Choose a minimal solid arrow $\alpha : i_0 \longrightarrow j_0$ of $\mathcal{A}$. In each of the following cases, use (25.19). If $e_{i_0} \in I$, delete e_{i_0}. Proceed similarly for e_{j_0}, if $e_{j_0} \in I$. Assume $e_{i_0}, e_{j_0} \notin I$. If $\delta(\alpha) = 0$ and $i_0 = j_0$, absorb the loop α to obtain $(\mathcal{A}^a, I^a)$ and then apply to it (25.18) or (25.20) to produce the required $(\mathcal{A}', I') \in \mathcal{R}$; if $\delta(\alpha) = 0$ with $i_0 \neq j_0$, consider the edge reduction of α; if $\delta(\alpha)$ is a dotted arrow of $\mathcal{A}$ apply regularization; finally, in the remaining case, reduce to the previous one by a suitable change of basis for the layer of $\mathcal{A}$.

Exercise 28.24. *Let $\mathcal{A}$ be a seminested ditalgebra over the algebraically closed field k. Suppose that I is a triangular ideal of $\mathcal{A}$ such that $(\mathcal{A}, I)$ is of finite representation type. Then, there is a trivial ditalgebra $\mathcal{A}'$, obtained from $\mathcal{A}$ by a finite sequence of basic operations of type a, d, r, e, u, s, as in (25.19), and a triangular ideal I' of $\mathcal{A}'$ such that the associated functor $F : \mathcal{A}'\text{-mod} \longrightarrow \mathcal{A}\text{-mod}$ induces an equivalence $\overline{F} : (\mathcal{A}', I')\text{-mod} \longrightarrow (\mathcal{A}, I)\text{-mod}$.*

Hint: Apply (25.20) to eliminate all marked points of $\mathcal{A}$ and then use (28.23).

Exercise 28.25. *Assume Λ is a finite-dimensional algebra over the algebraically closed field k. Assume that Λ has finite representation type. Show that there is an equivalence of categories $\mathcal{A}\text{-mod} \longrightarrow \Lambda\text{-mod}$, where $\mathcal{A}$ is a trivial ditalgebra over k.*

Hint: Assume that Λ is basic and write it as a quotient kQ/I of a path algebra kQ of a quiver Q over an admissible ideal I. Then, consider the associated seminested ditalgebra $\mathcal{A} = \mathcal{A}(\Lambda)$ with: trivial differential $\delta = 0$, layer (R, W), where R is a trivial algebra, W_0 is freely generated by the arrows of Q and $W_1 = 0$. Thus, I is triangular ideal of $\mathcal{A}$. Apply (28.24) and then delete the superfluous points.

Exercise 28.26. *Let $\mathcal{A}$ be a seminested ditalgebra. Assume that, for some dimension $d \in \mathbb{N}$, there are infinitely many non-isomorphic indecomposable $\mathcal{A}$-modules with dimension d. Then, there is a rational algebra $\Gamma = k[x]_{f(x)}$ and an $A_{\mathcal{A}}$-Γ-bimodule Z, finitely generated free as a right Γ-module such that the functor*

$$\Gamma\text{-mod} \xrightarrow{Z\otimes_\Gamma -} A_{\mathcal{A}}\text{-mod} \xrightarrow{L_{\mathcal{A}}} \mathcal{A}\text{-mod}$$

preserves isomorphism classes and indecomposables. If k is an infinite field, Γ admits infinitely many non-isomorphic indecomposables for each dimension d; hence, $\mathcal{A}$ admits infinitely many dimensions $d_1 < d_2 < \cdots$, each one of them with infinitely many non-isomorphic indecomposable d_i-dimensional $\mathcal{A}$-modules.

Hints: Work with dimension vectors $\ell \in \mathbb{L}(\mathcal{A})$ instead of dimensions $d \in \mathbb{N}$ and notice that we may assume that ℓ is sincere. The proof can be done by induction on the zero norm: for $\ell \in \mathbb{L}(\mathcal{A})$, $\|\ell\|^0_{\mathcal{A}} := \sum_{\alpha \in \mathbb{B}_0} \ell_{t(\alpha)}\ell_{s(\alpha)}$. If $\mathcal{A}$ has a marked point i_0, we can take $\Gamma := Re_{i_0}$ and consider the Γ-A-bimodule $Z = \Gamma$ as a right Γ-module, with left action of A given by its natural left action of R and defining $W_0 Z = 0$. At the base of the induction, $\|\ell\|^0_{\mathcal{A}} = 0$, our assumption implies that $\mathcal{A}$ admits a marked point. For the induction step, we can assume that $\mathcal{A}$ has no marked point, thus $\|\ell\|^0_{\mathcal{A}} = \|\ell\|_{\mathcal{A}}$ is the usual norm, and hence $\mathcal{A}$ is not minimal. Choose a minimal solid arrow α of $\mathcal{A}$ and consider the cases: if $\delta(\alpha) \neq 0$, apply regularization and the induction hypothesis; if $\delta(\alpha) = 0$ and $s(\alpha) = t(\alpha)$, apply absorption of α to get a ditalgebra with a marked point; finally, if $\delta(\alpha) = 0$ and $s(\alpha) \neq t(\alpha)$, apply reduction of the edge α and the induction hypothesis.

Exercise 28.27. *Assume that Λ is a finite-dimensional k-algebra over an infinite field k. Assume, furthermore, that Λ trivially splits over its radical or that the field k is algebraically closed. Show that if there is an infinite family of non-isomorphic indecomposable Λ-modules of some dimension d, then there is a rational algebra Γ and a Λ-Γ-bimodule Z finitely generated free by the right, such that the functor*

$$Z \otimes_\Gamma - : \Gamma\text{-mod} \longrightarrow \Lambda\text{-mod}$$

preserves isomorphism classes and indecomposables. Hence, there are infinitely many dimensions $d_1 < d_2 < \cdots$, *such that, for each* i, *there are infinitely many non-isomorphic indecomposable* d_i*-dimensional* Λ*-modules.*

Hints: Assume Λ *is basic and apply (28.26) to the Drozd's ditalgebra* $\mathcal{D}^{\Lambda}$. *Consider the functor* Ξ_{Λ}, *use (22.20) and (22.18)(1) to construct a rational algebra* Γ' *and a* Λ-Γ'*-bimodule* Z' *finitely generated by the right, such that* $Z' \otimes -$ *preserves indecomposability and isomorphism classes. Then, adjust* Γ' *and* Z', *as in the proof of (27.5), to obtain the wanted bimodule finitely generated free by the right.*

References. The content of this section is taken from [15]. Some primitive form of the underlying ideas of (28.8) are implicit in [28]. The series of exercises (8.22), (8.24), (14.11), (25.19), (25.20), (28.23), (28.24) and (28.25) evolved from [12]. Statements (28.26) and (28.27) came from [28].

29

Modules over non-wild algebras

This section is devoted to the proof of Theorem (29.9) below, which states that some part of the study of Λ-modules with bounded length, of a non-wild finite-dimensional algebra Λ over an algebraically closed field, can be reduced to the study of the modules over one minimal ditalgebra. We prepare to prove this with some preliminary lemmas.

Lemma 29.1. *Assume that* $F : \mathcal{B}\text{-Mod} \longrightarrow \mathcal{A}\text{-Mod}$ *is a reduction functor and that* K *is a field extension of* k. *If* $\widehat{F} : \mathcal{B}^K\text{-Mod} \longrightarrow \mathcal{A}^K\text{-Mod}$ *is the corresponding induced functor, as in (26.8), then, in the following, the left square commutes and the right square commutes up to isomorphism.*

$$\begin{array}{ccc} \mathcal{B}\text{-}K\text{-Mod}_p & \xrightarrow{F^K} & \mathcal{A}\text{-}K\text{-Mod}_p \\ {\scriptstyle \Psi}\downarrow & & \downarrow{\scriptstyle \Psi} \\ \mathcal{B}^K\text{-Mod} & \xrightarrow{\widehat{F}} & \mathcal{A}^K\text{-Mod} \end{array} \qquad \begin{array}{ccc} \mathcal{B}^K\text{-Mod} & \xrightarrow{\widehat{F}} & \mathcal{A}^K\text{-Mod} \\ {\scriptstyle (-)^K}\uparrow & & \uparrow{\scriptstyle (-)^K} \\ \mathcal{B}\text{-Mod} & \xrightarrow{F} & \mathcal{A}\text{-Mod}. \end{array}$$

Proof. It is enough to verify this for each basic reduction functor $F = F_z$, with $z \in \{a, r, d, e, u\}$. From (21.6)(1) and (21.7), we know that F^K restricts to proper bimodules. The following diagram satisfies the appropriate commutativities

$$\begin{array}{ccc} \mathcal{A}^z\text{-}K\text{-Mod}_p & \xrightarrow{F_z^K} & \mathcal{A}\text{-}K\text{-Mod}_p \\ {\scriptstyle \Psi}\downarrow & & \downarrow{\scriptstyle \Psi} \\ \mathcal{A}^{zK}\text{-Mod} & \xrightarrow{\widehat{F_z}} & \mathcal{A}^K\text{-Mod} \\ {\scriptstyle (-)^K}\uparrow & & \uparrow{\scriptstyle (-)^K} \\ \mathcal{A}^z\text{-Mod} & \xrightarrow{F_z} & \mathcal{A}\text{-Mod}. \end{array}$$

Indeed, this follows from (21.6)(2) and from (20.4), (20.5), (20.6), for $z \in \{a, r, d\}$. For the case $F = F_X$, which includes the cases $z = u$ and $z = e$, it follows from (21.7) and (20.11). □

Lemma 29.2. *Let Λ be a finite-dimensional k-algebra, which trivially splits over its radical and consider its Drozd's ditalgebra $\mathcal{D}^\Lambda$, as in (19.1). Assume there is a reduction functor $F : \mathcal{B}\text{-Mod} \longrightarrow \mathcal{D}^\Lambda\text{-Mod}$ with $\mathcal{B}$ a minimal ditalgebra. If the module $N \in \mathcal{B}\text{-Mod}$ is mapped by the composition*

$$\mathcal{B}\text{-Mod} \xrightarrow{F} \mathcal{D}^\Lambda\text{-Mod} \xrightarrow{\Xi_\Lambda} \mathcal{P}^1(\Lambda)$$

to an indecomposable object of the form $(P, 0, 0)$, then $N \cong S(i)$, for some non-marked point i of $\mathcal{B}$, see (23.8).

Proof. From (25.2), (25.3), (25.4) and (25.7), we know that the functor $F = F^k$ satisfies $\underline{\ell}_k F(N) = [F]\underline{\ell}_k N$, for some matrix $[F]$. A similar statement holds for the functor Ξ_Λ. More precisely: If $(P, Q, f) \in \mathcal{P}^1(\Lambda)$ with $P \cong \oplus_{i=1}^r n_i P_i$ and $Q \cong \oplus_{i=1}^r m_i P_i$, where $P_1, \ldots, P_r$ is a set of representatives of the isomorphism classes of projective indecomposable Λ-modules, then define

$$\widehat{\underline{\ell}}(P, Q, f) = (n_1, \ldots, n_r; m_1, \ldots, m_r).$$

Since Λ trivially splits over its radical, say $\Lambda = S \oplus J$, with S a trivial subalgebra of Λ, then every simple Λ-module is one dimensional. Then, having in mind the receipt of Ξ_Λ and (22.19), we have $\widehat{\underline{\ell}}\Xi_\Lambda(M) = \underline{\ell}_k M$, for all $M \in \mathcal{D}^\Lambda\text{-Mod}$. Then, if we call $G = \Xi_\Lambda F$, there is a matrix $[G]$ with $\widehat{\underline{\ell}}G(N') = [G]\underline{\ell}_k N'$, for all $N' \in \mathcal{B}\text{-mod}$.

Since P is indecomposable projective, hence with finite dimension, the same is true for N. Having in mind (23.8), all we have to see is that $G(J(i, n, \pi)) \cong (P, 0, 0)$ entails a contradiction. Indeed, in this case, for all $m \in \mathbb{N}$, we have

$$\begin{aligned} \widehat{\underline{\ell}}G(J(i, mn, \pi)) &= [G]\underline{\ell}_k J(i, mn, \pi) \\ &= [G]m\underline{\ell}_k J(i, n, \pi) \\ &= m[G]\underline{\ell}_k J(i, n, \pi) \\ &= m\widehat{\underline{\ell}}G(J(i, n, \pi)), \end{aligned}$$

and hence $G(J(i, mn, \pi))$ is also an indecomposable of the form $(Q, 0, 0)$. But there are only a finite number of such non-isomorphic objects. It follows that $G(J(i, mn, \pi)) \cong G(J(i, m'n, \pi))$ for some different numbers $m, m' \in \mathbb{N}$. We obtain the contradiction $J(i, mn, \pi) \cong J(i, m'n, \pi)$. □

Lemma 29.3. *Let Λ be a finite-dimensional k-algebra, which trivially splits over its radical. Let K be any field extension of k. Consider the K-algebra $\widehat{\Lambda} = \Lambda^K$ and its Drozd's ditalgebra $\mathcal{D}^{\widehat{\Lambda}}$. Assume there is a reduction functor*

$F:\mathcal{B}$-Mod$\longrightarrow \mathcal{D}^\Lambda$-Mod with $\mathcal{B}$ a minimal ditalgebra. Then, if the module $N\in \mathcal{B}^K$-Mod is mapped by the composition

$$\mathcal{B}^K\text{-Mod}\xrightarrow{\widehat{F}}(\mathcal{D}^\Lambda)^K\text{-Mod}\xrightarrow{F_{\xi_\Lambda}}\mathcal{D}^{\widehat{\Lambda}}\text{-Mod}\xrightarrow{\Xi_{\widehat{\Lambda}}}\mathcal{P}^1(\widehat{\Lambda})$$

to an indecomposable object of the form $(P,0,0)$, then $N\cong S(i)$, for some central primitive idempotent $\widehat{e}_i$ of the minimal K-ditalgebra $\mathcal{B}^K$.

Proof. By definition, $\widehat{F}$ is a finite composition of functors of the form F_ξ, where ξ is an isomorphism of layered ditalgebras, and basic reduction K-functors. As noticed in, (25.11), we know that F_ξ preserves dimension vectors: $\underline{\ell}_K(F_\xi(M))=\underline{\ell}_K(M)$. Thus, $\widehat{F}$ behaves linearly on dimension vectors, and we can apply the same argument we used for the proof of (29.2). □

Lemma 29.4. *Suppose that $F:\mathcal{B}$-Mod$\longrightarrow\mathcal{A}$-Mod is any full and faithful functor, where $\mathcal{A}$ and $\mathcal{B}$ are Roiter ditalgebras. Assume that $F(M)\cong N\oplus L$ in $\mathcal{A}$-Mod, then $M\cong N'\oplus L'$ in $\mathcal{B}$-Mod with $F(N')\cong N$.*

Proof. The existence of an isomorphism $F(M)\xrightarrow{\sigma}(N\oplus L)$ in $\mathcal{A}$-Mod implies the existence of the idempotent $\overline{e}=\sigma^{-1}\begin{pmatrix}0&0\\0&1\end{pmatrix}\sigma\in \mathrm{End}_{\mathcal{A}}(F(M))$, with kernel $\mathrm{Ker}\,\overline{e}\cong N$. Since F is full and faithful, there is an idempotent $e\in \mathrm{End}_{\mathcal{B}}(M)$ with $F(e)=\overline{e}$. From (5.12), we know that the idempotent e splits, and hence there is an isomorphism $M\xrightarrow{h}(N'\oplus L')$ in $\mathcal{B}$-Mod such that $e=h^{-1}\begin{pmatrix}0&0\\0&1\end{pmatrix}h$. Since F is additive, applying F gives an isomorphism $F(M)\xrightarrow{g}(F(N')\oplus F(L'))$ such that $\overline{e}=F(e)=g^{-1}\begin{pmatrix}0&0\\0&1\end{pmatrix}g$. Hence $F(N')\cong \mathrm{Ker}\,\overline{e}$ too, and $F(N')\cong N$. □

Lemma 29.5. *Let Λ be a finite-dimensional k-algebra and denote by J its radical. Take $M\in\Lambda$-Mod and let $P\xrightarrow{f}Q\longrightarrow M\longrightarrow 0$ be its minimal projective presentation. If $E:=\mathrm{End}_{\mathcal{P}^1(\Lambda)}(P,Q,f)^{op}$, then P, Q and M admit a natural structure of Λ-E-bimodules and the following inequality holds*

$$\ell_E(P/JP)+\ell_E(Q/JQ)\leq(1+\dim_k\Lambda)\times\ell_E(M).$$

Proof. Assume that $d:=\ell_E(M)$ is finite. First recall that given any algebra R and any filtration $0=M_m\subseteq\cdots\subseteq M_1\subseteq M_0=M$ of left (or right) R-modules, we always have that $\ell_R(M)=\sum_{i=0}^{m-1}\ell_R(M_i/M_{i+1})$. In particular, for our finite-dimensional k-algebra Λ, assuming that $J^m=0$, we know that $\dim_k\Lambda=\sum_i\dim_k(J^i/J^{i+1})$. Moreover, for any Λ-R-bimodule N we have that $\ell_R(N) = \sum_i \ell_R(J^i N/J^{i+1}N)$. Clearly, $\ell_E(Q/JQ) = \ell_E(M/JM) \leq \ell_E(M) = d$. If $t_i := \dim_k J^i/J^{i+1}$, then we have an epimorphism $[\Lambda/J]^{t_i} \longrightarrow J^i/J^{i+1} \longrightarrow 0$ of right Λ/J-modules. Hence, an epimorphism of right E-modules

$$([\Lambda/J] \otimes_\Lambda Q)^{t_i} \cong [\Lambda/J]^{t_i} \otimes_\Lambda Q \longrightarrow [J^i/J^{i+1}] \otimes_\Lambda Q \longrightarrow 0.$$

Thus, $\ell_E([J^i/J^{i+1}] \otimes_\Lambda Q) \leq t_i \times \ell_E([\Lambda/J] \otimes_\Lambda Q) \leq t_i \times \ell_E(Q/JQ) \leq t_i \times d$. Then

$$\begin{aligned} \ell_E(Q) &= \ell_E(\Lambda \otimes_\Lambda Q) \\ &= \textstyle\sum_i \ell_E([J^i \otimes_\Lambda Q]/[J^{i+1} \otimes_\Lambda Q]) \\ &= \textstyle\sum_i \ell_E([J^i/J^{i+1}] \otimes_\Lambda Q) \\ &\leq \textstyle\sum_i t_i \times d \\ &= \dim_k \Lambda \times d. \end{aligned}$$

Therefore, since $f : P \to Q$ is a morphism of Λ-E-bimodules

$$\ell_E(P/JP) = \ell_E(\mathrm{Im}\,(f)/J\mathrm{Im}\,(f)) \leq \ell_E(\mathrm{Im}\,(f)) \leq \ell_E(Q) \leq \dim_k \Lambda \times d,$$

and, finally, $\ell_E(Q/JQ) + \ell_E(P/JP) \leq d + \dim_k \Lambda \times d = (1 + \dim_k \Lambda) \times d$. □

Lemma 29.6. *Under the assumptions and notations of (28.22), the reduction functor* $F : \mathcal{B}\text{-Mod} \longrightarrow \mathcal{A}\text{-Mod}$ *constructed there satisfies that, for any field extension* K *of the algebraically closed field* k *and any* $M \in \mathcal{A}^K\text{-Mod}$ *with* $\dim_K M \leq d$, *there is some* $N \in \mathcal{B}^K\text{-mod}$ *with* $\widehat{F}(N) \cong M$ *in* $\mathcal{A}^K\text{-Mod}$.

Proof. We want to apply (28.22) to the trivial extension $k \leq k$ and the k-algebra $E = K$. From the second diagram of (29.1), we have an up to isomorphism commutative diagram

$$\begin{array}{ccc} \mathcal{B}^k\text{-Mod} & \xrightarrow{\overline{F}} & \mathcal{A}^k\text{-Mod} \\ {\scriptstyle (-)^k}\uparrow & & \uparrow{\scriptstyle (-)^k} \\ \mathcal{B}\text{-Mod} & \xrightarrow{F} & \mathcal{A}\text{-Mod}, \end{array}$$

where $\overline{F}$ is the induced reduction functor associated to F and the extension $k \leq k$, and, as noted at the end of (20.2), $(-)^k$ is an equivalence of categories.

Then, from (29.1), we have an up to isomorphism commutative diagram

$$\begin{array}{ccc}
\mathcal{B}^k\text{-}K\text{-Mod}_p & \xrightarrow{\overline{F}^K} & \mathcal{A}^k\text{-}K\text{-Mod}_p \\
{\scriptstyle ((-)^k)^K}\big\uparrow & & \big\uparrow{\scriptstyle ((-)^k)^K} \\
\mathcal{B}\text{-}K\text{-Mod}_p & \xrightarrow{F^K} & \mathcal{A}\text{-}K\text{-Mod}_p \\
{\scriptstyle \Psi}\big\downarrow & & \big\downarrow{\scriptstyle \Psi} \\
\mathcal{B}^K\text{-Mod} & \xrightarrow{\widehat{F}} & \mathcal{A}^K\text{-Mod},
\end{array}$$

where the vertical arrows are dimension-preserving equivalences. Then, it follows from (28.22), that $M \cong \widehat{F}(N)$, for some $N \in \mathcal{B}^K$-mod. □

Definition 29.7. *Given an algebra Λ and a Λ-module M, the rule $f \cdot m := f(m)$, for $m \in M$ and $f \in \text{End}_\Lambda(M)$ defines a structure of a left $\text{End}_\Lambda(M)$-module on M. The endolength of M, denoted by* endol (M), *is by definition the length of M as an $\text{End}_\Lambda(M)$-module.*

Lemma 29.8. *Assume that Λ' and Λ are finite-dimensional k-algebras and $F : \Lambda'\text{-Mod} \longrightarrow \Lambda\text{-Mod}$ is an equivalence of their module categories. Then:*

(1) For any $M' \in \Lambda'$-Mod, we have $\text{endol}(M') = \text{endol}(F(M'))$.

(2) Assume that $F = P \otimes_{\Lambda'} -$, for some Λ-Λ'-bimodule P. Take $M' \in \Lambda'$-Mod and let E' (resp. $\overline{E}$) denote the opposite endomorphism algebra of the object in the category $\mathcal{P}^2(\Lambda')$ (resp. in $\mathcal{P}^2(\Lambda)$) corresponding to the minimal projective presentation of M' (resp. of $P \otimes_{\Lambda'} M'$). Then, the functor F induces an isomorphism of algebras $\phi : E' \longrightarrow \overline{E}$. Thus, M' is naturally a right E'-module and $P \otimes_{\Lambda'} M'$ is naturally a right $\overline{E}$-module. Through ϕ^{-1}, M' is a right $\overline{E}$-module, then we have an equality of Λ-$\overline{E}$-bimodules

$$P \otimes_{\Lambda'} (M'_{\overline{E}}) = (P \otimes_{\Lambda'} M')_{\overline{E}}.$$

(3) There is a Λ-Λ'-bimodule P, which is finitely generated projective by the right and such that, for any field extension K of k, we have an equivalence of categories

$$P^K \otimes_{\Lambda'^K} - : \Lambda'^K\text{-Mod} \longrightarrow \Lambda^K\text{-Mod}.$$

Moreover, there is some $s \in \mathbb{N}$ such that, for any field extension K of k and any $M' \in \Lambda'^K$-mod, we have that $\dim_K M' \leq s \times \dim_K(P^K \otimes_{\Lambda'^K} M')$.

Proof. Consider the functor $G : \Lambda\text{-Mod} \longrightarrow \Lambda'\text{-Mod}$, with $FG \cong I_{\Lambda\text{-Mod}}$ and $GF \cong I_{\Lambda'\text{-Mod}}$. By Morita's theorem, $F \cong P \otimes_{\Lambda'} -$ and $G \cong Q \otimes_\Lambda -$, where

P is a Λ-Λ'-bimodule and Q is a Λ'-Λ-bimodule, both finitely generated and projective by the right.

(1) The equivalence $P \otimes_{\Lambda'} -$ induces an isomorphism of algebras $\mathrm{End}_{\Lambda'}(M') \cong \mathrm{End}_{\Lambda}(P \otimes_{\Lambda'} M')$; thus, every endomorphism of $P \otimes_{\Lambda'} M'$ has the form $1 \otimes f$, for some $f \in \mathrm{End}_{\Lambda'}(M')$. Then, every strictly descending chain $M'_n \subseteq M'_{n-1} \subseteq \cdots \subseteq M'_1 = M'$ of $\mathrm{End}_{\Lambda'}(M')$-submodules of M' determines the strictly descending chain $P \otimes_{\Lambda'} M'_n \subseteq P \otimes_{\Lambda'} M'_{n-1} \subseteq \cdots \subseteq P \otimes_{\Lambda'} M'_1 = P \otimes_{\Lambda'} M'$ of $\mathrm{End}_{\Lambda}(P \otimes_{\Lambda'} M')$-submodules of $P \otimes_{\Lambda'} M'$. It follows that $\mathrm{endol}\,(M') \leq \mathrm{endol}\,(P \otimes_{\Lambda'} M')$. With a similar argument applied to $P \otimes_{\Lambda'} M'$, we obtain $\mathrm{endol}\,(P \otimes_{\Lambda'} M') \leq \mathrm{endol}\,(Q \otimes_{\Lambda} P \otimes_{\Lambda'} M') = \mathrm{endol}\,(M')$.

(2) Since $F = P \otimes_{\Lambda'} -$ preserves minimal projective presentations, as well as commutative squares, we have an isomorphism of algebras $\phi : E' \longrightarrow \overline{E}$ such that $\phi(f_1, f_2) = (1 \otimes f_1, 1 \otimes f_2)$. Take $p \in P$, $m' \in M'$ and $(f_1, f_2) \in E'$; denote by $f \in \mathrm{End}_{\Lambda'}(M')$ the morphism induced by (f_1, f_2). Then, clearly we have $(p \otimes m')(f_1, f_2) = p \otimes f(m') = (p \otimes m')(1 \otimes f_1, 1 \otimes f_2)$. Then, $P \otimes_{\Lambda'} (M'_{E'}) = (P \otimes_{\Lambda'} M')_{E'}$. (2) follows from this.

(3) By Morita's theorem, $\Lambda' \cong \mathrm{End}_{\Lambda}(P)^{op}$, for some finitely generated projective generator P in Λ-Mod; P is naturally a right Λ'-module which is projective; and there is an equivalence of categories

$$P \otimes_{\Lambda'} - : \Lambda'\text{-Mod} \longrightarrow \Lambda\text{-Mod}.$$

If K is a field extension of k, then $\Lambda'^K \cong \mathrm{End}_{\Lambda}(P)^{opK} \cong \mathrm{End}_{\Lambda^K}(P^K)^{op}$, where P^K is a finitely generated projective generator in Λ^K-Mod. Therefore, again by Morita's theorem, we have an equivalence of categories

$$P^K \otimes_{\Lambda'^K} - : \Lambda'^K\text{-Mod} \longrightarrow \Lambda^K\text{-Mod}.$$

The moreover part follows from the same argument used to prove (27.11). Indeed, if $Q \otimes -$ is the quasi-inverse of $P \otimes -$, then, by Morita's theorem, $Q \cong {}^*P = \mathrm{Hom}_{\Lambda}(P, \Lambda)$; therefore, for any field extension K of k, $Q^K \cong ({}^*P)^K = (\mathrm{Hom}_{\Lambda}(P, \Lambda))^K \cong \mathrm{Hom}_{\Lambda^K}(P^K, \Lambda^K) = {}^*(P^K)$, and $Q^K \otimes -$ is the quasi inverse of $P^K \otimes -$. If Q is a direct summand of the right Λ-module Λ^s, then the module Q^K is a direct summand of the right Λ^K-module Λ^{Ks}. □

Theorem 29.9. *Let Λ be a finite-dimensional non-wild algebra over an algebraically closed field k and let $d \in \mathbb{N}$. Then, there is a minimal k-ditalgebra $\mathcal{B}$ and a Λ-B-bimodule T, where $B = A_{\mathcal{B}}$, which is finitely generated as right B-module, satisfying items (1) and (2) below, for any field extension K of k:*

(1) *Every Λ^K-module M of endolength $\leq d$ is isomorphic to $T^K \otimes_{\mathcal{B}^K} N$, for some B^K-E-bimodule N of finite E-length, where E is the opposite endomorphism K-algebra of the object in $\mathcal{P}^2(\Lambda^K)$ corresponding to M.*

(2) *There is a functor $F^{[K]} : \mathcal{B}^K\text{-Mod} \longrightarrow \Lambda^K\text{-Mod}$, which is full, reflects isomorphisms, preserves indecomposability and whose composition*

$$B^K\text{-Mod} \xrightarrow{L_{\mathcal{B}^K}} \mathcal{B}^K\text{-Mod} \xrightarrow{F^{[K]}} \Lambda^K\text{-Mod}$$

is naturally isomorphic to $T^K \otimes_{B^K} -$. Moreover, every Λ^K-module of dimension $\leq d$ over K is isomorphic to $F^{[K]}(N)$, for some $N \in \mathcal{B}^K$-mod.

Proof. We first show:

Claim I. We can assume that Λ is a basic algebra.

Proof of claim I: Assume the theorem holds for basic algebras, assume Λ is not basic and take $d \in \mathbb{N}$. Consider a basic finite-dimensional k-algebra Λ' Morita equivalent to Λ. From (22.15), we know that Λ' is not a wild algebra. Consider a Λ-Λ'-bimodule P as in (29.8)(3). Thus, there is some $s \in \mathbb{N}$ such that, for any field extension K of k and any $M' \in \Lambda'^K$-mod, we have that $\dim_K M' \leq s \times \dim_K(P^K \otimes_{\Lambda'^K} M')$. Define $d' := s \times d$. Then, by assumption, there is a minimal k-ditalgebra $\mathcal{B}$ and a Λ'-B-bimodule T', which is finitely generated as right B-module, satisfying items (1′) and (2′) below, for any field extension K of k:

(1′) *Every Λ'^K-module M' of endolength $\leq d'$ is isomorphic to $T'^K \otimes_{B^K} N$ for some B^K-E'-bimodule N of finite E'-length, where E' is the opposite endomorphism K-algebra of the object in $\mathcal{P}^2(\Lambda'^K)$ corresponding to M'.*

(2′) *There is a functor $F'^{[K]} : \mathcal{B}^K\text{-Mod} \longrightarrow \Lambda'^K\text{-Mod}$, which is full, reflects isomorphisms, preserves indecomposability and whose composition*

$$B^K\text{-Mod} \xrightarrow{L_{\mathcal{B}^K}} \mathcal{B}^K\text{-Mod} \xrightarrow{F'^{[K]}} \Lambda'^K\text{-Mod}$$

is naturally isomorphic to $T'^K \otimes_{B^K} -$. Moreover, every Λ'^K-module of dimension $\leq d'$ over K is isomorphic to $F'^{[K]}(N)$, for some $N \in \mathcal{B}^K$-mod.

Consider the Λ-B-bimodule $T := P \otimes_{\Lambda'} T'$. Since P is a finitely generated projective right Λ'-module, the bimodule T is finitely generated as a right B-module. Let us show that, for each field extension K of k, items (1) and (2) hold for Λ.

We know from (29.8)(3), that $P^K \otimes_{\Lambda'^K} - : \Lambda'^K\text{-Mod} \longrightarrow \Lambda^K\text{-Mod}$ is an equivalence. Thus, if M is a Λ^K-module of endolength $\leq d$ and $M' \in \Lambda'^K$-Mod satisfies $P^K \otimes_{\Lambda'^K} M' \cong M$, then, by (29.8)(1), $\text{endol}\,(M') = \text{endol}\,(M) \leq d \leq d'$, and $M' \cong T'^K \otimes_{B^K} N$, for some $N \in B^K$-E'-bimodule N of finite E'-length, where E' is the opposite endomorphism K-algebra of the object in $\mathcal{P}^2(\Lambda'^K)$ corresponding to M'.

Assume $\phi : P^K \otimes_{\Lambda'^K} M' \longrightarrow M$ is an isomorphism in Λ^K-Mod and let $\overline{E}$ (resp. E) denote the opposite endomorphism K-algebra of the object in $\mathcal{P}^2(\Lambda^K)$ corresponding to $P^K \otimes_{\Lambda'^K} M'$ (resp. to M). Then, ϕ induces an isomorphism of algebras $\overline{E} \longrightarrow E$ such that $\phi : P^K \otimes_{\Lambda'^K} M' \longrightarrow M$ is an isomorphism of Λ^K-E-bimodules. Then, from (29.8)(2), we obtain an isomorphism of algebras $E' \cong \overline{E}$ and isomorphisms of Λ^K-E-bimodules

$$\begin{aligned} T^K \otimes_{B^K} N_E &\cong P^K \otimes_{\Lambda'^K} T'^K \otimes_{B^K} N_E \cong P^K \otimes_{\Lambda'^K} (M'_E) \\ &\cong (P^K \otimes_{\Lambda'^K} M')_E \cong M_E, \end{aligned}$$

where N is a B^K-E-bimodule of finite E-length, and we have verified (1).

For (2), define $F^{[K]} : \mathcal{B}^K\text{-Mod} \longrightarrow \Lambda^K\text{-Mod}$, as the composition of $F'^{[K]}$ with $P^K \otimes_{\Lambda'^K} -$. Thus, $F^{[K]}$ is full, reflects isomorphisms and preserves indecomposability. Its composition

$$B^K\text{-Mod} \xrightarrow{L_{B^K}} \mathcal{B}^K\text{-Mod} \xrightarrow{F^{[K]}} \Lambda^K\text{-Mod}$$

is naturally isomorphic to $P^K \otimes_{\Lambda'^K} T'^K \otimes_{B^K} - \cong T^K \otimes_{B^K} -$. Moreover, if $M \in \Lambda^K$-mod satisfies $\dim_K M \leq d$ and $M \cong P^K \otimes_{\Lambda'^K} M'$, then by (29.8)(3), $\dim_K M' \leq s \times d = d'$, so $M' \cong F'^{[K]}(N)$ in Λ'^K-mod, for some $N \in B^K$-mod. Thus, $M \cong F^{[K]}(N)$, for the same N. Then, (2) holds too, and our Claim I is proved.

From now on, we assume that our given finite-dimensional algebra Λ is basic. Since k is an algebraically closed field, the algebra Λ trivially splits over its radical J and, by (23.25), its Drozd's ditalgebra $\mathcal{D}^\Lambda$ is nested.

From (22.21), we know that $\mathcal{D}^\Lambda$ is not wild. Apply (28.22) to the k-ditalgebra $\mathcal{D}^\Lambda$ and the integer $e := (1 + \dim_k \Lambda) \times d$, to obtain a minimal k-ditalgebra $\mathcal{C}$ and a reduction functor $G : \mathcal{C}\text{-Mod} \to \mathcal{D}^\Lambda\text{-Mod}$. Let Θ be the set of central primitive idempotents e_j of $\mathcal{C}$, such that $e_j \mathcal{C} \cong k$ and such that the representation $S(j) \in \mathcal{C}$-Mod is mapped by the composition

$$\mathcal{C}\text{-Mod} \xrightarrow{G} \mathcal{D}^\Lambda\text{-Mod} \xrightarrow{\Xi_\Lambda} \mathcal{P}^1(\Lambda)$$

to an object not in $\mathcal{P}^2(\Lambda)$. Thus, for $e_j \in \Theta$, the triple $\Xi_\Lambda G(S(j))$ has the form $(P, 0, 0)$, for some indecomposable projective Λ-module P.

Since the field k is algebraically closed, the finite-dimensional K-algebra $\widehat{\Lambda} := \Lambda^K$ trivially splits over its radical $\widehat{J} := J^K$. Denote by $\widehat{\Xi}_\Lambda$ the composition $(\mathcal{D}^\Lambda)^K\text{-Mod} \xrightarrow{F_{\xi_\Lambda}} \mathcal{D}^{\widehat{\Lambda}}\text{-Mod} \xrightarrow{\Xi_{\widehat{\Lambda}}} \mathcal{P}^1(\widehat{\Lambda})$, see (20.13), and consider the induced functors $\widehat{G} : \mathcal{C}^K\text{-Mod} \to (\mathcal{D}^\Lambda)^K\text{-Mod}$, corresponding to G. We have:

Claim II. Given $M \in \mathcal{C}^K$-Mod, then $\widehat{\Xi}_\Lambda \widehat{G}(M) \in \mathcal{P}^2(\widehat{\Lambda})$ iff $\widehat{e}_j M = 0$, for all $e_j \in \Theta$.

Proof of claim II: Let $M \in \mathcal{C}^K$-Mod. Then, if $\widehat{e}_j M \neq 0$, for some $e_j \in \Theta$, then if $\widehat{S}(j)$ denotes the one-dimensional representation of the ditalgebra $\mathcal{C}^K$ at $\widehat{e}_j$, $\widehat{S}(j)$ is a direct summand of M, because $\mathcal{C}^K$ is minimal and $\widehat{e}_j\mathcal{C}^K \cong K$. Then, $\Xi_{\widehat{\Lambda}} F_{\xi_\Lambda} \widehat{G}(\widehat{S}(j))$ is a direct summand of $\Xi_{\widehat{\Lambda}} F_{\xi_\Lambda} \widehat{G}(M)$. Notice that $\widehat{S}(j) \cong S(j)^K$, where $S(j)$ denotes the one-dimensional representation of the ditalgebra $\mathcal{C}$ at e_j. Then recall that the following diagram commutes up to isomorphism

$$\begin{array}{ccccccc}
\mathcal{C}^K\text{-Mod} & \xrightarrow{\widehat{G}} & (\mathcal{D}^\Lambda)^K\text{-Mod} & \xrightarrow{F_{\xi_\Lambda}} & \mathcal{D}^{\widehat{\Lambda}}\text{-Mod} & \xrightarrow{\Xi_{\widehat{\Lambda}}} & \mathcal{P}^1(\widehat{\Lambda}) \\
\uparrow{\scriptstyle (-)^K} & & \uparrow{\scriptstyle (-)^K} & & & & \uparrow{\scriptstyle (-)^K} \\
\mathcal{C}\text{-Mod} & \xrightarrow{G} & \mathcal{D}^\Lambda\text{-Mod} & & \xrightarrow{\Xi_\Lambda} & & \mathcal{P}^1(\Lambda).
\end{array}$$

Then, if $\Xi_\Lambda G(S(j)) \cong (P, 0, 0)$, we obtain $\Xi_{\widehat{\Lambda}} F_{\xi_\Lambda} \widehat{G}(\widehat{S}(j)) \cong \Xi_\Lambda G(S(j))^K \cong (P^K, 0, 0)$, which is not in $\mathcal{P}^2(\widehat{\Lambda})$. Thus, $\Xi_{\widehat{\Lambda}} F_{\xi_\Lambda} \widehat{G}(M) \notin \mathcal{P}^2(\widehat{\Lambda})$.

Now, if $\Xi_{\widehat{\Lambda}} F_{\xi_\Lambda} \widehat{G}(M) \cong (P, 0, 0) \oplus (Q, Q', \alpha)$ from (29.4), we get a decomposition $M = M_1 \oplus M_2$ in $\mathcal{C}^K$-Mod, with $\Xi_{\widehat{\Lambda}} F_{\xi_\Lambda} \widehat{G}(M_1) \cong (P, 0, 0)$. Thus, from (29.3), we get $M_1 \cong \widehat{S}(j)$, for some $e_j \in \Theta$, and $\widehat{e}_j M \neq 0$, and our claim II is proved.

Consider the minimal k-ditalgebra $\mathcal{B}$ obtained from $\mathcal{C}$ by deleting the idempotents of Θ, and the corresponding reduction functor $U : \mathcal{B}\text{-Mod} \longrightarrow \mathcal{C}\text{-Mod}$, and set $H := GU$ and $T := Z \otimes_A H(B)$, where $A = A_{\mathcal{D}^\Lambda}$ and Z is the transition bimodule associated to Λ (as in (22.18)). Then, T is finitely generated over B by construction. Denote by $F^{[K]}$ the composition represented in the first row of the following diagram

$$\begin{array}{ccccccccc}
\mathcal{B}^K\text{-Mod} & \xrightarrow{\widehat{H}} & (\mathcal{D}^\Lambda)^K\text{-Mod} & \xrightarrow{F_{\xi_\Lambda}} & \mathcal{D}^{\widehat{\Lambda}}\text{-Mod} & \xrightarrow{\Xi_{\widehat{\Lambda}}} & \mathcal{P}^1(\widehat{\Lambda}) & \xrightarrow{\text{Cok}} & \widehat{\Lambda}\text{-Mod} \\
\uparrow{\scriptstyle L_{\mathcal{B}^K}} & & & & \uparrow{\scriptstyle L_{\mathcal{D}^{\widehat{\Lambda}}}} & & & & \\
B^K\text{-Mod} & & \xrightarrow{F_{\xi_\Lambda}\widehat{H}(B^K)\otimes -} & & A_{\mathcal{D}^{\widehat{\Lambda}}}\text{-Mod}. & & & &
\end{array}$$

By (22.7), the square commutes up to isomorphism, thus if $\widehat{Z}$ denotes the transition bimodule associated to $\widehat{\Lambda}$, then by (22.18), the composition

$$B^K\text{-Mod} \xrightarrow{L_{\mathcal{B}^K}} \mathcal{B}^K\text{-Mod} \xrightarrow{F^{[K]}} \widehat{\Lambda}\text{-Mod}$$

is naturally isomorphic to $\widehat{Z} \otimes_{\widehat{A}} (F_{\xi_\Lambda}\widehat{H})(B^K) \otimes_{B^K} -$. Then, using successively (22.18), (29.1), (20.13), (20.12) and, again, (22.18), and the definition of T, we have

$$\begin{aligned}\widehat{Z}\otimes_{\widehat{A}}(F_{\xi_\Lambda}\widehat{H})(B^K) &\cong \mathrm{Cok}\Xi_{\widehat{\Lambda}}F_{\xi_\Lambda}\widehat{H}(B^K)\\ &\cong \mathrm{Cok}\Xi_{\widehat{\Lambda}}F_{\xi_\Lambda}(H(B)^K)\\ &\cong \mathrm{Cok}[(\Xi_\Lambda H(B))^K]\\ &\cong [\mathrm{Cok}(\Xi_\Lambda H(B))]^K\\ &\cong (Z\otimes_A H(B))^K = T^K.\end{aligned}$$

Therefore, $F^{[K]}L_{\mathcal{B}^K} = \mathrm{Cok}\Xi_{\widehat{\Lambda}}F_{\xi_\Lambda}\widehat{H}L_{\mathcal{B}^K} \cong T^K \otimes_{B^K} -$.

Proof of part (1): Let M be a $\widehat{\Lambda}$-module with endolength $\leq d$ and consider a minimal projective presentation $P\xrightarrow{f}Q\xrightarrow{g}M\longrightarrow 0$ in $\widehat{\Lambda}$-Mod. Let E be the opposite endomorphism K-algebra of (P, Q, f) in the category $\mathcal{P}^1(\widehat{\Lambda})$. Then, P, Q and M have natural structures of $\widehat{\Lambda}$-E-bimodules. Since $\mathrm{End}_{\widehat{\Lambda}}(M)^{op}$ is a quotient of E, we have that M is a $\widehat{\Lambda}$-E-bimodule with $\ell_E(M) \leq \mathrm{endol}\,(M) \leq d$.

Clearly, $(P, Q, f) \in \mathcal{P}^1(\widehat{\Lambda})^E$. From (21.9), we know that $(P, Q, f) \cong \Xi_{\widehat{\Lambda}}(L)$ in $\mathcal{P}^1(\widehat{\Lambda})^E$, for some $L \in \mathcal{D}^{\widehat{\Lambda}}$-$E$-Mod. Then, from (21.10), we know that P and Q admit a natural structure of $\widehat{\Lambda}$-E-bimodules and, moreover, $\ell_E(L) = \ell_E(P/\widehat{J}P) + \ell_E(Q/\widehat{J}Q)$. Combining this with an application of (29.5) to the K-algebra $\widehat{\Lambda}$ and the $\widehat{\Lambda}$-module L, we obtain $\ell_E(L) \leq (1 + \dim_K \widehat{\Lambda}) \times d = e$.

Applying (28.22) to the field extension $k \leq K$ and the K-algebra E, we get that $(F^E_{\xi_\Lambda})^{-1}(L) \cong \widehat{G}^E(N')$, for some $N' \in \mathcal{C}^K$-E-mod. From our claim II and (25.14), we get $N' \cong \widehat{U}^E(N'')$, for some $N'' \in \mathcal{B}^K$-E-mod. From the above remarks, we obtain $M \cong (\mathrm{Cok}\Xi_{\widehat{\Lambda}}F_{\xi_\Lambda}\widehat{H})(N'') \cong T^K \otimes_{B^K} N''$, an isomorphism of left Λ^K-modules. Here, we do not know that N'' is a proper bimodule. Since $\mathcal{B}^K$ is a minimal ditalgebra, we have the forgetful functor $U_{\mathcal{B}^K} : \mathcal{B}^K\text{-mod}\longrightarrow B^K\text{-Mod}$ such that $U_{\mathcal{B}^K}(g^0, g^1) = g^0$. Then, if we consider the B^K-E-bimodule $N := U^E_{\mathcal{B}^K}(N'')$, we have $\ell_E(N) = \ell_E(N'') < \infty$ and $M \cong T^K \otimes_{B^K} N$ in Λ^K-Mod, as required in the first part of the theorem.

Proof of part (2):. It is clear that $F^{[K]}$ is full. From our claim II, we have that the composition

$$\mathcal{B}^K\text{-Mod}\xrightarrow{\widehat{H}}(\mathcal{D}^\Lambda)^K\text{-Mod}\xrightarrow{F_{\xi_\Lambda}}\mathcal{D}^{\widehat{\Lambda}}\text{-Mod}\xrightarrow{\Xi_{\widehat{\Lambda}}}\mathcal{P}^1(\widehat{\Lambda})$$

sends $\mathcal{B}^K$-Mod into $\mathcal{P}^2(\widehat{\Lambda})$. Therefore, from (18.10), we know that $F^{[K]}$ reflects isomorphisms and preserves indecomposables.

Now, let $M \in \widehat{\Lambda}$-Mod satisfy $\dim_K(M) \leq d$ and consider a minimal projective presentation $P\xrightarrow{f}Q\xrightarrow{g}M\longrightarrow 0$ in $\widehat{\Lambda}$-Mod. Then, applying (27.13) to $\widehat{\Lambda}$ and M, we know that $\ell_{\widehat{\Lambda}}(Q/\widehat{J}Q) \leq d$ and $\ell_{\widehat{\Lambda}}(P/\widehat{J}P) \leq \dim_K \widehat{\Lambda} \times d$. Let

$L \in \mathcal{D}^{\widehat{\Lambda}}$-Mod with $\Xi_{\widehat{\Lambda}}(L) \cong (P, Q, f)$. Then, from (22.19), we know that

$$\dim_K(L) \leq \ell_{\widehat{\Lambda}}(P/\widehat{J}P) + \ell_{\widehat{\Lambda}}(Q/\widehat{J}Q) \leq \dim_K \widehat{\Lambda} \times d + d = e.$$

From (29.6), we know that for the functor $\mathcal{C}^K\text{-Mod} \xrightarrow{\widehat{G}} (\mathcal{D}^{\Lambda})^K\text{-Mod}$ we have $F_{\xi_\Lambda}^{-1}(L) \cong \widehat{G}(N')$, for some $N' \in \mathcal{C}^K$-mod. Since $(P, Q, f) \in \mathcal{P}^2(\widehat{\Lambda})$, it follows, from our claim II and (25.14), that $N' \cong \widehat{U}(N)$, for some $N \in \mathcal{B}^K$-mod. Therefore, $L \cong F_{\xi_\Lambda}\widehat{H}(N)$, and $M \cong F^{[K]}(N)$, as claimed. □

Corollary 29.10. *Let Λ be a finite-dimensional non-wild algebra over an algebraically closed field k and let $d \in \mathbb{N}$. Consider the minimal k-ditalgebra $\mathcal{B}$ and the Λ-B-bimodule T, which is finitely generated as right B-module, satisfying items (1) and (2) of the last theorem, for any field extension K of k. Write $\widehat{B} := B^K$, $\widehat{\Lambda} = \Lambda^K$, $\widehat{\mathcal{B}} = \mathcal{B}^K$ and $\widehat{T} = T^K$. Then, for any field extension $K \leq \Omega$, they satisfy also that there is a functor $F^{[\![\Omega]\!]} : \widehat{\mathcal{B}}\text{-}\Omega\text{-Mod}_p \longrightarrow \widehat{\Lambda}\text{-}\Omega\text{-Mod}$, which is full, reflects isomorphisms, preserves indecomposability and whose composition*

$$\widehat{B}\text{-}\Omega\text{-Mod} \xrightarrow{(L_{\widehat{\mathcal{B}}})^{\Omega}} \widehat{\mathcal{B}}\text{-}\Omega\text{-Mod}_p \xrightarrow{F^{[\![\Omega]\!]}} \widehat{\Lambda}\text{-}\Omega\text{-Mod}$$

is naturally isomorphic to $\widehat{T} \otimes_{\widehat{B}} -$. Moreover, every $\widehat{\Lambda}$-Ω-bimodule of dimension $\leq d$ over Ω is isomorphic to $F^{[\![\Omega]\!]}(N)$, for some $N \in \widehat{\mathcal{B}}\text{-}\Omega\text{-mod}_p$.

Proof. Consider the canonical isomorphisms $\psi : \widehat{\Lambda}^{\Omega} = \widehat{\Lambda} \otimes_K \Omega \longrightarrow \Lambda \otimes_k \Omega = \Lambda^{\Omega}$ and $\varphi : \widehat{B}^{\Omega} = \widehat{B} \otimes_K \Omega \longrightarrow B \otimes_k \Omega = B^{\Omega}$. Then, φ induces an isomorphism of ditalgebras $\zeta : \widehat{\mathcal{B}}^{\Omega} = \widehat{\mathcal{B}} \otimes_K \Omega \longrightarrow \mathcal{B} \otimes_k \Omega = \mathcal{B}^{\Omega}$. Define $F^{[\![\Omega]\!]}$ as the composition of functors making commutative the right square in the following diagram

$$\begin{array}{ccccc}
B^{\Omega}\text{-Mod} & \xrightarrow{L_{\mathcal{B}^{\Omega}}} & \mathcal{B}^{\Omega}\text{-Mod} & \xrightarrow{F^{[\Omega]}} & \Lambda^{\Omega}\text{-Mod} \\
\downarrow F_{\varphi} & & \downarrow F_{\zeta} & & \downarrow F_{\psi} \\
\widehat{B}^{\Omega}\text{-Mod} & \xrightarrow{L_{\widehat{\mathcal{B}}^{\Omega}}} & \widehat{\mathcal{B}}^{\Omega}\text{-Mod} & & \widehat{\Lambda}^{\Omega}\text{-Mod} \\
\uparrow \Psi' & & \uparrow \Psi & & \uparrow \Psi'' \\
\widehat{B}\text{-}\Omega\text{-Mod} & \xrightarrow{(L_{\widehat{\mathcal{B}}})^{\Omega}} & \widehat{\mathcal{B}}\text{-}\Omega\text{-Mod}_p & \xrightarrow{F^{[\![\Omega]\!]}} & \widehat{\Lambda}\text{-}\Omega\text{-Mod},
\end{array}$$

where Ψ', Ψ and Ψ'' denote the canonical isomorphism functors, see (21.2). By (29.9), we know that $F^{[\Omega]}$ is full, preserves indecomposables and reflects isomorphisms. Then, so does $F^{[\![\Omega]\!]}$. But (29.9) claims also that the functor defined by the upper row of the diagram is isomorphic to $T^{\Omega} \otimes_{B^{\Omega}} -$. Thus, in order to show that the functor defined by the lower row of the diagram is isomorphic to $\widehat{T} \otimes_{\widehat{B}} -$, it will be enough to show that the rest of the diagram commutes up to isomorphism, which is clear, and that the following one commutes up to isomorphism too

$$\begin{array}{ccc} B^{\Omega}\text{-Mod} & \xrightarrow{T^{\Omega}\otimes_{B^{\Omega}}-} & \Lambda^{\Omega}\text{-Mod} \\ {\scriptstyle F_{\varphi}}\big\downarrow & & \big\downarrow{\scriptstyle F_{\psi}} \\ \widehat{B}^{\Omega}\text{-Mod} & \xrightarrow{\widehat{T}^{\Omega}\otimes_{\widehat{B}^{\Omega}}-} & \widehat{\Lambda}^{\Omega}\text{-Mod} \\ {\scriptstyle \Psi'}\big\uparrow & & \big\uparrow{\scriptstyle \Psi''} \\ \widehat{B}\text{-}\Omega\text{-Mod} & \xrightarrow{\widehat{T}\otimes_{\widehat{B}}-} & \widehat{\Lambda}\text{-}\Omega\text{-Mod}. \end{array}$$

Given, $N \in B^{\Omega}$-Mod, we have the natural isomorphism

$$\theta_N : F_{\psi}(T^{\Omega} \otimes_{B^{\Omega}} N) \longrightarrow \widehat{T}^{\Omega} \otimes_{\widehat{B}^{\Omega}} F_{\varphi}(N)$$

given by $\theta_N([t \otimes \alpha] \otimes n) = [t \otimes 1 \otimes \alpha] \otimes n$, for $t \in T, \alpha \in \Omega$ and $n \in N$. Its inverse satisfies $\theta_N^{-1}([t \otimes \lambda \otimes \alpha] \otimes n) = [t \otimes \lambda\alpha] \otimes n$, for $t \in T$, $\lambda \in K$, $\alpha \in \Omega$ and $n \in N$. Moreover, for $M \in \widehat{B}$-Ω-Mod, we have the natural isomorphism

$$\sigma_M : (\Psi'')^{-1}(\widehat{T}^{\Omega} \otimes_{\widehat{B}^{\Omega}} \Psi'(M)) \longrightarrow \widehat{T} \otimes_{\widehat{B}} M$$

given by $\sigma_M([\widehat{t} \otimes \alpha] \otimes m) = \widehat{t} \otimes m\alpha$, for $\widehat{t} \in \widehat{T}$, $\alpha \in \Omega$ and $m \in M$. Its inverse satisfies $\sigma_M^{-1}(\widehat{t} \otimes m) = [\widehat{t} \otimes 1] \otimes m$, for $m \in M$ and $\widehat{t} \in \widehat{T}$. Therefore, the last diagram is commutative up to isomorphism. Since the isomorphisms of categories appearing in the previous diagrams preserve Ω-dimension, the last statement of this corollary follows from the corresponding statement for $F^{[\Omega]}$. □

Exercise 29.11. *Prove (29.2) using (22.7) and (22.20).*

Exercise 29.12. *If Λ is an algebra and $M, N \in \Lambda$-Mod, then*

$$\text{endol}\,(M), \text{endol}\,(N) \leq \text{endol}\,(M \oplus N) \leq \text{endol}\,(M) + \text{endol}\,(N).$$

Hint: We can identify $\text{End}_{\Lambda}(M \oplus N)$ *with the matrix algebra*

$$\begin{pmatrix} \text{End}_{\Lambda}(M) & \text{Hom}_{\Lambda}(N, M) \\ \text{Hom}_{\Lambda}(M, N) & \text{End}_{\Lambda}(N) \end{pmatrix}$$

and consider the canonical morphism of algebras ψ from the product $\Gamma :=$ $\text{End}_{\Lambda}(M) \times \text{End}_{\Lambda}(N)$ to $\text{End}_{\Lambda}(M \oplus N)$ which maps (f, g) onto the diagonal matrix determined by f and g. Then, by restriction of scalars through ψ, the space $M \oplus N$ has a natural structure of a Γ-module. For the second inequality, use that $\ell_{\Gamma}(M \oplus N) = \text{endol}\,(M) + \text{endol}\,(N)$. For the first inequality, show that: for any $\text{End}_{\Lambda}(M)$-submodule X of M, the $\text{End}_{\Lambda}(M \oplus N)$-submodule $\langle X \rangle$ of $M \oplus N$ generated by X satisfies that $X = \langle X \rangle \cap M$. For *this, notice that we have a vector space decomposition* $\langle X \rangle = X \oplus Z$, *where* $Z = \sum_{\theta \in \mathrm{Hom}_\Lambda(M,N)} \theta(X)$.

References. In this section, we proved a scalar extended version, reducing to only one minimal ditalgebra, of one of the main results of [20]. Statement (29.12) is taken from [20].

30

Absolute wildness

In this section, we explore the notion of absolute wildness, as defined below, which will be useful to us in Section 31. We will show that, over an algebraically closed field, absolute wildness is equivalent to wildness.

Definition 30.1. *Assume that $\mathcal{A}$ is a ditalgebra over the field k. We will say that $\mathcal{A}$ is absolutely wild iff there is an A-$k\langle x, y\rangle$-bimodule Z, which is a free right $k\langle x, y\rangle$-module of finite rank and such that, for any field extension K of k, the functor*

$$K\langle x, y\rangle\text{-Mod} \xrightarrow{Z^K \otimes_{K\langle x,y\rangle} -} A^K\text{-Mod} \xrightarrow{L_{\mathcal{A}^K}} \mathcal{A}^K\text{-Mod}$$

preserves isomorphism classes of indecomposables.

The k-algebra Λ is absolutely wild iff the regular ditalgebra $\mathcal{R}_\Lambda = (\Lambda, 0)$, with layer $(\Lambda, 0)$ is so.

Lemma 30.2. *Let R be the star k-algebra, that is the path k-algebra of the following quiver Q*

$$\begin{array}{ccccc} 1 & & & & 2 \\ & \searrow^{\alpha_1} & & {}^{\alpha_2}\swarrow & \\ & & 6 & & \\ & {}^{\alpha_3}\nearrow & \uparrow \alpha_4 & \nwarrow^{\alpha_5} & \\ 3 & & 4 & & 5. \end{array}$$

Then, R and its opposite algebra R^{op} are absolutely wild.

Proof. Consider the R-$k\langle x, y\rangle$-bimodule B given by the following representation B_0 of Q in the category of free right $k\langle x, y\rangle$-modules

$$\begin{array}{ccccc} k\langle x, y\rangle & & & & k\langle x, y\rangle \\ & \searrow^{t_1} & & {}^{t_2}\swarrow & \\ & & k\langle x, y\rangle^2 & & \\ & {}^{t_3}\nearrow & \uparrow t_4 & \nwarrow^{t_5} & \\ k\langle x, y\rangle & & k\langle x, y\rangle & & k\langle x, y\rangle \end{array}$$

where

$$t_1 = \begin{pmatrix} 1 \\ 0 \end{pmatrix}, t_2 = \begin{pmatrix} 0 \\ 1 \end{pmatrix}, t_3 = \begin{pmatrix} 1 \\ x \end{pmatrix}, t_4 = \begin{pmatrix} 1 \\ 1 \end{pmatrix}, t_5 = \begin{pmatrix} 1 \\ y \end{pmatrix}.$$

Thus, B is a free right $k\langle x, y\rangle$-module of rank 7 and the action of each arrow α_i of Q on B by the left is given by the corresponding matrix t_i. Then, the functor $F := B \otimes_{k\langle x,y\rangle} - : k\langle x, y\rangle\text{-Mod} \longrightarrow R\text{-Mod}$ preserves indecomposables and isomorphism classes. Indeed, for $M \in k\langle x, y\rangle$-Mod, the R-module $F(M)$ admits the following representation

$$\begin{array}{ccccc} M & & & & M \\ & \searrow^{T_1} & & {}^{T_2}\swarrow & \\ & & M^2 & & \\ & {}^{T_3}\nearrow & \uparrow T_4 & \nwarrow^{T_5} & \\ M & & M & & M \end{array}$$

where

$$T_1 = \begin{pmatrix} I \\ 0 \end{pmatrix}, T_2 = \begin{pmatrix} 0 \\ I \end{pmatrix}, T_3 = \begin{pmatrix} I \\ xI \end{pmatrix}, T_4 = \begin{pmatrix} I \\ I \end{pmatrix}, T_5 = \begin{pmatrix} I \\ yI \end{pmatrix}.$$

Then, a morphism $\Psi : F(M) \longrightarrow F(M')$ determines linear maps $\Psi_i : M \longrightarrow M'$, for $i \in [1, 5]$ and a matrix of linear maps $\Psi_6 = \begin{pmatrix} A & B \\ C & D \end{pmatrix} : M^2 \longrightarrow (M')^2$ satisfying, $\Psi_6 T_i = T_i' \Psi_i$, for $i \in [1, 5]$. This implies that $B = C = 0$, $A = D = \Psi_i$, for $i \in [1, 5]$, $Ax = xA$ and $Ay = yA$. Thus, $A : M \longrightarrow M'$ is a morphism of $k\langle x, y\rangle$-modules. Then, we have that Ψ is an isomorphism in R-Mod implies that A is an isomorphism in $k\langle x, y\rangle$-Mod, and, for $M = M'$, $\Psi \in \text{End}_R(F(M))$ is a non-trivial idempotent iff $A \in \text{End}_{k\langle x,y\rangle}(M)$ is so.

Now, if K is any field extension of k, we can identify $R^K = R \otimes_k K$ with the path K-algebra associated to the same quiver Q. Furthermore, we have an isomorphism of K-algebras $k\langle x, y\rangle \otimes_k K \cong K\langle x, y\rangle$, which induces an isomorphism of Λ^K-$K\langle x, y\rangle$-bimodules between B^K and the bimodule obtained from the representation B_0^K of Q in the category of free right $K\langle x, y\rangle$-modules, which can be defined by the same diagram used to define the representation B_0, where the field k is replaced by K. Then, the same argument given above, using K and B^K instead of k and B, proves that the functor $F := B^K \otimes_{K\langle x,y\rangle} - : K\langle x, y\rangle\text{-Mod} \longrightarrow R^K\text{-Mod}$ preserves indecomposables and isomorphism classes. Notice that the same argument applies to R^{op} (where we have to consider the appropriate representation obtained from B_0 transposing the matrices t_i). □

Definition 30.3. *A k-ditalgebra $\mathcal{A}$ is called an absolutely Roiter ditalgebra iff, for any field extension K of k, $\mathcal{A}^K$ is a Roiter ditalgebra.*

Lemma 30.4. *Let $\mathcal{A}'$ be a triangular proper subditalgebra of the absolutely Roiter ditalgebra $\mathcal{A}$. If $\mathcal{A}'$ is absolutely wild, so is $\mathcal{A}$.*

Proof. By assumption, there is an A'-$k\langle x, y\rangle$-bimodule Z, which is a free right $k\langle x, y\rangle$-module of finite rank and such that, for any field extension K of k, the functor

$$K\langle x, y\rangle\text{-Mod} \xrightarrow{Z^K \otimes_{K\langle x,y\rangle} -} A'^K\text{-Mod} \xrightarrow{L_{\mathcal{A}'^K}} \mathcal{A}'^K\text{-Mod}$$

preserves isomorphism classes of indecomposables. Consider the canonical projection of algebras $\pi : A \longrightarrow A'$, as in (12.3) and consider the A-$k\langle x, y\rangle$-bimodule $F_\pi(Z)$, which is a free right $k\langle x, y\rangle$-module of finite rank. Then, notice that for any field extension K of k, we have that $\mathcal{A}'^K$ is a triangular proper subditalgebra of the Roiter ditalgebra $\mathcal{A}^K$. Moreover, if $\pi^K : A^K \longrightarrow A'^K$ denotes the canonical projection of algebras, then we have a commutative diagram

$$\begin{array}{ccc} A'^K\text{-Mod} & \xrightarrow{F_{\pi^K}} & A^K\text{-Mod} \\ \uparrow{\scriptstyle (-)^K} & & \uparrow{\scriptstyle (-)^K} \\ A'\text{-Mod} & \xrightarrow{F_\pi} & A\text{-Mod}. \end{array}$$

Then, since Z^K produces the wildness of $\mathcal{A}'^K$, from (22.13) we know that $F_{\pi^K}(Z^K) \cong F_\pi(Z)^K$ produces the wildness of $\mathcal{A}^K$ and we are done. □

Lemma 30.5. *Assume that $H : \mathcal{B}'\text{-Mod} \longrightarrow \mathcal{B}\text{-Mod}$ is a functor obtained as a finite composition of functors of the form F_X, for some complete admissible module X as in (20.9), or F_{ϕ_z}, for some morphism of ditalgebras $\phi_z : \mathcal{A} \longrightarrow \mathcal{A}^z$, where $z \in \{d, r, a, p\}$ as in (20.4), (20.5), (20.6) or (20.7). Then, if $\mathcal{B}'$ is absolutely wild, so is $\mathcal{B}$.*

Proof. We adopt in this proof the same notation as in (25.11), although here we may be dealing with layered ditalgebras which are not necessarily seminested. Thus, if $\mathcal{A}^z$ is obtained from $\mathcal{A}$ by one of the operations $z \in \{d, r, a, p, X\}$ as in the statement of our lemma, we can consider the associated full and faithful functors $F_z : \mathcal{A}^z\text{-Mod} \longrightarrow \mathcal{A}\text{-Mod}$. Moreover, we can consider, for each field extension K of k and $z \in \{d, r, a, p, X\}$, the associated functor $F_{z^K} : \mathcal{A}^{Kz}\text{-Mod} \longrightarrow \mathcal{A}^K\text{-Mod}$ and the isomorphism $\xi_z : \mathcal{A}^{Kz} \longrightarrow \mathcal{A}^{zK}$ of layered ditalgebras, introduced in the above-listed lemmas and (20.11). We are assuming that X extends to K, for each field extension K of k. Consider the corresponding functor $\widehat{F}_z := F_{\xi_z} F_{z^K} : \mathcal{A}^{zK}\text{-Mod} \longrightarrow \mathcal{A}^K\text{-Mod}$. Moreover, if $H = F_{z_t} \cdots F_{z_2} F_{z_1}$, where $z_t, \ldots, z_1 \in \{d, r, a, p, X\}$, consider

$$\widehat{H} := \widehat{F}_{z_t} \cdots \widehat{F}_{z_2} \widehat{F}_{z_1} : \mathcal{B}'^K\text{-Mod} \longrightarrow \mathcal{B}^K\text{-Mod}.$$

Now, by assumption, there is a B'-$k\langle x, y\rangle$-bimodule Z_0, which is a free right $k\langle x, y\rangle$-module of finite rank and such that, for any field extension K of k, the functor

$$K\langle x, y\rangle\text{-Mod} \xrightarrow{Z_0^K \otimes_{K\langle x,y\rangle} -} B'^K\text{-Mod} \xrightarrow{L_{\mathcal{B}'^K}} \mathcal{B}'^K\text{-Mod}$$

preserves isomorphism classes of indecomposables.

Consider the B-$k\langle x, y\rangle$-bimodule $Z := H(Z_0)$. Since H is a composition of functors of the form either induced by a morphism of layered ditalgebras ϕ_z or of type F_X, then $H(Z_0)$ is finitely generated projective right module and, therefore, a free right $k\langle x, y\rangle$-module of finite rank. Now, fix any field extension K of k. Then, (22.7) implies the existence of an isomorphism of functors

$$\widehat{H} L_{\mathcal{B}'^K}(Z_0^K \otimes_{K\langle x,y\rangle} -) \cong L_{\mathcal{B}^K}(\widehat{H}(Z_0^K) \otimes_{K\langle x,y\rangle} -),$$

and, therefore, $Z^K = H(Z_0)^K \cong \widehat{H}(Z_0^K)$ is a B^K-$K\langle x, y\rangle$-bimodule such that $L_{\mathcal{B}^K}(Z^K \otimes_{K\langle x,y\rangle} -)$ preserves isomorphism classes of indecomposables. That is $\mathcal{B}$ is absolutely wild. □

Theorem 30.6. *Assume that $\mathcal{A}$ is a seminested wild ditalgebra over the algebraically closed field k. Then, $\mathcal{A}$ is an absolutely wild ditalgebra.*

Proof. From (27.10), we know there is a reduction functor $G : \mathcal{C}\text{-Mod} \longrightarrow \mathcal{A}\text{-Mod}$, where $\mathcal{C}$ is a critical ditalgebra.

The given proof that a critical ditalgebra $\mathcal{C}$ is wild required the performance of the following operations: first a suitable (generalized) unravelling $\mathcal{C} \mapsto \mathcal{C}^u$, then a suitable quotient $\mathcal{C}^u \mapsto \mathcal{C}^u/I$, which can also be attained as a finite sequence of basic steps of type (8.18), that is $\mathcal{C}^u \mapsto \mathcal{C}^{up_1} \mapsto \mathcal{C}^{up_1p_2} \mapsto \cdots \mapsto \mathcal{C}^{up_1p_2\cdots p_t} \cong \mathcal{C}^u/I$, one deletion of idempotents $\mathcal{C}^{up_1p_2\cdots p_t} \mapsto \mathcal{C}^{up_1p_2\cdots p_td}$, and one absorption $\mathcal{C}^{up_1p_2\cdots p_td} \mapsto \mathcal{C}^{up_1p_2\cdots p_tda} = \mathcal{B}$ to attain finally a Roiter ditalgebra $\mathcal{B}$ with layer of the form (R, W) or (R^{op}, W), where R is the star algebra. Thus, $\mathcal{B}$ admits an absolutely wild initial subditalgebra by (30.2). Consider any field extension K of k. Since $\mathcal{C}$ is seminested, by (23.7), $\mathcal{C}^K$ is seminested, thus a Roiter ditalgebra. From (20.4), (20.5), (20.6), (20.7) and (20.11), we have $\mathcal{B}^K = \mathcal{C}^{up_1\cdots p_tdaK} \cong \mathcal{C}^{Kup_1\cdots p_tda}$. This last ditalgebra is a Roiter ditalgebra by (9.3) and (16.3). Thus, $\mathcal{B}$ is an absolutely Roiter ditalgebra and, from (30.4), we get that $\mathcal{B}$ is absolutely wild too. We can denote by H the composition

$$\mathcal{B}\text{-Mod} \xrightarrow{F_dF_a} \mathcal{C}^{up_1p_2\cdots p_t}\text{-Mod} \xrightarrow{F_{p_1}\cdots F_{p_t}} \mathcal{C}^u\text{-Mod} \xrightarrow{F_X} \mathcal{C}\text{-Mod},$$

where F_a, F_d, F_{p_i} are induced by a morphisms of ditalgebras. Notice that, here, X satisfies (20.9) and apply (30.5) to obtain that $\mathcal{C}$ is absolutely wild. Now, (30.5) applied to the reduction functor $G : \mathcal{C}\text{-Mod} \longrightarrow \mathcal{A}\text{-Mod}$, implies that $\mathcal{A}$ is absolutely wild too. Here, again, we are using that every reduction of an edge or any unravelling are performed with admissible modules X satisfying (20.9). □

Lemma 30.7. *Let $h \in k[x, y]$ be a non-zero polynomial and $\Lambda := k[x, y]_h$ the localization of $k[x, y]$ at the powers of h. Assume furthermore that the field k is infinite. Then, Λ is absolutely wild.*

Proof. We start with the argument used in the proof of (22.16). First assume that $h \in k[x, y]$ is a polynomial satisfying that $h(0, 0) \neq 0$ and consider the $k[x, y]_h$-$k\langle x, y\rangle$-bimodule B, free of finite rank as a right $k\langle x, y\rangle$-module, introduced in the proof of (22.16) that realizes the wildness of $k[x, y]_h$. The same argument used there to show that the functor

$$B \otimes_{k\langle x,y\rangle} - : k\langle x, y\rangle\text{-Mod} \longrightarrow k[x, y]_h\text{-Mod}$$

preserves isomorphism classes and indecomposables shows that the functor

$$B^K \otimes_{K\langle x,y\rangle} - : K\langle x, y\rangle\text{-Mod} \longrightarrow K[x, y]_h\text{-Mod}$$

preserves isomorphism classes and indecomposables. Indeed, the matrices defining B^K are the same as the matrices defining B and the argument depends only on this. Moreover, if $0 \neq g \in k[x, y]$, the isomorphism of k-algebras $\psi : k[x, y]_g \longrightarrow k[x, y]_h$, where $h(0, 0) \neq 0$ is defined by a substitution of the form $x \mapsto x + \lambda$ and $y \mapsto y + \mu$, for some $\lambda, \mu \in k$. The same change of variables determines an isomorphism of K-algebras $\widehat{\psi} : K[x, y]_g \longrightarrow K[x, y]_h$. Moreover, if H denotes the $k[x, y]_g$-$k\langle x, y\rangle$-bimodule obtained by restriction of scalars by the left from B through ψ, then H^K coincides with the $K[x, y]_g$-$K\langle x, y\rangle$-bimodule obtained by restriction of scalars by the left from B^K through $\widehat{\psi}$. We have shown that, whenever $0 \neq g \in k[x, y]$, then, H is a

$k[x,y]_g$-$k\langle x,y\rangle$-bimodule, which is free of finite rank by the right, and such that, for any field extension K of k,

$$H^K \otimes_{K\langle x,y\rangle} - : K\langle x,y\rangle\text{-Mod} \longrightarrow K[x,y]_g\text{-Mod}$$

preserves isomorphism classes and indecomposables. □

Proposition 30.8. *Let $\mathcal{A}$ be a ditalgebra over the infinite field k. Write $A := A_{\mathcal{A}}$. Then, the following statements are equivalent:*

(i) $\mathcal{A}$ is absolutely wild.
(ii) There is an A-$k\langle x,y\rangle$-bimodule C, which is finitely generated as a right $k\langle x,y\rangle$-module, and such that the functor

$$K\langle x,y\rangle\text{-Mod} \xrightarrow{C^K \otimes_{K\langle x,y\rangle} -} A^K\text{-Mod} \xrightarrow{L_{\mathcal{A}^K}} \mathcal{A}^K\text{-Mod}$$

preserves isoclasses of indecomposables, for any field extension K of k.
(iii) There is an A-$k[x,y]$-bimodule D, which is finitely generated as a right $k[x,y]$-module, and such that the functor

$$K[x,y]\text{-Mod} \xrightarrow{D^K \otimes_{K[x,y]} -} A^K\text{-Mod} \xrightarrow{L_{\mathcal{A}^K}} \mathcal{A}^K\text{-Mod}$$

preserves isoclasses of indecomposables, for any field extension K of k.
(iv) There are an $h \in k[x,y]$ and an A-$k[x,y]_h$-bimodule E, which is finitely generated free as a right $k[x,y]_h$-module, and such that the functor

$$K[x,y]_h\text{-Mod} \xrightarrow{E^K \otimes_{K[x,y]_h} -} A^K\text{-Mod} \xrightarrow{L_{\mathcal{A}^K}} \mathcal{A}^K\text{-Mod}$$

preserves isoclasses of indecomposables, for any field extension K of k.
(v) There is an absolutely wild k-algebra Γ and an A-Γ-bimodule F, which is finitely generated free as a right Γ-module, and such that the functor

$$\Gamma^K\text{-Mod} \xrightarrow{F^K \otimes_{\Gamma^K} -} A^K\text{-Mod} \xrightarrow{L_{\mathcal{A}^K}} \mathcal{A}^K\text{-Mod}$$

preserves isoclasses of indecomposables, for any field extension K of k.

Proof. We will follow the strategy of the proof of (22.17).

(i) ⇒ (ii) is clear.
(ii) ⇒ (iii): Consider the $k\langle x,y\rangle$-$k[x,y]$-bimodule Z obtained from the regular $k[x,y]$-bimodule $k[x,y]$ by left restriction of scalars through the quotient $k\langle x,y\rangle \to k[x,y]$. Then, $D := C \otimes_{k\langle x,y\rangle} Z$ is an A-$k[x,y]$-bimodule. Since $C_{k\langle x,y\rangle}$ is finitely generated so is $D_{k[x,y]}$. Moreover, $Z^K \cong K[x,y]$ and, therefore, the functor $Z^K \otimes_{K[x,y]} - : K[x,y]\text{-Mod} \to K\langle x,y\rangle\text{-Mod}$ is full and faithful. Thus, it preserves indecomposables and isomorphism classes. By assumption, the functor $L_{\mathcal{A}^K}(C^K \otimes_{K\langle x,y\rangle} -)$:

$K\langle x, y\rangle$-Mod $\to \mathcal{A}^K$-Mod, preserves isomorphism classes of indecomposables, then, since $D^K \cong C^K \otimes_{K\langle x,y\rangle} Z^K$, so does the functor $L_{\mathcal{A}^K}(D^K \otimes_{K[x,y]} -) : K[x, y]$-Mod $\to \mathcal{A}^K$-Mod.

(iii) $\Rightarrow$ (iv): in the proof of (22.17), (iii) $\Rightarrow$ (iv), we found that for some $h \in k[x, y]$, the $k[x, y]$-$k[x, y]_h$-bimodule $E = D \otimes_{k[x,y]} k[x, y]_h$ is free of finite rank as a right $k[x, y]_h$-module. If $\Gamma := k[x, y]_h$, then $\Gamma^K \cong K[x, y]_h$ and then

$$\Gamma^K \otimes_{\Gamma^K} - : \Gamma^K\text{-Mod} \to K[x, y]\text{-Mod},$$

is full and faithful. By assumption, we know that the functor

$$L_{\mathcal{A}^K}(D^K \otimes_{K[x,y]} -) : K[x, y]\text{-Mod} \to \mathcal{A}^K\text{-Mod}$$

preserves isomorphism classes of indecomposables. Then, $E^K \cong D^K \otimes_{K[x,y]} \Gamma^K$ implies that

$$L_{\mathcal{A}^K}(E^K \otimes_{\Gamma^K} -) : \Gamma^K\text{-Mod} \to \mathcal{A}^K\text{-Mod}$$

also preserves isomorphism classes of indecomposables.

(iv) $\Rightarrow$ (v): By (30.7), $\Gamma := k[x, y]_h$ is absolutely wild. We can take $F := E$.

(v) $\Rightarrow$ (i): Let G be a Γ-$k\langle x, y\rangle$-bimodule, free of finite rank as a right $k\langle x, y\rangle$-module, that realizes the absolute wildness of Γ. Define $B := F \otimes_\Gamma G$, which is an A-$k\langle x, y\rangle$-bimodule free of finite rank by the right. Since $G^K \otimes_{K\langle x,y\rangle} - : K\langle x, y\rangle$-Mod $\to \Gamma^K$-Mod and $L_{\mathcal{A}^K}(F^K \otimes_{\Gamma^K} -) : \Gamma^K$-Mod $\to \mathcal{A}^K$-Mod preserve isomorphism classes of indecomposables, and $B^K \cong F^K \otimes_{\Gamma^K} G^K$ so does the functor $L_{\mathcal{A}^K}(B^K \otimes_{K\langle x,y\rangle} -) : K\langle x, y\rangle$-Mod $\to \mathcal{A}^K$-Mod, that is B realizes the absolute wildness of $\mathcal{A}$. □

Lemma 30.9. *If Λ and Λ' are Morita equivalent finite-dimensional k-algebras and Λ' is absolutely wild, then Λ is absolutely wild too.*

Proof. Assume Λ' is absolutely wild and take an Λ'-$k\langle x, y\rangle$-bimodule Z_0, which is a free right $k\langle x, y\rangle$-module of finite rank and such that, for any field extension K of k, the functor

$$K\langle x, y\rangle\text{-Mod} \xrightarrow{Z_0^K \otimes_{K\langle x,y\rangle} -} \Lambda'^K\text{-Mod}$$

preserves isomorphism classes of indecomposables. By (29.8)(3), there is a Λ-Λ'-bimodule P, which is finitely generated projective by the right and such that

$$P^K \otimes_{\Lambda'^K} - : \Lambda'^K\text{-Mod} \longrightarrow \Lambda^K\text{-Mod}$$

is an equivalence of categories, for any field extension K of k. Consider the Λ-$k\langle x, y\rangle$-bimodule $P \otimes_{\Lambda'} Z_0$, which is projective (hence free) as a right $k\langle x, y\rangle$-module, and let K be any field extension of k.

Then, $(P \otimes_{\Lambda'} Z_0)^K \otimes_{K\langle x,y\rangle} - \cong P^K \otimes_{\Lambda'^K} Z_0^K \otimes_{K\langle x,y\rangle} -$ are functors which preserve isomorphism classes of indecomposables. Thus, Λ is absolutely wild. □

Theorem 30.10. *Let Λ be a finite-dimensional wild algebra over an algebraically closed field k. Then, Λ is absolutely wild.*

Proof. By (30.9), we may assume that Λ is basic. Then, the Drozd's ditalgebra $\mathcal{D} := \mathcal{D}^\Lambda$ is nested. Since Λ is wild, from (27.14), we know that $\mathcal{D}$ is wild. Then, by (30.6), there is an A-$k\langle x, y\rangle$-bimodule Z, which is a free right $k\langle x, y\rangle$-module of finite rank and such that, for any field extension K of k, the functor

$$K\langle x, y\rangle\text{-Mod} \xrightarrow{Z^K \otimes_{K\langle x,y\rangle} -} A^K\text{-Mod} \xrightarrow{L_{\mathcal{D}^K}} \mathcal{D}^K\text{-Mod}$$

preserves isomorphism classes of indecomposables.

Let $B := \mathrm{Cok}\,\Xi_\Lambda(Z)$ be the A-$k\langle x, y\rangle$-bimodule given in (22.18). Then, B is finitely generated as a right $k\langle x, y\rangle$-module. Write $\widehat{\Lambda} := \Lambda^K$ and consider the following up to isomorphism commutative diagram (see (22.7))

$$\begin{array}{ccccccc}
K\langle x, y\rangle\text{-Mod} & \xrightarrow{Z^K \otimes_{K\langle x,y\rangle} -} & A^K\text{-Mod} & \xrightarrow{L_{\mathcal{D}^K}} & \mathcal{D}^K\text{-Mod} & & \\
\| & & \downarrow F_{\xi_\Lambda} & & \downarrow F_{\xi_\Lambda} & & \\
K\langle x, y\rangle\text{-Mod} & \xrightarrow{F_{\xi_\Lambda}(Z^K)\otimes_{K\langle x,y\rangle} -} & A_{\mathcal{D}^{\widehat{\Lambda}}}\text{-Mod} & \xrightarrow{L_{\mathcal{D}^{\widehat{\Lambda}}}} & \mathcal{D}^{\widehat{\Lambda}}\text{-Mod} & \xrightarrow{\Xi_{\widehat{\Lambda}}} & \mathcal{P}^1(\widehat{\Lambda}).
\end{array}$$

We have that $F_{\xi_\Lambda}(Z^K)$ is a bimodule which is free of finite rank as a right $K\langle x, y\rangle$-module. Then, the lower row of our diagram maps any indecomposable in $\mathcal{P}^2(\widehat{\Lambda})$, by (22.20). Then, by (22.18)(1), if we make $\widehat{B} := \mathrm{Cok}\,\Xi_{\widehat{\Lambda}} F_{\xi_\Lambda}(Z^K)$, the functor

$$\widehat{B} \otimes_{K\langle x,y\rangle} - \cong \mathrm{Cok}\,\Xi_{\widehat{\Lambda}}(F_{\xi_\Lambda}(Z^K) \otimes_{K\langle x,y\rangle} -) : K\langle x, y\rangle\text{-Mod} \longrightarrow \widehat{\Lambda}\text{-Mod}$$

preserves isomorphism classes of indecomposables. Finally, we just have to notice that, from (20.12) and (20.13), we obtain: $\widehat{B} = \mathrm{Cok}\,\Xi_{\widehat{\Lambda}} F_{\xi_\Lambda}(Z^K) \cong (\mathrm{Cok}\,\Xi_\Lambda(Z))^K = B^K$, and we are done. We used (30.8). □

31

Generic modules and tameness

In this section, we shall prove, for any finite-dimensional algebra Λ over an algebraically closed field k, that Λ is tame iff it is generically tame in the sense of the following definition. Moreover, we shall study generic modules of extended algebras Λ^K.

Definition 31.1. *If Λ is a k-algebra, over an abitrary field k, then a Λ-module G is called generic iff it is indecomposable, of infinite length as a Λ-module but with finite endolength. The algebra Λ is called generically tame iff, for each $d \in \mathbb{N}$, there are only finitely many isomorphism classes of generic Λ-modules of endolength d.*

We first recall a few elementary remarks on endolength and generic modules needed to give a scalar extended version of some important structure theorems on generic modules and their endomorphism algebras, and, then, to state and prove our Theorem (31.13).

Lemma 31.2. *Assume that Λ is a k-algebra and G is a $\Lambda^{k(x)}$-module with finite dimension over $k(x)$. Then, the endolength of the restriction ${}_{\Lambda}G$ of G to Λ is bounded by $\dim_{k(x)} G$. Thus, if furthermore ${}_{\Lambda}G$ is indecomposable and not finitely generated over Λ, then ${}_{\Lambda}G$ is a generic Λ-module.*

Proof. We have the following canonical embeddings of k-algebras

$$k(x) \subseteq \mathrm{End}_{\Lambda^{k(x)}}(G) \subseteq \mathrm{End}_{\Lambda}(G).$$

Thus, any chain $\cdots \subseteq G_{i+1} \subseteq G_i \subseteq \cdots \subseteq G_1 \subseteq G_0 = G$ of $\mathrm{End}_{\Lambda}(G)$-submodules of G is also a chain of $k(x)$-subvector spaces of G, and is, therefore, stationary. The length of any refining chain is bounded by $\dim_{k(x)} G$. □

Lemma 31.3. *Let k be any field. Then:*

(1) *If R is a minimal k-algebra with decomposition of unit as a sum of orthogonal central primitive idempotents $1 = \sum_{i=1}^{n} e_i$ and $I := \{i \in [1, n] \mid Re_i \not\cong k\}$, then R admits only $|I|$ non-isomorphic generic R-modules $\{Q_i\}_{i \in I}$. The R-module Q_i is obtained from the field of fractions $\mathrm{frac}(Re_i) \cong k(x)$ of Re_i by restriction of scalars through the inclusion $Re_i \longrightarrow \mathrm{frac}(Re_i)$. We call Q_i the principal generic R-module at the point i.*

(2) *Consider, for each $\lambda \in k$, the $k\langle x, y\rangle$-module $H_\lambda := k(x)[y]/(y - \lambda)$, where $k\langle x, y\rangle$ acts by restriction via the epimorphism of algebras*

$$k\langle x, y\rangle \longrightarrow k[x, y] \longrightarrow k[x, y]_{k[x]^*} \cong k(x)[y].$$

Here, $k[x, y]_{k[x]^*}$ denotes the localization of $k[x, y]$ at the multiplicative subset $k[x]^*$ formed by the non-zero elements in $k[x]$. Then, $\{H_\lambda\}_{\lambda \in k}$ is a family of non-isomorphic generic $k\langle x, y\rangle$-modules with endolength one.

Proof.

(1) It will be enough to show that, for the principal ideal domain $D = k[x]_{f(x)}$, the D-module given by its field of fractions Q is the unique generic D-module, up to isomorphism. An elementary arithmetical argument shows that Q is not a finitely generated D-module. Since $\mathrm{End}_D(Q) \cong \mathrm{End}_Q(Q)$, we know that Q is an indecomposable D-module. Moreover, $\mathrm{endol}\,(Q) = \dim_Q(Q) = 1$ and hence Q is generic. For the uniqueness, we first notice the following claim: if $M \in D$-Mod is indecomposable and has finite endolength, then, for any $d \in D^*$, either multiplication by d is an isomorphism or $d^i M = 0$, for some i. Indeed, notice that the map $d : M \longrightarrow M$ with receipt $d(m) = dm$ is a morphism of $\mathrm{End}_D(M)$-modules of finite length: if $f \in \mathrm{End}_D(M)$, then $d(f \cdot m) = d(f(m)) = f(dm) = f \cdot d(m)$. Then, by Fitting's Lemma, either d^i is an isomorphism for some i, or $d^i = 0$ for some i. This proves the claim.

Now assume that M is a generic D-module. Suppose that M is not faithful, that means that $dM = 0$ for some $d \in D \setminus \{0\}$. Then, M is a $D/(d)$-module with $\mathrm{End}_D(M) = \mathrm{End}_{D/(d)}(M)$. Therefore, M is an indecomposable not finitely generated $D/(d)$-module. Since the indecomposable finitely generated $D/(d)$-modules have the form $D/(\pi^\alpha)$, where $\pi^\alpha | d$ and $\pi \in D$ irreducible, the artinian ring $D/(d)$ has only a finite number of non-isomorphic indecomposable finitely generated modules. By Auslander's theorem (see [2]), every $D/(d)$-module is a direct sum of finitely generated indecomposable modules. Thus, there are no indecomposables in $D/(d)$-Mod which are not finitely generated: a contradiction.

All this shows that M is faithful and, from the claim, we obtain that, for any $d \in D^*$, multiplication by d gives an automorphism $\mu_d : M \longrightarrow M$ of the k-vector space M. Then, the algebra morphism $\mu : D \longrightarrow \mathrm{End}_k(M)$ extends to an algebra morphism $Q \longrightarrow \mathrm{End}_k(M)$ and M has a natural structure of a Q-module which restricts to our original ${}_DM$. Then, $\mathrm{End}_D(M) = \mathrm{End}_Q(M)$ and M is an indecomposable Q-module. Thus $M \cong Q$, as we wanted to show.

(2) We know that, for $\lambda \in k$, the $k[y]$-modules $S_\lambda := k[y]/(y - \lambda)$ are non-isomorphic one-k-dimensional $k[y]$-modules. By the same reason, the modules $H_\lambda = k(x)[y]/(y - \lambda)$ are non-isomorphic one-$k(x)$-dimensional modules over the algebra $k(x)[y] \cong k[x, y]_{k[x]^*}$. By Silver's theorem applied to the epimorphism of algebras $\psi : k\langle x, y\rangle \longrightarrow k(x)[y]$ given in the statement of our lemma, we know that the restriction functor $F_\psi : k(x)[y]\text{-Mod} \longrightarrow k\langle x, y\rangle\text{-Mod}$ is full and faithful. Then, $\{H_\lambda\}_{\lambda \in k}$ is a family of non-isomorphic indecomposable $k\langle x, y\rangle$-modules. Since $k(x) \subseteq \mathrm{End}_{k(x)[y]}(H_\lambda) \cong \mathrm{End}_{k\langle x,y\rangle}(H_\lambda)$ and H_λ is one-$k(x)$-dimensional, $\mathrm{endol}\,(H_\lambda) = 1$. It is not hard to see that H_λ is not a finitely generated $k\langle x, y\rangle$-module. Therefore, H_λ is a generic $k\langle x, y\rangle$-module. □

Lemma 31.4. *Let Δ and Λ be k-algebras and let B be any Λ-Δ-bimodule that is free of finite rank n as a right Δ-module. Consider the functor $F := B \otimes_\Delta - : \Delta\text{-Mod} \longrightarrow \Lambda\text{-Mod}$. Then, for any $M \in \Delta\text{-Mod}$*

$$\mathrm{endol}\,(F(M)) \leq \mathrm{endol}\,(M) \times n.$$

Proof. Fix $M \in \Delta\text{-Mod}$ and write $N := F(M) = B \otimes_\Delta M$. We can assume that $B = \Delta^n$, as right Δ-modules. The functor F induces the morphism of algebras $\varphi : \mathrm{End}_\Delta(M) \longrightarrow \mathrm{End}_\Lambda(N)$ with receipt $\varphi(f) = 1 \otimes f$. Thus, N admits by restriction via φ a natural structure of left $\mathrm{End}_\Delta(M)$-module. Then, the canonical isomorphism of vector spaces $\psi : N = B \otimes_\Delta M \longrightarrow M^n$ is in fact an isomorphism of $\mathrm{End}_\Delta(M)$-modules. Indeed, for $(b_i) \in B$, $m \in M$ and $f \in \mathrm{End}_\Delta(M)$, we have: $\psi(f \cdot [(b_i) \otimes m]) = \psi((1 \otimes f)[(b_i) \otimes m]) = \psi((b_i) \otimes f(m)) = (b_i f(m)) = (f(b_i m)) = f \cdot (b_i m) = f \cdot \psi[(b_i) \otimes m]$. Then, $\ell_{\mathrm{End}_\Delta(M)}(N) = \ell_{\mathrm{End}_\Delta(M)}(M^n) = n \times \mathrm{endol}\,(M) =: d$. Now, if $\cdots \subseteq N_{i+1} \subseteq N_i \subseteq \cdots \subseteq N_1 \subseteq N$ is a descending chain of $\mathrm{End}_\Lambda(N)$-submodules of N, then it is also a descending chain of $\mathrm{End}_\Delta(M)$-submodules of N. Then, if $\mathrm{endol}\,(M)$ is finite, this chain is stationary and has length $\leq d$. It follows that $\mathrm{endol}\,(N) \leq d$. □

Corollary 31.5. *A wild algebra Λ over an infinite field k is not generically tame.*

Proof. Assume that $F = Z \otimes_{k\langle x,y\rangle} - : k\langle x, y\rangle\text{-Mod} \longrightarrow \Lambda\text{-Mod}$ is a functor which preserves isomorphism classes of indecomposables, where Z is a Λ-$k\langle x, y\rangle$-bimodule which is free of finite rank by the right. Then, by (31.4), the family $\{H_\lambda\}_{\lambda \in k}$ of generic $k\langle x, y\rangle$-modules constructed in (31.3)(2) is mapped by F to an infinite family of non-isomorphic generic Λ-modules with bounded endolength. □

Remark 31.6. *Recall, for instance from* [1] 15, *that, for any ring* R

$$\text{rad}R = \{x \in R \mid 1 - axb \text{ is invertible, for all } a, b \in R\}.$$

Then, we have:

(1) If $\varphi : R \longrightarrow S$ *is a surjective ring homomorphism satisfying that* $x \in R$ *is invertible whenever* $\varphi(x)$ *is invertible in* S*, then* φ *induces an isomorphism* $\varphi : R/\text{rad}R \longrightarrow S/\text{rad}S$.

(2) If $\mathcal{B}$ *is a minimal ditalgebra and* $M \in \mathcal{B}\text{-Mod}$*, every* $f \in \text{End}_{\mathcal{B}}(M)$ *decomposes as* $f = (f^0, 0) + (0, f^1)$*, with* $(f^0, 0), (0, f^1) \in \text{End}_{\mathcal{B}}(M)$*. By (5.4), the second summand* $(0, f^1) \in \text{radEnd}_{\mathcal{B}}(M)$*. Thus, the canonical embedding functor* $L_{\mathcal{B}} : B\text{-Mod} \longrightarrow \mathcal{B}\text{-Mod}$ *induces a surjective morphism of algebras*

$$\varphi_{\mathcal{B}} : \text{End}_B(M) \longrightarrow \text{End}_{\mathcal{B}}(M)/\text{radEnd}_{\mathcal{B}}(M).$$

If, moreover, $\text{End}_R(M)$ *is a division algebra, then* $\text{radEnd}_{\mathcal{B}}(M)$ *is a nilpotent ideal (see (5.18)).*

Proposition 31.7. *Let* Λ *be a finite-dimensional non-wild algebra over the algebraically closed field* k *and take* $d \in \mathbb{N}$*. Then, if* $\mathcal{B}$ *and* T *are the minimal* k*-ditalgebra and bimodule obtained by applying Theorem (29.9), with the integer* d*, we have, for any field extension* K *of* k*:*

(1) The Λ^K*-modules of the form* $G = T^K \otimes_{B^K} \widehat{Q}_z$*, where* $\widehat{Q}_z$ *is some principal generic* B^K*-module, are generic, and satisfy*

$$\text{End}_{\Lambda^K}(G)/\text{radEnd}_{\Lambda^K}(G) \cong K(t).$$

Moreover, $\text{radEnd}_{\Lambda^K}(G)$ *is a nilpotent ideal.*

(2) Any generic Λ^K*-module of endolength* $\leq d$ *arises this way.*

Proof. Write $\widehat{\Lambda} = \Lambda^K$, $\widehat{B} = B^K$, $\widehat{R} = R^K$, $\widehat{T} = T^K$ and, as stated above, $\widehat{Q}_z$ is a principal generic $\widehat{B}$-module (at the point z, where $Re_z = k[t]_{f(t)}$, thus $\widehat{Re}_z = K[t]_{f(t)}$).

(1) The module $\widehat{Q}_z$ is naturally an object in $\widehat{\mathcal{B}}$-$K(t)$-mod$_p$, this implies that $G = \widehat{T} \otimes_{\widehat{B}} \widehat{Q}_z$ is a $\widehat{\Lambda}^{K(t)}$-module, finite dimensional over $K(t)$ (see (29.10)). By

(31.2), the $\widehat{\Lambda}$-module G has finite endolength. Also, since $F^{[K]}$ is full and reflects isomorphisms, by (29.9), it induces first a surjective algebra morphism $\psi : \mathrm{End}_{\widehat{B}}(\widehat{Q}_z) \longrightarrow \mathrm{End}_{\widehat{\Lambda}}(G)$ and then, by (31.6)(1), the third isomorphism of algebras in the following chain

$$K(t) \cong \mathrm{End}_{\widehat{B}}(\widehat{Q}_z) \cong \mathrm{End}_{\widehat{B}}(\widehat{Q}_z)/\mathrm{radEnd}_{\widehat{B}}(\widehat{Q}_z) \cong \mathrm{End}_{\widehat{\Lambda}}(G)/\mathrm{radEnd}_{\widehat{\Lambda}}(G),$$

where the first isomorphism is clear and the second one, $\varphi_{\widehat{B}}$, is defined in (31.6)(2). In particular, G is indecomposable and infinite dimensional over K, so that G is a generic $\widehat{\Lambda}$-module. From (31.6), $\mathrm{radEnd}_{\widehat{B}}(\widehat{Q}_z)$ is nilpotent. Since $\psi(\mathrm{radEnd}_{\widehat{B}}(\widehat{Q}_z)) = \mathrm{radEnd}_{\widehat{\Lambda}}(G)$, we obtain that $\mathrm{radEnd}_{\widehat{\Lambda}}(G)$ is nilpotent too.

(2) Let G be a generic $\widehat{\Lambda}$-module of endolength $\leq d$. By (29.9)(1), we know that $G \cong \widehat{T} \otimes_{\widehat{B}} N$, for some $N \in \widehat{B}$-E-mod, where E is the opposite endomorphism K-algebra of the object in $\mathcal{P}^2(\widehat{\Lambda})$ corresponding to G. Then, N is a $\widehat{B}$-module of finite endolength. Indeed, any chain of $\mathrm{End}_{\widehat{B}}(N)^{op}$-submodules of N is a chain of E-submodules of N by restriction through $E \to \mathrm{End}_{\widehat{B}}(N)^{op}$. Now N is an indecomposable module because $F^{[K]}$ reflects isomorphisms, and N is not in $\widehat{B}$-mod since G is infinite dimensional over K. Thus, N is a generic $\widehat{B}$-module, and so N is isomorphic to some $\widehat{Q}_z$ by (31.3). □

Corollary 31.8. *Let Λ be a non-wild finite-dimensional algebra over the algebraically closed field k, let K be any field extension of k and take $d \in \mathbb{N}$. Then, the isomorphism classes of generic Λ^K-modules of endolength $\leq d$ are represented by a finite collection $G_1, \ldots, G_n$ of non-isomorphic Λ^K-$K(t)$-bimodules, which are indecomposable as Λ^K-modules and satisfy $\dim_{K(t)} G_i = \mathrm{endol}\,(G_i)$, for all i.*

Proof. Using (31.7) and its notation, we only have to show that $\dim_{K(t)} G = \mathrm{endol}\,(G)$, for all $G = T^K \otimes_{B^K} \widehat{Q}_z$. We have seen that the K-algebra morphism

$$K(t) \cong \mathrm{End}_{\widehat{B}}(\widehat{Q}_z) \xrightarrow{L} \mathrm{End}_{\widehat{B}}(\widehat{Q}_z) \xrightarrow{F^{[K]}} \mathrm{End}_{\widehat{\Lambda}}(G),$$

induces an isomorphism $K(t) \cong \mathrm{End}_{\widehat{\Lambda}}(G)/\mathrm{radEnd}_{\widehat{\Lambda}}(G)$. Then, given a composition series $0 = G_e \subseteq \cdots \subseteq G_1 \subseteq G_0 = G$ of the $\mathrm{End}_{\widehat{\Lambda}}(G)$-module G, we know that each simple $\mathrm{End}_{\widehat{\Lambda}}(G)$-module G_i/G_{i+1} is one dimensional over $K(t)$. Thus, $\mathrm{endol}\,(G) = e = \sum_{i=0}^{e-1} \dim_{K(t)} G_i/G_{i+1} = \dim_{K(t)} G$. □

Theorem 31.9. *Let Λ be a finite-dimensional algebra over the algebraically closed field k. Then, the following statements are equivalent:*

(1) Λ is tame.
(2) Λ is not wild.
(3) There does not exists a Λ-$k\langle x,y\rangle$-bimodule Z which is finitely generated free by the right, such that the functor $Z\otimes_{k\langle x,y\rangle} - : k\langle x,y\rangle\text{-mod}\longrightarrow\Lambda\text{-mod}$ preserves isomorphism classes and indecomposables.
(4) Λ is generically tame.

Proof. We already know, by (22.5), that (3) $\Rightarrow$ (2). By (27.9), (1) $\Rightarrow$ (3). If Λ is not wild, from (31.7) applied to the trivial extension $K = k$, we obtain that Λ is generically tame. Thus, (2) $\Rightarrow$ (4). Finally, (4) $\Rightarrow$ (1) follows from (31.5) and (27.14). □

Definition 31.10. *Assume Λ is an algebra over the field k and let K be any field extension of k. Then, a Λ^K-module $\widehat{G}$ is called rationally induced from some Λ-$k(t)$-bimodule G iff $\widehat{G}\cong G\otimes_{k(t)} K(t)$.*

Theorem 31.11. *Let Λ be a finite-dimensional algebra over the algebraically closed field k and let K be any field extension of k. Then:*

(1) If Λ is not a wild algebra and d is a natural number, then the generic Λ^K-modules of endolength smaller than d are the Λ^K-modules rationally induced from the generic Λ-modules of endolength smaller than d (equipped with the Λ-$k(t)$-structure described in (31.8)).
(2) Λ is wild iff Λ^K is wild.

Proof.

(1) From (31.7), applied to the extension $K = k$, the generic Λ-modules with endolength $\leq d$ have the form $G\cong T\otimes_B Q_z$, for some principal generic B-module Q_z. From the same proposition, now applied to the extension $k\leq K$, the generic Λ^K-modules with endolength $\leq d$ have the form $\widehat{G}\cong T^K\otimes_{B^K}\widehat{Q}_z$, for some principal generic B^K-module $\widehat{Q}_z$. Notice that $\widehat{Q}_z\cong Q_z\otimes_{k(t)} K(t)$, then consider the isomorphism of Λ^K-modules

$$\widehat{G}\cong (T\otimes_k K)\otimes_{B^K}(Q_z\otimes_{k(t)} K(t)) \xrightarrow{\ \phi\ } (T\otimes_B Q_z)\otimes_{k(t)} K(t) \cong G\otimes_{k(t)} K(t)$$

determined by $\phi(b\otimes c\otimes q\otimes v) = b\otimes q\otimes cv$, with inverse map determined by $\phi'(b\otimes q\otimes v) = b\otimes 1\otimes q\otimes v$, for $b\in T$, $c\in K$, $q\in Q_z$ and $v\in K(t)$.

(2) If Λ is wild, by (30.10), it is absolutely wild and, in particular, Λ^K is wild. For the converse, assume Λ^K is wild, but Λ is not. Since Λ^K is wild, by (31.5), it is not generically tame. Since Λ is not wild, from (1) the generic

Λ^K-modules of endolength $\leq d$ are of the form $G \otimes_{k(t)} K(t)$, where G are generic Λ-modules of endolength $\leq d$. Since Λ is not wild, it is generically tame (31.9). Thus, Λ^K is generically tame: a contradiction. $\square$

Proposition 31.12. *Let Λ be a finite-dimensional non-wild algebra over the algebraically closed field k and let K be any field extension of k. Then, Λ^K is generically tame and if G is a generic Λ^K-module, then* $\mathrm{End}_{\Lambda^K}(G)$ *splits over its radical and any two splittings are conjugate.*

Proof. Write $\Gamma := \mathrm{End}_{\widehat{\Lambda}}(G)$, where $\widehat{\Lambda} := \Lambda^K$ and G is a fixed generic $\widehat{\Lambda}$-module. From (31.7), we know that Γ is a local K-algebra with $\Gamma/\mathrm{rad}\Gamma \cong K(x)$. Denote by $\nu : \Gamma \longrightarrow K(x)$ the corresponding quotient and choose $\gamma \in \Gamma$ such that $\nu(\gamma) = x$. Then, consider the algebra morphism $K[x] \xrightarrow{\varphi} \Gamma$ which maps x onto γ. If $p(x) \in K[x]$ is a non-zero polynomial, then $\nu\varphi(p(x)) = \nu(p(\gamma)) = p(x) \neq 0$, and, hence, $\varphi(p(x)) \notin \mathrm{Ker}\,\nu = \mathrm{rad}\Gamma$ is invertible. Thus, the map φ extends to a morphism of K-algebras $\overline{\varphi} : K(x) \longrightarrow \Gamma$, which determines a section for the canonical projection $\Gamma \longrightarrow \Gamma/\mathrm{rad}\Gamma$. Thus, Γ splits over its radical.

Now assume we have two splittings of Γ over its radical: $\Gamma = S_1 \oplus \mathrm{rad}\Gamma$ and $\Gamma = S_2 \oplus \mathrm{rad}\Gamma$. Then, S_1 and S_2 are isomorphic to $K(x)$. Denote by $s_j : K(x) \longrightarrow S_j$ the corresponding isomorphisms and consider the $\widehat{\Lambda}$-$K(x)$-bimodules G_1 and G_2, which are equal to G as $\widehat{\Lambda}$-modules and whose structure as right $K(x)$-modules is obtained by restriction of scalars through s_1 and s_2, respectively. Thus, for each j, we have a morphism of algebras $\psi_j : K(x) \longrightarrow \mathrm{End}_{\widehat{\Lambda}}(G_j) = \Gamma$ determined by the structure of $\widehat{\Lambda}$-$K(x)$-bimodule of G_j. The length of any descending chain of Γ-submodules of G_j is bounded by $\dim_{K(x)} G_j$, because this chain is also a descending chain of $K(x)$-subvector spaces of G_j, by restriction through ψ_j. Moreover, any Γ-composition series of G_j is a chain of Γ-submodules of G_j, with only one composition factor $K(x) \cong \Gamma/\mathrm{rad}\Gamma$. Then, the dimension of each of the $K(x)$-vector spaces G_1 and G_2 coincides with the endolength e of G.

Using the integer e in (29.9), we find a minimal k-ditalgebra $\mathcal{B}$ and a Λ-B-bimodule T satisfying the condition (2) stated there. Write $\widehat{T} = T^K$ and $\widehat{\mathcal{B}} = \mathcal{B}^K$. Using the field extension $\Omega = K(x)$ in (29.10), we find that $G_1 \cong F^{[\![K(x)]\!]}(N_1)$ and $G_2 \cong F^{[\![K(x)]\!]}(N_2)$, for some $N_1, N_2 \in \widehat{\mathcal{B}}$-$K(x)$-$\mathrm{mod}_p$. Since the functor $F^{[\![K(x)]\!]}$ reflects isomorphisms and G_1, G_2 are indecomposable $\widehat{\Lambda}$-modules, N_1, N_2 are indecomposables too. By (31.2), N_1 and N_2 are generic $\widehat{B}$-modules and, by (31.3), they are both isomorphic to the same principal generic $\widehat{B}$-module $Q := \widehat{Q}_z$. Each $N_j \in \widehat{\mathcal{B}}$-$K(x)$-$\mathrm{mod}_p$ is equipped with a bimodule structure $\phi_j : K(x) \longrightarrow \mathrm{End}_{\widehat{\mathcal{B}}}(N_j)^{op}$. We know there is an isomorphism of

K-algebras $\phi_0 : K(x) \longrightarrow \mathrm{End}_{K(x)}(Q)^{op} \cong \mathrm{End}_{\widehat{B}}(Q)^{op}$. Denote by Q_0 the corresponding $\widehat{\mathcal{B}}$-$K(x)$-bimodule and make $G_0 := F^{[\![K(x)]\!]}(Q_0) \in \widehat{\Lambda}$-$K(x)$-mod. Since ${}_{\widehat{B}}N_j \cong Q$ and $F^{[\![K(x)]\!]}(N_j) \cong G_j$, we have that

$$_{\widehat{\Lambda}}G_0 \cong \widehat{T} \otimes_{\widehat{B}} Q \cong \widehat{T} \otimes_{\widehat{B}} N_j \cong_{\widehat{\Lambda}} G_j = G.$$

Thus, as we have shown before for G_1 and G_2, we have that $\dim_{K(x)} G_0 = e$. Now, consider, for $j \in \{1, 2\}$, the morphism of fields ξ_j defined by the following diagram

$$\begin{array}{ccc} K(x) & \xrightarrow{\phi_j} & \mathrm{End}_{\widehat{B}}(N_j)^{op} \\ \xi_j \downarrow & & \downarrow \cong \\ K(x) & \xrightarrow{\phi_0} & \mathrm{End}_{\widehat{B}}(Q)^{op}. \end{array}$$

Thus, $N_j \cong (Q_0)_{\xi_j}$ in $\widehat{\mathcal{B}}$-$K(x)$-mod, and $G_j \cong \widehat{T} \otimes_{\widehat{B}} N_j \cong \widehat{T} \otimes_{\widehat{B}} (Q_0)_{\xi_j} \cong (G_0)_{\xi_j}$ in $\widehat{\Lambda}$-$K(x)$-mod. Denote by $L_j := \xi_j(K(x))$, the image subfield of $K(x)$ under the morphism ξ_j. Then

$$\dim_{K(x)} G_0 = e = \dim_{K(x)} G_j = \dim_{K(x)}(G_0)_{\xi_j} = \dim_{L_j} G_0.$$

Moreover, the L_j-vector space G_0, obtained from the $K(x)$-vector space G_0 by restriction, has dimension $\dim_{K(x)} G_0 \times \dim_{L_j} K(x)$. Then, $L_j = K(x)$ and ξ_j is an automorphism.

Now, from $G_j \cong (G_0)_{\xi_j}$, for $j \in \{1, 2\}$, we obtain $(G_1)_{\xi_1^{-1}} \cong G_0 \cong (G_2)_{\xi_2^{-1}}$, and, therefore, $G_2 \cong (G_1)_\xi$, where $\xi := \xi_1^{-1}\xi_2$ is an automorphism of the field $K(x)$. Denote by θ any isomorphism $G_2 \longrightarrow (G_1)_\xi$ in $\widehat{\Lambda}$-$K(x)$-Mod. Then, if we write $\cdot$, $\odot$ and $*$ for the action of $K(x)$ by the right on G corresponding to the $\widehat{\Lambda}$-$K(x)$-bimodules G_2, $(G_1)_\xi$ and G_1, respectively, we have

$$s_1(\xi(\alpha))[\theta(g)] = \theta(g) * \xi(\alpha) = \theta(g) \odot \alpha = \theta(g \cdot \alpha) = \theta(s_2(\alpha)[g]),$$

for $\alpha \in K(x)$ and $g \in G$. Hence, $\theta^{-1}s_1(\xi(\alpha))\theta = s_2(\alpha)$, for any $\alpha \in K(x)$, and then $\theta^{-1}S_1\theta = S_2$. □

Theorem 31.13. *Let k be an algebraically closed field, denote by $K = k(t)$ the field of fractions of the polynomial algebra $k[t]$, and let Λ be a finite-dimensional algebra over k. Then, Λ^K is not wild iff every generic Λ^K-module is rationally induced from a generic Λ-module.*

Proof. We only have to show the converse of theorem (31.11)(1) for the extension $k \leq K = k(t)$. Assume that Λ^K is wild, or equivalently, that Λ is wild. We want to expose a generic Λ^K-module, which is not rationally induced from a generic Λ-module. Consider the $K\langle x, y\rangle$-module $H_t := K(x)[y]/(y - t)$, as in (31.3)(2), where t is the canonical trascendental element of $K = k(t)$. Consider also the Λ-$k\langle x, y\rangle$-bimodule B, which is free of finite rank over $k\langle x, y\rangle$, such that for every field extension $k \leq K$, the functor

$$B^K \otimes_{K\langle x,y\rangle} - : K\langle x, y\rangle\text{-Mod} \longrightarrow \Lambda^K\text{-Mod}$$

preserves isomorphism classes of indecomposables, provided by (30.10).

From (31.4), we know that the Λ^K-module $G = B^K \otimes_{K\langle x,y\rangle} H_t$ is generic. We will show that it is not a rationally induced module. First, notice that the restriction of H_t to $k\langle x, y\rangle$, which we denote by $\text{Res}(H_t)$, is indecomposable. Indeed, since the actions of x and y on H_t commute, we have

$$\text{End}_{k\langle x,y\rangle}(\text{Res}(H_t)) = \text{End}_{k\langle x,y\rangle}\left(\frac{k(t)(x)[y]}{(y-t)}\right) \cong \text{End}_{k[x,y]}\left(\frac{k(t)(x)[y]}{(y-t)}\right).$$

Let us see that the action of any non-zero $f(x, y) \in k[x, y]$ on the $k[x, y]$-module $k(t)(x)[y]/(y - t)$ is invertible. The polynomial $f(x, y)$ acts on this space by multiplication of its class modulo the ideal $(y - t)$, thus we want to see that this class is not zero (because this space is a field). Assume that $0 \neq f(x, y) = \sum_{i=0}^{n} a_i(x)y^i$, with $a_i(x) \in k[x]$, and $f(x, y) = [y - t]g(y)$, where $g(y) = \sum_{j=0}^{m} b_j y^j \in k(t, x)[y]$, with $b_j \in k(t, x)$ and $b_m \neq 0$. Notice that $n \neq 0$ because, otherwise, $a_0(x) = [y - t]g(y)$ has the non-zero summand $b_m y^{m+1}$, which is impossible. Thus, $n > 0$ and the equality $f(x, y) = [y - t]g(y)$ implies that $n = m + 1$ and $a_0(x) = -tb_0$; for $i \in [1, m]$, $a_i(x) = b_{i-1} - tb_i$; and $a_{m+1}(x) = b_m$. This implies that $a_{m+1}(x) = b_m = -a_0(x)t^{-(m+1)} - a_1(x)t^{-m} - \cdots - a_m(x)t^{-1}$. Equivalently, $\sum_{i=0}^{m+1} a_i(x)t^i = 0$ and t is an algebraic element over $k(x)$, a contradiction.

Then, the morphism $k[x, y] \longrightarrow k(t, x)[y]/(y - t)$ extends to an isomorphism of fields $k(x, y) \cong k(t, x)[y]/(y - t)$, and, therefore

$$\begin{aligned}\text{End}_{k[x,y]}\left(\frac{k(t)(x)[y]}{(y-t)}\right) &\cong \text{End}_{k(x,y)}\left(\frac{k(t)(x)[y]}{(y-t)}\right) \cong \text{End}_{k(x,y)}(k(x, y)) \\ &\cong k(x, y).\end{aligned}$$

We have the following up to isomorphism commutative diagram of functors

$$\begin{array}{ccc} K\langle x, y\rangle\text{-Mod} & \xrightarrow{B^K \otimes_{K\langle x,y\rangle} -} & \Lambda^K\text{-Mod} \\ \downarrow{\scriptstyle \text{Res}} & & \downarrow{\scriptstyle \text{Res}} \\ k\langle x, y\rangle\text{-Mod} & \xrightarrow{B \otimes_{k\langle x,y\rangle} -} & \Lambda\text{-Mod}, \end{array}$$

where Res denotes restriction functors. The natural isomorphism is given, for each $M \in K\langle x, y\rangle$-Mod, by the map $\eta_M : \text{Res}(B^K \otimes_{K\langle x,y\rangle} M) \longrightarrow B \otimes_{k\langle x,y\rangle} \text{Res}(M)$, such that, $b \otimes c \otimes m \mapsto b \otimes cm$, for any $b \in B, c \in K, m \in M$.

Now, assume that the generic Λ^K-module G is rationally induced from some Λ-$k(z)$-bimodule G_0, thus $G \cong G_0 \otimes_{k(z)} K(z)$. Then

$$B \otimes_{k\langle x,y\rangle} \mathrm{Res}(H_t) \cong \mathrm{Res}\left(B^K \otimes_{K\langle x,y\rangle} H_t\right) = \mathrm{Res}(G) \cong \mathrm{Res}(G_0 \otimes_{k(z)} K(z)),$$

is a decomposable Λ-module. But we have seen that $\mathrm{Res} H_t$ is indecomposable and that $B \otimes_{k\langle x,y\rangle} -$, which may be identified with $B^k \otimes_{k\langle x,y\rangle} -$, preserves indecomposability, a contradiction. □

Exercise 31.14. *Statement (31.7)(1) uses a principal generic B^K-module $\widehat{Q}_z$. This module cannot be identified with the induced B^K-module $Q_z^K = Q_z \otimes_k K$, where Q_z is the corresponding principal generic B-module. Indeed, show that if $Q = k(x)$ and $K = k(y)$, then $Q^K = k(x) \otimes_k k(y) \ncong k(x, y) \cong K(x)$.*

Hint. Show that the inverse of $x + y$ in $k(x, y)$ does not belong to the image of the embedding map $k(x) \otimes_k k(y) \longrightarrow k(x, y)$. Thus, the element $x + y$ of the algebra $K[x] = k(x)[y]$ acts invertibly in the $K[x]$-module $\widehat{Q} = K(x)$ but not on the $K[x] = k[x] \otimes_k K$-module Q^K.

References. The notions of generic module and generic tameness are due to Crawley-Boevey. In [19] it is shown that generic tameness coincides with tameness, over an algebraic closed ground field. The study of generic modules has been pursued in [21]. (31.7) and (31.12) are scalar extended versions of important results from [20]. The examples in (31.3) are taken from [20] and [21].

32

Almost split sequences and tameness

In this section, we show Crawley-Boevey's Structure Theorem for the Auslander–Reiten sequences of tame finite-dimensional algebras Λ, over an algebraically closed field. We show first the corresponding statement for non-wild finite-dimensional seminested ditalgebras and, then, transfer it to Λ-mod.

We start with a series of lemmas which describe an exact structure with almost split sequences, for minimal ditalgebras, which will be fruitfully transferred later, with the help of suitable reduction functors to almost split sequences in the exact category of modules of a finite-dimensional tame ditalgebra.

Lemma 32.1. *Assume that $\mathcal{A}$ is a Roiter ditalgebra with layer (R, W). Assume that S is a k-subalgebra of R, $W_0 = 0$ and W_1 is freely generated by some S-S-subbimodule, as defined in (15.6). Then, $\mathcal{A}$ is S-acceptable, as in (6.4).*

Proof. We only prove the first condition of definition (6.4), since the proof of the second is dual. Fix a morphism $g : E \longrightarrow N$ in $\mathcal{A}$-Mod such that $g^0 : E \longrightarrow N$ is a retraction in S-Mod. Since $W_0 = 0$, we have $(g^0, 0), (0, g^1) \in \mathrm{Hom}_{\mathcal{A}}(E, N)$ and $g = (g^0, 0) + (0, g^1)$. Now, suppose that W_1 is freely generated by its S-S-subbimodule $\widetilde{W}_1$ and let $\rho : N \longrightarrow E$ be a right inverse for g^0 in S-Mod. Consider the composition of S-S-bimodule morphisms

$$\widetilde{W}_1 \xrightarrow{\widetilde{g}^1} \mathrm{Hom}_k(E, N) \xrightarrow{\rho^*} \mathrm{Hom}_k(E, E),$$

where ρ^* is induced by ρ and the map $\widetilde{g}^1$ is the restriction of the R-R-bimodule morphism $g^1 : W_1 \longrightarrow \mathrm{Hom}_k(E, N)$, and then extend this composition to a morphism of R-R-bimodules $\psi^1 : W_1 \longrightarrow \mathrm{Hom}_k(E, E)$. Then, $(0, \psi^1) \in \mathrm{End}_{\mathcal{A}}(E)$ and $(0, g^1) = (g^0, 0)(0, \psi^1)$ in $\mathcal{A}$-Mod, because

$$g^0\psi^1[w] = g^0(\rho^*[\widetilde{g}^1[w]]) = (g^0\rho)(\widetilde{g}^1[w]) = \widetilde{g}^1[w],$$

for any $w \in \widetilde{W}_1$. Then

$$g = (g^0, 0) + (0, g^1) = (g^0, 0) + (g^0, 0)(0, \psi^1) = (g^0, 0)[1_E + (0, \psi^1)].$$

From (5.4), we know that $(0, \psi^1) \in \mathrm{End}_{\mathcal{A}}(E)$ is nilpotent and, therefore, $h := 1_E + (0, \psi^1)$ is an isomorphism satisfying: $(g^0, 0)h = g$. Thus, $gh^{-1} = (g^0, 0)$. □

Lemma 32.2. *Let k be any field and consider the principal ideal domain $H := k[x]_{f(x)}$. Then, the indecomposable finite-dimensional H-modules are those of the form $H/H\pi^n$, where $n \geq 1$ and π is an irreducible element in H. The category H-mod of finite-dimensional H-modules has almost split sequences, which are the following*

$$E_1^\pi : 0 \longrightarrow H/H\pi \xrightarrow{\mu_\pi} H/H\pi^2 \xrightarrow{\mu_1} H/H\pi \longrightarrow 0$$

$$E_n^\pi : 0 \longrightarrow H/H\pi^n \xrightarrow{\begin{pmatrix}\mu_\pi \\ \mu_1\end{pmatrix}} H/H\pi^{n+1} \oplus H/H\pi^{n-1} \xrightarrow{(\mu_1, -\mu_\pi)} H/H\pi^n \longrightarrow 0,$$

where $n \geq 2$, π is an irreducible element of H and μ_1, μ_π are the morphism induced by multiplication by 1 and π, respectively.

Proof. Fix an irreducible element $\pi \in H$ and consider an exact sequence E_n^π, for $n \geq 1$ as above:

$$0 \longrightarrow H/H\pi^n \xrightarrow{f} E \xrightarrow{g} H/H\pi^n \longrightarrow 0.$$

We have to show that any non-retraction morphism $h : N \longrightarrow H/H\pi^n$ of finite-dimensional H-modules factors through g. It is enough to verify this for every indecomposable $N \in H$-mod and every non-isomorphism $h : N \longrightarrow H/H\pi^n$. Then, $N \cong H/H\pi_0^t$, for some $t \geq 1$ and some irreducible $\pi_0 \in H$, and we may assume that the isomorphism is an equality. Since $\mathrm{Hom}_H(H/H\pi_0^t, H/H\pi^n) = 0$, whenever π_0 and π are not associated elements in H, we can assume that $\pi_0 = \pi$.

Recall that, for $n \leq t$, multiplication by any $a \in H$ induces a morphism of H-modules $\mu_a : H/H\pi^t \longrightarrow H/H\pi^n$. Moreover, there is an isomorphism

$$\begin{array}{ccc} H/H\pi^n & \longrightarrow & \mathrm{Hom}_H(H/H\pi^t, H/H\pi^n) \\ \overline{a} & \longmapsto & \mu_a, \end{array}$$

which is an isomorphism of algebras if $t = n$. Similarly, for $n > t$, there is an isomorphism

$$\begin{array}{ccc} H\pi^{n-t}/H\pi^n & \longrightarrow & \mathrm{Hom}_H(H/H\pi^t, H/H\pi^n) \\ \overline{a} & \longmapsto & \mu_a. \end{array}$$

For the sequence E_1^π: if $t = 1 = n$, $\mathrm{End}_H(H/H\pi) \cong H/H\pi$ is a field, so h non-isomorphism means that $h = \mu_a = 0 = \mu_1 0$. Otherwise, $t \geq 2$, and $h = \mu_a = \mu_1\mu_a$.

For the sequence E_n^π with $n \geq 2$: if $n < t$, then $h = \mu_a = (\mu_1, -\mu_\pi)\begin{pmatrix}\mu_a\\0\end{pmatrix}$; if $n > t$, since $h = \mu_a$ with $\pi^{n-t} \mid a$, we can write $a = \pi b$ with $\pi^{n-t-1} \mid b$, and, therefore, $h = \mu_a = \mu_{\pi b} = (\mu_1, -\mu_\pi)\begin{pmatrix}0\\-\mu_b\end{pmatrix}$; finally, if $n = t$, $\mathrm{End}_H(H/H\pi^n) \cong H/H\pi^n$ is a local algebra with radical $\pi H/H\pi^n$, thus $h = \mu_a$ non-isomorphism means that $\pi \mid a$ and, therefore, we can proceed as in last case to factor h. □

Proposition 32.3. *Let $\mathcal{B}$ be a minimal k-ditalgebra and let $\widehat{\mathcal{E}}_{\mathcal{B}}$ be the class of composable pairs of morphisms $M\xrightarrow{f}E\xrightarrow{g}N$ in $\mathcal{B}$-*Mod *such that $gf = 0$ and*

$$0\longrightarrow M\xrightarrow{f^0}E\xrightarrow{g^0}N\longrightarrow 0$$

*is an exact sequence in B-*Mod*. Then, $\widehat{\mathcal{E}}_{\mathcal{B}}$ is an exact structure of $\mathcal{B}$-*Mod*. Moreover, in the exact category ($\mathcal{B}$-*mod*, $\widehat{\mathcal{E}}_{\mathcal{B}}$) each almost split conflation is isomorphic to the image of some almost split sequence in B-*mod*, under the embedding functor $L_{\mathcal{B}} : B$-*Mod*$\longrightarrow\mathcal{B}$-*Mod*. Then, they are of the form*

$$J(z,1,\pi)\xrightarrow{\phi_1}J(z,2,\pi)\xrightarrow{\psi_1}J(z,1,\pi)$$
$$J(z,n,\pi)\xrightarrow{\phi_n}J(z,n+1,\pi)\oplus J(z,n-1,\pi)\xrightarrow{\psi_n}J(z,n,\pi),\ \textit{for } n>1.$$

Proof. The fact that $\widehat{\mathcal{E}}_{\mathcal{B}}$ is an exact structure of $\mathcal{B}$-Mod is a consequence of (15.7)(2), (32.1) and (6.7), because if S denotes the trivial part of the algebra $B = R$, then $\widehat{\mathcal{E}}_{\mathcal{B}} = \mathcal{E}_{\mathcal{B},S}$. Then, it is clear that, relative to this exact structure, $L_{\mathcal{B}} : B$-Mod$\longrightarrow\mathcal{B}$-Mod is an exact functor. Moreover, $L_{\mathcal{B}}$ maps non-split exact sequences into non-split conflations. By (22.11), we also know that $L_{\mathcal{B}}$ preserves indecomposability.

Let us show that for any surjective right almost split morphism $p : E\longrightarrow N$ in B-Mod we have that $(p,0) : E\longrightarrow N$ is a right almost split morphism in $\mathcal{B}$-Mod. Suppose that $f = (f^0, f^1) : L\longrightarrow N$ is a non-retraction in $\mathcal{B}$-Mod. From (5.9), we know that f^0 is not a retraction in B-Mod. Since p is right almost split, there exists a B-module morphism $g^0 : L\longrightarrow E$ such that $pg^0 = f^0$. On the other hand, there is a commutative diagram of R-R-bimodule morphisms

$$\begin{array}{ccc} & & W_1 \\ & {\scriptstyle g^1}\swarrow & \big\downarrow{\scriptstyle f^1} \\ \mathrm{Hom}_k(L,E) & \xrightarrow{p_*} & \mathrm{Hom}_k(L,N) \longrightarrow 0, \end{array}$$

where p_* is surjective because p is so, and g^1 exists because W_1 is a freely generated R-R-bimodule, see (23.4)(2). Then, as remarked in (4.9), the pair (g^0, g^1) determines a morphism $g : L \longrightarrow E$ in $\mathcal{B}$-Mod such that $(p, 0)g = f$. Thus, $(p, 0)$ is a right almost split morphism in $\mathcal{B}$-Mod.

It follows, from (32.2), that $L_{\mathcal{B}}$ preserves almost split sequences, and we obtain the last part of the proposition. □

Lemma 32.4. *Let $\mathcal{A}$ and $\mathcal{B}$ be seminested ditalgebras and consider any reduction functor $H : \mathcal{B}$-Mod$\longrightarrow\mathcal{A}$-Mod. Assume that $\mathcal{B}$ is minimal and that the algebra R of the layer (R, W) of $\mathcal{A}$ is trivial. Consider any extension field K of k, make $\widehat{\mathcal{A}} := \mathcal{A}^K$ and $\widehat{\mathcal{B}} = \mathcal{B}^K$. Then*

$$\widehat{H} : (\widehat{\mathcal{B}}\text{-Mod}, \widehat{\mathcal{E}}_{\widehat{\mathcal{B}}}) \longrightarrow (\widehat{\mathcal{A}}\text{-Mod}, \mathcal{E}_{\widehat{\mathcal{A}}})$$

is an exact functor. Here, $\mathcal{E}_{\widehat{\mathcal{A}}}$ is the exact structure of (6.8) and $\widehat{\mathcal{E}}_{\widehat{\mathcal{B}}}$ is the exact structure of (32.3).

Proof. According to (22.7), if $\widehat{A} = (A_{\mathcal{A}})^K$ and $\widehat{B} = (A_{\mathcal{B}})^K$, there is an (up to isomorphism) commutative diagram

$$\begin{array}{ccccc}
\widehat{B}\text{-Mod} & \xrightarrow{\widehat{B}\otimes_{\widehat{B}}-} & \widehat{\mathcal{B}}\text{-Mod} & \xrightarrow{L_{\widehat{\mathcal{B}}}} & \widehat{\mathcal{B}}\text{-Mod} \\
\| & & \downarrow \underline{\widehat{H}} & & \downarrow \widehat{H} \\
\widehat{B}\text{-Mod} & \xrightarrow{\widehat{H}(\widehat{B})\otimes_{\widehat{B}}-} & \widehat{A}\text{-Mod} & \xrightarrow{L_{\widehat{\mathcal{A}}}} & \widehat{\mathcal{A}}\text{-Mod},
\end{array}$$

and $\underline{\widehat{H}} \cong \widehat{H}(\widehat{B}) \otimes_{\widehat{B}} -$ is an exact functor. Now take a conflation $\xi \in \widehat{\mathcal{E}}_{\widehat{\mathcal{B}}}$, so that it is isomorphic to an exact pair of the form

$$Y \xrightarrow{(f^0,0)} E \xrightarrow{(g^0,0)} X, \text{ where } 0 \longrightarrow Y \xrightarrow{f^0} E \xrightarrow{g^0} X \longrightarrow 0$$

is an exact sequence in $\widehat{B}$-Mod. Then, from the commutativity of the diagram, $\widehat{H}(\xi)$ is isomorphic to the image under the embedding functor $L_{\widehat{\mathcal{A}}}$ of the exact sequence of $\widehat{A}$-modules

$$0 \longrightarrow \widehat{H}(\widehat{B}) \otimes_{\widehat{B}} Y \xrightarrow{1\otimes f^0} \widehat{H}(\widehat{B}) \otimes_{\widehat{B}} E \xrightarrow{1\otimes g^0} \widehat{H}(\widehat{B}) \otimes_{\widehat{B}} X \longrightarrow 0,$$

which is a split exact sequence of R^K-modules, because R^K is semisimple. Thus, $\widehat{H}(\xi) \in \mathcal{E}_{\widehat{\mathcal{A}}}$, as claimed. □

Lemma 32.5. *Let $\mathcal{A}$ be a finite-dimensional seminested ditalgebra, with layer (R, W), over the field k (see (7.15)). Thus R is a trivial k-algebra. From (7.18), we know that the exact category $(\mathcal{A}\text{-mod}, \mathcal{E}_{\mathcal{A}})$ has almost split conflations. Thus, for any non $\mathcal{E}_{\mathcal{A}}$-projective indecomposable object $M \in \mathcal{A}$-mod, there is an almost split conflation $N \longrightarrow E \longrightarrow M$. By (7.11), the indecomposable object N is uniquely determined up to isomorphism. We call N the translate of*

M and denote it by $\tau(M)$. Under our assumptions, there is a number $c_{\mathcal{A}}$ such that, for any non $\mathcal{E}_{\mathcal{A}}$-projective indecomposable $M \in \mathcal{A}$-mod, the following inequality holds

$$\dim_k \tau(M) \leq c_{\mathcal{A}} \dim_k M.$$

Proof. Our assumptions imply that $\mathbb{E} = \mathrm{End}_{\mathcal{A}}(A)^{op}$ is a finite-dimensional k-algebra. Therefore, there is a number $c_{\mathbb{E}}$ such that, for any finite-dimensional indecomposable non-projective $\mathbb{E}$-module N, the inequality $\dim_k Dtr(N) \leq c_{\mathbb{E}} \dim_k N$ is satisfied. Here, D and tr denote the usual duality and transpose constructions, thus $Dtr(N)$ is the Auslander–Reiten translate of N in $\mathbb{E}$-mod.

Fix any non $\mathcal{E}_{\mathcal{A}}$-projective indecomposable object $M \in \mathcal{A}$-mod and consider the associated almost split conflation in $\mathcal{A}$-mod

$$\tau(M) \xrightarrow{i} E' \xrightarrow{d} M.$$

Let us recall the construction of this conflation, from the proof of (7.18). We considered the functor $F = \mathrm{Hom}_{\mathcal{A}}(A, -) : \mathcal{A}\text{-mod} \longrightarrow \mathbb{E}\text{-mod}$ and we showed first that $F(M)$ is an indecomposable $\mathbb{E}$-module and, therefore, there is a minimal right almost split map $g : E \longrightarrow F(M)$ in $\mathbb{E}$-mod. If $F(M)$ was a projective $\mathbb{E}$-module, there would be a retraction $F(A^n) \cong \mathbb{E}^n \longrightarrow F(M)$; hence, there would be a retraction $A^n \longrightarrow M$, because F is full and faithful. Therefore, M would be an $\mathcal{E}_{\mathcal{A}}$-projective object, a contradiction. Thus, the non-projective indecomposable module $F(M)$ admits an almost split sequence in $\mathbb{E}$-mod of the form

$$0 \longrightarrow Dtr\,F(M) \xrightarrow{f} E \xrightarrow{g} F(M) \longrightarrow 0.$$

Then, we considered the multiplication map $F(E) = \mathbb{E} \otimes_A E \xrightarrow{p} E$ and showed that the morphism $g' : E \longrightarrow M$ such that $F(g') = gp$ is right almost split in $\mathcal{A}$-mod. Then, we considered a minimal right almost split morphism $d : E' \longrightarrow M$ in $\mathcal{A}$-Mod, which turned out to be a deflation. Then, the conflation (i, d), where $i : \tau(M) \longrightarrow E'$ is the kernel of d is an almost split conflation. Now, since d is minimal right almost split, E' is a direct summand of E. Then, from the inequality

$$\dim_k E = \dim_k Dtr\,F(M) + \dim_k F(M) \leq (c_{\mathbb{E}} + 1) \dim_k F(M),$$

we obtain, $\dim_k \tau(M) \leq \dim_k E' \leq \dim_k E \leq (c_{\mathbb{E}} + 1) \dim_k F(M)$.

Moreover, $F(M) = \mathrm{Hom}_{\mathcal{A}}(A, M)$ can be identified with a subspace of $\mathrm{Phom}_{\mathcal{A}}(A, M) = \mathrm{Hom}_R(A, M) \times \mathrm{Hom}_{R\text{-}R}(W_1, \mathrm{Hom}_k(A, M))$. Therefore, $\dim_k F(M) \leq (\dim_k W_1 + 1) \times \dim_k A \times \dim_k M$. Then, if we make $c_{\mathcal{A}} := (c_{\mathbb{E}} + 1)(\dim_k W_1 + 1) \dim_k A$, we obtain

$$\dim_k \tau(M) \leq (c_{\mathbb{E}} + 1) \dim_k F(M) \leq c_{\mathcal{A}} \dim_k M.$$

□

Theorem 32.6. *Let $\mathcal{A}$ be a tame finite-dimensional seminested ditalgebra over the algebraically closed field k. Let K be any field extension of k and write $\widehat{\mathcal{A}} := \mathcal{A}^K$. Thus, we know that $\widehat{\mathcal{A}}$ is seminested and finite dimensional over K, in particular ($\widehat{\mathcal{A}}$-mod, $\mathcal{E}_{\widehat{\mathcal{A}}}$) has almost split conflations and a translation $\widehat{\tau}$. Then, for every dimension d, almost all d-dimensional indecomposable modules $M \in \widehat{\mathcal{A}}$-mod satisfy $\widehat{\tau}(M) \cong M$, see (27.3).*

Proof. Let e be a number such that if M is any non $\mathcal{E}_{\widehat{\mathcal{A}}}$-projective indecomposable object of $\widehat{\mathcal{A}}$-mod of K-dimension at most d, and

$$\widehat{\tau}(M) \xrightarrow{f} E \xrightarrow{g} M$$

is an almost split conflation ending at M, then $\dim_K \widehat{\tau}(M)$, $\dim_K E$, $\dim_K M \leq e$. We know that this number exists by (32.5). Since $\mathcal{A}$ is not wild, there are a minimal k-ditalgebra $\mathcal{B}$ and a reduction functor $F : \mathcal{B}\text{-Mod} \longrightarrow \mathcal{A}\text{-Mod}$, satisfying statement (2) of (28.22). We write $\widehat{\mathcal{B}} = \mathcal{B}^K$. Suppose that $M \in \widehat{\mathcal{A}}$-mod has K-dimension at most d and is not $\mathcal{E}_{\widehat{\mathcal{A}}}$-projective. By (6.20), we are excluding finitely many possible isomorphism classes of indecomposable modules with K-dimension at most d. Then, there are $N', E', M' \in \widehat{\mathcal{B}}$-mod with $\widehat{F}(N') \cong \widehat{\tau}(M)$, $\widehat{F}(E') \cong E$ and $\widehat{F}(M') \cong M$, where N' and M' are indecomposables, see (29.6). We can assume that these three isomorphisms are equalities. Since there are only finitely possible indecomposable objects in $\widehat{\mathcal{B}}$-mod of the form $S(z)$, we can assume that the module M' is not of this form, and hence we can assume that $M' = J(z, m, \pi)$, for some point z of $\widehat{\mathcal{B}}$ and some irreducible element $\pi \in \widehat{B}e_z$. Since the morphism $g : E \longrightarrow M$ is minimal right almost split and $\widehat{F}$ is full and faithful, the morphism $g' : E' \longrightarrow J(z, m, \pi)$ in $\mathcal{B}$-mod with $\widehat{F}(g') = g$ is minimal right almost split too. Similarly, the morphism $f' : N' \longrightarrow E'$ such that $\widehat{F}(f') = f$ is a minimal left almost split morphism in $\widehat{\mathcal{B}}$-mod.

Since the existence of a minimal right (resp. left) almost split morphism $G \longrightarrow M'$ (resp. $N' \longrightarrow G$) in $\widehat{\mathcal{B}}$-mod determines G up to isomorphism, we have, using (32.3) and (23.8), the following.

If $m = 1$, then $E' \cong J(z, 2, \pi)$ is indecomposable; therefore, $N' \ncong J(z, s, \pi)$, for $s > 1$. Thus, $N' \cong J(z, 1, \pi) = M'$.

If $m \geq 2$, $E' \cong J(z, m+1, \pi) \oplus J(z, m-1, \pi)$ splits. In particular, $N' \ncong J(z, 1, \pi)$. If $N' \cong J(z, s, \pi)$ with $s \geq 2$, then we also have that $E' \cong J(z, s+1, \pi) \oplus J(z, s-1, \pi)$. Since $\widehat{\mathcal{B}}$-mod is a Krull–Schmidt category, $m = s$ and, therefore, $N' \cong M'$.

Then, in any case, $\widehat{\tau}(M) \cong M$. □

Lemma 32.7. *Assume that $\mathcal{C}$ and $\mathcal{D}$ are Krull–Schmidt categories with exact structures which admit almost split conflations in the literal sense (that is, we* *do not require that they have minimal right almost split morphisms ending at projective indecomposable objects and minimal left almost split morphisms starting at injective indecomposable objects). Let* $F : \mathcal{C} \longrightarrow \mathcal{D}$ *be a full and faithful exact functor. Suppose that there is a sequence* $\{E_n\}_{n \in \mathbb{N}}$ *of non-isomorphic indecomposable objects in* $\mathcal{C}$ *and almost split conflations*

$$\zeta_1 : E_1 \longrightarrow E_2 \longrightarrow E_1,$$

$$\zeta_n : E_n \longrightarrow E_{n+1} \oplus E_{n-1} \longrightarrow E_n, \text{ for } n \geq 2.$$

Assume, furthermore, that $\tau F(E_1) \cong F(E_1)$. *Then, the image* $F(\zeta_n)$ *is an almost split conflation in* $\mathcal{D}$ *and* $\tau F(E_n) \cong F(E_n)$, *for all* $n \in \mathbb{N}$.

Proof. We first observe that if $\zeta : X \longrightarrow E \longrightarrow Y$ is a non-split conflation in $\mathcal{C}$, then $F(\zeta)$ is a non-split conflation of $\mathcal{D}$, because F is exact, full and faithful. In particular, each $F(\zeta_n)$ is a non-split conflation and $F(E_n)$ is a non-projective indecomposable object, for any $n \in \mathbb{N}$.

Now, we claim that if $\zeta : X \longrightarrow E \longrightarrow Y$ is an almost split conflation in $\mathcal{C}$ and we have an isomorphism $h : F(X) \longrightarrow \tau F(Y)$, then $F(\zeta)$ is an almost split conflation in $\mathcal{D}$. Indeed, $\zeta \in \mathrm{socExt}_{\mathcal{C}}(Y, X)$, where $\mathrm{Ext}_{\mathcal{C}}(Y, X)$ is considered as a right $\mathrm{End}_{\mathcal{C}}(Y)$-module. Since F is exact, full and faithful, $F(\zeta) \in \mathrm{socExt}_{\mathcal{D}}(F(Y), F(X))$, where $\mathrm{Ext}_{\mathcal{D}}(F(Y), F(X))$ is considered as a right $\mathrm{End}_{\mathcal{D}}(F(Y))$-module. But, then, the isomorphism

$$h^* : \mathrm{Ext}_{\mathcal{D}}(F(Y), F(X)) \longrightarrow \mathrm{Ext}_{\mathcal{D}}(F(Y), \tau F(Y))$$

maps $F(\zeta)$ onto a conflation $\widehat{\zeta}$ in the socle of $\mathrm{Ext}_{\mathcal{D}}(F(Y), \tau F(Y))$. Therefore, the isomorphic conflations $F(\zeta)$ and $\widehat{\zeta}$ are almost split conflations of $\mathcal{D}$.

The last claim implies that $F(\zeta_1) : F(E_1) \longrightarrow F(E_2) \longrightarrow F(E_1)$ is an almost split conflation. Then, there is an irreducible morphism $F(E_1) \longrightarrow F(E_2)$ and, if $\tau F(E_2) \longrightarrow G_2 \longrightarrow F(E_2)$ is an almost split conflation, then $F(E_1)$ is a direct summand of G_2 and there is an irreducible morphism $\tau F(E_2) \longrightarrow F(E_1)$. Since $F(\zeta_1)$ is an almost split conflation, we obtain $\tau F(E_2) \cong F(E_2)$. Again, our claim implies that $F(\zeta_2)$ is an almost split conflation.

Finally, assume we have shown, for a fixed $n \geq 2$, that

$$F(\zeta_n) : F(E_n) \longrightarrow F(E_{n+1}) \oplus F(E_{n-1}) \longrightarrow F(E_n)$$

is an almost split conflation, $\tau F(E_{n-1}) \cong F(E_{n-1})$ and $\tau F(E_n) \cong F(E_n)$, and let us show the corresponding statement for $F(\zeta_{n+1})$. By assumption, there is an irreducible morphism $F(E_n) \longrightarrow F(E_{n+1})$. It follows that in the almost split conflation $\tau F(E_{n+1}) \longrightarrow G_{n+1} \longrightarrow F(E_{n+1})$, the object $F(E_n)$ is a direct summand of G_{n+1}. Thus, there is an irreducible morphism $\tau F(E_{n+1}) \longrightarrow F(E_n)$. This implies that $\tau F(E_{n+1}) \cong F(E_{n+1})$ or $\tau F(E_{n+1}) \cong F(E_{n-1})$. If we are in the second case, our assumption implies that $\tau F(E_{n+1}) \cong F(E_{n-1}) \cong \tau F(E_{n-1})$. Then, $F(E_{n+1}) \cong F(E_{n-1})$ and $E_{n+1} \cong E_{n-1}$, which is impossible. Then, $\tau F(E_{n+1}) \cong F(E_{n+1})$ and, from our claim, $F(\zeta_{n+1})$ is an almost split conflation. □

Theorem 32.8. *Let Λ be a tame finite-dimensional algebra over the algebraically field closed k. Let K be any field extension of k and write $\widehat{\Lambda} := \Lambda^K$. Thus, $\widehat{\Lambda}$-mod has almost split sequences and a translation $\widehat{\tau}$. Then, for every dimension d, almost all d-dimensional indecomposable modules $M \in \widehat{\Lambda}$-mod satisfy $\widehat{\tau}(M) \cong M$.*

Proof. We may assume that Λ is basic. Thus, Λ trivially splits over its radical (and so does $\widehat{\Lambda}$). We know that $\mathcal{D} := \mathcal{D}^\Lambda$ and $\widehat{\mathcal{D}} := \mathcal{D}^K$ are nested finite-dimensional ditalgebras. Consider also the Drozd's ditalgebra $\mathcal{D}^{\widehat{\Lambda}}$ of $\widehat{\Lambda}$ and the functors

$$\widehat{\mathcal{D}}\text{-Mod} \xrightarrow{F_{\xi_\Lambda}} \mathcal{D}^{\widehat{\Lambda}}\text{-Mod} \xrightarrow{\Xi_{\widehat{\Lambda}}} \mathcal{P}^1(\widehat{\Lambda}) \xrightarrow{\text{Cok}} \widehat{\Lambda}\text{-Mod}.$$

Let d be a dimension and $M \in \widehat{\Lambda}$-mod with K-dimension d. Assume that $\overline{M} \in \widehat{\mathcal{D}}$-mod satisfies $\text{Cok}\,\Xi_{\widehat{\Lambda}} F_{\xi_\Lambda}(\overline{M}) \cong M$, where $(P, Q, \alpha) := \Xi_{\widehat{\Lambda}} F_{\xi_\Lambda}(\overline{M}) \in \mathcal{P}^2(\widehat{\Lambda})$. Then, if $\widehat{J}$ denotes the radical of $\widehat{\Lambda}$, from (22.19) and (27.13), we obtain

$$\dim_K \overline{M} = \ell_{\widehat{\Lambda}}(P/\widehat{J}P) + \ell_{\widehat{\Lambda}}(Q/\widehat{J}Q) \leq (1 + \dim_K \widehat{\Lambda}) \times \dim_K M.$$

Apply (32.6) to $\mathcal{D}$, K and $e := (1 + \dim_k \Lambda) \times d$ to obtain that almost all indecomposable modules $\overline{M} \in \widehat{\mathcal{D}}$-mod with $\dim_K \overline{M} \leq e$ satisfy $\widehat{\tau}\overline{M} \cong \overline{M}$. Using, (18.12), (18.13) and (19.10), we know that the composition $\text{Cok}\,\Xi_{\widehat{\Lambda}} F_{\xi_\Lambda}$ preserves almost split conflations, with the exception of those ending at $\overline{S}_P$ with $\Xi_{\widehat{\Lambda}} F_{\xi_\Lambda}(\overline{S}_P) = (P, 0, 0)$, where P is an indecomposable projective $\widehat{\Lambda}$-module. The result follows from this. □

Definition 32.9. *The Auslander–Reiten quiver $\Gamma_{\mathcal{A}}$ of the finite-dimensional Roiter ditalgebra $\mathcal{A}$, endowed with the exact structure $\mathcal{E}_{\mathcal{A}}$, is a valued translation quiver whose vertices are the isomorphism classes $[M]$ of indecomposable finite-dimensional objects in $\mathcal{A}$-mod; there is an arrow $[M] \longrightarrow [N]$ in $\Gamma_{\mathcal{A}}$ iff there is an irreducible morphism $M \longrightarrow N$ in $\mathcal{A}$-mod; the arrow $[M] \longrightarrow [N]$ has valuation (u, v) if there is a minimal right almost split morphism $uM \oplus X \longrightarrow N$, where M is not a direct summand of X, and if there is a minimal left almost split morphism $M \longrightarrow vN \oplus Y$, where N is not a direct summand of Y. The vertices corresponding to projective (resp. injective) objects are called projective (resp. injective) vertices. Then, there is a translation map τ from non-projective vertices to the non-injective vertices.*

A connected component $\mathcal{C}$ of $\Gamma_{\mathcal{A}}$ is called a homogeneous tube iff it contains no projective or injective vertex and is of the form $\mathbb{Z}\mathbb{A}_\infty/G$, where:

We recall that, given a valued quiver Δ, the translation quiver $\mathbb{Z}\Delta$ has as set of vertices $\mathbb{Z}\times\Delta_0$, where Δ_0 is the vertex set of Δ. The translation τ of $\mathbb{Z}\Delta$ is given by $\tau(i,z)=(i-1,z)$. The arrows in the quiver $\mathbb{Z}\Delta$ are given by the following rule: for each valued arrow $\alpha: x\xrightarrow{(u,v)}y$ in Δ, there are arrows $\alpha_i:(i,x)\xrightarrow{(u,v)}(i,y)$ and $\beta_i:(i-1,y)\xrightarrow{(v,u)}(i,x)$, for each $i\in\mathbb{Z}$. Recall also that $\mathbb{A}_\infty$ is the quiver which has vertex set $\{x_i\}_{i\in\mathbb{N}\cup\{0\}}$; it has one arrow $x_i\longrightarrow x_{i+1}$, with trivial valuation $(1,1)$, for any vertex x_i. Now, G denotes the group of automorphisms of $\mathbb{Z}\mathbb{A}_\infty$ generated by the automorphism of translation quivers determined by $(i,x)\mapsto\tau(i,x)$. Then, the translation quiver $\mathbb{Z}\mathbb{A}_\infty/G$ is by definition the quotient translation quiver of $\mathbb{Z}\mathbb{A}_\infty$ modulo G.

Corollary 32.10. *Let $\mathcal{A}$ be a tame finite-dimensional seminested ditalgebra over the algebraically closed field k. Let K be any field extension of k and write $\widehat{\mathcal{A}}:=\mathcal{A}^K$. Then, for every dimension d, almost all d-dimensional indecomposable modules $M\in\widehat{\mathcal{A}}$-mod lie in a homogeneous tube of $\Gamma_{\widehat{\mathcal{A}}}$.*

Moreover, for every $d\in\mathbb{N}$, there is a minimal K-ditalgebra $\mathcal{L}$ and reduction functor $G:\mathcal{L}\text{-Mod}\longrightarrow\widehat{\mathcal{A}}\text{-Mod}$ such that:

(1) G is full, faithful and controls the K-dimension.
(2) Almost all d-dimensional indecomposable $\widehat{\mathcal{A}}$-modules have the form $G(J(z,n,\pi))$, for some indecomposable $\mathcal{L}$-module $J(z,n,\pi)$.
(3) $G:(\mathcal{L}\text{-mod},\widehat{\mathcal{E}}_{\mathcal{L}})\longrightarrow(\widehat{\mathcal{A}}\text{-mod},\mathcal{E}_{\widehat{\mathcal{A}}})$ preserves almost split conflations.

Proof. We know that $\widehat{\mathcal{A}}$ is seminested and finite dimensional over K, in particular $(\widehat{\mathcal{A}}\text{-mod},\mathcal{E}_{\widehat{\mathcal{A}}})$ has almost split conflations and a translation $\widehat{\tau}$. Fix a dimension d. Since $\mathcal{A}$ is not wild, there are a minimal k-ditalgebra $\mathcal{B}$ and a reduction functor $F:\mathcal{B}\text{-Mod}\longrightarrow\mathcal{A}\text{-Mod}$, satisfying statement (2) of (28.22). We write $\widehat{\mathcal{B}}=\mathcal{B}^K$. From (29.6), the isoclasses $[M]$ of indecomposable objects $M\in\widehat{\mathcal{A}}$-mod with $\dim_K M\leq d$ are of the form $[\widehat{F}(J(z,n,\pi))]$ or $[\widehat{F}(S(z'))]$, where $J(z,n,\pi)$ and $S(z')$ run in the class of canonical representatives of indecomposable objects in $\widehat{\mathcal{B}}$-mod (see (23.8)). Denote by $\mathcal{C}^d$ the set of isoclasses $[\widehat{F}(J(z,n,\pi))]$ in $\mathcal{A}$-mod with $\dim_K\widehat{F}(J(z,n,\pi))\leq d$. Then, it will be enough to show that almost all $[X]\in\mathcal{C}^d$ is such that X lies in a homogeneous tube. We claim that

$$E^d:=\{[\widehat{F}(J(z,n,\pi))]\in\mathcal{C}^d\mid\widehat{\tau}\widehat{F}(J(z,1,\pi))\not\cong\widehat{F}(J(z,1,\pi))\}$$

is a finite set. Indeed, we know from (32.6), that Z^d, defined as

$$\{J(z,1,\pi)\mid\widehat{\tau}\widehat{F}(J(z,1,\pi))\not\cong\widehat{F}(J(z,1,\pi))\text{ and }\dim_K\widehat{F}(J(z,1,\pi))\leq d\}$$

is a finite set. From (22.7), the restriction $\underline{\widehat{F}} : B\text{-Mod}\longrightarrow A\text{-Mod}$ of $\widehat{F}$ is exact. Thus, $\dim_K \widehat{F}(J(z,1,\pi)) \leq \dim_K \widehat{F}(J(z,n,\pi))$. If $[\widehat{F}(J(z,n,\pi))] \in E^d$, then $\widehat{\tau}\widehat{F}(J(z,1,\pi)) \ncong \widehat{F}(J(z,1,\pi))$ and $J(z,1,\pi) \in Z^d$, which is a finite set. This means that there are only finitely many possible z and π such that $[\widehat{F}(J(z,n,\pi))] \in E^d$. Moreover, $\dim_K J(z,n,\pi) \leq \dim_K \widehat{F}(J(z,n,\pi)) \leq d$ imply there are only finitely many possibilities for n.

Now, we show that for every $[X] \in \mathcal{C}^d \backslash E^d$, the module X lies in a homogeneous tube. Thus, we can assume that X has the form $X = \widehat{F}(J(z,n,\pi))$, with $\widehat{\tau}\widehat{F}(J(z,1,\pi)) \cong \widehat{F}(J(z,1,\pi))$. In (32.3) we described the almost split conflations of $(\widehat{\mathcal{B}}\text{-mod}, \widehat{\mathcal{E}}_{\widehat{\mathcal{B}}})$. From (32.4), we know that the full and faithful functor

$$\widehat{F} : (\widehat{\mathcal{B}}\text{-mod}, \widehat{\mathcal{E}}_{\widehat{\mathcal{B}}})\longrightarrow(\widehat{\mathcal{A}}\text{-mod}, \mathcal{E}_{\widehat{\mathcal{A}}})$$

is exact, then, from (32.7), we obtain almost split conflations

$$\widehat{F}(J(z,1,\pi))\longrightarrow\widehat{F}(J(z,2,\pi))\longrightarrow\widehat{F}(J(z,1,\pi)) \text{ and}$$
$$\widehat{F}(J(z,n,\pi))\longrightarrow\widehat{F}(J(z,n+1,\pi)) \oplus \widehat{F}(J(z,n-1,\pi))$$
$$\longrightarrow\widehat{F}(J(z,n,\pi)), \text{ for } n > 1.$$

The first part of the corollary follows from this.

For the moreover part, let $\mathcal{L}$ be the minimal K-ditalgebra obtained from $\widehat{\mathcal{B}}$ by successively localizing at the marked points z of $\widehat{\mathcal{B}}$ such that $J(z,1,\pi) \in Z^d$, using the irreducible element $\pi \in Re_z = K[x]_{f_z}$, where we may assume, after replacing π by an associate irreducible element of Re_z if necessary, that the coefficients of π lie in K, see (26.4). By (26.5), there is a full and faithful functor $F^l : \mathcal{L}\text{-Mod}\longrightarrow\widehat{\mathcal{B}}\text{-Mod}$ which preserves the K-dimension. Then, the composition functor $G := \widehat{F}F^l : \mathcal{L}\text{-Mod}\longrightarrow\widehat{\mathcal{A}}\text{-Mod}$ satisfies (1). Moreover, we have $F^l(J(z,n,\pi)) \cong J(z,n,\pi)$, for any indecomposable $J(z,n,\pi) \in \widehat{\mathcal{B}}\text{-Mod}$ such that $J(z,1,\pi) \notin Z^d$. Then, since $\mathcal{C}^d \backslash E^d$ is finite, we obtain (2). Finally, (3) is a consequence of (32.4) and (32.7). □

Theorem 32.11. *Let Λ be a tame finite-dimensional algebra over the algebraically closed field k. Let K be any field extension of k and write $\widehat{\Lambda} := \Lambda^K$. Then, for every dimension d, almost all d-dimensional indecomposable modules $M \in \widehat{\Lambda}$-mod lie in a homogeneous tube of $\Gamma_{\widehat{\Lambda}}$.*

Moreover, for every $d \in \mathbb{N}$, there are rational K-algebras $\Gamma_1, \ldots, \Gamma_p$ and $\widehat{\Lambda}$-Γ_i-bimodules $T_1, \ldots, T_p$ such that:

(1) Each T_i is free of finite rank, as a right Γ_i-module.
(2) The functor $T_i \otimes_{\Gamma_i} - : \Gamma_i\text{-mod}\longrightarrow\widehat{\Lambda}\text{-mod}$ preserves isoclasses, indecomposables and almost split sequences.

(3) Almost all d-dimensional indecomposable $\widehat{\Lambda}$-modules are isomorphic to one of the form $T_i \otimes_{\Gamma_i} \Gamma_i/(\pi)^n$, for some irreducible element $\pi \in \Gamma_i$ and $n \in \mathbb{N}$.

Proof. First notice that it will be enough to prove theorem (32.11′), which is obtained from (32.11) replacing conditon (1) by:

(1′) Each T_i is finitely generated as a right Γ_i-module.

Indeed, the argument given in the proof of (27.5) applies here, to replace each $\Gamma_i = K[x]_{f_i}$ by a new rational algebra $\Gamma_i' = K[x]_{hf_i}$, for a suitable $h \in K[x]$, and determining new bimodules $T_i' = T_i \otimes_{\Gamma_i} \Gamma_i'$, which are free of finite rank as right Γ_i'-modules satisfying that almost all $\widehat{\Lambda}$-modules M with $\dim_K M \leq d$ are isomorphic to one of the form $T_i' \otimes_{\Gamma_i'} \Gamma_i'/(\pi)^n$, for some irreducible element $\pi \in \Gamma_i'$ and some $n \in \mathbb{N}$, and each functor $T_i' \otimes_{\Gamma_i'} -$ preserves isoclasses and indecomposables. Given the description of almost split sequences given in (32.2), it is clear that the canonical embedding

$$\Gamma_i' \otimes_{\Gamma_i'} - : \Gamma_i'\text{-mod} \longrightarrow \Gamma_i\text{-mod}$$

preserves almost split sequences. Thus, each $T_i' \otimes_{\Gamma_i'} - = T_i \otimes_{\Gamma_i} (\Gamma_i' \otimes_{\Gamma_i'} -)$ preserves almost split sequences.

We can assume that Λ is a basic algebra. Indeed, assume the theorem holds for basic algebras, assume Λ is not basic and take $d \in \mathbb{N}$. Consider a basic finite-dimensional k-algebra Λ_0 Morita equivalent to Λ. From (22.15), we know that Λ_0 is not a wild algebra. Consider a Λ-Λ_0-bimodule P as in (29.8)(3). Thus, there is some $s \in \mathbb{N}$ such that, for any field extension K of k and any $M' \in \Lambda_0^K$-mod, we have that $\dim_K M' \leq s \times \dim_K(P^K \otimes_{\widehat{\Lambda}_0} M')$. Define $d_0 := s \times d$. Apply theorem (32.11′) to the basic algebra Λ_0 with the integer d_0, to obtain a $\widehat{\Lambda}_0$-Γ_i-bimodules T_i' satisfying that $(T_i')_{\Gamma_i}$ is finitely generated; that $T_i' \otimes_{\Gamma_i} - : \Gamma_i\text{-mod} \longrightarrow \widehat{\Lambda}_0\text{-mod}$ preserve isoclasses, indecomposables and almost split sequences; and that almost all indecomposable $\widehat{\Lambda}_0$-modules M with $\dim_K M \leq d_0$ are of the form $M \cong T_i' \otimes_{\Gamma_i} \Gamma_i/(\pi)^n$, for some irreducible element $\pi \in \Gamma_i$ and $n \in \mathbb{N}$. Then, the family $(\Gamma_i, T_i := P^K \otimes_{\widehat{\Lambda}_0} T_i')$ satisfies (1′), (2) and (3).

Assume now that Λ is basic and consider the finite k-dimensional nested Drozd's ditalgebra $\mathcal{D} = \mathcal{D}^\Lambda$. Fix $d \in \mathbb{N}$, put $e := (1 + \dim_k \Lambda) \times d$ and construct, as in the proof of (29.9), a minimal k-ditalgebra $\mathcal{B}$ and a reduction functor $H : \mathcal{B}\text{-Mod} \longrightarrow \mathcal{D}\text{-Mod}$ such that, for the field extension K of k, making $E = K$:

(a) The functor $\widehat{H} : \widehat{\mathcal{B}}\text{-Mod} \longrightarrow \widehat{\mathcal{D}}\text{-Mod}$ controls the K-dimension.
(b) For any $M \in F_{\xi_\Lambda}^{-1}\Xi_{\widehat{\Lambda}}^{-1}(\mathcal{P}^2(\widehat{\Lambda}))$ with $\dim_K M \leq e$, there exists $N \in \widehat{\mathcal{B}}\text{-Mod}$ with $\widehat{H}(N) \cong M$ in $\widehat{\mathcal{D}}\text{-Mod}$.
(c) $\widehat{H}(N) \in F_{\xi_\Lambda}^{-1}\Xi_{\widehat{\Lambda}}^{-1}(\mathcal{P}^2(\widehat{\Lambda}))$, for any $N \in \widehat{\mathcal{B}}\text{-Mod}$.

Now, apply to the functor $H : \mathcal{B}\text{-Mod} \longrightarrow \mathcal{D}\text{-Mod}$ the same procedure applied to F in the proof of (32.10) to construct a minimal K-ditalgebra $\mathcal{L}$ and a functor

$$G := \widehat{H}F^l : \mathcal{L}\text{-Mod} \longrightarrow \widehat{\mathcal{D}}\text{-Mod}$$

which satisfy:

(α) The functor G is full, faithful and controls K-dimension.
(β) $G : (\mathcal{L}\text{-mod}, \widehat{\mathcal{E}}_{\mathcal{L}}) \longrightarrow (\widehat{\mathcal{D}}\text{-mod}, \mathcal{E}_{\widehat{\mathcal{D}}})$ preserves almost split conflations.
(γ) Almost any $M \in F_{\xi_\Lambda}^{-1}\Xi_{\widehat{\Lambda}}^{-1}(\mathcal{P}^2(\widehat{\Lambda}))$, with $\dim_K M \leq e$ has the form $M \cong G(J(z, n, \pi))$, for some indecomposable $\mathcal{L}$-module $J(z, n, \pi)$.

Consider also the functor $G^{[K]}$ defined as the composition

$$\mathcal{L}\text{-Mod} \xrightarrow{G} \widehat{\mathcal{D}}\text{-Mod} \xrightarrow{F_{\xi_\Lambda}} \mathcal{D}^{\widehat{\Lambda}}\text{-Mod} \xrightarrow{\Xi_{\widehat{\Lambda}}} \mathcal{P}^1(\widehat{\Lambda}) \xrightarrow{\text{Cok}} \widehat{\Lambda}\text{-Mod}.$$

Then, follow the proof of part (2) of (29.9) to see that $G^{[K]}$ is full, reflects isomorphisms and preserves indecomposability; that the composition

$$L\text{-Mod} \xrightarrow{L_{\mathcal{L}}} \mathcal{L}\text{-Mod} \xrightarrow{G^{[K]}} \widehat{\Lambda}\text{-Mod}$$

is naturally isomorphic to $\widehat{T} \otimes_{\widehat{A}} (F_{\xi_\Lambda}G)(L) \otimes_L -$, where $\widehat{T}$ is the transition bimodule associated to $\widehat{\Lambda}$ (see (22.18)), and that almost any indecomposable $\widehat{\Lambda}$-module M with $\dim_K M \leq d$ is isomorphic to $G^{[K]}(J(z, n, \pi))$, for some indecomposable $J(z, n, \pi)$ of $\mathcal{L}$-Mod. Furthermore

$$G^{[K]} : (\mathcal{L}\text{-mod}, \widehat{\mathcal{E}}_{\mathcal{L}}) \longrightarrow \widehat{\Lambda}\text{-mod}$$

preserves almost split conflations, because of (β), (19.10), (18.13) and (18.12).

Now, for each marked vertex z of $\mathcal{L}$, consider the rational K-algebra $\Gamma(z) := Le_z$. Then, the canonical embedding $\Gamma(z) \otimes_{\Gamma(z)} - : \Gamma(z)\text{-mod} \longrightarrow L\text{-mod}$ preserves indecomposables, isoclasses (in fact it is full and faithful), and almost split sequences. Clearly, each indecomposable $J(z, n, \pi)$ of L-mod is isomorphic to $\Gamma(z) \otimes_{\Gamma(z)} \Gamma(z)/(\pi)^n$. Finally, consider the $\widehat{\Lambda}$-$\Gamma(z)$-bimodules

$$T(z) := \widehat{T} \otimes_{\widehat{A}} (F_{\xi_\Lambda}G)(L) \otimes_L \Gamma(z).$$

Then, the finite family of rational K-algebras $\Gamma(z)$ and bimodules $T(z)$, for z a marked vertex of $\mathcal{L}$, satisfies the requirements of (32.11). □

Exercise 32.12. *Let $\mathcal{A} = (T, \delta)$ be a seminested ditalgebra. Assume $\alpha : i \longrightarrow j$ is a solid arrow between non-marked points of of $\mathcal{A}$ with $\delta(\alpha) = 0$. Suppose that $M \in \mathcal{A}$-mod satisfies $\mathrm{Ext}_{\mathcal{A}}(M, M) = 0$. Show that $i \neq j$ implies that the map $M_\alpha : M_i \longrightarrow M_j$ is injective or surjective, while $i = j$ implies that $M_i = 0$. As a consequence, if Λ is a basic hereditary algebra over an algebraically closed field and $M \in \Lambda$-mod satisfies that $\mathrm{Ext}_\Lambda(M, M) = 0$, then M_α is either surjective or injective, for any arrow α of the quiver of Λ.*

Exercise 32.13. *Let $\mathcal{A}$ be a seminested ditalgebra with layer (R, W). We will say that $\mathcal{A}$ is acceptable iff $\mathcal{A}$ is R_0-acceptable, where R_0 is the trivial part of the minimal algebra R. Thus, if $\mathcal{A}$ is acceptable, then the exact structure $\widehat{\mathcal{E}}_{\mathcal{A}} := \mathcal{E}_{\mathcal{A},R_0}$ consists of the composable pairs of morphisms $M \xrightarrow{f} E \xrightarrow{g} N$ in $\mathcal{A}$-Mod such that $gf = 0$ and*

$$0 \longrightarrow M \xrightarrow{f^0} E \xrightarrow{g^0} N \longrightarrow 0$$

is an exact sequence in R-Mod. Show that:

(1) Assume $\xi : \mathcal{A} \longrightarrow \mathcal{A}'$ is a morphism of seminested ditalgebras (not necessarily compatible with the layers) such that $\xi(V)$ generates V' and such that $F_\xi : \mathcal{A}'$-Mod$\longrightarrow \mathcal{A}$-Mod is full, faithful and its image is closed under isomorphisms. Then, $\mathcal{A}'$ is acceptable whenever $\mathcal{A}$ is so. In particular, $\mathcal{A}^d$, $\mathcal{A}^a$, $\mathcal{A}^l$ and $\mathcal{A}^s$ are acceptable if $\mathcal{A}$ is acceptable.
(2) $\mathcal{A}^a$ or $\mathcal{A}^r$ acceptable imply that $\mathcal{A}$ is acceptable too.

Exercise 32.14. *Let $\mathcal{A}$ be a seminested ditalgebra with layer (R, W). Consider the class $\widehat{\mathcal{E}}_{\mathcal{A}}$ of composable pairs of morphisms $M \xrightarrow{f} E \xrightarrow{g} N$ in $\mathcal{A}$-Mod such that $gf = 0$ and*

$$0 \longrightarrow M \xrightarrow{f^0} E \xrightarrow{g^0} N \longrightarrow 0$$

is an exact sequence in R-Mod. Assume that $\widehat{\mathcal{E}}_{\mathcal{A}}$ is an exact structure on $\mathcal{A}$-Mod. Notice that this is the case if $\mathcal{A}$ is acceptable, as in (32.13). Show that any finite-dimensional $\widehat{\mathcal{E}}_{\mathcal{A}}$-projective or $\widehat{\mathcal{E}}_{\mathcal{A}}$-injective $\mathcal{A}$-module M satisfies that $M_t = 0$, for all marked point t of $\mathcal{A}$.

Exercise 32.15. *Recall that, given a ditalgebra $\mathcal{A}$, the radical* $\mathrm{rad}_{\mathcal{A}}$ *of the category $\mathcal{A}$-mod is an ideal of the category $\mathcal{A}$-mod. For each $M, N \in \mathcal{A}$-mod, the space* $\mathrm{rad}_{\mathcal{A}}(M, N)$ *is by definition the collection of elements $f \in \mathrm{Hom}_{\mathcal{A}}(M, N)$ such that $1 - gf$ is invertible, for any $g \in \mathrm{Hom}_{\mathcal{A}}(N, M)$, see [32](3.2). For $m \geq 2$, the mth power $\mathrm{rad}^m_{\mathcal{A}}$ of $\mathrm{rad}_{\mathcal{A}}$ is defined, recursively, for $M, N \in \mathcal{A}$-mod, as the subspace $\mathrm{rad}^m_{\mathcal{A}}(M, N)$ of $\mathrm{rad}_{\mathcal{A}}(M, N)$ consisting of finite sums of the form $\sum_i h_i f_i$, where $f_i \in \mathrm{rad}_{\mathcal{A}}(M, N_i)$ and $h_i \in \mathrm{rad}^{m-1}_{\mathcal{A}}(N_i, N)$. We agree that*

$\mathrm{rad}^1_{\mathcal{A}} = \mathrm{rad}_{\mathcal{A}}$. *Finally, let us consider the intersection* $\mathrm{rad}^{\infty}_{\mathcal{A}} := \cap_{m=1}^{\infty}\mathrm{rad}^m_{\mathcal{A}}$. *For* $M, N \in \mathcal{A}$-mod, *show that:*

(1) If $\mathcal{A}$ *is a ditalgebra with layer* (R, W), *where* W *is a finitely generated* R-R*-bimodule, then* $\mathrm{rad}^{\infty}_{\mathcal{A}}(M, N) = \mathrm{rad}^{s}_{\mathcal{A}}(M, N)$, *for some* $s \geq 1$.
(2) If $\mathcal{A}$ *is a minimal ditalgebra and* $U_{\mathcal{A}}$ *denotes the forgetful functor, then* $U_{\mathcal{A}}[\mathrm{rad}_{\mathcal{A}}(M, N)] \subseteq \mathrm{rad}_R(M, N)$.

Exercise 32.16. *Let* $\mathcal{A}$ *be a minimal ditalgebra and* $M \in \mathcal{A}$-mod *an indecomposable module of type* $M \cong J(z, m, \pi)$ *(see (23.8), we say that another indecomposable module* M' *has the same type of* M *iff* $M' \cong J(z, m', \pi)$, *for some natural* m'*). Show that:*

(1) Given $f \in \mathrm{Hom}_{\mathcal{A}}(E, M)$, f *is an irreducible morphism in* $\mathcal{A}$-mod *iff there exists some morphism* $f' : E' \longrightarrow M$ *such that* $(f, f') : E \oplus E' \longrightarrow M$ *is a minimal right almost split morphism.*
(2) If N *is an indecomposable module in* $\mathcal{A}$-mod *and* $f \in \mathrm{rad}^n_{\mathcal{A}}(N, M)$, *with* $n \geq 2$, *then there is a natural number* $s \geq 1$, *indecomposable modules* $M_1, \ldots, M_s$ *of the same type of* M, *morphisms* $g_i \in \mathrm{rad}_{\mathcal{A}}(N, M_i)$ *and* $h_i \in \mathrm{Hom}_{\mathcal{A}}(M_i, M)$ *with each* g_i *a sum of compositions of* $n - 1$ *irreducible morphisms between indecomposable modules of the same type of* M *such that* $f = \sum_{i=1}^{s} h_i g_i$.
(3) Show that $\mathrm{rad}^{m+n+1}_{\mathcal{A}}(J(z, m, \pi), J(z, n, \pi)) = 0$.

Hint: Apply the arguments of [3]V(5.3) and [3]V(7.4) to the almost split conflations described in (32.3). Part (3) follows from (1), (2) and the fact that any composition of $m + n$ *irreducible maps from* $J(z, m, \pi)$ *to* $J(z, n, \pi)$ *between indecomposables of the same type in* $\mathcal{A}$-mod *is zero.*

Exercise 32.17. *Assume that* $\mathcal{A}$ *is a minimal ditalgebra and take any indecomposable modules* $M, N \in \mathcal{A}$-mod. *Show that whenever* $f = (f^0, f^1) \in \mathrm{rad}^{\infty}_{\mathcal{A}}(M, N)$, *we have that* $f^0 = 0$.

Exercise 32.18. *Let* $\mathcal{A}$ *be a minimal ditalgebra and let* $f = (f^0, 0) \in \mathrm{Hom}_{\mathcal{A}}(M, N)$ *be such that* f^0 *is surjective. Assume that* $h = (0, h^1) \in \mathrm{Hom}_{\mathcal{A}}(L, N)$. *Show that there is a morphism* $g \in \mathrm{Hom}_{\mathcal{A}}(L, M)$ *with* $h = fg$. *State and prove the dual statement.*

Exercise 32.19. *Let* $\mathcal{A}$ *be a minimal ditalgebra and let* z *be a marked point of* $\mathcal{A}$. *Assume that* $M, N \in \mathcal{A}$-mod *are indecomposable modules such that* $M_z \neq 0$ *or* $N_z \neq 0$. *Then, given any* $f \in \mathrm{Hom}_{\mathcal{A}}(M, N)$, *we have that* $f \in \mathrm{rad}^{\infty}_{\mathcal{A}}(M, N)$ *iff* $f^0 = 0$.

Hint. Use (32.17) and (32.18).

Exercise 32.20. *Let $\mathcal{A}$ be a minimal ditalgebra. Given non-marked vertices i_0, j_0 of $\mathcal{A}$, we can consider the one-dimensional $\mathcal{A}$-modules S_{i_0} and S_{j_0} and, for each arrow $v \in \mathbb{B}_1$ from i_0 to j_0, we will denote by f_v the morphism $f_v = (0, f_v^1) \in \mathrm{Hom}_{\mathcal{A}}(S_{i_0}, S_{j_0})$ defined by $f_v^1(v) = 1_k$ and $f_v^1(w) = 0$, for any other dotted arrow $w \in \mathbb{B}_1 \setminus \{v\}$. Show that:*

(1) If $v_1, \ldots, v_m$ are the different dotted arrows from i_0 to j_0 in $\mathcal{A}$, then the morphisms $f_{v_1}, \ldots, f_{v_m}$ generate the vector space $\mathrm{rad}_{\mathcal{A}}(S_{i_0}, S_{j_0})$.
(2) If $\delta(v) = 0$, then f_v is an irreducible morphism of $\mathcal{A}$-Mod.
(3) If $v_1, \ldots, v_m$ are different dotted arrows with zero differential, from i_0 to j_0 in $\mathcal{A}$, then the irreducible morphisms $f_{v_1}, \ldots, f_{v_m}$ are linearly independent modulo the radical squared $\mathrm{rad}^2_{\mathcal{A}}(S_{i_0}, S_{j_0})$.
(4) Consider the following example: $\mathcal{A}$ is the minimal ditalgebra with different points $i_0, i_1, \ldots, i_n, j_0$ with $n \geq 1$; dotted arrows $v_s : i_0 \dashrightarrow j_0$, for $s \in [1, m]$ with $m \geq 2$, $u_t : i_0 \dashrightarrow i_t$, for $t \in [1, n]$, and $w_t : i_t \dashrightarrow j_0$, for $t \in [1, n]$, where, $\delta(w_t) = 0$, $\delta(u_t) = 0$ and $\delta(v_1) = \delta(v_2) = \cdots = \delta(v_m) = \sum_t \mu_t w_t u_t$, for some $\mu_1, \ldots, \mu_n \in k$. Then, $f_{v_1}, \ldots, f_{v_m}$ are irreducible morphisms in $\mathcal{A}$-Mod, but they are linearly dependent modulo the radical squared whenever $(\mu_1, \ldots, \mu_n) \neq 0$. Moreover, if we consider the change of basis $v'_1 := v_1, v'_2 := v_2 - v_1, \ldots, v'_m := v_m - v_1$, then $f_{v'_2}, \ldots, f_{v'_m}$ determines a basis of $(\mathrm{rad}_{\mathcal{A}}/\mathrm{rad}^2_{\mathcal{A}})(S_{i_0}, S_{j_0})$.
(5) Suppose now that $\mathcal{A}$ is a trivial ditalgebra (that is minimal with trivial R in its layer). Given two dotted arrows v and v' from i_0 to j_0 in $\mathcal{A}$, we say that v and v' have disjoint differentials iff we can write $\delta(v) = \sum_i \mu_i w_i u_i$ and $\delta(v') = \sum_j \mu'_j w'_j u'_j$, where $w_i, u_i, w'_j, u'_j \in \mathbb{B}_1$ and $\mu_i, \mu'_j \in k$, with disjoint sets $\{(w_i, u_i) \mid i\}$ and $\{(w'_j, u'_j) \mid j\}$. Assume that any pair of dotted arrows from i_0 to j_0 in $\mathcal{A}$ have disjoint differentials. Then, the collection $\{f_v\}_v$, where v runs in the set of dotted arrows from i_0 to j_0 with zero differential induces a basis of $(\mathrm{rad}_{\mathcal{A}}/\mathrm{rad}^2_{\mathcal{A}})(S_{i_0}, S_{j_0})$.

Exercise 32.21. *Suppose that $\mathcal{A}$ is a trivial ditalgebra. Show that, for each indecomposable $S \in \mathcal{A}$-mod, there is some natural number n with $\mathrm{rad}^n_{\mathcal{A}}(S, -) = 0$ and $\mathrm{rad}^n_{\mathcal{A}}(-, S) = 0$.*

References. Statements (32.5) and (32.7) are taken from [13]. (32.19) comes from [15]. Statement (32.6) is a scalar extended version of [13](12.2). Statement (32.11) is a scalar extended version of one of the main results of [19].

33

Varieties of modules over ditalgebras

In this section, we present a resumé of the basic geometric notions needed for the proof of (27.9), (22.4) and a useful property of the quadratic forms of tame nested ditalgebras. For the first part, we can consider semipresented ditalgebras, but for the more refined geometric remarks on varieties of modules over ditalgebras, we will have to restrict ourselves to presented ditalgebras. We assume that our ground field k is algebraically closed and consider only algebraic varieties which are locally affine spaces which satisfy the Hausdorff Axiom, in the sense of [33]. The reader may use [40] as a general detailed reference for the rudiments on varieties of modules for algebras.

We start with the general setting.

Definition 33.1. *A left action of the irreducible variety $\mathbb{H}$ on the variety X is a triad of morphisms of varieties*

$$\begin{cases} \mathbb{H} \times X \longrightarrow X, & (h, x) \mapsto hx; \\ \mathbb{H} \times \mathbb{H} \times X \longrightarrow \mathbb{H}, & (h, g, x) \mapsto h \cdot_x g; \\ \mathbb{H} \times X \longrightarrow \mathbb{H}, & (h, x) \mapsto h^{-x}, \end{cases}$$

satisfying $(h \cdot_x g)x = h(gx)$ and $h^{-x}(hx) = x$, for all $x \in X$ and $h, g \in \mathbb{H}$. Abusing the language, we refer to the action with a reference to the first map, assuming that the second one, which "provides pointwise products", and the third one, which "provides pointwise inversions", are attached to the first.

Remark 33.2. *Some properties of actions of algebraic groups can be shown, using essentially the same argument, for general left actions. We have, for instance:*

(1) The relation in the set X defined by $x \sim y$ iff $hx = y$, for some $h \in \mathbb{H}$, is an equivalence relation in X. For each $x \in X$, its equivalence class $\mathbb{H}x = \{hx \mid h \in \mathbb{H}\}$ will be called the orbit of x under the action of $\mathbb{H}$.
(2) Each orbit $\mathbb{H}x$ is irreducible and constructible. The irreducible components of X are unions of orbits.
(3) If $W \subseteq X$ and U is an open subset of $\overline{\mathbb{H}W}$, which is contained in $\mathbb{H}W$, then $\mathbb{H}U$ is an open subset of $\overline{\mathbb{H}W}$ too.
(4) Each orbit $\mathbb{H}x$ is open in its closure $\overline{\mathbb{H}x}$.
(5) $\overline{\mathbb{H}x} \setminus \mathbb{H}x$ is a union of orbits and $\dim(\overline{\mathbb{H}x} \setminus \mathbb{H}x) < \dim \mathbb{H}x$.

Definition 33.3. *A left action $\mathbb{H} \times X \longrightarrow X$ of the irreducible variety $\mathbb{H}$ on the variety X is called a left grouplike action if it satisfies the following:*

(1) $h \cdot_x (g \cdot_x f) = (h \cdot_{fx} g) \cdot_x f$, for all $x \in X$ and all $f, g, h \in \mathbb{H}$.
(2) There is $I \in \mathbb{H}$ such that $I \cdot_x h = h = h \cdot_x I$, for all $x \in X$ and $h \in \mathbb{H}$.
(3) $h^{-x} \cdot_x h = I = h \cdot_{hx} h^{-x}$, for all $x \in X$ and $h \in \mathbb{H}$.

Lemma 33.4. *Given a left grouplike action $\mathbb{H} \times X \longrightarrow X$, $x \in X$ and $f, g, h \in \mathbb{H}$, we have:*

(1) If $h \cdot_x f = h \cdot_x g$ and $fx = gx$, then $f = g$.
(2) If $g \cdot_x h = f \cdot_x h$, then $f = g$.
(3) If $f^{-x} = g^{-x}$ and $fx = gx$, then $f = g$.
(4) If $\mathrm{Stab}_{\mathbb{H}}(x) := \{h \in \mathbb{H} \mid hx = x\}$, then $\dim \mathbb{H}x = \dim \mathbb{H} - \dim \mathrm{Stab}_{\mathbb{H}}(x)$.

Proposition 33.5. *Let $\mathbb{H} \times X \longrightarrow X$ be a left grouplike action of the irreducible variety $\mathbb{H}$ on the variety X. Then:*

(1) If $W \subseteq X$ is an irreducible subvariety, then:
  (1.a) $\dim \mathbb{H}W + \dim(W \cap \mathbb{H}w) \geq \dim W + \dim \mathbb{H}w$, for each $w \in W$.
  (1.b) $\dim \mathbb{H}W + \dim(W \cap \mathbb{H}u) \leq \dim W + \dim \mathbb{H}u$, for each $u \in U$, where U is some non-empty open subset of $\overline{W}$ contained in W.
(2) Let W and T be subvarieties of X such that $\mathbb{H}T = X$ and that there exists a non-negative integer t such that $\dim(W \cap \mathbb{H}w) \leq t$, for each $w \in W$, then

$$\dim T \geq \dim W - t.$$

Remark 33.6. *In the following we will construct two varieties $\mathbb{H}$ and X, together with right and left actions of $\mathbb{H}$ on X, such that*

$$h(xh) = x = (hx)h, \quad \text{for all } x \in X \text{ and } h \in \mathbb{H}.$$

This implies that $\mathbb{H}$ acts on X by automorphisms, this means that $x \mapsto hx$ defines an automorphism of the variety X, for all $h \in \mathbb{H}$.

Notation 33.7. *Throughout this section, we fix a semipresented ditalgebra $\mathcal{A} = (T, \delta)$, over our algebraically closed field k. Adopt the notation of (27.7). Hence, $\mathcal{A}$ is a Roiter ditalgebra, with layer (R, W), such that:*

(1) R is a finitely generated k-algebra, which admits a decomposition as a finite product of indecomposable k-algebras, hence its unit decomposes as a finite sum $1 = \sum_{i=1}^{n} e_i$ of central primitive orthogonal idempotents.
(2) There is a set filtration, of some finite directed subsets $\mathbb{B}_0 \subseteq W_0$,

$$\emptyset = \mathbb{B}_0^0 \subseteq \mathbb{B}_0^1 \subseteq \cdots \subseteq \mathbb{B}_0^{\ell_0} = \mathbb{B}_0,$$

such that each $\mathbb{B}_0^j$ generates the R-R-bimodule W_0^j, for $j \in [0, \ell_0]$, where

$$0 = W_0^0 \subseteq W_0^1 \subseteq \cdots \subseteq W_0^{\ell_0} = W_0$$

is the triangular filtration of W_0, as in (5.1).
(3) The R-R-bimodule W_1 is freely generated by the finite directed subset $\mathbb{B}_1$.

This notation, together with the notations introduced in (33.8), (33.9), (33.10) and so forth, will remain fixed until the end of this section.

In the following, whenever we add the assumption that $\mathcal{A}$ is presented, we will adopt all the notation of (27.7). Thus there will be a set filtration of the given directed subset $\mathbb{B}_1 \subseteq W_1$, with finite directed subsets

$$\emptyset = \mathbb{B}_1^0 \subseteq \mathbb{B}_1^1 \subseteq \cdots \subseteq \mathbb{B}_1^{\ell_1} = \mathbb{B}_1,$$

such that each $\mathbb{B}_1^j$ generates the R-R-bimodule W_1^j, for $j \in [0, \ell_1]$, where

$$0 = W_1^0 \subseteq W_1^1 \subseteq \cdots \subseteq W_1^{\ell_1} = W_1$$

is the triangular filtration of W_1, as in (5.1).

Definition 33.8. *Consider a directed set $\mathbb{L}$ of R such that $\mathbb{L} \cup \{e_1, \ldots, e_n\}$ generates the k-algebra R. Consider the quiver Q_A with n points $\{1, \ldots, n\}$ and arrows $\mathbb{L} \cup \mathbb{B}_0$. By definition, the arrow α has source vertex $s(\alpha)$ and target point $t(\alpha)$. Then there is a surjective morphism of k-algebras $\psi : kQ_A \longrightarrow A_\mathcal{A}$, with: $\psi(\tau_i) = e_i$, where τ_i denotes the trivial path of Q_A at the vertex i; $\psi(\alpha) = \alpha$, for $\alpha \in \mathbb{B}_0$; and $\psi(r) = r$, for all $r \in \mathbb{L}$. Then, we have a presentation of $A_\mathcal{A}$ as a quiver algebra with relations $\mathrm{Ker}\,\psi = I_\mathcal{A}$. Hence, each object in $\mathcal{A}$-mod can be seen as a representation of the quiver Q_A bounded by a directed set of relations $I_\mathcal{A}^\tau$. Here, we mean directed relative to the canonical decomposition of the unit $1 = \sum_{i=1}^{n} \tau_i$ of the quiver algebra kQ_A, and we can take $I_\mathcal{A}^\tau = \cup_{i,j} \tau_j I_\mathcal{A} \tau_i$.*

Given a dimension vector $\underline{d} \in \mathbb{N}^n$, *consider the affine space*

$$\mathbb{A}_{\underline{d}} = \prod_{r \in \mathbb{L}} k^{(d_{t(r)} \times d_{s(r)})} \times \prod_{\alpha \in \mathbb{B}_0} k^{(d_{t(\alpha)} \times d_{s(\alpha)})}.$$

In this section, $\mathbb{N}$ *denotes the set of natural numbers, including* 0, $\mathbb{N}^* = \mathbb{N} \setminus \{0\}$, *and we adopt the standard conventions on empty matrices with zero rows or zero columns. Having this in mind, we can define*

$$\operatorname{mod}_{\mathcal{A}}(\underline{d}) = \{M \in \mathbb{A}_{\underline{d}} \mid \rho(M) = 0, \text{ for all } \rho \in I^{\tau}_{\mathcal{A}}\}.$$

Remark 33.9. *There is a bijective correspondence between the set of modules of the ditalgebra* $\mathcal{A}$ *with dimension vector* $\underline{d}$ *and underlying vectorspaces* $\{k^{d_i}\}$ *(which satisfy the relations* $I^{\tau}_{\mathcal{A}}$*) and the closed affine space* $\operatorname{mod}_{\mathcal{A}}(\underline{d})$ *of* $\mathbb{A}_{\underline{d}}$. *We abuse of the language and denote with the same symbol each point* M *in* $\operatorname{mod}_{\mathcal{A}}(\underline{d})$ *and the representation (or module) it determines.*

If we proceed as before, considering only the set $\mathbb{L}$ *as set of arrows, we get a quiver* Q_R, *and a presentation of the algebra* R *as a quotient of the quiver algebra* kQ_R *by some ideal* I_R. *Thus we can consider the closed affine subspace* $\operatorname{mod}_R(\underline{d})$ *in* $\prod_{r \in \mathbb{L}} k^{(d_{t(r)} \times d_{s(r)})}$. *We can also consider the algebraic group* $\mathbb{G}_{\underline{d}} = \prod_{i=1}^n \mathbb{G}_{d_i}$, *where* $\mathbb{G}_{d_i}$ *denotes the general linear group of* $d_i \times d_i$ *invertible matrices over the field* k. *Then* $\mathbb{G}_{\underline{d}}$ *acts on* $\operatorname{mod}_R(\underline{d})$ *with receipt* $\{H_i\}\{M_r\} = \{H_{t(r)} M_r H^{-1}_{s(r)}\}$. *Moreover, we know that two* $\underline{d}$*-dimensional* R*-modules are isomorphic iff their corresponding points in* $\operatorname{mod}_R(\underline{d})$ *belong to the same orbit under the action of* $\mathbb{G}_{\underline{d}}$.

Definition 33.10. *In order to handle the differential* δ, *it is convenient to extend the quiver* $Q_{\mathcal{A}}$ *to a new quiver* $\widetilde{Q}_{\mathcal{A}}$ *by adding the new set of arrows* $\mathbb{B}_1$. *As before, by definition, the arrow* x *has source point* $s(x)$ *and target point* $t(x)$. *Then* $T \cong T_R(W)$ *is identified with a quotient of the quiver algebra* $k\widetilde{Q}_{\mathcal{A}}$ *modulo some ideal* $\widetilde{I}_{\mathcal{A}}$.

The triangularity of the ditalgebra at level 0 *means that there is an ordering "*$<$*" of the set* $\mathbb{B}_0$ *such that, for all* $\alpha \in \mathbb{B}_0$

$$\delta(\alpha) \equiv \sum_{x \in \mathbb{B}_1} f^{\alpha}_x(\{r\}_{r \in \mathbb{L}}, \{\beta\}_{\beta < \alpha}) x g^{\alpha}_x(\{r\}_{r \in \mathbb{L}}, \{\beta\}_{\beta < \alpha}),$$

modulo the ideal $\widetilde{I}_{\mathcal{A}}$, *where* f^{α}_x *and* g^{α}_x *are directed elements in* $kQ_{\mathcal{A}}$.

In case $\mathcal{A}$ *is presented, the triangularity of the ditalgebra at level 1 means that there is an ordering "*$<$*" of the set* $\mathbb{B}_1$ *such that, for all* $x \in \mathbb{B}_1$

$$\delta(x) \equiv \sum_{y,z<x} f^x_{y,z}(\{r\}_{r \in \mathbb{L}}, \{\beta\}_{\beta \in \mathbb{B}_0}) y g^x_{y,z}(\{r\}_{r \in \mathbb{L}}, \{\beta\}_{\beta \in \mathbb{B}_0}) z h^x_{y,z}(\{r\}_{r \in \mathbb{L}}, \{\beta\}_{\beta \in \mathbb{B}_0}),$$

modulo the ideal $\widetilde{I}_{\mathcal{A}}$, *where* $f^x_{y,z}$, $g^x_{y,z}$ *and* $h^x_{y,z}$ *are directed elements in* $kQ_{\mathcal{A}}$.

Given a dimension vector $\underline{d} \in \mathbb{N}^n$*, consider the affine space*

$$\mathbb{M}_{\underline{d}} = \prod_{i=1}^{n} k^{(d_i \times d_i)} \times \prod_{x \in \mathbb{B}_1} k^{(d_{t(x)} \times d_{s(x)})},$$

and its open subspace $\mathbb{H}_{\underline{d}} = \mathbb{G}_{\underline{d}} \times \prod_{x \in \mathbb{B}_1} k^{(d_{t(x)} \times d_{s(x)})}$*. Notice that*

$$h_{\mathcal{A}}(\underline{d}) := \dim \mathbb{M}_{\underline{d}} = \dim \mathbb{H}_{\underline{d}} = \sum_{i=1}^{n} d_i^2 + \sum_{x \in \mathbb{B}_1} (d_{t(x)} \times d_{s(x)}).$$

We will define the morphism of varieties $\phi : \mathbb{H}_{\underline{d}} \times \mathbb{A}_{\underline{d}} \longrightarrow \mathbb{A}_{\underline{d}}$ *as follows. Let* $h = (\{H_i\}, \{H_x\}) \in \mathbb{H}_{\underline{d}}$ *and* $M = (\{M_r\}, \{M_\alpha\}) \in \mathbb{A}_{\underline{d}}$*, then* $\phi(h, M) = (\{N_r\}, \{N_\alpha\})$*, where*

$$N_r = H_{t(r)} M_r H_{s(r)}^{-1}, \text{ for all } r \in \mathbb{L}, \text{ and recursively, for } \alpha \in \mathbb{B}_0,$$
$$N_\alpha = H_{t(\alpha)} M_\alpha H_{s(\alpha)}^{-1} + \sum_{x \in \mathbb{B}_1} f_x^\alpha(\{N_r\}_{r \in \mathbb{L}}, \{N_\beta\}_{\beta < \alpha}) H_x g_x^\alpha(\{M_r\}_{r \in \mathbb{L}}, \{M_\beta\}_{\beta < \alpha}) H_{s(\alpha)}^{-1}.$$

We shall denote $\phi(h, M)$ *simply by* hM*. Notice that the element* $I_{\underline{d}} = (\{I_{d_i}\}, \{0\})$ *of* $\mathbb{H}_{\underline{d}}$ *satisfies* $I_{\underline{d}} M = M$*, for all* $M \in \mathbb{A}_{\underline{d}}$*.*

Lemma 33.11. *For any semipresented ditalgebra* $\mathcal{A}$ *and any dimension vector* $\underline{d} \in \mathbb{N}^n$*, the map* $\phi : \mathbb{H}_{\underline{d}} \times \mathbb{A}_{\underline{d}} \longrightarrow \mathbb{A}_{\underline{d}}$ *restricts to a morphism of varieties* $\mathbb{H}_{\underline{d}} \times \mathrm{mod}_{\mathcal{A}}(\underline{d}) \longrightarrow \mathrm{mod}_{\mathcal{A}}(\underline{d})$*.*

Proof. Take $M = (\{M_r\}, \{M_\alpha\}) \in \mathrm{mod}_{\mathcal{A}}(\underline{d})$ and $h = (\{H_i\}, \{H_x\}) \in \mathbb{H}_{\underline{d}}$. Let $N = hM = (\{N_r\}, \{N_\alpha\})$. We need to show that $\rho(N) = 0$, for all $\rho \in I_{\mathcal{A}}^{\tau}$. The $A_{\mathcal{A}}$-module corresponding to M, which we denote also by M, can be considered as an R-module; hence, it determines an element $M_0 = \{M_r\} \in \mathrm{mod}_R(\underline{d})$, the variety of representations of the algebra R. The point $N_0 = \{N_r\}$ is, by definition of N, in the same orbit of M_0 in $\mathrm{mod}_R(\underline{d})$ under the action of $\mathbb{G}_{\underline{d}}$. Hence, it determines an R-module N_0. Moreover, $h^0 = \{H_i\} \in \mathbb{G}_{\underline{d}}$ determines an isomorphism $M_0 \to N_0$ of R-modules. Since our bimodule W_1 is freely generated by $\mathbb{B}_1$ and each H_x determines an element in $\mathrm{Hom}_k(e_{s(x)} M_0, e_{t(x)} N_0) \cong e_{t(x)} \mathrm{Hom}_k(M_0, N_0) e_{s(x)}$, we obtain a morphism of R-bimodules $h^1 : W_1 \longrightarrow \mathrm{Hom}_k(M_0, N_0)$. Then, since $\mathcal{A}$ is a Roiter ditalgebra, there is a structure of $A_{\mathcal{A}}$-module L on N_0 such that $(h^0, h^1) : M \longrightarrow L$ is an isomorphism in $\mathcal{A}$-mod. This means precisely that the point in $\mathrm{mod}_{\mathcal{A}}(\underline{d})$ corresponding to L is N. □

Remark 33.12. *Let $\mathcal{A}$ be a semipresented ditalgebra and Δ any k-algebra. Then, given $\underline{d} \in \mathbb{N}^n$, a $\underline{d}$-ranked representation of the ditalgebra $\mathcal{A}$ over Δ, is a family*

$$Z = (\{Z_r\}, \{Z_\alpha\}) \in \prod_{r \in \mathbb{L}} \Delta^{(d_{t(r)} \times d_{s(r)})} \times \prod_{\alpha \in \mathbb{B}_0} \Delta^{(d_{t(\alpha)} \times d_{s(\alpha)})},$$

such that $\rho(Z) = 0$, for all $\rho \in \mathcal{I}^\tau_{\mathcal{A}}$.

(1) Recall that a rational algebra Γ is, by definition, a k-algebra of the form $\Gamma = k[x]_f$, for some polynomial $f \in k[x]$. We denote by $D(\Gamma)$ the set of points $\lambda \in k$, with $f(\lambda) \neq 0$. Given any $\underline{d}$-ranked representation Z of the ditalgebra $\mathcal{A}$ over Γ, we can evaluate the components of Z at any $\lambda \in D(\Gamma)$, to obtain $Z(\lambda) = (\{Z_r(\lambda)\}, \{Z_\alpha(\lambda)\}) \in \mathrm{mod}_{\mathcal{A}}(\underline{d})$. Any A-Γ-bimodule, free and finitely generated by the right, with rank vector $\underline{d}$, determines a $\underline{d}$-ranked representation Z and hence a curve

$$\begin{array}{ccc} D(\Gamma) & \xrightarrow{\gamma} & \mathrm{mod}_{\mathcal{A}}(\underline{d}) \\ \lambda & \mapsto & Z(\lambda). \end{array}$$

Then, the ditalgebra $\mathcal{A}$ is tame iff, for each dimension vector $\underline{d}$, there is a finite number of $\underline{d}$-ranked representations $Z^1, \ldots, Z^s$ of $\mathcal{A}$ over the rational k-algebras $\Gamma_1, \ldots, \Gamma_s$, respectively, such that, for all indecomposable $M \in \mathrm{mod}_{\mathcal{A}}(\underline{d})$, there are $i \in [1, s]$ and $\lambda \in E_i := D(\Gamma_i)$ such that $Z^i(\lambda) \cong M$ in $\mathcal{A}$-mod. Thus, if $\mathcal{A}$ is tame, for each dimension vector $\underline{d}$, there is a finite number of curves $\{\gamma_i : E_i \longrightarrow \mathrm{mod}_{\mathcal{A}}(\underline{d})\}_{i=1}^s$, defined on cofinite subsets E_i of k, such that every $\underline{d}$-dimensional indecomposable $\mathcal{A}$-module is represented by a point in $\cup_{i=1}^s \mathbb{H}_{\underline{d}} \gamma_i(E_i)$.

(2) Now, from (22.6), we know that any A-$k\langle x, y\rangle$-bimodule Z, which is finitely generated free by the right, determines an $\underline{r}$-ranked representation Z of $\mathcal{A}$ over $k\langle x, y\rangle$, where $\underline{r} \in \mathbb{N}^n$ is such that r_i is the rank of the free right $k\langle x, y\rangle$-module $e_i Z$. Then, the functor $Z \otimes_{k\langle x,y\rangle} -$ can be identified with the functor $F_Z : k\langle x, y\rangle\text{-mod} \rightarrow \mathcal{A}_{\mathcal{A}}\text{-mod}$, sending each M to $F_Z(M) = (\{e_i Z \otimes_{k\langle x,y\rangle} M\}_{i=1}^n, \{Z_r \otimes I_M\}, \{Z_\alpha \otimes I_M\})$, which is in fact isomorphic to the functor G_Z which maps each M onto $G_Z(M) = (\{M^{r_i}\}_{i=1}^n, \{Z_r(X, Y)\}, \{Z_\alpha(X, Y)\})$, where $Z_r(X, Y)$ (resp. $Z_\alpha(X, Y)$) is the matrix $Z_r(x, y)$ (resp. $Z_\alpha(x, y)$) evaluated at the morphisms X, Y determined by the action of x, y over M, respectively.

For any natural number $c \in \mathbb{N}$, the composition functor $L_{\mathcal{A}}(Z \otimes_{k\langle x,y\rangle} -)$ induces a morphism of varieties

$$\begin{array}{ccc} \mathrm{mod}_{k\langle x,y\rangle}(c) & \xrightarrow{\varphi_c} & \mathrm{mod}_{\mathcal{A}}(c\underline{r}) \\ (X, Y) & \mapsto & (\{Z_r(X, Y)\}, \{Z_\alpha(X, Y)\}), \end{array}$$

where $\mathrm{mod}_{k\langle x,y\rangle}(c) = k^{(c\times c)} \times k^{(c\times c)}$ is an affine irreducible variety.

We can now proceed with one of the missing proofs of Section 27.

Proof of (27.9). We have that two $\underline{d}$-dimensional modules $M, N \in \mathcal{A}$-mod are isomorphic iff their corresponding points in $\mathrm{mod}_{\mathcal{A}}(\underline{d})$ satisfy $N = hM$, for some $h \in \mathbb{H}_{\underline{d}}$. It follows that $M \in \mathbb{H}_{\underline{d}}N$ iff $N \in \mathbb{H}_{\underline{d}}M$; and $\mathbb{H}_{\underline{d}}(\mathbb{H}_{\underline{d}}M) \subseteq \mathbb{H}_{\underline{d}}M$.

By assumption and (33.12)(2), for some dimension vector $\underline{d} \in \mathbb{N}^n$, there is a morphism of varieties $\varphi : k^2 = \mathrm{mod}_{k\langle x,y\rangle}(1) \longrightarrow \mathrm{mod}_{\mathcal{A}}(\underline{d})$, induced by the functor $F := L_{\mathcal{A}}(Z \otimes_{k\langle x,y\rangle} -) : k\langle x, y\rangle\text{-mod} \longrightarrow \mathcal{A}\text{-mod}$. Since F preserves isomorphism classes of indecomposables, each $\varphi(\lambda, \mu)$ represents an indecomposable $\mathcal{A}$-module and $\varphi(\lambda, \mu) \notin \mathbb{H}_{\underline{d}}\varphi(\lambda', \mu')$, for all different pairs $(\lambda, \mu), (\lambda', \mu') \in k^2$. Consider, for each $\lambda \in k$, the curve $\varphi_\lambda : k \longrightarrow \mathrm{mod}_{\mathcal{A}}(\underline{d})$, defined by $\varphi_\lambda(\mu) = \varphi(\lambda, \mu)$, for $\mu \in k$.

From (33.12)(1), if $\mathcal{A}$ is tame, there is a finite number of curves $\{\gamma_i : E_i \longrightarrow \mathrm{mod}_{\mathcal{A}}(\underline{d})\}_{i=1}^m$, defined on cofinite subsets E_i of k, such that every $\underline{d}$-dimensional indecomposable $\mathcal{A}$-module is represented by a point in $\cup_{i=1}^m \mathbb{H}_{\underline{d}}\gamma_i(E_i)$.

Then, for each $\lambda \in k$, $\varphi_\lambda(k) \subseteq \cup_{i=1}^m \mathbb{H}_{\underline{d}}\gamma_i(E_i)$. It follows that $\varphi_\lambda(D^\lambda) \subseteq \mathbb{H}_{\underline{d}}\gamma_i(E_i)$, for some cofinite subset $D^\lambda \subseteq k$ and some i depending on λ. Then, $\gamma_i(E_i^\lambda) \subseteq \mathbb{H}_{\underline{d}}\varphi_\lambda(D^\lambda)$, for some cofinite subset $E_i^\lambda \subset E_i$. Since we are dealing with finitely many curves $\gamma_1, \ldots, \gamma_m$, then there is $\lambda' \neq \lambda$ such that $\varphi_{\lambda'}(D^{\lambda'}) \subseteq \mathbb{H}_{\underline{d}}\gamma_i(E_i)$, for the same i. Since E_i^λ is cofinite in E_i and $D^{\lambda'}$ is infinite, there exists $\mu \in D^{\lambda'}$ such that $\varphi_{\lambda'}(\mu) \in \mathbb{H}_{\underline{d}}\gamma_i(E_i^\lambda) \subseteq \mathbb{H}_{\underline{d}}\varphi_\lambda(D^\lambda)$. This entails a contradiction. □

From now on we assume furthermore that $\mathcal{A}$ is a presented ditalgebra. This will enable us to enrich the structure of the morphism $\phi : \mathbb{H}_{\underline{d}} \times \mathrm{mod}_{\mathcal{A}}(\underline{d}) \longrightarrow \mathrm{mod}_{\mathcal{A}}(\underline{d})$ to a left grouplike action of $\mathbb{H}_{\underline{d}}$ on $\mathrm{mod}_{\mathcal{A}}(\underline{d})$.

Lemma 33.13. *We have a morphism of varieties* $\mathbb{H}_{\underline{d}} \times \mathbb{H}_{\underline{d}} \times \mathrm{mod}_{\mathcal{A}}(\underline{d}) \longrightarrow \mathbb{H}_{\underline{d}}$, *which maps each triad formed by* $f = (\{F_i\}, \{F_x\}), h = (\{H_i\}, \{H_x\}) \in \mathbb{H}_{\underline{d}}$ *and* $M = (\{M_r\}, \{M_\alpha\}) \in \mathrm{mod}_{\mathcal{A}}(\underline{d})$ *onto* $h \cdot_M f = (\{G_i\}, \{G_x\})$ *defined by*

$$
\begin{aligned}
G_i &= H_i F_i, \text{ for all } i \in [1, n], \text{ and} \\
G_x &= H_{t(x)} F_x + H_x F_{s(x)} \\
&\quad + \sum_{y,z<x} f^x_{y,z}(h(fM)) H_y g^x_{y,z}(fM) F_z h^x_{y,z}(M), \text{ for all } x \in \mathbb{B}_1.
\end{aligned}
$$

It satisfies the formula $h(fM) = (h \cdot_M f)M$, *for all* h, f *and* M. *We also have that* $g \cdot_M (h \cdot_M f) = (g \cdot_{fM} h) \cdot_M f$, *for all* $M \in \mathrm{mod}_{\mathcal{A}}(\underline{d})$ *and all* $g, h, f \in \mathbb{H}_{\underline{d}}$. *Moreover, the element* $I_{\underline{d}} \in \mathbb{H}_{\underline{d}}$ *satisfies* $I_{\underline{d}} \cdot_M f = f = f \cdot_M I_{\underline{d}}$, *for all* $M \in \mathrm{mod}_{\mathcal{A}}(\underline{d})$ *and* $f \in \mathbb{H}_{\underline{d}}$.

Proof. In the category of representations of the ditalgebra $\mathcal{A}$, the families of matrices f and h determine isomorphisms of representations $f : M \longrightarrow fM$ and $h : fM \longrightarrow h(fM)$. Hence, we have the composition $h \cdot f = (g^0, g^1) : M \longrightarrow h(fM)$ in the same category. The corresponding family of matrices, associated to the family of linear transformations $(\{(g^0)_{|k^{d_i}}\}, \{g^1(x)_{|k^{d_i}}\})$, is precisely $h \cdot_M f$. Then $h(fM) = (h \cdot_M f)M$. □

Lemma 33.14. *There is a morphism of varieties $\mathbb{H}_{\underline{d}} \times \mathrm{mod}_{\mathcal{A}}(\underline{d}) \longrightarrow \mathbb{H}_{\underline{d}}$ which maps each pair formed by $h = (\{H_i\}, \{H_x\}) \in \mathbb{H}_{\underline{d}}$ and $M = (\{M_r\}, \{M_\alpha\}) \in \mathrm{mod}_{\mathcal{A}}(\underline{d})$ onto $h^{-M} = (\{G_i\}, \{G_x\})$ defined by*

$$G_i = H_i^{-1}, \text{ for all } i \in [1, n], \text{ and, recursively, for } x \in \mathbb{B}_1,$$

$$G_x = -H_{t(x)}^{-1} H_x H_{s(x)}^{-1} - \sum_{y,z<x} f_{y,z}^x(M) G_y g_{y,z}^x(hM) H_z h_{y,z}^x(M) H_{s(x)}^{-1}.$$

It satisfies the formula $h^{-M} \cdot_M h = I_{\underline{d}} = h \cdot_{hM} h^{-M}$, for all $M \in \mathrm{mod}_{\mathcal{A}}(\underline{d})$ and all $h \in \mathbb{H}_{\underline{d}}$. Hence, $h^{-M}(hM) = M$, for all h and M.

Proof. As before, in the category of representations of the ditalgebra $\mathcal{A}$, the family of matrices h determines an isomorphism of representations $h : M \longrightarrow hM$. Then, it has an inverse $h^{-1} : hM \longrightarrow M$. Hence, we have the composition $h^{-1} \cdot h = (Id, \{0\}) : M \longrightarrow M$ in the same category. It follows that the family of matrices corresponding to the morphism h^{-1} is precisely $h^{-M} = (\{G_i\}, \{G_x\})$, and $h^{-M} \cdot_M h = I_{\underline{d}} = h \cdot_{hM} h^{-M}$. □

Then, all the remarks at the begining of this section apply to the left grouplike action of $\mathbb{H}_{\underline{d}}$ on the variety $\mathrm{mod}_{\mathcal{A}}(\underline{d})$.

Remark 33.15. *If $\delta(x)$ is linear for all $x \in \mathbb{B}_1$, that is if $\delta(x) \equiv \sum_{y,z<x} \lambda_{y,z}^x yz$ modulo the ideal $\widetilde{I}_A$, for some $\lambda_{y,z}^x \in k$, then $h \cdot_M f$ and h^{-M} do not depend on M. Then, the properties defining a grouplike action in (33.3) imply that $\mathbb{H}_{\underline{d}}$ is an algebraic group with product $h \cdot_M f$, inverse h^{-M} and unit $I_{\underline{d}}$. Thus, in these cases, we have a usual algebraic group action on the variety $\mathrm{mod}_{\mathcal{A}}(\underline{d})$. They include all linear matrix problems (as defined in [22]). Matrix problems associated to partially ordered sets are examples of this situation (see Section 34).*

There are many examples of ditalgebras with non-linear $\delta(x)$, for some $x \in \mathbb{B}_1$, such that the map $h \mapsto h^{-M}$ is not an involution $\mathbb{H}_{\underline{d}} \to \mathbb{H}_{\underline{d}}$. Hence the triple which constitutes the grouplike action of $\mathbb{H}_{\underline{d}}$ on $\mathrm{mod}_{\mathcal{A}}(\underline{d})$ is not of the form produced by the action of an algebraic group structure.

We will define a right action of the variety $\mathbb{H}_{\underline{d}}$ on $\mathrm{mod}_{\mathcal{A}}(\underline{d})$. First define the morphism of varieties $\phi' : \mathbb{A}_{\underline{d}} \times \mathbb{H}_{\underline{d}} \longrightarrow \mathbb{A}_{\underline{d}}$ as follows.

Let $h = (\{H_i\}, \{H_x\}) \in \mathbb{H}_{\underline{d}}$ and $N = (\{N_r\}, \{N_\alpha\}) \in \mathbb{A}_{\underline{d}}$, then $\phi'(N, h) = (\{M_r\}, \{M_\alpha\})$, where

$$\begin{aligned} M_r &= H_{t(r)}^{-1} N_r H_{s(r)}, \text{ for all } r \in \mathbb{L}, \text{ and recursively, for } \alpha \in \mathbb{B}_0, \\ M_\alpha &= H_{t(\alpha)}^{-1} N_\alpha H_{s(\alpha)} \\ &\quad - \sum_{x \in \mathbb{B}_1} H_{t(\alpha)}^{-1} f_x^\alpha(\{N_r\}_{r\in\mathbb{L}}, \{N_\beta\}_{\beta<\alpha}) H_x g_x^\alpha(\{M_r\}_{r\in\mathbb{L}}, \{M_\beta\}_{\beta<\alpha}). \end{aligned}$$

Again, we denote $\phi'(N, h)$ simply by Nh.

Lemma 33.16. *The map* $\phi' : \mathbb{A}_{\underline{d}} \times \mathbb{H}_{\underline{d}} \longrightarrow \mathbb{A}_{\underline{d}}$ *restricts to a right grouplike action* $\mathrm{mod}_{\mathcal{A}}(\underline{d}) \times \mathbb{H}_{\underline{d}} \longrightarrow \mathrm{mod}_{\mathcal{A}}(\underline{d})$. *Moreover*

$$h(Nh) = N = (hN)h, \textit{ for all } N \in \mathrm{mod}_{\mathcal{A}}(\underline{d}) \textit{ and } h \in \mathbb{H}_{\underline{d}}.$$

Proof. We proceed as we did in the proof of (33.11). Take $N = (\{N_r\}, \{N_\alpha\}) \in \mathrm{mod}_{\mathcal{A}}(\underline{d})$ and $h = (\{H_i\}, \{H_x\}) \in \mathbb{H}_{\underline{d}}$. Let $M = Nh = (\{M_r\}, \{M_\alpha\})$. Consider the R-modules $M_0 = \{M_r\}$, $N_0 = \{N_r\} \in \mathrm{mod}_R(\underline{d})$. Then we have an isomorphism $h^0 : M_0 \to N_0$ in R-mod determined by the matrices $\{H_i\}$. Since our bimodule W_1 is freely generated by $\mathbb{B}_1$, and each H_x determines an element in $\mathrm{Hom}_k(e_{s(x)}M_0, e_{t(x)}N_0) \cong e_{t(x)}\mathrm{Hom}_k(M_0, N_0)e_{s(x)}$, we obtain a morphism of R-bimodules $h^1 : W_1 \longrightarrow \mathrm{Hom}_k(M_0, N_0)$. Then, since $\mathcal{A}$ is a Roiter ditalgebra, there is a structure of $A_{\mathcal{A}}$-module L on M_0 such that $(h^0, h^1) : L \to N$ is an isomorphism in $\mathcal{A}$-mod. This means precisely that the point in $\mathrm{mod}_{\mathcal{A}}(\underline{d})$ corresponding to L is Nh.

The pointwise products and the pointwise inversions corresponding to this right action are obtained as in (33.13) and (33.14), working on the right side.

Finally, since $Nh \xrightarrow{h} N$ and $N \xrightarrow{h} hN$ are isomorphisms in $\mathcal{A}$-mod, we have $N = h(Nh)$ and $N = (hN)h$. □

Lemma 33.17. *Given* $M, N \in \mathrm{mod}_{\mathcal{A}}(\underline{d})$, *the space of morphisms between* M *and* N *in the category* $\mathcal{A}$-mod *determines a linear subspace* $\mathrm{Hom}_{\mathcal{A}}(M, N)$ *in* $\mathbb{M}_{\underline{d}}$. *The automorphisms of* M *in* $\mathcal{A}$-mod *determine a dense open subspace* $\mathrm{Aut}_{\mathcal{A}}(M)$ *of* $\mathrm{End}_{\mathcal{A}}(M)$. *They are both irreducible varieties and share dimension. Moreover,* $\mathrm{Stab}_{\mathbb{H}_{\underline{d}}}(M) = \mathrm{Aut}_{\mathcal{A}}(M)$.

Proof. Assume that $M = (\{M_r\}, \{M_\alpha\})$ and $N = (\{N_r\}, \{N_\alpha\})$, then an element $h = (\{H_i\}, \{H_x\}) \in \mathbb{M}_{\underline{d}}$ determines a morphism $M \longrightarrow N$ in $\mathcal{A}$-mod *iff*

$$\begin{aligned} N_r H_{s(r)} &= H_{t(r)} M_r, \text{ for all } r \in \mathbb{L}, \text{ and, for all } \alpha \in \mathbb{B}_0, \\ N_\alpha H_{s(\alpha)} &= H_{t(\alpha)} M_\alpha + \sum_{x \in \mathbb{B}_1} f_x^\alpha(N) H_x g_x^\alpha(M). \end{aligned}$$

Then, these families h are bounden by a linear system of equations. The automorphisms $h = (\{H_i\}, \{H_x\})$ of M are elements of $\mathrm{End}_{\mathcal{A}}(M)$ with $\{H_i\} \in \mathbb{G}_{\underline{d}}$, hence they constitute an open subspace. □

Lemma 33.18. *Given dimension vectors $\underline{d}, \underline{d}' \in \mathbb{N}^n$, $M \in \mathrm{mod}_{\mathcal{A}}(\underline{d})$ and $N \in \mathrm{mod}_{\mathcal{A}}(\underline{d}')$, define $\langle M, N\rangle = \dim_k(\mathrm{Hom}_{\mathcal{A}}(M, N))$. Then, for each $t \in \mathbb{N}$, the set $\{(M, N) \mid \langle M, N\rangle \geq t\}$ is closed in $\mathrm{mod}_{\mathcal{A}}(\underline{d}) \times \mathrm{mod}_{\mathcal{A}}(\underline{d}')$.*

Lemma 33.19. *If we fix $M, N, L \in \mathrm{mod}_{\mathcal{A}}(\underline{d})$, the composition in the category $\mathcal{A}$-mod determines a bilinear map, hence a morphism of varieties*

$$\mathrm{Hom}_{\mathcal{A}}(M, N) \times \mathrm{Hom}_{\mathcal{A}}(N, L) \longrightarrow \mathrm{Hom}_{\mathcal{A}}(M, L), \qquad (f, h) \mapsto hf.$$

Where, by definition, if $f = (\{F_i\}, \{F_x\})$ and $h = (\{H_i\}, \{H_x\})$, then the components $(\{G_i\}, \{G_x\})$ of hf are given by

$$G_i = H_i F_i, \text{ for all } i \in [1, n], \text{ and}$$
$$G_x = H_{t(x)} F_x + H_x F_{s(x)} + \sum_{y,z<x} f^x_{y,z}(L) H_y g^x_{y,z}(N) F_z h^x_{y,z}(M), \text{ for all } x \in \mathbb{B}_1.$$

They satisfy the associativity formula $h(gf) = (hg)f$, whenever we have $f \in \mathrm{Hom}_{\mathcal{A}}(M, N)$, $g \in \mathrm{Hom}_{\mathcal{A}}(N, L)$ and $h \in \mathrm{Hom}_{\mathcal{A}}(L, T)$. Notice that, if $f, g \in \mathbb{H}_{\underline{d}}$, $f \in \mathrm{Hom}_{\mathcal{A}}(M, fM)$ and $g \in \mathrm{Hom}_{\mathcal{A}}(fM, g(fM))$, then $g \cdot_M f = gf$. In particular, $\mathrm{End}_{\mathcal{A}}(M)$ is an algebraic ring.

Lemma 33.20. *Given $M \in \mathrm{mod}_{\mathcal{A}}(\underline{d})$, the inversion in the category $\mathcal{A}$-mod determines a morphism of varieties*

$$\mathrm{Aut}_{\mathcal{A}}(M) \longrightarrow \mathrm{Aut}_{\mathcal{A}}(M), \qquad h \mapsto h^{-1},$$

where if $h = (\{H_i\}, \{H_x\})$, then the components $(\{G_i\}, \{G_x\})$ of h^{-1} are given by

$$G_i = H_i^{-1}, \text{ for all } i \in [1, n], \text{ and recursively, for } x \in \mathbb{B}_1,$$
$$G_x = -H^{-1}_{t(x)} H_x H^{-1}_{s(x)} - \sum_{y,z<x} f^x_{y,z}(M) G_y g^x_{y,z}(M) H_z h^x_{y,z}(M) H^{-1}_{s(x)}.$$

It satisfies the formula $h^{-1}h = I_{\underline{d}}$, for all $h \in \mathrm{Aut}_{\mathcal{A}}(M)$. Notice that, if $h \in \mathrm{Stab}_{\mathbb{H}_{\underline{d}}}(M)$, then $h \in \mathrm{Aut}_{\mathcal{A}}(M)$, and $h^{-1} = h^{-M}$. In particular, $\mathrm{Aut}_{\mathcal{A}}(M)$ is an algebraic group, with unit $I_{\underline{d}}$.

Lemma 33.21. *If $\mathcal{A}$ is a tame presented ditalgebra, then, for any dimension vector $\underline{d}$, there is a closed subvariety $\mathbb{T}_{\underline{d}}$ of $\mathrm{mod}_{\mathcal{A}}(\underline{d})$ which intersects all $\mathbb{H}_{\underline{d}}$-orbits in $\mathrm{mod}_{\mathcal{A}}(\underline{d})$ and such that $\dim \mathbb{T}_{\underline{d}} \leq \sum_{i=1}^n d_i$.*

Proof. We can use the curves γ_i described in (33.12)(1) to construct $\mathbb{T}_{\underline{d}}$ as in [40](5.1), because $\mathcal{A}$-mod is a Krull–Schmidt category. □

Lemma 33.22. *Assume there is an A-$k\langle x,y\rangle$-bimodule B which is finitely generated free by the right, such that the functor*

$$k\langle x,y\rangle\text{-mod} \xrightarrow{B\otimes_{k\langle x,y\rangle}-} A\text{-mod} \xrightarrow{L_{\mathcal{A}}} \mathcal{A}\text{-mod}$$

preserves isomorphism classes. Let c be a natural number and let $\underline{r}$ be the vector of ranks of B. Then there is an irreducible subvariety W_c of $\text{mod}_{\mathcal{A}}(c\underline{r})$ *satisfying:*

(1) $\dim W_c = 2c^2$*; and*
(2) $\dim(W_c \cap \mathbb{H}_{c\underline{r}}M) \leq c^2 - 1$, for each $M \in \text{mod}_{\mathcal{A}}(c\underline{r})$.

Proof. *(Sketch)* Let V_c be the closure of the image of the morphism φ_c described in (33.12)(2), and let W_c be an open subset of V_c contained in the image of φ_c. Then, W_c is an irreducible variety that satisfies (1) and (2). See [40](5.2). □

Proposition 33.23. *Let $\mathcal{A}$ be a presented ditalgebra and assume there is an A-$k\langle x,y\rangle$-bimodule Z, which is finitely generated free by the right, such that the functor*

$$k\langle x,y\rangle\text{-mod} \xrightarrow{Z\otimes_{k\langle x,y\rangle}-} A\text{-mod} \xrightarrow{L_{\mathcal{A}}} \mathcal{A}\text{-mod}$$

preserves isomorphism classes. Then, $\mathcal{A}$ is not tame.

Proof. Suppose $\mathcal{A}$ is tame. Then, from (33.21) and (33.22), we obtain an $\underline{r} \in \mathbb{N}^n$ and, for each $c \in \mathbb{N}$, subvarieties $\mathbb{T}_{c\underline{r}}$ and W_c of $\text{mod}_{\mathcal{A}}(c\underline{r})$ as described before. Take $X = \text{mod}_{\mathcal{A}}(c\underline{r})$, $\mathbb{H} = \mathbb{H}_{c\underline{r}}$, $W = W_c$, $T = \mathbb{T}_{c\underline{r}}$, and $t = c^2 - 1$ in (33.5). Then, $c^2 + 1 = \dim W_c - (c^2 - 1) \leq \dim \mathbb{T}_{c\underline{r}} \leq c\left(\sum_{i=1}^n r_i\right)$. This is clearly false for big c. □

Now we can prove the characterization of wildness anounced in Section 22.

Proof of (22.4). Clearly, (1) implies (2). From (22.5), we have that (2) implies (3). It is also clear that (3) implies (4). From (33.23), we know that (4) implies that Λ is not tame and, hence, by (27.14), that Λ is wild. □

Now we give some properties of generic decompositions of representations of presented ditalgebras, which extend well-known facts for module varieties for algebras, using the pointwise products and pointwise inversions attached to the given action.

Definition 33.24. *If $\underline{d} \in \mathbb{N}^n$ is a dimension vector, then* $\text{ind}_{\mathcal{A}}(\underline{d})$ *denotes the constructible subspace of* $\text{mod}_{\mathcal{A}}(\underline{d})$ *defined by the indecomposable representations in* $\mathcal{A}$-mod *with dimension vector $\underline{d}$.*

Given a decomposition $\underline{d} = m_1\underline{d}^1 + \cdots + m_s\underline{d}^s$, with $\underline{m} = (m_1, \ldots, m_s) \in (\mathbb{N}^*)^s$ and $\underline{d}^j \in \mathbb{N}^n$, for all $j \in [1,s]$, we denote by $\text{dec}_{\mathcal{A}}(\underline{d}; \underline{d}^1, \ldots, \underline{d}^s; \underline{m})$ *the constructible subset of* $\text{mod}_{\mathcal{A}}(\underline{d})$ *determined by the representations M in*

$\mathcal{A}$-mod *with dimension vector* $\underline{d}$, *such that* $M \cong m_1 M^1 \bigoplus \cdots \bigoplus m_s M^s$, *for some indecomposable representations* M^j, *with* $\underline{\dim} M^j = \underline{d}^j$, *and* $M^i \not\cong M^j$ *for* $i \neq j$.

Given an irreducible component $\mathcal{C}$ *of* $\mathrm{mod}_{\mathcal{A}}(\underline{d})$, *a decomposition* $\underline{d} = m_1 \underline{d}^1 + \cdots + m_s \underline{d}^s$, *is a generic decomposition of* $\mathcal{C}$ *iff*

$$\mathrm{dec}_{\mathcal{A}}(\underline{d}; \underline{d}^1, \ldots, \underline{d}^s; \underline{m}) \cap \mathcal{C} \text{ is a dense subset of } \mathcal{C}.$$

We say that the decomposition of $\underline{d}$ *is indecomposable iff* $s = 1$ *and* $m_1 = 1$.

Proposition 33.25. *There is a unique generic decomposition for any irreducible component* $\mathcal{C}$ *of* $\mathrm{mod}_{\mathcal{A}}(\underline{d})$. *Assume that* $\underline{d} = m_1 \underline{d}^1 + \cdots + m_s \underline{d}^s$, *with* $s \in \mathbb{N}$, $\underline{d}^j \in \mathbb{N}^n$ *and* $m_j \in \mathbb{N}^*$, *is a generic decomposition of* $\mathcal{C}$. *Then, for each* i, *there is an irreducible component* $\mathcal{C}_i$ *of* $\mathrm{mod}_{\mathcal{A}}(\underline{d}^i)$ *such that its generic decomposition is indecomposable. Moreover, we have the inequality*

$$\dim \mathbb{H}_{\underline{d}} - \dim \mathcal{C} \geq \sum_{i=1}^{s} (\dim \mathbb{H}_{\underline{d}^i} - \dim \mathcal{C}_i).$$

Proof. For each decomposition $\underline{d} = m_1 \underline{d}^1 + \cdots + m_t \underline{d}^t$ with $\underline{d}^i \in \mathbb{N}^n$ and $m_j \in \mathbb{N}^*$, we have a morphism of varieties

$$\begin{aligned} \varphi_{\underline{d}^1,\ldots,\underline{d}^t;\underline{m}} : \mathbb{H}_{\underline{d}} \times \mathrm{mod}_{\mathcal{A}}(\underline{d}^1) \times \cdots \times \mathrm{mod}_{\mathcal{A}}(\underline{d}^t) &\to \quad \mathrm{mod}_{\mathcal{A}}(\underline{d}), \\ (h, \{M^i\}) \qquad\qquad &\mapsto h(\oplus_{i=1}^t m_i M^i). \end{aligned}$$

For $r \in [1, t]$, denote by $I_r = \{j \in [1, t] \mid \underline{d}^i = \underline{d}^r\}$. Then, if $|I_r| > 1$ and $i, j \in I_r$ are different indexes with $i < j$, we can consider the morphism of varieties

$$\begin{array}{c} \mathbb{H}_{\underline{d}^r} \times (\mathbb{H}_{\underline{d}} \times \mathrm{mod}_{\mathcal{A}}(\underline{d}^1) \times \cdots \times \widehat{\mathrm{mod}_{\mathcal{A}}(\underline{d}^j)} \times \cdots \times \mathrm{mod}_{\mathcal{A}}(\underline{d}^t)) \\ \big\downarrow \psi_{r;i,j} \\ \mathbb{H}_{\underline{d}} \times \mathrm{mod}_{\mathcal{A}}(\underline{d}^1) \times \cdots \times \mathrm{mod}_{\mathcal{A}}(\underline{d}^t) \end{array}$$

defined by

$$\begin{aligned} &\psi_{r;i,j}(h_r; h; N^1, \ldots, N^i, \ldots, \widehat{N^j}, \ldots, N^t) \\ &= (h; N^1, \ldots, N^i, \ldots, h_r N^i, \ldots, N^t). \end{aligned}$$

The terms with a hat are omitted and $h_r N^i$ appears in the $j + 1$ position of the vector at the right in the last formula. Then

$$G_{r;i,j} := \psi_{r;i,j}[\mathbb{H}_{\underline{d}^r} \times (\mathbb{H}_{\underline{d}} \times \mathrm{mod}_{\mathcal{A}}(\underline{d}^1) \times \cdots \times \widehat{\mathrm{mod}_{\mathcal{A}}(\underline{d}^j)} \times \cdots \times \mathrm{mod}_{\mathcal{A}}(\underline{d}^t))]$$

is constructible for all such r, i, j. Clearly

$$\mathrm{dec}_{\mathcal{A}}(\underline{d};\underline{d}^1,\ldots,\underline{d}^t;\underline{m})$$
$$= \varphi_{\underline{d}^1,\ldots,\underline{d}^t;\underline{m}}\left[\mathbb{H}_{\underline{d}} \times \mathrm{ind}_{\mathcal{A}}(\underline{d}^1) \times \cdots \mathrm{ind}_{\mathcal{A}}(\underline{d}^t)\backslash\left(\cup_{r;i,j} G_{r;i,j}\right)\right].$$

Moreover, $\mathrm{mod}_{\mathcal{A}}(\underline{d}) = \bigcup\{\mathrm{dec}_{\mathcal{A}}(\underline{d};\underline{d}^1,\ldots,\underline{d}^t;\underline{m}) \mid \sum_{i=1}^t m_i\underline{d}^i = \underline{d}\}$. Thus, there is a decomposition $\underline{d} = \sum_{i=1}^s m_i\underline{d}^i$ such that

$$\mathcal{C} = \overline{\mathrm{dec}_{\mathcal{A}}(\underline{d};\underline{d}^1,\ldots,\underline{d}^s;\underline{m}) \cap \mathcal{C}},$$

which is the wanted generic decomposition.

Now, let U be an open subset of $\mathcal{C}$ contained in $\mathrm{dec}_{\mathcal{A}}(\underline{d};\underline{d}^1,\ldots,\underline{d}^s;\underline{m}) \cap \mathcal{C}$. We may assume that U is an open subset of $\mathrm{mod}_{\mathcal{A}}(\underline{d})$. Now consider, for each i, the open set U_i of $\mathrm{mod}_{\mathcal{A}}(\underline{d}^i)$ obtained as the image of the projection $\pi_i : \mathbb{H}_{\underline{d}} \times \mathrm{mod}_{\mathcal{A}}(\underline{d}^1) \times \cdots \times \mathrm{mod}_{\mathcal{A}}(\underline{d}^s) \to \mathrm{mod}_{\mathcal{A}}(\underline{d}^i)$ of the open set $\varphi^{-1}_{\underline{d}^1,\ldots,\underline{d}^s;\underline{m}}(U)$. Choose an irreducible component $\mathcal{C}_i$ of $\mathrm{mod}_{\mathcal{A}}(\underline{d}^i)$ that intersects U_i. Then, the generic decomposition of $\mathcal{C}_i$ is indecomposable.

It is not hard to see that, for a suitable choice of the components $\mathcal{C}_i$, the following restriction of $\varphi_{\underline{d}^1,\ldots,\underline{d}^s;\underline{m}}$ is a dominant morphism

$$\varphi := \varphi_{\underline{d}^1,\ldots,\underline{d}^s;\underline{m}}| : \mathbb{H}_{\underline{d}} \times \mathcal{C}_1 \times \cdots \times \mathcal{C}_s \to \mathcal{C}.$$

Then, there is an $N \in \mathcal{C}$ such that $\dim(\varphi^{-1}(N)) = \dim \mathbb{H}_{\underline{d}} + \sum_{i=1}^s \dim \mathcal{C}_i - \dim \mathcal{C}$. Now, choose a point $(h, M^1,\ldots,M^s) \in \varphi^{-1}(N)$ and consider the embedding of varieties

$$\begin{array}{lll}\psi : \mathbb{H}_{\underline{d}^1} \times \cdots \times \mathbb{H}_{\underline{d}^s} \to & & \varphi^{-1}(N) \\ \qquad (h_1,\ldots,h_s) & \mapsto & (h \cdot_{(\oplus m_i h_i M^i)} (\oplus m_i h_i^{-M^i}), h_1 M^1,\ldots,h_s M^s).\end{array}$$

Then, $\sum_{i=1}^s \dim \mathbb{H}_{\underline{d}^i} \leq \dim(\varphi^{-1}(N)) = \dim \mathbb{H}_{\underline{d}} + \sum_{i=1}^s \dim \mathcal{C}_i - \dim \mathcal{C}$. □

Lemma 33.26. *If $\mathcal{A}$ is tame, then* $\dim(\mathrm{mod}_{\mathcal{A}}(\underline{d})) \leq h_{\mathcal{A}}(\underline{d})$, *for each* $\underline{d} \in \mathbb{N}^n$.

Proof. From the last result, it is enough to show that $\dim \mathbb{H}_{\underline{d}} - \dim \mathcal{C} \geq 0$, if $\mathcal{C}$ is an irreducible component of $\mathrm{mod}_{\mathcal{A}}(\underline{d})$ whose generic decomposition is indecomposable. Since $\mathcal{A}$ is tame, there is a finite number of curves $\kappa_i : D_i \to \mathrm{mod}_{\mathcal{A}}(\underline{d})$, $i \in [1, r]$, where D_i is a cofinite subset of k, such that each indecomposable representation of dimension vector $\underline{d}$ is isomorphic to some point on one of these curves. Then, for some i, there is a well-defined dominant morphism of varieties

$$\varphi : \mathbb{H}_{\underline{d}} \times D_i \to \mathcal{C}, \qquad (h, t) \mapsto h\kappa_i(t).$$

Choose a point $N \in \mathrm{Im}\,\varphi$ such that $\dim(\varphi^{-1}(N)) = \dim \mathbb{H}_{\underline{d}} - \dim \mathcal{C} + 1$, and $(f, t) \in \varphi^{-1}(N)$. Observe that we have an injective morphism $\mathrm{Aut}_{\mathcal{A}}(\kappa_i(t)) \to$

$\varphi^{-1}(N)$ defined by $h \mapsto (f \cdot_{(\kappa_i(t))} h, t)$. Comparing dimensions, we obtain the formula. □

Proposition 33.27. *Given is a seminested ditalgebra $\mathcal{A}$, assume that $Re_i = k[x_i]_{f_i(x_i)}$, for $i \in [1, m]$ and $Re_i = k$, for $i \in [m+1, n]$. Consider the quadratic form $q_{\mathcal{A}}$ of $\mathcal{A}$, that is the quadratic form $q_{\mathcal{A}} : \mathbb{Z}^n \to \mathbb{Z}$ defined, for $z \in \mathbb{Z}^n$, by*

$$q_{\mathcal{A}}(z) = \sum_{i=m+1}^{n} z_i^2 - \sum_{\alpha \in \mathbb{B}_0} z_{t(\alpha)} z_{s(\alpha)} + \sum_{x \in \mathbb{B}_1} z_{t(x)} z_{s(x)}.$$

Thus $q_{\mathcal{A}}(z) := (z, z)_{\mathcal{A}}$, see (23.30). Assume $\mathcal{A}$ is nested and tame. Then, the quadratic form $q_{\mathcal{A}}$ is weakly semipositive, that is $q_{\mathcal{A}}(\underline{d}) \geq 0$, for all $\underline{d} \in \mathbb{N}^n$.

Proof. From [33](3.3), we know that the difference $\dim \mathbb{A}_{\underline{d}} - \dim \operatorname{mod}_{\mathcal{A}}(\underline{d})$ is bounded by the number μ of polynomials whose set of zeros is $\operatorname{mod}_{\mathcal{A}}(\underline{d})$. We can choose the generators $\mathbb{L}$ of R as follows: $\mathbb{L} := \cup_{i=1}^{m} \{x_i, y_i\}$, where $y_i = f(x_i)^{-1}$. Thus, from (33.26), we have

$$\dim \mathbb{A}_{\underline{d}} - \mu \leq \dim \operatorname{mod}_{\mathcal{A}}(\underline{d}) \leq h_{\mathcal{A}}(\underline{d}).$$

We claim that the ideal $I_A = \operatorname{Ker} \psi$ of kQ_A is generated by the set of relations $P := \{f_i(x_i)y_i - \tau_i \mid i \in [1, m]\}$. Consider the canonical projection $\pi : kQ_A \longrightarrow kQ_A/\langle P \rangle =: A'$ and the induced morphism of algebras $\overline{\psi} : A' \longrightarrow A$ with $\overline{\psi}\pi = \psi$. It will be enough to show that $\overline{\psi}$ is an isomorphism because, then, $I_A = \operatorname{Ker} \psi = \operatorname{Ker} \overline{\psi}\pi = \operatorname{Ker} \pi = \langle P \rangle$. In order to exhibit an inverse for $\overline{\psi}$, recall that the algebra A is freely generated by the pair (R, W_0). Consider the morphism of algebras $\varphi_0 : R \longrightarrow A'$ defined by $e_i \mapsto \pi(\tau_i)$ and $x_i \mapsto \pi(x_i)$. There is such morphism of algebras because $\varphi(f_i(x_i)) = f_i(\pi(x_i)) = \pi(f_i(x_i))$ is invertible in the algebra $\pi(\tau_i)A'\pi(\tau_i)$, with inverse $\pi(y_i)$. Since, W_0 is an R-R-bimodule freely generated by $\mathbb{B}_0$, there is a morphism of R-R-bimodules $\varphi_1 : W_0 \longrightarrow A'$ such that $\varphi_1(\alpha) = \pi(\alpha)$, for all $\alpha \in \mathbb{B}_0$. Then, there is a morphism of algebras $\varphi : A \longrightarrow A'$ which extends φ_0 and φ_1. Clearly, $\varphi = (\overline{\psi})^{-1}$. Therefore, as claimed, we have $I_A = \langle P \rangle$.

Then, $\mu = \sum_{i=1}^{m} d_i^2$ and last inequality becomes

$$\sum_{\alpha \in \mathbb{B}_0} d_{t(\alpha)} \times d_{s(\alpha)} + \sum_{i=1}^{m} 2d_i^2 - \sum_{i=1}^{m} d_i^2 \leq \sum_{i=1}^{n} d_i^2 + \sum_{x \in \mathbb{B}_1} d_{t(x)} \times d_{s(x)}.$$

That is precisely $q_{\mathcal{A}}(\underline{d}) \geq 0$. □

Proposition 33.28. *Assume that $\mathcal{A}$ is a nested ditalgebra with finite representation type. Then, its quadratic form $q_{\mathcal{A}}$, as defined in (33.27), is weakly positive, that is $q_{\mathcal{A}}(\underline{d}) > 0$, for all $\underline{d} \in \mathbb{N}^n$.*

Proof. In our situation, each irreducible component of $\mathrm{mod}_{\mathcal{A}}(\underline{d})$ is a finite union of $\mathbb{H}_{\underline{d}}$-orbits. This is the case, in particular, for the irreducible component of maximal dimension. Then, this component coincides with the closure of one of the orbits $\mathbb{H}_{\underline{d}}M$ of $\mathbb{H}_{\underline{d}}$ in $\mathrm{mod}_{\mathcal{A}}(\underline{d})$. Then, from (33.4) and (33.17), we know that $\dim \mathrm{mod}_{\mathcal{A}}(\underline{d}) = \dim \mathbb{H}_{\underline{d}} - \dim \mathrm{Aut}_{\mathcal{A}}(M)$. Using this fact and [33](3.3), as in the proof of (33.27), we obtain $q_{\mathcal{A}}(\underline{d}) \geq \dim \mathrm{Aut}_{\mathcal{A}}(M) \geq 1$, as we wanted to show. □

Exercise 33.29. *Lt $\mathcal{A}$ be a nested ditalgebra with only one point, which is not marked, n solid arrows and m dotted arrows. Show that $\mathcal{A}$ is wild whenever $n > m + 1$.*

Exercise 33.30. *Assume that the nested ditalgebra $\mathcal{A}$ is of finite representation type. Show that, for any $M \in \mathcal{A}$-mod with $\dim \mathrm{End}_{\mathcal{A}}(M) = 1$, we have $\mathrm{Ext}_{\mathcal{A}}(M, M) = 0$.*

Hint: Use (33.28) and (23.30).

Exercise 33.31. *Given $\underline{d} \in \mathbb{N}^n$, from (33.18), we can consider, for $t \in [1, h_{\mathcal{A}}(\underline{d})]$, the following $\mathbb{H}_{\underline{d}}$-stable closed subspace of $\mathrm{mod}_{\mathcal{A}}(\underline{d})$*

$$\mathrm{mod}_{\mathcal{A}}(\underline{d}, t) := \{M \in \mathrm{mod}_{\mathcal{A}}(\underline{d}) \mid \dim_k(\mathrm{End}_{\mathcal{A}}(M)) \geq t\}.$$

Show that the following statements are equivalent for a presented ditalgebra $\mathcal{A}$:

(1) $\mathcal{A}$ is tame;
(2) For each dimension vector $\underline{d} \in \mathbb{N}^n$, there is a constructible subset C of $\mathrm{ind}_{\mathcal{A}}(\underline{d})$ with $\dim C \leq 1$ and such that $\mathbb{H}_{\underline{d}}C = \mathrm{ind}_{\mathcal{A}}(\underline{d})$;
(3) For each dimension vector $\underline{d} \in \mathbb{N}^n$, if C is a constructible subset of $\mathrm{ind}_{\mathcal{A}}(\underline{d})$ such that it intersects each $\mathbb{H}_{\underline{d}}$-orbit in at most a finite number of points, then $\dim C \leq 1$.
(4) For each dimension vector $\underline{d} \in \mathbb{N}^n$, there is a constructible subset C of $\mathrm{mod}_{\mathcal{A}}(\underline{d})$ with $\dim C \leq \sum_i d_i$ such that $\mathbb{H}_{\underline{d}}C = \mathrm{mod}_{\mathcal{A}}(\underline{d})$;
(5) For each dimension vector $\underline{d} \in \mathbb{N}^n$ and each $t \in [1, h_{\mathcal{A}}(\underline{d})]$, we have that

$$\dim(\mathrm{mod}_{\mathcal{A}}(\underline{d}, t)) \leq h_{\mathcal{A}}(\underline{d}) + \sum_i d_i - t.$$

Hint: Compare with [40](5.5) and [24](1.3).

Exercise 33.32. *Notice that if $\underline{d} \in \mathbb{N}^n$, for each $s \in [1, h_{\mathcal{A}}(\underline{d})]$, we obtain a constructible subspace of $\mathrm{mod}_{\mathcal{A}}(\underline{d})$ as follows*

$$\mathrm{ind}^s_{\mathcal{A}}(\underline{d}) := \{x \in \mathrm{ind}_{\mathcal{A}}(\underline{d}) \mid \dim \mathbb{H}_{\underline{d}}x = s\}.$$

Indeed, with the notation of (33.24) and (33.31) in mind, observe that

$$\mathrm{ind}^s_{\mathcal{A}}(\underline{d}) = \mathrm{ind}_{\mathcal{A}}(\underline{d}) \cap [\mathrm{mod}_{\mathcal{A}}(\underline{d}, h_{\mathcal{A}}(\underline{d}) - s) \setminus \mathrm{mod}_{\mathcal{A}}(\underline{d}, h_{\mathcal{A}}(\underline{d}) - s + 1)].$$

By definition, the number of parameters of $\mathcal{A}$ in dimension $\underline{d}$ is

$$p_{\mathcal{A}}(\underline{d}) := \max_s(\dim(\mathrm{ind}^s_{\mathcal{A}}(\underline{d})) - s).$$

Prove the following statements, for a presented ditalgebra $\mathcal{A}$:

(1) $\mathcal{A}$ is tame if and only if $p_{\mathcal{A}}(\underline{d}) \leq 1$, for all $\underline{d} \in \mathbb{N}^n$.
(2) If $\mathcal{A}$ is wild, then $\limsup_{|\underline{d}| \to \infty} p_{\mathcal{A}}(\underline{d}) / h_{\mathcal{A}}(\underline{d}) > 0$, where $|\underline{d}| = \sum_i d_i$.

Hint: Compare with [40](5.8) and [25].

References. The idea of transferring the Tits quadratic form, used for representations of quivers, to the context of differential graded categories is present in [46]. The fact that tame nested bocses have a weakly non-negative quadratic form is due to Drozd, see [28]. The presentation of the basics of varieties of modules for presented ditalgebras given here is developed from part of [10], it extends to varieties of modules over ditalgebras some well-known results for varieties of modules over algebras (see for instance [40]).

34

Ditalgebras of partially ordered sets

In this section, we examine an important family of nested ditalgebras. Namely, the ditalgebras associated to *posets* (that is finite partially ordered sets). We will provide an explicit conceptual relationship between the category of representations of a poset $\mathbb{S}$ and the category of modules over a ditalgebra $\mathcal{A}^{\mathbb{S}}$. This relation will be described without assumptions on finiteness of dimensions of representations. Moreover, if $\Lambda = \Lambda_{\mathbb{S}}$ denotes the suspension algebra associated to the poset $\mathbb{S}$, the relationship with the equivalence Ξ_Λ studied before will be exhibited.

Proposition 34.1. *Let $\mathbb{S} = (X, \leq)$ and $\mathbb{T} = (Y, \leq)$ be any posets. Consider the bigraph (with no marked points) $\mathbb{B} := (\mathcal{P}, \mathbb{B}_0, \mathbb{B}_1)$ defined by: $\mathcal{P}$ is the disjoint union of X and Y; there is a solid arrow $\alpha_{y,x} : x \to y$ for each $(x, y) \in X \times Y$; there is a dotted arrow $v_{x',x} : x \to x'$ iff $x, x' \in X$ satisfy $x' < x$; and there is a dotted arrow $w_{y',y} : y \to y'$ iff $y, y' \in Y$ satisfy $y' < y$. Then, for any solid arrow $\alpha_{y,x}$ of $\mathbb{B}$, define*

$$\delta(\alpha_{y,x}) := \sum_{y<y'} w_{y,y'}\alpha_{y',x} - \sum_{x'<x} \alpha_{y,x'}v_{x',x}$$

and, for the dotted arrows $v_{x',x}$ and $w_{y',y}$ of $\mathbb{B}$, define

$$\delta(v_{x',x}) := \sum_{x'<z<x} v_{x',z}v_{z,x} \quad \text{and} \quad \delta(w_{y',y}) := \sum_{y'<z<y} w_{y',z}w_{z,y} \, .$$

These data define a nested ditalgebra $\mathcal{A}(\mathbb{S}, \mathbb{T})$. Moreover, we have the formula

$$\mathcal{A}(\mathbb{S}, \mathbb{T})^{op} \cong \mathcal{A}(\mathbb{T}^{op}, \mathbb{S}^{op}),$$

where the posets $\mathbb{S}^{op}$ and $\mathbb{T}^{op}$ denote the posets obtained from $\mathbb{S}$ and $\mathbb{T}$ by reversing the order. We refer to $\mathcal{A}(\mathbb{S}, \mathbb{T})$ as the ditalgebra associated to the pair of posets $(\mathbb{S}, \mathbb{T})$.

Proof. We consider the product R of $|X \cup Y|$ copies of the field k, one for each element of the disjoint union $X \cup Y$. For each $z \in X \cup Y$, denote by e_z the canonical idempotent of R corresponding to the copy z of k. Thus, we have a decomposition $R = (\oplus_{x \in X} ke_x) \oplus (\oplus_{y \in Y} ke_y)$. Then, we consider, respectively, the R-R-bimodules freely generated by the solid and the dotted arrows

$$\begin{cases} W_0 = \bigoplus_{(x,y) \in X \times Y} R\alpha_{y,x} R \\ W_1 = [\bigoplus_{x'<x} Rv_{x',x} R] \bigoplus \left[\bigoplus_{y'<y} Rw_{y',y} R\right]. \end{cases}$$

Make $T = T_R(W)$, with $W = W_0 \oplus W_1$. By definition, we clearly have, for any solid arrow $\alpha_{y,x}$, that $\delta(\alpha_{y,x}) \in [T]_1$, and, for any dotted arrows $v_{x',x}$ and $w_{y',y}$, we have $\delta(v_{x',x}), \delta(w_{y',y}) \in [T]_2$. Then, we have linear maps

$$\begin{cases} W_0 = \bigoplus_{(x,y) \in X \times Y} k\alpha_{y,x} \xrightarrow{\delta} [T]_1 \\ W_1 = [\bigoplus_{x'<x} kv_{x',x}] \bigoplus \left[\bigoplus_{y'<y} kw_{y',y}\right] \xrightarrow{\delta} [T]_2 \end{cases}$$

which are, in fact, morphisms of R-R-bimodules and, therefore, can be extended to a differential δ over T using (4.4).

We want to show that $\delta^2(W) = 0$. For any solid arrow $\alpha_{y,x}$, we have

$$\begin{aligned} \delta^2(\alpha_{y,x}) &= \delta(\textstyle\sum_{y<y'} w_{y,y'}\alpha_{y',x}) - \delta(\sum_{x'<x} \alpha_{y,x'} v_{x',x}) \\ &= \textstyle\sum_{y<y'} [\delta(w_{y,y'})\alpha_{y',x} - w_{y,y'}\delta(\alpha_{y',x})] \\ &\quad - \textstyle\sum_{x'<x} [\delta(\alpha_{y,x'})v_{x',x} + \alpha_{y,x'}\delta(v_{x',x})] \\ &= \textstyle\sum_{y<y'} \left(\sum_{y<z<y'} w_{y,z}w_{z,y'}\right) \alpha_{y',x} \\ &\quad - \textstyle\sum_{y<y'} w_{y,y'} \left(\sum_{y'<y''} w_{y',y''}\alpha_{y'',x} - \sum_{x'<x} \alpha_{y',x'} v_{x',x}\right) \\ &\quad - \textstyle\sum_{x'<x} \left(\sum_{y<y'} w_{y,y'}\alpha_{y',x'} - \sum_{x''<x'} \alpha_{y,x''} v_{x'',x'}\right) v_{x',x} \\ &\quad - \textstyle\sum_{x'<x} \alpha_{y,x'} \left(\sum_{x'<z<x} v_{x',z}v_{z,x}\right) \\ &= \textstyle\sum_{y<z<y'} w_{y,z}w_{z,y'}\alpha_{y',x} - \sum_{y<y'<y''} w_{y,y'}w_{y',y''}\alpha_{y'',x} \\ &\quad \textstyle\sum_{x''<x'<x} \alpha_{y,x''}v_{x'',x'}v_{x',x} - \sum_{x'<z<x} \alpha_{y,x'}v_{x',z}v_{z,x} \\ &= 0 \end{aligned}$$

For the dotted arrow $v_{x',x}$, we have

$$\begin{aligned} \delta^2(v_{x',x}) &= \delta(\textstyle\sum_{x'<z<x} v_{x',z}v_{z,x}) \\ &= \textstyle\sum_{x'<z<x} (\delta(v_{x',z})v_{z,x} - v_{x',z}\delta(v_{z,x})) \\ &= \textstyle\sum_{x'<z<x} \sum_{x'<u<z} v_{x',u}v_{u,z}v_{z,x} - \sum_{x'<z<x} \sum_{z<t<x} v_{x',z}v_{z,t}v_{t,x} \\ &= \textstyle\sum_{x'<u<z<x} v_{x',u}v_{u,z}v_{z,x} - \sum_{x'<z<t<x} v_{x',z}v_{z,t}v_{t,x} \\ &= 0 \end{aligned}$$

Similarly, we have that $\delta^2(w_{y',y}) = 0$. Hence $\delta^2 = 0$ and $\mathcal{A}(\mathbb{S}, \mathbb{T}) = (T, \delta)$ is a ditalgebra with layer (R, W).

In order to show the triangularity of (R, W), we consider the partial order in the set of solid arrows determined by $\alpha_{y',x'} \leq \alpha_{y,x}$ iff $y \leq y'$ and $x \geq x'$; and the partial order in the set of dotted arrows determined by: $v_{x',x} \leq v_{t',t}$ iff $t' \leq x'$ and $t \geq x$; and, similarly, $w_{y',y} \leq w_{u',u}$ iff $u' \leq y'$ and $u \geq y$. Then, given a solid arrow $\alpha_{y,x}$ we can define: $\mathbb{h}(\alpha_{y,x}) = 1$, if $\alpha_{y,x}$ is minimal, and $\mathbb{h}(\alpha_{y,x}) = m$ iff m is the maximal number $m \geq 2$ such that there is a sequence $\alpha_{y_m,x_m} < \cdots < \alpha_{y_1,x_1} = \alpha_{y,x}$ of solid arrows. Then, $W_0^j := \bigoplus_{\mathbb{h}(\alpha_{y,x}) \leq j} R\alpha_{y,x}R$, defines the filtration of W_0 required in (5.1); the filtration required for W_1 is constructed similarly.

Having in mind (2.7), in order to prove the formula, we just have to replace the basis of the layer of $\mathcal{A}(\mathbb{S}, \mathbb{T})^{op}$: each $\alpha^{op}_{y,x}$ by $-\alpha^{op}_{y,x}$, each $v^{op}_{x',x}$ by $-v^{op}_{x',x}$, and each $w^{op}_{y',y}$ by $-w^{op}_{y',y}$. □

Notation 34.2. *With the notation of the previous proposition: if* $\mathbb{S} = (X, \leq)$ *consists of only one point,* $X = \{x_0\}$, *we write* $\mathcal{A}_{\mathbb{T}} := \mathcal{A}(\mathbb{S}, \mathbb{T})$. *Similarly, if* $\mathbb{T} = (Y, \leq)$ *consists of only one point,* $Y = \{y_0\}$, *we have the ditalgebra* $\mathcal{A}^{\mathbb{S}} := \mathcal{A}(\mathbb{S}, \mathbb{T})$.

Remark 34.3. *Given the poset* $\mathbb{S}$, *both ditalgebras* $\mathcal{A}_{\mathbb{S}}$ *and* $\mathcal{A}^{\mathbb{S}}$ *will be of interest to us later. We refer to them as ditalgebras associated to the poset* $\mathbb{S}$.

As a consequence of the above formula, we have

$$\mathcal{A}^{\mathbb{S}} \cong (\mathcal{A}_{\mathbb{S}^{op}})^{op} .$$

Lemma 34.4. *For any poset* $\mathbb{S} = (X, \leq)$, *consider the poset* $\overline{\mathbb{S}} = (\overline{X}, \leq)$ *with underlying set* $\overline{X} = X \cup \{m, M\}$, *where the order of* $\mathbb{S}$ *is extended to* $\overline{X}$ *making* $m \leq i$ *and* $i \leq M$, *for any* $i \in X$. *Consider the quiver* $Q_{\mathbb{S}}$ *with points* $\overline{X}$ *and such that, for* $i, j \in \overline{X}$, *there is an arrow* $\alpha : i \longrightarrow j$ *in* $Q_{\mathbb{S}}$ *iff* $i < j$ *and there is no* $t \in \overline{X}$ *with* $i < t < j$. *Given a field* k, *define the suspension algebra* $\Lambda = \Lambda_{\mathbb{S}}$ *over* k *as the quotient of the path algebra* $kQ_{\mathbb{S}}$ *modulo the ideal* I *generated by all differences of two paths in* $Q_{\mathbb{S}}$ *sharing their source point and their target point. Notice that* Λ *trivially splits over its radical. For any* Λ*-module* H, *consider the full subcategory* Sub(H) *of* Λ-Mod *formed by the objects* $N \in \Lambda$-Mod *which admit a monomorphism* $\varphi : N \longrightarrow H^{(Z)}$, *for some index set* Z. *Consider the canonical decomposition of the unit* $1 = \sum_{i \in X} e_i + e_m + e_M$ *of* Λ *as a sum of primitive orthogonal idempotents and the corresponding indecomposable projective* Λ*-modules* $P_i := \Lambda e_i$, *for* $i \in \overline{X}$. *Notice that* P_m *is an injective* Λ*-module. As usual, given* $N \in \Lambda$-Mod, *we make* $N_i := e_i N$, *for all* $i \in \overline{X}$. *Then, the following statements hold:*

(1) Sub(Λ) = Sub(P_m).
(2) Given $N \in \Lambda$-Mod, *we have that* $N \in$ Sub(Λ) *iff the multiplication by* α *map* $N_\alpha : N_i \longrightarrow N_j$ *is injective for each arrow* $\alpha : i \longrightarrow j$ *of* $Q_{\mathbb{S}}$.

(3) *Every $N \in \mathrm{Sub}(\Lambda)$ admits a decomposition $N \cong P_m^{(Z')} \oplus N'$, where P_m is not a direct summand of the projective cover of N'. Moreover, $N_m = 0$ iff P_m is not a direct summand of N.*

(4) *Every $N \in \mathrm{Sub}(\Lambda)$ admits a decomposition $N \cong P_M^{(Z'')} \oplus N''$, where P_M is not a direct summand of the projective cover of N''.*

Proof. Notice that, for each pair of different vertices r, s of $Q_{\mathbb{S}}$, there is at most one non-zero equivalence class in Λ of paths from r to s. This implies that, for $i \leq j$ in $\overline{\mathbb{S}}$, there is an inclusion of P_j into P_i. Thus, every indecomposable projective Λ-module is isomorphic to a submodule of P_m. This implies (1). Item (2) is easy to show.

Now assume that $N \in \mathrm{Sub}(\Lambda)$ and consider the non-zero path class $q_i : m \longrightarrow i$ in Λ corresponding to each $i \in \overline{X} \setminus \{m\}$. Then, consider the subrepresentation L of N defined by $L_m := N_m$ and, for each $i \in \overline{X} \setminus \{m\}$ by $L_i := N_{q_i}(N_m)$. Notice that if B is a vector space basis for N_m, then $L \cong P_m^{(B)}$. Then, consider the morphism of representations $\pi : N \longrightarrow L$ which maps every element $n \in N_i = L_i \oplus V_i$ onto its component in L_i. Since L is projective, π splits and we obtain the decomposition $N \cong P_m^{(B)} \oplus N'$ required for (3). Moreover, the projective cover P of N coincides with the projective cover of N/JN, but $N_m = 0$ implies that $(N/JM)_m = 0$. Hence P admits no direct summand of type P_m.

Now, assume that $N \in \mathrm{Sub}(\Lambda)$ and consider the non-zero path class $p_i : i \longrightarrow M$ in Λ corresponding to each $i \in \overline{X} \setminus \{M\}$. Consider the subrepresentation N'' of N defined by $N''_M := \sum_{i<M} N_{p_i}(N_i)$ and $N''_j := N_j$, for all $j \in \overline{X} \setminus \{M\}$. Then, consider a vector space decomposition $N_M = \sum_{i<M} N_{p_i}(N_i) \oplus V$ and choose a basis B for the vector space V. Then, we clearly have that $N \cong N'' \oplus P_M^{(B)}$. As before, $(JN'')_M = N''_M$ implies that $(N''/JN'')_M = 0$ and, hence, the projective cover of N'' admits no direct summand of type P_M. □

The following result is well known and its proof is left to the reader.

Proposition 34.5. *Given a poset $\mathbb{S} = (X, \leq)$ and a field k, the k-category of $\mathbb{S}$-spaces $\mathcal{S}(\mathbb{S})$ has as objects the families $(V, \{V_i\}_{i \in X})$ of vector spaces where each V_i is a subspace of V and $V_i \subseteq V_j$ whenever $i \leq j$. A morphism of $\mathbb{S}$-spaces $f : (V, \{V_i\}_i) \longrightarrow (W, \{W_i\}_i)$ is a linear map $f : V \longrightarrow W$ such that $f(V_i) \subseteq W_i$, for all i. The category $\mathcal{S}(\mathbb{S})$ is an additive category with kernels and cokernels, but it is not abelian. Consider the suspension algebra $\Lambda = \Lambda_{\mathbb{S}}$ associated to the poset $\mathbb{S}$ in (34.4). Using the notation introduced there, consider the full subcategory* $\mathrm{Sub}_{P_n}(\Lambda)$ *of* Λ-Mod *consisting of the modules in* $\mathrm{Sub}(\Lambda)$

without direct summands isomorphic to P_n. Then, the categories $\mathcal{S}(\mathbb{S})$ and $\mathrm{Sub}_{P_n}(\Lambda)$ *are equivalent.*

Lemma 34.6. *Let $\mathbb{S}$ be a poset, consider the suspension algebra $\Lambda = \Lambda_{\mathbb{S}}$ associated to the poset $\mathbb{S}$ in (34.4) and the notation introduced there. Recall that $\Lambda_{\mathbb{S}}$ trivially splits over its radical and consider the corresponding Drozd's ditalgebra $\mathcal{D}^{\Lambda_{\mathbb{S}}}$. Then the ditalgebra $\mathcal{A}^{\mathbb{S}}$ associated to the poset $\mathbb{S}$ in (34.2) can be realized as the ditalgebra obtained from $\mathcal{D}^{\Lambda_{\mathbb{S}}}$ by deletion of the idempotents $\{e''_x \mid x \in X\} \cup \{e''_M, e'_m, e'_M\}$ (see (23.25)).*

Proof. Recall that in $\Lambda_{\mathbb{S}}$, for each pair of different vertices r, s of $Q_{\mathbb{S}}$, there is at most one non-zero equivalence class of paths from r to s. Then, we can choose a dual basis $(p_j, \gamma_{p_j})_j$ of the radical J of $\Lambda_{\mathbb{S}}$, which is a projective right S-module, as in (23.25) with the additional property that $P = \{p_j\}_j$ is a *multiplicative basis* for J. This means that each product $p_r p_s$ is either zero or is some other basic element in P. Thus, for $p \in P$, we have the formula $\mu(\gamma_p) = \sum_{p_s p_r = p} \gamma_{p_r} \otimes \gamma_{p_s}$.

Then the formulas for the differential of $\mathcal{D}^{\Lambda_{\mathbb{S}}}$ are the following

$$\begin{aligned}
\delta^{\Lambda}\left(\begin{pmatrix} 0 & 0 \\ \gamma_p & 0 \end{pmatrix}\right) = {} & \sum_{p_s p_r = p} \left(\begin{pmatrix} 0 & 0 \\ 0 & \gamma_{p_r} \end{pmatrix} \otimes \begin{pmatrix} 0 & 0 \\ \gamma_{p_s} & 0 \end{pmatrix}\right) \\
& - \sum_{p_s p_r = p} \left(\begin{pmatrix} 0 & 0 \\ \gamma_{p_r} & 0 \end{pmatrix} \otimes \begin{pmatrix} \gamma_{p_s} & 0 \\ 0 & 0 \end{pmatrix}\right),
\end{aligned}$$

$$\delta^{\Lambda}\left(\begin{pmatrix} \gamma_p & 0 \\ 0 & 0 \end{pmatrix}\right) = \sum_{p_s p_r = p} \left(\begin{pmatrix} \gamma_{p_r} & 0 \\ 0 & 0 \end{pmatrix} \otimes \begin{pmatrix} \gamma_{p_s} & 0 \\ 0 & 0 \end{pmatrix}\right), \text{ and}$$

$$\delta^{\Lambda}\left(\begin{pmatrix} 0 & 0 \\ 0 & \gamma_p \end{pmatrix}\right) = \sum_{p_s p_r = p} \left(\begin{pmatrix} 0 & 0 \\ 0 & \gamma_{p_r} \end{pmatrix} \otimes \begin{pmatrix} 0 & 0 \\ 0 & \gamma_{p_s} \end{pmatrix}\right).$$

After eliminating the idempotents of $\mathcal{D} = \mathcal{D}^{\Lambda_{\mathbb{S}}}$ described above, we are only left with e''_m and $\{e'_i \mid i \in X\}$. These correspond to the points of $\mathcal{A}^{\mathbb{S}}$. Moreover, we have to consider only the arrows of $\mathcal{D}$ determined by non-zero path classes p of $\Lambda_{\mathbb{S}}$ determined by paths of $Q_{\mathbb{S}}$ starting at the point m and ending at some point i of X. Again, they correspond to the arrows of $\mathcal{A}^{\mathbb{S}}$. Then, the formulas for the differential simplify as

$$\delta^{\Lambda^d}\left(\begin{pmatrix} 0 & 0 \\ \gamma_p & 0 \end{pmatrix}\right) = - \sum_{p_s p_r = p} \left(\begin{pmatrix} 0 & 0 \\ \gamma_{p_r} & 0 \end{pmatrix} \otimes \begin{pmatrix} \gamma_{p_s} & 0 \\ 0 & 0 \end{pmatrix}\right), \text{ and}$$

$$\delta^{\Lambda^d}\left(\begin{pmatrix} \gamma_p & 0 \\ 0 & 0 \end{pmatrix}\right) = \sum_{p_s p_r = p} \left(\begin{pmatrix} \gamma_{p_r} & 0 \\ 0 & 0 \end{pmatrix} \otimes \begin{pmatrix} \gamma_{p_s} & 0 \\ 0 & 0 \end{pmatrix}\right).$$

Indeed, every solid arrow $\begin{pmatrix} 0 & 0 \\ \gamma_{p_s} & 0 \end{pmatrix}$ appearing in the first formula is annihilated when passing to the ditalgebra $\mathcal{D}^d$ because the path-class p_s does not start at the point m. Every arrow $\begin{pmatrix} 0 & 0 \\ 0 & \gamma_p \end{pmatrix}$ of $\mathcal{D}$ disappears in $\mathcal{D}^d$ too. □

Theorem 34.7. *Let $\mathbb{S} = (X, \leq)$ be a poset, consider the suspension algebra $\Lambda = \Lambda_{\mathbb{S}}$ associated to $\mathbb{S}$ in (34.4) and the notation introduced there. Consider the full subcategory $\mathcal{Q}(\Lambda_{\mathbb{S}})$ of $\mathcal{P}(\Lambda_{\mathbb{S}})$ formed by the morphisms between projectives of the form $\varphi : P \longrightarrow P_m^{(Z)}$, for some index set Z, such that P_M and P_m are not isomorphic to direct summands of P. Consider also the full subcategories $\mathcal{Q}^1(\Lambda_{\mathbb{S}})$ and $\mathcal{Q}^2(\Lambda_{\mathbb{S}})$ of $\mathcal{Q}(\Lambda_{\mathbb{S}})$ defined respectively by the classes of objects $\mathcal{Q}(\Lambda_{\mathbb{S}}) \cap \mathcal{P}^1(\Lambda_{\mathbb{S}})$ and $\mathcal{Q}(\Lambda_{\mathbb{S}}) \cap \mathcal{P}^2(\Lambda_{\mathbb{S}})$ (see (18.8)). Then, the following statements hold:*

(1) The equivalence $\Xi_{\Lambda_{\mathbb{S}}} : \mathcal{D}^{\Lambda_{\mathbb{S}}}\text{-Mod} \longrightarrow \mathcal{P}^1(\Lambda_{\mathbb{S}})$ induces an equivalence of categories $\Xi_{\mathbb{S}} : \mathcal{A}^{\mathbb{S}}\text{-Mod} \longrightarrow \mathcal{Q}^1(\Lambda_{\mathbb{S}})$ such that the following square of functors commutes

$$\begin{array}{ccc} \mathcal{D}^{\Lambda_{\mathbb{S}}}\text{-Mod} & \xrightarrow{\Xi_{\Lambda_{\mathbb{S}}}} & \mathcal{P}^1(\Lambda_{\mathbb{S}}) \\ \uparrow{\scriptstyle F_d} & & \uparrow{\scriptstyle I} \\ \mathcal{A}^{\mathbb{S}}\text{-Mod} & \xrightarrow{\Xi_{\mathbb{S}}} & \mathcal{Q}^1(\Lambda_{\mathbb{S}}). \end{array}$$

Here F_d is the functor associated to the deletion of idempotents $\mathcal{D}^{\Lambda_{\mathbb{S}}} \mapsto \mathcal{A}^{\mathbb{S}}$ described in (34.6) and I is the inclusion functor.

(2) Define $\mathcal{Q}^3(\Lambda_{\mathbb{S}})$ as the full subcategory of $\mathcal{Q}^2(\Lambda_{\mathbb{S}})$ with objects $\varphi : P \longrightarrow P_m^{(Z)}$ such that $\operatorname{soc}\operatorname{Im}\varphi = \operatorname{soc} P_m^{(Z)}$. Then, any object in $\mathcal{Q}^1(\Lambda_{\mathbb{S}})$ decomposes as a direct sum of the form

$$(P' \longrightarrow 0) \oplus (P \xrightarrow{\varphi} P_m^{(Z)}) \oplus (0 \longrightarrow P_m^{(Z')}),$$

where $\varphi \in \mathcal{Q}^3(\Lambda_{\mathbb{S}})$.

(3) Let $\operatorname{Sub}_{P_m,P_M}(\Lambda_{\mathbb{S}})$ be the full subcategory of $\Lambda_{\mathbb{S}}$-Mod formed by the objects in $\operatorname{Sub}(\Lambda_{\mathbb{S}})$ without direct summands of the form P_m or P_M. Consider the functor $\operatorname{Im} : \mathcal{Q}^3(\Lambda_{\mathbb{S}}) \longrightarrow \operatorname{Sub}_{P_m,P_M}(\Lambda_{\mathbb{S}})$ which maps every object $\varphi : P \longrightarrow P_m^{(Z)}$ onto $\operatorname{Im}\varphi$ and any morphism $f = (f_1, f_2) : \varphi \longrightarrow \varphi'$ onto the restriction $f_{2|} : \operatorname{Im}\varphi \longrightarrow \operatorname{Im}\varphi'$. Then, the functor Im is full, dense and reflects isomorphisms. Thus, from the representation theory point of view, the study of the category $\mathcal{S}(\mathbb{S})$ of $\mathbb{S}$-spaces, defined in (34.5), can be reduced to the study of the category of modules over the ditalgebra $\mathcal{A}^{\mathbb{S}}$.

Proof.

(1) Consider the splitting $\Lambda = \Lambda_{\mathbb{S}} = S \oplus J$ of the suspension algebra over its radical. Consider the canonical idempotents e_i of $\Lambda_{\mathbb{S}}$ and e'_i, e''_i of $R^{\Lambda_{\mathbb{S}}}$, as described in (23.25) (here i runs in $X \cup \{m, M\}$, as in (34.6)). Consider the idempotents $e' := \sum_i e'_i$ and $e'' := \sum_i e''_i$. Take $N \in \mathcal{A}^{\mathbb{S}}$-Mod, then $e''_i F_d(N) = 0$, for $i \in X$, $e'_M F_d(N) = 0$, $e'_m F_d(N) = 0$ and $e''_M F_d(N) = 0$.

By definition

$$\varphi := \Xi_{\Lambda_{\mathbb{S}}}(F_d(N)) : \Lambda_{\mathbb{S}} \otimes_S F_d(N)_1 \longrightarrow \Lambda_{\mathbb{S}} \otimes_S F_d(N)_2.$$

Thus

$$\begin{cases} P := \Lambda \otimes_S F_d(N)_1 = \Lambda \otimes_S e' F_d(N) = \Lambda \otimes_S (\oplus_{i \in X} e'_i F_d(N)) \\ \Lambda \otimes_S F_d(N)_2 = \Lambda \otimes_S e'' F_d(N) = \Lambda \otimes_S (e''_m F_d(N)) \cong P_m^{(Z)} \end{cases}$$

and P does not admit direct summands of the form P_m and P_M. Thus, $\varphi \in \mathcal{Q}^1(\Lambda_{\mathbb{S}})$. Since Ξ_Λ is an equivalence, so is $\Xi_{\mathbb{S}}$.

(2) Take $\varphi : P \longrightarrow P_m^{(Z)}$ in $\mathcal{Q}^1(\Lambda_{\mathbb{S}})$. From (18.9), we can eliminate a direct summand of type $P' \longrightarrow 0$ and assume that $\varphi \in \mathcal{Q}^2(\Lambda_{\mathbb{S}})$. We have that $\operatorname{soc} P_m^{(Z)} = \operatorname{soc}\operatorname{Im}\varphi \oplus T$ is a direct sum of copies of the socle of P_m, which is the simple module S_M, and $P_m = I_M$ is injective. Then, $\operatorname{Im}\varphi \subseteq P_m^{(Z)}$ is an injective envelope iff $\operatorname{soc}\operatorname{Im}\varphi = \operatorname{soc} P_m^{(Z)}$. Moreover, by the minimality of the injective envelope, the inclusion $\operatorname{Im}\varphi \subseteq P_m^{(Z)}$ splits as the direct sum $(\operatorname{Im}\varphi \subseteq P_m^{(Z')}) \oplus (0 \longrightarrow P_m^{(Z'')})$, where the first summand is the injective envelope of $\operatorname{Im}\varphi$. Statement (2) follows from this.

(3) Let us show that Im is well defined. Since $\operatorname{Im}\varphi \subseteq JP_m^{(Z)}$, we know that P_m is not a direct summand of $\operatorname{Im}\varphi$. Since $\operatorname{Ker}\varphi \subseteq JP$, $\varphi : P \longrightarrow \operatorname{Im}\varphi$ is a projective cover. But, P_M is not a direct summand of P; hence, P_M is not a direct summand of $\operatorname{Im}\varphi$.

Now, we show that Im is dense. Take $L \in \operatorname{Sub}_{P_m, P_M}(\Lambda_{\mathbb{S}})$. Consider the projective cover $P \xrightarrow{\pi} L$, the injective envelope $L \xrightarrow{\sigma} P_m^{(Z)}$ and the composition $\varphi := \sigma\pi$. By (34.4), $L = P_m^{(Z')} \oplus P_M^{(Z'')} \oplus \widehat{L}$, where the projective cover of $\widehat{L}$ does not admit direct summands of the form P_m or P_M. By assumption, $L = \widehat{L}$ and P_m, P_M are not direct summands of P. Since π is a projective cover, $\operatorname{Ker}\varphi \subseteq JP$. Again, (34.4) implies that $L_m = 0$ and, then, $\operatorname{Im}\varphi \cong L \subseteq JP_m^{(Z)}$. Thus, $\varphi \in \mathcal{Q}^3(\Lambda_{\mathbb{S}})$ and Im is a dense functor.

Given $\varphi : P \longrightarrow P_m^{(Z)} \in \mathcal{Q}^3(\Lambda_{\mathbb{S}})$, the inclusion $\sigma : \operatorname{Im}\varphi \longrightarrow P_m^{(Z)}$ is the injective envelope of $\operatorname{Im}\varphi$ and the restriction $\pi = \varphi^| : P \longrightarrow \operatorname{Im}\varphi$ is the projective cover. Thus, given another $\varphi' : P' \longrightarrow P_m^{(Z')} \in \mathcal{Q}^3(\Lambda_{\mathbb{S}})$ and a morphism $g : \operatorname{Im}\varphi \longrightarrow \operatorname{Im}\varphi'$, there is a commutative diagram in Λ-Mod

$$\begin{array}{ccccc} P & \xrightarrow{\pi} & \operatorname{Im}\varphi & \xrightarrow{\sigma} & P_m^{(Z)} \\ {\scriptstyle f_1}\downarrow & & {\scriptstyle g}\downarrow & & {\scriptstyle f_2}\downarrow \\ P' & \xrightarrow{\pi'} & \operatorname{Im}\varphi' & \xrightarrow{\sigma'} & P_m^{(Z')}. \end{array}$$

Then, $f = (f_1, f_2) : \varphi \longrightarrow \varphi'$ is a morphism in $\mathcal{Q}^3(\Lambda_{\mathbb{S}})$, which the functor Im maps onto g. Then, Im is a full functor. If g is an isomorphism, so will be f. Thus, the functor Im reflects isomorphisms. □

Exercise 34.8. *We say that the poset $\mathbb{S}'$ is contained in the poset $\mathbb{S}$ iff there is an order preserving injective map $\mathbb{S}' \to \mathbb{S}$. Show that if $\mathcal{A}_{\mathbb{S}'}$ is wild, then so is $\mathcal{A}_{\mathbb{S}}$ (see (34.2)). Then, show that if $\mathcal{A}_{\mathbb{S}}$ is wild, so is $\mathcal{A}(\mathbb{T}, \mathbb{S})$, for any poset $\mathbb{T}$.*

Hint: Show that $\mathcal{A}_{\mathbb{S}'}$ can be obtained from $\mathcal{A}_{\mathbb{S}}$ by deletion of a suitable set of idempotents of $\mathcal{A}_{\mathbb{S}}$. Thus, $\mathcal{A}_{\mathbb{S}'} \cong (\mathcal{A}_{\mathbb{S}})^d$ and we can apply (22.8).

Exercise 34.9. *Using (33.27) and the vectors $x \in \mathbb{N}^{\mathcal{P}}$ specified for each case, show that the ditalgebras $\mathcal{A}_{\mathbb{S}}$ associated to the following posets are wild (see (34.2)). Here, we assume that the ground field k is algebraically closed.*

(1) $\mathbb{S} = \{a_1; a_2; a_3; a_4; a_5\}$, *with* $x = (2, 1, 1, 1, 1, 1)$;
(2) $\mathbb{S} = \{a_1; a_2; a_3; a_4 < a_5\}$, *with* $x = (4, 2, 2, 2, 1, 1)$;
(3) $\mathbb{S} = \{a_1 < a_2; a_3 < a_4; a_5 < a_6 < a_7\}$, *with* $x = (6, 2, 2, 2, 2, 2, 1, 1)$;
(4) $\mathbb{S} = \{a_1; a_2 < a_3 < a_4; a_5 < a_6 < a_7 < a_8\}$, *with the vector* $x = (8, 4, 2, 2, 2, 2, 2, 1, 1)$;
(5) $\mathbb{S} = \{a_1 < a_2 > a_3 < a_4; a_5 < a_6 < a_7 < a_8 < a_9\}$, *with the vector* $x = (10, 4, 2, 2, 4, 2, 2, 2, 1, 1)$;
(6) $\mathbb{S} = \{a_1; a_2 < a_3; a_4 < a_5 < a_6 < a_7 < a_8 < a_9\}$, *with the vector* $x = (12, 6, 4, 4, 2, 2, 2, 2, 1, 1)$.

Exercise 34.10. *If $\mathbb{T}_2$ denotes poset with two uncomparable elements and $\mathbb{S}$ is any poset, consider the ditalgebra $\mathcal{B}^{\mathbb{S}} := \mathcal{A}(\mathbb{S}, \mathbb{T}_2)$, see (34.1). Thus, $\mathcal{B}^{\mathbb{S}}$ is a nested ditalgebra (with no marked points) which we call the ditalgebra with two sinks associated to the partially ordered set $\mathbb{S}$. Using (33.27) and (33.28), under the assumption that k is algebraically closed, show that:*

(1) $\mathcal{B}^{\mathbb{S}}$ *is of infinite representation type if* $\mathbb{S} = \{a_1; a_2\}$;
(2) $\mathcal{B}^{\mathbb{S}}$ *is wild if* $\mathbb{S} = \{a_1; a_1; a_3\}$ *or* $\mathbb{S} = \{a_1 < a_2; a_3 < a_4\}$.

References. Representations of posets were introduced by L. A. Nazarova and A. V. Roiter in [41]. Representations of pairs of partially ordered sets were introduced by M. Kleiner in [38]. Tame posets were classified in [42] and tame pairs of posets were classified in [39]. The formulation of the problem of pairs of posets in terms of differential graded categories is given in [46]. The idea of realizing categories of representations of posets as subcategories of module categories, as in (34.5), originated in [9]. Statement (34.9) is taken from [50].

35

Further examples of wild ditalgebras

This section consists of a series of lemmas which exhibit some interesting families of wild seminested ditalgebras with one or two points. The wildness is proved, as we did before in the case of critical ditalgebras, avoiding the explicit construction of a bimodule which produces wildness. Instead, we provide arguments with few computational verifications and which can be readily visualized in the appropriate diagrams. Since we want to use (33.27) (more precisely (34.9) and (34.10)), we will assume through the whole section that our ground field is algebraically closed.

For each one of the wild ditalgebras $\mathcal{A}$ presented here, by Theorem (27.10), there is a finite sequence of basic operations $\mathcal{A} \mapsto \mathcal{A}^{z_1} \mapsto \cdots \mapsto \mathcal{A}^{z_1 \cdots z_t} = \mathcal{C}$, with $z_1, \ldots, z_t \in \{a, d, r, e, u\}$ and $\mathcal{C}$ a critical ditalgebra. However, it is apparently simpler to verify the wildness of $\mathcal{A}$ by a different procedure.

Lemma 35.1. *Assume that $\mathcal{A} = (T, \delta)$ is a triangular ditalgebra with layer (R, W). Assume that we have a subset N of W_0 with a set filtration $\mathcal{F}(N)$: $\emptyset = N_0 \subseteq \cdots \subseteq N_\ell = N$. Furthermore, assume that there are R-R-bimodule decompositions $W_0 = W_0' \oplus W_0''$ and $W_1 = W_1' \oplus W_1''$ such that:*

(1) $W_0'' = \widetilde{N}$ is the R-R-subbimodule of W_0 generated by N.
(2) $\delta(N_{i+1}) \subseteq \langle N_i \rangle W_1' + W_1'' + W_1' \langle N_i \rangle$, where $\langle N_i \rangle$ denotes the ideal of $[T]_0$ generated by N_i;

Consider the ideal I of $\mathcal{A}$ generated by the set N and the canonical projection $\eta : \mathcal{A} \longrightarrow \mathcal{A}/I$. From (2), we know that N is a triangular subset of W_0 and hence, by (24.2), the quotient ditalgebra $\mathcal{A}/I = (\overline{T}, \overline{\delta})$ is a Roiter ditalgebra with layer $(R, \eta(W))$, where $\eta(W) = \eta(W_0) \oplus \eta(W_1)$. Here, we have more: the layer $(R, \eta(W))$ has the form $(R, W_0' \oplus W_1' \oplus \widehat{W}_1'')$, where $\widehat{W}_1''$ is some quotient of the R-R-bimodule W_1''.

Proof. From (24.2), we know that the layer of $\mathcal{A}/I$ has the form $(R, W_0' \oplus \eta(W_1))$. Thus we only have to examine the form that the 1-part of the bimodule of the layer takes at each step of the construction of $\mathcal{A}/I$ as a finite sequence of quotients $\mathcal{A}^{p_1 p_2 \cdots p_\ell}$ as in the proof of (24.2), where each $\mathcal{A}^{p_1 \cdots p_i}$ is obtained from $\mathcal{A}^{p_1 \cdots p_{i-1}}$ by factoring out the subbimodule $\widetilde{N}_i$ generated by the image of N_i in this quotient. Condition (2) implies that in $\mathcal{A}^{p_1}$ we obtain $W_1' \oplus W_1''(p_1)$, where $W_1''(p_1) = W_1''/\delta(\widetilde{N}_1)$; in $\mathcal{A}^{p_1 p_2}$ we obtain $W_1' \oplus W_1''(p_1 p_2)$, where $W_1''(p_1 p_2) = W_1''(p_1)/\delta^{p_1}(\widetilde{N}_2)$; in $\mathcal{A}^{p_1 p_2 p_3}$ we obtain $W_1' \oplus W_1''(p_1 p_2 p_3)$, where $W_1''(p_1 p_2 p_3) = W_1''(p_1 p_2)/\delta^{p_1 p_2}(\widetilde{N}_3)$; proceeding like this, in ℓ steps, in $\mathcal{A}^{p_1 p_2 \cdots p_\ell}$ we obtain $\eta(W_1) \cong W_1' \oplus W_1''(p_1 p_2 \cdots p_\ell)$, where $W_1''(p_1 p_2 \cdots p_\ell) = W_1''(p_1 \cdots p_{\ell-1})/\delta^{p_1 \cdots p_{\ell-1}}(\widetilde{N}_\ell)$. □

Lemma 35.2. *Let $\mathcal{A} = (T, \delta)$ be a seminested ditalgebra with only two points 1 and 2, with $Re_1 = k[x]_{f(x)}$ and $Re_2 = k$; at least one solid arrow $\alpha : 1 \to 2$ and at least one dotted arrow $v : 1 \to 2$. Moreover, assume that $\delta(\alpha) = vh(x)$, where $h(x) \in k[x]_{f(x)}$ has the form $0 \neq h(x) = (x - \lambda_1)^{m_1} \cdots (x - \lambda_q)^{m_q} g(x)$, where $\lambda_1, \ldots, \lambda_q \in k$ are pairwise different, $f(\lambda_j) \neq 0$ for all j, $g(x) \in k[x]_{f(x)}$ and $m_1, \ldots, m_q$ are non-negative integers. Then we have that $\mathcal{A}$ is wild if $\sum_{s=1}^{q} m_s \geq 4$.*

Proof. (Sketch) After replacing the basis of the layer of $\mathcal{A}$ (only replace v by vu, for some suitable unit $u \in k[x]_{f(x)}$), if necessary, we can assume that $g(x) \in k[x]$. We can also assume that $(x - \lambda_i)$ is not a factor of $g(x)$ for all i (indeed, we can increase the value of m_i if necessary). Then, by localizing at the point 1 using $g(x)$, we can assume that $g(x) = 1$ (because, once we know that this localization $\mathcal{A}^\ell$ is wild, we get that $\mathcal{A}$ is wild too). Thus, $h(x) = (x - \lambda_1)^{m_1} \cdots (x - \lambda_q)^{m_q}$ and $m_1 + \cdots + m_q \geq 4$. For any $s \in [1, q]$, we can write

$$h(x) = \sum_j c_{s,j}(x - \lambda_s)^j,$$

for some $c_{s,j} \in k$. Here, the root λ_s of $h(x)$ has multiplicity m in $h(x)$ iff $0 = c_{s,0} = c_{s,1} = \cdots = c_{s,m-1}$ and $c_{s,m} \neq 0$. We will consider five cases, according to the partitions of 4. In each case, we will consider an unravelling $\mathcal{A}^u$ at the point 1, using an appropriate $n \geq 2$ and the roots $\lambda_1, \ldots, \lambda_q$ of $h(x)$. In all five cases, we will fix $z := v_2^2$. Then, having in mind the τ-notation of (24.6), we get for any arrow $z \otimes \alpha \otimes w$ of $\mathcal{A}^u$ with $w \in \widetilde{Q}_{\lambda_s}$

$$\delta^u(z \otimes \alpha \otimes w) = \sum_{j \geq m_s} c_{s,j} z \otimes v \otimes \tau^j w - \sum_{w < w' \in \widetilde{Q}_{\lambda_s}} z \otimes \alpha \otimes w' \gamma_{w'},$$

for some $\gamma_{w'} \in P^*$. Let us examine in some detail the first case.

Case 1: $m_1 \geq 4$.

Consider the unravelling at the point 1 using $n = 8$ and $\lambda_1 \in k$. Make $z := v_2^2$. Then, we get for any arrow $z \otimes \alpha \otimes w$ of $\mathcal{A}^u$, with $w \in \widetilde{Q}_{\lambda_1}$

$$\delta^u(z \otimes \alpha \otimes w) = \sum_{j \geq m_1} c_{1,j} z \otimes v \otimes \tau^j w - \sum_{w < w' \in \widetilde{Q}_{\lambda_1}} z \otimes \alpha \otimes w' \gamma_{w'},$$

for some $\gamma_{w'} \in P^*$.

We shall write $z_a := x_{1,a}^{a-1}$, for $a \in [1, n]$. Recall that the "central axis" of the quiver $\widetilde{Q}_{\lambda_1}$ is the set of points $\tau^{-s} z_i$ in $\widetilde{Q}_{\lambda_1}$ determined by the equation $i = 2s + 1$. Consider the following subset N of arrows of W_0^u

$$N := \{z \otimes \alpha \otimes \tau^{-s} z_i \mid i \in [2, 8], s \in [0, (i-1)/2) \cap \mathbb{Z}\} \setminus \{z \otimes \alpha \otimes \tau^{-3} z_8\},$$

and consider the filtration $\mathcal{F}(N) : \emptyset = N_0 \subseteq N_1 \subseteq \cdots \subseteq N_\ell = N$, where $\ell = n - 1$ and each N_t is, by definition, the subset of N formed by the elements $z \otimes \alpha \otimes w \in N$ such that $\mathbb{h}(w) \leq t$. We want to factor out N, that is all the arrows of $\mathcal{A}^u$ produced by points in $\widetilde{Q}_{\lambda_1}$ below the central axis, with the only exception of $\tau^{-3} z_8$.

The formula for the differential δ^u stated above shows that N is triangular with filtration $\mathcal{F}(N)$. Moreover, $W_0^u = P^* \oplus (X^* \otimes_R W_1 \otimes_R X)$ and

$$\delta^u(N_i) \subseteq \langle N_{i-1} \rangle P^* + X^* \otimes_R W_1 \otimes_R X + P^* \langle N_{i-1} \rangle,$$

for all i. Consider the decomposition $W_0^u = W_0^c \oplus \widetilde{N}$, where $\widetilde{N}$ and W_0^c are the S-S-subbimodules of W_0^u generated, respectively, by the set of arrows N and $\mathbb{B}_0^u \setminus N$. Thus, we can apply (35.1). Therefore, the quotient ditalgebra $\mathcal{A}^u/I$, where I is the ideal of $\mathcal{A}^u$ generated by N, is a Roiter ditalgebra with layer $(S, W_0^c \oplus P^* \oplus \widehat{W}_1)$.

Now, select the five arrows $\alpha_i := z \otimes \alpha \otimes \tau^{-i} z_{2i+1}$, $i \in [0, 3]$, and $\alpha_4 := z \otimes \alpha \otimes \tau^{-3} z_8$. For simplicity, we denote by $\widehat{\delta}$ the differential of the quotient $\mathcal{A}^u/I$ and we write with the same symbol an arrow of $\mathcal{A}^u$ and its class modulo I. Notice that, in the quotient $\mathcal{A}^u/I$, we have

$$\begin{aligned}
\widehat{\delta}(\alpha_0) &= \widehat{\delta}(z \otimes \alpha \otimes z_1) = 0\\
\widehat{\delta}(\alpha_1) &= \widehat{\delta}(z \otimes \alpha \otimes \tau^{-1} z_3) = 0\\
\widehat{\delta}(\alpha_2) &= \widehat{\delta}(z \otimes \alpha \otimes \tau^{-2} z_5) = 0\\
\widehat{\delta}(\alpha_4) &= \widehat{\delta}(z \otimes \alpha \otimes \tau^{-3} z_8) = 0\\
\widehat{\delta}(\alpha_3) &= \widehat{\delta}(z \otimes \alpha \otimes \tau^{-3} z_7) = -(z \otimes \alpha \otimes \tau^{-3} z_8)\overline{\gamma} = -\alpha_4 \overline{\gamma},
\end{aligned}$$

where $\overline{\gamma}$ is the class, modulo I, of the element $\gamma_{\xi(\tau^{-3} z_7, \tau^{-3} z_8)} \in P^* \subseteq W_1^u$, associated to the class path $\xi(\tau^{-3} z_7, \tau^{-3} z_8)$ of Q given as the image of the path in $\widetilde{Q}_{\lambda_1}$ connecting $\tau^{-3} z_7$ with $\tau^{-3} z_8$ (see (23.24)). Consider the triangular proper subditalgebra $\mathcal{B}$ of $\mathcal{A}^u/I$ determined by the five solid arrows $\alpha_0, \ldots, \alpha_4$ and the dotted arrow $\overline{\gamma}$. Having in mind the formulas for the differentials of the arrows $\alpha_0, \ldots, \alpha_4$ given above and recalling from (23.22) that $\mathbb{h}(\overline{\gamma}) = \ell(\xi) = 1$ (thus $\delta^u(\overline{\gamma}) = 0$), we see that $\mathcal{B}$ is indeed a triangular proper subditalgebra of $\mathcal{A}^u/I$. By (22.13), we only have to show that $\mathcal{B}$ is wild.

We claim that after deleting some points of $\mathcal{B}$, we get a ditalgebra $\mathcal{B}^d$ which is isomorphic to the wild ditalgebra $\mathcal{A}^{\mathbb{S}}$ of the poset $\mathbb{S} = \{a_1; a_2; a_3; a_4 < a_5\}$ (see (34.9)(2)). And, therefore, $\mathcal{B}^d$ is wild and, so is $\mathcal{B}$, as we wanted to show. Indeed, delete from $\mathcal{B}$ all the points with the exception of: $a_0 := 2$, $a_1 := i_1$, $a_2 := i_3$, $a_3 := i_5$, $a_4 := i_8$ and $a_5 := i_7$.

Notice that the argument used in this case may be *codified* in the next figure. There, $\widetilde{Q}_{\lambda_1}$ is shown with some points surrounded by tiny circles (which indicate the points $w \in \widetilde{Q}_{\lambda_1}$ such that $z \otimes \alpha \otimes w \in N$; these are the arrows of $\mathcal{A}^u$ to be factored out) and some points surrounded by dotted rectangles (which indicate the points $w \in \widetilde{Q}_{\lambda_1}$ such that the arrow $z \otimes \alpha \otimes w$ is "selected"). Selected arrows, together with the dotted arrow γ_ξ of $\mathcal{A}^u$ corresponding to class paths inside dotted rectangles, determine the wild poset $\mathbb{S}$ sitting in $\mathcal{A}^u/I$. Notice that the points in the lower diagonal path of the picture are $z_1, z_2, \ldots, z_n$ (here $n = 8$). Notice also that $m_1 \geq 4$ implies that, for $j \geq m_1$, $\tau^j w = 0$, for any point w in a dotted rectangle: this geometric fact guarantees that the dotted arrows $z \otimes v \otimes \tau^j w$ do not appear in the differential of the corresponding selected arrow $z \otimes \alpha \otimes w$ of $\mathcal{A}^u$.

Case 1: $m_1 \geq 4$.

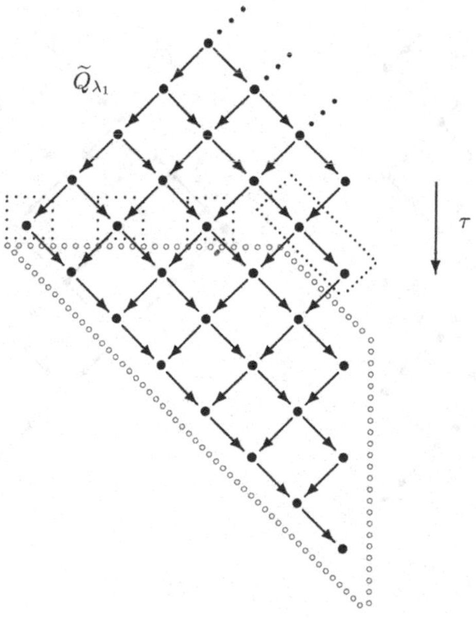

In each of the remaining four cases, we proceed similarly, with the argument codified in the corresponding figure as follows:

Case 2: $m_1, m_2 \geq 2$.

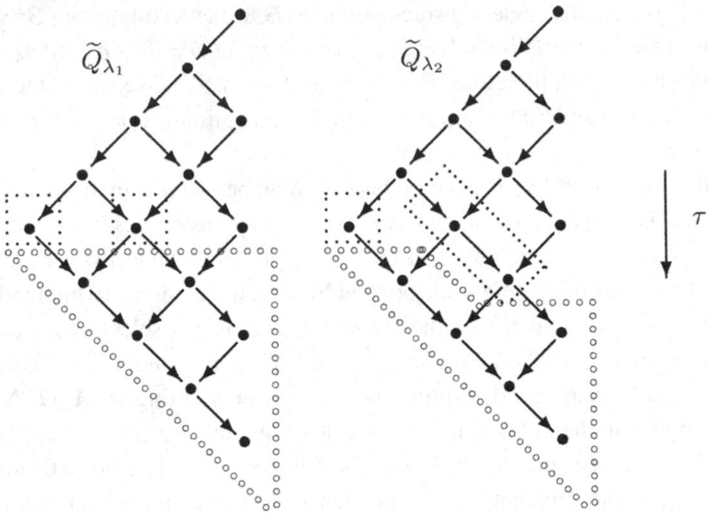

Case 3: $m_1 \geq 3$ *and* $m_2 \geq 1$.

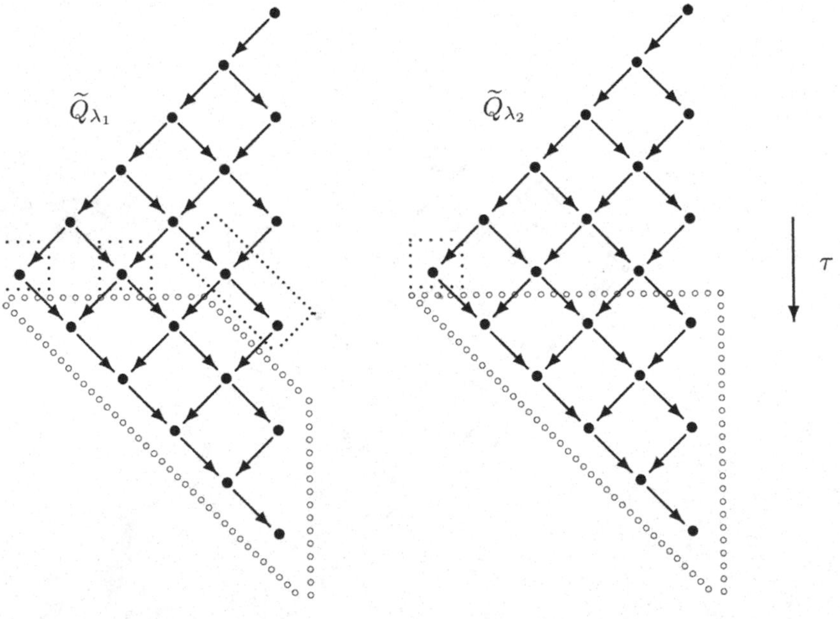

Case 4: $m_1 \geq 2$ *and* $m_2, m_3 \geq 1$.

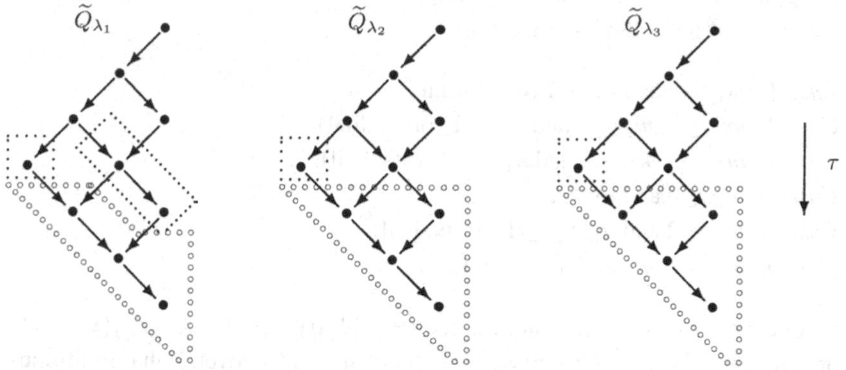

Case 5: $m_1, m_2, m_3, m_4 \geq 1$.

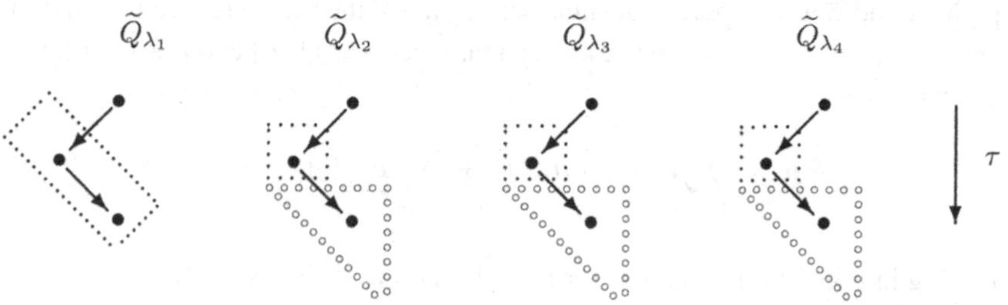

□

Lemma 35.3. *Assume that $\mathcal{A}$ is a seminested ditalgebra with layer (R, W) and differential δ. Assume $\mathcal{A}$ has only two points 1 and 2, with $Re_1 = k[x]_{f_1(x)}$ and $Re_2 = k[y]_{f_2(y)}$; at least one solid arrow $\alpha : 1 \to 2$ and at least two dotted arrows $v_1, v_2 : 1 \longrightarrow 2$. Moreover, assume that $\delta(\alpha) = h_2(y)v_2 + v_1h_1(x)$, for some non-zero elements $h_1(x) \in k[x]_{f_1(x)}$ and $h_2(y) \in k[y]_{f_2(y)}$, where*

$$\begin{cases} h_1(x) = (x-\lambda_1)^{m_1}(x-\lambda_2)^{m_2}\cdots(x-\lambda_q)^{m_q}g_1(x), \\ h_2(y) = (y-\mu_1)^{n_1}(y-\mu_2)^{n_2}\cdots(y-\mu_p)^{n_p}g_2(y), \end{cases}$$

where $\lambda_1, \lambda_2, \ldots, \lambda_q \in k$ are pairwise different and $f_1(\lambda_s) \neq 0$, for all $s \in [1, q]$; $\mu_1, \mu_2, \ldots, \mu_p \in k$ are pairwise different and $f_2(\mu_r) \neq 0$, for all $r \in [1, p]$; $m_1, m_2, \ldots, m_q, n_1, n_2, \ldots, n_p$ are non-negative integers; $g_1(x) \in k[x]_{f_1(x)}$ and $g_2(y) \in k[y]_{f_2(y)}$. Then we have that $\mathcal{A}$ is wild if ($\sum_{s=1}^q m_s \geq 2$ and $\sum_{r=1}^p n_r \geq 2$) or ($\sum_{s=1}^q m_s \geq 1$ and $\sum_{r=1}^p n_r \geq 3$) or, dually, ($\sum_{r=1}^p n_r \geq 1$ and $\sum_{s=1}^q m_s \geq 3$).

Proof. (Sketch) By a similar argument to the one exhibited at the beginning of the proof of (35.2), we may assume that $g_1(x) = 1$ and $g_2(y) = 1$. Then, we are in one of the following cases:

Case 1: $m_1 \geq 3$ and $n_1 \geq 1$ (or its dual).
Case 2: $m_1 \geq 2, m_2 \geq 1$ and $n_1 \geq 1$ (or its dual).
Case 3: $m_1, m_2, m_3 \geq 1$ and $n_1 \geq 1$ (or its dual).
Case 4: $m_1 \geq 2$ and $n_1 \geq 2$.
Case 5: $m_1 \geq 2$ and $n_1, n_2 \geq 1$ (or its dual).
Case 6: $m_1, m_2 \geq 1$ and $n_1, n_2 \geq 1$.

In every case, we can write, for each $s \in [1, q]$, $h_1(x) = \sum_j c_{s,j}(x - \lambda_s)^j$ and, for each $r \in [1, p]$, $h_2(y) = \sum_i c'_{r,i}(y - \mu_r)^i$. Moreover, λ_s has multiplicity m in $h_1(x)$ iff $0 = c_{s,0} = c_{s,1} = \cdots = c_{s,m-1}$ and $c_{s,m} \neq 0$; and, similarly, for $h_2(y)$. In every case, we will consider a simultaneous unwinding $\mathcal{A}^X$ at the points 1 and 2, using appropriate numbers m, n and the roots $\lambda_1, \ldots, \lambda_q$ of $h_1(x)$ and $\mu_1, \ldots, \mu_p$ of $h_2(y)$, see (24.8). Then, given $s \in [1, q]$ and $r \in [1, p]$, we can write

$$\delta(\alpha) = \sum_{i \geq n_r} c'_{r,i}(y - \mu_r)^i v_2 + \sum_{j \geq m_s} c_{s,j} v_1 (x - \lambda_s)^j.$$

Having in mind the τ-notation, for $z \in \widetilde{Q}^y_{\mu_r}$ and $w \in \widetilde{Q}^x_{\lambda_s}$, we have

$$\begin{aligned}\delta^X(z \otimes \alpha \otimes w) = & \textstyle\sum_{z > z' \in \widetilde{Q}^y_{\mu_r}} \gamma_{z'} z' \otimes \alpha \otimes w \\ & + \textstyle\sum_{i \geq n_r} c'_{r,i} \tau^{-i} z \otimes v_2 \otimes w + \sum_{j \geq m_s} c_{s,j} z \otimes v_1 \otimes \tau^j w \\ & - \textstyle\sum_{w < w' \in \widetilde{Q}^x_{\lambda_s}} z \otimes \alpha \otimes w' \gamma_{w'},\end{aligned}$$

for some $\gamma_{z'}, \gamma_{w'} \in P^*$.

We split the argument of the proof in two parts.

Part 1: If we are in one of the cases 1, 2 *or* 3.

Consider the ditalgebra $\mathcal{A}^X$ obtained from $\mathcal{A}$ by "simultaneous unwinding" at the point 1 with $m = 9$ and $\lambda_1, \ldots, \lambda_q$, and at the point 2 with $n = 1$ and μ_1, see (24.8). Fix $z := y^0_{1,1} \in \widetilde{Q}^y_{\mu_1}$. Then, since $n_1 \geq 1$, for any arrow $z \otimes \alpha \otimes w$ of $\mathcal{A}^X$ with $w \in \widetilde{Q}^x_{\lambda_s}$

$$\delta^X(z \otimes \alpha \otimes w) = \sum_{j \geq m_s} c_{s,j} z \otimes v_1 \otimes \tau^j w - \sum_{w < w' \in \widetilde{Q}^x_{\lambda_s}} z \otimes \alpha \otimes w' \gamma_{w'},$$

for some $\gamma_{w'} \in P^*$.

Then, we proceed in each one of the cases with the argument codified in the corresponding figure (see the argument in the proof of (35.2)) as follows. In all these cases, we can identify the wild ditalgebra $\mathcal{A}^{\mathbb{S}}$ associated to the poset $\mathbb{S} = \{a_1 < a_2; a_3 < a_4; a_5 < a_6 < a_7\}$ "sitting in" an appropriate quotient $\mathcal{A}^X/I$, for some ideal I of $\mathcal{A}^X$ generated by a triangular subset of W_0^X. Thus, the ditalgebra $\mathcal{A}$ is wild.

In Case 1, for instance, we consider the ideal I of $\mathcal{A}^X$ generated by the arrows $z \otimes \alpha \otimes w$, where w lies in the region surrounded by tiny circles. Then, consider the proper triangular subditalgebra $\mathcal{B}$ of $\mathcal{A}^X/I$ defined by the selected arrows $\alpha_w := z \otimes \alpha \otimes w$, where w is a point of $\widetilde{Q}^x_{\lambda_s}$ in a dotted rectangle for some $s \in [1, q]$, and the dotted arrows $\gamma_\xi \in P^*$ of $\mathcal{A}^X/I$ defined by a path in a dotted rectangle. Then, construct $\mathcal{B}^d$ eliminating all the points of $\mathcal{B}$ with the exception of $a_0 = i_1^y$, $a_1 = i_1^x$, $a_2 = i_2^x$, $a_3 = i_4^x$, $a_4 = i_5^x$, $a_5 = i_7^x$, $a_6 = i_8^x$ and $a_7 = i_9^x$. The fact that $\mathcal{B}^d$ is isomorphic to $\mathcal{A}^{\mathbb{S}}$ follows from the description of $\delta^X(\gamma_\xi)$ in (24.8).

Case 1: $m_1 \geq 3$ and $n_1 \geq 1$ (or its dual).

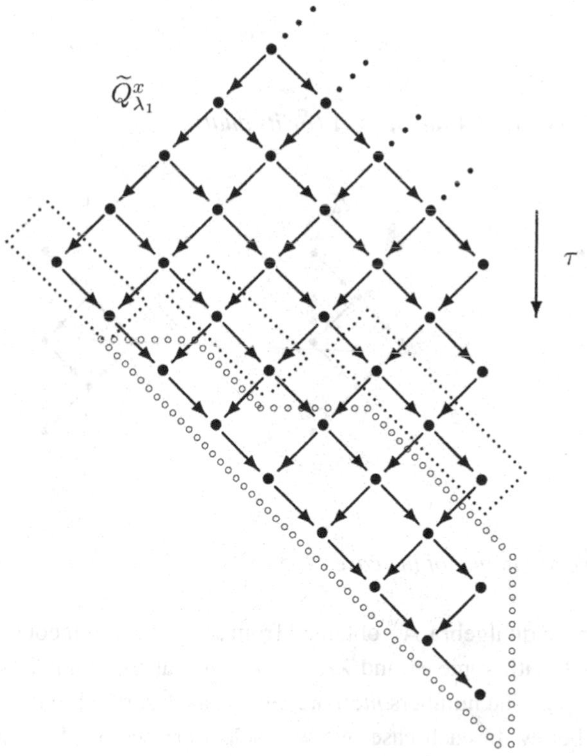

Case 2: $m_1 \geq 2$, $m_2 \geq 1$ *and* $n_1 \geq 1$ *(or its dual).*

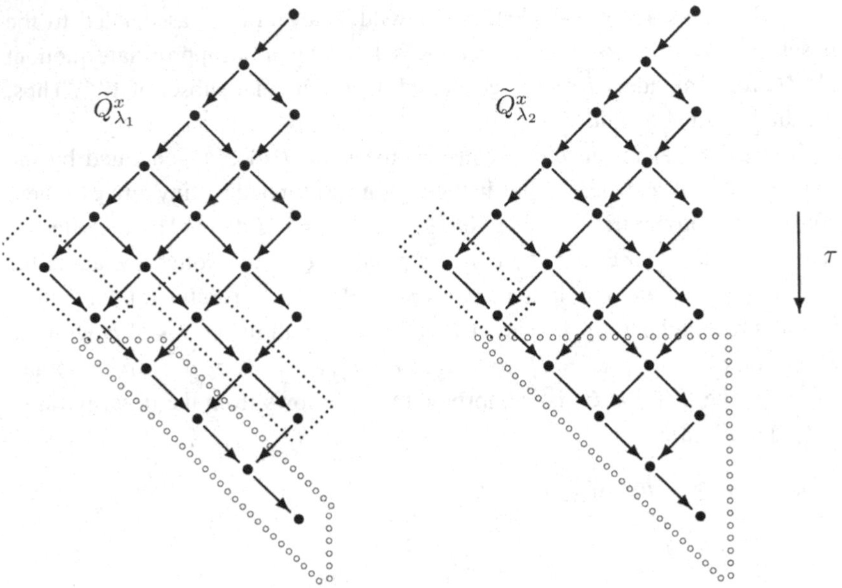

Case 3: $m_1, m_2, m_3 \geq 1$ *and* $n_1 \geq 1$ *(or its dual).*

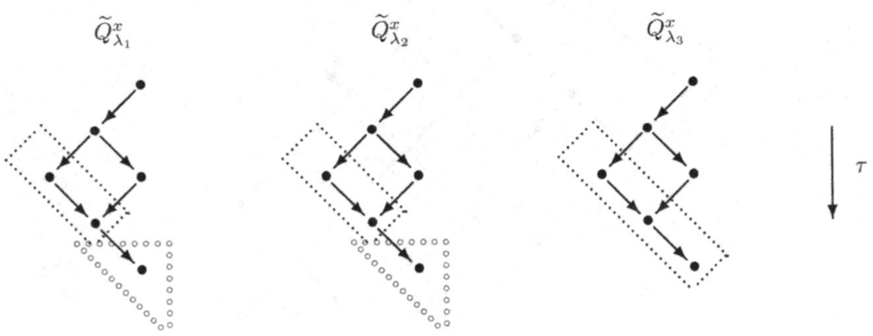

Part 2: If we are in one of the cases 4, 5 or 6.

Consider the ditalgebra $\mathcal{A}^X$ obtained from $\mathcal{A}$ by "simultaneous unwinding" at the point 1 with some m and $\lambda_1, \ldots, \lambda_q$, and at the point 2 with some n and $\mu_1, \ldots, \mu_p$. The numbers m, n are chosen as described in the corresponding figures below. In each case, we will select arrows of $\mathcal{A}^X$ determined by points in dotted rectangles in the components $\widetilde{Q}_{\mu_r}$ and $\widetilde{Q}_{\lambda_s}$, as shown in each picture. The regions surrounded by tiny circles determine the ideal I of $\mathcal{A}^X$ which has to be factored out. In these cases, we can locate in $\mathcal{A}^X/I$ the wild ditalgebra $\mathcal{B}^{\mathbb{S}}$ associated to the poset $\mathbb{S} = \{a_1 < a_2; a_3 < a_4\}$, considered in (34.10).

Now we describe the situation more precisely for Case 4. The argument is codified in the next picture. Thus, $n = 5$ and $m = 3$, and we are considering the ditalgebra $\mathcal{A}^X$ obtained from $\mathcal{A}$ by simultaneous unwinding at the points 1 and 2 using $(5; \lambda_1)$ and $(3; \mu_1)$, respectively. Then, select the arrows of $\mathcal{A}^X$ of the form $z \otimes \alpha \otimes w$, where w lies in some dotted rectangle of $\widetilde{Q}^x_{\lambda_1}$ and z lies in some dotted rectangle of $\widetilde{Q}^y_{\mu_1}$. Now, consider the triangular set N of solid arrows of $\mathcal{A}^X$ of the form $z \otimes \alpha \otimes w$, where $w \in \widetilde{Q}^x_{\lambda_1}$ is surrounded by tiny circles or $z \in \widetilde{Q}^y_{\mu_1}$ is surrounded by tiny circles. Then, consider the ideal I of $\mathcal{A}^X$ generated by N and the proper triangular subditalgebra $\mathcal{B}$ of $\mathcal{A}^X/I$ defined by the selected solid arrows, and by the dotted arrows $\gamma_\xi \in P^*$ of $\mathcal{A}^X/I$ defined by a path in a dotted rectangle. Then, delete the point i^x_3 to obtain $\mathcal{B}^d$. Thus, we are keeping the points $x_0 = i^y_1, x_1 = i^y_3, a_1 = i^x_1, a_2 = i^x_2, a_3 = i^x_4$ and $a_4 = i^x_5$. Then, $\mathcal{B}^d \cong \mathcal{B}^{\mathbb{S}}$ is a wild ditalgebra.

Case 4: $m_1 \geq 2$ *and* $n_1 \geq 2$.

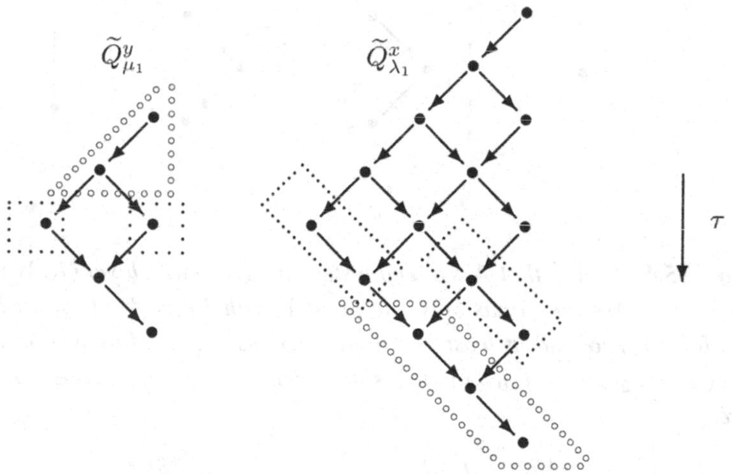

The procedures in the remaining cases are codified in the corresponding following figures.

Case 5: $m_1 \geq 2$ *and* $n_1, n_2 \geq 1$ *(or its dual).*

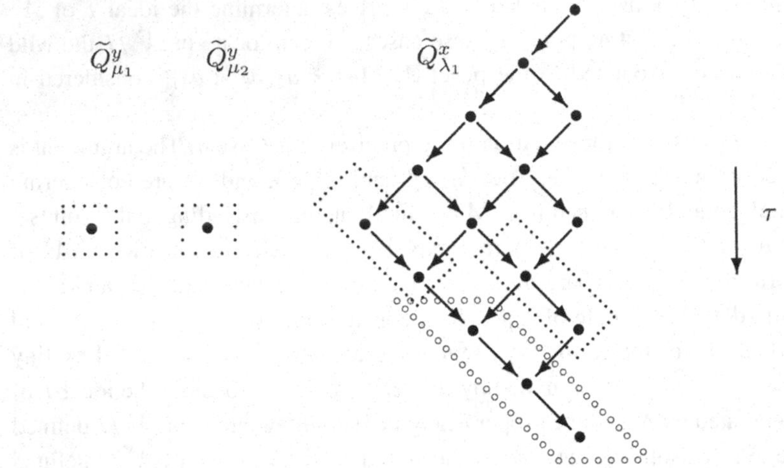

Case 6: $m_1, m_2 \geq 1$ *and* $n_1, n_2 \geq 1$.

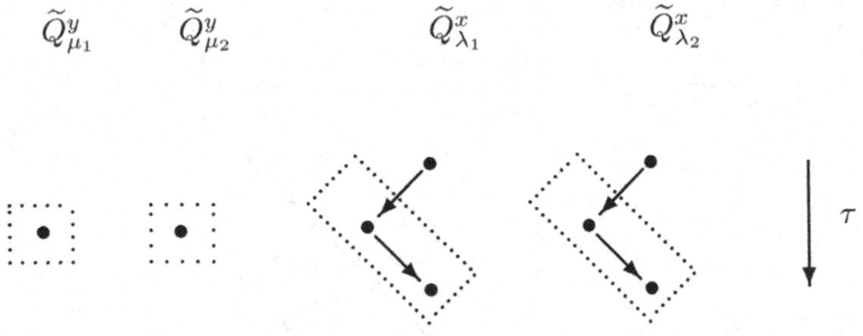

□

Lemma 35.4. *Assume that $\mathcal{A}$ is a seminested ditalgebra with layer (R, W) and differential δ. Assume $\mathcal{A}$ has only one point 1, with $Re_1 = k[x]_{f(x)}$; at least one solid arrow α and at least two dotted arrows v_1, v_2. Moreover, assume that $\delta(\alpha) = h_2(x)v_2 + v_1 h_1(x)$, for some elements $h_1(x), h_2(x) \in k[x]_{f(x)}$, where*

$$\begin{cases} h_1(x) = (x-\lambda_1)^{m_1}(x-\lambda_2)^{m_2}\cdots(x-\lambda_q)^{m_q} g_1(x), \\ h_2(x) = (x-\mu_1)^{n_1}(x-\mu_2)^{n_2}\cdots(x-\mu_p)^{n_p} g_2(x), \end{cases}$$

where $\lambda_1, \lambda_2, \ldots, \lambda_q \in k$ *are pairwise different and* $f(\lambda_s) \neq 0$, *for all* $s \in [1, q]$; $\mu_1, \mu_2, \ldots, \mu_p \in k$ *are pairwise different and* $f(\mu_r) \neq 0$, *for all* $r \in [1, p]$; $m_1, m_2, \ldots, m_q, n_1, n_2, \ldots, n_p$ *are non-negative integers;* $g_1(x), g_2(x) \in k[x]_{f(x)}$. *Then we have:*

(1) *If* $\{\lambda_1, \ldots, \lambda_q\} \cap \{\mu_1, \ldots, \mu_p\} = \emptyset$, *then* $\mathcal{A}$ *is wild whenever* ($\sum_{s=1}^{q} m_s \geq 2$ *and* $\sum_{r=1}^{p} n_r \geq 2$) *or* ($\sum_{s=1}^{q} m_s \geq 1$ *and* $\sum_{r=1}^{p} n_r \geq 3$) *or, dually,* ($\sum_{r=1}^{p} n_r \geq 1$ *and* $\sum_{s=1}^{q} m_s \geq 3$).
(2) *If* $\lambda_i = \mu_j$ *with* $m_i, n_j \geq 1$, *for some* $i \in [1, q]$ *and* $j \in [1, p]$, *then* $\mathcal{A}$ *is wild if* $\sum_{r=1}^{p} n_r \geq 2$ *or, dually,* $\sum_{s=1}^{q} m_s \geq 2$.

Proof. (Sketch) As we did at the beginning of the proof of (35.2), we may assume that $g_1(x) = 1$ and $g_2(x) = 1$.

Under the assumptions of (1), we are in one of the six cases considered at the beginning of (35.3). The pictures encoding the proof for each case given there, where we take $x = y$, can be used here too.

Under the assumptions of (2), we may assume that $i = 1 = j$ and we are in one of the following two cases (or their duals). In each case, there is a wild algebra sitting in the appropriate quotient $\mathcal{A}^X/I$ encoded in the corresponding given picture. In both cases, we have to select the solid arrows of the form $z \otimes \alpha \otimes w$, where w is in a dotted rectangle and z is the point in the right upper corner of $\widehat{Q}_{\lambda_1}$. Then, we can locate in $\mathcal{A}^X/I$ the wild algebra with two points formed by a loop based at the target point of an arrow.

Case 1: $\lambda_1 = \mu_1$, $m_1 \geq 2$ *and* $n_1 \geq 1$.

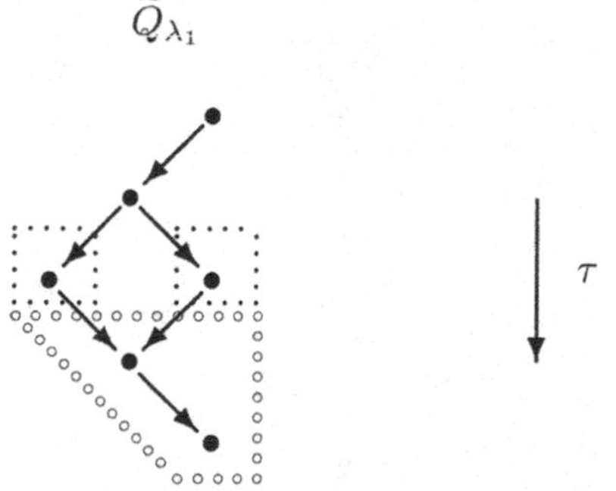

Case 2: $\lambda_1 = \mu_1$*;* $m_1, m_2 \geq 1$ *and* $n_1 \geq 1$.

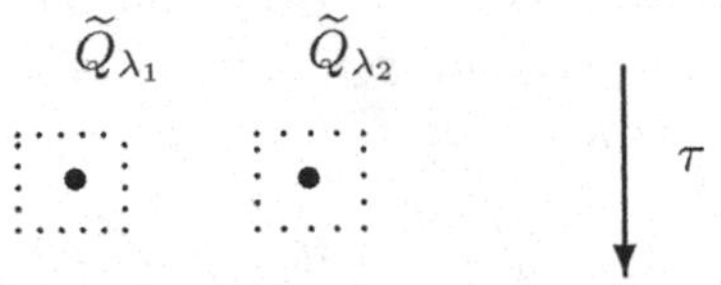

□

References. Lemmas (35.2), (35.3) and (35.4) are due to Yingbo Zhang and Xu Yunge (see [49]). Our proofs rest on the use of appropriate quotients and reductions using admissible modules instead of providing explicit bimodules producing wildness as in [49].

36

Answers to selected exercises

Section 2

Solution to Exercise (2.9): (1) Take any source $\{P' \xrightarrow{f_i} M_i\}_{i \in I}$ in $\mathcal{A}$-Mod. We are looking for some $f \in \mathrm{Hom}_{\mathcal{A}}(P', P)$ such that $\pi_i f = f_i$, for all $i \in I$. That is such that $\pi_i^0 f^0 = f_i^0$ and $\pi_i^0 f^1(v) = f_i^1(v)$, for all $i \in I$ and $v \in V = [T]_1$. Since $\{\pi_i^0 : P \longrightarrow M_i\}_{i \in I}$ is the product in k-Mod, there exists a unique linear map $f^0 : P' \longrightarrow P$ such that $\pi_i^0 f^0 = f_i^0$, for all $i \in I$. For the same reason, given $v \in V$, there exists a unique linear map $f^1(v) : P' \longrightarrow P$ such that $\pi_i^0 f^1(v) = f_i^1(v)$, for all $i \in I$. We claim that this rule defines a morphism of A-A-bimodules $f^1 : V \longrightarrow \mathrm{Hom}_k(P', P)$. Indeed, if $a, b \in A$, $v \in V$, $i \in I$ and $p' \in P'$, then

$$\begin{aligned} \pi_i^0[af^1(v)b][p'] &= \pi_i^0[af^1(v)(bp')] \\ &= a\pi_i^0[f^1(v)(bp')] \\ &= a(\pi_i^0 f^1(v))[bp'] \\ &= a(f_i^1(v))[bp'] \\ &= f_i^1(avb)[p']. \end{aligned}$$

Thus, by the uniqueness of the linear map $f^1(avb)$, we obtain that $f^1(avb) = af^1(v)b$, and that $f^1 \in \mathrm{Hom}_{A\text{-}A}(V, \mathrm{Hom}_k(P', P))$. Now, we show that $f := (f^0, f^1) \in \mathrm{Hom}_{\mathcal{A}}(P', P)$. Take $a \in A$ and $p' \in P'$, then we have to prove the equality $af^0(p') = f^0(ap') + f^1(\delta(a))[p']$ of elements in P. It is enough to show that $\pi_i^0(af^0[p']) = \pi_i^0(f^0[ap'] + f^1(\delta(a))[p'])$, for all $i \in I$. Given such an $i \in I$, we have

$$\begin{aligned} \pi_i^0(af^0[p']) &= a\pi_i^0(f^0[p']) \\ &= af_i^0[p'] \\ &= f_i^0[ap'] + f_i^1(\delta(a))[p'] \\ &= \pi_i^0(f^0[ap'] + f^1(\delta(a))[p']). \end{aligned}$$

Thus, indeed $f \in \mathrm{Hom}_{\mathcal{A}}(P', P)$. Moreover, by construction of f, we clearly have $\pi_i f = f_i$, for all $i \in I$, and its uniqueness.

(2) This is dual to the previous proof. □

Solution to Exercise (2.10): (1) Take $a, b \in A$, then

$$\begin{aligned}
\psi_g(a)\psi_g(b) &= \begin{pmatrix} \psi_M(a) & 0 \\ \psi_N(a)g^0 - g^0\psi_M(a) - g^1(\delta(a)) & \psi_N(a) \end{pmatrix} \\
&\quad \times \begin{pmatrix} \psi_M(b) & 0 \\ \psi_N(b)g^0 - g^0\psi_M(b) - g^1(\delta(b)) & \psi_N(b) \end{pmatrix} \\
&= \begin{pmatrix} \psi_M(a)\psi_M(b) & 0 \\ \Delta & \psi_N(a)\psi_N(b) \end{pmatrix} \\
&= \begin{pmatrix} \psi_M(ab) & 0 \\ \psi_N(ab)g^0 - g^0\psi_M(ab) - g^1(\delta(ab)) & \psi_N(ab) \end{pmatrix} \\
&= \psi_g(ab)
\end{aligned}$$

because ψ_M and ψ_N are morphisms of algebras and, by Leibniz rule

$$\begin{aligned}
\Delta &= \psi_N(a)g^0\psi_M(b) - g^0\psi_M(a)\psi_M(b) - g^1(\delta(a))\psi_M(b) \\
&\quad + \psi_N(a)\psi_N(b)g^0 - \psi_N(a)g^0\psi_M(b) - \psi_N(a)g^1(\delta(b)) \\
&= \psi_N(ab)g^0 - g^0\psi_M(ab) - g^1(\delta(a)b + a\delta(b)) \\
&= \psi_N(ab)g^0 - g^0\psi_M(ab) - g^1(\delta(ab))
\end{aligned}$$

(2) Take $f \in \mathrm{Hom}_{\mathcal{A}}(F_\eta(M), F_\eta(N))$ and assume that $f^1 = \eta_1^*(g^1)$. Take $g^0 = f^0$ and $g = (g^0, g^1)$. Notice that $\psi_M \eta_0 = \psi_{F_\eta(M)}$. Then, for any $a \in A$

$$\begin{aligned}
0 &= \psi_{F_\eta(N)}(a)f^0 - f^0\psi_{F_\eta(M)}(a) - f^1(\delta(a)) \\
&= \psi_N\eta(a)g^0 - g^0\psi_M\eta(a) - g^1(\eta\delta(a)) \\
&= \psi_N\eta(a)g^0 - g^0\psi_M\eta(a) - g^1(\delta'\eta(a)).
\end{aligned}$$

This implies that $\psi_g\eta_0 = \psi_0\eta_0$. Since η_0 is an epimorphism of algebras, $\psi_g = \psi_0$ and $g \in \mathrm{Hom}_{\mathcal{A}'}(M, N)$, as we wanted to show. □

Solution to Exercise (2.11): First we see that $gf = ((gf)^0, (gf)^1)$ is a morphism. Clearly, $(gf)^0 \in \mathrm{Hom}_k(M, L)$, because $\pi(g^1 \otimes f^1)(d) \in \mathrm{Hom}_k(M, L)$.

Let us verify that $(gf)^1 \in \mathrm{Hom}_{A\text{-}A}(V, \mathrm{Hom}_k(M, N))$. For this, take $v \in V$, $a \in A$ and $m \in M$, then

$$\begin{aligned}
[(gf)^1(av)](m) &= [g^0 f^1(av)](m) + [g^1(av)f^0](m) + [\pi(g^1 \otimes f^1)(\delta(av))](m) \\
&= ag^0[f^1(v)(m)] - g^1(\delta(a))[f^1(v)(m)] + [ag^1(v)f^0](m) \\
&\quad + [\pi(g^1 \otimes f^1)(\delta(a)v + a\delta(v))](m) \\
&= ag^0[f^1(v)(m)] - g^1(\delta(a))[f^1(v)(m)] + [ag^1(v)f^0](m) \\
&\quad + [g^1(\delta(a))f^1(v)](m) + a[\pi(g^1 \otimes f^1)(\delta(v))](m) \\
&= ag^0[f^1(v)(m)] + [ag^1(v)f^0](m) + a[\pi(g^1 \otimes f^1)(\delta(v))](m) \\
&= a[(gf)^1(v)](m).
\end{aligned}$$

Now, take $a \in A$, $v \in V$ and $m \in M$, then

$$
\begin{aligned}
[(gf)^1(va)](m) &= [g^0 f^1(va)](m) + [g^1(va) f^0](m) + [\pi(g^1 \otimes f^1)(\delta(va))](m) \\
&= [g^0(f^1(v)a)](m) + (g^1(v)a)[f^0(m)] \\
&\quad + [\pi(g^1 \otimes f^1)(\delta(v)a - v\delta(a))](m) \\
&= g^0[f^1(v)(am)] + (g^1(v))[af^0(m)] \\
&\quad + [\pi(g^1 \otimes f^1)(\delta(v)a - v\delta(a))](m) \\
&= g^0[f^1(v)(am)] + (g^1(v))[f^0(am) + f^1(\delta(a))(m)] \\
&\quad + [\pi(g^1 \otimes f^1)(\delta(v)a)](m) - [g^1(v) f^1(\delta(a))](m) \\
&= g^0[f^1(v)(am)] + (g^1(v))[f^0(am)] + [\pi(g^1 \otimes f^1)(\delta(v)a)](m) \\
&= [g^0 f^1(v)](am) + [g^1(v) f^0](am) + [\pi(g^1 \otimes f^1)(\delta(v))](am) \\
&= [(gf)^1(v)](am) \\
&= [(gf)^1(v)a](m).
\end{aligned}
$$

Finally, take $a \in A$ and $m \in M$, then $gf \in \operatorname{Hom}_A(M, L)$ because

$$
\begin{aligned}
[(gf)^0](am) &= [g^0 f^0](am) - [\pi(g^1 \otimes f^1)(d)](am) \\
&= [g^0 f^0](am) - [\pi(g^1 \otimes f^1)(da)](m) \\
&= g^0[af^0(m) - f^1(\delta(a))(m)] - [\pi(g^1 \otimes f^1)(\delta^2(a) + ad)](m) \\
&= g^0[af^0(m)] - g^0[f^1(\delta(a))](m) - [\pi(g^1 \otimes f^1)(\delta^2(a) + ad)](m) \\
&= ag^0[f^0(m)] - g^1(\delta(a))[f^0(m)] - g^0[f^1(\delta(a))](m) \\
&\quad - [\pi(g^1 \otimes f^1)(\delta^2(a) + ad)](m) \\
&= [a(g^0 f^0)](m) - [a\pi(g^1 \otimes f^1)(d)](m) \\
&\quad - [g^0 f^1(\delta(a)) + g^1(\delta(a)) f^0 + \pi(g^1 \otimes f^1)(\delta^2(a))][m] \\
&= [a(g^0 f^0)](m) - [a\pi(g^1 \otimes f^1)(d)](m) - (gf)^1(\delta(a))[m] \\
&= [a(gf)^0](m) - (gf)^1(\delta(a))[m].
\end{aligned}
$$

Now we see that $\mathcal{A}$-Mod is indeed a k-category. Clearly, $I_M = (I_M, 0)$ for all $M \in \mathcal{A}$-Mod. Now we show that the composition is associative (it is clearly bilinear). Consider the morphisms $f : M \longrightarrow N$, $g : N \longrightarrow L$ and $h : L \longrightarrow K$ in $\mathcal{A}$-Mod.

We show first that $[h(gf)]^0 = [(hg)f]^0$

$$
\begin{aligned}
[h(gf)]^0 &= h^0 (gf)^0 - \pi(h^1 \otimes (gf)^1)(d) \\
&= h^0[g^0 f^0 - \pi(g^1 \otimes f^1)(d)] - \pi(h^1 \otimes (gf)^1)(d) \\
&= h^0[g^0 f^0 - \pi(g^1 \otimes f^1)(d)] \\
&\quad - \pi\left(h^1 \otimes [g^0 f^1 + g^1 f^0 + \pi(g^1 \otimes f^1)\delta]\right)(d) \\
&= h^0 g^0 f^0 - [h^0 \pi(g^1 \otimes f^1)](d) \\
&\quad - \pi\left(h^1 \otimes g^0 f^1\right)(d) - \pi\left(h^1 \otimes g^1 f^0\right)(d) \\
&\quad - \pi\left(h^1 \otimes \pi(g^1 \otimes f^1)\delta\right)(d).
\end{aligned}
$$

Moreover,

$$\begin{aligned}
[(hg)f]^0 &= (hg)^0 f^0 - [\pi((hg)^1 \otimes f^1)](d)\\
&= h^0 g^0 f^0 - [\pi(h^1 \otimes g^1)(d)]f^0 - [\pi((hg)^1 \otimes f^1)](d)\\
&= h^0 g^0 f^0 - [\pi(h^1 \otimes g^1)(d)]f^0\\
&\quad - \pi\left([h^0 g^1 + h^1 g^0 + \pi(h^1 \otimes g^1)\delta] \otimes f^1\right)(d)\\
&= h^0 g^0 f^0 - [\pi(h^1 \otimes g^1)(d)]f^0 - \pi(h^0 g^1 \otimes f^1)[d]\\
&\quad - \pi(h^1 g^0 \otimes f^1)[d] - \pi(\pi(h^1 \otimes g^1)\delta \otimes f^1)[d].
\end{aligned}$$

We have to compare these lists of summands. Assume that $d = \sum_{i\in I} u_i \otimes w_i$, with $u_i, w_i \in V$. Similarly, write $\delta(u_i) = \sum_{j\in J_i} u^1_{ij} \otimes u^2_{ij}$, and $\delta(w_i) = \sum_{t\in T_i} w^1_{it} \otimes w^2_{it}$. Then, by Leibniz formula, $0 = \delta(d) = \sum_i [\delta(u_i)w_i - u_i\delta(w_i)]$, that is

$$\sum_{i,j} u^1_{ij} \otimes u^2_{ij} \otimes w_i = \sum_{i,t} u_i \otimes w^1_{it} \otimes w^2_{it}.$$

It implies that

$$\begin{aligned}
\pi(\pi(h^1 \otimes g^1)\delta \otimes f^1)[d] &= \textstyle\sum_i [\pi(h^1 \otimes g^1)(\delta(u_i))]f^1(w_i)\\
&= \textstyle\sum_{ij} h^1(u^1_{ij})g^1(u^2_{ij})f^1(w_i)\\
&= \textstyle\sum_{it} h^1(u_i)g^1(w^1_{it})f^1(w^2_{it})\\
&= \pi(h^1 \otimes \pi(g^1 \otimes f^1)\delta)[d].
\end{aligned}$$

We also have

$$\begin{aligned}
\pi(h^1 \otimes g^1 f^0)[d] &= \textstyle\sum_i h^1(u_i)g^1(w_i)f^0 = [\pi(h^1 \otimes g^1)(d)]f^0,\\
[h^0\pi(g^1 \otimes f^1)][d] &= \textstyle\sum_i h^0 g^1(u_i)f^1(w_i) = \pi(h^0 g^1 \otimes f^1)[d],\\
\pi(h^1 g^0 \otimes f^1)[d] &= \textstyle\sum_i h^1(u_i)g^0 f^1(w_i) = \pi(h^1 \otimes g^0 f^1)[d].
\end{aligned}$$

It follows that $[h(gf)]^0 = [(hg)f]^0$. In order to show that $[h(gf)]^1 = [(hg)f]^1$, consider $v \in V$, and let $\delta(v) = \sum_a u_a \otimes w_a$, and, for each a

$$\delta(u_a) = \sum_b u^1_{ab} \otimes u^2_{ab} \text{ and } \delta(w_a) = \sum_c w^1_{ac} \otimes w^2_{ac}.$$

Then

$$\begin{aligned}
[h(gf)]^1(v) &= h^0(gf)^1(v) + h^1(v)(gf)^0 + \textstyle\sum_a h^1(u_a)(gf)^1(w_a)\\
&= h^0[g^0 f^1(v) + g^1(v)f^0 + \textstyle\sum_a g^1(u_a)f^1(w_a)]\\
&\quad + h^1(v)[g^0 f^0 - \pi(g^1 \otimes f^1)(d)]\\
&\quad + \textstyle\sum_a h^1(u_a)[g^0 f^1(w_a) + g^1(w_a)f^0 + \sum_c g^1(w^1_{ac})f^1(w^2_{ac})],
\end{aligned}$$

and

$$\begin{aligned}[(hg)f]^1(v) &= (hg)^0 f^1(v) + (hg)^1(v) f^0 + \textstyle\sum_a (hg)^1(u_a) f^1(w_a) \\ &= [h^0 g^0 - \pi(h^1 \otimes g^1)(d)] f^1(v) \\ &\quad + [h^0 g^1(v) + h^1(v) g^0 + \textstyle\sum_a h^1(u_a) g^1(w_a)] f^0 \\ &\quad + \textstyle\sum_a [h^0 g^1(u_a) + h^1(u_a) g^0 + \sum_b h^1(u^1_{ab}) g^1(u^2_{ab})] f^1(w_a).\end{aligned}$$

Then, we have to show that

$$\sum_{a,c} h^1(u_a) g^1(w^1_{ac}) f^1(w^2_{ac}) - \sum_{i \in I} h^1(v) g^1(u_i) f^1(w_i)$$

and

$$\sum_{a,b} h^1(u^1_{ab}) g^1(u^2_{ab}) f^1(w_a) - \sum_{j \in I} h^1(u_j) g^1(w_j) f^1(v)$$

coincide. We have

$$\sum_a \delta(u_a) w_a + (-1)^{\deg(u_a)} u_a \delta(w_a) = \delta\Big(\sum_a u_a \otimes w_a\Big) = \delta^2(v) = dv - vd,$$

which implies, using that T is a t-algebra, that

$$\begin{aligned}\sum_{a,c} u_a \otimes w^1_{ac} \otimes w^2_{ac} - \sum_{i \in I} v \otimes u_i \otimes w_i &= \sum_{a,b} u^1_{ab} \otimes u^2_{ab} \otimes w_a \\ &\quad - \sum_{j \in I} u_j \otimes w_j \otimes v.\end{aligned}$$

Then, the following composition applied to last equality gives the desired result

$$\begin{aligned}V \otimes_A V \otimes_A V \xrightarrow{h^1 \otimes g^1 \otimes f^1} \mathrm{Hom}_k(L, K) \otimes_A \mathrm{Hom}_k(N, L) \otimes_A \mathrm{Hom}_k(M, N) \\ \xrightarrow{\pi} \mathrm{Hom}_k(M, K).\end{aligned}$$

□

Solution to Exercise (2.12): We first show that, whenever we have $(f^0, f^1) \in \mathrm{Hom}_{\mathcal{A}'}(M, N)$, then $F_\phi(f^0, f^1) \in \mathrm{Hom}_{\mathcal{A}}(F_\phi(M), F_\phi(N))$. Let $m \in M$ and $a \in A$. Then, we have

$$\begin{aligned}F_\phi(f)^0(am) &= f^0(am) \\ &= f^0[\phi_0(a)m] \\ &= \phi_0(a) f^0(m) - f^1(\delta'(\phi_0(a)))[m] \\ &= a f^0(m) - f^1 \phi_1(\delta(a))[m] \\ &= a F_\phi(f)^0[m] - F_\phi(f)^1(\delta(a))[m].\end{aligned}$$

In order to show that F_ϕ preserves the composition, take $f \in \mathrm{Hom}_{\mathcal{A}'}(M, N)$ and $g \in \mathrm{Hom}_{\mathcal{A}'}(N, L)$ in $\mathcal{A}'$-Mod. Assume that $d = \sum_{i \in I} u_i \otimes w_i$, thus

$d' = \phi(d) = \sum_{i \in I} \phi_1(u_i) \otimes \phi_1(w_i)$. Therefore

$$\begin{aligned}
\left[F_\phi(gf)\right]^0 &= (gf)^0 \\
&= g^0 f^0 - \pi(g^1 \otimes f^1)(d') \\
&= g^0 f^0 - \pi(g^1\phi_1 \otimes f^1\phi_1)(d) \\
&= \left[F_\phi(g)\right]^0 \left[F_\phi(f)\right]^0 - \pi\left[(F_\phi(g))^1 \otimes (F_\phi(f))^1\right](d) \\
&= \left[F_\phi(g)F_\phi(f)\right]^0.
\end{aligned}$$

On the other hand, for $v \in V$ with $\delta(v) = \sum_i v_i^1 \otimes v_i^2$, we have $\delta'(\phi(v)) = \phi\delta(v) = \sum_i \phi(v_i^1) \otimes \phi(v_i^2)$, and, therefore

$$\begin{aligned}
\left[F_\phi(g)F_\phi(f)\right]^1(v) &= F_\phi(g)^1(v)F_\phi(f)^0 + F_\phi(g)^0 F_\phi(f)^1(v) \\
&\quad + \pi(F_\phi(g)^1 \otimes F_\phi(f)^1)(\delta(v)) \\
&= g^1(\phi(v))f^0 + g^0 f^1(\phi(v)) + \textstyle\sum_i g^1(\phi(v_i^1))f^1(\phi(v_i^2)) \\
&= g^1(\phi(v))f^0 + g^0 f^1(\phi(v)) + \pi(g^1 \otimes f^1)(\delta'(\phi(v)) \\
&= (gf)^1(\phi(v)) \\
&= \left[F_\phi(gf)\right]^1(v).
\end{aligned}$$

We have seen that $F_\phi(gf) = F_\phi(g)F_\phi(f)$. Clearly, F_ϕ preserves identities. □

Section 3

Solution to Exercise (3.6): The first notice in the proof of (3.2) gives that $d \in V \otimes V$. The verification of the formula $\delta^2(u) = du - ud$, for any homogeneous $u \in T$, is done for $u = a \in A$ and $u = v \in V$ as in the proof of (3.2). For $u = v_1 v_2 \cdots v_n$, where $v_i \in V$, it follows from an easy induction argument using (1.7) and the fact that it already holds for each $v_i \in V$.

It remains to show that $\delta(d) = 0$. For this, make $d = \sum_i d_i^1 \otimes d_i^2$, with $d_i^1, d_i^2 \in V$. Thus

$$\begin{aligned}
\delta(d) &= \textstyle\sum_i \delta(d_i^1) \otimes d_i^2 - \sum_i d_i^1 \otimes \delta(d_i^2) \\
&= \textstyle\sum_i [\mu(d_i^1) - w \otimes d_i^1 - d_i^1 \otimes w] \otimes d_i^2 \\
&\quad - \textstyle\sum_i d_i^1 \otimes [\mu(d_i^2) - w \otimes d_i^2 - d_i^2 \otimes w] \\
&= \textstyle\sum_i \mu(d_i^1) \otimes d_i^2 - \sum_i w \otimes d_i^1 \otimes d_i^2 \\
&\quad - \textstyle\sum_i d_i^1 \otimes \mu(d_i^2) + \sum_i d_i^1 \otimes d_i^2 \otimes w.
\end{aligned}$$

We also have $\sum_i \mu(d_i^1) \otimes d_i^2 = (\mu \otimes 1)[w \otimes w - \mu(w)] = \mu(w) \otimes w - (\mu \otimes 1)\mu(w)$, and similarly $\sum_i d_i^1 \otimes \mu(d_i^2) = w \otimes \mu(w) - (1 \otimes \mu)\mu(w)$. It follows that

$$\begin{aligned}\delta(d) &= \mu(w) \otimes w - (\mu \otimes 1)\mu(w) - w \otimes [w \otimes w - \mu(w)] \\ &\quad - w \otimes \mu(w) + (1 \otimes \mu)\mu(w) + [w \otimes w - \mu(w)] \otimes w \\ &= (1 \otimes \mu)\mu(w) - (\mu \otimes 1)\mu(w) \\ &= 0.\end{aligned}$$

□

Solution to Exercise (3.7): We first show that μ is a morphism of right A-modules. For $a, b \in A$ and $v \in V$, we have

$$\begin{aligned}\mu[(wa + v)b] &= \mu[wab + vb] \\ &= w \otimes wab + \delta(vb) + vb \otimes w + w \otimes vb - dab \\ &= w \otimes wab + \delta(v)b - v \otimes \delta(b) + vb \otimes w + w \otimes vb - dab \\ &= w \otimes wab + \delta(v)b - v \otimes (bw - wb) \\ &\quad + vb \otimes w + w \otimes vb - dab \\ &= (\mu[wa + b])b\end{aligned}$$

Moreover

$$\begin{aligned}\mu(b[wa + v]) &= \mu(bwa + bv) \\ &= \mu(\delta(b)a + wba + bv) \\ &= w \otimes wba + \delta(\delta(b)a + bv) + (\delta(b)a + bv) \otimes w \\ &\quad + w \otimes (\delta(b)a + bv) - dba \\ &= w \otimes wba + \delta^2(b)a - \delta(b) \otimes \delta(a) \\ &\quad + \delta(b) \otimes v + b\delta(v) + \delta(b)a \otimes w + bv \otimes w \\ &\quad + w \otimes \delta(b)a + w \otimes bv - dba \\ &= w \otimes wba + (db - bd)a - (bw - wb) \otimes (aw - wa) \\ &\quad + (bw - wb) \otimes v + b\delta(v) + (bw - wb)a \otimes w + bv \otimes w \\ &\quad + w \otimes (bw - wb)a + w \otimes bv - dba \\ &= w \otimes wba - bda - bwa \otimes w + bw \otimes wa + wb \otimes aw \\ &\quad - wb \otimes wa + bw \otimes v - wb \otimes v + b\delta(v) + bwa \otimes w \\ &\quad - wba \otimes w + bv \otimes w + wb \otimes wa - w \otimes wba + w \otimes bv \\ &= -bda + bw \otimes wa + bw \otimes v + b\delta(v) + bv \otimes w \\ &= bw \otimes wa + b\delta(v) + bv \otimes w + bw \otimes v - bda \\ &= b\mu(wa + v).\end{aligned}$$

In order to show that $(1 \otimes \mu)\mu = (\mu \otimes 1)\mu$, it will be enough to verify the equality on generators of U. We have

$$\begin{aligned}(1 \otimes \mu)\mu(w) &= (1 \otimes \mu)[w \otimes w - d] \\ &= w \otimes \mu(w) - (1 \otimes \mu)(d) \\ &= w \otimes [w \otimes w - d] - (1 \otimes \mu)(d) \\ &= w \otimes w \otimes w - w \otimes d - (1 \otimes \mu)(d)\end{aligned}$$

and

$$\begin{aligned}(\mu \otimes 1)\mu(w) &= (\mu \otimes 1)[w \otimes w - d] \\ &= \mu(w) \otimes w - (\mu \otimes 1)(d) \\ &= [w \otimes w - d] \otimes w - (\mu \otimes 1)(d) \\ &= w \otimes w \otimes w - d \otimes w - (\mu \otimes 1)(d).\end{aligned}$$

Assume $d = \sum_i d_i^1 \otimes d_i^2$, with $d_i^1, d_i^2 \in V$. Then, $\delta(d) = 0$ implies that

$$\sum_i \delta(d_i^1) \otimes d_i^2 = \sum_i d_i^1 \otimes \delta(d_i^2).$$

Then

$$\begin{aligned}w \otimes d + (1 \otimes \mu)(d) &= w \otimes d + \textstyle\sum_i d_i^1 \otimes \mu(d_i^2) \\ &= w \otimes d + \textstyle\sum_i d_i^1 \otimes [\delta(d_i^2) + d_i^2 \otimes w + w \otimes d_i^2] \\ &= w \otimes d + \textstyle\sum_i \delta(d_i^1) \otimes d_i^2 + \sum_i d_i^1 \otimes d_i^2 \otimes w \\ &\quad + \textstyle\sum_i d_i^1 \otimes w \otimes d_i^2 \\ &= d \otimes w + \textstyle\sum_i [\delta(d_i^1) + d_i^1 \otimes w + w \otimes d_i^1] \otimes d_i^2 \\ &= d \otimes w + \textstyle\sum_i \mu(d_i^1) \otimes d_i^2 \\ &= d \otimes w + (\mu \otimes 1)(d).\end{aligned}$$

Moreover, if $v \in V$ and $\delta(v) = \sum_j v_j^1 \otimes v_j^2$ with $v_j^1, v_j^2 \in V$, we have that $dv - vd = \delta^2(v) = \sum_j \delta(v_j^1) \otimes v_j^2 - \sum_j v_j^1 \otimes \delta(v_j^2)$. Thus

$$\sum_j \delta(v_j^1) \otimes v_j^2 - dv = \sum_j v_j^1 \otimes \delta(v_j^2) - vd.$$

Then

$$\begin{aligned}(1 \otimes \mu)\mu(v) &= (1 \otimes \mu)[\delta(v) + v \otimes w + w \otimes v] \\ &= \textstyle\sum_j v_j^1 \otimes \mu(v_j^2) + v \otimes \mu(w) + w \otimes \mu(v) \\ &= \textstyle\sum_j v_j^1 \otimes [\delta(v_j^2) + v_j^2 \otimes w + w \otimes v_j^2] \\ &\quad + v \otimes [w \otimes w - d] + w \otimes [\delta(v) + v \otimes w + w \otimes v] \\ &= \textstyle\sum_j v_j^1 \otimes \delta(v_j^2) + \sum_j v_j^1 \otimes v_j^2 \otimes w + \sum_j v_j^1 \otimes w \otimes v_j^2 \\ &\quad + v \otimes w \otimes w - v \otimes d + w \otimes \delta(v) + w \otimes v \otimes w + w \otimes w \otimes v \\ &= \textstyle\sum_j [\delta(v_j^1) + w \otimes v_j^1 + v_j^1 \otimes w] \otimes v_j^2 \\ &\quad + [\delta(v) + w \otimes v + v \otimes w] \otimes w + [w \otimes w - d] \otimes v \\ &= \textstyle\sum_j \mu(v_j^1) \otimes v_j^2 + \mu(v) \otimes w + \mu(w) \otimes v \\ &= (\mu \otimes 1)[\delta(v) + v \otimes w + w \otimes v] \\ &= (\mu \otimes 1)\mu(v).\end{aligned}$$

Finally

$$\begin{aligned}(\epsilon \otimes 1)\mu(w) &= (\epsilon \otimes 1)[w \otimes w - d]\\ &= \epsilon(w) \otimes w - (\epsilon \otimes 1)(d)\\ &= 1 \otimes w\end{aligned}$$

and, for $v \in V$

$$\begin{aligned}(\epsilon \otimes 1)\mu(v) &= (\epsilon \otimes 1)[\delta(v) + v \otimes w + w \otimes v]\\ &= \textstyle\sum_j \epsilon(v_j^1) \otimes v_j^2 + \epsilon(v) \otimes w + \epsilon(w) \otimes v\\ &= 1 \otimes v.\end{aligned}$$

Thus, $(\epsilon \otimes 1)\mu$ is equivalent to 1_U. Similarly, $(1 \otimes \epsilon)\mu$ is equivalent to 1_U. □

Section 5

Solution to Exercise (5.15): Assume that f^0 is an epimorphism in R-Mod and let $g : M \longrightarrow E$ be a morphism in $\mathcal{A}$-Mod with $gf = 0$. Thus, $g^0 f^0 = 0$, and $g^0 = 0$. By assumption, there is a sequence of R-R-subbimodules $0 = W_1^0 \subseteq W_1^1 \subseteq \cdots \subseteq W_1^s = W_1$ such that $\delta(W_1^{j+1}) \subseteq AW_1^j AW_1^j A$, for all $j \in [0, s-1]$. Then, for any $w \in W_1^{j+1}$, say with $\delta(w) = \sum_i a_i w_i^1 b_i w_i^2 c_i$ and $w_i^1, w_i^2 \in W_1^j$, $a_i, b_i, c_i \in A$, we have that

$$0 = (gf)^1(w) = g^0 f^1(w) + g^1(w) f^0 + \sum_i a_i g^1(w_i^1) b_i f^1(w_i^2) c_i.$$

By triangular induction, we can assume that $g^1(w_i^1) = 0$, for all i, thus $g^1(w) f^0 = 0$. Then, using that f^0 is surjective, we obtain $g^1(w) = 0$ too. Then, $g = 0$. The statement on monomorphisms is dual. □

Solution to Exercise (5.16): Given $L, H \in \mathcal{A}$-Mod, an element

$$g = (g^0, g^1) \in \mathrm{Hom}_R(L, H) \times \mathrm{Hom}_{R\text{-}R}(W, \mathrm{Hom}_k(L, H))$$

is determined by the restriction linear maps $g_i^0 : e_i L \longrightarrow e_i H$, $i \in \{1, 2, 3\}$, and the linear map $g^1(v) : e_2 L \longrightarrow e_3 H$. This element g is a morphism $g : L \longrightarrow H$ in $\mathcal{A}$-Mod iff $g_2^0 L_\alpha = H_\alpha g_1^0$ and $H_\beta g_1^0 - g_3^0 L_\beta = g^1(v) L_\alpha$.

Now, assume that $g \in \mathrm{Hom}_{\mathcal{A}}(L, M)$ satisfies $fg = 0$. Then, $g_3^0 = 0$, $g_1^0 = 0$, $g^1(v) = 0$ because their codomains are zero. Since $fg = 0$, we also have that $0 = (fg)^1(v) = f_3^0 g^1(v) + f^1(v) g_2^0 = g_2^0$. Thus, $g = 0$ and f is a monomorphism.

Now assume that $f = \mathrm{Ker}\, g$, for some $g \in \mathrm{Hom}_{\mathcal{A}}(N, H)$. Then, $g_1^0 = 0$, $g_2^0 = 0$ and $g^1(v) = 0$ because their domains are zero. Thus, g is determined by the linear map g_3^0.

If $g_3^0 = 0$, then $g = 0$ and, therefore, $\mathrm{Ker}\, g = 1_N$. Then, $N \cong M$ in $\mathcal{A}$-Mod, which is not true. Then, $g_3^0 \neq 0$. Since its domain is k, g_3^0 is an injective map. Then, $g^0 : N \longrightarrow H$ is an injective map, which implies, for instance by (5.15), that g is a monomorphism. Again, since $gf = 0$, we obtain that $f = 0$, a contradiction. □

Solution to Exercise (5.17): The proof is similar to the proof of (5.3)(1). Assume that f^0 admits a left inverse $g^0 \in \mathrm{Hom}_R(N, M)$, thus $g^0 f^0 = id_M$.

We define inductively, for $j \in [1, r]$, morphisms $\phi_j : W_0^j \otimes_R N \longrightarrow N$ of left R-modules, using the next formula

$$(E_j): \qquad \phi_j(w \otimes n) = f^0[wg^0(n)] + f_{j-1}^1(\delta(w))[g^0(n)],$$

for $w \in W_0^j$ and $n \in N$, where $f_{j-1}^1 : A_{j-1}W_1A_{j-1} \longrightarrow \mathrm{Hom}_k(M, N)$ is the unique extension of f^1.

Here, f_j^1 makes sense whenever ϕ_j is defined.

The morphism ϕ_r determines the desired left A-module structure for N. Indeed, given any $m \in M, n := f^0(m) \in N$. Then, for any $w \in W_0$, $wf^0(m) \doteq wn = f^0(wg^0(n)) + f_r^1(\delta(w))[g^0(n)] = f^0(wm) + f_r^1(\delta(w))[m]$. □

Solution to Exercise (5.19): In order to show that $\delta^2 = 0$, it is enough to see that $\delta^2(W_0) = 0$ (since it is clear that $\delta^2(R \cup W_1) = 0$). $\delta^2(\alpha) = \delta(\alpha v \alpha) = \delta(\alpha)v\alpha + \alpha\delta(v\alpha) = \delta(\alpha)v\alpha + \alpha[\delta(v)\alpha - v\delta(\alpha)] = \alpha v \alpha v \alpha - \alpha v \alpha v \alpha = 0$. Then, $\mathcal{A}$ is a ditalgebra with (an obviously non-triangular) layer (R, W).

Consider any non-trivial R-module M. Consider the $T_R(W_0)$-module N with underlying R-module M and action map $\varphi : W_0 \otimes_R N \to N$ defined by $\varphi(\alpha \otimes n) = n$; that is, such that the action N_α of α on N is the identity map 1_N. Consider the pair $f = (f^0, f^1) \in \mathrm{Phom}_{R\text{-}W}(M, N)$ defined by $f^0 := 1_M$ and $f^1 \in \mathrm{Hom}_{R\text{-}R}(W_1, \mathrm{Hom}_k(M, N))$ given by $f^1(v) := -1_M$. Now, assume that there exists an $\mathcal{A}$-module $\widehat{M}$ with underlying R-module M such that $f \in \mathrm{Hom}_{\mathcal{A}}(\widehat{M}, N)$, and denote by $\widehat{M}_\alpha$ the action of α on $\widehat{M}$. This means that

$$N_\alpha f^0 = f^0 \widehat{M}_\alpha + f^1(\alpha v \alpha) = f^0 \widehat{M}_\alpha + N_\alpha f^1(v)\widehat{M}_\alpha = \widehat{M}_\alpha - \widehat{M}_\alpha = 0.$$

Thus, $1_N = N_\alpha = 0$. This is a contradiction, since $N \neq 0$. □

Solution to Exercise (5.20): We have that $\delta^2(W) = 0$. Indeed, $\delta^2(\alpha) = \delta(\alpha v) = \delta(\alpha)v + \alpha\delta(v) = \alpha v^2 - \alpha v^2 = 0$; and $-\delta^2(v) = \delta(v^2) = \delta(v)v - v\delta(v) = -v^3 + v^3 = 0$. Then, $\mathcal{A}$ is a ditalgebra with (an obviously non-triangular) layer (R, W).

Consider any non-trivial R-module H. Consider the $T_R(W_0)$-module M with underlying R-module H and action map $\varphi : W_0 \otimes_R H \to H$ defined by $\varphi(\alpha \otimes h) = 0$; that is, such that the action M_α of α on M is the zero map. Similarly, consider the $T_R(W_0)$-module N with underlying R-module H and action map $\varphi' : W_0 \otimes_R H \to H$ defined by $\varphi'(\alpha \otimes h) = h$; that is, such that the action N_α of α on N is the identity map 1_N. Then, consider the pair $f = (f^0, f^1) \in \mathrm{Phom}_{R\text{-}W}(M, N)$ defined by $f^0 := 1_H$ and $f^1 \in \mathrm{Hom}_{R\text{-}R}(W_1, \mathrm{Hom}_k(M, N))$ given by $f^1(v) := 1_H$. Then, since $N_\alpha f^0 = f^0 M_\alpha + N_\alpha f^1(v) = f^0 M_\alpha + f^1(\alpha v)$, we know that the pair f can be identified with an element in $\mathrm{Hom}_{\mathcal{A}}(M, N)$, which clearly satisfies that f^0 is an isomorphism in R-Mod. Now, assume that f is an isomorphism in $\mathcal{A}$-Mod and consider an inverse $g \in \mathrm{Hom}_{\mathcal{A}}(N, M)$. Then, g satisfies the relation $M_\alpha g^0 = g^0 N_\alpha + M_\alpha g^1(v)$, hence $g^0 = 0$, and $1_H = f^0 g^0 = 0$ is a contradiction because $H \neq 0$.

Now, we construct a non-split idempotent. Take $M = k$, where α acts as the zero map: $M_\alpha = 0$. Consider $g^0 := 0 \in \mathrm{Hom}_R(M, M)$ and $g^1 \in \mathrm{Hom}_{R\text{-}R}(W_1, \mathrm{Hom}_k(M, M))$ defined by $g^1(v) = -1_k$. Then, $M_\alpha g^0 = g^0 M_\alpha + M_\alpha g^1(v)$ holds and $g = (g^0, g^1)$ can be identified with an element of $\mathrm{Hom}_{\mathcal{A}}(M, M)$. The second component of the composition $g \circ g$ is determined by its value in v, but $(g \circ g)^1(v) = g^0 g^1(v) + g^1(v) g^0 - g^1(v) g^1(v) = -1_k = g^1(v)$. Then, $g^2 = g$ is a non-trivial idempotent which does not split because $\dim M = 1$. Indeed: If g was a split idempotent, by definition, this would imply that $1 f = f g$, with f isomorphism. Thus, $1 = g$, a contradiction. □

Solution to Exercise (5.21): Take $M := k \oplus k$, $f^0 := \begin{pmatrix} 0 & 0 \\ 1 & 0 \end{pmatrix}$ and $f^1(v) := \begin{pmatrix} 1 & 0 \\ 0 & 1 \end{pmatrix}$. Then, $f = (f^0, f^1)$ can be identified with an element of $\mathrm{Hom}_{\mathcal{A}}(M, M)$. Moreover, $(f^i)^0 = 0$, for $i \geq 2$, and

$$(f^2)^1(v) = \begin{pmatrix} 0 & 0 \\ 1 & 0 \end{pmatrix} + \begin{pmatrix} 0 & 0 \\ 1 & 0 \end{pmatrix} - \begin{pmatrix} 1 & 0 \\ 0 & 1 \end{pmatrix} = \begin{pmatrix} -1 & 0 \\ 2 & -1 \end{pmatrix},$$

$$(f^3)^1(v) = \begin{pmatrix} -1 & 0 \\ 2 & -1 \end{pmatrix} \begin{pmatrix} 0 & 0 \\ 1 & 0 \end{pmatrix} - \begin{pmatrix} -1 & 0 \\ 2 & -1 \end{pmatrix} = \begin{pmatrix} 1 & 0 \\ -3 & 1 \end{pmatrix},$$

$$(f^4)^1(v) = \begin{pmatrix} 1 & 0 \\ -3 & 1 \end{pmatrix} \begin{pmatrix} 0 & 0 \\ 1 & 0 \end{pmatrix} - \begin{pmatrix} 1 & 0 \\ -3 & 1 \end{pmatrix} = \begin{pmatrix} -1 & 0 \\ 4 & -1 \end{pmatrix},$$

$$(f^5)^1(v) = \begin{pmatrix} -1 & 0 \\ 4 & -1 \end{pmatrix} \begin{pmatrix} 0 & 0 \\ 1 & 0 \end{pmatrix} - \begin{pmatrix} -1 & 0 \\ 4 & -1 \end{pmatrix} = \begin{pmatrix} 1 & 0 \\ -5 & 1 \end{pmatrix},$$

and, in general, for $i \geq 2$, we have

$$(f^i)^1(v) = \begin{pmatrix} (-1)^{i+1} & 0 \\ (-1)^i i & (-1)^{i+1} \end{pmatrix} \neq 0.$$

Thus, f is not nilpotent.

Now, consider $N := k$, $g^0 := 1_k$ and $g^1(v) := 1_k$. Then, $g = (g^0, g^1)$ can be identified with an element in $\mathrm{Hom}_{\mathcal{A}}(N, N)$. Assume there is $f \in \mathrm{Hom}_{\mathcal{A}}(N, N)$ with $gf = 1_N$. Then, $f^0 = 1_k = g^0$ and $f^1(v) + g^1(v) - g^1(v)f^1(v) = 0$. Thus, $1_k = 0$, a contradiction. Hence g is not an isomorphism. □

Section 6

Solution to Exercise (6.21): If g is a retraction, there is a morphism $s : M \longrightarrow E$ with $gs = 1_M$. Since f is the kernel of g and $g(1_E - sg) = 0$, there is a morphism $p : E \longrightarrow N$ with $fp = 1_E - sg$. Hence, $fpf = (1_E - sg)f = f - sgf = f$. But f is a kernel, and hence a monomorphism, thus $pf = 1_N$ and f is a section. Dually, if we start assuming that f is a section, we get that g is a retraction. Thus, (2) and (3) are equivalent. If g is a retraction, and we proceed as before, we obtain the morphisms s and p which satisfy that $gs = 1_M$, $pf = 1_N$ and $1_E = fp + sg$, as required in (1). Thus, (2) is equivalent to (1). If we assume (1), we have a commutative diagram

$$\begin{array}{ccccc} N & \xrightarrow{(1_N,0)^t} & N \oplus M & \xrightarrow{(0,1_M)} & M \\ \| & & \downarrow{\scriptstyle (f,s)} & & \| \\ N & \xrightarrow{f} & E & \xrightarrow{g} & M, \end{array}$$

where (f, s) is an isomorphism with inverse $(p, g)^t$. Thus, ζ and ζ_0 are equivalent. Finally, (4) implies clearly (2). □

Solution to Exercise (6.22): The hypothesis on W_1 guarantees that all the Hom spaces we are considering are finite dimensional over k. We have the linear maps

$$\begin{array}{ccccc} \mathrm{Hom}_{\mathcal{A}}(M, N) & \xrightarrow{\eta} & \mathrm{Hom}_{\mathcal{A}}(E, E) & \xrightarrow{\rho} & \mathrm{Hom}_{\mathcal{A}}(N, M) \\ g & \mapsto & \sigma g \pi & & \\ & & f & \mapsto & \pi f \sigma \end{array}$$

Notice that η is injective (since π is epi and σ is mono, because π is a cokernel and σ is a kernel). Therefore, $\mathrm{Hom}_{\mathcal{A}}(M, N) \cong \mathrm{Im}\,\eta$. We also have that, $(\rho\eta)[g] = \pi(\sigma g \pi)\sigma = 0$. Thus, $\mathrm{Im}\,\eta \subseteq \mathrm{Ker}\,\rho$. We know that ρ induces an embedding $\overline{\rho}$ of $\mathrm{Hom}_{\mathcal{A}}(E, E)/\mathrm{Ker}\,\rho$ into $\mathrm{Hom}_{\mathcal{A}}(N, M)$.

Consider $f \in \operatorname{Ker}\rho$, then $\pi f \sigma = 0$. Since $\pi = \operatorname{Cok}\sigma$ and $\sigma = \operatorname{Ker}\pi$, there is a commutative diagram in $\mathcal{A}$-mod

$$\begin{array}{ccccc} N & \xrightarrow{\sigma} & E & \xrightarrow{\pi} & M \\ \downarrow{\scriptstyle f_N} & & \downarrow{\scriptstyle f} & & \downarrow{\scriptstyle f_M} \\ N & \xrightarrow{\sigma} & E & \xrightarrow{\pi} & M. \end{array}$$

Then, we can consider the map $\psi : \operatorname{Ker}\rho \longrightarrow \operatorname{End}_A(N) \times \operatorname{End}_A(M)$ with receipt $\psi(f) = (f_N, f_M)$. We claim that $\operatorname{Ker}\psi = \operatorname{Im}\eta$. Indeed, $\psi(f) = 0$ implies $f_N = 0$ and $f_M = 0$. Thus, $f\sigma = 0$ and $\pi f = 0$. Since $\pi f = 0$ and $\sigma = \operatorname{Ker}\pi$ imply there is $h \in \operatorname{Hom}_A(E, N)$ with $\sigma h = f$. Then, $0 = f\sigma = \sigma h \sigma$ and, hence, $h\sigma = 0$. But $\pi = \operatorname{Cok}\sigma$, and then there is $g \in \operatorname{Hom}_A(M, N)$ with $g\pi = h$. It follows that $f = \sigma h = \sigma g \pi \in \operatorname{Im}\eta$. Moreover, $f = \sigma g \pi$ obviously satisfies that $\pi f = 0$ and $f\sigma = 0$.

The equality $\operatorname{Ker}\psi = \operatorname{Im}\eta$ implies that ψ induces an embedding $\overline{\psi}$ of $\operatorname{Ker}\rho/\operatorname{Im}\eta$ into $\operatorname{End}_A(N) \times \operatorname{End}_A(M)$. From the above remarks, we get

$$\begin{aligned} \dim \operatorname{Ker}\rho - \dim \operatorname{Hom}_A(M, N) &= \dim \operatorname{Ker}\rho - \dim \operatorname{Im}\eta \\ &\leq \dim \operatorname{End}_A(N) + \dim \operatorname{End}_A(M) \end{aligned}$$

and $\dim \operatorname{End}_A(E) - \dim \operatorname{Ker}\rho \leq \dim \operatorname{Hom}_A(N, M)$. Adding these inequalities, we have that

$$\dim \operatorname{End}_A(E) \leq \dim \begin{pmatrix} \operatorname{End}_A(N) & \operatorname{Hom}_A(N, M) \\ \operatorname{Hom}_A(M, N) & \operatorname{End}_A(M) \end{pmatrix} = \dim \operatorname{End}_A(N \oplus M).$$

Assume we have an equality. Then, the embeddings $\overline{\rho}$ and $\overline{\psi}$ are isomorphisms. In particular, $(1_N, 0) \in \operatorname{Im}\overline{\psi}$. Therefore, there exists $f \in \operatorname{Ker}\rho$ such that the following diagram commutes

$$\begin{array}{ccccc} N & \xrightarrow{\sigma} & E & \xrightarrow{\pi} & M \\ \| & & \downarrow{\scriptstyle f} & & \downarrow{\scriptstyle 0} \\ N & \xrightarrow{\sigma} & E & \xrightarrow{\pi} & M. \end{array}$$

Then, $\pi f = 0$ implies the existence of $g \in \operatorname{Hom}_A(E, N)$ with $\sigma g = f$. Then, $\sigma g \sigma = f\sigma = \sigma = \sigma 1_N$. But σ is a monomorphism, thus $g\sigma = 1_N$. That is, σ is a section, and the original exact pair splits, by (6.21). □

Solution to Exercise (6.23): Otherwise, there is a non-split conflation $\xi : M_j \longrightarrow E \longrightarrow M_i$, for some $i \neq j$. Consider the injection σ_i :

$M_i \longrightarrow \oplus_{r\neq j} M_r$ and the projection $\pi_i : \oplus_{r\neq j} M_r \longrightarrow M_i$. Then, we have a pullback diagram

$$\begin{array}{ccccc} \xi\pi_i : M_j & \longrightarrow & E' & \longrightarrow & \oplus_{r\neq j} M_r \\ \| & & \downarrow & & \downarrow \pi_i \\ \xi \; : M_j & \longrightarrow & E & \longrightarrow & M_i, \end{array}$$

with $E' \in \mathcal{A}\text{-mod}_d$. Assume that $\xi\pi_i$ splits. Notice that $\pi_i\sigma_i = id$ implies that the maps they induce on the Ext groups satisfy $\sigma_i^*\pi_i^* = id$

$$\begin{array}{ccccc} \mathrm{Ext}_{\mathcal{A}}(M_i, M_j) & \xrightarrow{\pi_i^*} & \mathrm{Ext}_{\mathcal{A}}(\oplus_{r\neq j} M_r, M_j) & \xrightarrow{\sigma_i^*} & \mathrm{Ext}_{\mathcal{A}}(M_i, M_j) \\ \xi & \mapsto & \xi\pi_i & \mapsto & \xi. \end{array}$$

Then, ξ would split, which is not the case. Then, $\xi\pi_i$ is not split and we obtain, from (6.22), that

$$\dim \mathrm{End}_{\mathcal{A}}(E') < \dim \mathrm{End}_{\mathcal{A}}(\oplus_r M_r) = \dim \mathrm{End}_{\mathcal{A}}(M).$$

A contradiction, because $\dim \mathrm{End}_{\mathcal{A}}(M)$ was minimal. □

Solution to Exercise (6.24): Assume that (R, W) is the layer of $\mathcal{A}$. Since $G(\zeta)$ is split, by (6.21) there exist morphisms $s' : G(M) \longrightarrow G(E)$ and $p' : G(E) \longrightarrow G(N)$ such that $G(g)s' = 1_{G(M)}$, $p'G(f) = 1_{G(N)}$ and $1_{G(E)} = G(f)p' + s'G(g)$. Since G is full, then $p' = G(p)$ and $s' = G(s)$, for some $s : M \longrightarrow E$ and $p : E \longrightarrow N$. Since G is faithful, $gs = 1_M$, $pf = 1_N$ and $1_E = fp + sg$. Then, $g^0 s^0 = 1_M$, $p^0 f^0 = 1_N$ and $1_E = f^0 p^0 + s^0 g^0$ in R-Mod. Thus

$$0 \longrightarrow N \xrightarrow{f^0} E \xrightarrow{g^0} M \longrightarrow 0$$

is a split exact sequence in R-Mod. That the composition of $G(\zeta)$ is zero implies that the composition of ζ is zero too, because G is faithful. Then, $\zeta \in \mathcal{E}_{\mathcal{A}}$. Using (6.21) again, we get that ζ is split. □

Solution to Exercise (6.25): From (6.12), we can assume that $g^1 = 0$. By the naturality of ψ described in (6.13), we can consider the commutative diagram

$$\begin{array}{ccc} \mathrm{Hom}_R(W_0 \otimes_R Z, E) & \xrightarrow{\psi_A} & \mathrm{Ext}^1_{\mathcal{A}}(Z, E) \\ \downarrow (g^0)^* & & \downarrow g' \\ \mathrm{Hom}_R(W_0 \otimes_R Z, N) & \xrightarrow{\psi_A} & \mathrm{Ext}^1_{\mathcal{A}}(Z, N). \end{array}$$

Since g^0 is a retraction in R-Mod, $(g^0)^*$ is surjective. Since ψ_A is surjective too, so is g'. □

Section 7

Solution to Exercise (7.20): By (6.17), P is a projective object relative to the exact structure $\mathcal{E}_{\mathcal{A}}$. Hence, F is an exact functor.

Assume now that $h \in \mathrm{Hom}_{\mathcal{A}}(M, N)$ satisfies that $F(h) = 0$. This means that $hg = 0$, for any $g \in \mathrm{Hom}_{\mathcal{A}}(P, M)$, or, equivalently, $h^0 g^0 = 0$ and, for any $v \in V$, $h^1(v)g^0 + h^0 g^1(v) + \sum_i h^1(u_i)g^1(v_i) = 0$, where $\delta(v) = \sum_i u_i \otimes v_i$. We want to show that $h = 0$. Given $m_0 \in M$, there is a morphism of A-modules $g^0 : A \longrightarrow M$ such that $g^0(a) = am_0$. Since R is basic, R is a direct summand of Y, say $Y = R \oplus Y'$. Thus, $P = A \otimes_R Y = A \otimes_R (R \oplus Y') \cong A \oplus (A \otimes_R Y')$. Then, there is a morphism of A-modules $\hat{g}^0 : P \longrightarrow M$ such that $\hat{g}^0(1 \otimes 1) = m_0$.

Let $\hat{g} = (\hat{g}^0, 0)$, thus $\hat{g} \in \mathrm{Hom}_{\mathcal{A}}(P, M)$, and $h^0\hat{g}^0 = 0$ and $h^1(v)\hat{g}^0 = 0$. Evaluating at the element $1 \otimes 1 \in P$, we obtain $h^0(m_0) = 0$ and $h^1(v)[m_0] = 0$. Thus, $h = 0$ and F is a faithful functor.

Assume A is finite dimensional and M, N are finitely generated A-modules. We want to verify that $F : \mathrm{Hom}_{\mathcal{A}}(M, N) \longrightarrow \mathrm{Hom}_{\mathbf{E}}(F(M), F(N))$ is surjective. It is clear that $F : \mathrm{Hom}_{\mathcal{A}}(P, P) \longrightarrow \mathrm{Hom}_{\mathbf{E}}(F(P), F(P))$ is an isomorphism. Using that F is additive, for any natural numbers n and n', we obtain isomorphisms

$$F : \mathrm{Hom}_{\mathcal{A}}(P^n, P^{n'}) \longrightarrow \mathrm{Hom}_{\mathbf{E}}(F(P^n), F(P^{n'})).$$

From (6.17), we know that the $\mathcal{E}_{\mathcal{A}}$-projectives in $\mathcal{A}$-mod are direct summands of $\mathcal{A}$-modules the form $A \otimes_R X$, where X is a finite-dimensional R-module. For finite-dimensional R-modules X, we obtain direct summands of the objects P^n in $\mathcal{A}$-mod, for $n \in \mathbb{N}$. Assume that Q, Q' are objects in $\mathcal{A}$-mod of this type and consider a section $i : Q \longrightarrow P^n$ and a retraction $p : P^{n'} \longrightarrow Q'$. Then, the following diagram shows that the lower row is surjective too, hence an isomorphism

$$\begin{array}{ccc}
\mathrm{Hom}_{\mathcal{A}}(P^n, P^{n'}) & \xrightarrow{\cong} & \mathrm{Hom}_{\mathbf{E}}(F(P^n), F(P^{n'})) \\
\downarrow{\scriptstyle i^*} & & \downarrow{\scriptstyle F(i)^*} \\
\mathrm{Hom}_{\mathcal{A}}(Q, P^{n'}) & \xrightarrow{F} & \mathrm{Hom}_{\mathbf{E}}(F(Q), F(P^{n'})) \\
\downarrow{\scriptstyle p_*} & & \downarrow{\scriptstyle F(p)_*} \\
\mathrm{Hom}_{\mathcal{A}}(Q, Q') & \xrightarrow{F} & \mathrm{Hom}_{\mathbf{E}}(F(Q), F(Q')).
\end{array}$$

In our case, from (6.17), we know there is an exact sequence in A-mod, which splits in R-mod, of the form

$$0 \longrightarrow K \xrightarrow{u^0} A \otimes_R M \xrightarrow{p^0} M \longrightarrow 0.$$

Similarly, there is an exact sequence in A-mod, which splits in R-mod, of the form

$$0 \longrightarrow K' \xrightarrow{v^0} A \otimes_R K \xrightarrow{q^0} K \longrightarrow 0.$$

Make $Q_2 = A \otimes_R K$, $Q_1 = A \otimes_R M$, and $s^0 = u^0 q^0$. Then, (6.2) implies that their images under the canonical embedding in $\mathcal{A}$-Mod give the $\mathcal{E}_{\mathcal{A}}$-projective resolution

$$Q_2 \xrightarrow{s=(s^0,0)} Q_1 \xrightarrow{p=(p^0,0)} M,$$

and then, applying F, we get the exact sequence of $\mathbb{E}$-modules

$$F(Q_2) \xrightarrow{F(s)} F(Q_1) \xrightarrow{F(p)} F(M) \longrightarrow 0,$$

which is a projective resolution of $F(M)$ in $\mathbb{E}$-Mod. If we consider for N the corresponding construction $Q_2' \xrightarrow{s'} Q_1' \xrightarrow{p'} N$, then we have that any morphism $f \in \mathrm{Hom}_{\mathbb{E}}(F(M), F(N))$ can be lifted as follows

$$\begin{array}{ccccccc} F(Q_2) & \xrightarrow{F(s)} & F(Q_1) & \xrightarrow{F(p)} & F(M) & \longrightarrow & 0 \\ \downarrow{\scriptstyle f_2} & & \downarrow{\scriptstyle f_1} & & \downarrow{\scriptstyle f} & & \\ F(Q_2') & \xrightarrow{F(s')} & F(Q_1') & \xrightarrow{F(p')} & F(N) & \longrightarrow & 0. \end{array}$$

From the previous statements, $f_2 = F(h_2)$ and $f_1 = F(h_1)$ for some $h_1 \in \mathrm{Hom}_{\mathcal{A}}(Q_2, Q_2')$ and $h_2 \in \mathrm{Hom}_{\mathcal{A}}(Q_1, Q_1')$. Since $F(s'h_2) = F(s')f_2 = f_1 F(s) = F(h_1 s)$, we have $s'h_2 = h_1 s$, and hence there exists $h \in \mathrm{Hom}_{\mathcal{A}}(M, N)$ such that the following diagram commutes in $\mathcal{A}$-Mod

$$\begin{array}{ccccc} Q_2 & \xrightarrow{s} & Q_1 & \xrightarrow{p} & M \\ \downarrow{\scriptstyle h_2} & & \downarrow{\scriptstyle h_1} & & \downarrow{\scriptstyle h} \\ Q_2' & \xrightarrow{s'} & Q_1' & \xrightarrow{p'} & N. \end{array}$$

Applying F to this diagram, we obtain that $F(h) = f$, and F is full. □

Section 8

Solution to Exercise (8.21): By definition, I is an ideal of T. Since $\delta(S) \subseteq I$ and I is generated by S, we have that $\delta(I) \subseteq I$. I is a t-ideal of T because $I_A := I \cap A = 0$ and $I_V = ASA = I \cap V$, thus $I = \langle I_A \cup I_V \rangle$. Thus, I is an ideal of $\mathcal{A}$, and $\mathcal{A}/I$, as well as the canonical projection $\eta : \mathcal{A} \longrightarrow \mathcal{A}/I$, makes sense. Consider the morphism $\psi : T \longrightarrow T_A(V/I_V)$ determined by $\psi(a) = a$, for $a \in A$, and $\psi(v) = \widehat{v} := v + I_V$, for $v \in V$. Since $\psi(I) = \psi(TST) \subseteq \widehat{T}\psi(S)\widehat{T} = 0$, ψ induces a morphism of graded algebras $\overline{\psi} : T/I \longrightarrow \widehat{T}$ such that $\overline{\psi}\eta = \psi$. As in (8.4), we have that $\overline{\psi}$ is an isomorphism and we can transfer $\overline{\delta}$ onto a differential $\widehat{\delta}$ on $\widehat{T}$.

Consider the example: $R = k \times k$, where k is a field with char$k \neq 2$, $e_1 := (1, 0)$, $e_2 := (0, 1)$, $W_0 := ke_2 \otimes_k ke_1$ and $W_1 := ke_2 \otimes_k ke_1$. Write $\alpha := e_2 \otimes e_1 \in W_0$ and $v := e_2 \otimes e_1 \in W_1$. Consider the ditalgebra $\mathcal{A} := (T_R(W), \delta)$, where $\delta(\alpha) = v$ and $\delta(v) = 0$, and the subset $S = \{v\}$ of V. Take M associated to the representation $k \xrightarrow{1} k$ and N associated to the representation $k \xrightarrow{2} k$. Consider the morphism $g = (g^0, g^1) \in \mathrm{Hom}_{\mathcal{A}}(M, N)$ such that $g_1^0 = Id_k$, $g_2^0 = Id_k$ and $g^1(v) = Id_k$. g is indeed a morphism, because, $\alpha g^0(\mu) = 2\mu = \mu + \mu = g^0(\alpha\mu) + g^1(v)[\mu]$, for $\mu \in k$. But any morphism of the form $g = F_\eta(f)$ satisfies $g^1(v) = F_\eta(f)^1(v) = f^1\eta(v) = 0$. □

Solution to Exercise (8.22): Assume $f = (f^0, f^1) : M \longrightarrow N$ is an isomorphism in $\mathcal{A}$-Mod. Thus, $f^0 : M \longrightarrow N$ is an isomorphism of vector spaces. It is clear that $H_0 N = 0$, assume that $H_{i-1}N = 0$, for $i \in [1, \ell - 1]$ and let us see that $H_i N = 0$. Take $n \in N$ and $h \in H_i$, thus $n = f^0(m)$, for some $m \in M$. By assumption, $\delta(h) = \sum_r a_r h_r v_r + \sum_s v'_s h'_s a'_s$, where $a_r, a'_s \in A$, $h_r, h'_s \in H_{i-1}$ and $v_r, v'_s \in V$. Then

$$hn = hf^0[m] = f^0[hm] + \sum_r a_r h_r f^1(v_r)[m] + \sum_s f^1(v'_s)[h'_s a'_s m].$$

Since $H_{i-1}N = 0$ and $HM = 0$, $hn = 0$. □

Solution to Exercise (8.24): Since $\psi : T \longrightarrow T'$ is a morphism of graded algebras

$$\begin{aligned}\psi(\delta(H_i)) &\subseteq \psi(AH_{i-1}V + VH_{i-1}A)\\ &= \psi(A)\psi(H_{i-1})\psi(V) + \psi(V)\psi(H_{i-1})\psi(A)\\ &\subseteq A'\psi(H_{i-1})V' + V'\psi(H_{i-1})A'.\end{aligned}$$

Since $\psi\delta = \delta'\psi$, I' is indeed a triangular ideal of $\mathcal{A}'$. Now, assume that $M \in \mathcal{A}/I$-Mod and $N \in \mathcal{A}'$-Mod with $F_\psi(N) \cong F_\eta(M)$. Since I is triangular, by (8.22), $F_\psi(N)$ is annihilated by $I_{A'}$. Thus, $0 = I_A F_\psi(N) = I'_{A'}N$, and $N \cong F_{\eta'}(N')$ with $N' \in \mathcal{A}'/I'$-Mod. Thus, $F_\eta F_{\overline{\psi}}(N') \cong F_\psi F_{\eta'}(N') \cong F_\psi(N) \cong F_\eta(M)$. Since F_η is full and faithful, $F_{\overline{\psi}}(N') \cong M$. □

Section 9

Solution to Exercise (9.5): Since $I \cap R = 0$, $I \cap W_0 = 0$ and $I \cap W_1 = RSR$, we have $I = \langle I \cap R, I \cap W \rangle$, and then I is compatible with the layer (R, W). Make $\overline{W}_1 := W_1/RSR$, take $M \in \mathcal{A}/I$-Mod, $N \in R$-Mod and take $f = (f^0, f^1) \in \mathrm{Phom}_{R\text{-}\overline{W}_1}(M, N)$. Then, we have $g = (g^0, g^1) \in$

$\text{Phom}_{R\text{-}W_1}(M, N)$ defined by $g^0 = f^0$ and $g^1 := f^1\eta$, where $\eta : W_1 \longrightarrow \overline{W}_1$ is the canonical projection. Since $\mathcal{A}$ is a Roiter ditalgebra, there is a structure of A-module on N (which extends the action of R on N) such that $g = (g^0, \widehat{g}^1) \in \text{Hom}_{\mathcal{A}}(M, N)$. That is, such that, $\alpha g^0(m) = g^0(\alpha m) + \widehat{g}^1(\delta(\alpha))[m]$, for all $\alpha \in W_0$ and $m \in M$. We claim that $f = (f^0, \widehat{f}^1) \in \text{Hom}_{\mathcal{A}/I}(M, N)$. If $\delta(\alpha) = \sum_i a_i w_i b_i$, with $a_i, b_i \in A$ and $w_i \in W_1$, then, $\widehat{\delta}(\widehat{\alpha}) = \sum_i a_i \widehat{w}_i b_i$. Then, $\alpha f^0(m) = \alpha g^0(m) = g^0(\alpha m) + \widehat{g}^1(\delta(\alpha))[m] = f^0(\alpha m) + \sum_i a_i g^1(w_i) b_i[m] = f^0(\alpha m) + \sum_i a_i f^1(\widehat{w}_i) b_i[m] = f^0(\alpha m) + \widehat{f}^1(\widehat{\delta}(\widehat{\alpha}))[m]$. □

Solution to Exercise (9.6): Consider the conflation $N \xrightarrow{f} E \xrightarrow{g} M$ in $\mathcal{E}_{\mathcal{B}}$. Then, the conflation in $\mathcal{E}_{\mathcal{A}}$

$$F(N) \xrightarrow{F(f)} F(E) \xrightarrow{F(g)} F(M)$$

splits iff $F(g)$ is a retraction in $\mathcal{A}$-Mod (iff g is a retraction in $\mathcal{B}$-Mod, because F is full and faithful). □

Solution to Exercise (9.8): From (9.3), $\mathcal{A}^a$ is a Roiter ditalgebra. If $N \xrightarrow{f} E \xrightarrow{g} N \in \mathcal{E}_{\mathcal{A}^a}$, then the composition of the composable pair

$$F_a(N) \xrightarrow{F_a(f)} F_a(E) \xrightarrow{F_a(g)} F_a(M)$$

is zero. Moreover, since $F_a(f)^0 = f^0$ and $F_a(g)^0 = g^0$, then

$$0 \longrightarrow F_a(N) \xrightarrow{F_a(f)^0} F_a(E) \xrightarrow{F_a(g)^0} F_a(M) \longrightarrow 0$$

is a split exact sequence of R-modules. Thus, F_a is an exact functor.

Since F_a is an equivalence, from (9.6), F_a^* is a monomorphism. From (9.7), we know that G^* is an epimorphism. Finally

$$G^*([F_a(N) \xrightarrow{F_a(f)} F_a(E) \xrightarrow{F_a(g)} F_a(M)]) = 0,$$

iff $F_a(N) \xrightarrow{F_a(f)} F_a(E) \xrightarrow{F_a(g)} F_a(M)$ is a composable pair in $\mathcal{A}$-Mod with zero composition, whose "0 part" is a split exact sequence in R^a-Mod. This is also equivalent to

$$N \xrightarrow{f} E \xrightarrow{g} M \in \mathcal{E}_{\mathcal{A}^a}.$$

Therefore, the sequence of the statement is exact. □

Solution to Exercise (9.9): From (9.3), $\mathcal{A}^s$ is a Roiter ditalgebra. If $N \xrightarrow{f} E \xrightarrow{g} N \in \mathcal{E}_{\mathcal{A}^s}$, then the composition of the composable pair

$$F_s(N) \xrightarrow{F_s(f)} F_s(E) \xrightarrow{F_s(g)} F_s(M)$$

is zero. Moreover, since $F_s(f)^0 = f^0$ and $F_s(g)^0 = g^0$, then

$$0 \longrightarrow F_s(N) \xrightarrow{F_s(f)^0} F_s(E) \xrightarrow{F_s(g)^0} F_s(M) \longrightarrow 0$$

is a split exact sequence of R-modules. Thus, F_s is an exact functor.

Since F_s is full and faithful, from (9.6), F_s^* is a monomorphism.

Assume that $\zeta' : F_s(M) \xrightarrow{f} E \xrightarrow{g} F_s(N)$ is a conflation of $\mathcal{E}_{\mathcal{A}}$. From (5.14), we may assume that $f^1 = 0$ and $g^1 = 0$. Then

$$0 \longrightarrow F_s(M) \xrightarrow{f^0} E \xrightarrow{g^0} F_s(N) \longrightarrow 0$$

is a split exact sequence in R-Mod. Thus, as an R-module $E = M \oplus N$ is annihilated by the ideal I_R, which we factored out to obtain $\mathcal{A}^s$. This implies that $E = F_s(E)$. Then, the image of $\zeta : M \xrightarrow{f} E \xrightarrow{g} N$ under F_s is ζ'. The "0 part" of ζ is a split exact sequence of R^s-Mod because the "0 part" of ζ' is a split exact sequence of R-Mod. The composition of ζ is zero iff the composition of ζ' is zero because F_s is faithful. Thus, F_s^* is an epimorphism. □

Solution to Exercise (9.10): From (9.3), $\mathcal{A}^r$ is a Roiter ditalgebra.
If $N \xrightarrow{f} E \xrightarrow{g} N \in \mathcal{E}_{\mathcal{A}^r}$, then the composition of the composable pair

$$F_r(N) \xrightarrow{F_r(f)} F_r(E) \xrightarrow{F_r(g)} F_r(M)$$

is zero. Moreover, since $F_r(f)^0 = f^0$ and $F_r(g)^0 = g^0$, then

$$0 \longrightarrow F_r(N) \xrightarrow{F_r(f)^0} F_r(E) \xrightarrow{F_r(g)^0} F_r(M) \longrightarrow 0$$

is a split exact sequence of R-modules. Thus, F_r is an exact functor.

Since F_r is full and faithful, from (9.6), F_r^* is a monomorphism.

From (8.19) we know that $\delta_{|W_0'}$ injective implies that F_r is a dense functor (hence an equivalence). Thus, any element of $\mathrm{Ext}_{\mathcal{A}}(F_r(M), F_r(N))$ can be represented by a composable pair $F_r(N) \xrightarrow{F_r(f)} F_r(E) \xrightarrow{F_r(g)} F_r(M)$ in $\mathcal{E}_{\mathcal{A}}$. Thus, the composable pair $N \xrightarrow{f} E \xrightarrow{g} M$ has zero composition and its "0-part" satisfies that

$$0 \longrightarrow N \xrightarrow{f^0} E \xrightarrow{g^0} M \longrightarrow 0$$

is a split exact sequence in R-Mod. Then F_r^* is an epimorphism too. □

Section 10

Solution to Exercise (10.4): Assume that the Roiter ditalgebra $\mathcal{A}_i = (T_i, \delta_i)$ has layer (R_i, W^i), for $i \in \{1, 2\}$. We claim that the ditalgebra $\mathcal{A} := \mathcal{A}_1 \times \mathcal{A}_2$ is a Roiter ditalgebra with layer (R, W), where $R = R_1 \times R_2$ and $W =$

$W_0 \oplus W_1$, with $W_0 = W_0^1 \times W_0^2$ and $W_1 = W_1^1 \times W_1^2$. Take $M \in R$-Mod, $N \in \mathcal{A}$-Mod and $f = (f^0, f^1) \in \mathrm{Phom}_{R\text{-}W}(M, N)$. Then, we can consider $f_1 := (f_1^0, f_1^1) \in \mathrm{Phom}_{R_1\text{-}W^1}(e_1M, e_1N)$, defined by $f_1^0(e_1m) = f^0(e_1m)$ and $f_1^1(w_1) = f^1(w_1, 0)$ for $m \in M$ and $w_1 \in W_1^1$. Likewise, we can consider $f_2 := (f_2^0, f_2^1) \in \mathrm{Phom}_{R_2\text{-}W^2}(e_2M, e_2N)$. Since $\mathcal{A}_1$ and $\mathcal{A}_2$ are Roiter ditalgebras, there exist A_i-module structures $\widehat{e_iM}$ on the R_i-modules e_iM such that $f_i : \widehat{e_iM} \longrightarrow e_iN$ is a morphism in $\mathcal{A}_i$-Mod. Consider the functor $F = F_{\pi_1} \oplus F_{\pi_1}$ introduced in (10.3). Then, we have the isomorphism $\psi_N : F(e_1N, e_2N) \longrightarrow N$ of A-modules and the isomorphism $\psi_M : F(e_1M, e_2M) \longrightarrow M$ of R-modules. We can transfer the structure of A-module of $F(\widehat{e_1M}, \widehat{e_2M})$ over M using the R-isomorphism ψ_M. Then, we to obtain an A-module $\widehat{M}$, with underlying R-module M, and an isomorphism of A-modules $\varphi_M := (\psi_M)^{-1} : \widehat{M} \longrightarrow F(\widehat{e_1M}, \widehat{e_2M})$. Then, the composition of morphisms $(\psi_N, 0)F(f_1, f_2)(\varphi_M, 0) : \widehat{M} \longrightarrow N$ in $\mathcal{A}$-Mod is determined by f. A similar argument applies if we start with an R-module N and an A-module M. Thus, $\mathcal{A}$ is a Roiter ditalgebra. □

Section 12

Solution to Exercise (12.11): In the *Claim* of the proof of (12.9), this has already been proved for $t \in \langle \underline{W} \rangle$. If $t \in B$, our assumption implies that $\delta(t) = 0$. Moreover, $\delta^X(\sigma_{\nu,x}(t)) = \delta^X(\nu(tx)) = 0$. Notice that, for $x \in X$ and $p \in P$, we have $t(xp) = (tx)p$. Indeed, $t(xp) = tp^0(x) = p^0(tx) = (tx)p$, because $\delta(t) = 0$. Then

$$\begin{aligned}
\sigma_{\lambda(\nu),x}(t) - \sigma_{\nu,\rho(x)}(t) &= \textstyle\sum_{i,j} \nu[x_i p_j]\gamma_j \sigma_{\nu_i,x}(t) - \sum_j \sigma_{\nu,xp_j}(t)\gamma_j \\
&= \textstyle\sum_{i,j} \nu[x_i p_j]\gamma_j \nu_i(tx) - \sum_j \nu[t(xp_j)]\gamma_j \\
&= \textstyle\sum_{i,j} \nu[x_i p_j]\gamma_j \nu_i(tx) - \sum_j \nu[(tx)p_j)]\gamma_j \\
&= \textstyle\sum_{i,j} \nu[x_i p_j]\gamma_j \nu_i(tx) - \sum_{i,j} \nu[x_i \nu_i(tx) p_j]\gamma_j \\
&= \textstyle\sum_i \left[\sum_j \nu[x_i p_j]\gamma_j \nu_i(tx) - \sum_j \nu[x_i \nu_i(tx) p_j]\gamma_j\right] \\
&=: \Delta.
\end{aligned}$$

Then, $\Delta = 0$, because of the following: For $\nu \in X^*$ and $z \in X$, we can consider $\gamma_{\nu,z} \in P^*$ defined by $\gamma_{\nu,z}(p) := \nu(zp)$, for $p \in P$. Notice that $\gamma_{\nu,zs} = (\gamma_{\nu,z})s$, for $s \in S$. Indeed, for $p \in P$

$$\gamma_{\nu,zs}[p] = \nu[(zs)p] = \nu[z(sp)] = \gamma_{\nu,z}[sp] = (\gamma_{\nu,z})s[p].$$

Then, $\sum_j \nu(zp_j)\gamma_j s = \sum_j \gamma_{\nu,z}(p_j)\gamma_j s = (\gamma_{\nu,z})s = \gamma_{\nu,zs} = \sum_j \gamma_{\nu,zs}(p_j)\gamma_j = \sum_j \nu((zs)p_j)\gamma_j$. Then, taking $z = x_i$ and $s = \nu_i(tx)$ in the ith summand of Δ, we obtain that $\Delta = 0$. □

Section 14

Solution to Exercise (14.11): First recall that if (S, P) is the splitting associated to the algebra $\Gamma = \mathrm{End}_{\mathcal{A}'}(X)^{op}$ and $B = [T']_0$, then, we have a right additive filtration of S-S-subbimodules of P

$$0 = P^{(\ell_P+1)} \subseteq P^{(\ell_P)} \subseteq \cdots \subseteq P^{(1)} = P,$$

such that $P^{(i)}P^{(j)} \subseteq P^{(i+j)}$ for all $i, j \in [1, \ell_P]$ with $i + j \leq \ell_P$, and $P^{(i)}P^{(j)} = 0$, otherwise; and we have a right additive filtration of B-S-subbimodules of X

$$\mathcal{F}(X): \qquad 0 = X_0 \subseteq X_1 \subseteq \cdots \subseteq X_{\ell_X} = X$$

such that $X_j P \subseteq X_{j-1}$ for all $j \in [1, \ell_X]$.

(1) We show that $\delta^X(H_m^X) \subseteq A^X H_{m-1}^X V^X + V^X H_{m-1}^X A^X$, for all $m \in [1, 2\ell_X(\ell + 1)]$. Take $\nu \in X^*$, $h \in H$ and $x \in X$ such that $\mathbb{h}(\nu) + 2\ell_X \mathbb{h}(h) + \mathbb{h}(x) \leq m$. We want to study $\delta^X(\sigma_{\nu,x}(h))$. Since $h \in H \subseteq A$ has degree zero, by (12.11)

$$\delta^X(\sigma_{\nu,x}(h)) = \sigma_{\lambda(\nu),x}(h) + \sigma_{\nu,x}(\delta(h)) - \sigma_{\nu,\rho(x)}(h).$$

We look separately at each one of the summands. Since, by (12.8)(2), $\lambda(\nu) = \sum_{i,j} \nu(x_i p_j)\gamma_j \otimes \nu_i$, the first summand is $\sum_{i,j} \nu(x_i p_j)\gamma_j \sigma_{\nu_i,x}(h) \in P^* H_{m-1}^X \subseteq V^X H_{m-1}^X$, because from lemma (14.7), $\mathbb{h}(\nu_i) + 2\ell_X \mathbb{h}(h) + \mathbb{h}(x) < \mathbb{h}(\nu) + 2\ell_X \mathbb{h}(h) + \mathbb{h}(x) \leq m$, whenever $\nu(x_i p_j) \neq 0$.

Since $\rho(x) = \sum_j x p_j \otimes \gamma_j$, by (12.8)(2), we have that the third summand is $-\sum_j \sigma_{\nu,xp_j}(h)\gamma_j \subseteq H_{m-1}^X P^* \subseteq H_{m-1}^X V^X$ because, from (14.7), $\mathbb{h}(\nu) + 2\ell_X \mathbb{h}(h) + \mathbb{h}(xp_j) < \mathbb{h}(\nu) + 2\ell_X \mathbb{h}(h) + \mathbb{h}(x) \leq m$, whenever $x \neq 0$.

By the triangularity of I, we have $\delta(h) = \sum_r a_r h_r v_r + \sum_s v_s' h_s' a_s'$, where $a_r, a_s' \in A$, $h_r, h_s' \in H_{\mathbb{h}(h)-1}$ and $v_r, v_s' \in V$. Then, by (12.8)(3), we have

$$\sigma_{\nu,x}(\delta(h)) = \sum_{r,i,j} \sigma_{\nu,x_i}(a_r)\sigma_{\nu_i,x_j}(h_r)\sigma_{\nu_j,x}(v_r) + \sum_{s,i,j} \sigma_{\nu,x_i}(v_s')\sigma_{\nu_i,x_j}(h_s')\sigma_{\nu_j,x}(a_s'),$$

which belongs to $A^X H_{m-1}^X V^X + V^X H_{m-1}^X A^X$. Indeed, we can assume that $\nu \neq 0$ and $x \neq 0$. Thus, $2\ell_X \mathbb{h}(h) < m$. Then, $\mathbb{h}(\nu_i) + 2\ell_X \mathbb{h}(h_r) + \mathbb{h}(x_j) \leq 2\ell_X(\mathbb{h}(h_r) + 1) \leq 2\ell_X \mathbb{h}(h) < m$ and then $\sigma_{\nu_i,x_j}(h_r) \in H_{m-1}^X$. Similarly, $\sigma_{\nu_i,x_j}(h_s') \in H_{m-1}^X$.

(2) If $a \in I_A = \langle H \rangle$, then $a = \sum_r a_r h_r b_r$ with $a_r, b_r \in A$ and $h_r \in H$. Then, for $\nu \in X^*$ and $x \in X$, (12.8)(3) implies that

$$\sigma_{\nu,x}(a) = \sum_{r,i,j} \sigma_{\nu,x_i}(a_r)\sigma_{\nu_i,x_j}(h_r)\sigma_{\nu_j,x}(b_r).$$

Thus, $I_{A^X}^X = \langle H^X \rangle \subseteq \langle \{\sigma_{\nu,x}(a) \mid \nu \in X^*, x \in X \text{ and } a \in I\} \rangle \subseteq \langle H^X \rangle = I_{A^X}^X$.

(3) Assume that $N \in \mathcal{A}^X$-Mod satisfies that $I^X_{A^X} N = 0$. From (12.10), we have that, for $a \in A$, $x \in X$ and $n \in N$

$$a \cdot (x \otimes n) = \sum_i x_i \otimes \sigma_{\nu_i,x}(a) * n,$$

where $*$ denotes the action of A^X on N. Then, $I^X_{A^X} * N = 0$ implies that $I_A \cdot F^X(N) = 0$. Thus, F^X induces a functor $\overline{F}^X : (\mathcal{A}^X, I^X)$-Mod$\longrightarrow(\mathcal{A}, I)$-Mod. □

Section 15

Solution to Exercise (15.19): By assumption, we know the existence of an additive R_0-R_0-bimodule filtration $\mathcal{F}(\widetilde{W})$, for some R_0-R_0-subbimodule $\widetilde{W}$ of W, such that the product map $\rho : R \otimes_{R_0} \widetilde{W} \otimes_{R_0} R \longrightarrow W$ determines an isomorphism of R-R-bimodule filtrations $R \otimes_{R_0} \mathcal{F}(\widetilde{W}) \otimes_{R_0} R \xrightarrow{\cong} \mathcal{F}(W)$. Denote by ℓ, ℓ' and ℓ'' the lengths of $\mathcal{F}(W)$, $\mathcal{F}(W')$ and $\mathcal{F}(W'')$, respectively. Then, $\ell = \ell' + \ell''$ and $W_{\ell'} = W'$. Make $\widetilde{W}' := \widetilde{W}_{\ell'}$ and notice that $\mathcal{F}(\widetilde{W}') := \mathcal{F}(\widetilde{W})_{\widetilde{W}'}$ is an additive R_0-R_0-bimodule filtration such that the product map induces an isomorphism of filtrations $R \otimes_{R_0} \mathcal{F}(\widetilde{W}') \otimes_{R_0} R \xrightarrow{\cong} \mathcal{F}(W')$. Thus, $\mathcal{F}(W')$ is an R_0-R_0-free filtration.

Moreover, since $\mathcal{F}(\widetilde{W})$ is an additive filtration, for any $j \in [0, \ell]$, we have an R_0-R_0-bimodule filtration decomposition $\mathcal{F}(\widetilde{W}) = \mathcal{F}(\widetilde{W})_{\widetilde{W}_j} \oplus \mathcal{F}(\widetilde{V}''[j])$, for some additive R_0-R_0-bimodule filtration $\mathcal{F}(\widetilde{V}''[j])$. We want to consider the particular case $j = \ell'$, where we make $\widetilde{V}'' := \widetilde{V}''[\ell']$, and we have $\widetilde{W}' = \widetilde{W}_{\ell'}$. Since the product map induces an isomorphism

$$\begin{aligned}(R \otimes_{R_0} \mathcal{F}(\widetilde{W}') \otimes_{R_0} R) \oplus (R \otimes_{R_0} \mathcal{F}(\widetilde{V}'') \otimes_{R_0} R)\\ = R \otimes_{R_0} \mathcal{F}(\widetilde{W}) \otimes_{R_0} R \xrightarrow{\cong} \mathcal{F}(W),\end{aligned}$$

we know that the R-R-bimodule filtration

$$\mathcal{F}(V'') : 0 \subseteq R\widetilde{V}''_1 R \subseteq \cdots \subseteq R\widetilde{V}''_{\ell''} R = R\widetilde{V}'' R =: V''$$

is R_0-R_0-free and $\mathcal{F}(W) = \mathcal{F}(W') \oplus \mathcal{F}(V'')$.

Finally, since $W' \oplus V'' = W = W' \oplus W''$, there is an isomorphism of R-R-bimodules $\psi : V'' \longrightarrow W''$, which can be used to translate the R_0-R_0-free filtration $\mathcal{F}(V'')$ of V'' into the R_0-R_0-free filtration $\mathcal{F}'(W'') := \psi(\mathcal{F}(V''))$ of $W'' = \psi(V'')$. Indeed, the additive R_0-R_0-bimodule filtration

$$\psi(\mathcal{F}(\widetilde{V}'')) : 0 \subseteq \psi(\widetilde{V}''_1) \subseteq \cdots \subseteq \psi(\widetilde{V}''_{\ell''}) = \psi(\widetilde{V}'')$$

of $\psi(\widetilde{V}'')$ satisfies that the product map induces an isomorphism

$$R \otimes_{R_0} \psi(\mathcal{F}(\widetilde{V}'')) \otimes_{R_0} R \xrightarrow{\cong} \mathcal{F}'(W'').$$

□

Section 16

Solution to Exercise (16.4): The functor F_r is clearly exact because the first component of the layers of $\mathcal{A}$ and $\mathcal{A}'$ is R, and the first components of f and $F_r(f)$ coincide. The commutativity of the diagram is easy to verify. The induced morphism F_r is surjective because i_0 is a section, thus i_0^* is a retraction.

If $W_1' = 0$, then $\delta(W_0') = 0$, $\mathcal{A}'$-Mod is essentially the same as B-Mod; $B = R^a$ and we recover (9.7). □

Solution to Exercise (16.5): Adopt the notation of (16.4). Assume that $f = (f^0, f^1) : F_r(M) \longrightarrow N$ is an isomorphism in $\mathcal{A}'$-Mod. Consider the morphism $\widehat{f^1} \in \mathrm{Hom}_{R\text{-}R}(W_1' \oplus W_1'', \mathrm{Hom}_k(M, N))$ defined by its components $f^1 \in \mathrm{Hom}_{R\text{-}R}(W_1', \mathrm{Hom}_k(M, N))$ and $0 \in \mathrm{Hom}_{R\text{-}R}(W_1'', \mathrm{Hom}_k(M, N))$. Since $\mathcal{A}$ is a Roiter ditalgebra, there is $\overline{N} \in \mathcal{A}$-Mod with underlying R-module N such that $\widehat{f} = (f^0, \widehat{f^1}) : M \longrightarrow \overline{N}$ is a morphism in $\mathcal{A}$-Mod (hence an isomorphism).

It remains to show that $F_r(\overline{N}) = N$. Let us denote by $\cdot$ the action of W_0' over N provided by the original $\mathcal{A}'$-module structure of N. Denote by $\circ$ the action of W_0' over N obtained by restriction of the action of A over the $\mathcal{A}$-module $\overline{N}$. We will show that $w \cdot n = w \circ n$, for any $n \in N$ and $w \in W_0'$. Recall that the layer (R, W') of $\mathcal{A}'$ is triangular, and, therefore, there is a filtration

$$0 = W_0^0 \subseteq W_0^1 \subseteq \cdots \subseteq W_0^{\ell_0'} = W_0',$$

which is not necessarily an initial part of the corresponding filtration of W_0, for the triangular layer (R, W) of $\mathcal{A}$. We will show, by induction on $j \in [0, \ell_0']$, that $w \cdot n = w \circ n$, for any $n \in N$ and $w \in W_0^j$. This is clear for $j = 0$. Assume that it holds for a fixed $j \in [0, \ell_0' - 1]$. Take $w \in W_0^{j+1}$, $n \in N$ and write $n = f^0(m)$. Then, we have

$$\begin{cases} w \cdot n = w \cdot f^0(m) = f^0(wm) + f^1(\delta(w))[m] & \text{and} \\ w \circ n = w \circ f^0(m) = f^0(wm) + \widehat{f^1}(\delta(w))[m], \end{cases}$$

because $f = (f^0, f^1)$ is a morphism in $\mathcal{A}'$-Mod, $\widehat{f} = (f^0, \widehat{f^1})$ is a morphism in $\mathcal{A}$-Mod and the differential of $\mathcal{A}'$ is the restriction of the differential δ of $\mathcal{A}$.

We have to show that

$$\widehat{f}^1(\delta(w))[m] = f^1(\delta(w))[m].$$

From the triangularity hypothesis on $\mathcal{A}'$, we know that $\delta(W_0^{j+1}) \subseteq B_j W_1' B_j$, where B_j is the subalgebra of B generated by R and W_0^j. Thus, $\delta(w) = \sum_i b_i w_i b_i'$ where $b_i, b_i' \in B_j$ and $w_i \in W_1'$. Our induction hypothesis implies that $b \cdot n = b \circ n$, for all $b \in B_j$ and $n \in N$. Then, we have

$$\begin{aligned} f^1(\delta(w))[m] &= f^1(\textstyle\sum_i b_i w_i b_i')[m] \\ &= \textstyle\sum_i b_i \cdot f^1(w_i) b_i'[m] \\ &= \textstyle\sum_i b_i \cdot f^1(w_i)[b_i' m] \\ &= \textstyle\sum_i b_i \circ \widehat{f}^1(w_i)[b_i' m] \\ &= \textstyle\sum_i b_i \circ \widehat{f}^1(w_i) b_i'[m] \\ &= \widehat{f}^1(\textstyle\sum_i b_i w_i b_i')[m] \\ &= \widehat{f}^1(\delta(w))[m] \end{aligned}$$

□

Solution to Exercise (16.6): From (16.3), $\mathcal{A}^X$ is a Roiter ditalgebra. If $N \xrightarrow{f} E \xrightarrow{g} N \in \mathcal{E}_{\mathcal{A}^X}$, from (5.14) we may assume that $f = (f^0, 0)$ and $g = (g^0, 0)$. Then, the composition of the composable pair

$$F_X(N) \xrightarrow{F_X(f)} F_X(E) \xrightarrow{F_X(g)} F_X(M)$$

is zero. Moreover, since $F_X(f)^0 = 1_X \otimes f^0$ and $F_X(g)^0 = 1_X \otimes g^0$, then

$$0 \longrightarrow F_X(N) \xrightarrow{F_X(f)^0} F_X(E) \xrightarrow{F_X(g)^0} F_X(M) \longrightarrow 0$$

is a split exact sequence of R-modules. Thus, F_X is an exact functor.

From (13.5), F_X is full and faithful. Hence, from (9.6), F_X^* is a monomorphism. From (16.4), we know that F_r^* is an epimorphism.

Suppose that $N \xrightarrow{f} E \xrightarrow{g} M$ is a conflation in $\mathcal{E}_{\mathcal{A}^X}$, where we assume that $f = (f^0, 0)$ and $g = (g^0, 0)$, as above. Then we have the split exact sequence of S-modules

$$0 \longrightarrow N \xrightarrow{f^0} E \xrightarrow{g^0} M \longrightarrow 0,$$

which, tensoring by the left, by the B-S-bimodule X determines the split exact sequence in B-Mod

$$0 \longrightarrow X \otimes_S N \xrightarrow{1_X \otimes f^0} X \otimes_S E \xrightarrow{1_X \otimes g^0} X \otimes_S M \longrightarrow 0.$$

Then, the conflation $F_r F_X(N) \xrightarrow{F_r F_X(f)} F_r F_X(E) \xrightarrow{F_r F_X(g)} F_r F_X(M)$, which coincides with

$$X \otimes_S N \xrightarrow{(1_X \otimes f^0, 0)} X \otimes_S E \xrightarrow{(1_X \otimes g^0, 0)} X \otimes_S M,$$

is a split conflation in $\mathcal{E}_{\mathcal{A}'}$, that is $F_r^* F_X^* = 0$.

Now assume that $F_X(N) \xrightarrow{u} H \xrightarrow{v} F_X(M)$ is a conflation in $\mathcal{E}_{\mathcal{A}}$ which is mapped to zero by F_r^*. By (6.21), there is a commutative diagram in $\mathcal{A}'$-Mod

$$\begin{array}{ccccc} F_rF_X(N) & \xrightarrow{F_r(u)} & F_r(H) & \xrightarrow{F_r(v)} & F_rF_X(M) \\ \| & & \downarrow \cong & & \| \\ X \otimes_S N & \xrightarrow{(1,0)^t} & (X \otimes_S N) \oplus (X \otimes_S M) & \xrightarrow{(0,1)} & X \otimes_S M. \end{array}$$

Thus, $F_r(H) \cong X \otimes_S (N \oplus M)$ in $\mathcal{A}'$-Mod. By (16.5), there is $\overline{X \otimes_S (N \oplus M)} \in \mathcal{A}$-Mod with $\overline{X \otimes_S (N \oplus M)} \cong H$ and $F_r(\overline{X \otimes_S (N \oplus M)}) = X \otimes_S (N \oplus M)$. Then, from (16.1), we obtain the existence of some $E \in \mathcal{A}^X$-Mod with $F_X(E) = \overline{X \otimes_S (N \oplus M)} \cong H$. Since F_X is a full functor, we obtain that our original conflation in $\mathcal{E}_{\mathcal{A}}$ is equivalent to a conflation

$$F_X(N) \xrightarrow{F_X(f)} F_X(E) \xrightarrow{F_X(g)} F_X(M).$$

Now, consider the composable pair $\zeta : N \xrightarrow{f} E \xrightarrow{g} M$ in $\mathcal{A}^X$-Mod, and apply (6.24) to the full and faithful functor $G := F^{'X}$ (X is complete) and the split conflation $GF_{\hat{r}}(\zeta) = F_rF^X(\zeta)$ (by assumption) to get that $F_{\hat{r}}(\zeta)$ is a split conflation of $\mathcal{A}^{'X}$. Here, $\mathcal{A}'$ is a Roiter ditalgebra by (12.3). Since $F_{\hat{r}}(\zeta)$ is in particular a conflation, thus

$$0 \longrightarrow F_{\hat{r}}(N) \xrightarrow{F_{\hat{r}}(f)^0} F_{\hat{r}}(E) \xrightarrow{F_{\hat{r}}(g)^0} F_{\hat{r}}(M) \longrightarrow 0$$

is a split exact sequence in S-Mod, which implies that

$$0 \longrightarrow N \xrightarrow{f^0} E \xrightarrow{g^0} M \longrightarrow 0$$

is a split exact sequence in S-Mod too. Moreover, $F^X(\zeta)$ has zero composition, thus, since F_X is faithful, ζ has zero composition too. We have shown that ζ is a conflation in $\mathcal{E}_{\mathcal{A}^X}$. Therefore, the exactness of the sequence of the statement of the exercise is proved. □

Section 20

Solution to Exercise (20.14): Let $M, N \in \mathcal{A}$-Mod and denote by

$$\theta_{M,N} : \mathrm{Hom}_{\mathcal{A}}(M, N)^K \longrightarrow \mathrm{Hom}_{\mathcal{A}^K}(M^K, N^K)$$

the natural morphism defined by the same receipt of definition (20.8). Let (R, W) be the layer of $\mathcal{A}$ and recall that in (6.16) we described an isomorphism

$$\eta_{Y,N} : \mathrm{Hom}_{\mathcal{A}}(A \otimes_R Y, N) \longrightarrow \mathrm{Hom}_R(Y, N)\mathrm{Hom}_{R\text{-}R}(W_1, \mathrm{Hom}_k(A \otimes_R Y, N)),$$

natural in the variables $Y \in R$-Mod and $N \in A$-mod. Consider the corresponding isomorphism $\overline{\eta}$ for the layered ditalgebra $\mathcal{A}^K$, thus $\overline{\eta} = \overline{\eta}_{Y^K,N^K}$ goes from $\mathrm{Hom}_{\mathcal{A}^K}(A^K \otimes_{R^K} Y^K, N^K)$ to the product space $\mathrm{Hom}_{R^K}(Y^K, N^K) \times \mathrm{Hom}_{R^K\text{-}R^K}(W_1^K, \mathrm{Hom}_K(A^K \otimes_{R^K} Y^K, N^K))$. In the particular case $Y = R$, we have the canonical isomorphisms $\theta_0 : \mathrm{Hom}_R(R, N)^K \longrightarrow \mathrm{Hom}_{R^K}(R^K, N^K)$, and $\theta_1 : \mathrm{Hom}_{R\text{-}R}(W_1, \mathrm{Hom}_k(A \otimes_R R, N))^K \longrightarrow \mathrm{Hom}_{R^K\text{-}R^K}(W_1^K, \mathrm{Hom}_K(A^K \otimes_{R^K} R^K, N^K))$. Identifying $(A \otimes_R R)^K$ with $A^K \otimes_{R^K} R^K$, we see that $(\theta_0 \times \theta_1)\eta^K_{R,N} = \overline{\eta}_{R^K,N^K}\theta_{A\otimes R,N}$ and, hence, that $\theta_{A\otimes R,N}$ is an isomorphism. Then, $\theta_{A^n,N}$ is an isomorphism for all n. Since X is finitely presented, there is an exact sequence $A^n \to A^m \to X \to 0$ in A-Mod. Consider the image of this sequence in the exact category $\mathcal{A}$-Mod under the canonical embedding, apply the left exact functor $\mathrm{Hom}_{\mathcal{A}}(-, N)$ and then the extension functor to obtain the top sequence in the following exact commutative diagram

$$\begin{array}{ccccc}
0 \to \mathrm{Hom}_{\mathcal{A}}(X, N)^K & \to & \mathrm{Hom}_{\mathcal{A}}(A^m, N)^K & \to & \mathrm{Hom}_{\mathcal{A}}(A^n, N)^K \\
\downarrow{\scriptstyle \theta_M} & & \downarrow{\scriptstyle \theta_{A^m}} & & \downarrow{\scriptstyle \theta_{A^n}} \\
0 \to \mathrm{Hom}_{\mathcal{A}^K}(X^K, N^K) & \to & \mathrm{Hom}_{\mathcal{A}^K}(A^{Km}, N^K) & \to & \mathrm{Hom}_{\mathcal{A}^K}(A^{Kn}, N^K).
\end{array}$$

Then, X extends to K. □

Section 22

Solution to Exercise (22.23): If $M \in k\langle x, y\rangle$-Mod, $Z_t \otimes M \cong M^n$, where x and y act by the left via the matrices

$$\widehat{T}_x = \begin{pmatrix} 0 & I & \cdots & 0 & 0 & 0 & 0 \\ 0 & 0 & \ddots & \vdots & \vdots & \vdots & \vdots \\ 0 & 0 & \ddots & I & 0 & 0 & 0 \\ 0 & 0 & \ddots & 0 & I & 0 & 0 \\ 0 & 0 & \ddots & 0 & 0 & I & 0 \\ \vdots & \vdots & \ddots & 0 & 0 & 0 & I \\ 0 & 0 & \cdots & 0 & 0 & 0 & 0 \end{pmatrix} \text{ and } \widehat{T}_y = \begin{pmatrix} 0 & 0 & \cdots & 0 & 0 & 0 & 0 \\ I & 0 & \ddots & \vdots & \vdots & \vdots & \vdots \\ T_x & I & \ddots & 0 & 0 & 0 & 0 \\ T_y & T_x & \ddots & 0 & 0 & 0 & 0 \\ 0 & T_y & \ddots & I & 0 & 0 & 0 \\ \vdots & \vdots & \ddots & T_x & I & 0 & 0 \\ 0 & 0 & \cdots & T_y & T_x & I & 0 \end{pmatrix},$$

where T_x and T_y are the actions of x and y on the left module M. Consider $M' \in k\langle x, y\rangle$-Mod, with its corresponding matrices $\widehat{T}'_x$ and $\widehat{T}'_y$, and take $A = (A_{i,j}) \in \mathrm{Hom}_{k\langle x,y\rangle}(M^n, M'^n)$, a matrix with $n \times n$ k-linear components $A_{i,j} : M \to M'$. We have the equalities $A\widehat{T}_x = \widehat{T}'_x A$ and $A\widehat{T}_y = \widehat{T}'_y A$. From the first one, we derive that A has the form

$$A = \begin{pmatrix} A_1 & A_2 & A_3 & \cdots & A_n \\ 0 & A_1 & A_2 & \cdots & A_{n-1} \\ 0 & 0 & A_1 & \cdots & A_{n-2} \\ \vdots & \vdots & \vdots & \ddots & \vdots \\ 0 & 0 & 0 & \cdots & A_1 \end{pmatrix}.$$

From the second, we get first that $A_n = A_{n-1} = \cdots = A_2 = 0$, and then that $A_1 \in \mathrm{Hom}_{k\langle x,y\rangle}(M, M')$. Thus, A is an isomorphism iff A_1 is so. When $M = M'$, A is a non-trivial idempotent iff A_1 is so. Therefore, $Z_t \otimes_{k\langle x,y\rangle} -$ preserves indecomposability and isomorphism classes. The fact: $\dim Z_t \otimes M \geq t$, for any non-zero $M \in k\langle x, y\rangle$-Mod, is clear. □

Solution to Exercise (22.24): Consider the morphism of algebras $\varphi := (\psi_|)^{-1} : A \longrightarrow A'$ and notice that $\varphi\psi = 1_{A'}$. Then, apply (22.12). □

Solution to Exercise (22.25): By (9.5), we have that $\mathcal{A} = (T, \delta)$ and $\mathcal{A}/I = (\widehat{T}, \widehat{\delta})$ are both Roiter ditalgebras. The canonical projection $\eta : \mathcal{A} \longrightarrow \mathcal{A}/I$ is a morphism such that its restriction $\eta_| : A = [T]_0 \longrightarrow [\widehat{T}]_0$ is an isomorphism. Then, we can apply (22.24). □

Solution to Exercise (22.26): Denote by $\xi_A : A_{\mathcal{D}^{\Lambda^K}} \longrightarrow (A_{\mathcal{D}^\Lambda})^K = A^K$ the restriction of the morphism $\xi_\Lambda : \mathcal{D}^{\Lambda^K} \longrightarrow \mathcal{D}^{\Lambda K}$, as defined in (20.13). It induces the functor $F_{\xi_A} : A^K\text{-Mod} \longrightarrow A_{\mathcal{D}^{\Lambda^K}}\text{-Mod}$. Then, our claim follows from the commutativity of

$$\begin{array}{ccccccc} A^K\text{-Mod} & \xrightarrow{L_{\mathcal{D}^{\Lambda^K}} F_{\xi_A}} & \mathcal{D}^{\Lambda^K}\text{-Mod} & \xrightarrow{\Xi_{\Lambda^K}} & P^1(\Lambda^K) & \xrightarrow{\mathrm{Cok}^1} & \Lambda^K\text{-Mod} \\ \uparrow (-)^K & & \uparrow F_{\xi_\Lambda}(-)^K & & \uparrow (-)^K & & \uparrow (-)^K \\ A\text{-Mod} & \xrightarrow{L_{\mathcal{D}^\Lambda}} & \mathcal{D}^\Lambda\text{-Mod} & \xrightarrow{\Xi_\Lambda} & P^1(\Lambda) & \xrightarrow{\mathrm{Cok}^1} & \Lambda\text{-Mod}, \end{array}$$

stated in (20.12) and (20.13), and the definition of the transition bimodule. Indeed, $\xi_A^{-1} : F_{\xi_A}(A^K) \longrightarrow A_{\mathcal{D}^{\Lambda^K}}$ is an isomorphism of $A_{\mathcal{D}^{\Lambda^K}}$-modules. □

Solution to Exercise (22.28): Since $L_{\mathcal{A}}$ is faithful, in order to show that G is faithful, it will be enough to show that $F := Z \otimes_\Delta - : \Delta\text{-mod} \to A_{\mathcal{A}}\text{-mod}$ is so. Since $L_{\mathcal{A}}$ reflects isomorphisms, we only have to show that F is faithful and reflects isomorphisms.

Take a morphism $f : M \to N$ in Δ-mod and assume that $F(f) = 0$. Consider the cokernel C of f and an exact sequence in Δ-mod:

$$M \xrightarrow{f} N \xrightarrow{g} C \longrightarrow 0.$$

Then, $F(M)\xrightarrow{F(f)}F(N)\xrightarrow{F(g)}F(C)\longrightarrow 0$ is exact in $A_{\mathcal{A}}$-mod. By assumption, $F(f)=0$. Thus, $F(g)$ is an isomorphism, and so is $G(g):G(N)\to G(C)$. By assumption, $N\cong C$ in Δ-mod. Since g is a surjective morphism in Δ-mod, then g is an isomorphism. Hence, $f=0$. We have shown that F is faithful.

Now, assume that $f:M\to N$ in Δ-mod satisfies that $F(f)$ is an isomorphism. Thus, $G(f)$ is an isomorphism and, by assumption, $M\cong N$. Then, it will be enough to show that f is surjective. Consider the cokernel C of f and an exact sequence in Δ-mod:

$$M\xrightarrow{f}N\xrightarrow{g}C\longrightarrow 0.$$

As before, we have the exact sequence $F(M)\xrightarrow{F(f)}F(N)\xrightarrow{F(g)}F(C)\longrightarrow 0$ in $A_{\mathcal{A}}$-mod. Since $F(f)$ is an isomorphism in $A_{\mathcal{A}}$-mod, we have $F(g)=0$. By the first part, we obtain that $g=0$. Therefore, f is surjective and F reflects isomorphisms. □

Solution to Exercise (22.29): Consider the wild algebra

$$\Gamma := k[u,v,w]/\langle u^2,v^2,w^2,uv,vw,uw\rangle$$

and the functor $G:k\langle x,y\rangle\text{-Mod}\longrightarrow\Gamma\text{-Mod}$ given in (22.27). Note that, for $M\in k\langle x,y\rangle$-mod, $\dim GM=2\dim M$ and $(\dim_k M)^2\leq\dim_k \mathrm{End}_\Gamma(GM)$.

Assume that $\mathcal{A}$ is wild and the inequality in the statement of the exercise holds for almost all finite-dimensional indecomposable $\mathcal{A}$-module M (up to isomorphism). From (22.17)(v), we know the existence of an A-Γ-bimodule C, finitely generated free by the right, say of rank r, such that

$$\Gamma\text{-Mod}\xrightarrow{C\otimes -}A_{\mathcal{A}}\text{-Mod}\xrightarrow{L_A}\mathcal{A}\text{-Mod}$$

preserves isomorphism classes of indecomposables. With the same argument given in the proof of (22.5), we obtain that

$$F:=[\Gamma\text{-mod}\xrightarrow{C\otimes -}A_{\mathcal{A}}\text{-mod}\xrightarrow{L_A}\mathcal{A}\text{-mod}]$$

preserves isomorphism classes and indecomposables. From (22.28), we know that F is a faithful functor. Then, for almost all indecomposable $M\in k\langle x,y\rangle$-mod, we have, by assumption, $\frac{1}{4}(\dim_k GM)^2=(\dim_k M)^2\leq \dim_k\mathrm{End}_\Gamma(GM)\leq\dim_k\mathrm{End}_{\mathcal{A}}(FGM)\leq c\dim_k FGM=cr\dim_k GM$.

Then, $4(\dim_k M)^2=(\dim_k GM)^2\leq 4cr\dim_k GM=8cr\dim_k M$. That is, $\dim_k M\leq 2cr$. This is a contradiction because $k\langle x,y\rangle$ admits infinitely many isomorphism classes of indecomposables M of arbitrary big finite dimension. □

Section 23

Solution to Exercise (23.26): Assume $U \xrightarrow{g} V \longrightarrow 0$ is an exact sequence of R-R-bimodules and $f : W \longrightarrow V$ is any morphism. Denote by $\mathbb{B}$ the directed subset of W which freely generates it. Choose $u_w \in U$ with $g(u_w) = f(w)$, for any $w \in \mathbb{B}$. Thus, $g(e_{t(w)}u_w e_{s(w)}) = e_{t(w)}g(u_w)e_{s(w)} = e_{t(w)}f(w)e_{s(w)} = f(w)$. Since W is freely generated by $\mathbb{B}$, there is a morphism of R-R-bimodules $\overline{f} : W \longrightarrow U$ such that $\overline{f}(w) = e_{t(w)}u_w e_{s(w)}$, for each $w \in \mathbb{B}$. Then, $g\overline{f} = f$, and W is a projective left $R \otimes_k R$-module.

Consider now the example W freely generated by one element $1 \xrightarrow{w} 2$ over the algebra $R = k \times k$. Thus, $W = Re_2 \otimes_k e_1 R$ is a one-dimensional space, while the algebra $R \otimes_k R$ is four dimensional. Thus, W cannot be free.

The last statement follows from the proof of (15.7)(2) □

Solution to Exercise (23.27): Assume $\Gamma := \mathrm{End}_{\mathcal{A}'}(X)^{op}$ has splitting $\Gamma = S \oplus P$, denote by $1 = \sum_{j=1}^{s} e_j$ the decomposition of the unit of R, where (R, W) is the layer of $\mathcal{A}$, as a sum of central primitive orthogonal idempotents, and by $1 = \sum_{i=1}^{t} f_i$ the canonical corresponding decomposition of the unit of Γ; that is, where each f_i is obtained as the composition of the projection $X \to Y_i$ with the injection $Y_i \to X$. Consider the associated filtrations $\mathcal{F}(P) :\ 0 = P_0 \subseteq \cdots \subseteq P$, given by the powers of the radical P, and $\mathcal{F}(X) :\ 0 = XP_0 \subseteq \cdots \subseteq XP \subseteq X$. Recall that $\mathcal{A}^X$ has layer (S, W^X). Then, consider the finite dual basis $(x_i, \nu_i)_{i\in I}$ and $(p_j, \gamma_j)_{j\in J}$ for the right S-modules X and P, which are compatible with the given filtrations and such that each of the elements $x_i, \nu_i, p_j, \gamma_j$ is directed. According to (23.11), this can be done in such a way that $\{x_i\}_{i\in I}$ is a basis for the vector space X over k. Then, for any S-module M, we have $X \otimes_S M \cong \oplus_{i\in I} kx_i \otimes_S M \cong \oplus_{i\in I} kx_i \otimes_k f_{s(x_i)}M$. Thus any element of $X \otimes_S M$ can be written uniquely as a sum $\sum_{i\in I} x_i \otimes m_i \in X \otimes_S M$, where $m_i \in f_{s(x_i)}M$.

Assume that $f^0 : M \longrightarrow N$ is injective and that $F^X(f)^0(\sum_i x_i \otimes m_i) = 0$, where $m_i = f_{s(x_i)}m_i$, for all $i \in I$. Then

$$0 = \sum_{i\in I} x_i \otimes f^0(m_i) + \sum_{i,j} x_i p_j \otimes f^1(\gamma_j)[m_i].$$

Make $I_q := \{i \in I \mid \mathbb{h}(x_i) = q\}$, for $q \in [1, \ell_X]$, where ℓ_X is the length of the filtration $\mathcal{F}(X)$. We show by downward induction on q that $m_i = 0$, for all $i \in I_q$. If $q = \ell_X$, then

$$-\sum_{i\in I_q} x_i \otimes f^0(m_i) = \sum_{i\in I\setminus I_q} x_i \otimes f^0(m_i) + \sum_{j\in J; i\in I} x_i p_j \otimes f^1(\gamma_j)[m_i],$$

where $\mathbb{h}(x_i p_j) < \ell_X = q$. This implies that $\sum_{i \in I_q} x_i \otimes f^0(m_i) = 0$, and then each $x_i \otimes f^0(m_i) = 0$, for all $i \in I_{\ell_X}$. Thus, $f^0(m_i) = 0$ and, since f^0 is injective, $m_i = 0$, for $i \in I_{\ell_X}$. Now, assume that $q < \ell_X$ and $m_i = 0$, for all $i \in I_{q'}$, for $q' \in [q+1, \ell_X]$. Then, from the former formula, we have that

$$
\begin{aligned}
-\sum_{i \in I_q} x_i \otimes f^0(m_i) = & \sum_{i \in I \setminus I_{\ell_X} \cup \cdots \cup I_q} x_i \otimes f^0(m_i) \\
& + \sum_{j \in J; i \in I \setminus I_{\ell_X} \cup \cdots \cup I_{q+1}} x_i p_j \otimes f^1(\gamma_j)[m_i],
\end{aligned}
$$

where $\mathbb{h}(x_i p_j) < q$. This implies that $\sum_{i \in I_q} x_i \otimes f^0(m_i) = 0$, and then, as before, $f^0(m_i) = 0$, for all $i \in I_q$, which implies that $m_i = 0$, for $i \in I_q$. In a finite number of steps, we obtain that all $m_i = 0$. Thus $\sum_i x_i \otimes m_i = 0$, and $F^X(f)^0$ is injective.

Now, assume that f^0 is a surjective map, and consider the typical generators $x_i \otimes n \in X \otimes_S N$, with $n = f_{s(x_i)} n$. Notice that, for $i \in I_1$, $\mathbb{h}(x_i) = 1$ and $x_i p = 0$, for any $p \in P$. Thus, if we choose $m \in f_{s(x_i)} M$ with $f^0(m) = n$, then

$$x_i \otimes n = x_i \otimes f^0(m) = F^X(f)^0(x_i \otimes m).$$

Now assume that we have shown that $x_i \otimes n' \in \operatorname{Im} F^X(f)^0$, for any $i \in I_{q'}$, with $q' < q$, and $n' = f_{s(x_i)} n'$ in N. Then, for $i \in I_q$ and $n = f_{s(x_i)} n$ in N, if we choose $m = f_{s(x_i)} m$ in M with $f^0(m) = n$, we have

$$x_i \otimes n = F^X(f)^0(x_i \otimes m) - \sum_j x_i p_j \otimes f^1(\gamma_j)[m].$$

By induction, the right sum is already in the image of $F^X(f)^0$, because $\mathbb{h}(x_i p_j) < q$, for all the indexes in the considered range. Thus, $x_i \otimes n \in \operatorname{Im} F^X(f)^0$. At the end of this finite induction, we have that $F^X(f)^0$ is surjective. □

Solution to Exercise (23.28): If $R = Re_1 \times \cdots \times Re_{i_0} \times \cdots \times Re_n$, then the quotient algebra $R^s = R/I_R \cong Re_1 \times \cdots \times ke_{i_0} \times \cdots \times Re_n$. Thus, R^s is minimal and the statements on the marked points of $\mathcal{A}^s$ hold. The triangularity of (R^s, W^s) was shown in (8.12).

Then, we have to notice that whenever U is an R-R-bimodule freely generated by its subset $\mathbb{B}$, then $U/(I_R U + U I_R)$ is an R/I_R-R/I_R-bimodule freely generated by its subset $\nu(\mathbb{B})$, where $\nu : U \longrightarrow U/(I_R U + U I_R)$ is the canonical projection. This follows from (23.4)(2). Then, $\mathcal{A}^s$ is nested (resp. seminested) with the basis described in (1).

Part (2) follows from (8.10). □

Solution to Exercise (23.29): Make $\widehat{I} := \widehat{A} \cap I$, then, from (8.9), we can identify $[T']_0$ with $\widehat{A}/\widehat{I}$ and $[T']_1$ with $\widehat{V}/(\widehat{I}\widehat{V} + \widehat{\delta}(\widehat{I}) + \widehat{V}\widehat{I})$. In order to apply (2.10), we will show that $\eta_0 : A \longrightarrow A' = \widehat{A}/\widehat{I}$ is an epimorphism of algebras and that $\eta_1 : V \longrightarrow V' := [T']_1$ induces an isomorphism

$$\mathrm{Hom}_{A'\text{-}A'}(V', U) \xrightarrow{\eta_1^*} \mathrm{Hom}_{A\text{-}A}(V, U),$$

for any A'-A'-bimodule U.

If $\psi, \psi' : A' \longrightarrow B$ are morphisms of algebras such that $\psi\eta = \psi'\eta$, then $\psi(\eta(a)) = \psi'(\eta(a))$, for $a \in A$. Thus, in order to show that $\psi = \psi'$, we only have to show that $\psi(\pi(\alpha^t)) = \psi(\pi(\alpha^t))$, for any $\alpha \in S$. Indeed, the equality $\eta(\alpha) = \eta(\alpha)\pi(\alpha^t)\eta(\alpha)$ implies that

$$\begin{aligned}\psi(\eta(\alpha))\psi(\pi(\alpha^t))\psi(\eta(\alpha)) &= \psi(\eta(\alpha)) = \psi'(\eta(\alpha)) \\ &= \psi(\eta(\alpha))\psi'(\pi(\alpha^t))\psi(\eta(\alpha)).\end{aligned}$$

Then, multiplying this equality by $\psi(\pi(\alpha^t))$ at the left and at the right, we obtain that

$$\begin{aligned}\psi'(\pi(\alpha^t)) &= \psi'\eta(e_{s(\alpha)})\psi'(\pi(\alpha^t))\psi'\eta(e_{t(\alpha)}) \\ &= \psi\eta(e_{s(\alpha)})\psi'(\pi(\alpha^t))\psi\eta(e_{t(\alpha)}) \\ &= \psi(\pi(\alpha^t))\psi(\eta(\alpha))\psi'(\pi(\alpha^t))\psi(\eta(\alpha))\psi(\pi(\alpha^t)) \\ &= \psi(\pi(\alpha^t))\psi(\eta(\alpha))\psi(\pi(\alpha^t))\psi(\eta(\alpha))\psi(\pi(\alpha^t)) \\ &= \psi\eta(e_{s(\alpha)})\psi(\pi(\alpha^t))\psi\eta(e_{t(\alpha)}) \\ &= \psi(\pi(\alpha^t)).\end{aligned}$$

Thus, $\psi = \psi'$ and η_0 is an epimorphism of algebras.

Now, assume that U is an A'-A'-bimodule, hence by restriction through $\eta_0 : A \longrightarrow A'$ it is also an A-A-bimodule, and assume that $f^1 : V \longrightarrow U$ is a morphism of A-A-bimodules. We want to show the existence of a morphism $g^1 : V' \longrightarrow U$ of A'-A'-bimodules such that $f^1 = g^1\eta_1$, where $\eta_1 : V \longrightarrow V'$ is the restriction of the morphism η.

For this, we first construct a morphism $h^1 : \widehat{V} \longrightarrow U$ of $\widehat{A}$-$\widehat{A}$-bimodules such that $f^1 = h^1\sigma_1$. Again, U is considered here as an $\widehat{A}$-$\widehat{A}$-bimodule by restriction using $\pi_0 : \widehat{A} \longrightarrow A'$.

From (4.1)(4), we know that the A-A-bimodule V is freely generated by the subset $\mathbb{B}_1$. Similarly, the $\widehat{A}$-$\widehat{A}$-bimodule $\widehat{V}$ is freely generated by the set $\mathbb{B}_1 \cup S_1^t$. Then, in order to define h^1 it will be enough to give its values on this basis. Thus, define $h^1(\beta) := f^1(\beta)$, for $\beta \in \mathbb{B}_1$, and $h^1(v_\alpha^t) := -\pi(\alpha^t)f^1(\delta(\alpha))\pi(\alpha^t)$, for any $\alpha \in S$. Clearly, this determines a morphism $h^1 : \widehat{V} \longrightarrow U$ of $\widehat{A}$-$\widehat{A}$-bimodules such that $f^1 = h^1\sigma_1$.

Now, since $\widehat{I}$ annihilates U by the left and by the right, it is clear that $h^1(\widehat{I}\widehat{V}) = 0$ and that $h^1(\widehat{V}\widehat{I}) = 0$. In order to show that $h^1(\widehat{\delta}(\widehat{I})) = 0$, it is enough to show that $h^1\widehat{\delta}(\alpha\alpha^t - e_{t(\alpha)}) = 0$ and that $h^1\widehat{\delta}(\alpha^t\alpha - e_{s(\alpha)}) = 0$, for $\alpha \in S$. Fix $\alpha \in S$, then

$$\begin{aligned} h^1(\widehat{\delta}(\alpha\alpha^t - e_{t(\alpha)})) &= h^1(\widehat{\delta}(\alpha)\alpha^t) + h^1(\alpha\widehat{\delta}(\alpha^t)) - h^1(\widehat{\delta}(e_{t(\alpha)})) \\ &= h^1(\delta(\alpha)\alpha^t) + h^1(\alpha v^t_\alpha) - h^1(0) \\ &= h^1(\delta(\alpha)\alpha^t) + \alpha h^1(v^t_\alpha) \\ &= h^1(\delta(\alpha)\alpha^t) - \alpha\alpha^t f^1(\delta(\alpha))\alpha^t \\ &= h^1(\delta(\alpha))\alpha^t - \alpha\alpha^t f^1(\delta(\alpha))\alpha^t \\ &= f^1(\delta(\alpha))\alpha^t - f^1(\delta(\alpha))\alpha^t \\ &= 0. \end{aligned}$$

Similarly, $h^1\widehat{\delta}(\alpha^t\alpha - e_{s(\alpha)}) = 0$. Then, h^1 induces a morphism of $\widehat{A}$-$\widehat{A}$-bimodules $g^1 : V' \longrightarrow U$ such that $g^1\pi_1 = h^1$. Then, g^1 is a morphism of A'-A'-bimodules such that $g^1\eta_1 = g^1\pi_1\sigma_1 = h^1\sigma_1 = f^1$ as we wanted.

Now, given any morphism $g^1 \in \mathrm{Hom}_{A'\text{-}A'}(V', U)$ with $f^1 := \eta_1^*(g^1) = 0 \in \mathrm{Hom}_{A\text{-}A}(V, U)$, reversing the last argument, we know that, for $\alpha \in S$

$$\begin{aligned} 0 &= g^1(\pi(\widehat{\delta}(\alpha\alpha^t - e_{t(\alpha)}))) \\ &= g^1(\pi(\widehat{\delta}(\alpha)\alpha^t + \alpha\widehat{\delta}(\alpha^t))) \\ &= g^1(\pi(\delta(\alpha)\alpha^t + \alpha v^t_\alpha)) \\ &= g^1(\pi(\delta(\alpha)))\pi(\alpha^t) + \pi(\alpha)g^1(\pi(v^t_\alpha)). \end{aligned}$$

Thus, $\pi(\alpha)g^1(\pi(v^t_\alpha)) = -g^1(\pi(\delta(\alpha)))\pi(\alpha^t)$. Then, multiplying at the left by $\pi(\alpha^t)$, we obtain

$$\begin{aligned} g^1(\pi(v^t_\alpha)) &= \pi(\alpha^t)\pi(\alpha)g^1(\pi(v^t_\alpha)) \\ &= -\pi(\alpha^t)g^1(\pi(\delta(\alpha)))\pi(\alpha^t) \\ &= -\pi(\alpha^t)f^1(\delta(\alpha))\pi(\alpha^t) \\ &= 0 \end{aligned}$$

Therefore, $g^1 = 0$ and η_1^* is an isomorphism of vector spaces. Thus, the functor F_η is full and faithful by (2.10). □

Solution to Exercise (23.30): From (6.13), we have $\dim \mathrm{Hom}_A(M, N) - \dim \mathrm{Ext}_A(M, N) = \dim \mathrm{Phom}_{R\text{-}W}(M, N) - \dim \mathrm{Hom}_R(W_0 \otimes_R M, N)$. We also have $\mathrm{Phom}_{R\text{-}W}(M, N) = \mathrm{Hom}_R(M, N) \times \mathrm{Hom}_{R\text{-}R}(W_1, \mathrm{Hom}_k\ (M, N))$ and $\dim \mathrm{Hom}_R(M, N) = \sum_i \dim M_i \dim N_i$. Moreover, we also have the equality $\dim \mathrm{Hom}_{R\text{-}R}(W_1, \mathrm{Hom}_k(M, N)) = \sum_{v \in \mathbb{B}_1} \dim N_{t(v)} \quad \dim M_{s(v)}$, and the series of equalities $\dim \mathrm{Hom}_R(W_0 \otimes_R M, N) = \dim \mathrm{Hom}_{R\text{-}R} (W_0, \mathrm{Hom}_k(M, N)) = \sum_{\alpha \in \mathbb{B}_0} \dim N_{t(\alpha)} \dim M_{s(\alpha)}$. Combining these equalities, we obtain what we wanted. □

Solution to Exercise (23.31): First consider the R-R-bimodule morphism $\psi' : W \longrightarrow T$ defined by $\psi'(\alpha) = \alpha'$ and $\psi'(w) = w$, for every arrow $w \in \mathbb{B}_0 \cup \mathbb{B}_1$ different from α. Since T is freely generated by the pair (R, W), 1_R and ψ' extend to a morphism of algebras $\overline{\psi}' : T \longrightarrow T$. Similarly, if we define $\alpha'' := r^{-1}(\alpha - \sum_t r^t_{\lambda_t+1}\beta^t_{\lambda_t}r^t_{\lambda_t}\cdots r^t_2\beta^t_2 r^t_1\beta^t_1 r^t_0)$. Then, as in the previous construction, we obtain a morphism $\overline{\psi}'' : T \longrightarrow T$, which is the inverse of $\overline{\psi}'$. This implies that the image $\overline{\psi}'(W_0) = W'_0$ is freely generated by the directed set $\mathbb{B}'_0$ and also that the pair (R, W') is a layer for $\mathcal{A}$.

Since (R, W) is a triangular layer, we have R-R-bimodule filtrations $\mathcal{F}_0$ and $\mathcal{F}_1$ of W_0 and W_1, respectively, satisfying conditions (1) and (2) of (5.1). Now, for the layer (R, W'), we take $\mathcal{F}'_1 := \mathcal{F}_1$. For the R-R-bimodule filtration $\mathcal{F}_0$: $0 = W_0^0 \subseteq W_0^1 \subseteq \cdots \subseteq W_0^{\ell_0} = W_0$ of W_0, if we make $h := \mathrm{h}(\alpha)$, we know that $W_0^{h+j} = R\alpha R \oplus \left(\oplus_{\sigma\in\mathbb{B}_0^j} R\sigma R\right)$, for some $\mathbb{B}_0^j \subseteq \mathbb{B}_0$, for all $j \in [0, \ell_0 - h]$. Then, by definition, the filtration $\mathcal{F}'_0$ of W'_0 is given by $[W'_0]^i = W_0^i$, for all $i < h$, and $[W'_0]^{h+j} = R\alpha' R \oplus \left(\oplus_{\sigma\in\mathbb{B}_0^j} R\sigma R\right)$, for all $j \in [0, \ell_0 - h]$.

Condition (5.1)(2) holds trivially for $\mathcal{F}'_1$. In order to show that condition (5.1)(1) holds for $\mathcal{F}'_0$, it is enough to show that $\delta(\alpha') \in A_{h-1}W_1A_{h-1}$. But this is an easy consequence of the assumption on the heights of the solid arrows appearing in the definition of α' and the condition (5.1)(1) for $\mathcal{F}_0$. □

Solution to Exercise (23.32): We can construct, as in (23.31), an isomorphism of layered t-algebras $T \longrightarrow T$ mapping W onto W'. Thus, W'_1 is freely generated by the directed set $\mathbb{B}'_1$ and the pair (R, W') is a layer for $\mathcal{A}$.

Since (R, W) is a triangular layer, we have R-R-bimodule filtrations $\mathcal{F}_0$ and $\mathcal{F}_1$ of W_0 and W_1, respectively, satisfying conditions (1) and (2) of (5.1). Now, for the layer (R, W'), we take $\mathcal{F}'_0 := \mathcal{F}_0$. For the R-R-bimodule filtration $\mathcal{F}_1$ of W_1, if we make $h := \mathrm{h}(v)$, we know that $W_1^{h+j} = RvR \oplus \left(\oplus_{w\in\mathbb{B}_1^j} RwR\right)$, for some $\mathbb{B}_1^j \subseteq \mathbb{B}_1$, for all $j \in [0, \ell_1 - h]$. Then, by definition, the filtration $\mathcal{F}'_1$ of W'_1 is given by $[W'_1]^s = W_1^s$, for all $s < h$, and $[W'_1]^{h+j} = Rv'R \oplus \left(\oplus_{w\in\mathbb{B}_1^j} RwR\right)$, for all $j \in [0, \ell_1 - h]$. Now, $\delta(W_0^{s+1}) \subseteq A_s\overline{W}_1A_s \subseteq A_sW'_1A_s$, for all $s < \mu$; and, for all $s \geq \mu$, all β^t_i, α^t_j lie in A_s, then $A_sW'_1A_s$ contains v and, therefore, $\delta(W_0^{s+1}) \subseteq A_sW_1A_s = A_sW'_1A_s$. Thus, condition (5.1)(1) holds for $\mathcal{F}'_0$.

In order to show that condition (5.1)(2) holds for $\mathcal{F}'_1$, notice that $AW_1^iA = A[W'_1]^iA$, for all $i \in [0, \ell_1]$. Thus, $\delta(w) \in AW_1^iAW_1^iA = A[W'_1]^iA[W'_1]^iA$, whenever $w \in [W'_1]^{i+1} \cap \left(\mathbb{B}'_1\setminus\{v'\}\right)$. Finally, $\delta(v') = r\delta(v) + \delta(\xi) \in AW_1^{h-1}AW_1^{h-1}A$, because condition (5.1)(2) holds for $\mathcal{F}_1$, $\mathrm{h}(v) = \mathrm{h}(v')$ and, by assumption, $\mathrm{h}(\delta(\beta^t_i)), \mathrm{h}(\delta(\alpha^t_j)), \mathrm{h}(v_t) < \mathrm{h}(v)$, for all i, j, t. □

Section 24

Solution to Exercise (24.12): Consider the morphism of layered ditalgebras ψ : $\mathcal{A}_0 \longrightarrow \mathcal{A}_{p,q}$ determined by $\psi_{|R} = id_R$, $\psi(\alpha) = \alpha$ and $\psi(v_t) := q_t(y)v_tp_t(x)$, for all t. Notice that $A_0 = A_{p,q}$ and $\psi_| : A_0 \longrightarrow A_{p,q}$ is the identity map. Then, apply (22.24). □

Section 25

Solution to Exercise (25.17): First notice that for $N \in \mathcal{A}^X$-Mod with underlying free S-module, we have that $F^X(N) \cong X \otimes_S N \cong \oplus_i X f_i \otimes_{Sf_i} f_i N \cong \oplus_i \mathrm{rank}(N_i) X f_i$. Thus, the multiplicity of Xf_i as a direct summand of ${}_B F^X(N)$ is the rank of N_i as a left free Sf_i-module.

If $M \in \mathcal{A}$-Mod has the form ${}_B M \cong \oplus_i m_i X f_i$ in B-Mod, we can consider the left S-module $N := \oplus_i m_i S f_i$. Then, there is an isomorphism of left B-modules $X \otimes_S N \cong M$, and we can apply (25.5) to obtain an $\mathcal{A}^X$-module $\overline{N}$, with the same S-module structure that N, such that $F^X(\overline{N}) \cong M$ in $\mathcal{A}$-Mod. □

Solution to Exercise (25.18): It follows from (21.3) and (21.6)(1). Having in mind the description of the layer given in (23.28) and the definition of the norm, we get (4). □

Solution to Exercise (25.19): Since F_a, F_r, F_d, F_s are restriction functors, this follows from (8.24). For $z \in \{e, u\}$, the functor $F_z = F^X$, for a suitable admissible triangular module X over a proper subditalgebra $\mathcal{A}'$ of $\mathcal{A}$ with $\delta(W_0') = 0$. Thus, from (14.11), we already have a triangular ideal I^X of $\mathcal{A}^X$ and an associated functor

$$F_z = F^X : (\mathcal{A}^X, I^X)\text{-Mod} \longrightarrow (\mathcal{A}, I)\text{-Mod}.$$

We denote by H the filtered set of generators of I and by H^X the corresponding set of generators for I^X constructed in (14.11). Assume that $N \in \mathcal{A}^X$-Mod and $M \in (\mathcal{A}, I)$-Mod with $F_X(N) \cong M$. From (8.22), we know that $H \cdot M = 0$ implies that $H \cdot F^X(N) = 0$. It remains to prove that $H^X * N = 0$, since then $N \in (\mathcal{A}^X, I^X)$-Mod satisfies that $\overline{F}_X(N) \cong M$.

We consider only the case of the unravelling, since the argument for the edge reduction is similar. From (12.10), we have

$$\begin{aligned} h \cdot (x \otimes n) = & \sum_{s,a,t} x^t_{s,a} \otimes \sigma_{v_t^{s,a},x}(h) * n + x^{i_0}_{i_*} \otimes \sigma_{v^{i_*}_{i_0},x}(h) * n \\ & + \sum_{i \neq i_0} x^i_i \otimes \sigma_{v^i_i,x}(h) * n, \end{aligned}$$

where $h \in H$ and $x \in X$ and $n \in N$, and $(x^t_{s,a}, v_t^{s,a})_{s,a,t} \cup (x^i_i, v^i_i)_{i \neq i_0} \cup \{(x^{i_0}_{i_*}, v^{i_*}_{i_0})\}$ is the dual basis of X constructed in (23.23). If we denote by $\{f_j\}_j = (f_{i_{s,a}})_{s,a} \cup \{f_i\}_{i \neq i_0} \cup \{f_{i_*}\}$ the canonical idempotents of S, then we have

$$\begin{aligned} X\otimes_S N &\cong \oplus_j Xf_j \otimes_{Sf_j} f_jN \\ &= \left[\oplus_j X_1 f_j \otimes_{Sf_j} f_jN\right] \oplus \left[\oplus_j X_2 f_j \otimes_{Sf_j} f_jN\right] \\ &\quad \oplus \left[\oplus_j X_3 f_j \otimes_{Sf_j} f_jN\right] \\ &= \left[\oplus_{s,a} X_1 f_{i_{s,a}} \otimes_{Sf_{i_{s,a}}} f_{i_{s,a}}N\right] \oplus \left[X_2 f_{i_*} \otimes_{Sf_{i_*}} f_{i_*}N\right] \\ &\quad \oplus \left[\oplus_{i\neq i_0} X_3 f_i \otimes_{Sf_i} f_iN\right] \\ &\cong \left[\oplus_{s,a,t} kx^t_{s,a} \otimes_k f_{i_{s,a}}N\right] \oplus f_{i_*}N \oplus \left[\oplus_{i\neq i_0} f_iN\right], \end{aligned}$$

and the last isomorphism maps the element $0 = h\cdot(x\otimes n)$ onto

$$\sum_{s,a,t} x^t_{s,a} \otimes \sigma_{\nu_t^{s,a},x}(h) * n + \sigma_{\nu_{i_0}^{i*},x}(h) * n + \sum_{i\neq i_0} \sigma_{\nu_i^i,x}(h) * n.$$

We used that $\sigma_{\nu_t^{s,a},x}(h) * n \in f_{i_{s,a}}N$, $\sigma_{\nu_{i_0}^{i*},x}(h) * n \in f_{i_*}N$, and $\sigma_{\nu_i^i,x}(h) * n \in f_iN$. This implies that $\sigma_{\nu_i,x}(h) * n = 0$, for all $i \in I$, where now $(x_i, \nu_i)_{i\in I}$ denotes the chosen dual basis for X_S. But, for $\nu \in X^*$, $\sigma_{\nu,x}(h) = \sum_i \nu(x_i)\sigma_{\nu_i,x}(h)$, see (12.8)(4), and then $H^X * N = 0$. □

Solution to Exercise (25.20): Assume that (R, W) is the layer of $\mathcal{A}$ and $Re_{i_0} = k[x]_{f(x)}$. Notice that $I \cap Re_{i_0} \neq 0$. Otherwise, for each natural number m and each $\lambda \in k$ with $f(\lambda) \neq 0$, there is an $\mathcal{A}$-module $M(m,\lambda)$ with support $\{i_0\}$, where the action of x on it is given by the $m\times m$ Jordan block with eigenvalue λ. Since $I_{\mathcal{A}}M(m,\lambda) = 0$ and all these modules are not isomorphic, $(\mathcal{A}, I)$ is not of finite representation type; a contradiction.

Consider any monic polynomial $h(x) \in k[x]$ that generates the ideal $I \cap Re_{i_0}$ of the principal ideal domain $k[x]_{f(x)}$, and decompose it as a product of its linear factors (with distinct roots $\lambda_1, \dots, \lambda_q$)

$$h(x) = (x-\lambda_1)^{r_1}(x-\lambda_2)^{r_2}\cdots(x-\lambda_q)^{r_q}.$$

After multiplying $h(x)$ by the inverses of some linear factors of $f(x)$, if necessary, we can assume that $h(x)$ and $f(x)$ have no common root. Take

$$r \geq \max\{||M|| \mid M \text{ indecomposable in } (\mathcal{A}, I)\text{-mod}\}.$$

Such a maximum exists because $(\mathcal{A}, I)$ has finite representation type. Now, consider the ditalgebra $\mathcal{A}^u$ obtained from $\mathcal{A}$ by unravelling at i_0 using r and the scalars $\lambda_1, \lambda_2, \dots, \lambda_q$. From (25.19), we know that the associated reduction functor $F_u : \mathcal{A}^u\text{-mod} \longrightarrow \mathcal{A}\text{-mod}$ induces a full and faithful functor $\overline{F}_u : (\mathcal{A}^u, I^u)\text{-mod} \longrightarrow (\mathcal{A}, I)\text{-mod}$.

From (25.9)(1), we know that any $M \in \mathcal{A}$-mod, which is indecomposable in $(\mathcal{A}, I)$-mod, has norm $\leq r$ and hence is of the form $F_u(N) \cong M$, for some $N \in \mathcal{A}^u$-mod. From (25.19), we obtain that it has the form $\overline{F}_u(N) \cong M$, for some $N \in (\mathcal{A}^u, I^u)$-mod. This implies that $\overline{F}_u$ is an equivalence.

Define $g(x) := (x - \lambda_1)\cdots(x - \lambda_q)$. Observe that given any $M \in (\mathcal{A}, I)$-mod, $M(h(x)) = 0$ implies that the eigenvalues of $M(x)$ are included in $\{\lambda_1, \ldots, \lambda_q\}$. Since our field is algebraically closed, for this M, $M(g(x))$ is not invertible whenever $M(x) \neq 0$. Then, using (25.9)(3), we obtain that any $N \in (\mathcal{A}^u, I^u)$-mod with $F_u(N)(x) \neq 0$ and $F_u(N)(h(x)) = 0$, the inequality $||N|| \leq ||F_u(N)||$ is strict. Moreover, $||N|| \leq ||F_u(N)||$, for any $N \in (\mathcal{A}^u, I^u)$-mod.

With the notation used in the proofs of (23.23) and (25.9), we have the dual basis element $x_{i_*}^{i_0} = e_{i_0} \in (e_{i_0}R)_{g(x)} = X_2 f_{i_*}$ of X and its corresponding dual element $\nu_{i_0}^{i_*} \in X^*$, which satisfies $\nu_{i_0}^{i_*}(re_{i_0}) = rf_{i_*}$, for any $r \in R$. Then, $\sigma_{\nu_{i_0}^{i_*}, x_{i_*}^{i_0}}(h(x)e_{i_0}) = \nu_{i_0}^{i_*}([h(x)e_{i_0}]x_{i_*}^{i_0}) = h(x)f_{i_*} \in Sf_{i_*}$ which is identified with $k[x]_{f(x)g(x)}$. Therefore, the ideal I^u contains f_{i_*}, because $g(x) \in I^u \cap Sf_{i_*}$ is invertible in Sf_{i_*}. This implies that any object $N \in (\mathcal{A}^u, I^u)$-mod satisfies that $N_{i_*} = 0$. Now consider the finitely presented ditalgebra $\mathcal{A}' := \mathcal{A}^{ud}$ obtained from $\mathcal{A}^u$ by deletion of the idempotent f_{i_*}. From the previous argument and (25.19), the full and faithful reduction functor $F_d : \mathcal{A}^{ud}\text{-mod} \longrightarrow \mathcal{A}^u\text{-mod}$ induces an equivalence $\overline{F}_d : (\mathcal{A}^{ud}, I^{ud})\text{-mod} \longrightarrow (\mathcal{A}^u, I^u)\text{-mod}$. Then, the composition $F_u F_d$ induces an equivalence $\overline{F}_u \overline{F}_d : (\mathcal{A}^{ud}, I^{ud})\text{-mod} \longrightarrow (\mathcal{A}, I)\text{-mod}$. Then, (2) holds. (1) is clear. For (3), recall first that F_d is a norm preserving functor, then $||N|| = ||F_d(N)|| \leq ||F_u F_d(N)||$, for any $N \in (\mathcal{A}^{ud}, I^{ud})$-mod. Moreover, $F_u F_d(N)(x) \neq 0$ implies $||N|| = ||F_d(N)|| < ||F_u F_d(N)||$. Hence (3) holds. □

Section 26

Solution to Exercise (26.11): Take $M, M' \in B$-mod as in the statement of the exercise. Consider the decomposition $1 = \sum_{i=1}^n e_n$ of the unit of R as a sum of central primitive orthogonal idempotents. Write $s = s(\alpha)$ and $t = t(\alpha)$. Notice that B admits only the following non-isomorphic indecomposable modules: the simples S_i, for $i \in [1, n]$ and the projective-injective I. Notice that $e_s I = ke_s$, $e_t I = ke_t$, and $e_i I = 0$, for $i \notin \{s, t\}$. Notice also that $\mathrm{Ext}_B(S_s, S_t) \neq 0$, and all other possible extension groups between indecomposables are zero. Then, $M \cong [\oplus_{i=1}^n m_i S_i] \oplus mI$ and $M' \cong [\oplus_{i=1}^n m_i' S_i] \oplus m'I$. Assume $M \cong M'$ in R-mod. Then, we have that $m_i = m_i'$, for $i \notin \{s, t\}$, $m_s + m = m_s' + m'$ and $m_t + m = m_t' + m'$. We have to show that $m_s = m_s'$ (and, hence, $m = m'$ and $m_t = m_t'$). If $m_s \neq m_s'$, then we may assume that $m_t > m_t'$. Since $\mathrm{Ext}_B(M, M) = 0$ and $m_t > 0$, we have that $m_s = 0$. Then, $m_t + m_s' + m' = m_t' + m'$, a contradiction. Therefore, $M \cong M'$ in B-mod. □

Solution to Exercise (26.12): We may assume that M (and hence also N) is a *sincere $\mathcal{A}$-module*, that is such that $M_i \neq 0$, for each point i of $\mathcal{A}$. Indeed, if M is not sincere, we consider the ditalgebra $\mathcal{A}^d$ obtained from $\mathcal{A}$ by elimination of the idempotents e_i such that $M_i = 0 = N_i$. Then, the associated reduction functor $F_d : \mathcal{A}^d\text{-mod} \longrightarrow \mathcal{A}\text{-mod}$ maps two sincere $\mathcal{A}^d$-modules M' and N' onto $F_d(M') \cong M$ and $F_d(N') \cong N$. By (9.9), we know that $\mathrm{Ext}_{\mathcal{A}^d}(M', M') = 0$ and $\mathrm{Ext}_{\mathcal{A}^d}(N', N') = 0$. Since $\underline{\ell}(M') = \underline{\ell}(N')$, we have $M' \cong N'$ in $\mathcal{A}^d$-mod and, hence, $M \cong N$ in $\mathcal{A}$-mod.

From now on, we assume that M and N are sincere.

The proof will be done by induction on the common norm $||M||_{\mathcal{A}} = ||N||_{\mathcal{A}}$.

If $||M||_{\mathcal{A}} = 0$, then $W_0 = 0$ and, hence, $A_{\mathcal{A}} = R$. But, clearly, $\underline{\ell}(M) = \underline{\ell}(N)$ iff $M \cong N$ in R-mod. Thus, $M \cong N$ in $\mathcal{A}$-mod.

We have shown the statement of the exercise for the base of the induction case $||M||_{\mathcal{A}} = 0$. Now, assume that $M, N \in \mathcal{A}$-mod are sincere $\mathcal{A}$-modules with positive norm and no self-extensions, and that for any seminested ditalgebra $\mathcal{A}'$ with no marked points and any $M', N' \in \mathcal{A}'$-mod (not necessarily sincere) without self-extensions and satisfying that $\underline{\ell}(M') = \underline{\ell}(N')$ and $||M'||_{\mathcal{A}'} < ||M||_{\mathcal{A}}$, then $M' \cong N'$ in $\mathcal{A}'$-mod.

Consider an arrow $\alpha : i_0 \to j_0$ in $\mathbb{B}_0$ with minimal height in the filtration $\mathcal{F}(W_0)$ given by the triangularity requirement. Then, $\delta(\alpha) \in W_1$. Thus, $\delta(\alpha) = \sum_{j=1}^m c_j v_j$, for some $v_j \in \mathbb{B}_1$ and $c_j \in k$.

Case 1: If $c_t \neq 0$, for some $t \in [1, m]$. Make $v_t' := \sum_{j=1}^m c_j v_j$ and consider the change of basis for W_1 where $v_1, \ldots, v_t, \ldots, v_m$ is replaced by $v_1, \ldots, v_t', \ldots, v_m$ using (26.1). Consider the ditalgebra $\mathcal{A}^r$ obtained from $\mathcal{A}$ by regularization of α and the associated length vector preserving equivalence $F_r : \mathcal{A}^r\text{-mod} \longrightarrow \mathcal{A}\text{-mod}$. If $M', N' \in \mathcal{A}^r$-mod are such that $F_r(M') \cong M$ and $F_r(N') \cong N$, then $\underline{\ell}(M') = \underline{\ell}(N')$. Moreover, by (9.10), we know that $\mathrm{Ext}_{\mathcal{A}^r}(M', M') = 0$ and $\mathrm{Ext}_{\mathcal{A}^r}(N', N') = 0$. Since M and N are sincere, we know that $||M'||_{\mathcal{A}^r} < ||M||_{\mathcal{A}}$. By induction hypothesis, $M' \cong N'$ in $\mathcal{A}^r$-mod. Hence, $M \cong N$ in $\mathcal{A}$-mod.

Case 2: If $\delta(\alpha) = 0$. Assume $i_0 = j_0$ and consider the proper subditalgebra $\mathcal{A}'$ of $\mathcal{A}$ generated by R and α. Consider also the restriction functor $F_r : \mathcal{A}\text{-mod} \longrightarrow \mathcal{A}'\text{-mod}$. From (16.4), we know that $\mathrm{Ext}_{\mathcal{A}'}(F_r(M), F_r(M)) = 0$. By assumption, R is semisimple, therefore $\mathrm{Ext}_{A_{\mathcal{A}'}}(F_r(M), F_r(M))$ embeds in $\mathrm{Ext}_{\mathcal{A}'}(F_r(M), F_r(M))$. Moreover, $A_{\mathcal{A}'}$ is a non-trivial minimal algebra and, therefore, it admits no sincere module without self-extensions (see (32.2)); a contradiction.

This contradiction implies that $i_0 \neq j_0$. Consider the ditalgebra $\mathcal{A}^e$ obtained from $\mathcal{A}$ by reduction of the edge $\alpha : i_0 \longrightarrow j_0$, and the associated equivalence functor $F_e : \mathcal{A}^e\text{-mod} \longrightarrow \mathcal{A}\text{-mod}$. If $M', N' \in \mathcal{A}^e\text{-mod}$ are such that $F(M') \cong M$ and $F(N') \cong N$, then, by (16.6), we know that $\mathrm{Ext}_{\mathcal{A}^e}(M', M') = 0$ and $\mathrm{Ext}_{\mathcal{A}^e}(N', N') = 0$. Moreover, since M and N are sincere, we know that $||M'||_{\mathcal{A}^e} < ||M||_{\mathcal{A}}$.

Then, it will be enough to show that the $\mathcal{A}^e$-modules M' and N' satisfy that $\underline{\ell}(M') = \underline{\ell}(N')$ (or, equivalently, $M' \cong N'$ in S-mod). Because, then, by induction hypothesis, $M' \cong N'$ in $\mathcal{A}'$-mod. Hence, $M \cong N$ in $\mathcal{A}$-mod.

From (25.6), we know that, if B is the subalgebra of $A_{\mathcal{A}}$ generated by R and α, then the left B-module $M \cong F_e(M')$ decomposes in B-Mod as

$$(\dim M'_{i_*})Xf_{i_*} \oplus (\dim M'_{j_*})Xf_{j_*} \oplus (\dim M'_z)Xf_z \oplus \left[\oplus_{u\notin\{i_*,j_*,z\}}(\dim M'_u)Xf_u\right],$$

where $Xf_{i_*} \cong S_{i_0}$, $Xf_{j_*} \cong S_{j_0}$, $Xf_u \cong S_u$ and $Xf_z \cong I$ are, respectively, the indecomposable B-modules: simple at i_0, simple at j_0, simple at the point u, and the injective-projective I. We are using the notation of (23.18). Since $M \cong N$ as left R-modules, by (26.11), we have that in fact they are isomorphic as left B-modules. That is that the multiplicities of the finite-dimensional modules S_{i_0}, S_{j_0}, I and $Xf_u \cong S_u$ coincide with the corresponding numbers for N'. Then, $M' \cong M'$ in S-Mod. As claimed. □

Solution to Exercise (26.13): Assume that the modules $M, N \in \mathcal{A}$-mod satisfy that $\mathrm{Ext}_{\mathcal{A}}(M, M) = 0$, $\mathrm{Ext}_{\mathcal{A}}(M, M) = 0$ and ${}_RM \cong {}_RN$ in R-mod. Take a basic R-module X such that ${}_RM \in \mathrm{add}\,(X)$. Consider R as an initial subditalgebra $\mathcal{A}'$ of $\mathcal{A}$.

Denote by $1 = \sum_i f_i$ a decomposition (as a sum of central primitive orthogonal idempotents) of the unit of the algebra S, appearing in the splitting $\Gamma = \mathrm{End}_{\mathcal{A}'}(X)^{op} = \mathrm{End}_R(X)^{op} = S \oplus P$ of Γ by its radical P. There is such splitting because k is an algebraically closed field. Since X is basic, $X = \oplus_i Xf_i$ is the decomposition of X as a direct sum of indecomposable R-modules; X is a complete triangular admissible $\mathcal{A}'$-module by (17.4). $\mathcal{A}^X$ is a seminested ditalgebra with layer (S, W^X), by (23.13).

Consider the associated functor $F^X : \mathcal{A}^X\text{-mod} \longrightarrow \mathcal{A}\text{-mod}$. Since, ${}_RM, {}_RN \in \mathrm{add}\,(X)$ and the R-modules Xf_i are indecomposables, from (25.17), we know there exist $M', N' \in \mathcal{A}^X$-mod with $F^X(M') \cong M$ and $F^X(N') \cong N$. Since $M \cong N$ in R-mod, M' and N' can be chosen such that $M' \cong N'$ in S-mod (see the solution of (25.17)).

From (16.6), we know that $\mathrm{Ext}_{\mathcal{A}^X}(M', M') = 0$ and $\mathrm{Ext}_{\mathcal{A}^X}(N', N') = 0$. Finally, by (26.12), we get that $M' \cong N'$ in $\mathcal{A}^X$-mod and, therefore, $M \cong N$ in $\mathcal{A}$-mod, as claimed. □

Section 27

Solution to Exercise (27.15): We know that any Morita equivalence $F : \Lambda'\text{-mod} \longrightarrow \Lambda\text{-mod}$ preserves the number of non-isomorphic indecomposable projective modules and satisfies that $\underline{p}(M) = \underline{p}(F(M))$ then, for each $M \in \Lambda'$-mod, we can assume that Λ is basic. Then, the Drozd's ditalgebra $\mathcal{D} = \mathcal{D}^{\Lambda}$ is a nested ditalgebra without marked points, to which (26.12) can be applied. Consider tense modules $M, N \in \Lambda$-mod, with $\underline{p}(M) = \underline{p}(N)$. Consider the equivalence functor $\Xi_{\Lambda} : \mathcal{D}\text{-mod} \longrightarrow \mathcal{P}^1(\Lambda)$ and $M', N' \in \mathcal{D}$-mod with $\Xi_{\Lambda}(M') = \phi_M$ and $\Xi_{\Lambda}(N') = \phi_N$. Since M and N are tense modules, from (19.10), we know that $\mathrm{Ext}_{\mathcal{D}}(M', M') = 0$ and $\mathrm{Ext}_{\mathcal{D}}(N', N') = 0$. From the definition of Ξ_{Λ} and (22.19), we know that $\underline{\ell}(M') = \underline{p}(M)$ and $\underline{\ell}(N') = \underline{p}(N)$. Then, by (26.12), $M' \cong N'$. Therefore, $M \cong N$. □

Section 28

Solution to Exercise (28.23): Take a pair $(\mathcal{A}, I) \in \mathcal{R}$, say $\mathcal{A} = \mathcal{A}_0^{z_1 \cdots z_t}$ and $I = I_0^{z_1 \cdots z_t}$. Then, the associated reduction functor $F : \mathcal{A}\text{-mod} \longrightarrow \mathcal{A}_0\text{-mod}$ induces an equivalence $\overline{F} : (\mathcal{A}, I)\text{-mod} \longrightarrow (\mathcal{A}_0, I_0)\text{-mod}$.

Assume that $\mathcal{A}'$ is a seminested ditalgebra with no marked points and $\mathcal{A}' = \mathcal{A}^{z'_1 \cdots z'_s}$ is obtained from $\mathcal{A}$ by a finite sequence of basic operations $z'_1, \ldots, z'_s \in \{a, d, r, e, u, s\}$ and $I' = I^{z'_1 \cdots z'_s}$. Then the associated reduction functor $F : \mathcal{A}'\text{-mod} \longrightarrow \mathcal{A}\text{-mod}$ induces a functor $\overline{F} : (\mathcal{A}', I')\text{-mod} \longrightarrow (\mathcal{A}, I)\text{-mod}$, which preserves families of non-isomorphic indecomposable objects. Then, $(\mathcal{A}', I')$ has finite representation type and, if we define $M(\mathcal{A}', I')$ as the direct sum in $(\mathcal{A}', I')$-mod of a chosen set of representatives of all the indecomposable objects in $(\mathcal{A}', I')$-mod, we know that $F[M(\mathcal{A}', I')]$ is a direct summand of $M(\mathcal{A}, I)$. Thus, $||M(\mathcal{A}', I')|| \leq ||F[M(\mathcal{A}', I')]|| \leq ||M(\mathcal{A}, I)||$.

Assume that the given $\mathcal{A}$ is a non-trivial ditalgebra, say with layer (R, W) and differential δ. Then, there exists a minimal solid arrow $\alpha : i_0 \longrightarrow j_0$. Thus, $\delta(\alpha) \in W_1$.

Case 1: $e_{i_0} \in I$ or $e_{j_0} \in I$.

If $e_{i_0} \in I$, then $M_{i_0} = 0$, for any $M \in (\mathcal{A}, I)$-mod, and we consider the ditalgebra $\mathcal{A}^d$ obtained from $\mathcal{A}$ by deletion of the idempotent e_{i_0}. Then, by (25.19), $\overline{F}_d : (\mathcal{A}^d, I^d)\text{-mod} \longrightarrow (\mathcal{A}, I)\text{-mod}$ is a norm preserving equivalence. Thus, $\overline{FF_d}$ is an equivalence too and $(\mathcal{A}^d, I^d) \in \mathcal{R}$. Clearly, $c(\mathcal{A}^d, I^d) < c(\mathcal{A}, I)$. If $e_{j_0} \in I$, we proceed similarly.

From now on we assume that $e_{i_0}, e_{j_0} \notin I$.

Case 2: $\delta(\alpha) = 0$ and $i_0 = j_0$.

Consider the ditalgebra $\mathcal{A}^a$ obtained from $\mathcal{A}$ by absorption of the loop α. Then, we have a norm preserving equivalence $\overline{F}_a : (\mathcal{A}^a, I^a)\text{-mod} \longrightarrow (\mathcal{A}, I)\text{-mod}$. Notice that $e_{i_0} \notin I = I^a$, make $x := \alpha$, thus $R^a e_{i_0} = k[x]$, and consider the following subcases:

Subcase 2.1: $M(x) = 0$, for any $M \in (\mathcal{A}^a, I^a)$-mod.

This occurs if $x \in I^a$. Apply (23.28) to $(\mathcal{A}^a, I^a)$ to obtain a pair $(\mathcal{A}^{as}, I^{as})$ whose first component is a ditalgebra without marked points and an associated functor $F_s : \mathcal{A}^{as}\text{-mod} \longrightarrow \mathcal{A}^a\text{-mod}$ which induces an equivalence $\overline{F}_s : (\mathcal{A}^{as}, I^{as})\text{-mod} \longrightarrow (\mathcal{A}^a, I^a)\text{-mod}$. Then, $\overline{F}\overline{F}_a\overline{F}_s$ is an equivalence too, and $(\mathcal{A}^{as}, I^{as}) \in \mathcal{R}$. Moreover, (25.18) implies that $c(\mathcal{A}^{as}, I^{as}) < c(\mathcal{A}, I)$.

Subcase 2.2: $M(x) \neq 0$, for some $M \in (\mathcal{A}^a, I^a)$-mod.

Then, there is an indecomposable object N in $(\mathcal{A}^a, I^a)$-mod with $N(x) \neq 0$. Then apply (25.20) to $(\mathcal{A}^a, I^a)$ to obtain a pair $(\mathcal{A}^{a\prime}, I^{a\prime})$, where $\mathcal{A}^{a\prime}$ has no marked points and the associated reduction functor $F_a F_\prime : \mathcal{A}^{a\prime}\text{-mod} \longrightarrow \mathcal{A}\text{-mod}$ induces an equivalence $\overline{F}_a \overline{F}_\prime : (\mathcal{A}^{a\prime}, I^{a\prime})\text{-mod} \longrightarrow (\mathcal{A}, I)\text{-mod}$. Hence $(\mathcal{A}^{a\prime}, I^{a\prime}) \in \mathcal{R}$. Moreover, since $M(\mathcal{A}, I)(x) \neq 0$, then $||M(\mathcal{A}^{a\prime}, I^{a\prime})|| < ||M(\mathcal{A}, I)||$ and $c(\mathcal{A}^{a\prime}, I^{a\prime}) < c(\mathcal{A}, I)$.

Case 3: $\delta(\alpha) = 0$ and $i_0 \neq j_0$.

Consider the pair $(\mathcal{A}^e, I^e)$, where $\mathcal{A}^e$ is obtained from $\mathcal{A}$ by reduction of the edge α. From (25.19), we get that the associated reduction functor F_e induces an equivalence $\overline{F}_e : (\mathcal{A}^e, I^e)\text{-mod} \longrightarrow (\mathcal{A}, I)\text{-mod}$. Thus, $\overline{F F_e}$ is an equivalence too and $(\mathcal{A}^e, I^e) \in \mathcal{R}$. Clearly, $F_e[M(\mathcal{A}^e, I^e)] \cong M(\mathcal{A}, I)$ and, since $e_{i_0}, e_{j_0} \notin I$, there are indecomposable objects $N, N' \in (\mathcal{A}, I)$-mod with $N_{i_0} \neq 0$ and $N'_{j_0} \neq 0$. Thus, $M(\mathcal{A}, I)_{i_0} \neq 0$ and $M(\mathcal{A}, I)_{j_0} \neq 0$, therefore $||M(\mathcal{A}^e, I^e)|| < ||F_e[M(\mathcal{A}^e, I^e)]|| = ||M(\mathcal{A}, I)||$. Hence, $c(\mathcal{A}^e, I^e) < c(\mathcal{A}, I)$.

Case 4: $\delta(\alpha) \neq 0$.

If $\delta(\alpha)$ is a basis element of W_1, then consider the ditalgebra $\mathcal{A}^r$ obtained from $\mathcal{A}$ by regularization of the arrow α and the associated equivalence $F_r : \mathcal{A}^r\text{-mod} \longrightarrow \mathcal{A}\text{-mod}$. From (25.19), we get that F_r induces an equivalence $\overline{F}_r : (\mathcal{A}^r, I^r)\text{-mod} \longrightarrow (\mathcal{A}, I)\text{-mod}$. Thus, $\overline{F F_r}$ is an equivalence too and $(\mathcal{A}^r, I^r) \in \mathcal{R}$. As in *Case 3*, $M(\mathcal{A}, I)_{i_0} \neq 0$ and $M(\mathcal{A}, I)_{j_0} \neq 0$, imply that $c(\mathcal{A}^r, I^r) < c(\mathcal{A}, I)$.

Finally, if $\delta(\alpha) = \sum_i \lambda_i w_i$, where $w_1, \dots, w_s$ is a basis for W_1 as a freely generated R-R-bimodule, and $\lambda_1, \dots, \lambda_s \in k$ not all zero. Then, after a suitable change of basis of W_1, we are reduced to the previous situation.

□

Solution to Exercise (28.24): Consider the family $\mathcal{R} = \mathcal{R}(\mathcal{A}, I)$, defined in (28.23), and the numbers $c(\mathcal{A}', I')$ associated to each $(\mathcal{A}', I') \in \mathcal{R}$.

Notice that, after applying (25.20) to each marked point of the given $\mathcal{A}$, we get a pair $(\mathcal{A}', I') \in \mathcal{R}$. If $\mathcal{A}'$ is not trivial, (28.23) implies the existence of $(\mathcal{A}'', I'') \in \mathcal{R}$ with $c(\mathcal{A}'', I'') < c(\mathcal{A}, I)$. If $\mathcal{A}''$ is not trivial, we can repeat this construction. This procedure must stop at some step because $c(\mathcal{A}', I')$ is a positive integer. □

Solution to Exercise (28.25): Composing with a Morita equivalence, if necessary, we may indeed assume that Λ is basic, and hence isomorphic to a quotient kQ/I as in the hint. By (28.24), there is a trivial seminested ditalgebra $\mathcal{A}'$, obtained from $\mathcal{A} = \mathcal{A}(\Lambda)$ by a finite sequence of basic operations of type a, d, r, e, u, s, as in (25.19), and a triangular ideal I' of $\mathcal{A}'$ such that the associated functor $F : \mathcal{A}'\text{-mod} \longrightarrow \mathcal{A}\text{-mod}$ induces an equivalence

$$\overline{F} : (\mathcal{A}', I')\text{-mod} \longrightarrow (\mathcal{A}, I)\text{-mod}.$$

Since $\mathcal{A}'$ is a trivial ditalgebra, its layer has the form (R', W'), where R' is a finite product of copies of k and $I' \cap A'$ is an ideal of $A' = R'$. Then, $I' \cap A'$ is a product of some copies of k and $\{0\}$. Then, we can delete all idempotents e_i of $\mathcal{A}'$ corresponding to non-trivial factors k of the ideal $I' \cap A'$, to obtain a new trivial seminested ditalgebra $\mathcal{A}'^d$ and a triangular ideal $I'^d = 0$. Thus, an equivalence

$$\mathcal{A}'^d\text{-mod} \longrightarrow (\mathcal{A}, I)\text{-mod} \simeq \Lambda\text{-mod}.$$

□

Solution to Exercise (28.26): The existence of a dimension $d \in \mathbb{N}$, as in the statement of the exercise, is equivalent to the existence of some dimension vector $\ell \in \mathbb{L}(\mathcal{A})$ and infinitely many non-isomorphic indecomposable $\mathcal{A}$-modules with dimension vector ℓ. We may assume that ℓ is *sincere*, that is such that $\ell_i \neq 0$, for all point $i \in \{1, \dots, n\}$ of $\mathcal{A}$. Indeed, if ℓ was not sincere, we consider the ditalgebra $\mathcal{A}^d$ obtained from $\mathcal{A}$ by elimination of the idempotents e_i such that $\ell_i = 0$. Then, the associated reduction functor $F_d : \mathcal{A}^d\text{-mod} \longrightarrow \mathcal{A}\text{-mod}$ maps an infinite family of non-isomorphic indecomposable $\mathcal{A}^d$-modules with the same sincere dimension vector ℓ^d onto the given infinite family of non-isomorphic indecomposable $\mathcal{A}$-modules with dimension vector ℓ. If the statement of the exercise holds for $\mathcal{A}^d$ and the sincere dimension vector ℓ^d, we have an $\mathcal{A}^d$-Γ bimodule Z, finitely generated projective by the right and such that the composition functor of the upper row in the following diagram preserves isomorphism classes and indecomposables

$$\begin{array}{ccccc} \Gamma\text{-mod} & \xrightarrow{Z\otimes_\Gamma -} & \mathcal{A}^d\text{-mod} & \xrightarrow{L_{\mathcal{A}^d}} & \mathcal{A}^d\text{-mod} \\ \| & & \downarrow{F_d} & & \downarrow{F_d} \\ \Gamma\text{-mod} & \xrightarrow{F_d(Z)\otimes_\Gamma -} & \mathcal{A}\text{-mod} & \xrightarrow{L_\mathcal{A}} & \mathcal{A}\text{-mod}. \end{array}$$

From (22.7), we know that the diagram is commutative, up to isomorphism, and $F_d(Z)$ is finitely generated projective by the right. Since F_d is full and faithful, the composition functor of the lower row of the diagram preserves isomorphism classes and indecomposables.

From now on, we assume ℓ is sincere. The proof will be done by induction on the *zero norm*: for $\ell \in \mathbb{L}(\mathcal{A})$, by definition

$$||\ell||^0_\mathcal{A} := \sum_{\alpha\in\mathbb{B}_0} \ell_{t(\alpha)}\ell_{s(\alpha)}.$$

If $\mathcal{A}$ has a marked point i_0, say with $Re_{i_0} = k[x]_{f(x)} =: \Gamma$, then we can consider the A-Γ-bimodule Z defined by $Z := Re_{i_0}$ as a right Γ-module and the left action of A on Z is determined by the natural action of R and by making each element of W_0 act on Z annihilating everything. Then, clearly the composition

$$\Gamma\text{-mod} \xrightarrow{Z\otimes_\Gamma -} A_\mathcal{A}\text{-mod} \xrightarrow{L_A} \mathcal{A}\text{-mod}$$

preserves isomorphism classes and indecomposables. If $||\ell||^0_\mathcal{A} = 0$, then $\mathcal{A}$ is a minimal ditalgebra and, since $\mathcal{A}$ is not of finite representation type, $\mathcal{A}$ admits at least one marked point. We have shown the statement of the exercise for the base of the induction case $||\ell||^0_\mathcal{A} = 0$. Now, assume that $\ell \in \mathbb{L}(\mathcal{A})$ is a sincere dimension vector of $\mathcal{A}$ and that the assertion has been proved for all ditalgebras and all (not necessarily sincere dimension vectors) $\ell' \in \mathbb{L}(\mathcal{A}')$ with $||\ell'||^0_{\mathcal{A}'} < ||\ell||^0_\mathcal{A}$.

Now, we can also assume that $\mathcal{A}$ has no marked point (hence $||\cdot||^0_\mathcal{A} = ||\cdot||$, the usual norm). Consider a minimal solid arrow $\alpha \in \mathbb{B}_0$. Then, $\delta(\alpha) \in W_1$. Thus, $\delta(\alpha) = \sum_{j=1}^m c_j v_j$, for some $v_j \in \mathbb{B}_1$ and $c_j \in k$.

Case 1: If $c_t \neq 0$, make $v'_t := \sum_{j=1}^m c_j v_j$, and consider the change of basis for W_1 where $v_1, \ldots, v_t, \ldots, v_m$ is replaced by $v_1, \ldots, v'_t, \ldots, v_m$ using (26.1). Consider the ditalgebra $\mathcal{A}^r$ obtained from $\mathcal{A}$ by regularization of α and the associated length preserving equivalence

$F_r : \mathcal{A}^r\text{-mod} \longrightarrow \mathcal{A}\text{-mod}$. If $M \in \mathcal{A}^r$-mod is mapped by F_r to a member of the given family of non-isomorphic indecomposable $\mathcal{A}$-modules of fixed sincere dimension vector ℓ, then $\ell(M) = \ell$ and $||\ell||_{\mathcal{A}^r} < ||\ell||_{\mathcal{A}}$. Then, we apply our induction hypothesis to $\mathcal{A}^r$ and ℓ, to obtain an appropriate A^r-Γ-bimodule Z such that the functor

$$\Gamma\text{-mod} \xrightarrow{Z\otimes_\Gamma -} A^r\text{-mod} \xrightarrow{L_{\mathcal{A}^r}} \mathcal{A}^r\text{-mod}$$

preserves isomorphism classes and indecomposables. Proceeding with F_r as we did before with F_d, we obtain what we wanted for $\mathcal{A}$.

Case 2: If $\delta(\alpha) = 0$.

Subcase 2.1: If $s(\alpha) = t(\alpha)$, we consider the ditalgebra $\mathcal{A}^a$ obtained from $\mathcal{A}$ by absorption of the loop α and the associated length preserving equivalence $F_a : \mathcal{A}^a\text{-mod} \longrightarrow \mathcal{A}\text{-mod}$. Then, $\mathcal{A}^a$ has a marked point and, as we have already seen, there is a functor

$$\Gamma\text{-mod} \xrightarrow{Z\otimes_\Gamma -} A^a\text{-mod} \xrightarrow{L_{\mathcal{A}^a}} \mathcal{A}^a\text{-mod}$$

preserving isomorphism classes and indecomposables. Proceeding with F_a as we did before with F_d, we obtain what we wanted for $\mathcal{A}$.

Subcase 2.2: If $s(\alpha) \neq t(\alpha)$. Consider the ditalgebra $\mathcal{A}^e$ obtained from $\mathcal{A}$ by reduction of the edge $\alpha : i_0 \longrightarrow j_0$, and the associated reduction functor $F_e : \mathcal{A}^e\text{-mod} \longrightarrow \mathcal{A}\text{-mod}$. There is only a finite number of dimension vectors $\ell' \in \mathbb{L}(\mathcal{A}^e)$ such that $t_e(\ell') = \ell$. Then, there is $\ell' \in \mathbb{L}(\mathcal{A}^e)$ and an infinite family of non-isomorphic indecomposable finite-dimensional $\mathcal{A}^e$-modules of dimension vector ℓ', which are mapped by F_e onto infinitely members of the given family of non-isomorphic ℓ-dimensional $\mathcal{A}$-modules. Since ℓ is sincere, $||\ell'||_{\mathcal{A}^e} < ||\ell||_{\mathcal{A}}$. Then, we can apply our induction hypothesis to $\mathcal{A}^e$ and ℓ', to obtain a suitable composition functor

$$\Gamma\text{-mod} \xrightarrow{Z\otimes_\Gamma -} A^e\text{-mod} \xrightarrow{L_{\mathcal{A}^e}} \mathcal{A}^e\text{-mod}$$

preserving isomorphism classes and indecomposability. Proceeding with F_e, as we did before with F_d, we obtain what we wanted for $\mathcal{A}$.

□

Solution to Exercise (28.27): By (27.11), we may assume that Λ is basic. Then, Λ trivially splits over its radical and we can consider the Drozd's nested ditalgebra $\mathcal{D} := \mathcal{D}^\Lambda$ (see (23.25)). By (28.26) there is a rational algebra $\Gamma = k[x]_{f(x)}$ and an $A_{\mathcal{A}}$-Γ-bimodule Z, finitely generated free as a right Γ-module such that the functor

$$\Gamma\text{-mod} \xrightarrow{Z\otimes_\Gamma -} A\text{-mod} \xrightarrow{L_{\mathcal{D}}} \mathcal{D}^\Lambda\text{-mod}$$

preserves isomorphism classes and indecomposables

Consider the functor $\Xi_\Lambda : \mathcal{D}^\Lambda\text{-Mod} \longrightarrow \mathcal{P}^1(\Lambda)$. By (22.20), the objects of the form $\Xi_\Lambda(Z \otimes_\Gamma M)$, with $M \in \Gamma\text{-Mod}$, cannot be of the form $(P, 0, 0)$. Thus, the image of any indecomposable under the functor $\Xi_\Lambda L_{\mathcal{D}}(Z \otimes_\Gamma -)$ lies in $\mathcal{P}^2(\Lambda)$, and its composition with the cokernel functor $\mathcal{P}^2(\Lambda) \longrightarrow \Lambda\text{-Mod}$ gives a functor which preserves isomorphism classes of indecomposables. By (22.18)(1), the mentioned composition is naturally isomorphic to $Z' \otimes_\Gamma -$, where $Z' = \mathrm{Cok}\,\Xi_\Lambda(Z)$ is a Λ-Γ-bimodule finitely generated by the right. Then, we can apply the argument in the proof of (27.5) to obtain a rational algebra Γ' and a Λ-Γ'-bimodule Z'' finitely generated free by the right such that $Z'' \otimes_{\Gamma'} - : \Gamma'\text{-mod} \longrightarrow \Lambda\text{-mod}$ preserves isomorphism classes of indecomposables. By Krull–Schmidt theorem for Λ-mod, the functor $Z'' \otimes_{\Gamma'} - : \Gamma'\text{-mod} \longrightarrow \Lambda\text{-mod}$ preserves indecomposables and isomorphism classes. □

Section 29

Solution to Exercise (29.12): We can identify $\mathrm{End}_\Lambda(M \oplus N)$ with the matrix algebra

$$\begin{pmatrix} \mathrm{End}_\Lambda(M) & \mathrm{Hom}_\Lambda(N, M) \\ \mathrm{Hom}_\Lambda(M, N) & \mathrm{End}_\Lambda(N) \end{pmatrix}$$

and consider the canonical morphism of algebras

$$\Gamma := \mathrm{End}_\Lambda(M) \times \mathrm{End}_\Lambda(N) \xrightarrow{\psi} \mathrm{End}_\Lambda(M \oplus N)$$

which maps (f, g) onto the diagonal matrix determined by f and g. Then, by restriction of scalars through ψ, the space $M \oplus N$ has a natural structure of a Γ-module.

For the second inequality, we can assume that $\mathrm{endol}\,(M) = s$ and $\mathrm{endol}\,(N) = t$ are finite. Consider a composition series $0 = M_s \subseteq \cdots \subseteq M_1 \subseteq M_0 = M$ of the $\mathrm{End}_\Lambda(M)$-module M and a composition series $0 = N_t \subseteq \cdots \subseteq N_1 \subseteq N_0 = N$ of the $\mathrm{End}_\Lambda(N)$-module N. Then, the filtration

$$0 = M_s \subseteq \cdots \subseteq M_1 \subseteq M \subseteq M \oplus N_{t-1} \subseteq \cdots \\ \subseteq M \oplus N_1 \subseteq M \oplus N_0 = M \oplus N$$

is a composition series of length $s+t$ of the Γ-module $M \oplus N$ (since each simple Γ-module is either a simple $\mathrm{End}_\Lambda(M)$-module or a simple $\mathrm{End}_\Lambda(N)$-module). Then, we have that $\ell_\Gamma(M \oplus N) = \mathrm{endol}\,(M) + \mathrm{endol}\,(N)$. Moreover each chain of $\mathrm{End}_\Lambda(M \oplus N)$-submodules of $M \oplus N$ is a chain of Γ-submodules of $M \oplus N$, by restriction through ψ. Then, $\mathrm{endol}\,(M \oplus N) \leq \ell_\Gamma(M \oplus N) = \mathrm{endol}\,(M) + \mathrm{endol}\,(N)$.

For the first inequality, we shall establish the claim: for any $\mathrm{End}_\Lambda(M)$-submodule X of M, the $\mathrm{End}_\Lambda(M \oplus N)$-submodule $\langle X \rangle$ of $M \oplus N$ generated by X satisfies that $X = \langle X \rangle \cap M$. Indeed, every element of $\langle X \rangle$ is a finite sum of the form

$$\sum_i \begin{pmatrix} f_i & \theta_i' \\ \theta_i & g_i \end{pmatrix} \begin{pmatrix} x_i \\ 0 \end{pmatrix} = \begin{pmatrix} \sum_i f_i(x_i) \\ \sum_i \theta_i(x_i) \end{pmatrix},$$

where the square matrices belong to the matrix algebra described above and $x_i \in X$. It follows that we have a vector space decomposition $\langle X \rangle = X \oplus Z$, where $Z = \sum_{\theta \in \mathrm{Hom}_\Lambda(M,N)} \theta(X)$. Hence, $X = \langle X \rangle \cap M$ as claimed. Now, assume that the $\mathrm{End}_\Lambda(M \oplus N)$-module $M \oplus N$ admits a composition series of finite length ℓ. Any chain of $\mathrm{End}_\Lambda(M)$-submodules $\cdots \subseteq X_j \subseteq \cdots \subseteq X_1 \subseteq M$ determines the chain $\cdots \subseteq \langle X_j \rangle \subseteq \cdots \subseteq \langle X_1 \rangle \subseteq \langle M \rangle \subseteq M \oplus N$ of $\mathrm{End}_\Lambda(M \oplus N)$-submodules of $M \oplus N$. Thus, this last chain must stabilize and refine to a composition series of length $\leq \ell$. From the claim, the chain $\cdots \subseteq X_j \subseteq \cdots \subseteq X_1 \subseteq M$ stabilizes and refines to a composition series of length $\leq \ell$. □

Section 32

Solution to Exercise (32.12): Consider the subalgebra T' of T generated by R and α. Then, the restriction of δ to T' is zero and $\mathcal{A}' := (T', 0)$ is a proper triangular subditalgebra of $\mathcal{A}$. From (16.4), we know that $\mathrm{Ext}_{\mathcal{A}'}(M, M)$ is a quotient of $\mathrm{Ext}_{\mathcal{A}}(M, M)$. Thus, $\mathrm{Ext}_{\mathcal{A}'}(M, M) = 0$.

Assume that $i = j$, then $M = \oplus_{s \in \mathcal{P}} M_s$ is a decomposition in T'-mod, and, if $M_i \neq 0$, then M_i admits a direct summand $N \cong H/H\pi^n$, where $H = k[\alpha]$ and $\pi \in k[\alpha]$ is an irreducible element. From, (32.2), we know that $\mathrm{Ext}_H(N, N) \neq 0$ and, hence, also $\mathrm{Ext}_{\mathcal{A}'}(M, M) \neq 0$; a contradiction.

Now suppose that $i \neq j$. Then, $N := M_i \oplus M_j$ is a module over the path k-algebra H of the quiver $i \xrightarrow{\alpha} j$. Denote by S_i and S_j the simple H-modules associated to the vertices i and j, respectively. Denote by I the projective injective indecomposable H-module. Then, if N_α is not surjective, S_j is a direct summand of N; if N_α is not injective, S_i is a direct summand of N. But there is a non-split exact sequence $S_j \longrightarrow I \longrightarrow S_i$, that is $\mathrm{Ext}_H(S_i, S_j) \neq 0$. Therefore, we obtain that N_α is either surjective or injective, as claimed. □

Solution to Exercise (32.13):

(1) Let $f = (f^0, f^1) : M \longrightarrow N$ be a morphism in $\mathcal{A}'$-Mod such that f^0 is surjective. Then, clearly, $F_\xi(f) = (f^0, f^1\xi) : F_\xi(M) \longrightarrow F_\xi(N)$ is a morphism in $\mathcal{A}$-Mod with f^0 surjective. Since $\mathcal{A}$ is acceptable, there is a commutative square in $\mathcal{A}$-Mod

$$\begin{array}{ccc} F_\xi(M) & \xrightarrow{(f^0, f^1\xi)} & F_\xi(N) \\ {\scriptstyle s}\downarrow & & \downarrow{\scriptstyle r} \\ X & \xrightarrow{(\varphi^0, 0)} & Y, \end{array}$$

where r, s are isomorphisms. Since $\mathrm{Im}\, F_\xi$ is closed under isomorphisms, $X = F_\xi(M')$ and $Y = F_\xi(N')$, for some $M', N' \in \mathcal{A}'$-Mod. Since F_ξ is full and faithful, there is a commutative square in $\mathcal{A}'$-Mod

$$\begin{array}{ccc} M & \xrightarrow{(f^0, f^1)} & N \\ {\scriptstyle s'}\downarrow & & \downarrow{\scriptstyle r'} \\ M' & \xrightarrow{h} & N', \end{array}$$

which is mapped by F_ξ onto the previous one. Thus, r', s' are isomorphisms. Since $\xi(V)$ generates V' and $F(h) = (h^0, h^1\xi) = (\varphi^0, 0)$, we have that $h^1 = 0$, and $\mathcal{A}'$ is acceptable.

(2) Let $z \in \{a, r\}$, and take a morphism $f : M \longrightarrow N$ be in $\mathcal{A}$-Mod such that f^0 is surjective. Since $F_z : \mathcal{A}^z\text{-Mod} \longrightarrow \mathcal{A}\text{-Mod}$ is dense and full, there is a commutative square in $\mathcal{A}$-Mod

$$\begin{array}{ccc} M & \xrightarrow{(f^0, f^1)} & N \\ {\scriptstyle s'}\downarrow & & \downarrow{\scriptstyle r'} \\ F_z(M') & \xrightarrow{F_z(h)} & F_z(N'), \end{array}$$

where r', s' are isomorphisms. Then, the first component h^0 of $F_z(h)$ is surjective. Therefore, there is a commutative square in $\mathcal{A}^z$-Mod

$$\begin{array}{ccc} M' & \xrightarrow{(h^0, h^1)} & N' \\ {\scriptstyle s}\downarrow & & \downarrow{\scriptstyle r} \\ M'' & \xrightarrow{(\varphi^0, 0)} & N'', \end{array}$$

where r, s are isomorphisms. Then, consider the commutative diagram in $\mathcal{A}$-Mod

$$\begin{array}{ccc}
M & \xrightarrow{(f^0,f^1)} & N \\
\downarrow{\scriptstyle s'} & & \downarrow{\scriptstyle r'} \\
F_z(M') & \xrightarrow{F_z(h^0,h^1)} & F_z(N') \\
\downarrow{\scriptstyle F_z(s)} & & \downarrow{\scriptstyle F_z(r)} \\
F_z(M'') & \xrightarrow{F_z(\varphi^0,0)} & F_z(N''),
\end{array}$$

where $F_z(\varphi^0, 0) = (\varphi^0, 0)$, because $F_z = F_\xi$ is induced by a morphism $\xi : \mathcal{A} \longrightarrow \mathcal{A}^z$.

□

Solution to Exercise (32.14): Let t be a marked point of $\mathcal{A}$, say with $H := Re_t = k[x]_{f(x)}$, and assume that $M_t \neq 0$. For instance, from (32.2), we know there is a non-split epimorphism $\pi : E_t \longrightarrow M_t$ in H-mod. Choose any linear map $\sigma : M_t \longrightarrow E_t$ with $\pi\sigma = 1_{M_t}$. Consider the R-module $E = \oplus_{i \in \mathcal{P}} E_i$ defined by its components: the Re_t-module E_t is the H-module fixed above and $E_i := M_i$, for all point $i \neq t$. The action E_α of a solid arrow $\alpha : i \longrightarrow j$ of $\mathcal{A}$ on E is given as follows

$$\begin{cases}
E_\alpha := M_\alpha & \text{if } i, j \neq t; \\
E_\alpha := \sigma \circ M_\alpha & \text{if } i \neq t = j; \\
E_\alpha := M_\alpha \circ \pi & \text{if } i = t \neq j; \\
E_\alpha := \sigma \circ M_\alpha \circ \pi & \text{if } i = t = j.
\end{cases}$$

Since A is freely generated by R and W_0 (freely generated by the solid arrows), we obtain an A-module E. The map $f^0 := \oplus_i f_i^0$, with $f_i^0 := 1_{M_i}$, for all $i \neq t$, and $f_t^0 := \pi$, is clearly an epimorphism of R-mod. In fact $f^0 : E \longrightarrow M$ is an epimorphism of A-modules, because

$$\begin{array}{ccc}
E_i & \xrightarrow{E_\alpha} & E_j \\
\downarrow{\scriptstyle f_i^0} & & \downarrow{\scriptstyle f_j^0} \\
M_i & \xrightarrow{M_\alpha} & M_j
\end{array}$$

commutes for all solid arrow $\alpha : i \longrightarrow j$ of $\mathcal{A}$. Moreover, f^0 is not split in A-mod because f_t^0 is not split in Re_t-mod. Consider its kernel $g^0 : N \longrightarrow E$ in A-mod. Then, applying the canonical embedding functor, we have the following non-split exact pair in $\widehat{\mathcal{E}}_{\mathcal{A}}$

$$N \xrightarrow{(g^0,0)} E \xrightarrow{(f^0,0)} M.$$

It follows that, M is not $\widehat{\mathcal{E}}_{\mathcal{A}}$-projective. Similarly, we show that M is not $\widehat{\mathcal{E}}_{\mathcal{A}}$-injective. □

Solution to Exercise (32.17): Assume $f = (f^0, f^1) \in \mathrm{rad}^{\infty}_{\mathcal{A}}(M, N)$ satisfies that $f^0 \neq 0$. Then, having in mind the description of the indecomposable $\mathcal{A}$-modules, there exists a unique point z of $\mathcal{A}$ with $M_z \neq 0$ and $N_z \neq 0$. Since $f \in \mathrm{rad}_{\mathcal{A}}(M, N)$, f does not split and, therefore, z is a marked point. Moreover, $M \cong J(z, m, \pi)$ and $N \cong J(z, n, \pi)$, for some $m, n \geq 1$. Then, from (32.16), we get $\mathrm{rad}^s_{\mathcal{A}}(J(z, m, \pi), J(z, n, \pi)) = 0$, for some s. Therefore, we have the contradiction $f = 0$. □

Solution to Exercise (32.19): We already have (32.17), thus we only have to take $f \in \mathrm{Hom}_{\mathcal{A}}(M, N)$ with $f^0 = 0$ and show that $f \in \mathrm{rad}^{\infty}_{\mathcal{A}}(M, N)$. Assume that $M_z \neq 0$, thus $M \cong J(z, m, \pi)$. Write $H := Re_z$, then, for any natural number ℓ, we have the injective morphism of H-modules $\mu_{\pi^\ell} : H/(\pi^m) \longrightarrow H/(\pi^{m+\ell})$. It induces the morphism $h_\ell = (\mu_{\pi^\ell}, 0) : J(z, m, \pi) \longrightarrow J(z, m + \ell, \pi)$ in $\mathcal{A}$-mod. In fact, $h_\ell \in \mathrm{rad}^{\ell}_{\mathcal{A}}(J(z, m, \pi), J(z, m + \ell, \pi))$, because $(\mu_{\pi^\ell}, 0)$ is the composition of ℓ radical morphisms of the form $(\mu_\pi, 0)$. From (32.19), for any $\ell \geq 1$, there is a morphism $t_\ell : J(z, m + \ell, \pi) \longrightarrow N$ in $\mathcal{A}$-mod such that $t_\ell h_\ell = f$. Thus, $f \in \mathrm{rad}^{\ell}_{\mathcal{A}}(M, N)$, for any ℓ. Then, $f \in \mathrm{rad}^{\infty}_{\mathcal{A}}(M, N)$. The case $N_z \neq 0$ is discussed dually. □

Solution to Exercise (32.20): First notice that any morphism $g : S_{i_0} \longrightarrow M$ in $\mathcal{A}$-Mod with $g^0 \neq 0$ is a section, see (5.9). Dually, any morphism $h : M \longrightarrow S_{j_0}$ in $\mathcal{A}$-Mod with $h^0 \neq 0$ is a retraction.

(1) If $f : S_{i_0} \longrightarrow S_{j_0}$ is a radical morphism, from the previous remark, we must have that $f^0 = 0$. Each $f^1(v_s)$ is a scalar multiple, say by $a_s \in k$, of 1_k. Thus, $f = \sum_s a_s f_{v_s}$.
(2) Since $f^0_v = 0$, f_v is not a section or a retraction. Assume we have a factorization $f_v = hg$ in $\mathcal{A}$-Mod. If, under the assumption $\delta(v) = 0$, we have that $g^0 = 0$ and $h^0 = 0$, then we obtain the contradiction $1_k = f^1_v(v) = h^0 g^1(v) + h^1(v) g^0 + 0 = 0$. Thus, $g^0 \neq 0$ or $h^0 \neq 0$ and, therefore, g is a section or h is a retraction.
(3) Suppose that $\sum_i \lambda_i f_{v_i} = \sum_j h_j g_j$, where all $\lambda_i \in k$, and all h_j, g_j are radical morphisms. From the remark previous to the proof of (1), $g^0_j = 0$ and

$h_j^0 = 0$. Then, if $s \in [1, m]$, since the dotted arrow v_s has zero differential, we obtain $\lambda_s 1_k = \left(\sum_i \lambda_i f_{v_i}\right)^1 [v_s] = 0 + 0 + 0$ and, therefore, $\lambda_s = 0$.

(4) Assume that $f_{v_s} = hg$ in $\mathcal{A}$-Mod, with $g^0 = 0$ and $h^0 = 0$. Then, for $s \neq r$, $1_k = f_{v_s}^1(v_s) = \sum_t \mu_t h^1(w_t) g^1(u_t) = f_{v_s}^1(v_r) = 0$, which is impossible. Then, $g^0 \neq 0$ or $h^0 \neq 0$ and, hence, g is a section or h is a retraction. Thus, each f_{v_s} is irreducible. Now, assume that $(\mu_1, \ldots, \mu_n) \neq 0$ and choose $(\theta_1, \ldots, \theta_n) \in k^n$ such that $\lambda := \sum_t \theta_t \mu_t \neq 0$. Then, $\lambda \sum_s f_{v_s} = \sum_t \theta_t f_{w_t} f_{u_t}$. Finally, after the change of basis, since $\lambda f_{v_1'} = \sum_t \theta_t f_{w_t} f_{u_t}$, we can apply (1) and (3).

(5) If v is a dotted arrow from i_0 to j_0 in $\mathcal{A}$ and $\delta(v) = \sum_{t=1}^m \mu_t w_t u_t \neq 0$, then, choosing $\theta_1, \ldots, \theta_m$ such that $\lambda := \sum_t \theta_t \mu_t \neq 0$, we obtain that $\lambda f_v = \sum_t \theta_t f_{w_t} f_{u_t} \in \text{rad}^2_{\mathcal{A}}(S_{i_0}, S_{j_0})$. Then, apply (1) and (3).

□

References

[1] Anderson, F. W. and Fuller, K. R. *Rings and Categories of Modules.* Springer Verlag GTM **13** (1974).

[2] Auslander, M. Representation theory of artin algebras II. *Comm. Algebra* **1**(4) (1974) 269–310.

[3] Auslander, M., Reiten, I. and Smalø, S. O. Representation theory of artin algebras. *Cambridge Studies in Advanced Mathematics* **36** (1995).

[4] Auslander, M. and Smalø, S. O. Almost split sequences in subcategories. *J. Algebra* **69** (1981) 426–454. *Addendum J. Algebra* **71** (1981) 592–594.

[5] Bautista, R. The category of morphisms between projective modules. *Comm. Algebra* **32**(11) (2004) 4303–4331.

[6] Bautista, R., Boza, J. and Pérez, E. Reduction functors and exact structures for bocses. *Bol. Soc. Mat. Mexicana* (3) **9**(1) (2003) 21–60.

[7] Bautista, R. and Kleiner, M. M. Almost split sequences for relatively projective modules. *J. Algebra* **135**(1) (1990) 19–56.

[8] Bautista, R. and Zuazua, R. Exact structures for lift categories. *Fields Institute Communications* **45** (2005) 37–56.

[9] Bautista, R. and Martínez-Villa, R. *Representations of Partially Ordered Sets and 1-Gorenstein Artin Algebras: Proceedings, Conference on Ring Theory Antwerp* (1978) New York, Marcel Dekker (1979) 385–433.

[10] Bautista, R., Raggi-Cárdenas, A. G. and Salmerón, L. On varieties of representations of bocses. *Bol. Soc. Mat. Mexicana* (3) **8**(1) (2002) 5–30.

[11] Bautista, R. and Salmerón, L. On discrete and inductive algebras. *Fields Institute Communications* **45** (2005) 17–35.

[12] Bautista, R. and Salmerón, L. The Kiev algorithm for bocses applies directly to representation-finite algebras. *Manuscripta Math.* **65** (1989) 281–287.

[13] Bautista, R. and Zuazua, R. One-parameter families of modules for tame algebras and bocses. *Algebras and Representation Theory* **8** (2005) 635–677.

[14] Bautista, R. and Zhang Yingbo Representations of a k-algebra over the rational functions field over k. *J. of Algebra* **267** (2003) 342–358.

[15] Bautista, R., Drozd, Y., Zeng, X. and Zhang, Y. On hom-spaces of tame algebras. *Central European J. of Maths.* **5**(2) (2007) 215–263.

[16] Burt, W. L. and Butler M. C. R. Almost split sequences for bocses: representations of finite dimensional algebras. *Canadian Math. Soc. Conference Proceedings* **11** (1990) 89–121.

[17] Cohn, P. M. Free rings and their relations. *London Math. Soc. Monograph* 19, Academic Press (1985).

[18] Cohn, P. M. Noncommutative unique factorization domains. *Trans. Amer. Math. Soc.* **109**(2) (1963) 313–331. Corr. *ibid.* **119** (1965) 552.

[19] Crawley-Boevey, W. W. On tame algebras and bocses. *Proc. London Math. Soc.* **56**(3) (1988) 451–483.

[20] Crawley-Boevey, W. W. Tame algebras and generic modules. *Proc. London Math. Soc.* **63**(3) (1991) 241–265.

[21] Crawley-Boevey, W. W. Modules of finite length over their endomorphism rings. *Representations of Algebras and Related Topics*, eds. S. Brenner and H. Tachikawa, *London Math. Lect. Notes Series* **168** (1992) 127–184.

[22] Crawley-Boevey, W. W. *Matrix Problems and Drozd's Theorem: Topics in Algebra.*, Vol. 26, Part 1, Banach Center Publ., PWN-Polish Science Publ. (1990) 199–222.

[23] Curtis, Ch. W. and Reiner, I. *Methods of Representation Theory Vol I.* New York, John Wiley & Sons (1981).

[24] De la Peña, J. A. Functors preserving tameness. *Fundamenta Math.* **137** (1991) 177–185.

[25] De la Peña, J. A. Sur les degrés de liberté des indécomposables. *C.R. Acad. Sci. Paris* **412**(I) (1991) 545–548.

[26] Donovan, P. and Freislich, M. R. The representation theory of finite graphs and associated algebras. *Carleton Mathematical Lecture Notes*, No. 5. Carleton University, Ottawa, Ont. (1973).

[27] Dräxler P., Reiten I., Smalø, S. O. and Solberg Ø. Exact categories and vector space categories. With an appendix by B. Keller. *Trans. Amer. Math. Soc.* **351**(2) (1999) 647–682.

[28] Drozd, Yu. A. *Tame and wild matrix problems.* Representations and quadratic forms. [Institute of Mathematics, Academy of Sciences, Ukranian SSR, Kiev (1979) 39–74] *Trans. Amer. Math. Soc.* **128** (1986) 31–55.

[29] Drozd, Yu. A. and Kirichenko, V. V. *Finite-Dimensional Algebras.* Springer-Verlag, Berlin (1994).

[30] Drozd, Yu. A. and Ovsienko, A. *Coverings of Tame Boxes and Algebras.* Preprint (2000).

[31] Gabriel, P. Unzerlegbare Darstellungen I. *Manuscrpita Math.* (1972) 71–103.

[32] Gabriel, P. and Roiter, A.V. *Representations of Finite-Dimensional Algebras.* Springer Verlag (1977).

[33] Humphreys, J. E. *Linear Algebraic Groups.* GTM 21, Springer Verlag (1981).

[34] Jacobson, N. *The Theory of Rings: Math. Surveys II.* Providence R.I., Amer. Math. Soc. (1943).

[35] Jacobson, N. *Basic Algebra II.* New York, W. H. Freeman & Company (1980).

[36] Kasjan, S. Auslander–Reiten sequences under field extensions. *Proc. of the Amer. Math. Soc.* **128**(10) (2000) 2885–2896.

[37] Kasjan, S. Base field extensions and generic modules over finite dimensional algebras. *Arch. Math.* **77** (2001) 155–162.

[38] Kleiner, M. Partially ordered sets of finite type. *Zap. Naucn. Sem. Lomi.* **28** (1972), 32–41. Engl. transl. *J. Soviet Math.* **3** (1975) 607–615.

[39] Kleiner, M. Pairs of partially ordered sets of tame representation type. *Linear Algebra and Its Applications* **104** (1988) 103–115.

[40] Larrión, F., Raggi-Cárdenas, A. G. and Salmerón, L. Rudimentos de mansedumbre y salvajismo en teoría de representaciones. *Aportaciones Matemáticas* (Textos) **5** Soc. Mat. Mex. (1995).

[41] Nazarova, L. A. and Roiter, A. V. Representations of partially ordered sets. *Zap. Naucn. Sem. Lomi.* **28** (1972) 5–31. Engl. transl. *J. Soviet Math.* **3** (1975) 585–606.

[42] Nazarova, L. A. Partially ordered sets of infinite type. *Math. USSSR–Izv.* **9** (1975) 911–938.

[43] Ovsienko, S. A. *Generic Representations of Free Bocses.* Preprint 93–010, Universität Bielefeld.

[44] Pérez, E. Representaciones con grupo de auto-extensiones simple. Tesis Doctoral, Facultad de Ciencias, Universidad Nacional Autónoma de México (2001).

[45] Roiter, A. V. Matrix problems and representations of bocses. *Springer Lect. Notes in Math.* **831** (1980) 288–324.

[46] Roiter, A. V. and Kleiner, M. M. Representations of differential graded categories. *Matrix Problems*, Inst. Mat. Akad Nauk. Ukrain SSR, Kiev, 1977, 5–70. abriged English version: *Springer Lect. Notes in Math.* **488** (1975) 316–339.

[47] Rotman, J. J. *An Introduction to Homological Algebra.* Academic Press Inc. (1979).

[48] Silver, L. Noncommutative localisations and applications. *J. Algebra* **7** (1967) 44–76.

[49] Yingbo, Z. and Xu Yunge On tame and wild bocses. *Science in China Series A-Mathematics* **48**(4) (2005) 456–468.

[50] Zavadski, A. G. and Nazarova, L. A. Partially ordered sets of tame type. *Matrix Problems, Kiev* (1977) 122–143.

[51] Zuazua, R. Arboles Algorítmicos en la Teoría de Bocses. Tesis Doctoral. Facultad de Ciencias, Universidad Nacional Autónoma de México (1999).

Index

Printed in the United States
by Baker & Taylor Publisher Services